Microbial Genetics

Editors

Sylwia Okoń
Institute of Plant Genetics, Breeding and Biotechnology
University of Life Sciences in Lublin
Poland

Beata Zimowska
Department of Plant Protection
Subdepartment of Plant Pathology and Mycology
University of Life Sciences in Lublin
Poland

Mahendra Rai
Department of Biotechnology
SGB Amravati University
Amravati, Maharashtra
India

CRC Press
Taylor & Francis Group
Boca Raton London New York

CRC Press is an imprint of the
Taylor & Francis Group, an **informa** business
A SCIENCE PUBLISHERS BOOK

First edition published 2024
by CRC Press
2385 NW Executive Center Drive, Suite 320, Boca Raton FL 33431

and by CRC Press
4 Park Square, Milton Park, Abingdon, Oxon, OX14 4RN

© 2024 Sylwia Okoń, Beata Zimowska and Mahendra Rai

CRC Press is an imprint of Taylor & Francis Group, LLC

Library of Congress Cataloging-in-Publication Data (applied for)

ISBN: 978-1-032-35841-3 (hbk)
ISBN: 978-1-032-35842-0 (pbk)
ISBN: 978-1-003-32893-3 (ebk)

DOI: 10.1201/9781003328933

Typeset in Times New Roman
by Radiant Productions

Preface

Microorganisms constitute a very diverse group of organisms inhabiting various environments. They are characterized by various processes occurring at the genetic level and require authentic identification and differentiation. In the past, the traditional taxonomy was based on morphological markers which were not stable and changed according to environmental conditions. However, recent DNA-based technology has generated a revolution in determining the identity and taxonomy of microbes, particularly in fungi. In addition, the emergence of metagenomics has opened new avenues for the determination of the identity and taxonomy of non-culturable microorganisms.

The proposed book presents, on the one hand, fundamental issues related to the broadly understood genetics of microorganisms, the organization of genomes, and their differentiation. On the other hand, it presents detailed issues related to the methods of genetic research used in various groups of microorganisms, taking into account their taxonomy, methods of identification, and assessment of their diversity. The book is divided into four sections on genetic research of bacteria, fungi, viruses, phytoplasma, and protozoa. Section I deals with the basics of microbial genetics and begins with an introduction to microbial genetics, organization of microbial genomes, its variability, and diversity; basic principles of microbial genetics; metagenomics, genome sequencing, molecular phylogeny of microorganisms and mobile genetic elements and the evolution of microbes; Section II describes the molecular basis of viruses and phytoplasma including the genetic variation of viruses and how phytoplasmas manipulate host plants'-molecular mechanisms?; Section III discusses bacterial genetics, particularly genetic methods of identification, classification, and differentiation, and novel approaches for the genetic characterization of the secretome in the genus *Ralstonia;* finally, fungal genetics have been described in Section IV in which DNA-based techniques for studying the genetic diversity of fungi, transcriptomics, proteomics, and metabolomics, and the novel application of CRISPR/Cas technology in studying plant pathogenic oomycetes have been incorporated.

There is a diversity of representation of contributors from Europe, Brazil, Mexico, Egypt, India, and Iran. The proposed book will be very useful for postgraduate and research students of microbiology, genetics, biotechnology, and biology. It will be essential reading for the students owing to the inclusion of basic and applied aspects.

We are thankful to the authors for their up-to-date chapters. Their sincere efforts will update the readers' knowledge of Microbial Genetics. Further, we would like to sincerely thank everyone on the CRC/Taylor Francis team who uninterruptedly cooperated with us during the editing process and also in the publication of the book. We especially thank Mr. Raju Primlani, the publishing manager, who always motivated us to complete the book.

Sylwia Okoń, Poland
Beata Zimowska, Poland
Mahendra Rai, India

Contents

Section IV: Fungal and Protozoan Genetics

Section I

Basics of Microbial Genetics

CHAPTER 1

An Introduction to Microbial Genetics

Sylwia Okoń,[1,*] *Beata Zimowska*[2] and *Mahendra Rai*[3]

Introduction

Microbial genetics is the study of the mechanisms of heritable information in microorganisms. The main focus of this field of science is understanding microbial genome organization, evolution, and functioning. The genetics of microorganisms is also based on the study of genome plasticity in response to changes in the environment or adaptation to new living conditions. The development of genetics, molecular biology and genetic engineering introduced the study of microorganisms into the stage of intentional manipulation of genomes, learning about gene functions and metabolic pathways. Much of our understanding of mutation, genome rearrangement, DNA repair, DNA recombination mechanisms, DNA replication, DNA transfer, gene regulation, and much more has come from fundamental research in microbes.

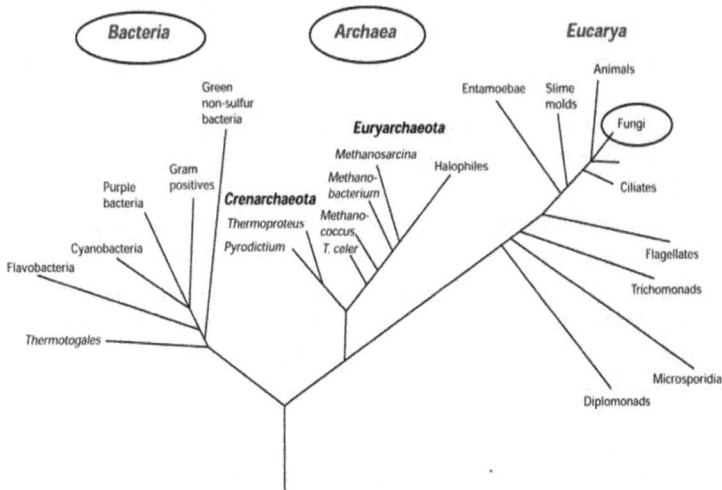

Figure 1. The universal tree of life-based on phylogenetic analysis of 16S and 18S rDNA of various microbial groups and plants and animals. Reprinted from Woese (1994) under a Creative Commons 4.0 International License.

[1] Institute of Plant Genetics, Breeding and Biotechnology, University of Life Sciences in Lublin, Akademicka 15, 20–950 Lublin.
[2] Department of Plant Protection, Subdepartment of Plant Pathology and Mycology, University of Life Sciences in Lublin, 7 K. St. Leszczyńskiego Street, 20-068 Lublin, Poland.
[3] Department of Biotechnology, SGB Amravati University, Amravati-444 602, Maharashtra, India.
* Corresponding author: sylwia.okon@up.lublin.pl

The term "microorganism" refers to a group of microscopic organisms, including bacteria, archaea, viruses, fungi, and protists. These organisms can be found in each of the major phyla (Figure 1). This very diverse group of organisms has been developing on Earth for billions of years. Microorganisms live in all environments on Earth that are occupied by macroscopic organisms and are the only forms of life in other environments, such as deep subsurface and "extreme" environments. Microorganisms play an important role in the functioning of ecosystems. They participate in carbon, nitrogen, sulfur, and other biogeochemical cycles. Microbes are also involved in bioremediation, decomposition of organic matter, degradation, and removal of pollutants, reclamation of degraded areas, etc. (Gupta et al. 2017, Kumar and Verma 2019).

Why Microorganisms are of Interest to Genetics?

The habitat of microorganisms is constantly changing, which forces microorganisms to have extraordinary abilities to adapt to environmental and evolutionary challenges. The adaptability of microorganisms stems from the fact that their genomes are extremely dynamic (Russell 1990, Mitchell et al. 2009, Bleuven and Landry 2016). The acquisition of foreign DNA combined with intragenomic rearrangements and duplication events may explain the remarkable ability of microorganisms to continuously explore new ecological niches (Sicheritz-Pontén and Andersson 2001). Microorganisms show remarkable diversity in the structure and organization of their genomes. Their genetic material can be DNA, RNA, in the form of single-stranded and double-stranded particles. It can be linear or circular, composed of one molecule or divided into several (Galagan et al. 2005, Dziewit et al. 2014, Campillo-Balderas et al. 2015, O'Carroll and Rein 2016, Stajich 2017). Several internal and external mechanisms are responsible for shaping the genomes of microorganisms. Among the internal mechanisms affecting the variability of microbial genomes, point mutations (substitutions, insertions, and deletions) should be distinguished, which result in the modification, inactivation or differential regulation of existing genes. Internal mechanisms also include the lengthening, loss, fusion, and duplication of genes. Among the external mechanisms shaping the structure of microorganisms genomes, the most important are fusion and horizontal gene transfer (del Duca et al. 2022).

Due to their size, microorganisms became an excellent object of genetic research, which formed the basis for further research. In particular, they were of great importance in the discovery that genetic material is DNA, that genes have a simple linear structure, that the genetic code is a triplet code, and that gene expression is regulated by specific genetic processes (Jacob and Monod 1961, Crick et al. 1961, Avery et al. 1979).

Microorganisms show remarkable diversity in terms of metabolic properties, cellular structures, and lifestyle. Even in relatively narrow taxonomic groups, phenotypic diversity between species is remarkable (Artuso et al. 2021). Their genomes contain clues as to how they started and travelled this evolutionary path. Therefore, they can provide a lot of valuable information not only for scientists involved in the study of microorganisms, but also for those who study climate change on Earth, or metabolic pathways of living organisms, adaptive possibilities or interactions between different groups of organisms. The current structure and organisation of genes and genomes are increasingly characterized by modern and efficient molecular techniques, allowing for a better understanding of the mechanisms underlying the evolution, expansion, and formation of organisms. The development of Next Generation Sequencing (NGS) has provided an unparalleled opportunity to obtain a sequential view of these evolutionary processes with larger, more complex genomes that encode a more complex life history and metabolism (Forde and O'Toole 2013, Brockhurst et al. 2011). In addition, thanks to the development of research on DNA sequencing, we have the opportunity to better understand the world of microorganisms, and the species composition of various ecosystems and discover new, previously unknown species (Feng et al. 2018, Acharya et al. 2020, Kwok et al. 2020, Azli et al. 2022).

A key step in studying the evolution of an organism and its genome is to determine its phylogeny. DNA sequencing techniques and the increasing availability of fully sequenced genomes have made it possible to carry out comparative analyses of a huge number of genes and genomes in recent years. The combination of genomic and evolutionary data has given rise to phylogenomic approaches that use phylogenetic principles to make sense of genomic data by using molecular data to infer species relationships and using available information about the evolutionary history of a species to gain insight into the mechanisms of molecular evolution (Philippe et al. 2005, Delsuc et al. 2005, del Duca et al. 2022).

Due to the short generation time and large population size, microbes evolve very quickly, making them excellent models for studying evolution in action and its genetic basis (Good and Hallatschek 2018, Brockhurst et al. 2011, Callahan et al. 2014). The diversity observed in the microbial world allows us to understand the processes behind genetic diversity. Bacteria, archaea, viruses, and yeasts are the subject of laboratory and *in silico* studies to study the evolution of their genomes and the relationship between genotypes and phenotypes (Achtman and Wagner 2008, Truong et al. 2017).

Genetic studies of microbes play a key role in our understanding of the flow of genetic information in biological systems and recent advances in molecular biology. Moreover, the hereditary processes of unicellular eukaryotic microorganisms are similar to those occurring in multicellular organisms, which allows scientists to collect information about this process and use it for deeper studies of plants, animals, and humans (Joyce and Palsson 2006, Yilmaz and Walhout 2017, Allaf and Peerhossaini 2022).

Genetic recombination in bacteria is an important process which is the rearrangement of genomes of donor and recipient to form new hybrid genomes. This process also occurs in viruses and that is why there is a diversity of genomes among viruses. For example, recombination is a very important evolutionary tool for RNA viruses such as coronaviruses. Both homologous and nonhomologous recombination have been reported in coronaviruses (Focosi and Maggi 2022).

Understanding the Genetic Basis of Microbial Metabolism Allows for Their Better Use

Understanding metabolic pathways and the possibilities of their regulation allows microorganisms to be used on an industrial scale, e.g., for the production of various types of enzymes (Adrio and Demain 2014, Patel et al. 2017), biofuels (Stephanopoulos 2007, Parisutham et al. 2014) or supporting the work of biorefineries (Sarkar et al. 2022). Engineering of metabolic pathways and their modelling through recombinant DNA allows the use of microorganisms to produce various types of natural compounds (Trenchard et al. 2015, Gottardi et al. 2017) (Figure 2).

Microorganisms play a significant role in increasing the productivity of agricultural crops and improving the condition of agricultural soils. Application of beneficial microbes may be a potential alternative to harmful chemical fertilizers and pesticides. Plant-associated soil microbes play a crucial role in plant growth and development such as nutrient cycling and crop productivity (Yan et al. 2008, Kumar and Verma 2019, Lopes et al. 2021).

Microorganisms are also the main cause of many plant, animal, and human diseases. Understanding the genetic basis of pathogenesis allows for better health protection and disease prevention strategies (Meena et al. 2019, Thakur et al. 2019).

Understanding the genetic and ecological processes associated with different microorganisms and their genomes and studying their metabolic potential may also open up new intriguing prospects for their possible biotechnological use. Artuso et al. (Artuso et al. 2021) proposed the use of *Aminobacter* species in bioaugmentation and bioremediation processes, while Gammuto et al. (2022) suggested the possibility of developing new tools based on the use of the azurine domain p28 *P. aeruginosa* for the treatment of cancer.

Research on microorganisms contributed to the dynamic development of genetics and had a huge impact on the development of modern genetic engineering and biotechnology.

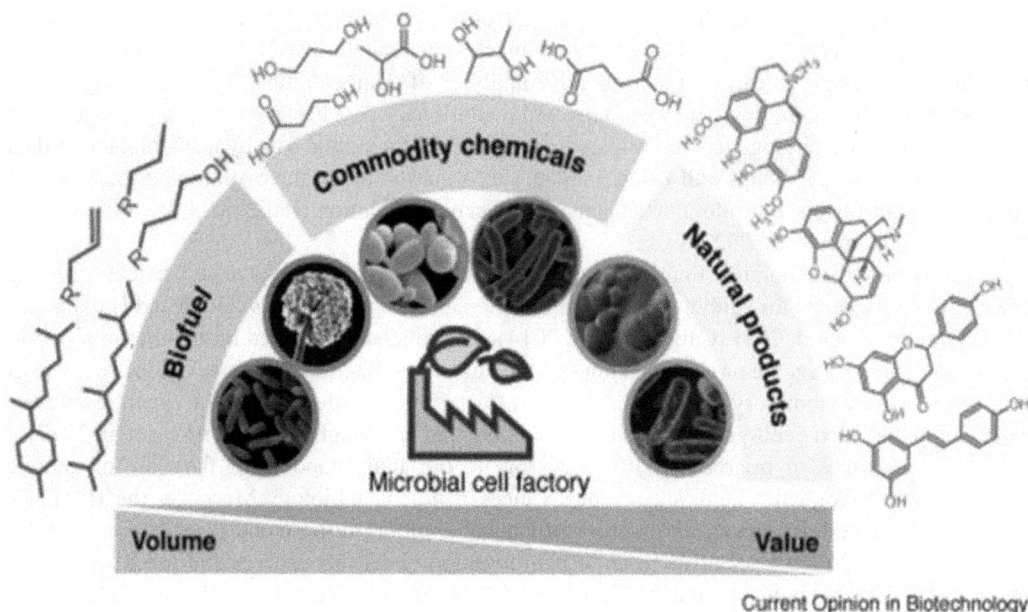

Figure 2. Chemicals produced by microbial cell factories. Different categories of chemicals such as biofuels, commodity chemicals, and natural products, from high-volume chemicals to high-value compounds which can be produced by microbial cell factories. Reprinted from Liu and Nielsen (2019) under a Creative Commons 4.0 International License.

The first genetic modifications in microorganisms was made by Cohen et al. (1973). Since then, the technology of genetic manipulation of organisms has had remarkable progress with a variety of microorganisms. Taking advantage of their rapid reproduction, genetically modified microbes were used to create biofactories for the gene of interest (Xie 2022). These biofactories are typically much less expensive to operate and maintain than alternative manufacturing procedures and can be used in almost any industry (Thodey et al. 2014, Ram et al. 2020, Alvarez et al. 2021, Mandal and Majumdar 2023). Genetic modifications of microorganisms have also contributed to the development of nanotechnology. Genetically engineered microbial cell factories are functional platforms contributing to nanobiosynthesis and related nanobiotechnology applications. A deep understanding of such biosynthetic pathways and genetic engineering techniques opens the door to wider horizons for the industrial exploitation of microbial nanofactories (Zhu et al. 2021).

Microorganisms are a Model for Studying Genome Editing Possibilities

Microorganisms are also excellent subjects for gene editing research. Genome editing (also called gene editing) is a group of technologies that give scientists the ability to change an organism's DNA. These technologies allow genetic material to be added, removed, or altered at particular locations in the genome. This has been made possible by the discovery and characteristic of the bacterial CRISPR-Cas antiviral defence system. CRISPR (Clustered Regularly Interspaced Short Palindromic Repeats) are RNA molecules that are like a guide through an extremely long DNA chain. They can find a precisely designated place in the genome and bring there the appropriate enzyme-Cas9. The Cas9 protein, on the other hand, works like scissors- it cuts the DNA in the precisely indicated place. Thanks to this, it is possible to activate or deactivate a specific DNA fragment and, for example, regulate the operation of genes. At the next stage, another, precisely designed DNA fragment can be inserted into this selected place of DNA (Figure 3) (Hsu et al. 2014, Bak et al. 2018).

The natural CRISPR system has been used for various biotechnological applications, including the generation of phage-resistant dairy cultures (Quiberoni et al. 2010) and the phylogenetic classification of bacterial strains (Horvath et al. 2008, 2009). Currently, the

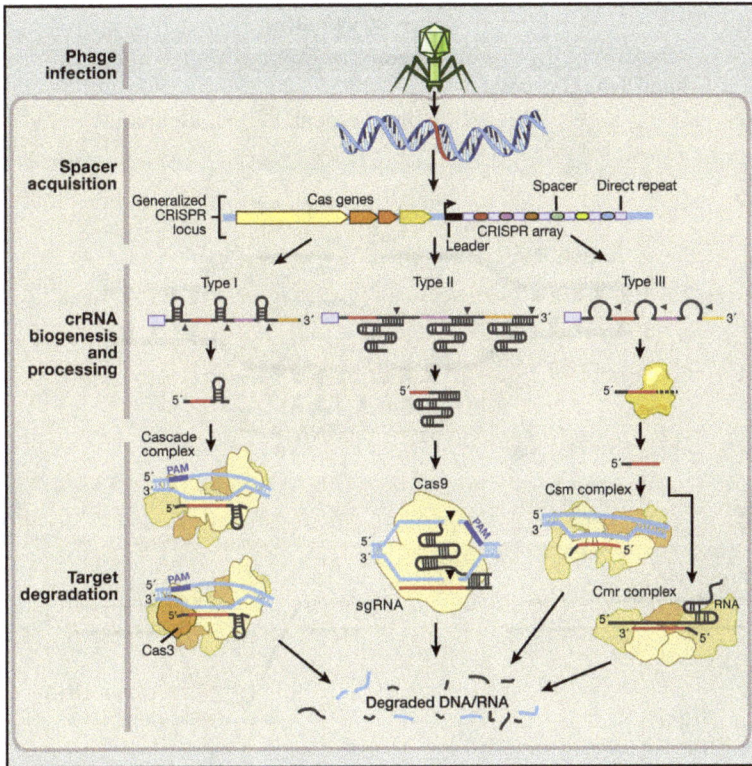

Figure 3. Natural mechanisms of microbial CRISPR systems in adaptive immunity. Reprinted from Hsu et al. (2014) under a Creative Commons 4.0 International License.

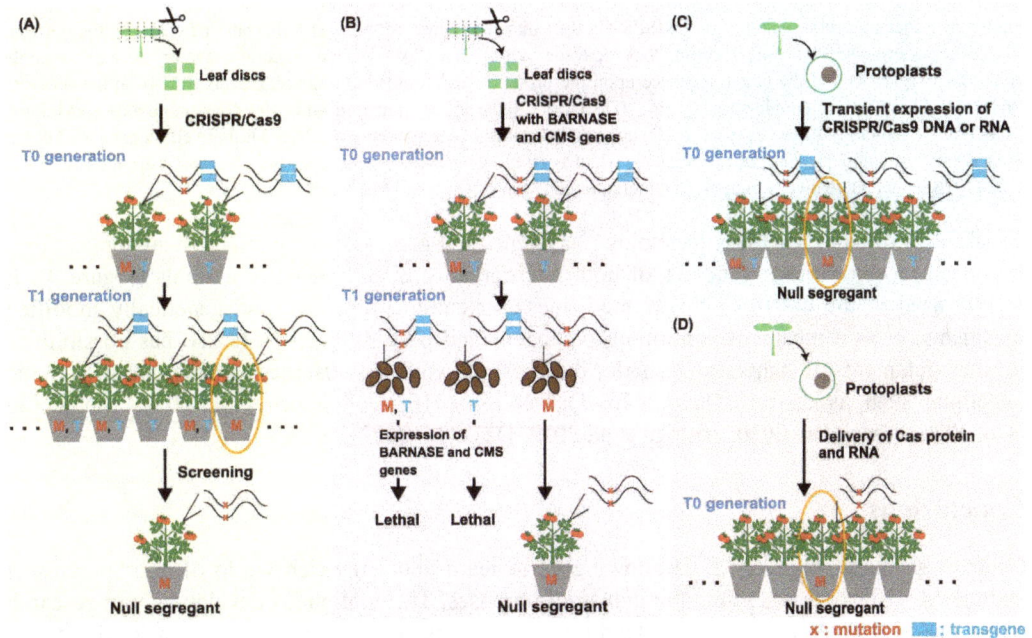

Figure 4. Generation of null segregants in plants by CRISPR/Cas9 technology. Representative methods for the production of null segregants are shown: isolation of null segregants by Mendelian segregation (a); programmed self-elimination of transgenic plants (b); transient expression of CRISPR/Cas9 (c); and ribonucleoprotein-mediated genome editing (d). Please refer to the text for detailed explanations. "M" and "T", plants with mutation and transgene insertion, respectively. Reprinted from Wada et al. (2020) under a Creative Commons 4.0 International License.

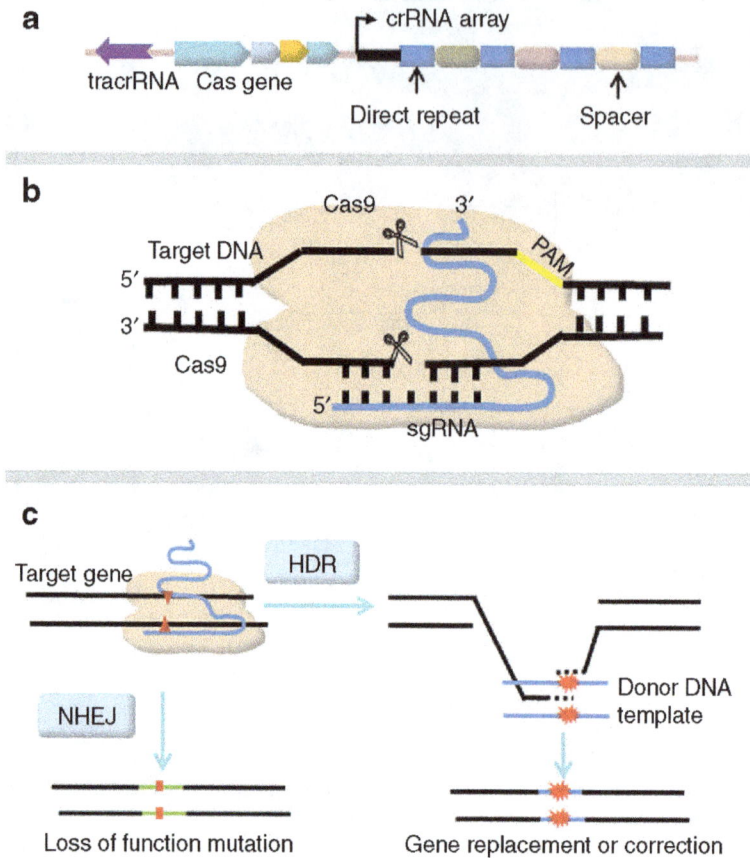

Figure 5. Schematic representation of CRISPR-Cas9-mediated genome editing. (a) Schematic of CRISPR locus (from *Streptococcus pyogenes*). (b) Site-specific DNA cleavage by nuclease Cas9 directed by complementary between a single guide RNA (sgRNA) and the target sequence upon the presence of a protospacer-adjacent motif (PAM) on the opposite strand. (c) The resultant double-strand breaks (DSBs) are subsequently repaired either by nonhomologous end-joining (NHEJ) or by homology-directed repair (HDR) upon the existence of a donor template. NHEJ is more efficient than HDR but is error-prone and may produce indel mutations, whereas HDR can provide precise gene modification. Reprinted from Dai et al. (2016) under a Creative Commons 4.0 International License.

possibilities of using CRISPR technology has generated great interest in biomedical applications. It is used to edit various genomes of both microorganisms and plants or animals (Figure 4). It can be used in the creation of new medicines, agricultural products, and genetically modified organisms, or as a means of controlling pathogens and pests (Figure 5). It also has possibilities in the treatment of inherited genetic diseases as well as diseases arising from somatic mutations such as cancer (Dow 2016, Dai et al. 2016, Donohoue et al. 2018, Montecillo et al. 2020, Wada et al. 2020, Nishiga et al. 2022, Das et al. 2022).

Conclusions

Genetics seems to be a key field allowing us to learn about the rich world of microorganisms and their use for both scientific and industrial purposes. The continuous development of research techniques allows us to better understand microorganisms not only at the molecular level but also at the metabolic level, which gives the possibility of modelling them according to human needs.

References

Acharya, K., A. Blackburn, J. Mohammed, A.T. Haile, A.M. Hiruy and D. Werner. 2020. Metagenomic water quality monitoring with a portable laboratory. Water Res. 184: 116112.

Achtman, M. and M. Wagner. 2008. Microbial diversity and the genetic nature of microbial species. Nat Rev Microbiol 6: 431–440.

Adrio, J.L. and A.L. Demain. 2014. Microbial enzymes: tools for biotechnological processes. Biomolecules 4: 117–139.

Allaf, M.M. and H. Peerhossaini. 2022. Cyanobacteria: model microorganisms and beyond. Microorganisms 10: 696.

Alvarez H.M., M.A. Hernández, M.P. Lanfranconi, R.A. Silva and M.S. Villalba. 2021. *Rhodococcus* as biofactories for microbial oil production. Molecules 26: 4871.

Artuso, I., P. Turrini, M. Pirolo, G.A. Lugli, M. Ventura and P. Visca. 2021. Phylogenomic reconstruction and metabolic potential of the genus *Aminobacter*. Microorganisms 9(6): 1332.

Avery, O.T., C.M. Macleod and M. McCarty. 1979. Studies on the chemical nature of the substance inducing transformation of *Pneumococcal* Types. Inductions of Transformation by a Desoxyribonucleic Acid Fraction Isolated from *Pneumococcus* Type III. J Exp Med 149(2): 297–326.

Azli, B., M.N. Razak, A.R. Omar, N.A. Mohd Zain, F. Abdul Razak and I. Nurulfiza. 2022. Metagenomics insights into the microbial diversity and microbiome network analysis on the heterogeneity of influent to effluent water. Front Microbiol 13: 715.

Bak R.O., N. Gomez-Ospina, M.H. Porteus. 2018.Gene editing on center stage. Trends Genet. 34(8): 600–611.

Bleuven, C. and C.R. Landry. 2016. Molecular and cellular bases of adaptation to a changing environment in microorganisms. P. Roy Soc. B Biol. Sci. 283: 1841.

Brockhurst, M.A., N. Colegrave and D.E. Rozen. 2011. Next-generation sequencing as a tool to study microbial evolution. Mol. Ecol. 20(5): 972–980.

Callahan, B.J., T. Fukami and D.S. Fisher. 2014. Rapid Evolution of adaptive niche construction in experimental microbial populations. Evolution 68(11): 3307–3316.

Campillo-Balderas, J.A., A. Lazcano and A. Becerra. 2015. Viral genome size distribution does not correlate with the antiquity of the host lineages. Front Ecol. Evol. 3: 143.

Cohen, S.N., A.C. Chang, H.W. Boyer and R.B. Helling. 1973. Construction of biologically functional bacterial plasmids *in vitro*. Proc. Natl. Acad. Sci. U S A 70(11): 3240–4.

Crick, F.H.C., L. Barnett, S. Brenner and R.J. Watts-Tobin. 1961. General nature of the genetic code for proteins. Nature 192(4809): 1227–1232.

Dai, W-J., L-Y. Zhu, Z-Y. Yan, Y. Xu, Q-L. Wang and X-J Lu. 2016. CRISPR-Cas9 For I*n Vivo* gene therapy: Promise and hurdles. Mol Ther Nucleic Acids. 5: e349.

Das, S., S. Bano, P. Kapse and G.C. Kundu. 2022. CRISPR based therapeutics: A new paradigm in cancer precision medicine. Mol Cancer. 21: 85.

Delsuc, F., H. Brinkmann and H. Philippe. 2005. Phylogenomics and the reconstruction of the tree of life. Nat. Rev. Genet. 6(5): 361–375.

del Duca, S., A. Vassallo, A. Mengoni and R. Fani. 2022. Microbial genetics and evolution. Microorganisms 10: 1274.

Donohoue, P.D., R. Barrangou, and A.P. May. 2018. Advances in industrial biotechnology using CRISPR-Cas systems. Trends Biot. 36(2): 134–146.

Dow, L.E. 2015. Modeling disease *in vivo* with CRISPR/Cas9. Trends Mol. Med. 21(10): 609–621.

Dziewit, L., J. Czarnecki, D. Wibberg, M. Radlinska, P. Mrozek, M. Szymczak et al. 2014. Architecture and functions of a multipartite genome of the methylotrophic bacterium *Paracoccus aminophilus* JCM 7686, containing primary and secondary chromids. BMC Genomics 15(1): 1–16.

Feng, G., T. Xie, X. Wang, J. Bai, L. Tang, H. Zhao et al. 2018. Metagenomic analysis of microbial community and function involved in Cd-Contaminated Soil. BMC Microbiol 18(1): 1–13.

Focosi, D. and F. Maggi. 2022. Recombination in Coronaviruses, with a Focus on SARS-CoV-2. Viruses 14(6): 1239. https://doi.org/10.3390/v14061239.

Forde, B.M. and P.W. O'Toole. 2013. Next-generation sequencing technologies and their impact on microbial genomics. Brief. Func. Genomics 12(5): 440–453.

Galagan, J.E., M.R. Henn, L.J. Ma, C.A. Cuomo and B. Birren. 2005. Genomics of the fungal kingdom: Insights into eukaryotic biology. Genome Res. 15(12): 1620–1631.

Gammuto, L., C. Chiellini, M. Iozzo, R. Fani and G. Petroni. 2022. The azurin coding gene: Origin and phylogenetic distribution. Microorganisms 10(1): 9.

Good, B.H. and O. Hallatschek. 2018. Effective models and the search for quantitative principles in microbial evolution. Curr. Opin. Microbiol. 45: 203–212.

Gottardi, M., J.D. Knudsen, L. Prado, M. Oreb, P. Branduardi and E. Boles. 2017. *De novo* biosynthesis of trans-cinnamic acid derivatives in *Saccharomyces cerevisiae*. Appl. Microbiol. Biot. 101(12): 4883–4893.

Gupta, A., R. Gupta, R.L. Singh, A. Gupta, R. Gupta and R.L. Singh. 2017. Microbes and environment. Principles and Applications of Environmental Biotechnology for a Sustainable Future, 43–84.

Horvath, P., D.A. Romero, A.C. Coûté-Monvoisin, M. Richards, H. Deveau, S. Moineau et al. 2008. Diversity, activity, and evolution of CRISPR Loci In *Streptococcus thermophilus*. J. Bacteriol. 190: 1401–1412.

Horvath, P., A.C. Coûté-Monvoisin, D.A. Romero, P. Boyaval, C. Fremaux and R. Barrangou. 2009. Comparative analysis of CRISPR loci in lactic acid bacteria genomes. Int. J. Food Microbiol. 131: 62–70.

Hsu, P.D., E.S. Lander and F. Zhang. 2014. Development and applications of CRISPR-Cas9 for genome engineering. Cell. 157(6): 1262–1278.

Jacob, F. and J. Monod. 1961. Genetic regulatory mechanisms in the synthesis of proteins. J. Mol. Biol. 3(3): 318–356.

Joyce, A.R. and B. Palsson. 2006. The model organism as a system: Integrating "omics" Data Sets. Nat. Rev. Mol. Cell Biol. 7(3): 198–210.

Kumar, A. and J.P. Verma. 2019. The role of microbes to improve crop productivity and soil health. Ecological Wisdom Inspired Restoration Engineering: 249–265.

Kwok, K.T.T., D.F. Nieuwenhuijse, M.V.T. Phan and M.P.G. Koopmans. 2020. Virus metagenomics in farm animals: A systematic review. Viruses 12: 107.

Liu, Y. and J. Nielsen. 2019. Recent trends in metabolic engineering of microbial chemical factories. Curr. Opin. Biotechnol. 60: 188–197.

Lopes, M.J.S., M.B. Dias-Filho and E.S.C. Gurgel. 2021. Successful plant growth-promoting microbes: Inoculation methods and abiotic factors. Front. Sustain. Food Syst. 5: 606454.

Mandal, D.D. and S. Majumdar. 2023. Bacteria as biofactory of pigments: evolution beyond therapeutics and biotechnological advancements. J. Biosci. Bioeng. 135(5): 349–358.

Meena, M., P. Swapnil, A. Zehra, M. Aamir, M.K. Dubey, C.B. Patel et al. 2019. Virulence Factors and Their Associated Genes in Microbes. New and Future Developments in Microbial Biotechnology and Bioengineering: Microbial Genes Biochemistry and Applications: 181–208.

Mitchell, A., G.H. Romano, B. Groisman, A. Yona, E. Dekel, M. Kupiec et al. 2009. Adaptive prediction of environmental changes by microorganisms. Nature 460(7252): 220–224.

Montecillo, J.A.V., L.L. Chu and H. Bae. 2022. CRISPR-Cas9 system for plant genome editing: Current approaches and emerging developments. Agronomy 10: 1033.

Nishiga, M., C. Liu, L.S. Qi and J.C. Wu. 2022. The use of new CRISPR tools in cardiovascular research and medicine. Nat. Rev. Cardiol. 19: 505–521.

O'Carroll, I.P. and A. Rein. 2016. Viral nucleic acids. Enc. Cell Biol. 1: 517–524.

Parisutham, V., T.H. Kim and S.K. Lee. 2014. Feasibilities of consolidated bioprocessing microbes: From pretreatment to biofuel production. Bioresource Technol. 161: 431–440.

Patel, A.K., R.R. Singhania and A. Pandey. 2017. Production, purification, and application of microbial enzymes. Biotechnology of Microbial Enzymes: Production, Biocatalysis and Industrial Applications, 13–41.

Philippe, H., F. Delsuc, H. Brinkmann and N. Lartillot. 2005. Phylogenomics. Annu. Rev. Ecol., Evol. S. 36: 541–562.

Quiberoni, A., S. Moineau, G.M. Rousseau, J. Reinheimer and H.W. Ackermann. 2010. *Streptococcus thermophilus* Bacteriophages. Int. Dairy J 20: 657–664.

Ram, S., M. Mitra, F. Shah, S.R. Tirkey and S. Mishra. 2020. Bacteria as an alternate biofactory for carotenoid production: A review of its applications, opportunities and challenges. J. Funct. Foods. 67: 103867.

Russell, N.J. 1990. Cold adaptation of microorganisms. Philos. Trans. R. Soc. Lond B. Biol. Sci. 326(1237): 595–611.

Sarkar, D., R. Hansdah, A. Kar and A. Sarkar. 2022. Microbial bioprospecting in development of integrated biomass based biorefineries. Bioprospecting of Microbial Diversity: Challenges and Applications in Biochemical Industry, Agriculture and Environment Protection, 257–275.

Sicheritz-Pontén, T. and S.G.E. Andersson. 2001. A phylogenomic approach to microbial evolution. Nucleic Acids Res. 29(2): 545–552.

Stajich, J.E. 2017. Fungal genomes and Insights into the evolution of the Kingdom. The Fungal Kingdom 5: 619–633.

Stephanopoulos, G. 2007. Challenges in engineering Microbes for Biofuels Production. Science 315(5813): 801–804.

Thakur, A., H. Mikkelsen and G. Jungersen. 2019. Intracellular pathogens: Host immunity and microbial persistence strategies. J. Immunol. Res. 1356540.

Thodey, K., S. Galanie and C. Smolke. 2014. A microbial biomanufacturing platform for natural and semisynthetic opioids. Nat. Chem. Biol. 10: 837–844.

Trenchard, I.J., M.S. Siddiqui, K. Thodey and C.D. Smolke. 2015. *De novo* production of the key branch point benzylisoquinoline alkaloid reticuline in yeast. Metab. Eng. 31: 74–83.

Truong, D.T., A. Tett, E. Pasolli, C. Huttenhower and N. Segata. 2017. Microbial strain-level population structure and genetic diversity from metagenomes. Genome Res. 27(4): 626–638.

Wada, N., R. Ueta, Y. Osakabe and K. Osakabe. 2022. Precision genome editing in plants: state-of-the-art in CRISPR/Cas9-based genome engineering. BMC Plant Biol. 20: 234.

Woese, C.R. 1994. There must be a prokaryote somewhere: microbiology's search for itself. Microbiol. Rev. 58(1): 1326(1237): 595–6119.

Xie, D. 2022. Continuous biomanufacturing with microbes—upstream progresses and challenges. Cur. Op. Biot. 78: 102793.

Yan, Y., J. Yang, Y. Dou, M. Chen, S. Ping, J. Peng et al. 2008. Nitrogen fixation island and rhizosphere competence traits in the genome of root-associated *Pseudomonas stutzeri* A1501. P. Natl. Acad. Sci. USA 105(21): 7564–7569.

Yilmaz, L.S. and A.J. Walhout. 2017. Metabolic network modeling with model organisms. Curr. Opin. Chem. Biol. 36: 32–39.

Zhu, C., Z. Ji, J. Ma, Z. Ding, J. Shen and Q. Wang. 2021. Recent advances of nanotechnology-facilitated bacteria-based drug and gene delivery systems for cancer treatment. Pharmaceutics. 13(7): 940.

CHAPTER 2

Genetic Organization of Microbial Genomes

Sylwia Okoń

Introduction

The genome is the complete genetic information of a living organism or virus. In eukaryotes, the term usually refers to the genetic material contained in a basic, single (haploid) set of chromosomes. In the case of prokaryotes, it refers to the single chromosome they contain, and in the case of viruses, to their genetic material molecule.

The genome is the set of all genes and other DNA sequences. It is all of the genetic material that an organism possesses. Each genome is a resource of information that is needed to build an organism, ensure its development and growth. The diversity of the world of microorganisms is related to the structure of their organisms, the living environment and the way they live, which means that their genomes are organized in a variety of ways.

The dynamic development of genomics, in particular related to the development of high-throughput nucleic acid sequencing techniques, provides more and more data on the structure of the genomes of microorganisms. This allows us to better understand the structure and organization of the genomes of different groups of microorganisms and shows the extraordinary diversity in the organization of genetic information. Analyses of metagenomes allowed to discover new, so far unidentified groups of microorganisms and expand our knowledge about the diversity of their genomes (Koonin et al. 2021). Viruses are the smallest group of particles belonging to the world of microorganisms. They show the greatest diversity in the organization of genetic material depending on different types of hosts. Their genetic material can be both DNA and RNA, which takes a single or double-stranded form. It can be composed of one or more nucleic acid molecules (Campillo-Balderas et al. 2015, O'Carroll and Rein 2016). Bacteria also show great variation in the structure and organization of their genomes. Their genetic material is DNA, but it can be organized in various ways. We observe diversity both in terms of the size of genomes, their structure and the type of genetic information carried (Dziewit et al. 2014). Microscopic fungi are representatives of eukaryotic organisms. These microorganisms have genomes organized into chromosomes and their genetic material is DNA. Mainly, the diversity of genomes relates to the different numbers of chromosomes and the size of the genetic material carrying the encoded information (Galagan et al. 2005, Stajich 2017).

This chapter presents and summarizes the variations in the organization of the genomes of different groups of microorganisms. It shows the enormous diversity of the genome structure of the most diverse group of organisms on earth.

Institute of Plant Genetics, Breeding and Biotechnology, University of Life Sciences in Lublin, Akademicka 15, 20–950 Lublin.
Email: sylwia.okon@up.lublin.pl

Viruses and Viroids

Viruses are small, non-viable infectious particles, obligate parasites that reproduce only within the host cell and must subjugate at least part of the host's genetic machinery in order to replicate and express their genes. A single virus particle called a virion, consists of a protein coat (capsid) and a nucleic acid that constitutes the genome.

Viral genomes are grouped according to the type and structure of nucleic acids. They can be in the form of double-stranded DNA (dsDNA), single-stranded DNA (ssDNA), double-stranded RNA (dsRNA) or single-stranded RNA (ssRNA). In most viruses, the genome is a single DNA or RNA molecule in a circular or linear form. Some viruses have split genomes, meaning that their genes are carried by several different molecules (Table 1).

The size of viral genomes is very diverse and depends on the type of viral nucleic acid and the type of host cell. The genome of DNA viruses is generally larger than that of RNA viruses. In addition, dsDNA viruses have a larger genome compared to ssDNA viruses. Similarly, dsRNA viruses have larger genomes than ssRNA viruses, which is associated with greater stability of double-stranded structures and fragility of RNA particles, especially in the single-stranded form (Campillo-Balderas et al. 2015, Chaitanya 2019).

The development of sequencing techniques allowed for a better understanding of the sequence of entire genomes of viruses, which confirmed the great diversity in the size and structure of the genomes of different groups of viruses. Studies have shown that the genome size of DNA viruses ranges from 1.8 kb (*Circoviridae*) to 2500 kb (*Poxiviridae*). The genome length of RNA viruses ranges from 1.7kb (*Deltavirus*) to 32kb (*Coronaviridae*). In addition, the size of genomes within individual groups may vary from a few to even several dozen kb (Campillo-Balderas et al. 2015).

dsDNA viruses are the most diverse in terms of size and genome complexity, ranging from 5kbp in Polyomaviruses typically to 130–365kbp in Poxviruses. Relatively recently, dsDNA viruses with much larger genomes have been identified and are currently being studied by many scientists (Guglielmini et al. 2019). The size of their genomes reaches up to 2.5Mbp (*Pandoraviruses*) (Philippe et al. 2013). The genomes of dsDNA viruses can be linear (*Poxiviridae*) or circular (*Polyomaviridae*). dsDNA can be in the form of a single molecule (Adenoviridae) or in the form of several segments (*Polydnaviridae*) (O'Carroll and Rein 2016).

Table 1. Types and structures of viral nucleic acids (Modrow et al. 2013, Campillo-Balderas et al. 2015, O'Carroll and Rein 2016, Fermin 2018, Malathi and Deví 2019).

Nucleid acid	Genome structure	Segments	Genome length	Example
dsDNA	Linear	1	15–2500 kb	*Poxiviridae*
	Linear	10	150–200 kb	*Polydnaviridae*
	Circular	1	4.5–300 kb	*Polyomaviridae*
ssDNA(-)	Circular	1	1.6–3.9 kb	*Anelloviruses*
ssDNA(+)	Circular	2–8	3–9 kb	*Nanoviridae*
	Circular	1	1.8–2.9 kb	*Circoviridae*
	Linear	1	4–6 kb	*Parvoviridae*
	Linear	2	12.5 kb	*Bidnaviridae*
dsRNA	Linear	1–12	18–32 kb	*Reoviridae*
	Linear	1	4.6–7 kb	*Totiviridae*
ssRNA (+)	Linear	1	4–9.7 kb	*Picornaviridae*
	Linear	2–3	6.3–12 kb	*Virgaviridae*
ssRNA(-)	Linear	1	11–19 kb	*Paramyxoviridae*
	Linear	2–8	11–25 kb	*Orthomyxoviridae*
	Circular	1	1.7 kb	*Deltavirus*
Ambisense RNA	Linear	2–3	11–19 kb	*Arenaviridae*

ssDNA viruses are a group of the smallest and simplest viruses, whose genome size is 2–6 kb. They encode only single structural proteins and proteins involved in the replication of their DNA. Some ssDNA viruses have a more complex structure. For example, the *Nanoviridae* genome consists of 6–8 spherical individually packed segments of about 1 kb each, encoding 5–7 proteins (Thomas et al. 2021). In *Pleolipoviridae*, the genome is in the form of an ssDNA spherical particle of about 10 kb. In contrast, in *Bidnaviridae*, the genome consists of two segments of about 6kb and 6.5 kb packed separately (Fermin 2018).

Among the ssDNA viruses, there are viruses whose DNA molecule has the positive (+), negative (-) and mixed (+/-) polarity. The only family of ssDNA(-) viruses is the *Anelloviridae* family. The genome, represented by a single circular ssDNA molecule of negative polarity, contains four potential ORFs. Proteins are expressed by alternative splicing of a single premRNA (Fermin 2018, Malathi and Devi 2019).

The ssDNA(+) viruses are characterized by a circular genome (de Sales Lima et al. 2015), usually in the form of a single molecule; in *Nanoviruses*, the genome is in the form of 6–8 separate packed segments, each of which encodes a single protein (Gronenborn 2004).

In the case of ssDNA(+/-) viruses, the genome is ambisense. This means that within the same or different molecules, the genome consists of DNA sequences coding in opposite directions. Most of these viruses have a single-piece genome, but *Geminiviruses* may have single-piece or bi-part genomes (Sun et al. 2020). The genome is circular, except for members of the animal virus family *Parvoviridae*, which present a linear genome (Fermin 2018, Pénzes et al. 2020).

dsRNA viruses form a large and diverse group of viruses. The length of their genomes ranges from 3.7 kb (*Cystoviridae*) to about 32 kb (*Reoviridae*). The dsRNA molecule has a linear form, very often consisting of 1–12 segments packed together in a virion. Each of these segments encodes information for one, sometimes two proteins (Fermin 2018, Arnold et al. 2021).

ssRNA viruses are viroids classified into two families; *Avsunviroidae* (four species) and *Pospiviroidae* (28 species). All infect only plant hosts. Viroids have a circular genome that is not surrounded by a protein coat. Biologically, they behave like viruses, but their genomes do not encode proteins. Viroid replication depends on host proteins (via a rolling circle model) (Sanger et al. 1976, Flores et al. 2004).

Most (+) ssRNA viruses have small genomes of 4–6 kb, but there are viruses whose genome is much larger, reaching 27–32 kb. The genome of these viruses takes a linear form and occurs in the form of a single molecule or in the form of several segments (Sawicki et al. 2007, Comas-Garcia 2019). Positive-sense RNA viruses, thanks to the appropriate polarization of the RNA molecule, are immediately translated, giving rise to proteins encoded in the viral genome, and also serve as templates for generating new viral genome particles (Modrow et al. 2013).

The genome of (-)ssRNA viruses is mostly linear, but there are viruses whose genome is circular (*Deltavirus*). It is made of one or more RNA segments. These segments encode 1 or 2 protein molecules. The size of the genomes of this group of viruses ranges from about 11 kb to about 25 kb. The smallest viruses belonging to this group, *Daltavirus*, have a much smaller genome of about 1.7 kb (Cho et al. 2013, de la Peña and Gago-Zachert 2022).

Several RNA viruses have an ambisense ssRNA (+/-) genome. That is, the genome contains all genetic information encoded in segments in sense or antisense orientation. Two families (*Tospoviridae* and *Arenaviridae*) and the genus *Phlebovirus* from the family *Phenuiviridae* have 57 species with ssRNA(+/-) genomes. They all form enveloped, spherical virions that surround bi- or tripartite linear genomes (Fermin 2018).

The number of genes encoded by viral genomes is also diverse and ranges from several to several hundred. This diversity is mainly due to the type of virus nucleic acid and the structure of the virus particle, e.g., head and tail phages have over 200 genes equal to proteins.

Bacteria

Bacterial genomes are very diverse both in terms of size, organization and type of genetic information carried. The genetic material of bacteria is DNA. The total DNA contained in a cell is called a replicon. A bacterial replicon can be organized in different ways depending on the bacterial species. Generally, it is divided into the primary replicon, which is the bacterial chromosome, and the secondary replicons, which include the "second chromosome", chromids, megaplasmids and plasmids (Figure 1) (diCenzo and Finan 2017).

The bacterial chromosome is the largest DNA molecule containing housekeeping genes necessary for the proper functioning of the bacterial cell. The genomes of about 59.5% of bacterial species do not contain secondary replicons, and their entire genetic material is represented by a

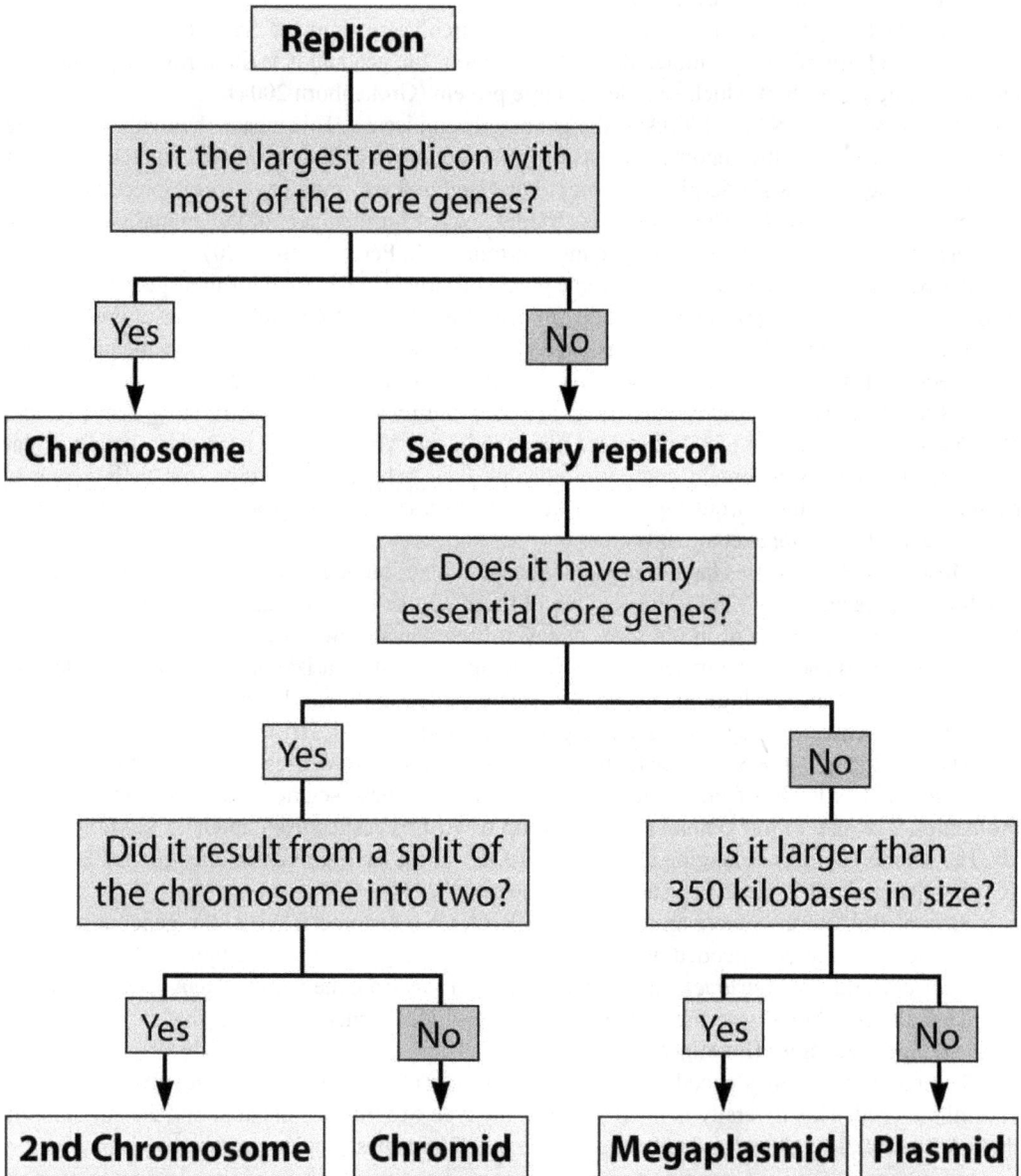

Figure 1. Classification of bacterial replicons. This flow chart illustrates the decisions involved in the classification of bacterial replicons. Reprinted from diCenzo and Finan (2017), open access, under a Creative Commons CC BY 4.0 license.

chromosome. However, complex genomes have been identified in many bacteria. In the case of some species of bacteria, the chromosome constitutes about 50% of the total DNA of the cell, the remaining DNA is contained in secondary replicons (Galibert et al. 2001, Chain et al. 2006). About 11% of bacterial species (whose genomes have been sequenced) have housekeeping genes located in various secondary replicons.

Secondary replicons found in a bacterial cell can be classified and divided depending on whether they encode basic genes related to the proper functioning of cells. Replicons that play a significant role in the functioning of cells with housekeeping genes include the second chromosome and the chromid (Harrison et al. 2010, diCenzo and Finan 2017). The second chromosome evolved as a result of the ancestral chromosome splitting into two replicons. It is extremely rare, however, that accurate finds related to sequencing and recording of genomes allow it to be distinguished from chromids. The second chromosome shows high synteny with the chromosomes of related species, and the distribution of housekeeping genes between the primary chromosome and the second chromosome is random. Second chromosomes have been identified in Salmonella enterica *Nocardia farcinica* cells. They are 0.73 Mb and 2.66 Mb, respectively (diCenzo and Finan 2017).

Chromids are structures formed from the combination of a chromosome fragment and a plasmid (Harrison et al. 2010). These replicons contain at least one housekeeping gene necessary for the proper functioning of the cell; however, their replication system is similar to the replication of plasmids. Some chromids carry only single essential genes (*Synorzibium meliloti* - chromid pSymB carries two genes) while others contain large DNA segments of chromosomal origin (*Paracoccus denitrificans* PD1222). Chromids, due to the coding of housekeeping genes, cannot be removed from bacterial cells due to the lethal effect. However, this "indispensability" condition raises some controversy. Some genes, which are considered essential in some bacteria, may have minor functions in others, and conversely, genes considered non-essential in some hosts may play key roles in housekeeping in others. In addition, some chromids may not meet the criterion of essentiality due to the fact that they carry housekeeping genes that are present in cells in several copies located in different replicons. The observations led to the division of chromids into two groups: primary chromids (chromids sensu stricto), which are obligatory because they contain at least one housekeeping gene that occurs in one copy in the genome. The second group consists of secondary chromids (chromids sensu lato), which are facultatively necessary, so they can be removed from the cell without causing a lethal effect (Petersen et al. 2013, Dziewit et al. 2014).

Plasmids and megaplasmids are secondary replicons that do not encode genes involved in housekeeping. However, they play an important role in bacterial evolution by allowing exogenous DNA to be introduced into cells by horizontal gene transfer. These replicons often determine various phenotypic features that enable bacteria to quickly adapt to changing environmental conditions and provide them with an advantage in competition with other microorganisms inhabiting the same ecological niche. Their division is determined by the size of the replicon- the average size of a plasmid is about 78.9 kb, and of a megaplasmid about 772 kb (Harrison et al. 2010, diCenzo and Finan 2017).

Globally, the bacterial chromosome adopts a conformation that depends on the size and shape of the cell. In terms of physical organization, the genome of a typical bacterial cell is a circular DNA molecule located in a nucleoid. The DNA double helix undergoes supercoiling, which allows the chromatin to be packed tightly together. According to the commonly accepted model, DNA in a bacterial cell is bound to a protein core from which about 40–50 supercoiled loops radiate. Each loop contains about 100 kb of DNA. Such a model of bacterial chromosome construction was developed on the basis of research on *E. coli,* and it is present in most species of bacteria.

In some species of bacteria, e.g., *Borelia*, the chromosome occurs in a linear form (Baril et al. 1989). These chromosomes evolved from circular chromosomes as a result of their breakage. At their ends, covalent bonds are formed between the 5' and 3' ends of the polynucleotides in the DNA double helix. Binding proteins at the ends of the DNA double helix have been identified in *Streptomyces* (Volff and Altenbuchner 2000).

Some bacteria have so-called multipart genomes, i.e., genomes that are divided into 2 or more pieces of DNA (White et al. 1999). In these genomes, apart from the primary replicon which is the chromosome, there are secondary replicons. Different species of bacteria represent a different level of organization and a different structure of multipart genomes, e.g., *Vibrio cholerae* bacteria have one main chromosome and a megaplasmid (Heidelberg et al. 2000).

The bacterial genome is very compact, e.g., in *E. coli*, non-coding DNA is only about 11% of the total and is scattered in the genome in small sections (Blattner et al. 1997). There is a theory that such a compact organization enables rapid replication, but this hypothesis has never been supported by reliable research. A characteristic feature of the bacterial genome is the lack of introns in the genes. Some species, e.g., *E. coli*, do not have discontinuous genes at all. A second characteristic of bacterial genomes is the low frequency of repetitive sequences. Some species have certain sequences that may be repeated in different parts of the genome. An example is insertion sequences (IS) having the ability to transpose between replicons (Hawkey et al. 2015).

One of the characteristics of bacterial genomes is the presence of specific structural and functional units called operons. An operon is a group of structural genes separated by one or two nucleotides and operator and promoter sequences that control their transcriptional activity. All genes in an operon are expressed collectively as a unit. In most cases, the genes in an operon are functionally related, encoding a set of proteins that are involved in a single biochemical activity, such as the use of sugar as an energy source (lactose operon) or the synthesis of an amino acid (tryptophan operon) (Osbourn and Field 2009, Koonin 2009).

In some species of bacteria, e.g., *Aquifex aeolicus*, operons contain genes that often have no biochemical connections. For example, one operon of this species contains six genes that encode proteins involved in DNA recombination, an enzyme involved in protein synthesis, a protein necessary for cell motility, an enzyme involved in nucleotide biosynthesis, and an enzyme involved in lipid synthesis (Swanson 2001).

The average sizes of bacterial genomes range from 0.16 to 13 Mb (Moran 2002, Batut et al. 2014). In bacteria, a strong correlation was observed between the size of the genome and the number of functional genes, it is assumed that there are about 950 genes per 1 Mb on average. The number of genes is therefore varied and reflects the nature of the ecological niches in which the bacteria live. The largest genomes belong to free-living species. These bacteria live in an environment that provides a wide range of physical and biological conditions to which the genomes of these species must respond (Qin et al. 2019). The smallest genomes belong to those bacterial species that are obligate parasites, e.g., *Mycoplasma genitalium*, whose genome is 0.58 Mb and contains 470 genes (Butt et al. 2012). The small number of genes means that these species are unable to synthesize many of the compounds they must obtain from their hosts, while the small size of the genome allows them to multiply quickly.

Archaea

Archaea are small, single-celled organisms initially classified as bacteria. However, the development of molecular techniques, including sequencing, allowed them to be clearly classified as a separate systematic group (Allers et al. 2005, Pace 2006).

Archaea differ from bacteria in many caches, including organization, size, and complexity of genetic material. However, there is very little information about the structure of archaeal genomes due to the relatively low number of cultivated archaeal species, their small size and often extreme growth conditions. Studies of archaeal chromosome biology have focused mainly on members of the euryarchaeal and crenarchaeal lineages and have shown that they typically possess multiple copies of their genomes, a condition known as oligoploidy. For example, *Methanococcus maripaludis* possesses up to 55 copies of the genome and *Haloferax volcanii* possesses ~20 copies. This is probably related to the extreme living conditions of these microorganisms (Hildenbrand et al. 2011, Zerulla et al. 2014).

The genetic material of archaea is circular DNA in the form of a single chromosome, the core of which is formed by histone proteins. It ranges in size from 0.5 to 5.8 Mbp. The archaeal genome smaller than 1 Mbp is found in a symbiont that derives nutrients from a host, and the small size of this genome reflects the loss of unnecessary genes (Galagan et al. 2001, Waters et al. 2003, Allers et al. 2005, French et al. 2007, Bell 2022).

Archaeal genomes resemble bacterial genomes in terms of the number, length and density of genes.

Archaeal genes average about 1 kbp in length and are either adjacent to neighbouring genes or separated by less than about 200 bp; this results in a high gene density with minimal non-coding regions. Only a few archaeal protein-coding genes have introns, and these introns are short. It is assumed that about 15% of proteins encoded by archaeal genes are unique to this group of microorganisms; however, their function is not yet fully understood (Graham et al. 2000).

In addition to the main chromosome, Archaea also have plasmids. So far, about 60 plasmids have been isolated. We can divide them into four groups: sulfolobus, haloarchaeal, thermococcal, and methanogenic, depending on the group of archaea from which they were isolated. They encode proteins that allow archaea to survive in extreme environments (Wang et al. 2015).

Fungi

Fungi are eukaryotic organisms. The organization of the genome is typical of all eukaryotes. Their genetic material is a linear DNA molecule forming complexes with proteins. The double helix of DNA, 140–150 bp long, is wound on a set of 8 positively charged core proteins called histones. The core is formed by histones H2A, H2B, H3 and H4. The DNA helix together with the core proteins form complexes called nucleosomes. The tight binding of proteins to DNA is possible due to the negatively charged phosphate groups in the DNA. Nucleosomes are separated from each other by DNA segments of 50–70 bp. In addition to the core octamer proteins, the so-called connecting histones (H1) are attached to the nucleosome, the interaction of which allows the formation of more compact structures called a solenoid. Further condensation of chromatin fibres leads to the formation of chromosomes. Each eukaryotic species has a certain number of DNA molecules organized into chromosomes located in the cell nucleus. All eukaryotic organisms studied so far have at least two chromosomes, and DNA molecules are always linear. The only variation at this level of eukaryotic genome structure is in the number of chromosomes (Saitou 2013).

There is a large variation in the number of chromosomes in fungi. Their number varies in the haploid system, from 3 in *Schizosaccharomyces pombe* to 23 in *Ustilago maydis* (Zolan 1995, Wood et al. 2002, Basse and Steinberg 2004).

Fungal genomes are the smallest eukaryotic genomes. Their size varies from 2 Mb in *Microsporidia* to 2 Gb in *Pucciniales*, the average size of the genome is about 35 Mb (Stajich 2017). The free-living yeasts in the *Ascomycota* and the *Cryptomycota* parasite *Rozella* typically have genomes in the 7 to 12 Mb range, and the *Basidiomycota Cryptococcus* yeasts are around 20 Mb (Goffeau et al. 1996, Loftus et al. 2005, Pombert et al. 2013, Toome et al. 2014).

At the upper end of the range of currently sequenced species are the 150- to 175-Mb genomes of *Cenococcum geophilum* (*Dothidiomycetes; Ascomycota*), *Sphaerobolus stellatus* (*Agaricomyoctina; Basidiomycota*), *Tuber melanosporum* (*Pezizales; Ascomycota*), and *Blumeria graminis* (*Leotiomycetes; ascomycota*) (Dujon et al. 2004, Peter et al. 2016).

Genomes of *Entomophthoromycotina* (*Zoopagomycotina*) are also extremely large. The genome of *Entomophaga aulicae* is estimated to be as large as 8 Gb based on nuclear staining approaches (Spanu et al. 2010).

Genome size estimation using flow cytometry has indicated that many of the rust fungi (*Pucciniales; Pucciniomycotina*) also have large genomes with estimated sizes of 300 to 900 Mb (Tavares et al. 2014). Variation in genome size is mainly due to the length and number of introns, varying amounts of repetitive sequences, including satellite DNA, transposable elements (TEs), and ribosomal genes.

The genome of fungi is characterized by high complexity. For example, the *S. cerevisiae* genome is 12 Mb long and contains about 6,000 genes. This is the result of the economical use of space, the genes in the genomes of fungi lie close to each other, most of the genes have a continuous structure and are characterized by a smaller number of repeating sequences than other eukaryotic organisms.

Fungal genomes display coding densities ranging from 37% to 61% and, as with other sequenced eukaryotes, gene density is inversely correlated with genome size. The length of the coding sequence is on average 1.3 to 1.9 kb. The number and length of introns varies greatly in particular fungal species (Bon et al. 2003, Galagan et al. 2005). Intron densities in fungi range from 5–6 introns per gene in basidiomycetes such as *Cryptococcus neoformans* (Loftus et al. 2005) to an average of one to two introns per gene for many recently sequenced ascomycetes (e.g., *Neurospora, Magnaporthe*) (Galagan et al. 2005, Dean et al. 2005) to < 300 total introns in the hemiascomycete yeast *S. cerevisae* (Goffeau et al. 1996). Introns are usually short, averaging 80 to 150 bp in many ascomycetes, the average intron length in *P. fructicola* was 50 bp, in *Pneumocystis jirovecii* (45 bp) and *Pseudocercospora fijiensis* (45 bp), and in *S. cerevisiae* 111 bp (Wang et al. 2020).

DNA repetitive elements, microsatellites or simple repeats, minisatellites, mobile transposing elements at the DNA level, retrotransposons and their various derivatives are ubiquitous components of all fungal genomes. Transposable elements (TEs) represent a significant but under-studied fraction of eukaryotic genomes. They are mobile genetic units that multiply and expand to distant regions of the genome. TEs are divided into two classes based on their transposition mechanism. Class I elements are transposed via RNA intermediates and include five orders (LTR, DIRS, PLE, LINE, and SINE) that are differentiated based on their structure and transposition system. Class II includes elements that are directly transposed from DNA to DNA. Their role as strong regulatory elements, genomic parasites and near-neutral sequences is constantly under review (Wessler 2006, Wicker et al. 2007, Raffaele and Kamoun 2012, Castanera et al. 2016).

Fungal genomes are extremely variable in TE content, ranging from 0.02 to 29.8% of their genome. Comprehensive TE analyzes of fungi have described exceptional variation in repeat content where amplification events are more related to fungal lifestyle than to phylogenetic proximity (Kent et al. 2017, Muszewska et al. 2019).

Some fungal plant pathogens have genomes with a distinctly dualistic architecture described by the two-way model of evolution. The core of the genome is densely packed with ordering genes, while the lifestyle adaptive part contains effector and TE genes (Kapitonov and Jurka 2001). This genomic architecture has been described for versatile fungal pathogens, among them *Fusarium, Leptosphaeria* and *Verticillium* (Ma et al. 2010, Riley et al. 2014, Larray et al. 1999). The lifestyle-specific genome is expected to be enriched for TEs as they may play a role in host shifting and adaptation to new ecological niches (Ellinghaus et al. 2008, Castanera et al. 2016). Mobile elements have played a key role in shaping the architecture and gene content of fungal genomes. High levels of genetic diversity associated with repetitive sequences often characterize the genomes of pathogenic species and this diversity has been proposed to be adaptive (Neafsey et al. 2010, de Jonge et al. 2013, Manning et al. 2013, Stukenbrock and Croll 2014).

The 60 Mb genome of *F. oxysporum* f. sp. *lycopersici* is characterized by a high content of repetitive sequences (17%) and transposable elements (4%) (Ma et al. 2010). In comparison, the genome of the related species *F. graminearum* is noticeably smaller (36 Mb) with only 0.24% repetitive sequences and 0.03% transposable elements. Also in terms of genome structure, the two *Fusarium* species are fundamentally different: the genome of *F. oxysporum* f. sp. *lycopersici* consists of 14 chromosomes, while the genome of *F. graminearum* contains only 4. The genome of the pathogen *Blumeria graminis* was strongly influenced by the activity of mobile genetic elements (Spanu et al. 2010). The large genome (~120 Mb) of *B. graminis* consists of 64% repetitive sequences, including both transposable and retro elements. Interestingly, the expansion of the *B. graminis* genome was accompanied by massive gene loss, which reflects an adaptation to the obligate biotrophy of the pathogen. Adaptive gene loss was directly mediated by the activity of mobile elements in the fungal genome.

Changes in the chromosomal copy number (i.e., ploidy) constitute a major process shaping eukaryote genome evolution (Otto and Whitton 2000). Increases in ploidy can occur either through the duplication of the chromosomal set(s) of an organism (autopolyploidy) or by the hybridization between different species (allopolyploidy).

Some fungi exist as stable haploid, diploid, or polyploid cells, others change ploidy in response to environmental conditions and aneuploidy is also observed in novel environments or during periods of stress (Todd et al. 2017). Many fungi undergo ploidy changes when adapting to unfavourable or new environments. Some fungi exist as stable haploid, diploid, or polyploid (triploid, tetraploid) hyphae, while others change ploidy under certain conditions and return to their original level of ploidy under other conditions. In *S. cerevisiae*, tetraploid strains have been found in food processing strain collections and from clinical sources (Albertin et al. 2009, al Safadi et al. 2010). While in most cases the polyploids in *Saccharomyces* were likely generated by autopolyploids, the *S. pastorianus* used for lager brewing has a complex origin of hybrid speciation (Libkind et al. 2011). *S. pastorianus* arose from the union of *S. cerevisiae* and *S. eubayanus*. Strong selective pressure during domestication in the brewing process led to numerous chromosomal rearrangements in extant *S. pastorianus*. The human pathogen *C. neoformans* is generally haploid, but strains have been shown to undergo ploidy changes during infection, including aneuploidy and transitions from haploid to diploid, and even higher states of ploidy, such as during giant cell formation in the lungs (Zaragoza and Nielsen 2013).

Aneuploidy (an abnormal number of chromosomes) is sometimes seen in fungi exposed to new or stressful environments and due to a previous change in ploidy. An organism's ability to replicate and segregate its genome with high fidelity is essential for its survival and generation of future generations. Errors in replication or segregation can lead to a change in ploidy or chromosome number. Several fungi tend to carry unequal numbers of chromosome copies due to duplication or loss of individual chromosomes. Unequal chromosome copy numbers are generally created by defects in the separation of chromatids during meiotic or mitotic cell divisions. Aneuploidy is generally considered harmful to the organism as it possibly disrupts gene regulation due to an unbalanced number of gene copies, but also is a prevalent strategy in fungal adaptation (Tsai and Nelliat 2019). Rapid aneuploidization was observed in *C. albicans* as a response to changing environmental conditions (Sah et al. 2021, Bouchonville et al. 2009). Also, *Cryptococcus* isolates have been found to develop higher levels of drug resistance by duplicating specific chromosomes (Sionov et al. 2010, Zaragoza and Nielsen 2013).

Conclusions

Complete sequencing of microbial genomes has enabled comparisons of the dynamic genome size and gene content across a range of times in microbial evolution. The growing number of complete microbial genomes provides an unprecedented opportunity to study biology and evolution. The diverse structure and organization of the genomes of microorganisms confirm their great diversity and provide grounds for further research. More detailed information about genome size and organization is also useful in microbial taxonomy.

References

Albertin, W., P. Marullo, M. Aigle, A. Bourgais, M. Bely, C. Dillmann et al. 2009. Evidence for autotetraploidy associated with reproductive isolation in *Saccharomyces cerevisiae*: towards a new domesticated species. J. Evolution Biol. 22(11): 2157–2170.

Allers, T. and M. Mevarech. 2005. Archaeal genetics—the third way. Nat. Rev. Genet. 6: 58–73.

Arnold, M.M., A. van Dijk and S. López. 2021. Double-stranded RNA viruses. In Virology, 33–67. John Wiley and Sons, Ltd.

Baril, C., C. Richaud, G. Baranton and I. saint Girons. 1989. Linear chromosome of *Borrelia burgdorferi*. Res. Microbiol. 140(8): 507–516.

Basse, C.W. and G. Steinberg. 2004. Ustilago Maydis, model system for analysis of the molecular basis of fungal pathogenicity. Mol. Plant Pathol. 5(2):83–92.

Batut, B., C. Knibbe, G. Marais and V. Daubin. 2014. Reductive genome evolution at both ends of the bacterial population size spectrum. Nat. Rev. Microbiol. 12(12): 841–850.

Bell, S.D. 2022. Form and function of archaeal genomes. Biochem. Soc. 50(6): 1931–1939.

Blattner, F.R., G. Plunkett, C.A. Bloch, N.T. Perna, V. Burland, M. Riley et al. 1997. The complete genome sequence of *Escherichia coli* K-12. Science 277(5331): 1453–1462.

Bon, E., S. Casaregola, G. Blandin, B. Llorente, C. Neuvéglise, M. Munsterkotter et al. 2003. Molecular evolution of eukaryotic genomes: Hemiascomycetous yeast spliceosomal introns. Nucleic Acids Res 31(4): 1121–1135.

Bouchonville, K., A. Forche, K.E.S. Tang, A. Selmecki and J. Berman. 2009. Aneuploid chromosomes are highly unstable during DNA transformation of *Candida albicans*. Eukaryot Cell 8(10): 1554.

Butt, A.M., S. Tahir, I. Nasrullah, M. Idrees, J. Lu and Y. Tong. 2012. *Mycoplasma genitalium:* A comparative genomics study of metabolic pathways for the identification of drug and vaccine targets. Infection, Genet. Evol. 12(1): 53–62.

Campillo-Balderas, J.A., A. Lazcano and A. Becerra. 2015. Viral genome size distribution does not correlate with the antiquity of the host lineages. Front Ecol. Evol. 3: 143.

Castanera, R., L. López-Varas, A. Borgognone, K. LaButti, A. Lapidus, J. Schmutz et al. 2016. Transposable elements versus the fungal genome: Impact on whole-genome architecture and transcriptional profiles. PLOS Genet 12(6): e1006108.

Chain, P.S.G., V.J. Denef, K.T. Konstantinidis, L.M. Vergez, L. Agulló, V.L. Reyes et al. 2006. *Burkholderia xenovorans* LB400 Harbors a Multi-Replicon, 9.73-Mbp Genome Shaped for Versatility. P. Natl. A. Sci. USA 103(42): 15280–15287.

Chaitanya, K.V. 2019. Structure and organization of virus genomes. Genome. Genom. 18: 1–30.

Cho, W.K., S. Lian, S.M. Kim, S.H. Park and K.H. Kim. 2013. Current insights into research on rice stripe virus. Plant Pathol. J. 29(3): 223.

Comas-Garcia, M. 2019. Packaging of genomic RNA in positive-sense single-stranded RNA viruses: A complex story. Viruses 11(3): 253.

Dean, R.A., N.J. Talbot, D.J. Ebbole, M.L. Farman, T.K. Mitchell, M.J. Orbach et al. 2005. The genome sequence of the rice blast fungus *Magnaporthe grisea*. Nature 434(7036): 980–986.

diCenzo, G.C. and T.M. Finan. 2017. The divided bacterial genome: structure, function, and evolution. Microbiol. Mol. Biol. R. 81(3): e0001917

Dujon, B., D. Sherman, G. Fischer, P.D.- Nature and undefined 2004. 2004. Genome Evolution in Yeasts. Nature 430: 35–44.

Dziewit, L., J. Czarnecki, D. Wibberg, M. Radlinska, P. Mrozek, M. Szymczak et al. 2014. Architecture and functions of a multipartite genome of the methylotrophic bacterium *Paracoccus aminophilus* JCM 7686, Containing Primary and Secondary Chromids. BMC Genomics 15(1): 1–16.

Ellinghaus, D., S. Kurtz and U. Willhoeft. 2008. LTR harvest, an efficient and flexible software for *de Novo* detection of LTR retrotransposons. BMC Bioinformatics 9: 18.

Fermin, G. 2018. Virion structure, genome organization, and taxonomy of viruses. Viruses : 17–54.

Flores, R., S. Delgado, M.E. Gas, A. Carbonell, D. Molina, S. Gago et al. 2004. Viroids: The minimal non-coding RNAs with autonomous replication. FEBS Letters 567(1): 42–48.

French, S.L., T.J. Santangelo, A.L. Beyer and J.N. Reeve. 2007.Transcription and translation are coupled in archaea. Mol. Biol. Evol. 24: 893–895.

Galagan, J.E., M.R. Henn, L.J. Ma, C.A. Cuomo and B. Birren. 2005. Genomics of the fungal kingdom: Insights into eukaryotic biology. Genome Res. 15(12): 1620–1631.

Galagan, J.E., C. Nusbaum, A. Roy, M.G. Endrizzi, P. Macdonald, W. FitzHugh et al. 2001. The composite genome of the legume symbiont *Sinorhizobium Meliloti*. Science 293(5530): 668–672.

Goffeau, A., G. Barrell, H. Bussey, R.W. Davis, B. Dujon, H. Feldmann et al. 1996. Life with 6000 genes. Science 274(5287): 546–567.

Graham, D.E., R. Overbeek, G.J. Olsen and C.R. Woese. 2000. An archaeal genomic signature. P Natl. A. Sci. USA 97(7): 3304–08.

Gronenborn, B. 2004. Nanoviruses: genome organisation and protein function. Vet. Microbiol. 98(2): 103–109.

Guglielmini, J., A.C. Woo, M. Krupovic, P. Forterre and M. Gaia. 2019. Diversification of giant and large eukaryotic DsDNA viruses predated the origin of modern eukaryotes. P. Natl. A Sci. USA 116(39): 19585–19592.

Harrison, P.W., R.P.J. Lower, N.K.D. Kim and J.P.W. Young. 2010. Introducing the bacterial "Chromid": not a chromosome, not a plasmid. Trends Microbiol. 18(4): 141–148.

Hawkey, J., M. Hamidian, R.R. Wick, D.J. Edwards, H. Billman-Jacobe, R.M. Hall et al. 2015. ISMapper: Identifying transposase insertion sites in bacterial genomes from short read sequence data. BMC Genomics 16(1): 1–11.

Heidelberg, J.F., J.A. Elsen, W.C. Nelson, R.A. Clayton, M.L. Gwinn, R.J. Dodson et al. 2000. DNA sequence of both chromosomes of the cholera pathogen *Vibrio cholerae*. Nature 406(6795): 477–483.

Hildenbrand, C., T. Stock, C. Lange, M. Rother and J. Soppa. 2011. Genome copy numbers and gene conversion in methanogenic archaea. J. Bact. 193: 734–743.

de Jonge, R., M.D. Bolton, A. Kombrink, G.C.M. van den Berg, K.A. Yadeta and B.P.H.J. Thomma. 2013. Extensive chromosomal reshuffling drives evolution of virulence in an asexual pathogen. Genome Res. 23(8): 1271–1282.

Kapitonov, V.V. and J. Jurka. 2001. Rolling-circle transposons in eukaryotes. P Natl. A Sci. USA 98(15): 8714–8719.

Kent, T. V., J. Uzunović and S.I. Wright. 2017. Coevolution between transposable elements and recombination. Philos. T R Soc. B. 372: 20160458.

Koonin, E.V. 2009. Evolution of genome architecture. Int. J. of Biochem. Cell B 41(2): 298–306.

Koonin, E.V., K.S. Makarova and Y.I. Wolf. 2021. Evolution of microbial genomics: Conceptual shifts over a quarter century. Trends Microbiol. 29(7): 582–592.

de la Peña, M. and S. Gago-Zachert. 2022. A life of research on circular RNAs and Ribozymes: Towards the origin of viroids, deltaviruses and life. Virus Res. 314: 198757.

Larray, L.M., G. Pérez, M.M. Peñas, J.J.P. Baars, T.S.P. Mikosch, A.G. Pisabarro et al. 1999. Molecular karyotype of the white rot fungus *Pleurotus ostreatus*. Appl. Environ. Microb. 65(8): 3413–3417.

Libkind, D., C.T. Hittinger, E. Valeŕio, C. Gonçalves, J. Dover, M. Johnston et al. 2011. Microbe domestication and the identification of the wild genetic stock of lager-brewing yeast. P. Natl. A. Sci. USA 108(35): 14539–14544.

Loftus, B.J., E. Fung, P. Roncaglia, D. Rowley, P. Amedeo, D. Bruno et al. 2005. The genome of the basidiomycetous yeast and human pathogen *Cryptococcus neoformans*. Science 307(5713): 1321–1324.

Ma, L., H. van der Does, K. Borkovich, J. Coleman, M. Daboussi, A. Di Pietro et al. 2010. 2010. Comparative Genomics reveals mobile pathogenicity chromosomes in *Fusarium*. Nature 464: 367–373.

Malathi, V.G. and P.R. Devi. 2019. SsDNA viruses: key players in global virome. Virus Dis. 30(1): 3.

Manning, V.A., I. Pandelova, B. Dhillon, L.J. Wilhelm, S.B. Goodwin, A.M. Berlin et al. 2013. Comparative genomics of a plant-pathogenic fungus, *Pyrenophora tritici-repentis*, reveals transduplication and the impact of repeat elements on pathogenicity and population divergence. G3: Genes Genom. Genet. 3(1): 41–63.

Modrow, S., D. Falke, U. Truyen and H. Schätzl. 2013. Viruses with single-stranded, positive-sense RNA genomes. Mol. Virol. 12: 185–349.

Moran, N.A. 2002. Microbial minimalism: Genome reduction in bacterial pathogens. Cell 108(5): 583–586.

Muszewska, A., K. Steczkiewicz, M. Stepniewska-Dziubinska and K. Ginalski. 2019. Transposable elements contribute to fungal genes and impact fungal lifestyle. Sci. Rep. 9(1): 1–10.

Neafsey, D.E., B.M. Barker, T.J. Sharpton, J.E. Stajich, D.J. Park, E. Whiston et al. 2010. Population genomic sequencing of coccidioides fungi reveals recent hybridization and transposon control. Genome Res. 20(7): 938–946.

O'Carroll, I.P. and A. Rein. 2016. Viral Nucleic acids. Enc. Cell Biol. 1: 517–524.

Osbourn, A.E. and B. Field. 2009. Operons. Cell Mol. Life Sci. 66(23): 3755.

Otto, S.P. and J. Whitton. 2000. Polyploid Incidence and Evolution. Annu. Rev. Genet. 34: 401–437.

Pace, N. 2006. Time For A Change. Nature 441: 289.

Pénzes, J.J., M. Söderlund-Venermo, M. Canuti, A.M. Eis-Hübinger, J. Hughes, S.F. Cotmore et al. 2020. Reorganizing the family *Parvoviridae*: a revised taxonomy independent of the canonical approach based on host association. Arch. Virol. 165(9): 2133–2146.

Peter, M., A. Kohler, R.A. Ohm, A. Kuo, J. Krützmann, E. Morin et al. 2016. Ectomycorrhizal ecology is imprinted in the genome of the dominant symbiotic fungus *Cenococcum geophilum*. Nat. Commun. 7: 12662.

Petersen, J., O. Frank, M. Göker and S. Pradella. 2013. Extrachromosomal, extraordinary and essential–the plasmids of the Roseobacter Clade. Appl. Microbiol. Biotechnol. 97(7): 2805–2815.

Philippe, N., M. Legendre, G. Doutre, Y. Couté, O. Poirot, M. Lescot et al. 2013. *Pandoraviruses*: Amoeba viruses with genomes up to 2.5 Mb reaching that of parasitic eukaryotes. Science 341(6143): 281–286.

Pombert, J.-F., J. Xu, D.R. Smith, D. Heiman, S. Young, C.A. Cuomo et al. 2013. Complete genome sequences from three genetically distinct strains reveal high intraspecies genetic diversity in the microsporidian *Encephalitozoon cuniculi*. Am. Soc. Microbiol. 12(4): 503–511.

Qin, Q.L., Y. Li, L.L. Sun, Z. bin Wang, S. Wang, X.L. Chen et al. 2019. Trophic specialization results in genomic reduction in free-living marine Idiomarina bacteria. MBio 10(1): e0254518.

Raffaele, S. and S. Kamoun. 2012. Genome evolution in filamentous plant pathogens: Why bigger can be better. Nat. Rev. Microbiol. 10: 417–430.

Riley, R., A.A. Salamov, D.W. Brown, L.G. Nagy, D. Floudas, B.W. Held et al. 2014. Extensive sampling of *Basidiomycete* genomes demonstrates inadequacy of the white-rot/brown-rot paradigm for wood decay fungi. P. Natl. A. Sci. USA 111(27): 9923–9928.

al Safadi, R., M. Weiss-Gayet, J. Briolay and M. Aigle. 2010. A polyploid population of *Saccharomyces cerevisiae* with separate sexes (Dioecy). FEMS Yeast Res. 10(6): 757–768.

Sah, S.K., J.J. Hayes and E. Rustchenko. 2021. The role of aneuploidy in the emergence of Echinocandin resistance in human fungal pathogen *Candida albicans*. PLOS Pathog 17(5): e1009564.

Saitou, N. 2013. Eukaryote genomes. Introduction to Evolutionary Genomics 17: 193.

de Sales Lima, F.E., S.P. Cibulski, H.F. dos Santos, T.F. Teixeira, A.P.M. Varela, P.M. Roehe et al. 2015. Genomic characterization of novel circular ssDNA viruses from insectivorous bats in Southern Brazil. PLOS ONE 10(2): e0118070.

Sanger, H.L., G. Klotz, D. Riesner, H.J. Gross and A.K. Kleinschmidt. 1976. Viroids are single stranded covalently closed circular RNA molecules existing as highly base paired rod like structures. P. Natl. A. Sci. USA 73(11): 3852–3856.

Sawicki, S.G., D.L. Sawicki and S.G. Siddell. 2007. A contemporary view of coronavirus transcription. J. Virol. 81(1): 20.

Sionov, E., H. Lee, Y.C. Chang and K.J. Kwon-Chung. 2010. *Cryptococcus neoformans* overcomes stress of Azole drugs by formation of disomy in specific multiple chromosomes. PLOS Pathog 6(4): e1000848.

Spanu, P.D., J.C. Abbott, J. Amselem, T.A. Burgis, D.M. Soanes, K. Stüber et al. 2010. Genome expansion and gene loss in powdery mildew fungi reveal tradeoffs in extreme parasitism. Science 330(6010): 1543–1546.

Stajich, J.E. 2017. Fungal genomes and insights into the evolution of the Kingdom. The Fungal Kingdom 5: 619–633.

Stukenbrock, E.H. and D. Croll. 2014. The evolving fungal genome. Fungal Biol. Rev. 28(1): 1–12.

Sun, S., Y. Hu, G. Jiang, Y. Tian, M. Ding, C. Yu et al. 2020. Molecular characterization and genomic function of grapevine *Geminivirus* A. Front Microbiol. 11: 2121.

Swanson, R.V. 2001. Genome of *Aquifex aeolicus*. Method Enzymol. 330: 158–169.

Tavares, S., A.P. Ramos, A.S. Pires, H.G. Azinheira, P. Caldeirinha, T. Link et al. 2014. Genome size analyses of *Pucciniales* reveal the largest fungal genomes. Front Plant Sci. 5: 422.

Thomas, J.E., B. Gronenborn, R.M. Harding, B. Mandal, I. Grigoras, J.W. Randles et al. 2021. ICTV virus taxonomy profile: Nanoviridae. J. Gen. Virol. 102(3): 001544.

Todd, R.T., A. Forche and A. Selmecki. 2017. Ploidy variation in fungi–polyploidy, aneuploidy, and genome evolution. Microbiol Spectrum 5(4): 10.1128/microbiolspec.FUNK-0051–2016.

Toome, M., R.A. Ohm, R.W. Riley, T.Y. James, K.L. Lazarus, B. Henrissat et al. 2014. Genome sequencing provides insight into the reproductive biology, nutritional mode and ploidy of the fern pathogen *Mixia osmundae*. New Phytol. 202(2): 554–564.

Tsai, H.J. and A. Nelliat. 2019. A double-edged sword: Aneuploidy is a prevalent strategy in fungal adaptation. Genes 10(10): 787.

Volff, J.-N. and J. Altenbuchner. 2000. A new beginning with new ends: Linearisation of circular chromosomes during bacterial evolution. FEMS Microbiol. Let. 186(2): 143–150.

Wang, B., X. Liang, M.L. Gleason, T. Hsiang, R. Zhang and G. Sun. 2020. A chromosome-scale assembly of the smallest *Dothideomycete* genome reveals a unique genome compaction mechanism in filamentous fungi. BMC Genomics 21(1): 1–13.

Wang, H., N. Peng, S.A. Shah, L. Huang and Q. She 2015. Archaeal extrachromosomal genetic elements. Microbiol. Mol. Biol. Rev. 79(1): 117–152.

Waters, E., M.J. Hohn, I. Ahel, D.E. Graham, M.D. Adams, M. Barnstead et al. 2003. The genome of *Nanoarchaeum equitans*: Insights into early archaeal evolution and derived parasitism. P. Natl. A. Sci. USA. 100(22): 12984–88.

Wessler, S.R. 2006. Transposable elements and the evolution of eukaryotic genomes. P. Natl. A Sci. USA 103(47): 17600–17601.

White, O., J.A. Eisen, J.F. Heidelberg, E.K. Hickey, J.D. Peterson, R.J. Dodson et al. 1999. Genome sequence of the radioresistant bacterium *Deinococcus radiodurans* R1. Science 286(5444): 1571.

Wicker, T., F. Sabot, A. Hua-Van, J.L. Bennetzen, P. Capy, B. Chalhoub et al. 2007. A unified classification system for eukaryotic transposable elements. Nat. Rev. Genet. 8: 973–982.

Wood, V., R. Gwilliam, M.A. Rajandream, M. Lyne, R. Lyne, A. Stewart et al. 2002. The genome sequence of *Schizosaccharomyces pombe*. Nature 415: 871–880.

Zaragoza, O. and K. Nielsen. 2013. Titan cells in *Cryptococcus neoformans*: Cells with a giant impact. Curr. Opin. Microbiol. 16(4): 409–413.

Zerulla, K., S. Chimileski, D. Nather, U. Gophna, R.T. Papke and J. Soppa. 2014. DNA as a phosphate storage polymer and the alternative advantages of polyploidy for growth or survival. PLoS ONE 9: e94819.

Zolan, M.E. 1995. Chromosome-length polymorphism in Fungi. Microbiol. Rev. 59(4): 686–698.

CHAPTER 3

Basic Principles of Microbial Replication, Transcription and Translation

Aleksandra Nucia

Introduction

The biological processes of a cell result from the genetic information it possesses. One of the most important discoveries of modern biology is the universality of biological processes occurring in every living cell. This concerns both the biosynthesis of basic building compounds, such as amino acids, sugars, fatty acids or nucleotides, but also the construction of proteins, starches, fats and nucleic acids from these compounds. These processes are conserved in the evolutionary sequence and are common to all cells. Knowledge of the structure of DNA has enabled understanding of the detailed molecular mechanisms of these processes (Kitadai and Maruyama 2018).

One of the most important biological processes is the flow of genetic information. According to the central dogma of molecular biology, the flow of genetic information is from DNA to RNA to protein (Figure 1) (Crick 1970, Cobb 2017). Later discoveries showed that the flow of genetic information is more complicated (Shapiro 2009, Bustamante et al. 2011, Koonin 2012, Camacho 2021). The dogma is only a framework for understanding the flow of information between the biopolymers present in every living cell: DNA, RNA and proteins. So there are 9 imaginable direct information transfers that can occur between them. The dogma divides these transfers into 3 groups (Crick 1970):

- general transfers (believed to occur normally in most cells),
- special transfers (known to occur but only under certain conditions for some viruses or in the laboratory),
- unknown transfers (believed never to occur),

General transfers describe the normal flow of biological information: DNA can be copied into DNA (DNA replication), DNA information can be copied into mRNA (transcription), and proteins can be synthesized using the information in mRNA as a template (translation). Such a flow of genetic information is observed in archaeal bacteria and fungi (Leonard and Mechali 2013, Kent and Dixon 2020). Special transfers describe: copying RNA from RNA (RNA replication), DNA synthesis using an RNA template (reverse transcription) which very often occurs in viruses (Lai 2005, Menéndez-Arias et al. 2017, V'kovski et al. 2020, Zhang et al. 2023). Unknown transfers describe: protein synthesis directly from a DNA template without using mRNA, protein copied from a protein, RNA synthesis using a protein's primary structure as a template, DNA synthesis using a protein's primary structure as a template. The first studies assumed that such a flow of information

Institute of Plant Genetics, Breeding Biotechnology, University of Life Sciences in Lublin, Akademicka 15, 20–950 Lublin, Poland.
Email: aleksandra.nucia@up.lublin.pl

DNA $\xrightarrow{\text{transcription}}$ RNA $\xrightarrow{\text{translation}}$ protein

reverse   *reverse*

Figure 1. The Central Dogma of Molecular Biology. Reprinted from Koonin (2012) under a Creative Commons Attribution 4.0 International License.

does not occur naturally; however, the development of molecular research allowed to discover that information can also be transferred between proteins (Bussard 2005).

Our knowledge of the molecular basis of life is constantly being developed and refined, but some basic elements of cellular functioning remain constant.

The chapter presents a summary of the most important information on the basic transfers taking place in the cells of microorganisms and viruses in the field of transferring genetic information that is the basis of the diversity of the world of microorganisms.

Replication

Replication is the process of DNA synthesis or copying genetic information contained in the nucleotide sequence by the cell. The process of accurate copying of genetic information is present in each cell undergoing division and it is the basis of biological inheritance (Alberts 2003). However, the exact replication mechanism is different between *Prokaryotes*, *Eukaryotes* and viruses due to genetic diversity of the genomes. The exact mechanism of replication depends on the type and shape of the genetic material, and the conformation of the chromosomes is particularly important (Worning et al. 2006, Maga 2013, O'Donnell et al. 2013). Bacterial genomes are organised on a single chromosome, and the genetic material is a circular molecule of double-stranded DNA. Viruses may have double-stranded DNA, single-stranded DNA or RNA. Fungi are eukaryotic organisms, so they have a linear chromosome having double-stranded DNA (Antolin and Black 2007, Mohanta and Bae 2015, Rampersad and Tennant 2018, Galli et al. 2019, Koonin et al. 2020). Therefore, genetic information can also be transferred from RNA to RNA, for example in the replication of some RNA viruses, and from RNA to DNA, such as in retroviruses. Their genome is made of RNA, and copies of it are transcribed into DNA during the infectious cycle (Chinchar 1999, Cann 2008).

Replication is an extremely complex process involving dozens of enzymes. The process must be very precise and carefully planned at the right place and time. At the enzymatic level, it involves not only the mechanism of creating progeny DNA strands programmed in the parental process but also determines the high efficiency and fidelity of the genome duplication process and DNA replication in full synchronization with the cell cycle (only once per cell division cycle) (Leonard and Grimwade 2015). The accuracy and fidelity of the replication process is affected by the precise mechanism of polymerization of new DNA strands, as well as the presence of protective systems. This process is critically dependent on DNA polymerases and other enzymes responsible for the replication reaction. These enzymes are responsible for the efficient selection of the correct nucleotides for the polymerization reaction, the detection and repair of DNA errors, and the excision of any misplaced nucleotides by their internal exonucleases (Bębenek and Ziuzia-Graczyk 2018).

According to the theory of DNA structure, the existence of a DNA double helix ensures that nucleic acids can reproduce themselves accurately. The elongation of the polynucleotide chain can only occur in one direction (5'–3'), which is a consequence of the attachment of new nucleotides to the 3' end of the DNA strand. Therefore, it is necessary to explain the fact that replication of the genome requires replication of both DNA strands, which run in opposite directions. Therefore, a feature of the replication process is its semi-conservative nature. This means that during replication, both DNA strands are templates for the synthesis of new strands. Nucleotides are added to them according to the rule of base complementarity, which allows the formation of new, identical DNA molecules. Both molecules contain one parent strand and another a progeny strand

(Matsunaga et al. 2003, Sclafani and Holzen 2007, Masai et al. 2010). For easier understanding, genome replication has been divided into three main phases (Solar et al. 1998, Firshein 2003, Rampersad and Tennant 2018):

1. Initiation – start of replication.
2. Elongation – elongation of polypeptide chains.
3. Termination – completion of replication and final assembly of new DNA strands.

Initiation begins in a precisely defined position called *ori* (*origin of replication*). In bacteria (having a double-stranded DNA molecule), replication usually starts from a single replication origin (*oriC*) and involves the whole chromosome. Therefore, a single origin controls replication of the entire genome. The initiation of replication of viruses proceeds similarly to the initiation of replication in bacteria and begins at a single specific origin (Méchali 2001, Robinson and Bell 2005, Rampersad and Tennant 2018, Trojanowski et al. 2018). However, compared to viruses having double-stranded DNA, in the genomes of viruses having a single-stranded DNA molecule, replication proceeds in two stages. First, a single DNA strand is transformed into a circular double-stranded molecule after the complementary strand is synthesized. In the second stage, large numbers of strands are synthesized, for which the newly formed strand is the matrix (Rampersad and Tennant 2018). In eukaryotic genomes (fungi), replication initiates at the range of hundreds to tens of thousands of origins of replications, so the process begins simultaneously at many sites along the chromosomal DNA (Robinson and Bell 2005, Gao et al. 2012, Katayama 2017). This site contains specific nucleotide sequences used primarily to bind initiator proteins or other proteins supporting the replication process. These proteins cause large changes in the DNA structure within which DNA strands are separated (O'Donnell et al. 2013). More specifically, DNA replication begins with the breaking of hydrogen bonds between DNA strands. At this point, replication forks are formed that move in two opposite directions, although DNA synthesis can only occur in the 5'–3' direction (Chesnokov and Svitin 2013). DNA polymerases cannot start the synthesis of new DNA strands, they can only elongate them. Therefore, DNA synthesis begins with the synthesis of RNA molecules. At the beginning of replication, the enzyme RNA polymerase, called primase, synthesizes short RNA fragments - primers on the DNA template (Figure 2). To these primers, DNA polymerase adds further triphosphonucleotides. At the end of the whole process, the primers are excised and removed, and the resulting gaps are removed (Van Der Endet et al. 1985, Guilliam et al. 2015).

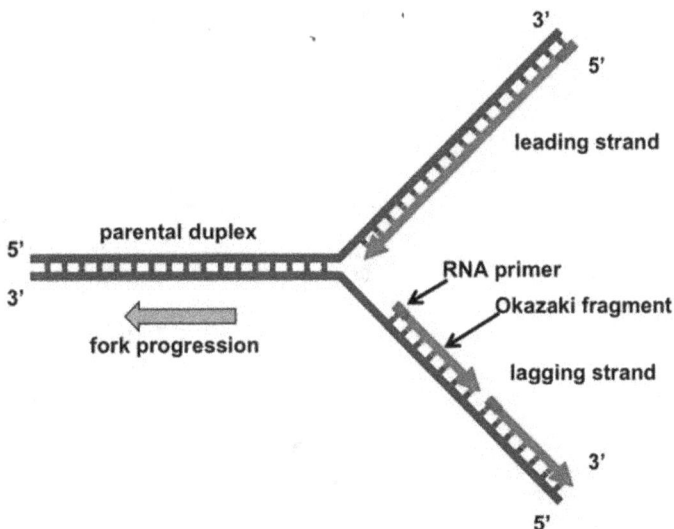

Figure 2. The replication fork. Reprinted from Leman and Noguchi (2013) under a Creative Commons Attribution 4.0 International License.

DNA replication is closely linked to the transcription process. The relationship between these processes occurs through the synthesis of primer RNA. Transcription is also necessary to activate DNA replication initiation sites, both in bacterial and eukaryotic cells. It is also worth noting that replication, regardless of the type of microorganism, can only be controlled at the initiation stage (Danis et al. 2004).

Elongation of polypeptide chain requires the participation of many enzyme proteins needed to synthesize new DNA molecules. This process occurs by DNA polymerases, which add nucleotides to the primer. All DNA polymerases (prokaryotic and eukaryotic) (Table 1) catalyze the chain elongation reaction of the synthesized DNA chain by forming a bond between the internal phosphate of the nucleotide and the 3'-OH group of the primer. In this way, a new DNA strand is formed. In addition to DNA polymerase, the activity of topoisomerase and helicase, which are responsible for the process of separating the two DNA strands and unwinding the matrix chains, is necessary (Duguet 1997, Vos et al. 2011, Wu et al. 2017).

Elongation of the chain occurs in the 5'–3' direction. Due to the anti-parallel nature of the DNA duplex, DNA synthesis occurs continuously on one leading strand and discontinuously on the other strand (lagging strand). Because DNA polymerases can only move along their templates in one direction (5'–3'), DNA synthesis on the lagging strand produces short DNA fragments–called Okazaki fragments, which are joined by DNA ligase (Zheng and Shen 2011, Balakrishnan and Bambara 2013, Prioleau and Macalpine 2016). Polynucleotide ligases are proteins that support replication. They catalyze the synthesis of a phosphodiester bond between adjacent 3'-OH and 5'-P residues of nucleotides hydrogen-bonded to the template DNA strand. During replication, they combine the Okazaki fragments into one long strand. Bacterial ligases require the presence of NAD as a coenzyme and energy source. The ligases of some viruses and all eukaryotes use ATP for this purpose. Bacterial and viral ligases are made of one polypeptide chain. In bacterial cells, the number of ligase molecules corresponds to the number of DNA polymerase I molecules, while in eukaryotic cells there are at least two ligases that differ in size (Tomkinson et al. 2006, Ellenberger and Tomkinson 2008, Shuman 2009). Proteins supporting replication also include: proteins stabilizing the single-stranded structure of DNA-SSB (single-strand binding proteins), helicases unwinding the double helix and topoisomerases causing relaxation of double helix twists. The newly formed strand is checked by exonucleases, which can cleave nucleotides from the ends of the DNA (Marceau 2012, Oakley 2019).

Termination is the final stage of replication. In the DNA of bacterial cells, the site of replication termination is determined by specific *ter* sequences, where the replication fork stops. The termination site on the circular bacterial chromosome is located exactly opposite the origin of replication (Mirkin and Mirkin 2007, Dewar and Walter 2017). In the DNA of eukaryotic cells, terminal sequences or any systems regulating termination are absent. Replication ends when replication forks travelling from opposite directions meet each other. Unlike circular bacterial chromosomes, eukaryotic chromosomes are made of linear DNA molecules that are terminated by very distinctive nucleotide sequences called telomeres. Telomeres are involved in the termination of chromosomal DNA replication, which ensures that the length of the respective genomes remains constant. Telomeres also stabilize and protect chromosomal DNA from the action of nucleases and from joining with other chromosomes (Dewar and Walter 2017, Maestroni et al. 2017, Bonnell et al. 2021).

Table 1. DNA polymerases involved in replication elongation in *Procaryota* and *Eukaryota* (Bębenek and Ziuzia-Graczyk 2018).

Organism	DNA polymerase
Bacterial DNA polymerases	I, II, III
DNA polymerases of eukaryotic cells	α, δ, ε, β, γ

Transcription

Transcription is the first and key step of gene expression. It is an essential step in using the information from genes in our DNA to make proteins. This process is similar to DNA replication and involves the creation of a polynucleotide RNA strand on a DNA template with the participation of specific enzymes. Transcription involves copying DNA fragments and converting them to RNA (Clancy 2008, Chen et al. 2020). Genes may be subdivided into two major groups: those whose final product is an RNA molecule (e.g., tRNA, rRNA, assorted regulatory RNAs—see below) and those whose final product is protein. The transcribed DNA molecules into protein-coding RNAs are called messenger RNAs (mRNAs). Generally, in this process, a single-stranded DNA template is used to synthesize a complementary strand of RNA by an enzyme called RNA polymerase. In particular, RNA polymerase builds strands of RNA from 5' to 3', adding each new nucleotide at the 3' end (Clancy 2008, Barba-Aliaga et al. 2021). Transcription will first be described in bacteria because it is simpler than in eukaryotes (Lehman 2008). The principles of transcription are similar in higher organisms, but the details are more complicated. The major differences between prokaryotes and eukaryotes occur in the initiation and regulation of transcription, rather than in the actual synthesis of RNA (Clancy 2008). In most organisms, transcription proceeds in three main stages, including initiation, elongation, and termination. These steps differ among different groups of microorganisms mainly due to the structure and complexity of the genomes (Figure 3, Figure 4).

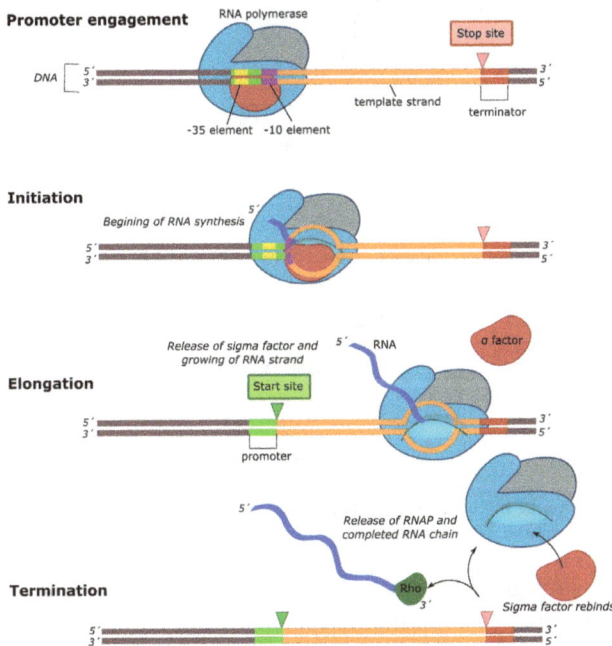

Figure 3. Graphic representation of the bacterial transcription process. Reprinted from Abril et al. (2020) under a Creative Commons Attribution 4.0 International License.

Initiation

The first step in transcription is initiation. In this stage, RNA polymerase binds to the DNA of the gene at a region called the promoter. A promoter is a regulatory sequence that determines the level of expression of a gene or group of genes. RNA polymerase in conjunction with a number of regulatory molecules recognizes the promoter sequence and starts the transcription process (Browning and Busby 2016, Barba-Aliaga et al. 2021).

In different groups of microorganisms, the initiation of transcription proceeds slightly differently. It is most diverse in viruses, which is related to the greatest diversity of their genomes.

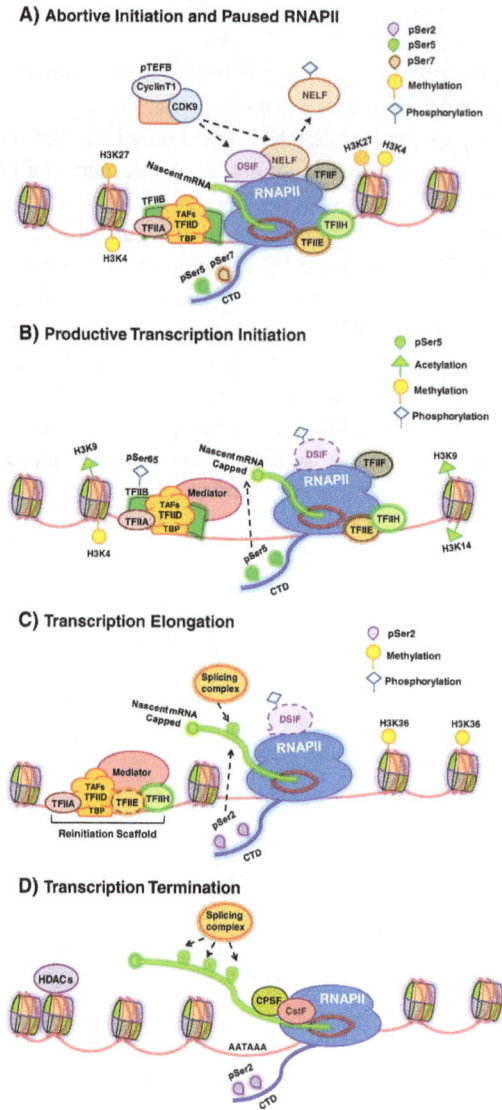

Figure 4. Graphic representation of the eukaryotes transcription process. Reprinted from Shandilya and Roberts (2012) under a Creative Commons Attribution 4.0 International License.

Transcription of a virus genome is one of the most important stages in its life cycle. It is the method by which viral genetic information is communicated to the host cell's protein synthesis machinery, enabling it to produce the viral proteins required for replication and the assembly of progeny viruses (Hull 2002). Viral transcription takes place in the context of the life cycle of a given virus and occurs in a time-regulated cascade of gene expression: as immediate-early, delayed-early and late. Early genes are associated with the generation of infection and the suppression of host antiviral responses; the middle genes are related to genome replication or latency fixing; and late genes are associated with the production of structural elements of new virions. Therefore, viral genomes are often arranged in such a way that the order of transcription matches the temporal requirements of gene expression. This basic feature was first observed in bacteriophages and is a common feature of many viruses that roughly divide their genes into those that are expressed during the early, middle, and late stages of infection (Broyles 2003, Yang et al. 2014).

Viruses use the host cell's transcription machinery to copy and rewrite their genetic information. Depending on the type and structure of the genetic material of viruses, transcription proceeds in a slightly different way. According to the Baltimore classification, viruses are divided into seven groups according to the mechanism of the relationship between the viral genome and mRNA formation (Baltimore 1971). In viruses, depending on the structure of their genome, transcription takes place in the presence of different polymerases, i.e., RNA-dependent RNA polymerases, RNA-dependent DNA polymerases, DNA-dependent RNA polymerases and DNA-dependent DNA polymerases. The genome of DNA viruses can be double-stranded (dsDNA) or single-stranded (ssDNA). DNA viruses transcribe mRNA using DNA-dependent RNA polymerases (also called RNA polymerases). RNA viruses can also be double-stranded (dsRNA) or single-stranded (ssRNA). Single-stranded RNA genomes can have a plus (+) or minus (-) sense. RNA viruses transcribe their genomes using RNA-dependent RNA polymerases. Reverse transcription viruses replicate using reverse transcription, the process of making DNA from RNA templates. Their genomes can be RNA or DNA. RNA-dependent RNA polymerases and reverse transcriptases are unique to viruses because the host cell does not require RNA replication or reverse transcription. Viral polymerases are generally active as a single protein capable of carrying out multiple functions related to viral genome synthesis (Choi 2012).

In bacteria, all transcription is performed by a single type of RNA polymerase. Bacterial RNA polymerase consists of 4 catalytic subunits involved in the transcription process. The RNA polymerase also includes the sigma regulatory unit, which determines the initiation of transcription, it is detached before the elongation step. Several different sigma factors have been identified, and each oversees the transcription of a unique set of genes. The sigma factors are therefore discriminatory as each binds a distinct set of promoter sequences (Murakami and Darst 2003, Feklístov et al. 2014, Paget 2015, Helmann 2019, Chen et al. 2020). An example of specialization of sigma factors for different gene promoters is bacterial sporulation of the species *Bacillus subtilis*. This bacterium exists in two states: vegetative (growing) and sporulating. The genes involved in spore formation are not normally expressed during vegetative growth. Interestingly, expression of the gene encoding the new sigma factor turns the first genes into sporulation. Subsequent expression of various sigma factors then turns on new sets of genes needed later in the sporulation process (Losick and Stragier 1992). Bacterial cells are able to tune their transcription programs to changing environments through numerous mechanisms regulating the activity of RNA polymerase or changing the set of promoters to which RNA polymerase can bind.

Fungi are eukaryotic organisms in which the transcription initiation process is more complex than in bacteria. First, eukaryotic cells contain three distinct nuclear RNA polymerases that transcribe different classes of genes. Protein-coding genes are transcribed by RNA polymerase II to form mRNA; ribosomal RNA (rRNA) and transfer RNA (tRNA) are transcribed by RNA polymerases I and III. RNA polymerase I is specifically dedicated to the transcription of the three largest types of rRNA, which are designated as 28S, 18S and 5.8S according to the sedimentation rate during high-speed centrifugation. RNA polymerase III transcribes tRNA genes and the smallest species of ribosomal RNA (5S rRNA). Some small RNAs involved in splicing and transporting proteins (snRNAs and scRNAs) are also transcribed by RNA polymerase III, while others are transcripts of polymerase II (Cooper 2000, Barba-Aliaga et al. 2021, Girbig et al. 2022).

In addition, a number of transcription factors (TFs) involved in the initiation of transcription and regulation of gene expression have been identified in fungi (Kensche et al. 2008, Shelest 2017). The transcription initiation stage is regulated by cis factors, i.e., regular sequences within promoters and enhancers, as well as trans factors, i.e., proteins binding to regulatory sequences (Gasch et al. 2004, Thompson and Regev 2009, García-Estrada et al. 2018, Siddiq and Wittkopp 2022). Wang et al. (2021) found that fungal gene transcription is modulated in response to changing conidia growth conditions and depending on their developmental stage. It is also adapted to the respective developmental stages of the fungus. Transcription factors also control the virulence of phytopathogenic fungi (John et al. 2021).

Elongation

Transcription elongation is a regulated process in which an RNA chain complementary to the template strand of DNA is synthesized as RNA polymerase moves along DNA. The underlying mechanism for transcription elongation is highly conserved between bacteria and eukaryotes. The elongation involves a gene-scale DNA tracking system accompanied by the formation of bonds at each base pair. Thus, specific DNA conformations/flexibility, DNA damage and DNA binding proteins are of particular importance in elongation regulation (Imashimizu et al. 2014, Couvillion et al. 2022). One of the main mechanisms regulating elongation in all kingdoms of life is the sequence-dependent pause (Landick 2006, Adelman and Lis 2012). In bacteria, the pause is important for coupling transcription with translation and for ensuring that regulatory factors are able to bind the elongation complex (Roberts et al. 1998, Artsimovitch and Landick 2002, Landick 2006, Kang et al. 2019). In eukaryotes, slow elongation, achieved by a pause around the exon-intron junction, allows spliceosome assembly, which increases the efficiency of alternative splicing (Pelechano et al. 2009, Agapov et al. 2022).

In bacteria, the elongation phase of the RNA chain begins with the release of the sigma subunit from the polymerase. RNA polymerase synthesizes mRNA complementary to the DNA template in the 5′ to 3′ direction. The mRNA chain elongation stage, similarly to the initiation stage, is subject to modifications and control by numerous factors (Borukhov et al. 2005, Belogurov and Artsimovitch 2015, Washburn et al. 2015). Some of these factors (GreA, NusA and Mfd) are evolutionarily conserved and affect RNA polymerase (RNAP) processivity by modulating transcription pausing, arrest, termination or anti-termination.

In fungi, the elongation phase is analogous to that in bacteria; however, this process is not as well understood as the regulation of transcription initiation. It is known, however, that the transition from transcriptional initiation to elongation is associated with a change in the factors that are associated with RNAPII. However, several factors have been discovered that stimulate elongation on nonchromatin, "naked" DNA templates (e.g., TFIIS and TFIIF). Recently, other factors have been identified that are thought to affect RNAPII transcript elongation via effects on chromatin. These include for example the yeast factors Spt4, Spt5 and Spt16/Pob3 (Formosa et al. 2001, Schweikhard et al. 2014, Blythe et al. 2016, Decker 2021, Linzer et al. 2021).

Termination

Termination is the final step in RNA synthesis. It defines the boundaries of transcription units, but also affects the stability and cellular localization of the transcripts produced (Porrua and Libri 2015). Transcription termination occurs when the transcribing RNA polymerase finds a sequence called the terminator and releases the DNA template and nascent RNA. Termination is required to prevent inappropriate transcription of downstream genes and to recycle the polymerase (Rosonina et al. 2006).

Little is known about how viral polymerases complete nucleic acid synthesis at the 5' end of the genome. Terminating exactly at the 5' end of the genome would be particularly difficult for viruses with a linear genome, such as single-stranded RNA or DNA viruses. If the end of the genome is not copied during successive rounds of replication, the genome will get shorter or longer. The exceptions are DNA-dependent RNA polymerases, which terminate RNA transcription at a tail signal. The newly synthesized RNA forms a GC-rich hairpin followed by a stretch of uridine nucleotides at the tail signal, and the RNA polymerase complex is released from the DNA template (Choi 2012).

Termination by prokaryotic RNA polymerases, as well as eukaryotic RNA polymerase (Pol) I and Pol III, occurs at specific positions, requiring DNA or RNA cis elements that can directly or indirectly destabilize the elongating polymerase–DNA–RNA ternary complex (Henkin 2000, Paule and White 2000).

In bacteria, two main transcription termination strategies have been described, Rho-dependent and Rho-independent (Peters et al. 2023). Rho is a transcription termination protein which binds to C-rich unstructured stretch of the nascent RNA and translocates along the RNA; eventually, it catches up to the elongation complex (EC) and dissociates it (Grylak-Mielnicka et al. 2016, Mitra et al. 2017, Banerjee et al. 2023). Several models for Rho-dependent transcription termination have been proposed, but the molecular details of the mechanism of Rho translocation and RNA polymerase dissociation on Rho-dependent terminators are not yet fully understood (Ray-Soni et al. 2016, Chen et al. 2019, Jain et al. 2019, Peters et al. 2023).

Rho-independent termination also called intrinsic termination is independent of termination factors. Essential elements of a Rho-independent terminator consist of a GC-rich dyad repeat that forms a stem-loop (hairpin) structure followed by a T-rich stretch, generating a U-rich tail in the RNA after termination (Adhya and Gottesman 1978). Rho-independent termination is achieved by formation of the stem-loop structure, which is facilitated by RNA polymerase pausing during transcription of the T-rich tract (Ray-Soni et al. 2016).

Most eukaryotic genes are transcribed by RNA polymerase II (Pol II), including those that produce mRNAs and many noncoding functional RNAs. Proper expression of these genes requires efficient termination by Pol II to avoid transcriptional interference and synthesis of extended, non-functional RNAs. In fungi, there are two main transcription termination pathways for RNA polymerase II. The first pathway depends on a macromolecular complex including cleavage and polyadenylation factor (CPF) and is generally responsible for transcription termination in protein-coding genes. The second pathway Nrd1–Nab3–Sen1 (NNS) involves recognition of an element in the nascent transcript by an essential RNA-binding protein and acts specifically on short genes transcribed by Pol II (Ansari and Hampsey 2005, Porrua and Libri 2013, Larochelle et al. 2018, Han et al. 2020).

Translation

Translation is the process of protein biosynthesis and the final step in the expression of genetic information in a cell. Gene expression proceeds in a two-step manner, beginning with transcription. In view of this fact, translation is closely related to transcription. At first, the information contained in the nucleotide sequence in the DNA is transcribed into the complementary sequence of nucleotides in the messenger RNA (mRNA), during transcription. Thereafter, the genetic information, encoded in the mRNA, is converted into a specific chain of amino acids that form a polypeptide. It is the sequence of amino acids in the protein. This process is called translation and takes place in the cytoplasm on the ribosomes in the 5' to 3' direction (Figure 5) (Chekulaeva and Landthaler 2016, Webster and Weixlbaumer 2021). In eukaryotes, translation and transcription occur in different cellular compartments, whereas in bacteria, these two major processes of gene expression are coupled in time and space (Proshkin et al. 2010).

The basis of translation is genetic code, more specifically, a triplet base code. This means that the DNA sequence of a gene is divided into a series of units consisting of three nucleotides, while three consecutive nucleotides form one amino acid. Each such unit is called a codon (Yarus et al. 2005, Chavence and Hughes 2017).

Initiation begins when the codon initiating the synthesis of the protein - AUG (encoding methionine) by the ribosomes and the initiator tRNA - is found on the mRNA. Ribosomes are made up of two types of subunits. In bacteria, ribosomes consist of a small subunit (30S) and a large subunit (50S), while in eukaryotes, ribosomes are also composed of two subunits (40S and 60S), which together form

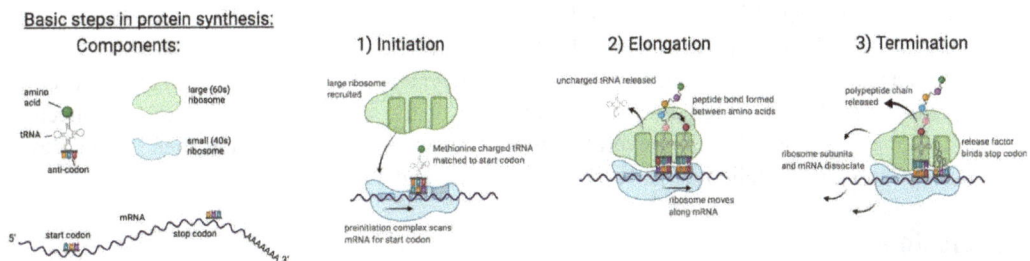

Figure 5. Basic schematic of translation. Reprinted from Marchingo and Cantrell (2022) under a Creative Commons Attribution 4.0 International License.

the 80S molecule (Lafontaine and Tollervey 2001, Preiss and Hentze 2003). Translation initiation factors are also involved in initiation of translation. In bacteria, there are only three initiation factors: IF-1, IF-2, IF-3. In Eukaryotes, a number of protein factors (eIF - Eukaryotic Initiation Factors) and PABP (Poly-A-binding Protein) are involved in initiation (Green and Noller 1997, Sonenberg and Dever 2003 Myasnikov et al. 2009, Sonneveld et al. 2020).

In Prokaryota, the process of translation starts with a group of initiating factors attaching to the 30S subunit. Once the initiator tRNA and mRNA have been attached, these protein factors are released, and the 50S subunit joins in. This results in the formation of a 70S ribosome on which translation can take place. The binding of the 50S subunit and the ejection of the IFs mark the irreversible transition to the elongation phase (Simonetti et al. 2008, Myasnikov et al. 2009). Eukaryotes also attach protein factors to the small subunit of the ribosome (40S). The initiator tRNA and mRNA are then attached to the 40S subunit. After the release of the protein factors, the 60S subunit attaches to the 40S subunit to form the 80S ribosome. Ribosomes formed in this way take part in the process of elongation of the peptide chain (Myasnikov et al. 2009, Hinnebusch 2014).

Elongation is the extension of the polypeptide chain. The amino acids are brought to the ribosome by the tRNA and joined together to form a chain. This stage takes place with the participation of elongation factors. Elongation factors (EF) bind the large subunit of the ribosome to the complex and move the ribosome along the mRNA. This allows the addition of further amino acids to the end of the polypeptide being formed (Agirrezabala and Frank 2009, Hummels and Kearns 2020).

During translation elongation, attention should be paid to two active RNA binding sites located in the ribosome: A-site and P-site. An amino-acyl-tRNA (aa-tRNA) carrying the amino acid is stably matched to the A-site of the ribosome. Amino acids are added to all tRNAs by enzymes. Then, the tRNA molecules, bound at the peptidyl site (P-site) and the A-site, oscillate between these two configurations. The next stage of elongation is the synthesis of the peptide bond, that is, the attachment of an amino acid the resulting polypeptide chain. The rapid formation of peptide bonds transfers the peptide attached to the tRNA at the P site to the amino acid at the A site. This amino acid still remains bound to the tRNA. Individual enzymes catalyze the reaction of attaching amino acids to individual tRNAs, having appropriate anticodons. During the shift from the acceptor site to the peptidyl site. After the formation of the peptide bond, the mRNA is shifted relative to the ribosome. The free (previous) tRNA is removed (Baranov et al. 2004, Blanchard et al. 2004).

Termination is the final step of polypeptide chain formation. Completion of polypeptide synthesis occurs when terminating factors - RFs - (Release Factors) - encounter one of the termination codons: UAA, UAG, UGA in mRNA. These codons are otherwise called nonsense codons and there is no corresponding tRNA for them, which gives a signal to terminate protein biosynthesis (Korostelev 2011, Rodnina 2018).

In bacteria, there are two termination factors: RF-1, which recognizes the triplets, UAA and UAG, and RF-2, which recognizes UAA and UGA. A third termination factor, RF-3, facilitates turnover of RF-1 and RF-2, but it is not necessary for peptidyl-tRNA hydrolysis. Termination of translation consists of three steps: recognizing the stop codon, hydrolyzing the ester bond of the peptidyl-tRNA, and dissociating RF1/RF2 with RF3 (Rodnina 2018). In Eukaryotes, termination is mediated by eRF1 and eRF3. Only one RF factor (RF-1) recognizes all nonsense codons at the A site of the ribosome. eRF-1 is also responsible for peptidyl-tRNA hydrolysis. eRF3 mediates release of the nascent peptide (Dever and Green 2012, Hellen 2018).

After translation is completed, the synthesized polypeptide is detached from the ribosome, and the ribosome itself disintegrates into subunits, with the mRNA strand being disconnected. This strand may be a template for several or more translational iterations and then breaks down (Dever and Green 2012, Hellen 2018, Nürenberg-Goloub and Tampé 2019).

Conclusions

The basic processes associated with the flow of genetic information are universal for all living organisms. However, considering the issue in more detail, the genetic diversity of microorganisms means that, depending on the microorganism, these processes proceed in slightly different ways. The development of molecular biology and biotechnology allows us to learn more and more accurately about the course of these processes and the ways of regulating these processes at various levels. Thanks to this, it is possible to control the basic life processes of microorganisms and use them for human needs. Accurate knowledge of the processes related to gene expression and regulation of expression allows, for example, to study the pathogenicity of microorganisms or responses to various types of stress. It also enables the study of adaptive abilities and the conduct of work related to the study of evolution.

References

Abril, A.G., J.L.R. Rama, A. Sánchez-Pérez and T.G. Villa. 2020. Prokaryotic sigma factors and their transcriptional counterparts in archaea and eukarya. Appl. Microbiol. Biotechnol. 104(10): 4289–4302.

Adelman, K. and J.T. Lis. 2012. Promoter-proximal pausing of RNA polymerase II: Emerging roles in metazoans. Nat. Rev. Genet. 13(10): 720–731.

Adhya, S. and M. Gottesman. 1978. Control of transcription termination. Annu. Rev. Biochem. 47: 967–996.

Agapov, A., A. Olina and A. Kulbachinskiy. 2022. RNA polymerase pausing, stalling and bypass during transcription of damaged DNA: From molecular basis to functional consequences. Nucleic Acids Res. 50(6): 3018–3041.

Agirrezabala, X. and Frank J. 2009. Elongation in translation as a dynamic interaction among the ribosome, tRNA, and elongation factors EF-G and EF-Tu. Q. Rev. Biophys. 42(3): 159–200.

Alberts, B. 2003. DNA replication and recombination. Nature 421(6921): 431–435.

Ansari, A. and M. Hampsey. 2005. A role for the CPF 3′-End processing machinery in RNAP II-dependent gene looping. Genes and Dev. 19(24): 2969–2978.

Antolin, M.F. and W.C. Black. 2007. Genes, Description of Encyclopedia of Biodiversity, 1–11.

Artsimovitch, I. and R. Landick. 2002. The transcriptional regulator RfaH stimulates RNA chain synthesis after recruitment to elongation complexes by the exposed nontemplate DNA Strand. Cell 109(2): 193–203.

Balakrishnan, L. and R.A. Bambara. 2013. Okazaki fragment metabolism. Cold spring Harb. Perspect. Biol. 5(2): a010173.

Baltimore, D. 1971. Expression of animal virus genomes. Bacteriol. Rev. 35(3): 235–241.

Banerjee, S., J. Chalissery, I. Bandey and R. Sen. 2006. Rho-dependent transcription termination: More questions than answers. J Microbiol 44(1): 11–22

Baranov, P.V., R.F. Gesteland and J.F. Atkins. 2004. Cold Spring Harb. Perspect. Biol. 10: 221–230.

Barba-Aliaga, M., P. Alepuz and J.E. Pérez-Ortín. 2021. Eukaryotic RNA polymerases: The many ways to transcribe a gene. Front. Mol. Biosci. 8: 663209.

Bębenek, A. and I. Ziuzia-Graczyk. 2018. Fidelity of DNA replication—a matter of proofreading. Curr. Genet. 64: 985–996.

Belogurov, G.A. and I. Artsimovitch. 2015. Regulation of transcript elongation. Ann. Rev. Microbiol. 69(1): 49–69.

Blanchard, S.C., H.D. Kim, R.L. Gonzalez, J.D. Puglisi and S. Chu. 2004. tRNA dynamics on the ribosome during translation. Proc. Natl. Acad. Sci. U.S.A. 101(35): 12893–12898.

Blythe, A.J., B. Yazar-Klosinski, M.W. Webster, E. Chen, M. Vandevenne, K. Bendak et al. 2016. The yeast transcription elongation factor Spt4/5 is a sequence-specific RNA binding protein. Protein Sci. A Publication of the Protein Society 25(9): 1710–1721.

Bonnell, E., E. Pasquier and R.J. Wellinger. 2021. Telomere replication: Solving multiple end replication problems. Front. Cell Devel. Biol. 9: 668171.

Borukhov, S., J. Lee and O. Laptenko. 2005. Bacterial transcription elongation factors: New insights into molecular mechanism of action. Mol. Microbiol. 55(5): 1315–1324.

Browning, D.F. and S.J.W. Busby. 2016. Local and global regulation of transcription initiation in bacteria. Front. Cell Devel. Biol. 14(10): 638–650.

Broyles, S.S. 2003. Vaccinia virus transcription. J. Gen. Virol. 84(9): 2293–2303.

Bussard, A.E. 2005. A scientific revolution? EMBO Reports 6(80): 691–694.

Bustamante, C., W. Cheng and Y.X. Meija. 2011. Revisiting the Central dogma one molecule at a time. Cell 144(4): 480–497.

Camacho, M.P. 2021. Beyond descriptive accuracy: the central dogma of molecular biology in scientific practice. Stud. His. Philos. Sci. Part A 86: 20–26.

Cann, A.J. 2008. Replication of viruses. Encyclopedia of Virology: 406–412.

Chekulaeva, M. and M. Landthaler. 2016. Eyes on translation. Mol. Cell. Cell Press. 63: 918–925.

Chen, J., T. Morita and S. Gottesman. 2019. Regulation of transcription termination of small RNAs and by small RNAs: Molecular mechanisms and biological functions. Front. Cell. Infect. Microbiol. 9: 201.

Chen, J., H. Boyaci and E.A. Campbell. 2020. Diverse and unified mechanisms of transcription initiation in bacteria. Nat. Rev. Microbiol. 19(2): 95–109.

Chesnokov, I. and A. Svitin. 2013. DNA replication: Eukaryotic origins and the origin recognition complex. In Encyclopaedia of Biological Chemistry: Second Edition, 108–113. Elsevier Inc.

Chevance, F.F.V. and K.T. Hughes. 2017. Case for the genetic code as a triplets. Proc. Natl. Acad. Sci. U.S.A. 114(18): 4745–4750.

Chinchar, V.G. 1999. Replication of viruses. Encyclopaedia of Virology: 1471–1478.

Choi, K.H. 2012. Viral polymerases. Adv. Exp. Med. Biol. 726: 267–304.

Clancy, S. 2008. RNA transcription by RNA polymerase: Prokaryotes vs Eukaryotes. Nature Education 1(1): 125.

Cobb, M. 2017. 60 Years ago, francis crick changed the logic of biology. PLOS Biology 15(9): e2003243.

Cooper, G.M. 2000. The Cell: A Molecular Approach. 2nd edition. Sunderland (MA): Sinauer Associates; 2000. Eukaryotic RNA Polymerases and General Transcription Factors.

Couvillion, M., K.M. Harlen, K.C. Lachance, K.L. Trotta, E. Smith, C. Brion et al. 2022. Transcription elongation is finely tuned by dozens of regulatory factors. ELife 11.

Crick, F. 1970. Central dogma of molecular biology. Nature 227(5258): 561–563.

Danis, E., K. Brodolin, S. Menut, D. Maiorano, C. Girard-Reydet and M. Méchali. 2004. Specification of a DNA replication origin by a transcription complex. Nat. Cell Biol. 6(8): 721–730.

Decker, T.M. 2021. Mechanisms of transcription elongation factor DSIF (Spt4-Spt5). J. Mol. Biol. 433(14).

Dever, T.E. and R. Green. 2012. The elongation, termination, and recycling phases of translation in eukaryotes. Cold Spring Harb. Perspect. Biol. 4(7): a013706.

Dewar, J.M. and J.C. Walter. 2017. Mechanisms of DNA replication termination. Nat. Rev. Mol. Cell Biol. 18: 507–516.

Duguet, M. 1997. When Helicase and topoisomerase meet! Journal of Cell Science 110(12): 1345–1350.

Ellenberger, T. and A.E. Tomkinson. 2008. Eukaryotic DNA Ligases: Structural and functional insights. Annual Review of Biochemistry. 77: 313–338.

Feklístov, A., B. Sharon, S.A. Darst and C.A. Gross. 2014. Bacterial sigma factors: A historical, structural, and genomic perspective. Ann. Rev. Org. 68: 357–376.

Firshein, W. 2002. Prokaryotic DNA replication. Modern Microbial Genetics: 1–26.

Formosa, T., P. Eriksson, J. Wittmeyer, J. Ginn, Y. Yu, and D.J. Stillman. 2001. Spt16–Pob3 and the HMG Protein Nhp6 combine to form the nucleosome-binding factor SPN. The EMBO Journal 20(13): 3506.

Galli, E., J.L. Ferat, J.M. Desfontaines, M.E. Val, O. Skovgaard, F.X. Barre et al. 2019. Replication termination without a Replication Fork Trap. Sci. Rep. 9(1): 8315.

García-Estrada, C., R. Domínguez-Santos, K. Kosalková and J.F. Martín. 2018. Transcription factors controlling primary and secondary metabolism in filamentous Fungi: The β-Lactam Paradigm. Ferment. 4(2): 47.

Gasch, A.P., A.M. Moses, D.Y. Chiang, H.B. Fraser, M. Berardini and M.B. Eisen. 2004. Conservation and evolution of Cis-Regulatory systems in ascomycete fungi. PLoS Biol. 2(12).

Girbig, M., A.D. Misiaszek and C.W. Müller. 2022. Structural insights into nuclear transcription by eukaryotic DNA-dependent RNA polymerases. Nature Rev. Mol. Cell Biol. 23(9): 603–622.

Green, R. and H.F. Noller. 1997. Ribosomes and translation. Ann. Rev. Biochem. 66: 679–716.

Grylak-Mielnicka, A., V. Bidnenko, J. Bardowski and E. Bidnenko. 2016. Transcription termination factor rho: A hub linking diverse physiological processes in bacteria. Microbiol. Res. Org. 162(3): 433–447.

Guilliam, T.A., B.A. Keen, N.C. Brissett and A.J. Doherty. 2015. Primase-polymerases are a functionally diverse superfamily of replication and repair enzymes. Nucleic Acids Res. 43(14): 6651–6664.

Han, Z., O. Jasnovidova, N. Haidara, A. Tudek, K. Kubicek, D. Libri et al. 2020. Termination of non-coding transcription in yeast relies on both an RNA Pol II CTD interaction domain and a CTD-Mimicking region in Sen1. The EMBO Journal 39(7): e101548.

Hellen, C.U.T. 2018. Translation termination and rimosome recycling in eukaryotes. Cold Spring Harb. Perspect. Biol. 10(10): a032656.

Helmann, J.D. 2019. Where to begin? Sigma factors and the selectivity of transcription initiation in bacteria. Mol. Microbiol. 112(2): 335–347.

Henkin, T.M. 2000. Transcription termination control in bacteria. Curr. Opin. Microbiol. 3(2): 149–153.

Hinnebush, A.G. 2014. The scanning mechanism of eukaryotic translation initiation. Annu. Rev. Biochem. 83: 779–812.

Hull, R. 2002. Expression of viral genomes. Matthews' Plant Virology 225–292.

Hummels, K.R. and D. Kearns. 2020. Translation elongation factor P (EF-P). FEMS Microbiol. Rev. 44: 208–2018.

Imashimizu, M., N. Shimamoto, T. Oshima and M. Kashlev. 2014. Transcription elongation. Transcription 5(1): e28285.

Jain, S., R. Gupta and R. Sen. 2019. Rho-dependent transcription termination in bacteria recycles RNA polymerases Stalled at DNA lesions. Nat. Commun. 10(1): 1–12.

John, E., K.B. Singh, R.P. Oliver and K.C. Tan. 2021. Transcription factor control of virulence in Phytopathogenic fungi. Mol. Plant Pathol. 22(7): 858–881.

Kang, J.Y., T.V. Mishanina, R. Landick and S.A. Darst. 2019. Mechanisms of transcriptional pausing in bacteria. J. Mol. Biol. 431(20): 4007.

Kensche, P.R., M. Oti, B.E. Dutilh and M.A. Huynen. 2008. Conservation of divergent transcription in fungi. Trends Genet. 24(5): 207–211.

Kent, R. and N. Dixon. 2020. Contemporary tools for regulating gene expression in bacteria. Trends Biotechnol. 38(3): 316–333.

Kitadai, N. and S. Maruyama. 2018. Origins of building blocks of life: A review. Geosci. Front. 9(4): 1117–1153.

Koonin, E.V. 2012. Does the central dogma still stand? Biology Direct 7(1): 1–7.

Koonin, E.V., V.V Dolja, M. Krupovic, A. Varsani, Y.I. Wolf, N. Yutin et al. 2020. Global Organization and Proposed Megataxonomy of the Virus World.

Korostelev, A.A. 2011. Structural aspects of translation termination on the ribosome. Cold Spring Harb. Perspect. Biol. 17: 1409–1421

Lafontaine, D.L.J. and Tollervey, D. 2001. The function and synthesis of ribosomes. Nat. Rev. Mol. Cell Biol. 2(7): 514–520.

Lai, M.M.C. 2005. RNA replication without RNA-dependent RNA polymerase: Surprises from hepatitis delta virus. J. Virol. 79(13): 7951.

Landick, R. 2006. The regulatory roles and mechanism of transcriptional pausing. Biochem. Soc. Trans. 34(6): 1062–1066.

Larochelle, M., M.A. Robert, J.N. Hébert, X. Liu, D. Matteau, S. Rodrigue et al. 2018. Common mechanism of transcription termination at coding and noncoding RNA genes in fission yeast. Nat. Comm. 9(1): 1–15.

Lehman, I.R. 2008. Historical perspective: arthur kornberg, a giant of 20th century biochemistry. Trends in Bioch. Sci. 33(6): 291–296.

Leman, A.R. and E. Noguchi. 2013. The replication fork: understanding the eukaryotic replication machinery and the challenges to genome duplication. Genes, 4(1): 1–32.

Leonard, A.C. and J.E. Grimwade. 2015. The orisome: Structure and function. Front. Microbiol. 6: 545.

Leonard, A.C. and M. Mechali. 2013. DNA replication origins. Cold Spring Harb. Perspect. Biol. 5(10): a010116.

Linzer, N., A. Trumbull, R. Nar, M.D. Gibbons, D.T. Yu, J. Strouboulis et al. 2021. Regulation of RNA polymerase II transcription initiation and elongation by transcription factor TFII-I. Front. Mol. Biosci. 8: 681550.

Losick, R. and P. Stragier. 1992. Crisscross regulation of cell-type-specific gene expression during development in *B. Subtilis*. Nature 355(6361): 601–604.

Maestroni, L., S. Matmati and S. Coulon. 2017. Solving the telomere replication problem. Genes. MDPI AG. 8(55): 1–16.

Maga, G. 2013. DNA Replication. In Brenner's Encyclopedia of Genetics: Second Edition, 392–394. Elsevier Inc.

Marceau, A.H. 2012. Functions of single-strand DNA-binding proteins in DNA replication, recombination, and repair. In, 922: 1–21.

Marchingo, J.M. and D.A. Cantrell. 2022. Protein synthesis, degradation, and energy metabolism in T cell immunity. Cell. Mol. Immunol. 19: 303–315.

Masai, H., S. Matsumoto, Z. You, N. Yoshizawa-Sugata and M. Oda. 2010. Eukaryotic chromosome DNA replication: Where, When, and How? Annu. Rev. Biochem. 79: 89–130

Matsunaga, F., C. Norais, P. Forterre and H. Myllykallio. 2003. Identification of short "eukaryotic" Okazaki Fragments synthesized from a prokaryotic replication origin. EMBO Reports 4(2): 154–158.

Méchali, M. 2001. DNA Replication Origins: From Sequence Specificity to Epigenetics. Nat. Rev. Genet. 2(8): 640–645.

Menéndez-Arias, L., A. Sebastián-Martín and M. Álvarez. 2017. Viral Reverse Transcriptases. Virus Res. Elsevier B.V.

Mirkin, E.V. and S.M. Mirkin. 2007. Replication fork stalling at natural impediments. Microbiol. Mol. Biol. Rev. 71(1): 13–35.

Mitra, P., G. Ghosh, M. Hafeezunnisa and R. Sen. 2017. Rho Protein: Roles and mechanisms. Annu. Rev. Microbiol. 71: 687–709.

Mohanta, T.K. and H. Bae. 2015. The diversity of fungal genome. Biol.Proced. Online 17(8): 1–9.

Murakami, K.S. and S.A. Darst. 2003. Bacterial RNA polymerases: The Wholo Story. Curr. Opin. Struct. Biol. 13(1): 31–39.

Myasnikov, A.G.M., A. Simonetti, S. Marzi and B.P. Klaholz. 2009. Structure–function insights into prokaryotic and eukaryotictranslation initiation. Struct. Biol. 19: 300–309.

Nürenberg-Goloub, E. and R. Tampé. 2019. Ribosome recycling in mRNA translation, quality control, and homeostasis. Biol. Chem. 401(1): 47–61.

Oakley, A.J. 2019. A structural view of bacterial DNA replication. Protein Sci. Blackwell Publishing Ltd. 28: 990–1004.

O'Donnell, M., L. Langston and B. Stillman. 2013. Principles and concepts of DNA Replication in Bacteria, Archaea, and Eukarya. Cold Spring Harb. Perspect. Biol. 5(7): a010108.

Paget, M.S. 2015. Bacterial sigma factors and anti-sigma factors: structure, function and distribution. Biomolecules 5(3): 1245.

Paule, M.R. and R.J. White. 2000. Survey and Summary Transcription by RNA Polymerases I and III. Nucleic Acids Res. 28(6): 1283–1298.

Pelechano, V., S. Jimeno-González, A. Rodríguez-Gil, J. García-Martínez, J.E. Pérez-Ortín and S. Chávez. 2009. Regulon-specific control of transcription elongation across the yeast genome. PLoS Genet. 5(8): e1000614.

Peters, J.M., A.D. Vangeloff and R. Landick. 2011. Bacterial transcription terminators: the RNA 3'-end chronicles. J. Mol. Biol. 412: 793–813.

Porrua, O. and D. Libri. 2013. A bacterial-like mechanism for transcription termination by the Sen1p Helicase in Budding Yeast. Nat. Struct. Mol. Biol. 20(7): 884–891.

Porrua, O. and D. Libri. 2015. Transcription termination and the control of the transcriptome: why, where and how to stop. Nat. Rev. Mol. Cell Biol. 16(3): 190–202.

Preiss, T. and M.W. Hentze. 2003. Starting the protein synthesis machine: eukaryotic translation initiation. BioEssays 25: 1201–1211.

Prioleau, M.-N. and D.M. Macalpine. 2016. DNA replication origins-where do we begin? Genes Dev. 30(15): 1683–97.

Proshkin, S., A.R. Rahmouni, A. Mironov and E. Nudler. 2010. Cooperation between translating ribosomes and RNA polymerase in transcroption elongation. Science 328: 504–508 (5977).

Rampersad, S. and P. Tennant. 2018. Replication and Expression Strategies of Viruses. In Viruses: Molecular Biology, Host Interactions, and Applications to Biotechnology, 55–82. Elsevier.

Ray-Soni, A., M.J. Bellecourt and R. Landick. 2016. Mechanisms of bacterial transcription termination: All good things must End. Annu. Rev. Biochem. 85: 319–347.

Roberts, J.W., W. Yarnell, E. Bartlett, J. Guo, M. Marr, D.C. Ko et al. 1998. Antitermination by bacteriophage lambda Q protein. Cold Spring Harb. Symp. Quant. Biol. 63: 319–325.

Robinson, N.P. and S.D. Bell. 2005. Origins of DNA replication in the three domains of Life. FEBS Journal. 272: 3757–3766.

Rodnina, M.V. 2018. Translation in prokaryotes. Cold Spring Harb. Perspect. Biol. 10(9): a032664.

Rosonina, E., S. Kaneko and J.L. Manley. 2006. Terminating the transcript: Breaking up is hard to Do. Genes and Development 20(9): 1050–1056.

Schweikhard, V., C. Meng, K. Murakami, C.D. Kaplan, R.D. Kornberg and S.M. Block. 2014. Transcription factors TFIIF and TFIIS promote transcript elongation by RNA polymerase II by synergistic and independent mechanisms. Proc. Natl. Acad. Sci. U.S.A. 111(18): 6642–6647.

Sclafani, R.A. and T.M. Holzen. 2007. Cell cycle regulation of DNA replication. Annual Rev. Genet. 41: 237–280.

Shandilya, J. and S.G.E. Roberts. 2012. The transcription cycle in eukaryotes: From productive initiation to RNA Polymerase II Recycling. Biochim. Biophys. Acta Gene Regul. Mech. 1819(5): 391–400.

Shapiro, J.A. 2009. Revisiting the central dogma in the 21st Century. Ann. N.Y. Acad. Sci. 1178(1): 6–28.

Shelest, E. 2017. Transcription factors in fungi: TFome Dynamics, Three Major Families, and Dual-Specificity TFs. Front. Genet. 8: 53.

Shuman, S. 2009. DNA Ligases: Progress and prospects. J. Biol. Chem. 284(26): 17365–17369.

Siddiq, M.A. and P.J. Wittkopp. 2022. Mechanisms of regulatory evolution in yeast. Curr. Opin. Genet. Dev. 77: 101998.

Simonetti, A., S. Marzi, L. Jenner, A. Myasnikov, P. Romby, G. Yusupova et al. 2009. A structural view of translation initiation in bacteria. Cell. Mol. Life Sci. 66(3): 423–436.

Solar, G., R. Giraldo, M.J. Ruiz-Echevarría, M. Espinosa and R. Díaz-Orejas. 1998. Replication and control of circular bacterial plasmids. Microbiol. Mol. Biol. Rev. 62(2): 434–464.

Sonenberg, N. and T.E. Dever. 2003. Eukaryotic translation initiation factors and regulators. Curr. Opin. 13: 56–63.

Sonneveld, S., B.M.P. Verhagen and M.E. Tanenbaum. 2020. Heterogeneity in mRNA translation. Trends Cell Biol. 30(8): 606–618.

Thompson, D.A. and A. Regev. 2009. Fungal regulatory evolution: Cis and Trans in the Balance. FEBS Letters 583(24): 3959–3965.

Tomkinson, A.E., S. Vijayakumar, J.M. Pascal and T. Ellenberger. 2006. DNA Ligases: structure, reaction mechanism, and function. Chem. Rev. 106: 687–699.

Trojanowski, D., J. Hołówka and J. Zakrzewska-Czerwińska. 2018. Where and when bacterial chromosome replication starts: a single cell perspective. Front. Microbiol. 9: 2819.

V'kovski, P., A. Kratzel, S. Steiner, H. Stalder and V. Thiel. 2020. Coronavirus biology and replication: Implications for SARS-CoV-2. Nat. Revi. Microbiol. 19(3): 155–170.

Van Der Endet, A., T.A. Baker, T. Ogawa and A. Kornberg. 1985. Initiation of enzymatic replication at the origin of the *Escherichia coli* chromosome: Primase as the sole priming enzyme (DNA/OrC/Plasmids). Proc. Nati. Acad. Sci. USA. 82(12): 3954–3958.

Vos, S.M., E.M. Tretter, B.H. Schmidt and J.M. Berger. 2011. All tangled up: how cells direct, manage and exploit topoisomerase function. Nat. Rev. Mol. Cell Biol. 12: 827–841.

Wang, F., P. Sethiya, X. Hu, S. Guo, Y. Chen, A. Li et al. 2021. Transcription in fungal conidia before dormancy produces phenotypically variable conidia that maximize survival in different environments. Nat. Microbiol. 6(8): 1066–1081.

Washburn, R.S., M.E. Gottesman, D.M. Hinton and S. Wigneshweraraj. 2015. Regulation of transcription elongation and termination. Biomolecules 5(2): 1063–1078.

Webster, M.W. and A. Weixbaumer. 2021. The intricate relationship between transcription and translation. Proc. Natl. Acad. Sci. U.S.A., 118(21): e2106284118.

Worning, P., L.J. Jensen, P.F. Hallin, H.-H. Staerfeldt and D.W. Ussery. 2006. Origin of replication in circular Prokaryotic chromosomes. Environ. Microbiol. 8(2): 353–361.

Wu, W.J., W. Yang and M.D. Tsai. 2017. How DNA polymerases catalyse replication and repair with contrasting fidelity. Nat. Rev. Chem. 1: 0068.

Yang, H., Y. Ma, Y. Wang, H. Yang, W. Shen and X. Chen. 2014. Transcription Regulation mechanisms of bacteriophages. 5(5): 300–304.

Yarus, M., J.G. Caporaso and R. Knight. 2005. Origins of the genetic code: The escaped triplet theory. Ann. Rev. Biochem. 74: 179–98.

Zhang, L., P. Bisht, A. Flamier, M.I. Barrasa, M. Friesen, A. Richards et al. 2023. LINE1-mediated reverse transcription and genomic integration of SARS-CoV-2 mRNA detected in virus-infected but not in viral mRNA-transfected cells. Viruses 15(3): 629.

Zheng, L. and B. Shen. 2011. Okazaki fragment maturation: Nucleases take centre stage. J. Mol. Cell Biol. 3(1): 23–30.

CHAPTER 4

Variability and Diversity in the Microbial Genomes

Sylwia Okoń

Introduction

Genetic diversity is usually defined as the number of genetic characteristics (alleles and genotypes) in a species. Genetic polymorphism varies among species and within genomes, and has important implications for the evolution and conservation of species (Ellegren and Galtier 2016). The diversity of microbes presently living on Earth is known to be high and is thought to be enormous, but the true extent of microbial diversity is largely unknown. New molecular tools are now permitting the diversity of microbes to be explored rapidly and their evolutionary relationships and history to be defined (Chivian et al. 2023, Bornemann et al. 2023). Next-generation sequencing combined with bioinformatics analysis capabilities allows for the analysis of entire ecosystem metagenomes. This gives the opportunity to learn about a whole range of new species of microorganisms (Garud and Pollard 2020, Shu and Huang 2022). The possibilities of sequencing and comparative analysis of both entire genomes and gene fragments gave the opportunity to reliably analyse the level of diversity of microorganisms of the same species and to search for differences at the genome level of organisms from different environments or under stress (Estrada et al. 2022, Morita et al. 2023, Sengupta and Azad 2023).

Various mechanisms are responsible for creating and maintaining the genetic diversity of microorganisms. They mainly concern various mutations appearing in nucleic acids, whether during replication or genome editing. Genomic variability also results from gene transfer between organisms as well as gene movement within the genome. This diversity is then modulated by natural selection and random genetic drift, the effects of which in turn depend on many factors, including genetic architecture, demographics and ecology (Arber 2000, Sanjuán and Domingo-Calap 2021). The variability that appears in the genomes of microorganisms can have a variety of effects that determine survival in a particular environment. It may be the result of adaptation to changed conditions and enables the adaptation of a given species. Genetic variability can also have negative effects that prevent survival and adaptation to new conditions (Taylor et al. 2017, Shroeder et al. 2018). Genetic variability is the driving force of evolution; different groups of microorganisms, due to different mechanisms of generating genetic variability, evolve in different ways and at different speeds.

The aim of this chapter is to present the genetic variability of various groups of microorganisms. Both variability at the level of nucleic acid sequences resulting from various types of DNA mutations, as well as variability resulting from chromosome rearrangements, gene transfer and the acquisition of new features related, for example, to adaptation to changing environmental conditions.

Institute of Plant Genetics, Breeding and Biotechnology, University of Life Sciences in Lublin, Akademicka 15, 20–950.
Email: Lublin. sylwia.okon@up.lublin.pl

Viruses

Viruses are a highly heterogeneous group of molecular parasites showing varied replication strategies, genome organizations, mutability, and evolutionary properties. Different mechanisms are responsible for creating and maintaining genetic diversity in viruses, including error-prone replication, repair avoidance, and genome editing, among others. This diversity is subsequently modulated by natural selection and random genetic drift, whose action in turn depends on multiple factors including viral genetic architecture, viral demography, and ecology (Sanjuán and Domingo-Calap 2021).

Viruses are the biological systems with the greatest variation in mutation rates, with the greatest differences between RNA and DNA viruses. Viral mutation rates are approximately between 10^{-8} and 10^{-4} substitutions per nucleotide per cell infection (s/n/c), with DNA viruses in the range of $10^{-8} - 10^{-6}$ and RNA viruses in the range of $10^{-6} - 10^{-4}$. These differences have several mechanistic bases. First, the polymerases of the vast majority of RNA viruses lack 3' exonuclease proofreading activity and hence are more error-prone than those of DNA viruses (Roberts and Weintraub 1988, Steinhauer et al. 1992). Single-stranded DNA viruses tend to mutate faster than double-stranded DNA viruses. The mechanisms underlying these differences are not well understood. Single-stranded nucleic acids may be more susceptible to oxidative deamination and other types of chemical damage than double-stranded nucleic acids. Single-strand DNA viruses and double-strand DNA viruses may also differ in their ability to repair themselves after replication (Seronello et al. 2011).

The genome size of DNA-based microorganisms, such as viruses, bacteria and unicellular eukaryotes, is generally inversely related to the mutation rate. As a result of this rule, the mutation rate per genome remains relatively constant at approximately 0.003 per copy. RNA viruses also seem to exhibit similar negative relationships, but because their genome sizes differ less, it is more difficult to detect such trends (Sanjuán and Domingo-Calap 2016).

The main sources of genetic variability of viruses include various mechanisms, the action of which depends on the type and structure of the viral nucleic acid.

Nucleotide mismatches can occur during replication of the viral genetic material. These are mainly due to the level of fidelity of the polymerase and the rate of base misincorporation.

Most nucleotide misincorporations can be corrected by polymerases that display 3' exonuclease activity. However, this activity is generally absent from RNA virus polymerases. Currently, only coronavirus RNA polymerases have been shown to perform 3' exonuclease proofreading. Lack of proofreading is a major factor responsible for the higher rates of spontaneous mutation of RNA viruses compared to DNA viruses. Although DNA virus replicases have 3' exonuclease activity, some amino acid substitutions in their polymerases inactivate proofreading. Some viruses, e.g., polyomaviruses, which use host polymerases for replication, may encode proteins that inactivate the 3' exonuclease proofreading domain of the host polymerase. Therefore, correction avoidance may also be a mechanism by which some DNA viruses increase their population diversity (Steinhauer et al. 1992, Sanjuán et al. 2010, Sanjuán and Domingo-Calap 2021).

Post-replicative repair is a highly efficient system for removing replication errors and repairing DNA damage. RNA viruses are not substrates of cellular repair systems, whereas some DNA viruses have evolved specific features to avoid repair. Eukaryotic viruses might also avoid repair or dysregulate repair pathways to increase their mutation rates. Eukaryotic viruses are known to interact with DNA damage response pathways. DNA damage response pathways comprise error-prone DNA polymerases for re-synthesis of excised strands, which might contribute to elevating viral mutation rates (Weitzman and Fradet-Turcotte 2018, Sanjuán and Domingo-Calap 2021, Lopez et al. 2022).

Diversity-generating retroelements are genetic cassettes found in different bacteria, plasmids, and DNA bacteriophages. They contain two short repeated sequences, called the template and the variable repeats. The repeat of the template is transcribed and then reverse transcribed by the RT encoded by the cassette. During this process, adenines are systematically replaced with random bases. The resulting cDNA is integrated, replacing the old repeat with a new, highly mutated sequence.

This process enables the hypermutation of specific phage genes involved in host attachment, a trait that is often subject to rapidly changing selection pressures (Guo et al. 2014, Benler et al. 2018).

Viral mutations can also arise in response to host enzymes, which include cellular cytidine deaminases and adenosine deaminases. Cellular cytidine deaminases are catalytic polypeptide-like enzymes (APOBECs) that massively introduce C-to-U base substitutions in retroviral cDNA leading to a characteristic pattern of G-to-A hypermutations in retroviral genomes. They may contribute to viral evolution and even promote immune escape and drug resistance. In addition to retroviruses, APOBECs can edit hepatitis B virus as well as other non-reverse-transcribed DNA viruses such as papillomaviruses, herpesviruses, and human polyomaviruses (Stavrou and Ross 2015, Sadeghpour et al. 2021, Sanjuán and Domingo-Calap 2021). RNA-dependent double-stranded adenosine deaminases (ADARs) are another type of host-encoded enzymes capable of editing and hypermutating the genomes of RNA viruses. ADARs deaminate adenosines in long regions of double-stranded RNAs, converting them to inosines, leading to A-G base substitutions (Ringlander et al. 2022).

Recombination plays an important role in viral evolution. Recombination is a pervasive process generating diversity in most viruses (Figure 1). It joins variants that arise independently within the same molecule, creating new opportunities for viruses to overcome selective pressures and to adapt to new environments and hosts (Pérez-Losada et al. 2015). Recombination differs from

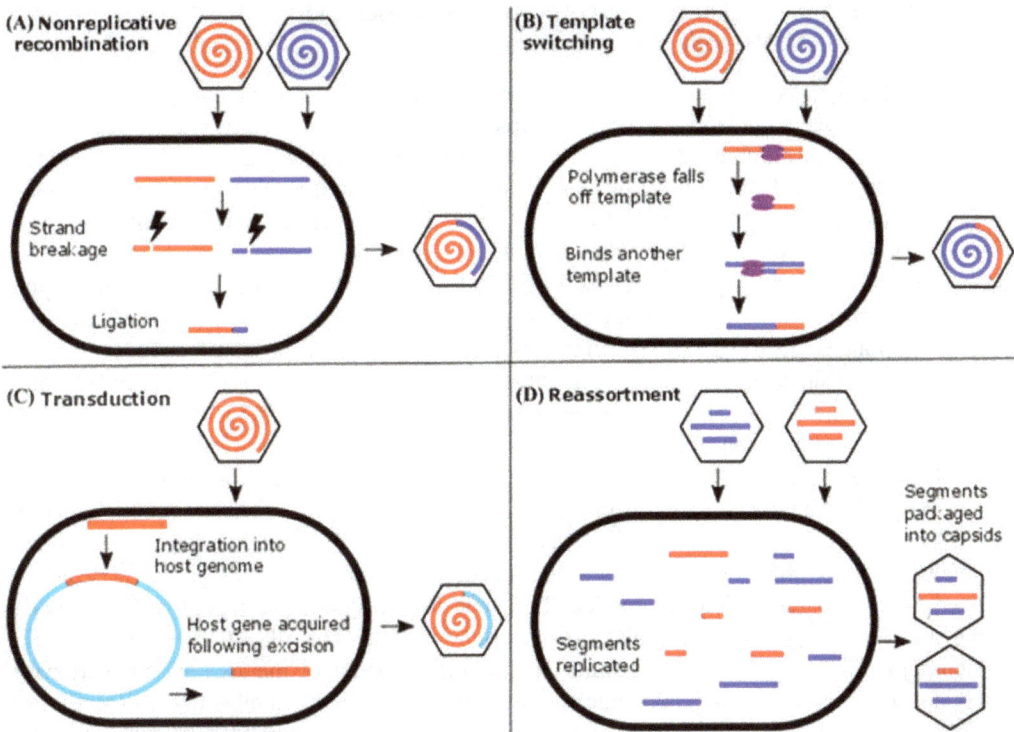

Figure 1. Major mechanisms of virus genetic recombination. (A) In non-replicative recombination, nucleic acid strand breakage and repair permit the recombination of genetic material from different sources into the same viral genome. Recombination can occur between homologous or nonhomologous sequences and between coinfecting viruses or between virus and foreign nucleic acid strands. (B) In replicative recombination or template switching, a polymerase molecule changes template during the process of replicating a nucleic acid strand. If the templates are derived from different sources, then novel genetic material can be introduced into the virus genome. (C) During the process of virus integration and excision from a host genome, viruses can acquire genetic material from the host. These genes can increase infectivity or aid in host suppression. (D) Reassortment occurs following coinfection of a host cell by multiple segmented viruses. Replicated genome segments are packaged into procapsids irrespective of the parent of origin. In this manner, segments from two or more parents can be packaged into the same procapsid, giving rise to progeny that are genetically different from either parent. Copyright © 2016 New York Academy of Sciences (Dennehy 2017).

mutation in the type and abundance of the genetic changes produced. Recombination between distantly related viruses has the ability to introduce major genetic changes in a single event, such as gene transfer, leading to the emergence of new viral subtypes or even new species. In segmented and multicomponent viruses, recombination is greatly facilitated by the ability to re-sort genome segments. Reassortment between viruses of the same subtype may also be an important source of diversity, may promote adaptation and accelerate immune escape. Even if reassortment is not possible, recombination can still take place through a template-switching mechanism whereby the viral polymerase and the nascent chain dissociate from one template and associate with another (Vijaykrishna et al. 2015, Lowen 2018).

Recombination is a more complex process in DNA viruses than in RNA viruses and involves specific host or virus recombinases. In addition, some DNA viruses encode their own recombination systems (Pérez-Losada et al. 2015).

Bacteria

On the basis of established knowledge of microbial genetics, one can distinguish three major natural strategies in the spontaneous generation of genetic variations in bacteria. These strategies are:

1. small local changes in the nucleotide sequence of the genome,
2. intragenomic reshuffling of segments of genomic sequences, and
3. the acquisition of DNA sequences from another organism.

Relatively often, a new mutation is explained by a local change in the DNA sequence, affecting one or several adjacent nucleotides. Some of these changes are explained by limited chemical or structural stability of the nucleotides, while others are attributed to the activities of the enzyme complex that mediates DNA replication. The main causes of spontaneous mutations occurring in the genome of bacteria relate primarily to errors made by polymerases during DNA replication. The DNA polymerases that replicate the genome are extremely accurate in their ability to pair nucleotides correctly, but they make errors that include pairing of unrelated dNTPs (Li and Lynch 2020). Mutations can also be the result of base pairing occurring in the form of rare tautomers (Pray 2008, Wang et al. 2011). Error correction by DNA polymerases is also influenced by the local sequence context (Sinha 1987). The high content of GC 5' or 3' relative to the mismatched nucleotide in nascent DNA makes correction less likely (Petruska and Goodman 1985, Carver et al. 1994). The reason for the formation of point mutations during the replication process may be the use of 8-oxo-dGTP as a substrate. Of the bases in DNA, guanine is the most common target for oxidation, forming 8-oxo-7,8-dihydroguanine (8-oxo-G) (Cadet et al. 2008, Kanvah et al. 2010). Mutations caused by 8-oxo-G are A·T → C·G transversions or G·C → T·A transversions. The oxidized guanine (GO) system limits mutagenesis by sanitizing nucleotide pools of 8-oxo-dGTP, removing oxidized 8-oxo-G bases in double-stranded DNA, and removing adenine paired with 8-oxo-G in double-stranded DNA (Michaels and Miller 1992, Lenhart et al. 2012). Ribonucleotide misincorporation is by far the most common error made by DNA polymerases, with the error rate largely due to the inequality of cellular nucleotide pools as NTPs far outnumber dNTPs (McElhinny et al. 2010, Reijns et al. 2012, Yao et al. 2013). Ribonucleotide excision repair (RER) is a biochemical pathway responsible for replacing ribonucleotides in DNA. It has been very well described in yeast (Lujan et al. 2014); despite the increasing number of studies on bacterial RER, the general mechanism of RER and the mechanisms of increased mutagenesis due to unrepaired ribonucleotides in bacterial DNA are poorly understood (Vaisman et al. 2013, 2014). Such errors generate mutations if they are not corrected before the next round of DNA replication by the corrective action of the replicative DNA polymerase or the DNA mismatch repair (MMR) system. Base pairing error generation, evasion of proofreading activity, and evasion of MMR each contribute to mutagenesis, but the mechanisms and factors affecting each are somewhat poorly understood.

Mutations may also arise as a result of incorrect alignment of the primer-template pair (Lovett 2004). During normal DNA replication, DNA polymerase can switch from using one template to another. This type of strand misalignment is called template switching and can lead to a variety of mutations, including base pair substitutions, insertions and deletions, and large structural rearrangements (Yoshiyama et al. 2001, Dutra and Lovett 2006, Anand et al. 2014, Tsaponina and Haber 2014). Template switching is facilitated by quasipalindrome sequences by forming a hairpin structure during DNA replication such that one arm of the quasipalindrome serves as the template for the other arm. Quasipalindromes are mutation hotspots in bacteria due to the increased likelihood of template switching events at hairpins (Viswanathan et al. 2000, Dutra and Lovett 2006, Seier et al. 2011, 2012). Long runs of homopolymers are hot spots for small insertions and deletions in bacteria due to some kind of misalignment of primers and templates referred to as "strand slippage" during DNA replication (Fujii et al. 1999, Lee et al. 2012, Schroeder et al. 2016).

The cause of point mutations in bacteria is also DNA damage by deamination of bases in nucleotides. Nucleobases in single-stranded DNA are more susceptible to deamination than those in double-stranded DNA (Frederico et al. 1990). Therefore, by transiently generating regions of single-stranded DNA, DNA transcription and replication greatly increase the likelihood of nucleobase deamination (Sung et al. 2015, Bhagwat et al. 2016, Schroeder et al. 2016).

DNA replication and transcription must use the same DNA template. DNA replication forks move much faster than transcription in bacteria, leading to inevitable conflicts between these two fundamental processes. These conflicts can lead to replication fork stops, homologous recombination and genomic instability and can also cause point mutations (Mirkin and Mirkin 2005, Mirkin et al. 2006, Srivatsan et al. 2010, Dutta et al. 2011). Replication-transcription conflicts strongly contribute to insertions and deletions in both the gene coding sequence and the promoter sequence (Dutta et al. 2011, Sankar et al. 2016).

Point mutations can arise spontaneously as a result of errors in DNA replication, or they can be induced by environmental mutagens that act directly on DNA. Chemicals that damage DNA can also indirectly cause mutations (Nicolette et al. 2018). Bacteria encode several genes to help repair DNA damage. If the amount of damage is significant, repair genes that are normally suppressed become active. This phenomenon is known as the SOS response. One of the induced genes reduces the correction of DNA polymerase and leads to an increase in the frequency of mutations (Radman 1975, Fijalkowska et al. 1997). This system may have evolved as a mechanism of hyperevolution to increase the ability to generate mutants that can survive in a hazardous environment, so it is a source of adaptive mutations (McKenzie et al. 2000, Foster 2007).

The mutation effect may be deleterious (inactivation or lower activity) or beneficial (enhanced or new activity) for bacterial cell. However, they appear at very different rates. For example, for *Escherichia coli* K-12, the rate of deleterious mutations per genome per replication is, at least, $2–8 \times 10^{-4}$, while that of beneficial mutations is, at least, 2×10^{-9} (Kibota and Lynch 1996, Boe et al. 2000, Imhof and Schlötterer 2001). Changes in the targets for several antibiotics can result in resistance or tolerance to the antibiotic (Levin-Reisman et al. 2017, Windels et al. 2019). In noncoding regions, point mutations can affect a variety of signals for expression and regulation of a gene (Yoshiyama et al. 2001, Sankar et al. 2016). Spontaneous single nucleotide changes can result in the generation of a functional gene. Such a process may account for the relationship of cholera toxin of Vibrio cholerae and enterotoxin of *E. coli*. The toxin proteins are highly homologous, but the genes are only expressed in the bacterial species from which they were isolated (Kuo et al. 2003, Shamini et al. 2011).

The second level of genetic variation in bacteria is changes caused by the intracellular transfer of genetic information. Global mutations (genome rearrangements) include inversion (also known as reversal), translocation, duplication, and transposition (Lara-Ramírez et al. 2011). Increasing number of prokaryotic genomes and their comparison have revealed the presence of a large number of genomic differences (Srivatsan et al. 2008, Skovgaard et al. 2011). These changes are caused by mobile genetic elements called transposable elements (TEs), which include transposons (Tn) and insertion sequences (IS). An insertion sequence (IS) is a small transposable element that encodes the proteins required for its own transposition, capable of jumping to other regions of the chromosome. Transposition rates vary between ISs and host species, but are often on the order of nucleotide substitution rates, making IS activity one of the more dynamic evolutionary forces found in many bacterial genomes. IS movement may also have functional implications for bacterial genomes. IS insertions before protein coding sequences may result in their increased expression, leading to different phenotypes depending on the function of the overexpressed gene (Olliver et al. 2005, Hamidian and Hall 2013).

Bacterial transposons belong to the DNA and Tn family transposons, which usually carry additional antibiotic resistance genes. They are made of insertion sequences and gene or genes located between them. Transposons can transfer from a plasmid to other plasmids or from a DNA chromosome to a plasmid and *vice versa*, resulting in the transmission of antibiotic resistance genes in bacteria (Babakhani and Oloomi 2018).

The third level of variation seen in bacterial genomes is variation due to horizontal gene flow. Horizontal Gene Transfer (HGT), also known as Lateral Gene Transfer (LTG), involves the transfer of genetic material from one organism to another, where the transferred genes are fixed in the genome. Intercellular transfer of genetic information enables bacteria to enrich their own gene pool with additional genetic information, often from taxonomically unrelated species of bacteria. This leads to a huge genetic variability of microorganisms within a single species and provides them with a rapid rate of evolution (Heuer and Smalla 2007).

HGT of genetic modules that allowed adaptation to rapidly evolving biotic interactions was frequently observed. Such interactions are, e.g., the production of antibiotics by microbes or their use by humans resulting in the spread of antibiotic resistance, the release of xenobiotics or new

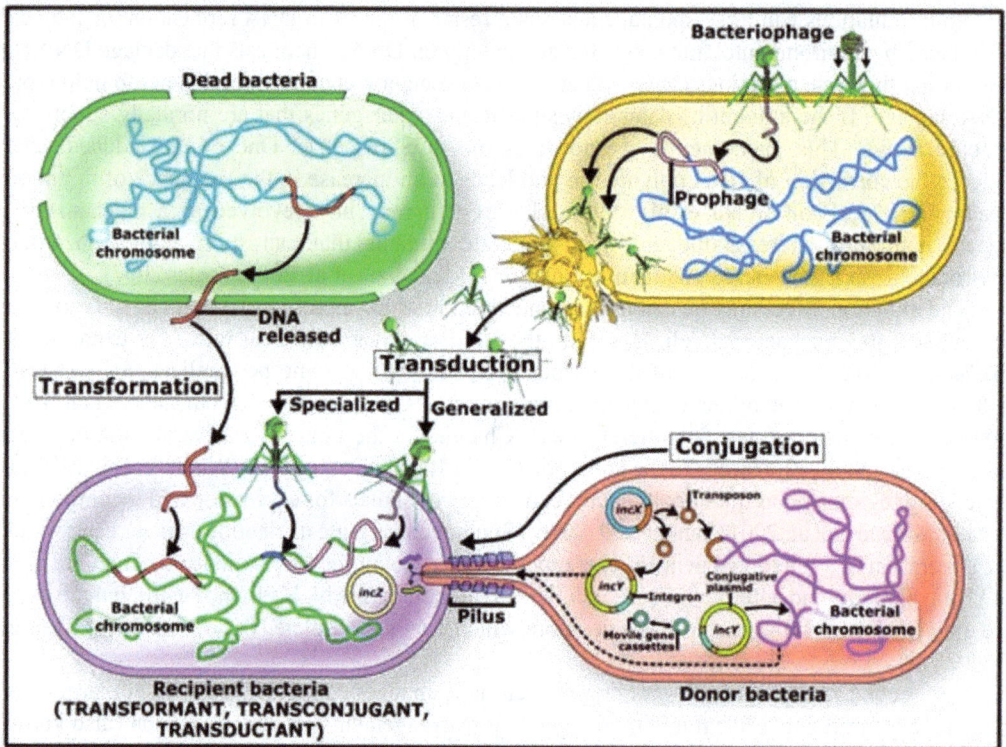

Figure 2. Involved mechanisms in horizontal gene transfer. Transduction, conjugation, and transformation are the main mechanisms by which bacterial species can mobilize and share genetic material with both related and non-related species. Reprinted from Bello-Lopez et al. (2019) under a Creative Commons Attribution (CC BY) license.

secondary metabolites and the spread of degradative genes and pathway assembly, or pathogenic and symbiotic interactions and the spread of genomic islands (Witte 1998, McManus et al. 2002, Top and Springael 2003, Hacker and Kaper 2003, Larraín-Linton et al. 2006).

Horizontal gene transfer (HGT) between bacteria is driven by three main processes: transformation (uptake of free DNA), transduction (bacteriophage-mediated gene transfer), and conjugation (gene transfer using plasmids or integrative conjugation elements) (Figure 2). Mobile genetic elements (MGEs) such as plasmids, bacteriophages, integrative conjugation elements, transposons, IS elements, integrons, gene cassettes and genomic islands are important carriers in the latter two processes (Ochman et al. 2000, Brüssow et al. 2004, Dobrindt et al. 2004).

Natural transformation is generally understood as the uptake of free DNA by competent bacteria (Chen and Dubnau 2004). DNA uptake can be followed by integration into the bacterial genome by homologous recombination, illicit homology-facilitated recombination, or by creating an autonomously replicating element (de Vries and Wackernagel 2002). Natural transformation provides a gene transfer mechanism that enables competent bacteria to generate genetic variation by utilizing the DNA present in their environment (Nielsen et al. 2000). The condition for natural transformation is the availability of free DNA, the development of competence, and the collection and stable integration of the captured DNA. The development of bacterial competence depends on many factors. It can be affected by, for example, environmental stress, changed growth conditions or lack of nutrients (Berka et al. 2002). Archaea and bacteria belonging to various groups of gram-positive bacteria, gram-negative bacteria, cyanobacteria, and green sulfur bacteria are capable of natural transformation. Many human pathogenic bacteria also have the ability to take up free DNA from the environment, including representatives of the genera *Campylobacter, Haemophilus, Helicobacter, Neisseria, Pseudomonas, Staphylococcus* and *Streptococcus* (Lorenz and Wackernagel 1994, O'Connell et al. 2022).

Transduction is a DNA harvesting mechanism by which non-viral DNA can be transferred from an infected host bacterium to a new host via infectious or non-infectious viral particles. The host DNA is mistakenly inserted into the empty phage head when the phage particle is produced. Defective phage particles that are released from the lysed host cells can adsorb to the new host cells and deliver the DNA encapsidated to the new host. The injected bacterial DNA can be integrated into the recipient genome. Although most bacteriophages only infect a narrow range of hosts, this gene transfer mechanism has the advantage that transducing phages can be rather stable under environmental conditions, do not require cell-cell contact, and the DNA in the transducing phage particles is protected (Wommack and Colwell 2000).

Most of the sequenced bacterial genomes contain prophage sequences (Canchaya et al. 2003). Many pathogenicity determinants (toxins) have been acquired via phages, e.g., by *Corynebacterium diphtheriae, Clostridium botulinum, Streptococcus pyogenes, Staphylococcus aureus* and Shiga toxin producing *E. coli* (Brüssow et al. 2004). Pathogenicity islands (PAI) are large genomic islands that carry one or more virulence gene, which often evolved from lysogenic bacteriophages, and are assumed to be more frequent in pathogenic strains than non-pathogenic strains (Hacker and Kaper 2003, Dobrindt et al. 2004). However, complete annotation of genome sequences revealed that some nonpathogenic strains can also carry PAIs encoding traits such as adhesins, iron uptake systems or proteases, which contribute to general adaptability, fitness and competitiveness, but lack prominent virulence factors (Grozdanov et al. 2004).

Conjugation is a multi-stage process of gene transfer by direct physical contact of donor and recipient cells with saws facilitating the process (Cabezón et al. 2015). The transfer of these conjugative genes requires a sophisticated machinery that ensures DNA and mating pair formation mobilization (Garcillán-Barcia et al. 2009, de La Cruz et al. 2010). These genes can be encoded by an autonomous replicating plasmid or by integrative conjugative elements (ICE) inserted in the chromosome (Wozniak and Waldor 2010, Smillie et al. 2010, Guglielmini et al. 2011). When DNA passes through the pile between bacteria of the same species, it is referred to as vertical transfer, when between different species, it is horizontal transfer.

The most important mobile elements involved in the conjugation process are plasmids, conjugative transposons and conjugative gene islands. Often traits conferring better fitness or the ability to colonize environmental niches are located on conjugative MGEs. Conjugative plasmids can mediate the lateral transfer of antibiotic resistance or virulence determinants between bacteria, allowing bacteria to adapt to otherwise hostile environments (Ochman et al. 2000, Ochman and Moran 2001). Conjugative plasmids can also induce biofilm development (Ghigo 2001). A biofilm niche is often found in an aquatic environment or at sites of bacterial infection. At these sites, conjugative plasmids may also play an important role in extensive lateral gene transfer, where virulence genes are mobilized between bacterial species.

Archeae

Archaea remains the least studied and least characterized domain of life, despite its importance not only to the ecology of our planet, but also to the evolution of eukaryotes. Members of the Archaea domain are diverse and widespread, found not only in extreme environments such as hot springs and deep hydrothermal vents, but also in a range of temperate and aerobic environments such as soils, marine and freshwater plankton. The development of DNA analysis technology has enabled the study of samples coming directly from the natural environment, which allowed the discovery of the microbial diversity of these ecosystems, including the diversity of archaea (Takai and Horikoshi 1999, Schleper et al. 2005).

The primary source of variability in Archaea, as in bacteria, is mutation. The accumulation of mutations in archaea has been studied on the example of *H. volcanii*, a model organism for archaea, and is $3.15 \times 10-10$ per site per generation or 0.0012 per genome per generation, and is similar to the value found in mesophilic prokaryotes (Lynch 2010, Kucukyildirim 2020). In principle, mutation rates can be influenced by a number of factors including growth conditions, growth medium, number of generations, temperature, and genome copy number (ploidy level). Extremely high temperatures in which *Hyperthermophilic* archaea grow are stressful and may result in very low growth/survival rates and have mutagenic effects. Also, multiple copies of the genome allow for the identification of more mutations than in haploid organisms (Kucukyildirim 2020).

Horizontal gene flow is also a source of archaeal variation. Gene transfer between organisms can occur via cytoplasmic bridges as in *H. volcanii,* which allow DNA transfer in both directions (Rosenshine et al. 1989, Sivabalasarma 2021) Gene transfer can also occur through cell aggregation as in *S. solfataricus* and *S. acidocaldarius*. Cellular aggregation is believed to enhance species-specific DNA transfer between *Sulfolobus* cells to provide enhanced repair of damaged DNA by homologous recombination (Fröls et al. 2008).

Fungi

The basis of the genetic variability of fungi is sexual recombination, which ensures the ability to generate offspring with genotypes other than parents and siblings. Random matings and mutations give an almost infinite number of gene combinations, which makes fungi genetically very diverse (Stukenbrock 2016, Möller and Stukenbrock 2017, Feurtey and Stukenbrock 2018). Sexual recombination in fungi is typically sporadic, and sexual stages have not been identified for many fungal lineages. The existence of sexual and parasexual cycles opens the possibility for recombinant lineages, including the formation of inter-species hybrids (Peter et al. 2018).

Mutations in fungal cells arise spontaneously under the influence of nucleotide mismatches during replication or under the influence of external factors such as UV light or chemical substances (Ma and Michailides 2005, Deising et al. 2008). Single-nucleotide polymorphisms (SNPs), usually caused by errors in DNA replication and repair, are a major source of genomic variation (Roberts and Weintraub 1988). They are responsible, among others, for increases in pathogenicity, and adaptation to new conditions. Genome duplication is a major evolutionary process that has shaped the genomes of several eukaryotic lineages including fungi. Duplications may concern individual genes, genomic segments or entire genomes (Wolfe and Shields 1997, Llorente et al. 2000). Such duplications are routinely acquired and lost in fungi (Koszul et al. 2004, 2006). Duplications occur more frequently in genes encoding proteins involved in the response to controls than in genes encoding housekeeping proteins. Most duplicates are rapidly lost, and where duplicated regions are retained, natural selection is likely to be responsible. Conserved duplications result in the differential expression of paralogs rather than the emergence of new gene functions (Wapinski et al. 2007).

Gene loss also contributes to the genetic variability of fungi. It can be an important factor contributing to, for example, the infection of new hosts through the loss of genes whose products engage host defence mechanisms (Daverdin et al. 2012, Hartmann et al. 2017).

Gene families are groups of non-allelic genes with high sequence similarity and encoding proteins with similar functions. Comparative phylogenetics of fungal genomes indicated that gene family expansion and contraction can be correlated with the adaptation of fungi to a specific environment. Pathogenesis-related genes are an example of such families (Dean et al. 2005). Such families expand or contract depending on the type of host (Baroncelli et al. 2016). Shrinkage and expansion of gene families has also been observed between plant and animal pathogens, e.g., expansion of genes encoding proteinases and contractions in genes encoding cellulases and other enzymes deconstructing plant cell walls have been documented in animal pathogens (Sharpton et al. 2009).

Horizontal gene transfer is also a source of fungal variability. For example, genes involved in adaptation or pathogenesis are transferred via HGT (Kurland et al. 2003, Friesen et al. 2006). Gene clusters involved in one metabolic pathway may also be transferred (Slot and Rokas 2011, Reynolds et al. 2017). Comparative and phylogenetic studies of fungal genomes allowed the discovery of

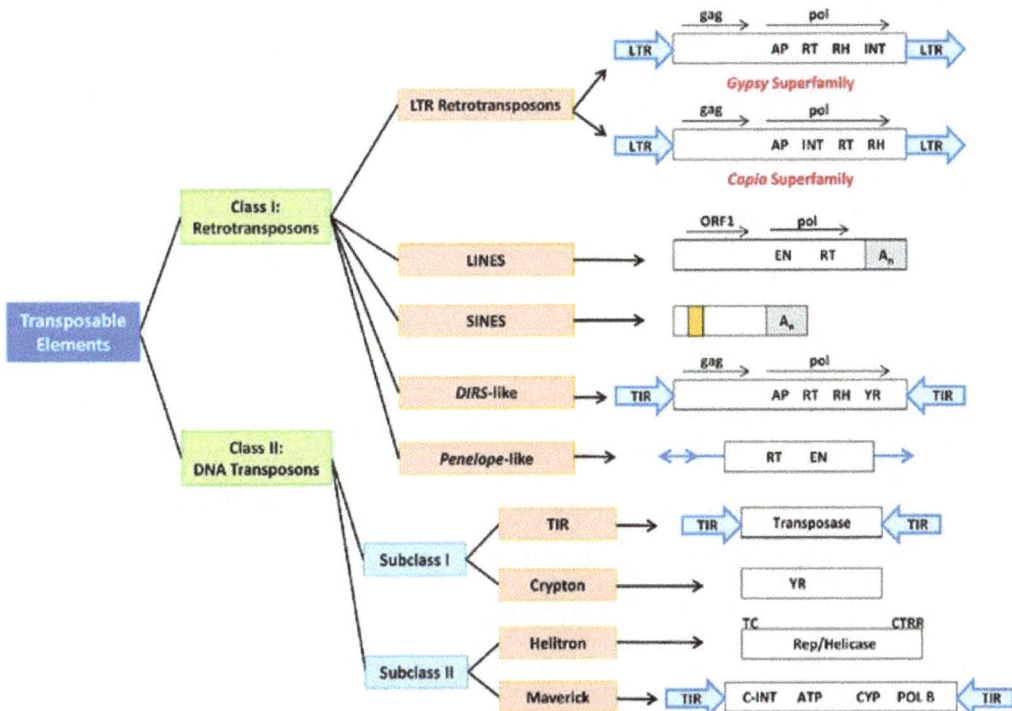

Figure 3. The orders and superfamilies of transposable elements present in fungi (Note: LTR = long terminal repeats, TIR = terminal inverted repeats, AP = aspartic protease, RT = reverse transcriptase, RH = RNase H, INT = integrase, EN = endonuclease, An = poly(A) tail, YR = tyrosine recombinase, ATP = packaging ATPase, CYP = cysteine protease, POL B = DNA polymerase B. Internal RNA polymerase III promoter region of SINEs is indicated by orange box). Reprinted from Razali et al. (2019) under a Creative Commons Attribution (CC BY 4.0) license.

genes from other species (kingdoms) obtained by means of horizontal gene flow. Plant-fungal HGTs have been shown to work in both directions, and although rare and apparently ancient, the exchange of transport proteins and siderophores is thought to have aided invasion into the terrestrial environment by both fungi and plants (Richards et al. 2009). Far more common are HGT events between the Fungi and Oomycota kingdoms; here it is safe to say that HGT from fungi, mainly the Ascomycota, enabled the Oomycota to become plant parasites through the transfer of genes related to cell wall destruction, nutrient uptake, and pathogenicity genes (Richards et al. 2011). An example of HGT is also the acquisition of bacterial glycoside hydrolase genes by fungi of the genus *Chytridiomycota*, which helped them adapt to feeding on plant cells in ruminant stomachs (Garcia-Vallvé et al. 2000). Horizontal gene transfer between fungi and other kingdoms is well documented; however, its mechanism is not well understood (Soanes and Richards 2014, Taylor et al. 2017, Husnik and McCutcheon 2018).

The variability in fungal genomes can also be caused by the presence of transposable elements (TEs), which can be transferred both within one genome and between genomes (Figure 3) (Daboussi et al. 2002, Muszewska et al. 2011, Castanera et al. 2017). Transposable elements (TEs) provide genetic variation by disrupting existing genes and gene sequences, stimulating rearrangements and duplications, and simply adding genetic material that can then evolve into new functions (Koufopanou et al. 2002, Vanheule et al. 2016, Horns et al. 2017).

Chromosome rearrangement can also occur in the fungal genome, which is also a source of fungal genetic variation, and provides a mechanism for gene family expansion and contraction through gene duplication, deletion, or disruption (Ohm et al. 2012, Vakirlis et al. 2016). Rearrangements can contribute to host specialization (Hartmann et al. 2017).

Conclusions

The development of sequencing techniques allowed us not only to learn the sequence of nucleotides in nucleic acids, but also to accurately study the genetic diversity of living organisms. Such research gave us the opportunity to determine at what levels we can identify the variability of microorganisms and what is the impact of genetic variability on their evolution. A more precise characterization of mutations makes it possible to study the adaptive mechanisms of microorganisms and opens the way to further work related to their more precise characterization.

References

Anand, R.P., O. Tsaponina, P.W. Greenwell, C.S. Lee, W. Du, T.D. Petes et al. 2014. Chromosome rearrangements via template switching between diverged repeated sequences. Gene Dev. 28(21): 2394–2406.

Arber, W. 2000. Genetic variation: molecular mechanisms and impact on microbial evolution. FEMS Microbiol. Rev. 24(1): 1–7.

Babakhani, S. and M. Oloomi. 2018. Transposons: The agents of antibiotic resistance in bacteria. J. Basic Microb. 58(11): 905–917.

Baroncelli, R., D.B. Amby, A. Zapparata, S. Sarrocco, G. Vannacci, G. le Floch et al. 2016. Gene family expansions and contractions are associated with host range in plant pathogens of the genus *Colletotrichum*. BMC Genom. 17(1): 1–17.

Bello-López J.M., O.A Cabrero-Martínez, G. Ibáñez-Cervantes, C. Hernández-Cortez, L.I. Pelcastre-Rodríguez, L.U. Gonzalez-Avila et al. 2019. Horizontal gene transfer and its association with antibiotic resistance in the genus *Aeromonas* spp. Microorganisms 7: 363.

Benler, S., A.G. Cobián-Güemes, K. McNair, S.H. Hung, K. Levi, R. Edwards et al. 2018. A diversity-generating retroelement encoded by a globally ubiquitous *Bacteroides* Phage. Microbiome 6(1): 1–10.

Berka, R.M., J. Hahn, M. Albano, I. Draskovic, M. Persuh, X. Cui et al. 2002. Microarray analysis of the *Bacillus subtilis* K-state: Genome-wide expression changes dependent on ComK. Mol. Microbiol. 43(5): 1331–1345.

Bhagwat, A.S., W. Hao, J.P. Townes, H. Lee, H. Tang and P.L. Foster. 2016. Strand-biased cytosine deamination at the replication fork causes cytosine to thymine mutations in *Escherichia coli*. P Natl A Sci USA 113(8): 2176–2181.

Boe, L., M. Danielsen, S. Knudsen, J.B. Petersen, J. Maymann and P.R. Jensen. 2000. The frequency of mutators in populations of *Escherichia coli*. Mut. Res. 448(1): 47–55.

Bornemann T., S.P. Esser, T.L. Stach, T. Burg and A.J. Probst. 2023. uBin: A manual refining tool for genomes from metagenomes. Envinron. Microbiol. 1–7: doi.org/10.1111/1462-2920.16351.

Brüssow, H., C. Canchaya and W.-D. Hardt. 2004. Phages and the evolution of bacterial pathogens: from genomic rearrangements to lysogenic conversion. Microbiol. Mol. Biol. R. 68(3): 560.

Cabezón, E., J. Ripoll-Rozada, A. Peña, F. de la Cruz and I. Arechaga. 2015. Towards an integrated model of bacterial conjugation. FEMS Microbiol. Rev. 39(1): 81–95.

Cadet, J., T. Douki and J. Ravanat. 2008. Oxidatively generated damage to the guanine moiety of DNA: Mechanistic aspects and formation in cells. ACS Pub. 41(8): 1075–1083.

Canchaya, C., C. Proux, G. Fournous, A. Bruttin and H. Brüssow. 2003. Prophage genomics. Microbiol Mol. Biol. R 67(2): 238–276.

Carver, T.E., R.A. Hochstrasser and D.P. Millar. 1994. Proofreading DNA: Recognition of aberrant DNA termini by the klenow fragment of DNA Polymerase I. P Natl. A Sci. USA 91(22): 10670–10674.

Castanera, R., A. Borgognone, A.G. Pisabarro and L. Ramírez. 2017. Biology, dynamics, and applications of transposable elements in *Basidiomycete* Fungi. Appl. Microbiol. Biot. 101(4): 1337–1350.

Chen, I. and D. Dubnau. 2004. DNA uptake during bacterial transformation. Nat. Rev. Microbiol. 2(3): 241–249.

Chivian, D., S.P. Jungbluth, P.S. Dehal, E.M. Wood-Charlson, R.S. Canon, B.H. Allen et al. 2023. Metagenome-assembled genome extraction and analysis from microbiomes using KBase. Nat. Protoc. 18: 208–238.

Daboussi, M.J., J.M. Davière, S. Graziani and T. Langin. 2002. Evolution of the Fot1 transposons in the genus *Fusarium*: Discontinuous distribution and epigenetic inactivation. Mol. Biol. Evol. 19(4): 510–520.

Daverdin, G., T. Rouxel, L. Gout, J.N. Aubertot, I. Fudal, M. Meyer et al. 2012. Genome structure and reproductive behaviour influence the evolutionary potential of a fungal phytopathogen. PLOS Pathog 8(11): e1003020.

Dean, R.A., N.J. Talbot, D.J. Ebbole, M.L. Farman, T.K. Mitchell, M.J. Orbach et al. 2005. The genome sequence of the rice blast fungus *Magnaporthe grisea*. Nature 434(7036): 980–986.

Deising, H.B., S. Reimann and S.F. Pascholati. 2008. Mechanisms and significance of fungicide resistance. Braz. J. Microbiol. 39(2): 286.

Dennehy, J.J. 2017. Evolutionary ecology of virus emergence. Ann. N Y Acad. Sci. 1389(1): 124-146.

Dobrindt, U., B. Hochhut, U. Hentschel and J. Hacker. 2004. Genomic islands in pathogenic and environmental microorganisms. Nat. Rev. Microbiol. 2(5): 414–424.

Dutra, B.E. and S.T. Lovett. 2006. Cis and trans-acting effects on a mutational hotspot involving a replication template switch. J. Mol. Biol. 356(2): 300–311.

Dutta, D., K. Shatalin, V. Epshtein, M.E. Gottesman and E. Nudler. 2011. Linking RNA polymerase backtracking to genome instability in *E. coli*. Cell 146(4): 533–543.

Ellegren, H. and N. Galtier. 2016. Determinants of genetic diversity. Nat. Rev. Genet. 17(7): 422–433.

Estrada, A.A., M. Gottschalk, C.J. Gebhart and D.G. Marthaler. 2022. Comparative analysis of *Streptococcus suis* genomes identifies novel candidate virulence-associated genes In North American isolates. Vet. Res. 53: 23.

Feurtey, A. and E.H. Stukenbrock. 2018. Interspecific gene exchange as a driver of adaptive evolution in fungi. Annu. Rev. Microbiol. 72: 377–398.

Fijalkowska, I.J., R.L. Dunn and R.M. Schaaper. 1997. Genetic requirements and mutational specificity of the *Escherichia coli* SOS mutator activity. J. Bacteriol. 179(23): 7435–7445.

Foster, P.L. 2007. Stress-induced mutagenesis in bacteria. Crit. Rev. Biochem. Mol. 42(5): 373.

Frederico, L.A., B.R. Shaw and T.A. Kunkel. 1990. A sensitive genetic assay for the detection of cytosine deamination: Determination of Rate constants and the activation energy. Biochemistry 29(10): 2532–2537.

Friesen, T.L., E.H. Stukenbrock, Z. Liu, S. Meinhardt, H. Ling, J.D. Faris et al. 2006. Emergence of a new disease as a result of interspecific virulence gene transfer. Nat. Genet. 38(8): 953–956.

Fröls, S., M. Ajon, M. Wagner, D. Teichmann, B. Zolghadr, M. Folea et al. 2008. UV-inducible cellular aggregation of the hyperthermophilic archaeon *Sulfolobus solfataricus* is mediated by pili formation. Mol. Microbiol. 70(4): 938–52.

Fujii, S., M. Akiyama, K. Aoki, Y. Sugaya, K. Higuchi, M. Hiraoka et al. 1999. DNA replication errors produced by the replicative apparatus of *Escherichia coli*. J. Mol. Biol. 289(4): 835–850.

Garcia-Vallvé, S., A. Romeu and J. Palau. 2000. Horizontal gene transfer of glycosyl hydrolases of the rumen fungi. Mol. Biol. Evol. 17(3): 352–361.

Garcillán-Barcia, M.P., M.V. Francia and F. de La Cruz. 2009. The diversity of conjugative relaxases and its application in plasmid classification. FEMS Microbiol. Rev. 33(3): 657–687.

Garud, N.R. and K.S. Pollard. 2020. Population genetics in the human microbiome. Trends Genet. 36(1): 53–67.

Ghigo, J.-M. 2001. Natural conjugative plasmids induce bacterial biofilm development. Nature 412(6845): 442–445.

Grozdanov, L., C. Raasch, J. Schulze, U. Sonnenborn, G. Gottschalk, J. Hacker et al. 2004. Analysis of the genome structure of the nonpathogenic probiotic *Escherichia coli* Strain Nissle 1917. J. Bacteriol. 186(16): 5432–5441.

Guglielmini, J., L. Quintais, M.P. Garcillán-Barcia, F. de la Cruz and E.P.C. Rocha. 2011. The repertoire of ICE in prokaryotes underscores the unity, diversity, and ubiquity of conjugation. PLOS Genet 7(8): e1002222.

Guo, H., D. Arambula, P. Ghosh and J.F. Miller. 2014. Diversity-generating retroelements in phage and bacterial genomes. Microbiol. Spectrum. 2(6): 10.1128/microbiolspec.MDNA3–0029-2014.

Hacker, J. and J.B. Kaper. 2003. Pathogenicity Islands and the evolution of microbes. Annu. Rev. Microbiol. 54: 641–679.

48 *Microbial Genetics*

Hamidian, M. and R. Hall. 2013. ISAba1 targets a specific position upstream of the intrinsic AmpC gene of *Acinetobacter baumannii* leading to cephalosporin resistance. J.Antimicrob. Chemoth. 68(11): 2682–2683.

Hartmann, F.E., A. Sánchez-Vallet, B.A. McDonald and D. Croll. 2017. A fungal wheat pathogen evolved host specialization by extensive chromosomal rearrangements. ISME J. 11(5): 1189–1204.

Heuer, H. and K. Smalla. 2007. Horizontal gene transfer between bacteria. Environ. Biosafety Res. 6(1-2): 3–13.

Horns, F., E. Petit and M.E. Hood. 2017. Massive expansion of gypsy-like retrotransposons in *Microbotryum* fungi. Genome Biol. Evol. 9(2): 363–371.

Husnik, F. and J. McCutcheon. 2018. Functional horizontal gene transfer from bacteria to Eukaryotes. Nat. Rev. Microbiol. 16: 67–79.

Imhof, M. and C. Schlötterer. 2001. Fitness effects of advantageous mutations in evolving *Escherichia coli* populations. P Natl. A Sci. USA 98(3): 1113–1117.

Kanvah, S., J. Joseph, G.B. Schuster, R.N. Barnett, C.L. Cleveland and U. Landman. 2010. Oxidation of DNA: Damage to nucleobases. ACS Publ. 43(2): 53.

Kibota, T.T. and M. Lynch. 1996. Estimate of the genomic mutation rate deleterious to overall fitness in *E. coli*. Nature 381(6584): 694–696.

Koszul, R., S. Caburet, B. Dujon and G. Fischer. 2004. Eucaryotic genome evolution through the spontaneous duplication of large chromosomal segments. EMBO J. 23(1): 234–243.

Koszul, R., B. Dujon and G. Fischer. 2006. Stability of large segmental duplications in the yeast genome. Genetics 172(4): 2211–2222.

Kucukyildirim S., M. Behringer, E. M. Williams, T.G. Doak and M. Lynch. 2020. Estimation of the genome-wide mutation rate and spectrum in the *Archaeal* species *Haloferax volcanii*, Genetics. 215(4): 1107–1116.

Koufopanou, V., M.R. Goddard and A. Burt. 2002. Adaptation for horizontal transfer in a homing Endonuclease. Mol. Biol. Evol. 19(3): 239–246.

Kuo, M.M.C., Y. Saimi and C. Kung. 2003. Gain-of-function mutations indicate That *Escherichia coli* Kch Forms a functional K+ Conduit *in vivo*. EMBO J. 22(16): 4049.

Kurland, C.G., B. Canback and O.G. Berg. 2003. Horizontal gene transfer: A critical view. PNAS 100(17): 9658–9662.

de La Cruz, F., L.S. Frost, R.J. Meyer and E.L. Zechner. 2010. Conjugative DNA metabolism in gram-negative bacteria. FEMS Microbiol. Rev. 34(1): 18–40.

Lara-Ramírez, E.E., A. Segura-Cabrera, X. Guo, G. Yu, C.A. García-Pérez and M.A. Rodríguez-Pérez. 2011. New implications on genomic adaptation derived from the *Helicobacter pylori* genome comparison. PLOS ONE 6(2): e17300.

Larraín-Linton, J., R. de La Iglesia, F. Melo and B. González. 2006. Molecular and population analyses of a recombination event in the catabolic plasmid PJP4. J. Bacteriol. 188(19): 6793–6801.

Lee, H., E. Popodi, H. Tang and P.L. Foster. 2012. Rate and molecular spectrum of spontaneous mutations in the bacterium *Escherichia coli* as determined by whole-genome sequencing. P Natl. A Sci. USA 109(41): E2774–E2783.

Lenhart, J.S., J.W. Schroeder, B.W. Walsh and L.A. Simmons. 2012. DNA repair and genome maintenance in *Bacillus subtilis*. Microbiol. Mol. Biol. R 76(3): 530–564.

Levin-Reisman, I., I. Ronin, O. Gefen, I. Braniss, N. Shoresh and N.Q. Balaban. 2017. Antibiotic tolerance facilitates the evolution of resistance. Science 355(6327): 826–830.

Li, W. and M. Lynch. 2020. Universally high transcript error rates in bacteria. ELife 9: 1–15.

Llorente, B., A. Malpertuy, C. Neuvéglise, J. de Montigny, M. Aigle, F. Artiguenave et al. 2000. Genomic exploration of the Hemiascomycetous Yeasts: 18. Comparative analysis of chromosome maps and synteny with *Saccharomyces cerevisiae*. FEBS Lett. 487(1): 101–112.

Lopez, A., R. Nichols Doyle, C. Sandoval, K. Nisson, V. Yang and O.I. Fregoso. 2022. Viral modulation of the DNA damage response and innate immunity: Two sides of the same coin. J. Mol. Biol. 434(6): 167327.

Lorenz, M.G. and W. Wackernagel. 1994. Bacterial gene transfer by natural genetic transformation in the environment. Microbiol. Rev. 58(3): 563–602.

Lovett, S.T. 2004. Encoded errors: Mutations and rearrangements mediated by misalignment at repetitive DNA sequences. Mol. Microbiol. 52(5): 1243–1253.

Lowen, A.C. 2018. It's in the Mix: Reassortment of segmented viral genomes. PLOS Pathog 14(9): e1007200.

Lujan, S.A., A.R. Clausen, A.B. Clark, H.K. MacAlpine, D.M. MacAlpine, E.P. Malc et al. 2014. Heterogeneous polymerase fidelity and mismatch repair bias genome variation and composition. Genome Res. 24(11): 1751–1764.

Lynch, M. 2010. Evolution of the mutation rate. Trends Genet. 26: 345–352.

Ma, Z. and T.J. Michailides. 2005. Advances in understanding molecular mechanisms of fungicide resistance and molecular detection of resistant genotypes in phytopathogenic fungi. Crop Prot. 24(10): 853–863.

McElhinny, S., D. Kumar, A. Clark, D.L. Watt, B.E. Watts, E. Ludstrom et al. 2010. Genome instability due to ribonucleotide incorporation into DNA. Nat. Chem. Biol. 6: 774–781.

McKenzie, G.J., R.S. Harris, P.L. Lee and S.M. Rosenberg. 2000. The SOS response regulates adaptive mutation. P Natl. A Sci. 97(12): 6646–6651.

McManus, P.S., V.O. Stockwell, G.W. Sundin and A.L. Jones. 2002. Antibiotic use in plant agriculture. Annu. Rev. Phyto. 40(1): 443–465.

Michaels, M.L. and J.H. Miller. 1992. The GO system protects organisms from the mutagenic effect of the spontaneous Lesion 8-Hydroxyguanine (7,8-Dihydro-8-Oxoguanine). J Bacteriol. 174(20): 6321–6325.

Mirkin, E.v. and S.M. Mirkin. 2005. Mechanisms of transcription-replication collisions in bacteria. Mol. Cell Biol. 25(3): 888–895.

Mirkin, E.v., D.C. Roa, E. Nudlet and S.M. Mirkin. 2006. Transcription regulatory elements are punctuation marks for DNA replication. P Natl. A Sci USA 103(19): 7276–7281.

Morita D., H. Arai, J. Isobe, E. Maenishi, T. Kumagai, F. Maruyama et al. 2023. Whole-genome and plasmid comparative analysis of *Campylobacter jejuni* from human patients in Toyama, Japan, from 2015 to 2019. Microbiol. Spectrum. 11(1): e02659–22

Möller, M. and E.H. Stukenbrock. 2017. Evolution and genome architecture in fungal plant pathogens. Nat. Rev. Microbiol. 15(12): 756–771.

Muszewska, A., M. Hoffman-Sommer and M. Grynberg. 2011. LTR retrotransposons in fungi. PLOS ONE 6(12): e29425.

Nicolette, J., E. Dakoulas, K. Pant, M. Crosby, A. Kondratiuk, J. Murray et al. 2018. A comparison of 24 chemicals in the six-well bacterial reverse mutation assay to the standard 100-Mm petri plate bacterial reverse mutation assay in two laboratories. Regul Toxicol Pharm. 100: 134–160.

Nielsen, K.M., K. Smalla and J.D. van Elsas. 2000. Natural transformation of *Acinetobacter* sp. strain BD413 with cell Lysates of *Acinetobacter* sp., *Pseudomonas fluorescens*, and *Burkholderia cepacia* in soil microcosms. Appl. Environ. Microb. 66(1): 206–212.

Ochman, H., J.G. Lawrence and E.A. Grolsman. 2000. Lateral gene transfer and the nature of bacterial innovation. Nature 405(6784): 299–304.

Ochman, H. and N.A. Moran. 2001. Genes lost and genes found: evolution of bacterial pathogenesis and symbiosis. Science 292(5519): 1096–1098.

O'Connell, L.M., P. Kelleher, I.M.H. van Rijswijck, P. de Waal, N.N.M.E. van Peij, J. Mahony et al. 2022. Natural Transformation in gram-positive bacteria and its biotechnological relevance to lactic acid bacteria. Annu. Rev. Food Sci. T 13: 409–431.

Ohm, R.A., N. Feau, B. Henrissat, C.L. Schoch, B.A. Horwitz, K.W. Barry et al. 2012. Diverse lifestyles and strategies of plant pathogenesis encoded in the genomes of eighteen *Dothideomycetes* Fungi. PLOS Pathog 8(12): e1003037.

Olliver, A., M. Vallé, E. Chaslus-Dancla and A. Cloeckaert. 2005. Overexpression of the multidrug efflux operon AcrEF by insertional activation with IS1 or IS10 elements in *Salmonella enterica* Serovar Typhimurium DT204 AcrB mutants selected with Fluoroquinolones. Antimicrob. Agents Ch 49(1): 289–301.

Pérez-Losada, M., M. Arenas, J.C. Galán, F. Palero and F. González-Candelas. 2015. Recombination in viruses: Mechanisms, methods of study, and evolutionary consequences. Infect. Genet. Evol. 30: 296.

Peter, J., M. de Chiara, A. Friedrich, J.X. Yue, D. Pflieger, A. Bergström et al. 2018. Genome evolution across 1,011 *Saccharomyces cerevisiae* isolates. Nature 556(7701): 339–344.

Petruska, J. and M.F. Goodman. 1985. Influence of neighboring bases on DNA polymerase insertion and proofreading fidelity. J. Biol. Chem. 260(12): 7533–7539.

Pray, L. 2008. DNA replication and causes of mutation. Nat. Edu. 1(1): 214.

Radman, M. 1975. SOS Repair hypothesis: phenomenology of an inducible DNA repair which is accompanied by Mutagenesis. Basic Life Sci. 5: 355–367.

Razali, M.N., B.H. Cheah and K. Nadarajah. 2019. Transposable elements adaptive role in genome plasticity, pathogenicity and evolution in fungal phytopathogens. Int. J. Mol. Sci. 20: 3597.

Reijns, M.A.M., B. Rabe, R.E. Rigby, P. Mill, K.R. Astell, L.A. Lettice et al. 2012. Enzymatic removal of ribonucleotides from DNA is essential for mammalian genome integrity and development. Cell 149(5): 1008–1022.

Reynolds, H.T., J.C. Slot, H.H. Divon, E. Lysøe, R.H. Proctor and D.W. Brown. 2017. Differential retention of gene functions in a secondary metabolite cluster. Mol. Biol. Evol. 34(8): 2002–2015.

Richards, T.A., D.M. Soanes, P.G. Foster, G. Leonard, C.R. Thornton and N.J. Talbot. 2009. Phylogenomic analysis demonstrates a pattern of rare and ancient horizontal gene transfer between plants and fungi. Plant Cell 21(7): 1897–1911.

Richards, T.A., D.M. Soanes, M.D.M. Jones, O. Vasieva, G. Leonard, K. Paszkiewicz et al. 2011. Horizontal gene transfer facilitated the evolution of plant parasitic mechanisms in the oomycetes. P Natl. A Sci. USA 108(37): 15258–15263.

Ringlander, J., J. Fingal, H. Kann, K. Prakash, G. Rydell, M. Andersson et al. 2022. Impact of ADAR-induced editing of minor viral RNA populations on replication and transmission of SARS-CoV-2. P Natl. A Sci. USA 119(6): e2112663119.

Roberts, J. and H. Weintraub. 1988. Cis-acting negative control of DNA replication in Eukaryotic cells. Cell 52(3): 397–404.

Rosenshine I, R. Tchelet and M. Mevarech. 1989. The mechanism of DNA transfer in the mating system of an archaebacterium. Science. 245(4924): 1387–89.

Sadeghpour, S., S. Khodaee, M. Rahnama, H. Rahimi and D. Ebrahimi. 2021. Human APOBEC3 variations and viral infection. Viruses 13(7): 1366.

Sanjuán, R. and P. Domingo-Calap. 2016. Mechanisms of viral mutation. Cell Mol. Life Sci. 73(23): 4433–4448.

Sanjuán, R. and P. Domingo-Calap. 2021. Genetic diversity and evolution of viral populations. Encyclopedia of Virology: 53–61.

Sanjuán, R., M.R. Nebot, N. Chirico, L.M. Mansky and R. Belshaw. 2010. Viral mutation rates. J. Virol. 84(19): 9733–9748.

Sankar, T.S., B.D. Wastuwidyaningtyas, Y. Dong, S.A. Lewis and J.D. Wang. 2016. The nature of mutations induced by replication–transcription collisions. Nature 535(7610): 178–181.

Schleper, C., G. Jurgens and M. Jonuscheit. 2005. Genomic studies of uncultivated archaea. Nat. Rev. Microbiol. 3: 479–488.

Schroeder, J.W., W.G. Hirst, G.A. Szewczyk and L.A. Simmons. 2016. The effect of local sequence context on mutational bias of genes encoded on the leading and lagging strands. Curr. Biol. 26(5): 692–697.

Seier, T., D.R. Padgett, G. Zilberberg, V.A. Sutera, N. Toha and S.T. Lovett. 2011. Insights into mutagenesis using *Escherichia coli* chromosomal LacZ strains that enable detection of a wide spectrum of mutational events. Genetics 188(2): 247–262.

Seier, T., G. Zilberberg, D.M. Zeiger and S.T. Lovett. 2012. Azidothymidine and other chain terminators are mutagenic for template-switch-generated genetic mutations. P Natl. A Sci. USA 109(16): 6171–6174.

Sengupta, S. and R.K. Azad. 2023. Leveraging comparative genomics to uncover alien genes in bacterial genomes. Microbial. Genomics 9: 000939.

Seronello, S., J. Montanez, K. Presleigh, M. Barlow, S.B. Park and J. Choi. 2011. Ethanol and reactive species increase basal sequence heterogeneity of *Hepatitis C* virus and produce variants with reduced susceptibility to antivirals. PLoS ONE 6(11): 27436.

Shamini, G., M. Ravichandran, J.T. Sinnott, C. Somboonwit, H.S. Siddhu, P. Shapshak et al. 2011. Structural inferences for cholera toxin mutations in *Vibrio cholerae*. Bioinformation 6(1): 1.

Sharpton, T.J., J.E. Stajich, S.D. Rounsley, M.J. Gardner, J.R. Wortman, V.S. Jordar et al. 2009. Comparative genomic analyses of the human fungal pathogens *Coccidioides* and their relatives. Genome Res. 19(10): 1722–1731.

Shu, W.S. and L.N. Huang. 2022. Microbial diversity in extreme environments. Nat. Rev. Microbiol. 20: 219–235.

Sinha, N.K. 1987. Specificity and efficiency of editing of mismatches involved in the formation of base-substitution mutations by the $3' \rightarrow 5'$ Exonuclease Activity of Phage T4 DNA Polymerase. P Natl. A Sci. USA 84: 915–919.

Sivabalasarma, S., H. Wetzel, P. Nußbaum, C. van der Does, M. Beeby and S-V. Albers. 2021. Analysis of cell–cell bridges in *Haloferax volcanii* using electron cryo-tomography reveal a continuous cytoplasm and S-Layer. Front. Microbiol. 11: 612239.

Skovgaard, O., M. Bak, A. Løbner-Olesen and N. Tommerup. 2011. Genome-wide detection of chromosomal rearrangements, indels, and mutations in circular chromosomes by short read sequencing. Genome Res. 21(8): 1388–1393.

Slot, J. and A. Rokas. 2011. Horizontal Transfer of a large and highly toxic secondary metabolic gene cluster between fungi. Currt. Biol. 21(2): 134–139.

Smillie, C., M.P. Garcillán-Barcia, M.V. Francia, E.P.C. Rocha and F. de la Cruz. 2010. Mobility of plasmids. Microbiol. Mol. Biol. R 74(3): 434–452.

Soanes, D. and T. Richards. 2014. Horizontal gene transfer in eukaryotic plant pathogens. Annu. Rev. Phyto. 52: 583–614.

Srivatsan, A., Y. Han, J. Peng, A.K. Tehranchi, R. Gibbs, J.D. Wang and R. Chen. 2008. High-precision, whole-genome sequencing of laboratory strains facilitates genetic studies. PLOS Genet 4(8): e1000139.

Srivatsan, A., A. Tehranchi, D.M. MacAlpine and J.D. Wang. 2010. Co-orientation of replication and transcription preserves genome integrity. PLOS Genet 6(1): e1000810.

Stavrou, S. and S.R. Ross. 2015. APOBEC3 proteins in viral immunity. J. Immunol. 195(10): 4565.

Steinhauer, D.A., E. Domingo and J.J. Holland. 1992. Lack of evidence for proofreading mechanisms associated with an RNA virus polymerase. Gene 122(2): 281–288.

Stukenbrock, E.H. 2016. The role of hybridization in the evolution and emergence of new fungal plant pathogens. Phytopathology 106(2): 104–112.

Sung, W., M.S. Ackerman, J.F. Gout, S.F. Miller, E. Williams, P.L. Foster and M. Lynch. 2015. Asymmetric context-dependent mutation patterns revealed through mutation–accumulation experiments. Mol. Biol. Evol. 32(7): 1672–1683.

Takai, K. and K. Horikoshi. 1999. Genetic diversity of Archaea in deep-sea hydrothermal vent environments. Genetics 152(4): 1285–1297.

Taylor, J.W., S. Branco, C. Gao, C. Hann-Soden, L. Montoya, I. Sylvain and P. Gladieux. 2017. Sources of fungal genetic variation and associating it with phenotypic diversity. Microbiol. Spectrum. 5(5): 10.1128/microbiolspec.FUNK-0057-2016.

Top, E.M. and D. Springael. 2003. The role of mobile genetic elements in bacterial adaptation to xenobiotic organic compounds. Curr. Opin. Biotech. 14(3): 262–269.

Tsaponina, O. and J.E. Haber. 2014. Frequent interchromosomal template switches during gene conversion in *S. cerevisiae*. Mol. Cell 55(4): 615–625.

Vaisman, A., J.P. McDonald, D. Huston, W. Kuban, L. Liu, B. van Houten et al. 2013. Removal of misincorporated ribonucleotides from prokaryotic genomes: An unexpected role for nucleotide excision repair. PLOS Genet 9(11): e1003878.

Vaisman, A., J.P. McDonald, S. Noll, D. Huston, G. Loeb, M.F. Goodman et al. 2014. Investigating the mechanisms of ribonucleotide excision repair in *Escherichia coli*. Mutat. Res-Fund Mol. M 761: 21–33.

Vakirlis, N., V. Sarilar, G. Drillon, A. Fleiss, N. Agier, J.P. Meyniel et al. 2016. Reconstruction of ancestral chromosome architecture and gene repertoire reveals principles of genome evolution in a model yeast genus. Genome Res. 26(7): 918–932.

Vanheule, A., K. Audenaert, S. Warris, H. van de Geest, E. Schijlen, M. Höfte et al. 2016. Living apart together: Crosstalk between the core and supernumerary genomes in a fungal plant pathogen. BMC Genomics 17(1): 1–18.

Vijaykrishna, D., R. Mukerji and G.J.D. Smith. 2015. RNA virus reassortment: An evolutionary mechanism for host jumps and immune evasion. PLOS Pathog. 11(7): e1004902.

Viswanathan, M., J.J. Lacirignola, R.L. Hurley and S.T. Lovett. 2000. A novel mutational hotspot in a natural Quasipalindrome in *Escherichia coli*. J. Mol. Biol. 302(3): 553–564.

de Vries, J. and W. Wackernagel. 2002. Integration of foreign DNA during natural transformation of *Acinetobacter* sp. by homology-facilitated illegitimate recombination. P Natl. A Sci. USA 99(4): 2094–2099.

Wang, W., H.W. Hellinga and L.S. Beese. 2011. Structural evidence for the rare tautomer hypothesis of spontaneous mutagenesis. P Natl. A Sci. USA 108(43): 17644–17648.

Wapinski, I., A. Pfeffer, N. Friedman and A. Regev. 2007. Natural history and evolutionary principles of gene duplication in fungi. Nature 449(7158): 54–61.

Weitzman, M.D. and A. Fradet-Turcotte. 2018. Virus DNA replication and the host DNA damage response. Ann. Rev. Virol. 5(1): 141.

Windels, E.M., J.E. Michiels, M. Fauvart, T. Wenseleers, B. van den Bergh and J. Michiels. 2019. Bacterial persistence promotes the evolution of antibiotic resistance by increasing survival and mutation rates. ISME J. 13(5): 1239–1251.

Witte, W. 1998. Medical consequences of antibiotics use in agriculture. Science 279(5353): 996–997.

Wolfe, K.H. and D.C. Shields. 1997. Molecular evidence for an ancient duplication of the entire yeast genome. Nature 387(6634): 708–713.

Wommack, K.E. and R.R. Colwell. 2000. Virioplankton: Viruses in aquatic ecosystems. Microbiol. Mol. Biol. R. 64(1): 69–114.

Wozniak, R.A.F. and M.K. Waldor. 2010. Integrative and conjugative elements: Mosaic mobile genetic elements enabling dynamic lateral gene flow. Nat. Rev. Microbiol. 8(8): 552–563.

Yao, N.Y., J.W. Schroeder, O. Yurieva, L.A. Simmons and M.E. O'Donnell. 2013. Cost of RNTP/DNTP pool imbalance at the replication fork. P Natl. A Sci. USA 110(32): 12942–12947.

Yoshiyama, K., K. Higuchi, H. Matsumura and H. Maki. 2001. Directionality of DNA replication fork movement strongly affects the generation of spontaneous mutations in *Escherichia coli*. J. Mol. Biol. 307(5): 1195–1206.

CHAPTER 5

Metagenomics
A Road to Novel Microbial Genes and Genomes

Tomasz Ociepa

Introduction

Nucleic acid sequencing from environmental samples holds great potential for the discovery and identification of a wide variety of microorganisms. This type of research is known under various terms, e.g., random, agnostic or shotgun high-throughput sequencing. The goal of metagenomic research is the possible identification and genomic characterization of all microorganisms present in a sample using a laboratory procedure. This approach gained popularity with the development of next-generation sequencing methods and the constantly decreasing costs of analysis. Metagenomics is a method of analysing the genome composed of all microorganisms inhabiting a given microbiome (e.g., soil). By sequencing variable fragments in genomes, it is possible to understand the genetic diversity, population composition and ecological impact of microorganisms in the studied environment. The hypervariable region sequencing technique provides insights into genetic diversity without the need for cell culture. Information obtained from sequencing, interpreted in terms of the microbiome and maintaining its homeostasis, can be valuable knowledge both in environmental protection and in the broadly understood industry (including the pharmaceutical industry).

Traditional methods of cultivating microorganisms are time-consuming and do not always bring the expected results. Some bacteria cannot be cultured in the laboratory, while others require special culture conditions. For the first time in 1985, Pace and colleagues presented a new culture-independent identification strategy for microorganisms by sequencing the 5S and 16S rRNA subunits, isolated directly from environmental samples (Lane et al. 1985a, 1985b). This innovative project definitely disturbed the previous achievements of Woese who, by conducting research on pure cultures of various cultured bacteria, proved that the 5S and 16S rRNA gene sequences are molecular determinants of their phylogeny (Woese 1987). There is no doubt that on the basis of the results of many works, based on a detailed analysis of these subunits at the level of nucleotide sequences, individual species and types can be identified, as well as a natural phylogenetic tree of bacteria can be constructed. In 1991, the concept of cloning bacterial DNA isolated directly from environmental samples was implemented. Schmidt and colleagues were the first to clone a mixture of bacterial DNA isolated directly from seawater samples in a phage vector (Schmidt et al. 1991). This was the way to construct the first environmental DNA library. Another success was in 1995 the cloning of DNA isolated from a mixture of thermophilic microorganisms that were grown in the laboratory on dried grass (lignocellulose). As a result of this study, a large DNA library was constructed, as well as clones expressing high cellulolytic activity were selected from it

Institute of Plant Genetics, Breeding and Biotechnology, University of Life Sciences in Lublin, Akademicka 15, 20–950, Lublin, Poland.
Email: tomasz.ociepa@up.lublin.pl

(Healy et al. 1995). However, in DeLong's laboratory in 1996, the construction of a huge environmental DNA library of prokaryotic organisms of a freshwater reservoir was characterized (Stein et al. 1996). A 40 kb clone was selected there, and then its genome fragment was identified. At the end of the 1990s, many laboratories achieved success in constructing a DNA library of microorganisms from various soil samples in various vectors.

In 2004, the work of Venter and colleagues was adopted, which described the construction of a metagenome with a billion base pair size of microorganisms from the Sargasso Sea in Bermuda, as well as the decipherment of their nucleotide sequences (Venter et al. 2004). The creation of a huge, metagenomic DNA library of marine microorganisms, in which more than a billion base pairs have been sequenced and about 1.2 million hitherto unknown genes have been identified, gave new opportunities for multidirectional cognitive and application research of marine bacteria. In 2006, Gill and colleagues analysed approximately 78 million base pairs of unique DNA sequence and 2,062 polymerase chain-amplified 16S ribosomal DNA sequences obtained from the fecal DNAs of two healthy adults. They found that the human gut microbiota contains at least 100 times as many genes as the human genome and provides important human physiological functions such as glycan biosynthesis and metabolism (Gill et al. 2006).

Reducing the sequencing costs of metagenomic DNA libraries and the use of new achievements in bioinformatics to analyse the created, huge databases allowed for acceleration and cognitive and application orientation of research in this new field of microbiology. This chapter presents the approaches used in this field and the state of the art in the form of databases and bioinformatics tools used to analyse the results of metagenome experiments.

Metagenomic Approaches

Metagenomics is the culture-independent genomic analysis of a community of microorganisms. It provides a community-wide assessment of metabolic function and bypasses the need for the isolation and laboratory cultivation of individual species. The analysis of metagenomic data provides a way to identify new organisms and isolate complete genomes from unculturable species that are present within an environmental sample. Metagenomics is a comprehensive field of study with

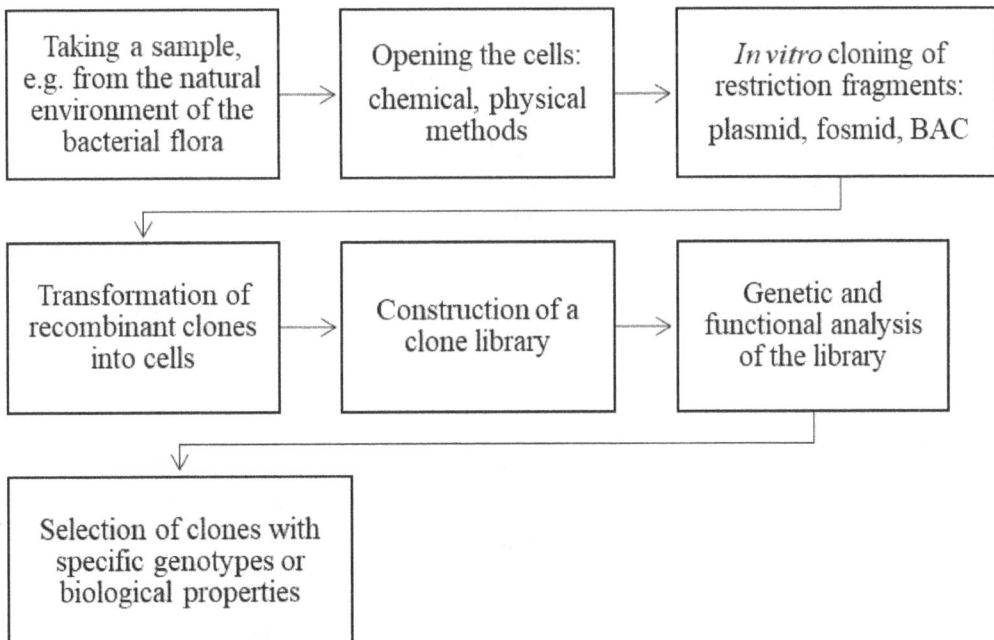

Figure 1. Stages of the experiment with the use of microorganisms.

two basic approaches: taxonomic application (sequence-based analysis) and functional application (function-driven analysis) or a combination of both, depending on the requirements of the objective (Navgire et al. 2022). Both the sequencing mentioned above are performed by any one method of next-generation sequencing technologies: nanopore technology (Baker 2022), sequencing by synthesis (Azli et al. 2022), pyrosequencing (Lazarevic et al. 2009), sequencing by ligation (Mitra et al. 2013), ion torrent sequencing (Cao et al. 2014) or single-molecule real-time sequencing (Wang et al. 2021). Metagenomics includes the following stages of experimental work (Figure 1).

In metagenomics, the most commonly used vectors for DNA cloning and library construction are the low-copy plasmid BAC (bacterial artificial chromosomes) and fosmids, because these vectors enable cloning of large DNA fragments (40–100 kb). Therefore, their stable replication is obtained in the host- recipient cells; most often, they are cells of the appropriate *E. coli* strains. In addition, phasmids, and above all derivatives of the BAC plasmid, as shuttle vectors (shuttle BACs) create the possibility of conjugative transfer of DNA from *E. coli* to another host, for example, *Streptomyces lividans* or *Pseudomonas putida*. In a situation where the purpose of the research is the selection of clones capable of expressing given proteins, enzymes or other bioproducts, then it is recommended to transfer the clone library from *E. coli* to the above-mentioned hosts. So far, in the analysis of metagenomic DNA libraries, serious methodological difficulties are caused by the low frequency of appearance of specific clones, mainly clones with sought biological activities. For this purpose, various activities and treatments are offered. One of the methods is to enrich environmental samples from which DNA for cloning is isolated with microorganisms with the desired biological properties. Another way is to enrich the cloned DNA mixture isolated from environmental samples into fractions that may contain the expected sequences or searched genes (Schloss and Handelsman 2003).

Sequence-Based Metagenomics

This approach is used to establish the phylogenetic relationships of a sequenced gene with taxonomic groups of microorganisms known in bioinformatics databases. In this case, phylogenetic clusters such as the 16S rRNA gene sequence are the target where operational taxonomic units (OTUs) (Franzén et al. 2015) are compared with their amplitude to estimate the abundance of microbial species in that particular environment (Edgar 2018).

16S rRNA gene sequencing is commonly used to identify, classify and quantify microbes in complex biological mixtures such as environmental samples and human gut microbiome samples. The 16S rRNA gene is a highly conserved component of the transcriptional machinery of all DNA-based life forms and is therefore well suited as a target gene for DNA sequencing in samples containing up to thousands of different species. Universal PCR primers can be designed to target a specific conserved 16S region, allowing the gene to be amplified in a wide range of different microorganisms from a single sample. Conveniently, the 16S rRNA gene consists of both conservative and variable regions. There are about 1500 base pairs (bp) in size, with nine highly conserved regions and nine variable regions (V1–V9) throughout the 16S rRNA. While the conserved region allows for universal amplification, sequencing of the variable regions enables the distinction between specific different microorganisms such as bacteria, archaea, and eukaryotic microbes. Identification of viruses requires metagenomic sequencing (direct sequencing of all DNA extracted from a microbial community) due to the lack of the 16S phylogenetic marker gene (Cox et al. 2013). Platforms such as Illumina sequencing use V3 and V4 regions to derive taxonomic classification by comparing these regions with regions already known and available in large public databases such as NCBI (Schoch et al. 2020), SILVA (Quast et al. 2013), GreenGene (DeSantis et al. 2006) or RDP (Cole et al. 2005). The output of 16S rRNA sequencing in the form of reads can be analyzed through a series of basic bioinformatics steps, which in combination are known as "pipes". These bioinformatics' pipelines remove sequencing errors or questionable reads to "clean up" the data, which can then be matched against microbial genome databases to accurately identify and profile the bacteria (and archaea) present in the samples. Multiple 16S sequencing pipelines are available, i.e., QIIME (Bolyen et al. 2019), Kraken (Wood et al. 2019), and UPARSE (Edgar 2013).

The internal transcribed spacer (ITS) region of rDNA is the most used barcode to study fungal diversity (Schoch et al. 2012). ITS region contains three partitions: ITS1, 5.8S and ITS2. The length of ITS sequence is highly variable from one fungal species to another and it is strongly dependent on the primers used to target the DNA sequence. For example, in *Ascomycota* and *Basidiomycota* the sequence lengths range between 600 and 900 bp (Toju et al. 2012). Amplicon-based high-throughput sequencing approaches include an enrichment step prior to sequencing which involves the use of PCR amplification of the target barcode. The most commonly used high-throughput sequencers have a limit on the maximum read length (e.g., MiSeq Illumina – 2 × 300 bp). This limitation forces the use of only one of the two sub-regions (ITS1 or ITS2) when using the amplicon-based HTS approach to determine fungal diversity. The generated sequenced data is analysed using PIPITS. The first pipeline with full bioinformatics automation is entirely dedicated to the sequencing of ITS regions of fungal origin. Part of the PIPITS_PROCESS pipeline uses the VSEARCH tool to group sequences into OTUs (Gweon et al. 2015). The 5.8S region does not contain enough information sites to be used for phylogenetic and DNA barcoding studies (Kohout et al. 2014).

The 18S rRNA gene is equivalent to the 16S rRNA gene in prokaryotes and is found in eukaryotes. It contains both conserved and variable regions (V1 through V9 with the exception of V6, which does not exist due to relative conservation), which may reflect differences between eukaryotic species such as fungi. 18S rRNA gene sequencing is commonly used to identify, classify, and quantify microbes within complex biological mixtures such as samples collected from the environment and gut. The ITS region was found to include ITS1, which has rDNA between 18S and 5.8S, and the ITS2 region, which has rDNA between 5.8S and 28S. The variability of ITS1 and ITS2 has been observed to be greater than that of the 18S rDNA gene, so it has an advantage over 18S in identifying fungi and lower eukaryotes at the species and subspecies level (Wang et al. 2014).

In general, metataxonomic research is more targeted, less expensive and less problematic in terms of the amount of data generated than whole genome sequencing. However, they provide only a limited amount of information about the test sample. This technique quantifies the composition of bacteria, fungi or plants in samples, for example from humans or animals, food, feed, water or wastewater. Unfortunately, in this case viruses, due to their enormous genetic diversity, do not have sequences conserved enough to be used in metataxonomic analysis. In turn, significant limitations of bacterial metataxonomy result from the fact that the differences in the 16S rRNA gene between some organisms are small. As a consequence, reliable analyses are possible at the genus level at most, and species identification must be confirmed by other methods. Therefore, the key is proper preparation of the material, selection of nucleic acid extraction methods, the use of high-fidelity enzymes (HiFi) and multi-stage control of the analytical process (Nagai et al. 2022).

Functional Metagenomics

This approach is used to find a sequence with a functional gene with a specific activity or if the gene is new and has a specific function in a functional pathway. Functional analysis of metagenomes aims to identify those functional abilities that are most significant for organisms living in the studied environment. Typically, the analysis is performed at the single gene level, focusing on the abundance of gene families, or at the pathway level where the occurrence of genes in pathways is considered. These processes begin with the identification of genes in the data and their function prediction, where function prediction is done by matching the data with function-oriented databases (Sharon et al. 2011). Annotating and assigning functions solely on the basis of homology is often confusing, as genes with similar sequences may have completely different functions in different microorganisms. This is especially true for regulatory genes; their specificity and biological function can rarely be predicted from sequence alone. Therefore, proper annotation of at least model organisms should include a functional analysis (Kozińska et al. 2019). This is achieved by shotgun metagenomics, which involves whole-genome sequencing (WGS), which additionally involves functional annotation of the gene (Nayfach et al. 2015, Brown et al. 2019). Functional annotation is divided into two steps: gene prediction and gene annotation, where gene prediction helps to find potential protein sequences. Once identified, these protein coding sequences are compared to protein families in databases and functionally described by family function matching (Gill et al. 2006).

The shotgun sequencing approach involves direct sequencing of all genetic material that is present in the sample. A very large amount of data is generated during shotgun sequencing – bacterial, viral, eukaryotic and archaeal sequences are obtained from a metagenomic sample, as well as a large fraction of unidentified reads. A typical shotgun metagenomics study comprises five steps, after the initial study design: (1) the collection, processing and sequencing of the samples; (2) preprocessing of the sequencing reads; (3) sequence analysis to profile taxonomic, functional and genomic features of the microbiome; (4) statistical and biological post-processing analysis, and (5) validation (Quince et al. 2017). Since shotgun metagenomics is PCR-independent, and therefore not biased by primers designed to target gene sequences that are expected to be conserved within prokaryotes or small eukaryotes, the method is able to detect microorganisms that may not be detected using targeted amplicon-based NGS methods (Boers et al. 2019). The main advantage of WGS metagenomics compared to marker gene sequencing is the ability to characterize the genetic and genomic diversity of the analysed sample and identify potential and new functions present in the studied community (Kim et al. 2022). Furthermore, with adequate sequencing depth, it is possible to assemble complete genomes from metagenomic data. Compared to the marker gene approach, WGS metagenomics is generally less affected by PCR-related errors necessary for marker gene amplification, such as the number of cycles used or the selection of primers and hypervariable regions (Sze and Schloss 2019). However, WGS metagenomic sequencing can also be affected by errors in metagenomic results, mainly due to the use of whole-genome amplification protocols that are used when working with low-concentration DNA samples (Sabina and Leamon 2015). Shotgun sequencing readouts require more complex bioinformatics methods to analyse the results. Pipelines in shotgun metagenomics also perform quality filtering steps, after which the purified sequencing data can either be assembled to create partial or complete microbial genomes, using pipelines such as Megahit (Li et al. 2015) or matched against microbial marker gene databases (using pipelines such as MetaPhlAn and HUMAnN). The results provide detailed information on the relative abundance of bacteria, fungi, viruses and other microbes in the sample, as well as the relative abundance of selected lists of microbial genes (e.g., metabolic genes or antibiotic resistance genes).

Applications of Metagenomics

The strategy of studying non-cultivated microorganisms has been extended to include cloning of DNA obtained directly from natural environments, as well as sequencing of extremely large environmental genomic libraries of bacteria from soils, waters, sewage, sediments and the digestive tracts of animals and humans. The goal of scientists is to decipher the wealth of genetic information contained in huge metagenomes for specific environments. It is also important to identify previously unknown genes, which increases knowledge about bacteria that have not been cultured so far and their role in natural environments. This brings desirable results in biotechnological solutions, which concern, among others, the production of innovative drugs, enzymes and other bioproducts. In addition, clinical metagenomics is developing rapidly, which sheds new light on the diagnosis and treatment of many infectious diseases.

Industrial Enzymes

Enzymes of microbial origin are of particular interest to the food and pharmaceutical industries due to the catalysis of reactions that can be difficult or expensive to maintain. This is due to the fact that there are often difficulties in the synthesis of chemical catalysts. Replacing the traditional chemical processes used to produce certain compounds or molecules with naturally sourced enzymatic pathways is a more environmentally friendly approach to large-scale production. Since microorganisms can catalyze a wide range of reactions, they are an obvious source of enzymes for industrial applications (Coughlan et al. 2015). Metagenomic studies allowed for the recognition and better characterization of obtaining enzymes such as amylase (Xu et al. 2014), ß-galactosidase (Vester et al. 2014), ß-glucuronidase (Neun et al. 2022), chitinase (Raimundo et al. 2021), esterase (Ouyang et al. 2013), glucosidase (Jiang et al. 2009), lipase (López-López et al. 2014), protease (Biver et al. 2013), and xylanase (Cheng et al. 2012).

Antibiotics and Bioactive Compounds

As in the food industry, the use of microbial enzymes is of particular interest in the biosynthesis of pharmaceutical products that have previously been synthesized by chemical means. Functional metagenomics can be used to discover genes capable of carrying out reactions of interest for obtaining bioactive substances or synthesizing intermediates in the pharmaceutical industry. There are a number of scientific reports describing the identification of many new genes, including genes encoding drug resistance such as ampicillin, amoxicillin (Martiny et al. 2011), β-lactamase (Forsberg et al. 2012), bleomycin (Mori et al. 2008), ceftazidime (Donato et al. 2010), chloramphenicol (Lang et al. 2010), kanamycin, gentamicin, rifampin (McGarvey et al. 2012), and tetracycline (Diaz-Torres et al. 2003). Some novel bioactive and biosynthetic pathways of industrial importance have been discovered, such as acid resistance genes (Guazzaroni et al. 2013), biotin- vitamin H (Streit and Entcheva 2003), prebiotic degradation (Cecchini et al. 2013), salt tolerance genes (Culligan et al. 2013), serine protease inhibitor– serpin gene (Jiang et al. 2011), and vibrioferrin (Fujita et al. 2011). Also, the relationship between environmental conditions and the content and abundance of membrane proteins was investigated (Patel et al. 2010).

Clinical Metagenomics

Metagenomics is also used to identify an unknown pathogen in disease outbreaks (Blauwkamp et al. 2019). The shotgun approach can also be used to discover and detect pathogens in clinical samples. For example, 16S/18S/ITS sequencing is a powerful and inexpensive tool for clinical microflora analysis (Church et al. 2020). It can be used to determine the species of intestinal microbes and their abundance, and allows you to monitor human health and well-being (Zhang et al. 2015). Metagenomics sheds light on the development of probiotics (Cecchini et al. 2013). By monitoring human-associated bacterial communities, it is possible to identify ways to modulate them to optimize human health. Next-generation clinical metagenomic sequencing (mNGS) is a comprehensive analysis of microbial and host genetic material (DNA and RNA) in patient samples (Liu et al. 2022). This new approach is changing the way doctors diagnose and treat infectious diseases, with applications spanning a wide range of fields, including antimicrobial resistance, the microbiome, host gene expression (transcriptomics) and oncology (Chiu and Miller 2019).

Conclusions

Metagenomics is a rapidly developing field of microbiology, giving the opportunity to study a much wider range of biodiversity of many environments and to use them for human benefit. It sets new directions for research into the hitherto unknown world of environmental bacteria, shaping the methodical possibilities of their identification and characterization at the level of genotypes and phenotypes. Metagenomics has created opportunities for thorough cognitive research in the field of taxonomy and phylogeny of bacteria living in various natural environments. It allowed for the analysis of their biological properties based on reading the genetic information of cloned genomes and genes.

References

Azli, B., M.N. Razak, A.R. Omar, N.A. Mohd Zain, F. Abdul Razak and I. Nurulfiza. 2022. Metagenomics insights into the microbial diversity and microbiome network analysis on the heterogeneity of influent to effluent water. Front Microbiol. 13: 715.

Baker, J.L. 2022. Using nanopore sequencing to obtain complete bacterial genomes from saliva samples. MSystems. 7(5): e0049122.

Biver, S., D. Portetelle and M. Vandenbol. 2013. Characterization of a new oxidant-stable serine protease isolated by functional metagenomics. Springerplus. 2(1): 1–10.

Blauwkamp, T.A., S. Thair, M.J. Rosen, L. Blair, M.S. Lindner, I.D. Vilfan et al. 2019. Analytical and clinical validation of a microbial cell-free DNA sequencing test for infectious disease. Nat. Microbiol. 4(4): 663–674.

Boers, S.A., R. Jansen and J.P. Hays. 2019. Understanding and overcoming the pitfalls and biases of Next-Generation Sequencing (NGS) methods for use in the routine clinical microbiological diagnostic laboratory. Eur. J. Clin. Microbiol. Infect. Dis. 38(6): 1059.

Bolyen, E., J.R. Rideout, M.R. Dillon, N.A. Bokulich, C.C. Abnet, G.A. Al-Ghalith et al. 2019. Reproducible, interactive, scalable and extensible microbiome data science using QIIME 2. Nat. Biotechnol. 37(8): 852–857.

Brown, S.M., H. Chen, Y. Hao, B.P. Laungani, T.A. Ali, C. Dong et al. 2019. MGS-Fast: Metagenomic shotgun data fast annotation using microbial gene catalogs. Gigascience 8(4): 1–9.

Cao, C., I. Lou, C. Huang and M.Y. Lee. 2014. Metagenomic sequencing of activated sludge filamentous bacteria community using the ion torrent platform. Desalination Water Treat. 57(5): 2175–2183.

Cecchini, D.A., E. Laville, S. Laguerre, P. Robe, M. Leclerc, J. Doré et al. 2013. Functional metagenomics reveals novel pathways of prebiotic breakdown by human gut bacteria. PLoS ONE. 8(9): e72766.

Cheng, F., J. Sheng, R. Dong, Y. Men, L. Gan and L. Shen. 2012. Novel Xylanase from a holstein cattle rumen metagenomic library and its application in Xylooligosaccharide and Ferulic acid production from wheat straw. J. Agric. Food Chem. 60(51): 12516–12524.

Chiu, C.Y. and S.A. Miller. 2019. Clinical metagenomics. Nat. Rev. Genet. 20(6): 341–355.

Church, D.L., L. Cerutti, A. Gürtler, T. Griener, A. Zelazny and S. Emler. 2020. Performance and application of 16S RRNA gene cycle sequencing for routine identification of bacteria in the clinical microbiology laboratory. Clin. Microbiol. Rev. 33(4): 1–74.

Cole, J.R., B. Chai, R.J. Farris, Q. Wang, S.A. Kulam, D.M. McGarrell et al. 2005. The ribosomal database project (RDP-II): sequences and tools for high-throughput RRNA analysis. Nucleic Acids Res. 33: 294–296.

Coughlan, L.M., P.D. Cotter, C. Hill and A. Alvarez-Ordóñez. 2015. Biotechnological applications of functional metagenomics in the food and pharmaceutical industries. Front Microbiol. 6: 672.

Cox, M.J., W.O.C.M. Cookson and M.F. Moffatt. 2013. Sequencing the human microbiome in health and disease. Hum. Mol. Genet. 22(1): 88–94.

Culligan, E.P., R.D. Sleator, J.R. Marchesi and C. Hill. 2013. Functional environmental screening of a metagenomic library identifies StlA; A Unique Salt Tolerance Locus from the Human Gut Microbiome. PLoS ONE. 8(12): e82985.

DeSantis, T.Z., P. Hugenholtz, N. Larsen, M. Rojas, E.L. Brodie, K. Keller et al. 2006. Greengenes, a Chimera-Checked 16S RRNA gene database and workbench compatible with ARB. Appl. Environ. Microbiol. 72(7): 5069–5072.

Diaz-Torres, M.L., R. McNab, D.A. Spratt, A. Villedieu, N. Hunt, M. Wilson et al. 2003. Novel tetracycline resistance determinant from the oral metagenome. Antimicrob. Agents Chemother. 47(4): 1430–1432.

Donato, J.J., L.A. Moe, B.J. Converse, K.D. Smart, E.C. Berklein, P.S. McManus et al. 2010. Metagenomic analysis of apple orchard soil reveals antibiotic resistance genes encoding predicted bifunctional proteins. Appl. Environ. Microbiol. 76(13): 4396–4401.

Edgar, R.C. 2013. UPARSE: Highly Accurate OTU sequences from microbial amplicon reads. Nat. Methods. 10(10): 996–998.

Edgar, R.C. 2018. Accuracy of taxonomy prediction for 16S RRNA and fungal ITS sequences. PeerJ. 6: e4652.

Forsberg, K.J., A. Reyes, B. Wang, E.M. Selleck, M.O.A. Sommer and G. Dantas. 2012. The shared antibiotic resistome of soil bacteria and human pathogens. Science. 337(6098): 1107–1111.

Franzén, O., J. Hu, X. Bao, S.H. Itzkowitz, I. Peter and A. Bashir. 2015. Improved OTU-Picking using long-read 16S RRNA gene amplicon sequencing and generic hierarchical clustering. Microbiome 3(43): 1–15.

Fujita, M.J., N. Kimura, A. Sakai, Y. Ichikawa, T. Hanyu and M. Otsuka. 2011. Cloning and heterologous expression of the vibrioferrin biosynthetic gene cluster from a marine metagenomic library. Biosci. Biotechnol. Biochem. 75(12): 2283–2287.

Gill, S.R., M. Pop, R.T. DeBoy, P.B. Eckburg, P.J. Turnbaugh, B.S. Samuel et al. 2006. Metagenomic analysis of the human distal gut microbiome. Science. 312(5778): 1355–1359.

Guazzaroni, M.E., V. Morgante, S. Mirete and J.E. González-Pastor. 2013. Novel acid resistance genes from the metagenome of the tinto river, an extremely acidic environment. Environ. Microbiol. 15(4): 1088–1102.

Gweon, H.S., A. Oliver, J. Taylor, T. Booth, M. Gibbs, D.S. Read et al. 2015. PIPITS: An automated pipeline for analyses of fungal internal transcribed spacer sequences from the illumina sequencing platform. Methods Ecol. Evol. 6(8): 973–980.

Healy, F.G., R.M. Ray, H.C. Aldrich, A.C. Wilkie, L.O. Ingram and K.T. Shanmugam. 1995. Direct isolation of functional genes encoding cellulases from the microbial consortia in a thermophilic, anaerobic digester maintained on lignocellulose. Appl. Microbiol. Biotechnol. 43(4): 667–674.

Jiang, C., G. Ma, S. Li, T. Hu, Z. Che, P. Shen et al. 2009. Characterization of a novel beta-glucosidase-like activity from a soil metagenome. J. Microbiol. 5: 542–548.

Jiang, C.J., Z.Y. Hao, R. Zeng, P.H. Shen, J.F. Li and B. Wu. 2011. Characterization of a novel serine protease inhibitor gene from a marine metagenome. Mar Drugs. 9: 1487–1501.

Kim, C.Y., J. Ma and I. Lee. 2022. HiFi metagenomic sequencing enables assembly of accurate and complete genomes from human gut microbiota. Nat. Commun. 13(1): 1–11.

Kohout, P., R. Sudová, M. Janoušková, M. Čtvrtlíková, M. Hejda, H. Pánková et al. 2014. Comparison of commonly used primer sets for evaluating *Arbuscular mycorrhizal* fungal communities: Is there a universal solution? Soil Biol. Biochem. 68: 482–493.

Kozińska, A., P. Seweryn and I. Sitkiewicz. 2019. A crash course in sequencing for a microbiologist. J. Appl. Genet. 60(1): 103–111.

Lane, D.J., B. Pace, G.J. Olsen, D.A. Stahl, M.L. Sogin and N.R. Pace. 1985a. Rapid determination of 16S ribosomal RNA sequences for phylogenetic analyses. Proc. Natl. Acad. Sci. U.S.A. 82(20): 6955–6959.

Lane, D.J., D.A. Stahl, G.J. Olsen, D.J. Heller and N.R. Pace. 1985b. Phylogenetic analysis of the genera *Thiobacillus* and *Thiomicrospira* by 5S RRNA sequences. J. Bacteriol. 163(1): 75–81.

Lang, K.S., J.M. Anderson, S. Schwarz, L. Williamson, J. Handelsman and R.S. Singer. 2010. Novel florfenicol and chloramphenicol resistance gene discovered in alaskan soil by using functional metagenomics. Appl. Environ. Microbiol. 76(15): 5321–5326.

Lazarevic, V., K. Whiteson, S. Huse, D. Hernandez, L. Farinelli, M. Østerås et al. 2009. Metagenomic study of the oral microbiota by illumina high-throughput sequencing. J. Microbiol. Methods. 79(3): 266–271.

Li, D., C.M. Liu, R. Luo, K. Sadakane and T.W. Lam. 2015. MEGAHIT: An ultra-fast single-node solution for large and complex metagenomics assembly via succinct de bruijn graph. Bioinformatics. 31(10): 1674–1676.

Liu, J., Q. Zhang, Y.Q. Dong, J. Yin and Y.Q. Qiu. 2022. Diagnostic accuracy of metagenomic next-generation sequencing in diagnosing infectious diseases: a meta-analysis. Sci. Rep. 12(1): 1–10.

López-López, O., M.E. Cerdán and M.I.G. Siso. 2014. New Extremophilic lipases and esterases from metagenomics. Curr. Protein Pept. Sci. 15(5): 445.

Martiny, A.C., J.B.H. Martiny, C. Weihe, A. Field and J.C. Ellis. 2011. Functional metagenomics reveals previously unrecognized diversity of antibiotic resistance genes in gulls. Front Microbiol. 2(238): 1–11.

McGarvey, K.M., K. Queitsch and S. Fields. 2012. Wide variation in antibiotic resistance proteins identified by functional metagenomic screening of a Soil DNA library. Appl. Environ. Microbiol. 78(6): 1708–1714.

Mitra, S., K. Förster-Fromme, A. Damms-Machado, T. Scheurenbrand, S. Biskup, D.H. Huson et al. 2013. Analysis of the intestinal microbiota using SOLiD 16S RRNA gene sequencing and SOLiD shotgun sequencing. BMC Genomics. 14(5): 1–11.

Mori, T., S. Mizuta, H. Suenaga and K. Miyazaki. 2008. Metagenomic screening for bleomycin resistance genes. Appl. Environ. Microbiol. 74(21): 6803–6805.

Nagai, S., S. Sildever, N. Nishi, S. Tazawa, L. Basti, T. Kobayashi et al. 2022. Comparing PCR-generated artifacts of different polymerases for improved accuracy of DNA metabarcoding. Metabarcoding Metagenom. 6: e77704.

Navgire, G.S., N. Goel, G. Sawhney, M. Sharma, P. Kaushik, Y.K. Mohanta et al. 2022. Analysis and interpretation of metagenomics data: An approach. Biol Proced Online. 24(1): 1–22.

Nayfach, S., P.H. Bradley, S.K. Wyman, T.J. Laurent, A. Williams, J.A. Eisen et al. 2015. Automated and accurate estimation of gene family abundance from shotgun metagenomes. PLoS Comput. Biol. 11: e1004573.

Neun, S., P. Brear, E. Campbell, T. Tryfona, K. El Omari, A. Wagner et al. 2022. Functional metagenomic screening identifies an unexpected β-glucuronidase. Nat. Chem. Biol. 18(10): 1096–1103.

Ouyang, L.M., J.Y. Liu, M. Qiao and J.H. Xu. 2013. Isolation and biochemical characterization of two novel metagenome-derived esterases. Appl. Biochem. Biotechnol. 169(1): 15–28.

Patel, P.v., T.A. Gianoulis, R.D. Bjornson, K.Y. Yip, D.M. Engelman and M.B. Gerstein. 2010. Analysis of membrane proteins in metagenomics: Networks of correlated environmental features and protein families. Genome Res. 20(7): 960.

Quast, C., E. Pruesse, P. Yilmaz, J. Gerken, T. Schweer, P. Yarza et al. 2013. The SILVA ribosomal RNA gene database project: Improved data processing and web-based tools. Nucleic Acids Res. 41: 590–596.

Quince, C., A.W. Walker, J.T. Simpson, N.J. Loman and N. Segata. 2017. Shotgun metagenomics, from Sampling to analysis. Nat. Biotechnol. 35(9): 833–844.

Raimundo, I., R. Silva, L. Meunier, S.M. Valente, A. Lago-Lestón, T. Keller-Costa et al. 2021. Functional metagenomics reveals differential chitin degradation and utilization features across free-living and host-associated marine microbiomes. Microbiome 9(1): 1–18.

Sabina, J. and J.H. Leamon. 2015. Bias in whole genome amplification: causes and considerations. Methods Mol. Biol. 1347: 15–41.

Schloss, P.D. and J. Handelsman. 2003. Biotechnological prospects from metagenomics. Curr. Opin. Biotechnol. 14(3): 303–310.

Schmidt, T.M., E.F. DeLong and N.R. Pace. 1991. Analysis of a marine picoplankton community by 16S RRNA gene cloning and sequencing. J. Bacteriol. 173(14): 4371–4378.

Schoch, C.L., K.A. Seifert, S. Huhndorf, V. Robert, J.L. Spouge, C.A. Levesque et al. 2012. Nuclear ribosomal Internal Transcribed Spacer (ITS) region as a universal DNA barcode marker for fungi. Proc. Natl. Acad. Sci. U.S.A. 109(16): 6241–6246.

Schoch, C.L., S. Ciufo, M. Domrachev, C.L. Hotton, S. Kannan, R. Khovanskaya et al. 2020. NCBI Taxonomy: A comprehensive update on curation, resources and tools. Database (Oxford). 1–21.

Sharon, I., S. Bercovici, R.Y. Pinter and T. Shlomi. 2011. Pathway-based functional analysis of metagenomes. J. Comput. Biol. 18(3): 495–505.

Stein, J.L., T.L. Marsh, K.Y. Wu, H. Shizuya and E.F. Delong. 1996. Characterization of uncultivated prokaryotes: Isolation and analysis of a 40-Kilobase-pair genome fragment from a planktonic marine Archaeon. J. Bacteriol. 178(3): 591–599.

Streit, W.R. and P. Entcheva. 2003. Biotin in microbes, the genes involved in its biosynthesis, its biochemical role and perspectives for biotechnological production. Appl. Microbiol. Biotechnol. 61(1): 21–31.

Sze, M.A. and P.D. Schloss. 2019. The impact of DNA polymerase and number of rounds of amplification in PCR on 16S RRNA gene sequence data. MSphere. 4(3): 1–13.

Toju, H., A.S. Tanabe, S. Yamamoto and H. Sato. 2012. High-coverage ITS primers for the DNA-based identification of ascomycetes and basidiomycetes in environmental samples. PLoS ONE. 7(7): e40863.

Venter, J.C., K. Remington, J.F. Heidelberg, A.L. Halpern, D. Rusch, J.A. Eisen et al. 2004. Environmental genome shotgun sequencing of the sargasso sea. Science 304(5667): 66–74.

Vester, J.K., M.A. Glaring and P. Stougaard. 2014. Discovery of novel enzymes with industrial potential from a cold and alkaline environment by a combination of functional metagenomics and culturing. Microb. Cell Fact. 13(1): 72.

Wang, L., J. Zhang, M. Zhou, Q. Chen, X. Yang, Y. Hou et al. 2021. Evaluation of the effect of antibiotics on gut microbiota in early life based on culturomics, SMRT sequencing and metagenomics sequencing methods. Anal Methods. 13(43): 5144–5156.

Wang, Y., R.M. Tian, Z.M. Gao, S. Bougouffa and P.Y. Qian. 2014. Optimal eukaryotic 18S and universal 16S/18S ribosomal RNA primers and their application in a study of symbiosis. PLoS ONE 9(3): e90053.

Woese, C.R. 1987. Bacterial evolution. Microbiol. Rev. 51(2): 221–271.

Wood, D.E., J. Lu and B. Langmead. 2019. Improved metagenomic analysis with Kraken 2. Genome Biol. 20(1): 1–13.

Xu, B., F. Yang, C. Xiong, J. Li, X. Tang, J. Zhou et al. 2014. Cloning and characterization of a novel α-amylase from a fecal microbial metagenome. J. Microbiol. Biotechnol. 24(4): 447–452.

Zhang, Y.J., S. Li, R.Y. Gan, T. Zhou, D.P. Xu and H. bin Li. 2015. Impacts of gut bacteria on human health and diseases. Int. J. Mol. Sci. 16(4): 7493–7519.

CHAPTER 6

Genome Sequencing as a Tool for Understanding Microorganisms

Tomasz Ociepa

Introduction

Knowledge of a complete and highly covered nucleotide sequence often opens up new possibilities in the field of understanding the functioning or evolution of various groups of microorganisms. The constant progress in the field of sequencing technology and the systematic reduction of costs per sample related to this type of analysis means that every year more research centers can perform this type of research on their own. Each year, the National Center for Biotechnology Information (NCBI) presents a report on the current state of records in the databases overseen by this institution. One of the most important databases from the point of view of bioinformatics experience is the Genomes database. The table and figure (Table 1, Figure 1) present part of the data from the report as of August 12, 2022, including data from the Genomes database with a description of individual sub-bases and a percentage increase in the number of records compared to the previous year (Sayers et al. 2022).

The largest percentage increase in records over the previous year is in the SRA database, which is where raw sequencing data and alignment information are stored. The archive accepts data from all branches of life as well as metagenomics and environmental surveys. In the mentioned archive, most data comes from reading the sequence of the human genetic material, followed by vertebrates, plants, bacteria, invertebrates, fungi, and the least data on archaea and viruses. The amount of data deposited in this archive increased from 40 Tb in 2010 to 29,482 Tb in 2020 (Stephan et al. 2022).

The growing amount of data from sequencing experiments require interpretation using bioinformatics tools, i.e., a combination of computer, mathematical or statistical analyses. The

Table 1. Number of records for the Genomes database deposited by the NCBI as of 12, 2022 (Sayers et al. 2022).

Database	Records	Description
Nucleotide	503 629 990	DNA and RNA sequences from GenBank and RefSeq
BioSample	28 001 796	Descriptions of biological source materials
SRA	23 813 19	High-throughput DNA/RNA sequence read archive
Taxonomy	2 571 112	Taxonomic classification and nomenclature catalogue
Assembly	1 388 980	Genome assembly information
BioProject	614 936	Biological projects providing data to NCBI
Genome	71 826	Genome sequencing projects by organism

Institute of Plant Genetics, Breeding and Biotechnology, University of Life Sciences in Lublin, Akademicka 15, 20–950, Lublin, Poland.
Email: tomasz.ociepa@up.lublin.pl

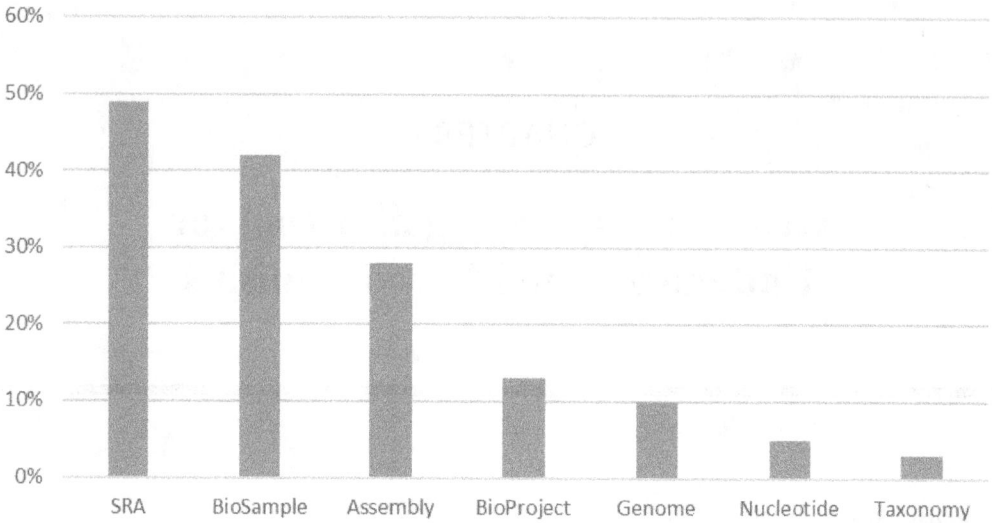

Figure 1. Percentage increase in the number of records in 2022 compared to 2021 (Sayers et al. 2022).

purpose of this chapter is to present the most important theoretical information about the sequencing process using various platforms. Various approaches to the analysis of the obtained results have also been described.

Sequencing Generations

Sequencing is a method of reading the order of nucleotide pairs in a DNA or RNA molecule. For over 50 years, many researchers have focused on developing techniques and technologies to facilitate this process. From the sequencing of short oligonucleotides to millions of bases, from the struggle to deduce the coding sequence of a single gene to the rapid and widely available whole genome sequencing (Heather and Chain 2016). This section outlines the different generations of sequencing technology, highlighting some of the key discoveries, researchers, and sequences along the way, as well as presenting the pros and cons of each generation.

First Generation Sequencing

Some groundbreaking discoveries in biology, chemistry and physics laid the foundations for the development of a method for reading the sequence of nucleic acids. Many scientists emphasize that the breakthrough date was 1953 when James Watson and Francis Crick proposed the three-dimensional structure of the helix and DNA using crystallographic data produced by Rosalind Franklin and Maurice Wilkins (Watson and Crick 1953). Since then, intensive work has been carried out to create a method for reading nucleotide sequences in the genome. Initial efforts focused on sequencing the most readily available populations of relatively pure RNA species, such as microbial ribosomal or transfer RNA or single-stranded RNA bacteriophage genomes. The first success appeared in 1965 thanks to Robert Holley, who managed to read the sequence of nucleotides in a tRNA molecule isolated from yeast (Holley et al. 1965). An important achievement in this field was the recognition of DNA sequence of bacteriophage λ by Ray Wu in 1968 (Wu and Kaiser 1968). They used the observation that the phage possessed 5′ overhanging 'cohesive' end and used DNA polymerase to fill the ends in with radioactive nucleotides, supplying each nucleotide one at a time and measuring incorporation to deduce sequence.

Sanger Method

The breakthrough came in 1977 when Frederick Sanger developed a technique commonly known as the dideoxy or simply the Sanger method. The first complete DNA genome to be sequenced by this method was that of the bacteriophage φX174 (Sanger et al. 1977a, 1977b). In this method, only single-stranded DNA (ssDNA) can be used as a template. A primer is then added to the single-stranded DNA to mark the starting point for sequencing. The whole process takes place with the participation of DNA polymerase, which during DNA synthesis incorporates a simple deoxynucleotide (dNTP) or dideoxynucleotide (ddNTP) marked with an appropriate dye into the resulting chain, and in the original version of the method with an isotope. The reaction is carried out using a mixture of dNTPs and ddNTPs, with the concentration of each of the ddNTPs (ddATP, ddTTP, ddGTP, ddCTP) selected to be statistically considered once per nascent DNA strand. The incorporation of ddNTP into the DNA strand completes the synthesis of the strand because no further nucleotide can be incorporated into it by the DNA polymerase. Labeling individual ddNTPs with various fluorescent markers allows for the identification of the nucleotide where DNA strand synthesis has been completed. The result of the reaction is the formation of a large number of DNA fragments of different lengths, depending on which nucleotide was attached to the DNA strand at its end. In the last stage of sequencing, electrophoresis is performed, the purpose of which is to show the terminal nucleotides, i.e., those located at the end of each DNA strand. After irradiation of the gel in which the nucleotides from each of the four tubes have been separated, only the newly added nucleotides can be imaged and shown on the chromatogram (Figure 2).

Due to the great interest in sequencing, over time this method began to be refined, enabling the automation of the process with the development of the technique used today, based on the use of fluorescently labeled nucleotides in combination with capillary electrophoresis and digital signal reading. The introduction of these improvements enabled the development in 1987 by Leroy Hood and Michael Hunkapiller, of the first commercially available sequencer capable of sequencing fragments of up to 1000 base pairs, which was called ABI 370 (Hood et al. 1987). With the development of automation of the Sanger sequencing technique, the stage of sequencing of full genomes began (Table 2). This period culminated in 2001 when the first draft of the human genome sequence was published using first-generation sequencing (Lander et al. 2001).

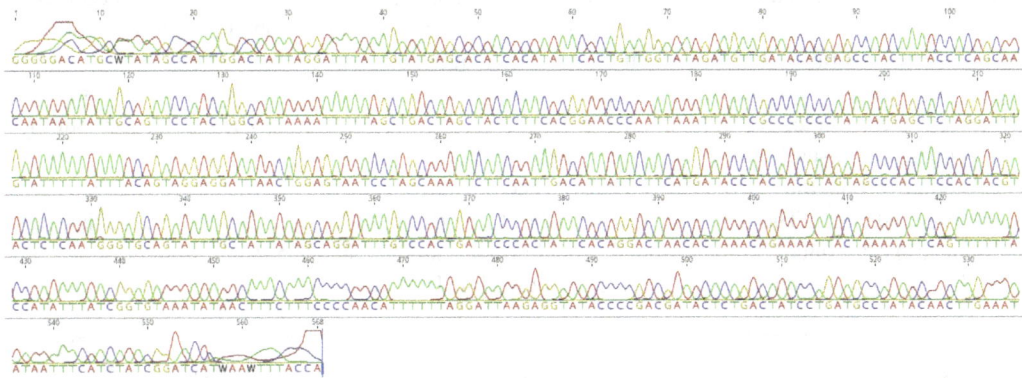

Figure 2. An example image of a chromatogram from Sanger sequencing (own research).

Table 2. List of microorganisms with known genome sequence based on first generation sequencing.

Organism	Genome size [M bp]	References
Haemophilus influenzae	1.8	(Fleischmann et al. 1995)
Mycoplasma pneumoniae	0.8	(Himmelreich et al. 1996)
Escherichia coli	4.6	(Blattner et al. 1997)
Saccharomyces cerevisiae	12.1	(Mewes et al. 1997)

Maxam-Gilbert Method

The second, alternative method of DNA sequencing is the Maxam-Gilbert chemical degradation method (Maxam and Gilbert 1977). This method is based on radioactive labeling of the 5' end of DNA and then degradation of the molecule in 4 separate reactions using 4 mixtures of reagents. In the first stage of the reaction, the nitrogenous bases of a given nucleotide are chemically modified. At the abasic site, the strand breaks behind the modified nucleotide. For the modification of G (guanine), a mixture of piperidine and dimethyl sulphate is used, for pairs A+G (adenine and guanine) - dimethyl sulphate, piperidine and formic acid, for C (cytosine) - hydrazine/sodium chloride and piperidine and for C+T (cytosine and thymine) - hydrazine and piperidine. The chemical reaction produces DNA fragments labeled at the 5' end. Since a given base occurs many times in the chain, we get a mixture of many fragments of different lengths in the test tube. As in the Sanger method, fragments from four separate reactions are separated by gel electrophoresis and DNA strands are visualized using autoradiography. The bands shown correspond to DNA fragments having a specific base at the 3' end (Maxam and Gilbert 1980).

Both methods were groundbreaking and invented at a similar time, but they had their general differences (Table 3) that later influenced their further development.

Table 3. Differences between the sequencing proposed by Sanger et al. (1977) and Maxam and Gilbert (1977).

Sanger	Maxam-Gilbert
Enzymatic method	Chemical method
Uses less hazardous chemicals	Uses more hazardous chemicals
Uses radioactively or fluorescently labeled ddNTPs	Uses radioactive isotope ^{32}P for labeling the ends of the DNA fragments
Routinely used for sequencing	Rarely used for sequencing

Second Generation Sequencing

Over time, the increasing demand for nucleic acid sequencing has contributed to the emergence and development of high-throughput sequencing techniques. In 2005 and in the following years, a new generation of sequencers appeared that broke the limitations of the first generation. New generation sequencing- NGS (next generations sequencing)- thus allows obtaining millions of reads in a time incomparably shorter than in the classic Sanger sequencing. To increase throughput and significantly reduce analysis time, a number of sequencing methods have been developed to avoid the tedious steps of cloning DNA fragments (Metzker 2009, Kchouk et al. 2017). The methodology in this generation usually has some general steps (Figure 3).

Library preparation involves random genome fragmentation followed by ligation with appropriate adapters. The approaches are related to platforms developed by companies that have introduced their sequencers over time. Roche/454 was launched in 2005, Illumina/Solexa in 2006, ABI/SOLiD in 2007 and in 2010 Life Technologies/Ion Torrent.

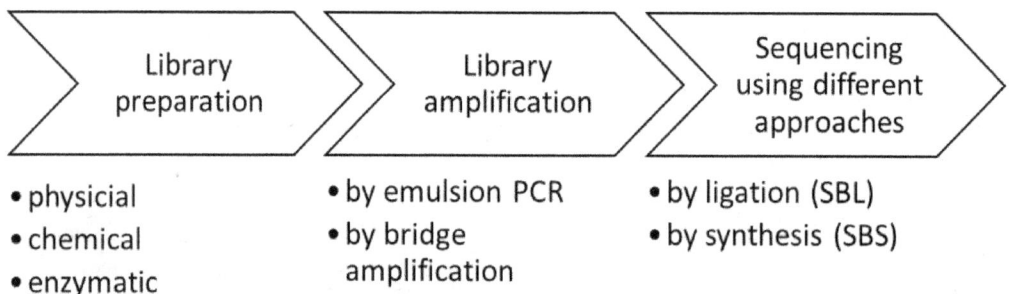

Figure 3. General flowchart of next-generation sequencing steps.

Roche/454 Sequencing

Roche/454 sequencing appeared in the market in 2005, using the pyrosequencing technique. The first microorganism sequenced with this platform was *Mycoplasma genitalium* (Margulies et al. 2005). Pyrosequencing is a "sequencing by synthesis" method, the results of which can be observed in real-time. Its origins date back to the second half of the 90s of the twentieth century (Ronaghi et al. 1996). This technique differs significantly from the previously described ones - it does not require the separation of strand fragments, e.g., by electrophoresis, making it much faster than sequencing by chemical degradation or chain termination. Pyrosequencing takes place in several, cyclically repeated stages. The procedure is based on 4 enzymes: Klenow fragment of DNA polymerase I, ATP sulfurylase, luciferase and an enzyme that degrades unincorporated deoxynucleotides, usually apyrase. The sequencing template, which is single-stranded DNA, is incubated with all four enzymes. Individual nucleotides are added to the reaction mixture one after the other. Attachment of subsequent nucleotides to the synthesized DNA strand in the reaction catalyzed by the Klenow fragment is accompanied by the release of pyrophosphate (PPi). The released PPi is converted to ATP by ATP sulfurylase. The next stage is the release of a quantum of light in the reactions of ATP to AMP conversion catalyzed by luciferase. The light reaction is the result of adding the correct nucleotide to the reaction mixture - complementary to the matrix. The last stage is the degradation of the remaining, unattached deoxynucleotides before the next one is added to the reaction so that they do not interfere with the result of subsequent cycles of DNA strand synthesis. It is important that in order to prevent the light reaction in the presence of dATP, α - thio - ATP is added as a substrate for DNA synthesis (Ahmadian et al. 2006).

Pyrosequencing is a simple technique to use and suitable for quantitative analysis of DNA sequences, but only a few dozen base pairs can be read in one experiment. Unlike other sequencing techniques, however, this method does not require the use of electrophoresis or any other method of separating DNA fragments, which significantly increases its speed. The sequence acquisition rate is 100 times faster than with the string termination method. At room temperature, the entire reaction from polymerization to light detection takes 3–4 seconds. The use of chemiluminescence has a positive effect on the sensitivity of the method. Each reaction can be carried out in very small volumes down to one picoliter. Another unquestionable advantage of this technique is also its high specificity obtained thanks to properly labeled primers and the use of a single-stranded template obtained in the PCR reaction. This ensures that the tested sample contains only the analyzed DNA fragment (Margulies et al. 2005, Gharizadeh et al. 2006). The main constraint is the problem in sequencing same nucleotide repeat (> 8 bp), that is, homopolymer sequencing (Mardis 2013). It is worth noting that since 2016, Roche has not supported the further development of this platform.

Pyrosequencing has been used in the sequencing of microbial genomes, which can be used to identify bacterial species, distinguish between bacterial strains, and detect genetic mutations that confer antimicrobial resistance (Cummings et al. 2013).

Illumina/Solexa Sequencing

Solexa developed the sequencing method while Illumina acquired the company and technology and began commercializing the sequencer, which was initially known as Genome Analyze (Voelkerding et al. 2009). In this method, short double-stranded adapters are attached to the fragmented DNA. After denaturation, a mixture of single-stranded DNA fragments and oligonucleotides (with a significant excess of the latter in relation to DNA fragments) are immobilized on a solid surface (a microplate placed in a flow cell). During the next step, each sequence attached to the fixed plate is amplified by "bridge PCR amplification", which creates several identical copies of each sequence (Adessi et al. 2000). A set of sequences formed from the same original sequence is called a cluster. After denaturation, the cycle repeats itself until the appropriate number of copies of the fragment is generated in the vicinity of a given fragment for sequencing. In this way, sectors are obtained on the microplate, each of which represents a statistically different amplified DNA fragment. The next step is the simultaneous sequencing of DNA (via synthesis) in all sectors. DNA is sequenced using specially designed nucleotides equipped with removable fluorescent tags, which each time end DNA synthesis in a given cycle- each of the 4 nucleotides is marked with a different fluorophore. The CCD camera records signals in individual sectors from complementary nucleotides newly attached in a given cycle, after which the fluorophores from all sectors are removed so that subsequently labeled nucleotides can be attached in the next sequencing cycle (Reuter et al. 2015).

The first sequencer developed with this technology was able to read short reads of around 35 bp. However, an important advantage is the use of paired-end (PE) reads, in which the sequence at both ends of each DNA cluster is recorded. Currently, the most advanced company's sequencer (NovaSeq X) can generate output range ~165 Gb - 16 Tb and read lengths 2 × 151 bp. For example, the longest reads can be made on MiSeq using SBS Kit Version, up to 2 × 301 bp. One of the main disadvantages of the Illumina/Solexa platform is the high requirements for sample quantification, because in the case of pooled samples, one sample may have a higher concentration than the others, which may result in cluster overlap and poor sequencing quality. The overall error rate of this sequencing technology is approximately 1% (Stoler and Nekrutenko 2021). Another limitation is the guanine-cytosine (GC) bias, introduced during bridge amplification (Mardis 2013). Another problem is dephasing, which means that different copies of the DNA in a cluster become out of sync (inconsistency). In other words, improper nucleotide deprotection results in fragments of different lengths in the cluster. This reduces the accuracy of base calling at the 3' end of DNA fragments, especially in the sequence of inverted repeats (Nakamura et al. 2011).

ABI/SOLiD Sequencing

In 2007, Applied Biosystems introduced a second-generation sequencing method, SOLiD, based on ligation. In this method, similar to "454" sequencing, the tested material is amplified by means of emulsion PCR. After the PCR reaction, the 3' end of one of the DNA strands is modified, which enables its covalent attachment to a solid support (glass microplate). In this way, as in the Solex method, a microplate with sectors is obtained, each of which represents a statistically different duplicated DNA fragment. In the next step, as a result of hybridization, a short oligonucleotide complementary to one of the primers (adapters) used for emulsion PCR is added. In the next step, short fluorescently labeled oligonucleotides are ligated in which the first two nucleotides are known; for example, a CA dinucleotide searches for a complementary GT sequence on a single-stranded DNA template. Then the fluorescent tag is removed, and in subsequent cycles, the next dinucleotides are found so that in total we get a double coverage of each position in the analyzed sequence (Shendure and Ji 2008).

This technology offers the highest accuracy of ~99.99% because each nucleotide has been sequenced twice, so there is much less chance of confusion with two adjacent colors. One limitation is that the time it takes per run is too long (6–7 days) along with producing less data compared to Illumina platforms (Voelkerding et al. 2009).

Ion Torrent Sequencing

Life Technologies presented the Ion Torrent semiconductor sequencing technology in 2010. The principle of sequencing is similar to pyrosequencing, but it does not use fluorescently labeled nucleotides (Rothberg et al. 2011). Amplification of libraries, as in the case of pyrosequencing and SOLiD, takes place in the form of emulsion PCR. Droplets containing the bead, DNA polymerase and other reagents necessary for the reaction are formed in the oil-water emulsion. DNA amplification is carried out on beads according to the principle: one DNA fragment - one bead. Then the uncoated, hollow beads are largely enriched and the full beads are transferred to a special plate called a chip. Ball complexes - amplified DNA fragments match the size of the wells of the chip, under which there are sensors measuring changes in pH. After manually inserting the chip into the sequencer, the sequencing process takes place. Nucleotides are added individually, and in the case of incorporation of a complementary nucleotide to the template strand, the released hydrogen ions are measured. The apparatus records the signal and its strength depends on the number of nucleotides attached in the cycle. In wells where the next nucleotide is not complementary to the one currently present in the solution, no incorporation or pH change occurs. The challenge in this method is the correct assessment of the number of built-in bases in long homopolymer strings, and the advantage is the elimination of some recording equipment (laser, camera), simplification of the process, shortening of reading time and reduction of sequencing costs (Loman et al. 2012).

Third Generation Sequencing

Second generation sequencing technologies revolutionized DNA analysis and were the most widely used compared to the first generation. However, second-generation technologies generally require an amplification step, which is a lengthy procedure in execution time. In addition, a large proportion of organisms have genomes that are very complex with many repetitive areas. *De novo* genome assembly based on short reads is a very challenging task. To solve the second-generation problems, scientists developed a new generation of sequencing called 'third-generation sequencing' (TGS). Third generation sequencing is referred to as sequencing of single molecules - nucleic acid molecules (Braslavsky et al. 2003). It allows to obtain readings with a length much longer than technologies of previous generations, while reducing the financial outlays necessary to carry out the analysis. There are two main approaches that characterize TGS: a single molecule real-time sequencing (SMRT) approach and a synthetic approach that builds on existing short read technologies used by Illumina and 10xGenomics for constructing long reads (Harris et al. 2008). The most widely used third-generation technology approach is SMRT, and sequencers that have used this approach include Pacific Biosciences (Eid et al. 2009) and Oxford Nanopore sequencing (Clarke et al. 2009).

Pacific Biosciences (PacBio)

Pacific Biosciences developed the first genomic sequencer using the SMRT approach in 2010. The template, called a SMRTbell, is a closed, single-stranded circular DNA that is created by ligating hairpin adaptors to both ends of a target double-stranded DNA (dsDNA) molecule. Single polymerase molecules are immobilized at the bottom of ZMWs (Zero-Mode Waveguides) reaction wells-holes with a diameter of several dozen nm made in a 100 nm thick metal layer on a glass surface (Levene et al. 2003). By illuminating the plate from the bottom with a laser beam, the contents of individual wells can be observed. Since the diameter of the pinholes is smaller than the wavelength, the actual volume that light reaches is 20 zeptolitres (20×10^{-21} litres). The signal from each incorporated nucleotide is recorded. When pyrophosphate is decoupled from the fluorophore, the signal disappears. Although all nucleotides are present in a high concentration in the reaction mixture, the background level is very low because the reaction time is several orders of magnitude longer than the random appearance of free nucleotides in the field of observation. Whenever a nucleotide is incorporated, it releases a luminous signal that is recorded by sensors. The detection of the labeled nucleotides makes it possible to determine the DNA sequence (Rhoads and Au 2015).

The company's first science instrument, named "PacBio RS", then a new version of the sequencer named "PacBio RS II", were released in April 2013, whereas PacBio Sequel is the latest platform released by Pacific BioSciences in autumn 2015 in collaboration with Roche Diagnostics.

The advantage of this technology is that the reaction can be monitored in real time which permits to gather the data related to base composition or sequence of the DNA template as well as the enzyme kinetics. The difference in enzyme kinetics provides us the clue about different modifications present in the DNA like methylation (Fang et al. 2012). Also, a big advantage is the fact that the sample for analysis can be prepared quickly, taking 4 to 6 hours instead of days.

Sequence reads are long- on average ~10 kbp, but individual, very long reads can be as long as 60 kbp. *De novo* sequencing can be easily performed because of the longer read length. Hence, a short read allows error in the assembly of fragments and formation of scaffolds in repeat and GC-rich regions (Bahassi and Stambrook 2014). The downside is the high error rate of around 13%, mostly insertion and deletion errors. These errors are randomly distributed along the long read (Koren et al. 2012).

Oxford Nanopore Technologies (ONT)

In 2014, Oxford Nanopore Technologies (ONT) introduced nanopore sequencing (Jain et al. 2015), which is based on the idea of using nanopores in a membrane to sequence single-stranded DNA or RNA molecules that was first proposed at the end of the 1980s (Deamer et al. 2016). In this sequencing technology, the first strand of the DNA molecule is connected by a hairpin to a complementary strand. The DNA fragment is passed through a protein nanopore (a nanopore is a nanoscale hole made of proteins or synthetic materials). When a DNA fragment is moved through a pore by a motor, a protein attached to the pore generates a change in ionic current caused by differences in the moving nucleotides occupying the pore. This change in ion current is recorded gradually and then interpreted to identify the sequence. The forward or template strand generates a so-called "1D template" read (D stands for direction), while the reverse or complementary strand generates a "1D complement" read. The consensus sequence obtained from the joint analysis of template and complement reads is called a "2D" read. It should be noted that not all fragments passing through pores are effective in generating 2D reads. Some only generate template reads, others generate template and completions, and finally only a small minority are of sufficient quality to obtain a 2D consensus from template reads and completions (Lu et al. 2016).

Among the advantages offered by this sequencer is the low cost and small device size. The sample is loaded into the device's port and data is displayed on the screen and generated without waiting for the process to complete. The MinION engine can provide very long reads in excess of 150 kbp, which can improve *de novo* assembly continuity. Nanopore sequencing devices demonstrate unequivocal features, which include the type of the starting material (DNA, cDNA or native RNA), the option for multiplex sequencing and the wide range of the initial amount per TGS library that ranges from 10 pg^{-1} mg (Athanasopoulou et al. 2021). However, this technology generates a high error rate of ~12% - scattered around ~3% mismatches, ~4% insertions and ~5% deletions (Ip et al. 2015).

Although both generations belong to high-throughput methods, they have general differences (Table 4).

Table 4. Comparison of two high-throughput sequencing generations.

Second generation	Third generation
Short reads, up to 700 bp	Long reads, well over 1000 bp
Many copies of the molecules	Single molecule
High accuracy	Lower accuracy
More difficult preparation of the library	Easier preparation of the library
More difficult *de novo* genome assembly	Easier *de novo* genome assembly

Approaches of Sequencing Experiments

Whole genome sequencing (WGS) has improved the study of complex biological phenomena in microorganisms such as population dynamics, host adaptation or infection outbreaks (McAdam et al. 2014). Bioinformatic analysis is an important component of the analytical scheme of whole genome studies. After the sequencing stage is completed, the computer-aided readings are submitted. The assembly of fragments can be done using two methods: resequencing and *de novo* assembly.

De novo sequencing allows you to know the genome of a selected organism (species, structure) without referring to a reference sequence. *De novo* assembly of genomes is based on bioinformatics methods that assemble a gene map from scratch, thus providing insight into the organization of the genetic information of the tested organism. *De novo* sequencing of the genome can be performed for animal, plant and microorganism, as well as a plasmid, artificial chromosome, etc. (Pérez-Cobas et al. 2020).

Genome resequencing consists in sequencing short fragments of genetic material, aligning them with the reference genome, and then extracting genetic variants. This technique allows to obtain information on variants stored in databases (concerning mainly coding regions), as well as on new variants and variants located outside the coding regions, which are not detectable by matrix methods or exome sequencing (Petrie and Xie 2021).

Considering the scale of sequencing and the area whose sequence is analysed, there are other sequencing technologies. Next-generation targeted sequencing focuses on specific regions of the genome. This technology is ideal for studying genes in specific pathways. It is faster and cheaper than WGS and also allows for deeper sequencing. Targeted sequencing is a particularly sensitive and powerful method for identifying variants and mutations (Benjamino et al. 2021).

Conclusions

Sequencing has come a long way from a labour-intensive method to high-throughput methods that analyse many genomes at once. Choosing the right technology depends largely on the application. In some experiments, Sanger sequencing is sufficient to show small differences. To look more broadly at a given problem, it is worth using new-generation platforms. A large amount of data is a challenge for researchers. There are several approaches to sequencing, and it is not always necessary to sequence the entire genome to study an organism. Microorganisms have played a key role in the development of sequencing methods. They were one of the first models for sequence studies. Currently, more and more sequences in bioinformatics databases come from experiments on microorganisms. Scientists from all over the world can thus compare the results of other teams' research with their own. This may contribute to a better understanding of many mechanisms or the process of evolution in a large group of microorganisms.

References

Adessi, C., G. Matton, G. Ayala, G. Turcatti, J.J. Mermod, P. Mayer et al. 2000. Solid phase DNA amplification: Characterisation of primer attachment and amplification mechanisms. Nucleic Acids Res. 28(20): 87.

Ahmadian, A., M. Ehn and S. Hober. 2006. Pyrosequencing: History, biochemistry and future. Clin. Chim. Acta. 363(1–2): 83–94.

Athanasopoulou, K., M.A. Boti, P.G. Adamopoulos, P.C. Skourou and A. Scorilas. 2021. Third-generation sequencing: The spearhead towards the radical transformation of modern genomics. Life 12(1): 30.

Bahassi, E.M. and P.J. Stambrook. 2014. Next-generation sequencing technologies: Breaking the sound barrier of human genetics. Mutagenesis 29(5): 303–310.

Benjamino, J., B. Leopold, D. Phillips and M.D. Adams. 2021. Genome-based targeted sequencing as a reproducible microbial community profiling assay. MSphere. 6(2): e01325–20.

Blattner, F.R., G. Plunkett, C.A. Bloch, N.T. Perna, V. Burland, M. Riley et al. 1997. The complete genome sequence of *Escherichia coli* K-12. Science. 277(5331): 1453–1462.

Braslavsky, I., B. Hebert, E. Kartalov and S.R. Quake. 2003. Sequence information can be obtained from single DNA molecules. Proc. Natl. Acad. Sci. U.S.A. 100(7): 3960–3964.

Clarke, J., H.C. Wu, L. Jayasinghe, A. Patel, S. Reid and H. Bayley. 2009. Continuous base identification for single-molecule nanopore DNA sequencing. Nat. Nanotechnol. 4(4): 265–270.

Cummings, P.J., R. Ahmed, J.A. Durocher, A. Jessen, T. Vardi and K.M. Obom. 2013. Pyrosequencing for microbial identification and characterization. J. Vis. Exp. 78: e50405.

Deamer, D., M. Akeson and D. Branton. 2016. Three decades of nanopore sequencing. Nat. Biotechnol. 34(5): 518–524.

Eid, J., A. Fehr, J. Gray, K. Luong, J. Lyle, G. Otto et al. 2009. Real-time DNA sequencing from single polymerase molecules. Science 323(5910): 133–138.

Fang, G., D. Munera, D.I. Friedman, A. Mandlik, M.C. Chao, O. Banerjee et al. 2012. Genome-wide mapping of methylated adenine residues in pathogenic *Escherichia coli* using single-molecule real-time sequencing. Nat. Biotechnol. 30(12): 1232–1239.

Fleischmann, R.D., M.D. Adams, O. White, R.A. Clayton, E.F. Kirkness, A.R. Kerlavage et al. 1995. Whole-genome random sequencing and assembly of haemophilus influenzae Rd. Science 269(5223): 496–512.

Gharizadeh, B., Z.S. Herman, R.G. Eason, O. Jejelowo and N. Pourmand. 2006. Large-scale pyrosequencing of synthetic DNA: A comparison with results from sanger dideoxy sequencing. Electrophoresis 27(15): 3042.

Harris, T.D., P.R. Buzby, H. Babcock, E. Beer, J. Bowers, I. Braslavsky et al. 2008. Single-molecule DNA sequencing of a viral genome. Science 320(5872): 106–109.

Heather, J.M. and B. Chain. 2016. The sequence of sequencers: The history of sequencing DNA. Genomics 107(1): 1.

Himmelreich, R., H. Hubert, H. Plagens, E. Pirkl, B.C. Li and R. Herrmann. 1996. Complete sequence analysis of the genome of the bacterium mycoplasma Pneumoniae. Nucleic Acids Res. 24(22): 4420.

Holley, R.W., J. Apgar, G.A. Everett, J.T. Madison, M. Marquisee, S.H. Merrill et al. 1965. Structure of a ribonucleic acid. Science. 147(3664): 1462–1465.

Hood, L.E., M.W. Hunkapiller and L.M. Smith. 1987. Automated DNA sequencing and analysis of the human genome. Genomics 1(3): 201–212.

Ip, C.L.C., M. Loose, J.R. Tyson, M. de Cesare, B.L. Brown, M. Jain et al. 2015. MinION analysis and reference consortium: Phase 1 data release and analysis. F1000Res. 4: 1075.

Jain, M., I.T. Fiddes, K.H. Miga, H.E. Olsen, B. Paten and M. Akeson. 2015. Improved data analysis for the MinION nanopore sequencer. Nat. Methods 12(4): 351–356.

Kchouk, M., J.-F. Gibrat and M. Elloumi. 2017. Generations of sequencing technologies: From first to next generation. Biol. Med. (Aligarh). 9: 395.

Koren, S., M.C. Schatz, B.P. Walenz, J. Martin, J.T. Howard, G. Ganapathy et al. 2012. Hybrid error correction and *de Novo* assembly of single-molecule sequencing reads. Nat. Biotechnol. 30(7): 693–700.

Lander, E.S., L.M. Linton, B. Birren, C. Nusbaum, M.C. Zody, J. Baldwin et al. 2001. Initial sequencing and analysis of the human genome. Nature 409(6822) 860–921.

Levene, H.J., J. Korlach, S.W. Turner, M. Foquet, H.G. Craighead and W.W. Webb. 2003. Zero-mode waveguides for single-molecule analysis at high concentrations. Science 299(5607): 682–686.

Loman, N.J., R. v. Misra, T.J. Dallman, C. Constantinidou, S.E. Gharbia, J. Wain et al. 2012. Performance comparison of benchtop high-throughput sequencing platforms. Nat. Biotechnol. 30(5): 434–439.

Lu, H., F. Giordano and Z. Ning. 2016. Oxford nanopore MinION sequencing and genome assembly. Genomics Proteomics Bioinformatics. 14(5): 265–279.

Mardis, E.R. 2013. Next-generation sequencing platforms. Annu. Rev. Anal. Chem. (Palo Alto Calif). 6: 287–303.

Margulies, M., M. Egholm, W.E. Altman, S. Attiya, J.S. Bader, L.A. Bemben et al. 2005. Genome sequencing in open microfabricated high density picoliter reactors. Nature. 437(7057): 376.

Maxam, A.M. and W. Gilbert. 1977. A new method for sequencing DNA. Proc. Natl. Acad. Sci. U.S.A. 74(2): 560.

Maxam, Allan M. and W. Gilbert. 1980. Sequencing end-labeled DNA with base-specific chemical cleavages. Meth. Enzymol. 65(1): 499–560.

McAdam, P.R., E.J. Richardson and R.R. Fitzgerald. 2014. High-throughput sequencing for the study of bacterial pathogen biology. Curr. Opin. Microbiol. 19(1): 106–113.

Metzker, M.L. 2009. Sequencing technologies—the next Generation. Nat. Rev. Genet. 11(1): 31–46.

Mewes, H.W., K. Albermann, M. Bahr, D. Frishman, A. Gielssner, J. Hani et al. 1997. Overview of the yeast genome. Nature. 387(6632): 737.

Nakamura, K., T. Oshima, T. Morimoto, S. Ikeda, H. Yoshikawa, Y. Shiwa et al. 2011. Sequence-specific error profile of illumina sequencers. Nucleic Acids Res. 39(13): e90.

Pérez-Cobas, A.E., L. Gomez-Valero and C. Buchrieser. 2020. Metagenomic approaches in microbial ecology: An update on whole-genome and marker gene sequencing analyses. Microb. Genom. 6(8): 1–22.

Petrie, K.L. and R. Xie. 2021. Resequencing of microbial isolates: A lab module to introduce novices to command-line bioinformatics. Front Microbiol. 12: 462.

Reuter, J.A., D. v. Spacek and M.P. Snyder. 2015. High-throughput sequencing technologies. Mol. Cell. 58(4): 586–597.

Rhoads, A. and K.F. Au. 2015. PacBio sequencing and its applications. Genomics Proteomics Bioinformatics. 13(5): 278–289.

Ronaghi, M., S. Karamohamed, B. Pettersson, M. Uhlén and P. Nyrén. 1996. Real-time DNA sequencing using detection of pyrophosphate release. Anal. Biochem. 242(1): 84–89.

Rothberg, J.M., W. Hinz, T.M. Rearick, J. Schultz, W. Mileski, M. Davey et al. 2011. An integrated semiconductor device enabling non-optical genome sequencing. Nature. 475(7356): 348–352.

Sanger, F., G.M. Air, B.G. Barrell, N.L. Brown, A.R. Coulson, J.C. Fiddes et al. 1977a. Nucleotide sequence of bacteriophage ΦX174 DNA. Nature. 265(5596): 687–695.

Sanger, F., S. Nicklen and A.R. Coulson. 1977b. DNA sequencing with chain-terminating inhibitors. Proc. Natl. Acad. Sci. U.S.A. 74(12): 5463.

Sayers, E.W., E.E. Bolton, J.R. Brister, K. Canese, J. Chan, D.C. Comeau et al. 2022. Database resources of the national center for biotechnology information in 2023. Nucleic Acids Res. 51: 29–38.

Shendure, J. and H. Ji. 2008. Next-generation DNA sequencing. Nat. Biotechnol. 26(10): 1135–1145.

Stephan, T., S.M. Burgess, H. Cheng, C.G. Danko, C.A. Gill, E.D. Jarvis et al. 2022. Darwinian genomics and diversity in the tree of life. Proc. Natl. Acad. Sci. U.S.A. 119(4): e2115644119.

Stoler, N. and A. Nekrutenko. 2021. Sequencing error profiles of illumina sequencing instruments. NAR Genom. Bioinform. 3(1): 1–9.

Voelkerding, K.v., S.A. Dames and J.D. Durtschi. 2009. Next-generation sequencing: From basic research to diagnostics. Clin Chem. 55(4): 641–658.

Watson, J.D. and F.H.C. Crick. 1953. Molecular structure of nucleic acids; A structure for deoxyribose nucleic acid. Nature. 171(4356): 737–738.

Wu, R. and A.D. Kaiser. 1968. Structure and base sequence in the cohesive ends of bacteriophage lambda DNA. J. Mol. Biol. 35(3): 523–537.

CHAPTER 7

Mobile Genetic Elements and
the Evolution of Microbes

*Sarah Alharbi,[1] Esraa Aldawood,[1] Norashirene Mohamed Jamil,[2]
Fatimah AlShehri,[3] Mohamed Lotfy Ashour[4,5]
and Mahmoud Abdelkhalek Elfaky[6,7],**

Introduction

Prokaryotic genomes consist of core genes, which are inherited vertically and encode vital cellular processes, and non-core genes, which are inherited horizontally (Touchon et al. 2009). Altogether, the core and non-core genes are referred to as pan-genome (Rankin et al. 2011). Interestingly, in *E. coli* the non-core genes account for approximately 90% of the pan-genome (Touchon et al. 2009). These accessory genes encode adaptive traits that enhance the fitness of bacteria according to the encountered niches. Horizontal gene transfer (HGT) of these accessory genes is facilitated mainly by mobile genetic elements (MGEs), which are fragments of DNA that can move around within a genome (intra-genomes) or across different genomes (inter-genomes). Example of MGEs includes plasmids, transposable elements, phages and their counterparts (gene transfer agents (GTAs)). Intra-genome mobility of MGEs includes the transposition of these elements from the host genomic DNA to a replicon and from one replicon to another replicon. Inter-genomes transfers are mediated by Horizontal gene transfer (HGT) between species that are more or less distantly related among bacteria or even between bacteria and eukaryotic cells (Burmeister 2015). However, the transfer of MGEs comes at a cost to the host which varies according to the type of mobile element (Starikova et al. 2013). Generally, due to the life or death dilemma imposed by bacteriophages, their cost is much more than other MGEs such as plasmids, integrative conjugative elements (ICEs), mobilizable islands or transposable elements (Rankin et al. 2011). Also, the integration of MGEs into the host chromosome can deactivate important functions in the cell. MGEs not only can affect the host but also may affect the host's neighbours (Rankin et al. 2011). This social effect can be positive by providing fitness to the host neighbours or negative such as the production of bacteriocins that harm the host's neighbours (West et al. 2007). The selective pressure exerted on both MGEs and the host is what defines the positive or negative effect of MGEs on bacteria (West et al. 2007).

[1] Department of Clinical Laboratory Sciences, College of Applied Medical Sciences, King Saud University, Riyadh, Saudi Arabia.
[2] Faculty of Applied Sciences, Universiti Teknologi MARA, 40450 Shah Alam, Selangor, Malaysia.
[3] Pharmaceutical Care Department, International Medical Center, Jeddah 23214, Saudi Arabia.
[4] Department of Pharmacognosy, Faculty of Pharmacy, Ain-Shams University, Abbasia, Cairo 11566, Egypt.
[5] Department of Pharmaceutical Sciences, Pharmacy Program, Batterjee Medical College, Jeddah 21442, Saudi Arabia.
[6] Department of Natural Products, Faculty of Pharmacy, King Abdulaziz University, Jeddah 21589, Saudi Arabia.
[7] Centre for Artificial Intelligence in Precision Medicines, King Abdulaziz University, Jeddah 21589, Saudi Arabia.
* Corresponding author: melfaky@kau.edu.sa

MGEs possess adaptations for moving genes not only between individual cells but also between strands of DNA (Rankin et al. 2011); thus, MGEs have a direct impact on genome evolution. Additionally, MGEs can facilitate homologous and non-homologous recombination events affecting the bacterial chromosome as a recent study showed that the presence of numerous IS elements in the *Shigella* genome leads to the pseudogenisation of affected genes and an increase in non-homologous recombination (Seferbekova et al. 2021). Therefore, MGEs and specifically IS elements have a direct effect on chromosomal rearrangements and genomic evolution (Frost et al. 2005). This chapter will discuss the diversity of MGEs, adaptive traits encoded by MGE accessory genes, and how the rearrangement caused by MGEs contributed to the evolution of bacteria.

Traits and Accessory Genes Carried by MGEs

MGEs can provide their host with genes that confer adaptive traits such as antibiotics resistance genes, metabolic traits, toxins or other virulence factors that help bacteria to exploit a specific niche (Hacker et al. 2001, Norman et al. 2009).

Antibiotics Resistance Genes

Although plasmid's main function was originally known to be carrying antibiotic multi-resistance (Hughes and Datta 1983), it is acknowledged that the resistance occurs due to extensive use of antibiotics. Resistance to antibiotics is concerning as it is predicted that due to the continuous development of resistance, we could return to the pre-antibiotic era (Frost et al. 2005). Plasmid- mediated resistance gene originally derived from point mutation that occurred in susceptible bacteria. Additionally, plasmids may originate from genes that encode protective mechanisms which are present in antibiotic-producing bacteria. Such genes can be mobilized if they are flanked by insertion sequences when they are excised by transposons of the Tn3 family, or as mobile cassettes by integrons (Frost et al. 2005). Tetracycline resistance can be acquired by two copies of IS10 flank tetracycline-resistant gene and regulatory gene to form transposon Tn10 (Ochman et al. 2000). Furthermore, the resistance to kanamycin, bleomycin, and streptomycin is achieved through the presence of two copies of IS50 elements that border a three-gene operon responsible for conferring resistance to these antibiotics, constructing transposon Tn5 (Ochman et al. 2000). Examples of some multiresistance plasmids and transposons associated with resistance in *S. aureus* and *Enterococci* are summarized in Table 1. These configurations facilitate the mobilization of groups of genes that resist different antibiotic classes and/or disinfectants in one conjugation event along with a promoter to express the genes in case of integrons (Frost et al. 2005). A study had investigated the distribution of antibiotic resistance genes across plasmids and chromosomes in three groups of Enterobacteriaceae: *Escherichia, Klebsiella* and *Salmonella*. It was found that genes responsible for antibiotic inactivation, replacement and target protection were usually present in plasmids rather than chromosomes, but resistance genes involved in efflux pumps were usually present in the chromosome (Hall et al. 2022a). Therefore, inhibiting horizontal gene transfer of MGEs in the first place can be the perfect strategy to tackle the antibiotic resistance issue.

Toxins and Virulence

Genes encoding toxins, virulence factors and cellular structures required for colonisation in animals and plants are carried by the phages and conjugative transposons (Frost et al. 2005). Bacterial-genome sequencing revealed that plasmids and phages are the keys to the divergence of closely related bacterial strains and species and a fundamental reason behind developing new pathogens (Brüssow et al. 2004, Frost et al. 2005). Phages exhibit an important role in converting avirulent strains into virulent ones that can produce toxins, as in the case of cholera and diphtheria (Faruque et al. 2005, Frost et al. 2005) and exotoxin A in *Streptococcus pyogenes* and Shiga toxin in enterohaemorrhagic strains of *E. coli* (Ochman et al. 2000). Moreover, the presence of a plasmid can be the only difference between commensal or soil bacterium and a deadly pathogen such as the case of anthrax (Okinaka et al. 1999). Additionally, the presence of horizontally acquired pathogenicity islands (PIs), which is chromosomally encoded clusters of virulence gene, can transform a commensal organism into a pathogen. PIs are usually present in tRNA and tRNA-like loci. In enteric bacteria, the integration site for PIs is tRNA^{selC} (Table 2). Examples of such enteric bacteria are enteropathogenic *E. coli* possessing 35-kb LEE island, uropathogenic *E. coli* possessing 70-kb PAI-1, *Shigella fexneri* possessing 24-kb SHI-2 island and *Salmonella enterica* possessing 17-kb SPI-3 island (Table 3) (Ochman et al. 2000). Interestingly, sequence analysis revealed that PIs are transferred and acquired through phage-mediated events. For example, in *Staphylococcus aureus* a phage promotes the excision, replication and mobilization of a PI holding the gene for toxic shock toxin (Ochman et al. 2000). Therefore, horizontal gene transfer can be a reason behind pathogenicity and tackling such events can be a way to prevent pandemics problem.

Metabolic Traits

Plasmids are important in establishing mutualisms between prokaryotes and eukaryotes. For example, symbiotic nitrogen fixation is mediated by a very large (> 250 kb) plasmid-encoded genes of the genus *Rhizobium* (López-Guerrero et al. 2012). These genes are important in the invasion of host-plant root cells and conversion of atmospheric dinitrogen to ammonia (a source of nitrogen) needed by the plant. On the other hand, Rhizobium benefits from carbohydrates generated photosynthetically by the plant. In other genera such as *Mesorhizobium,* symbiotic nitrogen fixation can be mediated by chromosomal symbiosis islands, which is very large (> 500 kb) and carry genes for Nod factor synthesis, nitrogen fixation, and island transfer (Sullivan et al. 2002). In addition to nitrogen fixation, MGEs are also found to contribute to the carbon elemental cycle (Frost et al. 2005). Moreover, in cyanobacteria, the phages which are called cyanophages provide the bacteria with genes involved in photosynthesis and genes that help the bacteria to survive inside the oceans (Frost et al. 2005). Natural selection plays an important role in acquiring new metabolic traits by horizontal transfer. In *E. coli,* acquiring the *lac* operon allows the bacterium to ferment the lactose in the mammalian colon, hence establishing a commensal relationship (Ochman et al. 2000). Therefore, horizontal gene transfer is essential in determining mutualism between the bacterium and the niches it may encounter.

Traits Involved in Bioremediation of Toxic Chemicals

Bioremediation involves using microorganisms or their enzymes to promote the degradation or removal of contaminants from the environment (Azubuike et al. 2016). Some naturally occurring plasmids that encode genes for the biotransformation of hydrocarbons have been found and received a patent for this discovery (Frost et al. 2005). Later, extensive progress in both genetic and biochemical basis have been made that enriched the understanding of plasmid-mediated remediation processes in polluted environments (Frost et al. 2005). Strains having plasmids carrying genes for organic and metal bioremediation have been used in industrial settings (Azubuike et al. 2016). These plasmids can be organized in large operons or present on genomic islands and some of them are conjugative (Lindstrom et al. 1991).

Role of MGEs in Genomic Rearrangement

Rearrangement events such as duplications, inversions, deletions, and translocations can have a fundamental effect on bacterial phenotype either by manipulating gene regulation or directly disrupting the gene (Darling et al. 2008). Recombination between inverted repeats causes inversions, recombination between direct repeats causes deletions, and recombination between

Table 1. Multiresistance plasmids in *S. aureus* and *Enterococci* (Partridge et al. 2018).

Plasmid	Size (kb)	Accessory gene	Tn	IS	Trait	Original source
pI258	28	*arsBC* *cadA* *merAB* *erm (B)* *blaZ*	Tn4004 ΔTn552 Tn551	IS257	Arsenic resistance, cadmium resistance, mercury resistance, resistance to MLS antibiotics, Penicillin's resistance	*S. aureus*
pSK1	28	*dfrA* *qacA* *aacA-aphD*	Tn4001	IS257 IS256	Trimethoprim resistance, antiseptic/disinfectant resistance, Gentamicin/kanamycin/tobramycin resistance	*S. aureus*
pUSA300-HOU-MR	27	*msrA* and *mphC* *bcrA* *sat4* *aphA-3* *blaZ* *cadD*	ΔTn552	IS257	Macrolide resistance, bacitracin resistance, streptothricin resistance, Kanamycin/neomycin resistance, Penicillin's resistance, cadmium resistance	*S. aureus*
pLW1043	58	*dfrA* *qacC* *aacA* *aphD* *vanA* *blaZ*	Tn4001-hybrid, Tn1545, Tn552	IS257 IS256	Trimethoprim resistance, antiseptic/disinfectant resistance, Gentamicin/kanamycin/tobramycin resistance, vancomycin resistance, Penicillin's resistance	*S. aureus*
pRE25	50	*cat* *erm(B)* *aadE* *aphA-3* *sat4*	Tn5405-like	IS216	Chloramphenicol resistance, resistance to MLS antibiotics, Streptomycin Kanamycin/neomycin, streptothricin resistance	*Enterococci*
pEF-01	32.4	*fexB, cfr*	none	IS216 IS1182	Chloramphenicol/florfenicol resistance, Phenicols/lincosamides/oxazolidinones/pleuromutilins/streptogramin A resistance	*Enterococci*

Table 2. Transfer RNA loci targeted by horizontally acquired DNA sequences tRNA (Ochman et al. 2000).

tRNA locus	Organism	Horizontally acquired DNA	Trait	Size (kb)
selC	*Escherichia coli*	Phage QR73	Phage genome	13
selC	Uropathogenic *E. coli*	PAI-1	Haemolysin	70
selC	Enteropathogenic *E. coli*	LEE	Type III secretion system; intimin receptor protein Tir	35
selC	*Salmonella enterica*	SPI-3	Macrophage survival protein MgtC; Mg2+ transporter MgtB	17
leuX	Uropathogenic *E. coli*	PAI-2	Haemolysin; prf Fmbriae	190
serT	*S. enterica*	SPI-5	Inositol phosphate phosphatase	7
ssrA	*Vibrio cholerae*	VP1	TcpA colonization factor and receptor for CTX phage	45
arg	*Corynebacterium diphtheriae*	Corynephages	Phage genome; toxin	36
glyV	*Pseudomonas* sp.	*clc* element	Chlorocatechol-degradation	105

direct repeats during replication results in duplication (Darling et al. 2008). A large scale of such rearrangements over time led to organismal evolution which was detected in various organisms such as *Escherichia coli, Salmonella enterica* var. Typhi, *Yersinia pestis, Pseudomonas stutzeri, Pseudomonas aeruginosa, Francisella tularensis, Helicobacter pylori, Mycobacterium leprae, Neisseria gonorrhoeae* and *Staphylococcus aureus* (Raeside et al. 2014).

MGEs such as transposons, IS elements and prophage play a direct role in recombination events and contribute to bacterial evolution. For example, *Shigella* is considered to be polyphyletic relative to *E. coli* that acquired virulence plasmid assisting intracellular lifestyle (Pupo et al. 2000). Moreover, *Shigella* gained a variety of mutations and deletions in some virulence genes and acquired IS elements as an adaptation to intracellular niche and as a result of lacking competitors compared to the large intestine niche faced by *E. coli* (Seferbekova et al. 2021). It has been found that *Shigella* gained extremely high rates of intragenomic rearrangements and had a decreased rate of homologous recombination compared to pathogenic and non-pathogenic *E. coli* (Seferbekova et al. 2021). It has been suggested that the accumulation of IS elements influences many traits related to genome evolution and play an essential role in the evolution of intracellular pathogens (Seferbekova et al. 2021). Moreover, in a recent investigation focusing on *Helicobacter pylori* genomes, researchers examined the correlation between genetic elements and rearrangements. The study revealed that the positioning of genomic islands played a vital role in determining the frequency of these rearrangements (Noureen et al. 2021). Therefore, MGEs play an essential role in the bacterial genome rearrangement and facilitate bacterial adaptation to various environmental niches within or outside the host.

Mobile Genetic Elements and Horizontal Gene Transfer

Horizontal gene transfer (HGT) is one of the adaptive mechanisms in bacteria that causes genome plasticity and promotes bacterial evolution (speciation and sub-speciation). There are three classical modes of HGT, which are transduction, conjugation, and transduction (Figure 1). Additionally, there are other mechanisms of HGT, which are Gene Transfer Agent (GTAs) trafficking (Solioz et al.

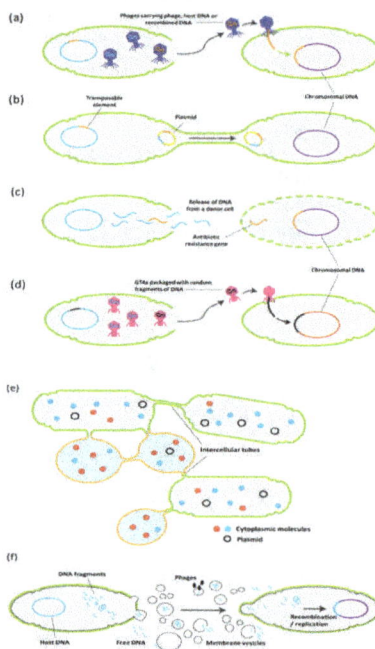

Figure 1. Various modes of Horizontal gene transfer (HGT).

1975), nanotubes transfer (Dubey and Ben-Yehuda 2011) and Membrane Vesicles (MVs) transfer (Domingues and Nielsen 2017, Mashburn-Warren and Whiteley 2006) (Figure 1).

(a) Transduction is a process where bacteriophages act as a vector to transfer bacterial DNA from a previously infected donor to recipient cell, (b) Conjugation is a DNA transfer process from a donor to recipient cell via cell surface pili or adhesins, (c) Transformation is the uptake, integration, and expression of naked DNA from the cytoplasm, (d) Gene transfer agents (GTAs) are bacteriophage-like particles that package unspecific segments of the bacterial genome and incomplete copies of their own genome. GTA particles are released through cell lysis, (e) Nanotubes are membranous intercellular bridges that allow cytoplasmic molecules to be transferred between bacteria of the same or different species, (f) Membrane vesicles (MVs) are spheres of lipid bilayers generated from the cell surface that contain DNA, RNA, proteins, and polysaccharides. MVs carrying genetic material can act as a vector of HGT by blebbing out from the outer membrane of donor cells into the surrounding environment and subsequently merge with the outer membrane of the recipient cells. Figure is based on and adapted from (Chiang et al. 2019, Dubey and Ben-Yehuda 2011, Lang et al. 2012, Von Wintersdorff et al. 2016).

Transposable Elements

Transposable elements (TEs) are fragments of DNA that are able to move themselves (and associated genes) from one site to another site within a genome or inter-genomes. Because these TEs are frequently found in numerous copies in various parts of a genome, they can aid homologous recombination (the interchange of sequences between segments that are identical or related). Therefore, TEs can be thought of as naturally occurring 'genetic engineers,' as they promote diverse DNA rearrangements via deletions, inversions, and replicon fusions, which contribute to genomic diversity. Class 1 Integrons, Insertion Sequences (IS), Composite Transposon, Integrative Conjugative Elements (ICEs), Integrative Mobilizable Elements (IMEs), and Integrated or Transposable bacteriophages are the examples of TEs. Additionally, a novel superfamily of mobile genetic element was recently identified, namely Casposon, which is also described here (Table 3).

Mechanisms of DNA transposition

Transposable elements move from one location (donor site) to another (target site) within or between bacterial genomes by a mechanism known as transposition (Hickman and Dyda 2015). This mechanism is mainly catalysed by the TEs-encoded transposases, which are sequence-specific DNA binding proteins. Most transposable elements encode their own transposases (autonomous elements). However, some TEs lack a transposase gene (non-autonomous elements), but they can be mobilised by autonomous elements present in the genome (Siguier et al. 2015). Transposases are classified based on their catalytic domain into four main classes: the DDE transposases (have two aspartates (D) and one glutamate (E) at their active sites), the HUH transposases (have two histidine residues (H) flanking a hydrophobic amino acid (U)), tyrosine-transposases, and serine-transposases (Curcio and Derbyshire 2003, Hickman and Dyda 2015). In casposons, a homolog of Cas 1 integrase (the casposase) functions as the transposase (Krupovic et al. 2014).

Transposition involves two main chemical steps, DNA breakage and ligation (insertion), which are mediated by the transposase's catalytic domain (Figure 2). During DNA breakage, the transposase recognises a specific DNA sequence (usually an inverted repeat (IRs)) flanking the integrated transposable elements and then excises the DNA element out of the donor DNA (Figure 2a). The transposase also recognises a specific DNA sequence in the target DNA, makes a cleavage and then inserts the excised elements into this new target site (Figure 2b). Most integrated TEs are flanked by direct repeats (DRs) of a characteristic length (2 to 9 bp or more) (Siguier et al. 2017). These DRs are generated when the host DNA repair system repairs the gap arising from staggered cuts of the target DNA (Figure 2b) (Curcio and Derbyshire 2003, Hickman

Table 3. Types of Transposable Elements.

Types of Transposable Element	Description	Examples	References
Integron and Gene Cassette	Integron is an element that can capture, and express open reading frames embedded in gene cassettes. It consists of three components: (1) a gene that encodes for a site-specific recombinase (*intI* encoding for integrase) which enables the insertion and integration activities of a particular gene cassette, (2) the proximal recombination site (*attI* which is the recognition site for integrase) and (3) a promoter (*Pc* which is responsible for the transcription of gene cassettes) Gene cassette is a small mobile element that often carries antibiotic resistance genes. It is a promoter-less and non-self-transmissible element residing in integron.	Class 1 integron (capture gene cassettes carrying vast antibiotic resistance genes)	(Domingues et al. 2012, Ghaly et al. 2017)
Insertion Sequence (IS)	A simple element, consisting of a single or two transposase gene (*tnp*) flanked by inverted repeats and is usually shorter than 2 kb. Transposition mechanism can be the conservative "cut-and-paste" (IS element is excised from the original site and inserted in a new target site) or replicative "copy-and-paste" (during the transposition reaction, the IS element is replicated, resulting in a transposing element that is a copy of the original element).	IS6 family elements (encode a single transposase)	(Varani et al. 2021)
Compound or Composite Transposon	An element carrying additional genes (often antibiotic re-sistance genes) in the middle, flanked by two copies of an IS element. This element depends on the transposase carried by the IS element for their insertion and excision.	Tn9 (associated with IS1) carrying chloramphenicol resistance gene, Tn10 (associated with IS10) carrying tetracycline resistance gene, Tn5 (associated with IS50) carrying streptomycin, bleomycin, and kanamycin resistance genes.	(Reznikoff 1993)
Transposable Prophage	Transposable prophage is a phage genome that has been integrated into the bacterium's chromosome. This prophage may excise and insert from and into the genome.	Lambda (Tyrosine recombinase with specific target site), Mu (DDE recombinase with multiple target sites)	(Boram and Abelson 1971, Weil and Signer 1968)
Conjugative Transposons (CTns) or Integrative Conjugative Elements (ICEs)	The conjugative transposons (CTns) are self-transferable elements that can integrate into and excise from bacterial chromosomes, as well as transmit themselves from one bacterium to another via conjugation. CTns are often modular in structure. Each module carries a set of genes responsible for excision, conjugative transfer, and integration into the new host genome. They carry accessory genes that confer virulence factors and/or resistance to antimicrobial agents. They are also known as integrative conjugative elements (CTns).	Tn916/Tn1545 family, ICEBs1 of *Bacillus subtilis*, pSAM2 of Actinomycete integrative and conjugative elements (AICEs)	(Auchtung et al. 2016, Franke and Clewell 1981, Johnson and Grossman 2015, Mullany et al. 1996)

Table 3 contd. ...

...Table 3 contd.

Types of Transposable Element	Description	Examples	References
Mobilisable Transposons (MTns) or Integrative Mobilisable Elements (IMEs)	The mobilisable transposons (MTns) and integrative mobilisable elements (IMEs) are non-conjugative MGEs that rely on other conjugative elements for the formation of conjugation pore or mating bridge for their mobilisation. MTns and IMEs carry a limited number of *mob* genes (only for DNA processing) and therefore are not self-transmissible.	Tn*4451* (Serine recombinase with specific target site), Tn*4555* (Tyrosine recombinase with variable target sites)	(Adams et al. 2004, Smith and Parker 1993)
Casposons	A novel superfamily of MGEs present in various archaeal and bacterial genomes. It encodes casposase (a homolog of Cas1 endonuclease used in CRISPR-Cas system). Casposons are variable in size ranging from approximately 8 to 20 kb and they are self-replicating MGEs. It can be divided into four families based on the phylogenetic analysis of their casposases and taxonomic distribution. Family 1 elements (restricted to the archaeal phylum *Thaumarchaeota*), Family 2 (prevalent in the Euryarchaeota archaeal order, notably in methanogens), Family 3 (bacteria-specific and found in proteobacteria and actinobacteria) and Family 4 (only found in some species of the euryarchaeon *Methanosarcina mazei*)	NitSJ-C1 (Ni-trosopumilus sp. SJ), AciBoo-C1 (*Acid-uliprofundum boonei* T469); HenMar-C1 (*Henriciella marina* DSM 19595); MetMaz1FA1A3-C1 (*Methanosarci-na mazei* strain 1.F.A.1A.3).	(Hickman and Dyda 2014, Krupovic et al. 2014, 2017)

and Dyda 2015). The catalytic mechanism of DNA breakage and re-joining varies according to the type of the transposase. These differences lead to various transposition pathways. For example, some transposases, such as serine and tyrosine transposases, cut out (excise) the transposon from the donor DNA and paste (insert) it into the target DNA (non-replicative, cut-out-paste in transposition). Other transposases, including some DDE transposases, copy out the element from the donor DNA and paste it into the target DNA. As a result, both the donor and the target DNA have a copy of the transposon (replicative, copy-and-paste transposition) (Siguier et al. 2017). HUH transposes of the IS 91 family mediate a rolling circle (RC) transposition (Mendiola et al. 1994).

Plasmids

Plasmids are autonomous, self-replicating extra-chromosomal DNA elements found in all three major domains: *bacteria, archaea,* and *eukaryota* (Smillie et al. 2010). Most plasmids are circular molecules but can also be linear (Hinnebusch and Tilly 1993, Ravin 2011). Plasmids vary in size from around 700 bp to 2.5 MB (mega plasmids) (Hall et al. 2022b, Norman et al. 2009). The plasmid backbone consists of core and accessory genes. The core genes encode functions essential for plasmid survival and propagation, including plasmid replication, maintenance, and mobility (Acman et al. 2020, Garcillán-Barcia et al. 2011). The accessory genes confer adaptive genetic traits that are clinically significant, such as antibiotic resistance, virulence, and pathogenicity. These adaptive traits enable the host to survive in stressful environments (Partridge et al. 2018, Pilla and Tang 2018). Plasmids can also carry other MGEs such as transposons and insertion sequences (ISs) (Partridge et al. 2018). Several criteria have been used to classify plasmids into different types. For example, plasmids have been categorized based on their mobility, replication, copy number, or incompatibility (Carattoli et al. 2005, Garcillán-Barcia et al. 2009, Shintani et al. 2015, Smillie et al. 2010).

Plasmids are essential drivers of HGT, disseminating adaptive genes among bacteria, thus enabling bacterial evolution (Frost et al. 2005). Of great medical importance is the evolution of plasmid-mediated antibiotic resistance. Plasmids are responsible for the–rapid dissemination of clinically significant antibiotic resistance genes (ARGs) such as *mcr-1* (mobile colistin resistance

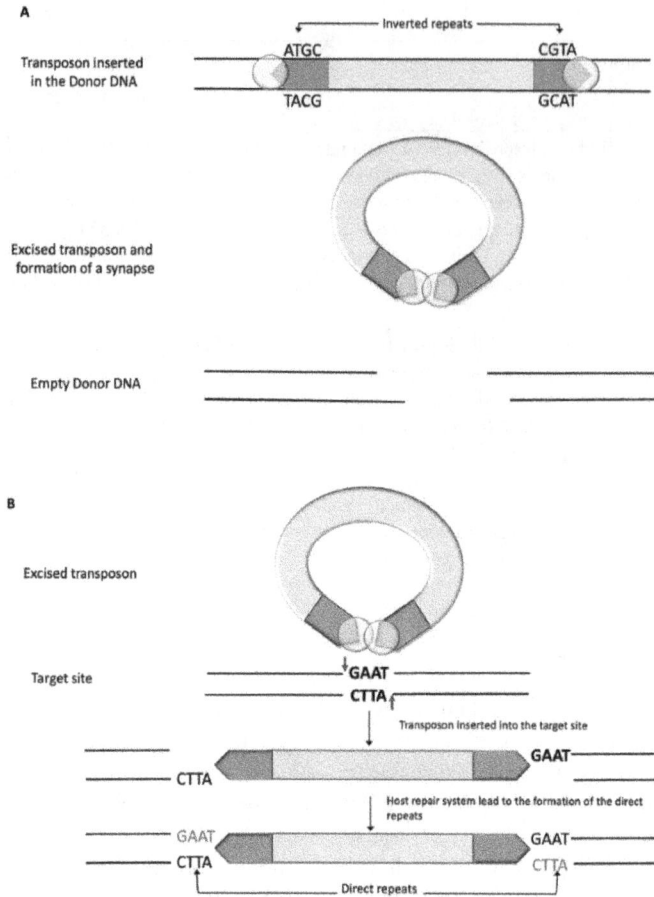

Figure 2. (a) In tansposon excision, the inverted repeats (IRs) flanking the integrated transposon are recognized and cleaved by the transposon encoded transposases. Transposases bound to the IRs are also bound to each other to form a synapse. (b) The excised transposon is then inserted into the target site via a staggered cut mediated by the transposases. The staggered cut and the subsequent repair of the resulting single-strand gaps result in the formation of direct repeats (DRs) at both ends of the inserted transposon.

gene), which confers resistance to colistin, the last resort treatment for severe infections caused by multidrug-resistant (MDR) Gram-negative bacteria (Liu et al. 2016). Also, the multicopy nature of plasmids accelerates the evolution of antibiotic resistance by increasing the expression of resistance genes, resulting in a high resistance level (Millan et al. 2016, Rodríguez-Beltrán et al. 2021). Unlike chromosomally encoded ARGs, which can persist only with the presence of positive selection, resistance genes carried on MGEs such as plasmids do not require positive selection to persist in the environment (Stevenson et al. 2017). Therefore, antibiotic reduction is insufficient to limit the spread of resistance genes. One potential strategy for combating the acquisition and transmission of ARGs is to eliminate the antibiotic resistance plasmids alongside treatment with antibiotics (Buckner et al. 2018, Vrancianu et al. 2020).

Plasmid Replication

Plasmids replicate independently of the chromosome. Replication initiates at the plasmid origin of replication *(ori)* site, which is known as *oriV* for origin of vegetative replication to differentiate it from *oriT*, the site of conjugation initiation (discussed later) (del Solar et al. 1998). Some plasmids have multiple origins of replication, such as plasmids of the phylum Firmicutes

(Lanza et al. 2015). The *ori* region determines several plasmid properties such as host range, copy number, and incompatibility. In addition to the *ori* region, plasmids encode Rep proteins located very close to the *ori* sequence. These proteins are necessary to initiate replication at this site. Plasmids also rely on host replication machinery such as helicases, primases, and polymerases (del Solar et al. 1998). Three mechanisms of plasmid replication have been described: theta replication, strand displacement, and rolling circle (RC) replication.

Theta replication and strand displacement–replication–involve opening the double-stranded DNA at the origin of replication followed by RNA priming, which initiates replication. Many theta-type replicating plasmids, such as pPS10 and R6K, contain at their origin of replication repeated DNA sequences known as iterons where Rep protein binds, leading to opening of the DNA duplex and generating a structure resembling the Greek letter theta (θ). In the Theta type of replication, the RNA primer proceeds unidirectionally or bidirectionally (Lilly and Camps 2015). In the strand-displacement plasmid replication type, the RNA primer proceeds bidirectionally. This type of replication is used by the incompatibility group Q (IncQ) plasmids (del Solar et al. 1998) Rolling Circle Replication is unidirectional and is initiated by cleavage of one of the plasmid DNA strands to generate a free 3'-OH end. Plasmids that use RC replication include pMV158, pC221, pUB110, and pHPK255 (Garcillán-Barcia et al. 2022).

Major Plasmid Properties Determined by the Ori Region

One of the major plasmid properties that is controlled by the *ori* region is the plasmid host range, which is the type of bacteria in which plasmids can stably replicate and be maintained. Plasmid host range is determined by the ability of the plasmid to replicate independently from the chromosomes or by the presence of multiple origins of replication (Norman et al. 2009). Plasmids are classified based on their host range into narrow or broad host plasmids. Narrow host range plasmids, such as ColE1-derived bacterial plasmids (e.g., pBR322 and pUC), can only replicate in closely related bacteria, whereas broad host range plasmids, such as RK2, RSF1010, and pBBR1MCS, can replicate in multiple bacterial species (Jain and Srivastava 2013, Kües and Stahl 1989). Broad host range plasmids encode all the machinery needed for the initiation of replication, and they do not depend on the host replication machinery. Other determinants that can influence plasmid host range include plasmid conjugation, plasmid fitness cost, or the ability of the plasmid to overcome host restriction systems (Lacroix and Citovsky 2016, Loftie-Eaton et al. 2016, Millan and MacLean 2017, Thomas and Nielsen 2005).

Another important plasmid property that is controlled by the *ori* region is the plasmid copy number, which is the average number of copies of a particular plasmid per cell. Plasmids are classified based on copy number into low and high copy number plasmids. Low copy number plasmids such as F and R plasmids are maintained at one to two copies per cell, whereas high copy number plasmids are maintained at tens to several hundred copies per cell and include many cloning vectors such as pUC and pBR322-derived plasmids (Nordström 2006, Projan et al. 1987, Vieira and Messing 1982). Generally, low copy number plasmids are large and carry functional conjugative systems, while high copy number plasmids are usually small and nonconjugative, but can be mobilized by conjugative elements coexisting in the same cell (Norman et al. 2009, Ramsay et al. 2016). Plasmid copy number has an essential role in the rapid evolution of bacteria. The multicopy nature of plasmids can increase the expression of specific adaptive traits such as antibiotic resistance. Also, increasing the number of plasmids per cell can increase the frequency of beneficial mutations (Rodríguez-Beltrán et al. 2021).

Plasmid stability depends on the plasmid copy number. Therefore, plasmid replication should keep up with the replication of the cell to maintain a stable copy number, avoiding their loss or accumulation in the cell during cell division. This means that a plasmid will replicate more if the copy number is lower than the required level. On the other hand, the plasmid replication will stop if the copy number reaches too high levels. Regulation of plasmid replication is achieved by three different mechanisms that inhibit the initiation of replication in a "dose dependent way," such that excess plasmid copies lead to suppression of plasmid replication. First, the anti-sense RNA (Counter transcribed RNA, ctRNA) hybridizes to and inhibits the function of an essential RNA required for DNA synthesis priming. Second, the antisense RNA inhibits the translation of a plasmid encoded protein that is essential for the initiation of plasmid replication (Rep protein). Third, regulation by coupling: at high plasmid concentration, (Rep) initiator proteins bind to repeat DNA sequences called iterons found in the *Ori* region and inhibit the plasmid replication (del Solar et al. 1998, Norman et al. 2009).

The phenomenon of plasmid incompatibility is another major plasmid characteristic that is controlled by the *ori* region. Incompatibility is the inability of two plasmids to co-exist stably in the same host cell in the absence of external selection pressure. It occurs when two plasmids have the same replication control or the same partitioning system. If two plasmids cannot co-exist in the same cell, they are in the same incompatibility group and are incompatible. If the two plasmids can coexist stably together, they are compatible (Hyland et al. 2014, Novick 1987).

Plasmid Maintenance

Stable inheritance of plasmids during cell division is crucial to ensure plasmid transmission to daughter cells and prevent plasmid loss. Plasmid stability can be achieved by plasmid segregation, which distributes plasmids equally between daughter cells at the time of cell division (Salje 2010). Plasmids have evolved several mechanisms that promote their segregational stability, thus preventing their loss from the bacterial population. These systems include multimer resolution (*res*), partitioning (*par*), and postsegregational killing systems.

Multimer resolution (*res*) systems convert plasmid dimers or multimers into monomers that can be stably inherited. Plasmid dimers arise from the fusion of two plasmids head to tail, resulting in a larger circular molecule, while plasmid multimers result from the fusion of more than two monomers. The generation of plasmid dimers and multimers is a result of homologous recombination events that may occur during cell division, or of improper termination of plasmid replication. Accumulation of these multimers can decrease the number of plasmid copies available for segregation and stable inheritance to daughter cells. To resolve plasmid multimers, site-specific recombinases (also called resolvases), such as tyrosine or serine recombinases, bind to recombination sites in the plasmids and resolve plasmid multimers into monomers. These site-specific recombination systems are either encoded by the host (e.g., the cerXerC/D system used by the ColE1 plasmid) or by the plasmid itself (e.g., the CreloxP system encoded by phage P1). Transposable elements such as transposons or ICEs carried on larger plasmids also contain resolvase genes for multimer resolution (Crozat et al. 2014, Midonet and Barre 2014, Norman et al. 2009).

Another system that ensures plasmid maintenance is the partitioning (*par*) system, which is usually encoded by low copy number plasmids. High copy number plasmids lack active partitioning systems. Partitioning systems ensure proper plasmid segregation after cell division, so each cell has at least one copy of the plasmid. These systems consist of three components: a cis-acting centromere like DNA site, a DNA-binding adaptor protein, and a motor protein which is a nucleotide triphosphatase. The adaptor protein recognizes and binds to the plasmid centromere, forming a nucleoprotein complex which interacts with the motor protein. The motor protein then pushes or pulls plasmids apart, resulting in plasmid separation (Baxter and Funnell 2014, Million-Weaver and Camps 2014).

Some plasmids avoid being lost during cell division by killing cells that do not inherit a copy of the plasmid. This mechanism is achieved by post-segregational cell killing systems, also known as addiction systems, because the host bacteria depend on these systems for their survival and thus become addicted to these systems. Two types of post-segregational cell killing systems are widely present in prokaryotes: toxin-antitoxin (TA) systems and restriction-modification (RM) systems. TA systems consist of two components encoded by the plasmid: a stable toxin and an unstable antitoxin. When a plasmid is present in the cell, both the toxin and the antitoxin are produced. The continuous presence of the antitoxin is required to inactivate the toxin and thus prevent cell death. However, when the plasmid is lost, the unstable antitoxin is degraded faster than the toxin (Hernández-Arriaga et al. 2014), resulting in cell death due to the function of the toxin. RM systems are often considered as bacterial defense systems against invading MGEs such as bacteriophages or plasmids, which impose fitness costs on their hosts. This system consists of restriction endonuclease and methyltransferase. The methyltransferase modifies the DNA host and thus protects the DNA from digestion by the endonuclease. RM systems can also act as post-segregational cell killing systems. The loss of the R-M encoding plasmid due to improper segregation can result in the loss of the protective activity of the methyltransferase, leaving the host DNA susceptible to digestion by restriction enzymes, which leads to cell death (Handa and Kobayashi 1999, Kulakauskas et al. 1995).

Plasmid Mobility

Plasmids move between different hosts via several mechanisms, including conjugation, natural transformation, transduction, and vesiduction (transfer via membrane vesicles) (Ammann et al. 2008, Smillie et al. 2010, Soler and Forterre 2020). Conjugation is a highly evolved mechanism that moves plasmids efficiently into other bacteria. Plasmids are classified on the basis of their conjugative mobility to conjugative (self-transmissible) plasmids, mobilizable plasmids, and non-mobilizable plasmids (Smillie et al. 2010).

Self-transmissible conjugative plasmids carry transfer (*tra*) genes which encode proteins involved in DNA transfer and replication (the Dtr component) and mating pair formation (the Mpf component). The Dtr component is responsible for processing the DNA via the relaxase, which is a site specific endonuclease that recognizes and binds to the origin of transfer sequence *oriT*, a DNA segment (usually a hundred base pairs (bp) in length) at which the conjugative transfer initiates. The relaxase bound to *oriT* generates a large nucleoprotein complex called the relaxosome. The Mpf component is responsible for bringing the two cells together. It consists of type IV coupling protein (T4CP) and type 4 secretion system (T4SS), a multiprotein secretion apparatus that extends from the cytoplasm to the cell's surface and serves as a channel for translocating single-stranded DNA (ssDNA) during conjugation (Norman et al. 2009, Smillie et al. 2010). The conjugative transfer of dsDNA was described for the Streptomyces conjugative plasmid, which encodes a single pore-forming ATPase responsible for translocating dsDNA from the donor cell (Thoma and Muth 2016). In Gram negative bacteria, the connection between the donor and the recipient cell is mediated by a pilus, which is assembled on the donor cell surface by T4SS (De La Cruz et al. 2010), while in Gram positive bacteria, this attachment is promoted by surface adhesins (Goessweiner-Mohr et al. 2014).

Mobilizable plasmids carry only *oriT* and relaxase and lack the MPF components. In order to be mobilized by conjugation, these plasmids depend on the mating-pair apparatus formed by self-transmissible elements such as conjugative plasmids or ICE coexisting in the same donor cell. The process by which the mobilizable plasmids are transferred is called mobilization (Smillie et al. 2010). Non-mobilizable plasmids are incapable of transfer by conjugation or mobilization, and they spread by natural transformation, transduction, or conduction. In conduction, non-mobilizable plasmids recombine with a mobilizable or conjugative plasmid to form a cointegrated plasmid which is then transferred to a recipient cell (Clark and Warren 2003).

During conjugation, the pilus extends from the surface of the donor cell and connects to the recipient cell, bringing the cells together. The cells are then connected via a pore. The coupling proteins then signal the relxase to initiate DNA transfer. In the donor cell, the relaxase creates a nick in one of the donor DNA strands at the *oriT* site. Subsequently, the conjugative plasmid encoded helicase unwinds the strands. The relaxase remains covalently attached to the 5' end of the single-stranded DNA (ssDNA) and is then transported from the donor cell to the recipient cell via the conjugation pore, carrying the attached ssDNA with it. In the recipient cell, relaxase facilitates the recyclization of ssDNA. A primase encoded by either the host or the conjugative element primes the complementary strand replication. The 3' end of the donor cell's remaining strand serves as a primer for DNA replication. As a result, the donor and recipient bacteria both contain a circular double-stranded copy of the element (Smillie et al. 2010).

Bacteriophages

Bacteriophage, which is also known as phage, is a type of virus that infects bacteria and archaea. It consists of proteins and DNA encapsulated in a structure called capsid. A temperate bacteriophage is a type of phage that can undergo both lytic and lysogenic cycles. In the lytic cycle, phages exploit a host cell to produce additional progeny virions, which are subsequently released out of the cell. In lysogenic cycles, the injected phage DNA is integrated into the host chromosome, which does not immediately result in bacterial death (Cieślik et al. 2021). Two very well studied temperate bacteriophages, Mu and Lambda, can integrate their genome into bacterial host chromosome via random transposition (Mu) or site-specific recombination (Lambda) (Casjens and Hendrix 2015). Temperate bacteriophages play a role in the spread of genetic material (including antibiotic resistance genes) from one bacterium to another via a process called transduction. There are three types of

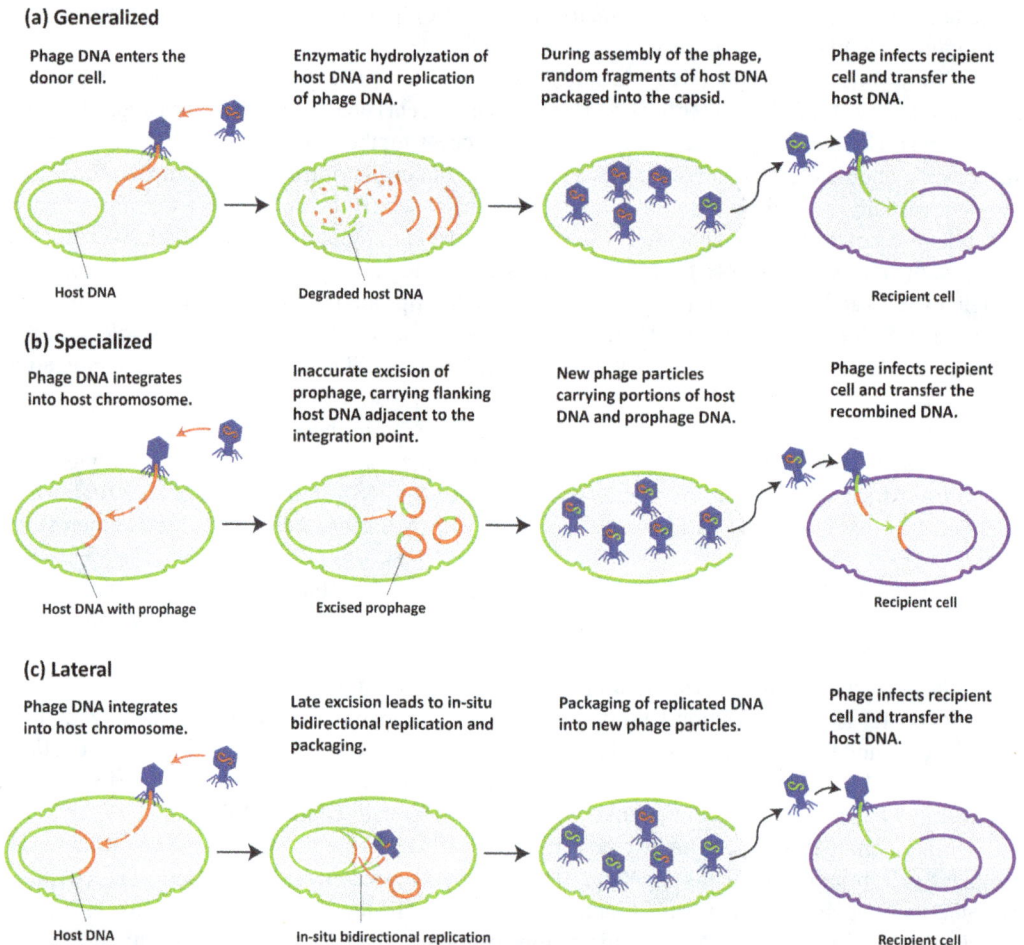

(a) Generalized

Phage DNA enters the donor cell.

Enzymatic hydrolyzation of host DNA and replication of phage DNA.

During assembly of the phage, random fragments of host DNA packaged into the capsid.

Phage infects recipient cell and transfer the host DNA.

Host DNA   Degraded host DNA   Recipient cell

(b) Specialized

Phage DNA integrates into host chromosome.

Inaccurate excision of prophage, carrying flanking host DNA adjacent to the integration point.

New phage particles carrying portions of host DNA and prophage DNA.

Phage infects recipient cell and transfer the recombined DNA.

Host DNA with prophage   Excised prophage   Recipient cell

(c) Lateral

Phage DNA integrates into host chromosome.

Late excision leads to in-situ bidirectional replication and packaging.

Packaging of replicated DNA into new phage particles.

Phage infects recipient cell and transfer the host DNA.

Host DNA   In-situ bidirectional replication   Recipient cell

Figure 3. Types of transductions Diagram showing three different types of transductions; (a) Generalised, (b) Specialised and (c) Lateral.

transduction: (A) Generalised transduction - the transmission of random bacterial chromosomal segments packaged by phage, (B) Specialised transduction - the transmission of phage DNA together with flanking chromosomal DNA in which the phage was inserted into due to imprecise prophage excision, (C) Lateral transduction - as a result of anomalous late excision, *in situ* replication, and packaging, lateral transduction involves the transfer of phage DNA plus a considerable amount of nearby chromosomal DNA (Figure 3) (Chiang et al. 2019).

Gene Transfer Agents

Gene Transfer Agents (GTAs) are phage-like particles that contain random DNA fragments of the cell producing it. It is capable of transferring genetic content and can be found in certain bacteria and archaea (Grüll et al. 2018). GTA genes are linked to phage genes, but they have unique characteristics that set them apart from phage. Unlike phage which can generally package their whole genomes and reproduce in infected host cells, GTA carries a smaller quantity of DNA (4–14 kbp) in their small-sized capsid, which is inadequate to encode for the phage-like structural form (Lang et al. 2012). The GTA-encoding genes are located in the host cell's genome and cannot be transferred to another cell. These genes are hypothesised to have originated from bacteriophage DNA that encoded defective phages that could not generate phage particles but had some phage-like structural properties (head and tail) (Casjens and Hendrix 2015, Yen et al. 1979). GTA particles are generated, then released via bacterial cell lysis and attached itself to target cells, most likely by tail-receptor contact (Lang et al. 2012, 2017).

Interestingly, GTAs have been suggested to possess several advantages over the previously discovered HGT processes. In contrast to naked DNA in the transformation process, GTA offers protection of the genetic material encapsulated in their capsid. Furthermore, unlike conjugation, the transfer of GTA does not rely on bacterial cell-to-cell contact that requires genes encoding the conjugation machinery. It is also thought to be more efficient in transferring random fragments of host DNA in comparison to most phages (Von Wintersdorff et al. 2016).

RcGTA, the most well-studied GTA, is derived from the purple photosynthetic bacteria *Rhodobacter capsulatus* (Marrs 1974, Solioz and Marrs 1977, Yen et al. 1979). It was the first GTA discovered as a result of co-culturing two strains of *R. capsulatus* with different antibiotic resistance profiles. The genes that code for RcGTA are separated into two clusters: a 14-kb gene cluster that codes for the protein required for head and tail morphogenesis, and a structural gene cluster that codes for the proteins that make up the head spikes and tail fibres. In their approximately 30-nm spiked-capsid, RcGTA particle carries about 4 kb of DNA (Grüll et al. 2018).

Antibacterial Defense Against Mobile Genetic Elements

Evolution of Bacterial Immune Mechanisms Against Mges

Overview

Bacteria are constantly attacked by viruses (bacteriophages) and have developed a variety of defensive responses to resist phage infections. The immune systems against bacteriophages can function at many stages of the phage life cycle, such as receptor binding, genome injection into the host cell, and intracellular genome replication. First, surface modification immunity arises from either masking, mutation, or deletion of host receptor proteins, which serve as phage entrance sites. Second, superinfection exclusion, which allows the host to prevent phage DNA injection or replication. Third, three intracellular defense systems can attack injected phage genomes: restriction-modification (RM), the CRISPR-Cas (clustered regularly interspaced short palindromic repeat–CRISPR-associated gene) system, and prokaryotic Argonaute (pAgo). Finally, infection failure can result during phage infection, through bacterial initiation of a suicide response known as abortive infection (Abi) (van Houte et al. 2016a). The molecular mechanisms underlying these immunological tactics have been illustrated in numerous studies, which will be summarized below (Bickle' and Kruger 1993, Labrie et al. 2010a).

Restriction–modification System

Restriction-modification (RM) systems are a group of immune systems that work by cleaving nonself, unmodified DNA while leaving modified self DNA alone. The majority of RM systems include two parts: a methyltransferase (MT) and a restriction endonuclease (RE) that recognizes certain DNA sequences and cleaves unmethylated DNA (restriction sites). The MT-catalyzed methylation of chromosomal restriction sites prevents self-cleavage of the host DNA (Roberts et al. 2003).

RM systems are categorized into four kinds based on subunit composition and biochemical properties such as protein structure, restriction site recognition, cofactor requirements, and substrate specificity (types I to IV). The restriction (HsdR), modification (HsdM), and specificity (HsdS) subunits of Type I systems encode a protein complex. HsdR's DNA translocation activity is triggered by unmodified target sequences, while HsdS remains coupled to the recognition sequence, resulting in loop creation in the DNA. When two complexes contact, cleavage occurs, and the cleavage site can be thousands of base pairs away from the actual recognition site. The restriction (HsdR), modification (HsdM), and specificity (HsdS) subunits of Type I systems encode a protein complex. HsdR's DNA translocation activity is triggered by unaltered target sequences, whereas HsdS stays attached to the recognition sequence, resulting in DNA loop formation. Cleavage happens when two complexes come into contact, and the cleavage site can be thousands of base pairs away from the recognition site (Roberts et al. 2003).

Restriction endonucleases can break down DNA that is detected as alien because it lacks the same sequence-specific chemical fingerprints. There's a good probability that small plasmids and single genes won't contain any of the more uncommon restriction sequences, avoiding the effect of enzymic cleavage (Thomas and Nielsen 2005).

Furthermore, the fact that DNA entering through conjugative transfer or natural transformation is single-stranded rather than double-stranded may provide some protection, and indeed, a comparison of transformation frequencies of double-stranded DNA confirms that the former mechanism has a greater ability to avoid destruction (Lacks and Springhorn 1984). Nonetheless, it is widely known that if the recipient contains a restriction system to which the incoming plasmid is susceptible, the number of transconjugants can be lowered (Moser et al. 1993, Pinedo and Smets 2005, Purdy et al. 2002).

The IncP-1 plasmids with a large host range appear to have adapted to the presence of such barriers by selecting versions that have lost the majority of their sites (Wilkins et al. 1996).

The barrier is roughly proportional to the number of target sites in the plasmid when a cloned DNA fragment with extra restriction sites is introduced into the plasmid, resulting in a reduced frequency of transfer and confirming that the barrier is roughly proportional to the number of target sites in the plasmid (Purdy et al. 2002, Wilkins et al. 1996).

This is a common challenge in the genetic manipulation of bacteria, and solutions range from utilizing a non-methylating *E. coli* host in which the receiver detects methylated DNA to cloning appropriate modifying enzymes into *E. coli* so that the DNA is already protected on entry (Bassett and Janisiewicz 2003).

Abortive Infection Systems

Abi is not a defense system in and of itself; rather, it is an immunological strategy that is expressed in a variety of bacterial defense systems. The basic idea behind this method is that after a bacterial cell detects an infection, it goes into survival mode. Before the phage can complete its reproduction cycle, it commits suicide. This prevents mature phage particles from emerging from the infected cell, preventing the phage epidemic from spreading to other cells and ensuring the colony's survival. Abi is an altruistic attribute in which one cell sacrifices itself for the benefit of the community. Given that bacteria usually reside in colonies of isogenic or nearly isogenic cells, and that phages are generally very specific, infecting only one species or subspecies of bacteria, this altruism is likely to protect only very closely related kin (Fukuyo et al. 2012, Hamilton 1964, van Houte et al. 2016b).

For many years, *Escherichia coli* has been the most frequently utilized model organism for studying phage infection, and most mechanistic understandings of Abi systems have been obtained using *E. coli* model systems (Keen 2015). *Lactococcus lactis* has also been intensively examined for its ability to limit phage infections, with more than 20 separate defensive genes or systems identified as Abi systems (named AbiA to AbiZ) in that bacterium. However, the mechanism of phage sensing and/or cell killing is only known for a tiny percentage of them (Barrangou and Horvath 2011, Chopin et al. 2005, Keen 2015, Sing and Klaenhammer 1990).

Rex was one of the first Abi systems to be discovered in *E. coli* (Dy et al. 2014, Gaddum et al. 1955). The Rex system is made up of two genes, rexA and rexB, which are both essential for defensive function and are expressed from a suppressed lambda prophage (Landsmann et al. 1982, Snyder and McWilliams 1989).

To safeguard the surrounding bacterial population, infected bacterial cells sacrifice themselves, thus taking one for the team. Mobile genetic elements like prophages and plasmids are commonly used to encode abortive systems (Abi) (Raza et al. 2021a). The rex system, which consists of the proteins rexA and rexB, is the best-known abortive mechanism thus far (Seed 2015). Phages form protein-DNA complexes that activate rexA at the start of an infection. The membrane-anchored rexB, an ion channel that produces a reduction in membrane potential, is activated by RexA (Labrie et al. 2010b). This reduces cellular ATP levels and, as a result, prevents cell multiplication (Labrie et al. 2010b, Seed 2015). As a result, phage infection fails because both ATP and ATP-dependent cellular components are in short supply (Labrie et al. 2010b).

One of the various mechanisms employed by bacteria to plan a premature cell death is the cyclic oligonucleotide-based anti-phage signaling system (CBASS), which is a form of Abi system (Lau et al. 2020). When phages infect cells, CBASS causes the creation of signaling cyclic oligonucleotides, which causes an effector to be activated, causing cell death (Lau et al. 2020, Millman et al. 2020).

In *S. epidermidis*, the action of a serine/threonine kinase (Stk) provides another example. Stk that has been activated phosphorylates proteins that are involved in all important cellular functions such as translation, transcription, metabolism, and repair. Changes in phosphoprotein levels cause bacterial cell death, preventing phage infection from spreading (Hampton et al. 2020a).

HORMA proteins, which are present in a variety of bacteria including *E. coli*, are also important in abortive adaptation. These proteins recognize phage products and trigger the production of cyclic tri-AMP via a cGAS/DncV-like nucleotidyltransferase. This second messenger causes ds DNA to be cleaved, resulting in cell death (Hampton et al. 2020a).

Due to a variety of predators attacking it, *Lactococcus lactis* exhibits a variety of Abi systems (Seed 2015). As a result, Abi systems develop a resistance mechanism that promotes cell death, restricting phage replication in a bacterial community.

CRISPR–Cas

The discovery of bacterial CRISPR-Cas systems that prevent bacterial viruses (phages) from infecting cells has revolutionized bacterial immunity while also providing intriguing tools for targeted genome editing. CRISPR systems wipe off phage genomes, prompting phages to produce anti-CRISPR (Acr) proteins that block Cas effectors (Bondy-Denomy et al. 2012, 2015). Although there are six different types of CRISPR systems (I–VI) found throughout the bacterial world (Koonin et al. 2017), Acr proteins have only been discovered for type I and II CRISPR systems

(Bondy-Denomy et al. 2012, Borges et al. 2017a, Koonin et al. 2017, Pawluk et al. 2016, Rauch et al. 2017).

Inhibiting type I-F and I-E CRISPR systems, Acr proteins were first found in *Pseudomonas aeruginosa*. *P. aeruginosa* strains also have a third CRISPR subtype (type IC) that isn't known to have any inhibitors. To screen for new acr candidates, we modified *P. aeruginosa* to target phage JBD30 with type I-C CRISPR-Cas (fig. S1A) and used it in parallel with existing type I-E (strain SMC4386) and I-F (strain PA14) CRISPR strains (Bondy-Denomy et al. 2012, Pawluk et al. 2014).

CRISPR-Cas systems are bacteria and archaea's adaptive immune systems. Their molecular mechanism has already been thoroughly examined elsewhere (Levy et al. 2017, Marraffini 2015, Reeks et al. 2013, Van Der Oost et al. 2014, Wiedenheft et al. 2012). CRISPR-Cas systems are made up of CRISPR loci and CRISPR-associated (cas) genes. The protein machinery that carries out the immune response is encoded by cas genes. Invader-derived sequences are separated by direct repeats in CRISPR loci, which serve as a genetic memory of prior infections. Based on phylogeny, cas gene composition, and CRISPR sequences, CRISPR-Cas systems are currently categorized into two different classes, six kinds, and sixteen subtypes. Regardless of differences, all systems follow the same basic idea of acquiring spacers from foreign DNA and integrating them into CRISPR loci on the host genome (adaptation). The CRISPR loci are then transcribed, and the resultant RNA molecule is processed to produce mature CRISPR RNA (crRNA), which binds to one or more Cas proteins (expression). Finally, crRNA-Cas complexes bind and cleave complementary nucleic acids (with the help of additional Cas nucleases in some situations), resulting in host immunity (interference) (Makarova et al. 2015, Shmakov et al. 2015).

Temperate bacteriophages generate prophage by introducing their genome into bacterial cells. Some DNA fragments are left behind, which aid in the acquisition of immunity (from bacteriophages) by surviving bacteria during subsequent infections. These fragments could potentially be horizontally transferred to other bacterial cells, conferring protection to them. Clustered regularly short palindromic repeats is the name given to this type of DNA sequence (CRISPR) (Raza et al. 2021b).

CRISPR contains an accompanying protein, an endonuclease that helps change the genome by generating cuts in double-stranded DNA. Cas9 is the name of the protein, and the CRISPR-Cas9 system is the name of the relationship between Cas9 and a guide RNA that matches the target gene. RNA-spacers (flanked by repeats) direct the Cas protein to target and break DNA (Gholizadeh et al. 2020). These spacers are DNA fragments gathered in the cell from phages that have attacked the bacterial cell. The insertion of spacers into CRISPR loci on the host genome eventually leads to phage infection avoidance (Barrangou et al. 2007).

CRISPR-Cas9 is widely used in the treatment of infectious diseases such as HIV (Cui et al. 2020). Other applications include detecting and cleaving antibiotic resistance-causing DNA (Gholizadeh et al. 2020). Certain bacteria's CRISPR-Cas immune systems, such as P. aeruginosa, are thought to be controlled by quorum sensing (Broniewski et al. 2021). Bacteria are protected by the CRISPR-Cas9 system from horizontally transmitted mobile elements. MDR bacteria, which lack this mechanism, can quickly acquire new genes and adapt to new medications (Wang et al. 2018). CRISPR-Cas9, on the other hand, prevents the phage life cycle from being completed.

Toxin–antitoxin Modules

The first TA system, ccdA/ccdB, was published 30 years ago as a module that improved the stability of the F plasmid by killing plasmid-free daughter cells after segregation. Following that, plasmid encoded TA systems were identified based on their ability to improve plasmid stability, although homology searches on bacterial chromosomes also revealed a number of TA systems (Ogura and Hiraga 1983).

A poisonous protein and an antitoxin, which is either an RNA (type I and III) or a protein, make up a toxin-antitoxin (TA) system (type II). Type II systems are prevalent in bacterial genomes, where they are transferred horizontally (Guglielmini and Melderen 2011). They usually consist of two genes encoding a toxin and a labile antitoxin, which are grouped in an operon. These tiny modules, when carried by mobile genetic components, contribute to their stability through a phenomenon known as addiction (Guglielmini and Melderen 2011).

Phage Growth Limitation and Bacteriophage Exclusion

Bacteria use a variety of strategies to protect themselves from bacteriophage infection, the most frequent of which are restriction-modification (R-M) systems (Bickle and Krüger 1993, Murray 2002). The HsdS subunit is necessary for target sequence recognition in type I R-M systems, and the sequence specificity of such subunits can be changed by natural processes in some circumstances (Dybvig et al. 1998, Schouler et al. 1998, Sitaraman and Dybvig 1997). When the products of distinct hsdS alleles interact with the other subunits of the system, HsdR (needed for restriction) and HsdM, they can confer unique R-M specificity (required for modification) (Dybvig et al. 1998, Schouler et al. 1998, Sitaraman and Dybvig 1997). High-frequency phase fluctuation can occur in R-M systems on occasion. The mod gene in the *Haemophilus influenzae* type III R-M system, for example, has 40 tetranucleotide (AGTC) repeats that expand and contract, causing frameshifts (De Bolle et al. 2000). While changing the sequence specificity of an R-M system may be seen as a response to a changing phage population, the benefit provided to bacteria by phase-variable R-M systems is unclear (Dybvig et al. 1998). The phage growth limitation (Pgl) system in *Streptomyces coelicolor* is an uncommon phage resistance mechanism (Sumby and Smith 2003). The phenotype is defined by the ability of the infecting phage (e.g., C31) to go through a single phage burst, which results in progeny that are severely attenuated in a second infectious cycle (Sumby and Smith 2003).

Argonaute-Centered RNA/DNA-Guided Defense

Argonautes are big proteins with around 800–1200 amino acids that contain non-catalytic domains such as the PAZ (PIWI-Argonaute-Zwille), MID (Middle), and N domains, as well as two domain linkers, L1 and L2, in addition to the catalytic PIWI domain (Cerutti et al. 2000, Parker et al. 2005, Parker and Barford 2006, Swarts et al. 2014a). The MID domain is found in all Ago proteins and is required for binding the 5'-end of the guide. The PAZ domain, which has an OB-fold core like those seen in many nucleic acid-binding proteins, isn't required for guide binding but helps to maintain the guide from the 3'end. The N domain is not essential for guide loading, but it does play a role in the dissociation of the loaded dsRNA's second passenger strand and target cleavage.

Although Argonautes were first identified as highly conserved eukaryote-specific proteins, prokaryotic homologs of eukaryotic Ago (hence referred to as pAgo and eAgo, respectively) were quickly discovered in a wide range of bacteria and archaea (Bohmert et al. 1998, Tabara et al. 1999). However, the distribution of pAgo is limited, with only approximately a third of archaeal genomes and roughly 10% of bacterial genomes encoding a member of this family (Swarts et al. 2014a).

The idea of pAgo's defense function has been investigated in the lab, with promising results, while the scope of the tests is still limited. The ability of pAgos from the bacteria *Aquifex aeolicus* (Yuan et al. 2005) and *Thermus thermophilus* (Sheng et al. 2014), as well as the archaea *Methanocaldococcus jannaschii* (Willkomm et al. 2016) and *Pyrococcus furiosus* (Swarts et al. 2015) to cleave target nucleic acids *in vitro* has been established. Interestingly, all three catalytically active pAgos use ssDNA guides, although their capacity to cleave RNA or DNA differs. In contrast, the RNA-binding pAgo of the bacteria *Rhodobacter sphaeroides*, which was predicted to be inactive due to mutations in the catalytic core of the PIWI domain, has shown no nuclease activity (Olovnikov et al. 2013).

The defensive functions of the pAgo from *R. sphaeroides* (Olovnikov et al. 2013) and *T. thermophilus* (Swarts et al. 2014b) have been demonstrated. By cleaving the plasmid DNA with plasmid-derived short ssDNA guides, the *T. thermophilus* Ago limits plasmid replication. The process of guide production is unknown; however, it has been established that the PIWI domain's catalytic residues are essential (Swarts et al. 2014b). As a result, it appears that pAgo initially shreds the plasmid DNA in a guide- (and presumably, sequence) independent manner before acquiring the guides and becoming a target-specific nuclease. It's unclear what causes the self/non-self discrimination in the first place (Olovnikov et al. 2013). Association with short RNAs that represent much of the bacterial transcriptome has been observed for the *R. sphaeroides* pAgo.

Furthermore, this Ago is coupled with tiny RNA-complementing ssDNA molecules, and this DNA population is enriched in "foreign" sequences, such as those from plasmids and mobile elements inserted into the bacterial chromosome. pAgo appears to sample degradation products of the bacterial transcriptome in *R. sphaeroides* and then selectively creates complementary DNAs for foreign sequences that are employed to suppress the expression of corresponding elements via yet undiscovered processes. It's unclear whether the operation of this catalytically inactive pAgo necessitates the involvement of other nucleases. Nonetheless, the presence of pAgo within evolutionarily conserved operons containing nuclease and helicase genes suggests a complicated organization of prokaryotic Ago-centered defense systems that needs to be examined further. Such studies should shed light on the mechanisms used by prokaryotic pAgo-centered defense systems to create guide RNA and DNA molecules and distinguish parasite genomes from host genomes (Makarova et al. 2009, Swarts et al. 2014b).

Unlike its prokaryotic counterparts, the Ago-centered molecular machinery involved in RNAi in eukaryotes has been extensively explored. The eukaryotic Ago family is incredibly diverse, with various catalytically active (slicers) and even more numerous inactivated forms (Azlan et al. 2016, Höck and Meister 2008, Hutvagner and Simard 2008, Meister 2013).

Eukaryotes have a range of regulatory mechanisms in the micro(mi)RNA branch, in addition to the defense function of the small interfering (si) RNA branch of RNAi (Iwakawa and Tomari 2015, Jonas and Izaurralde 2015, Reis 2017).

In most circumstances, RNAi's defensive function involves active eAgo cleavage of foreign (virus) dsRNAs, whereas miRNA pathways involve binding and reversible deactivation of mRNA by inactive eAgo variants (although in certain cases, mRNA breakdown by other nucleases is favored). RNAi's antiviral and regulatory branches appear to be intertwined: infection with a virus causes the creation of endogenous siRNA, which silences many host genes (Cao et al. 2014).

Anti-Defense Gene Clustering in Mobile Genetic Elements

Foreign genetic elements are always threatening to infiltrate all cellular life. Prokaryotes are outnumbered by a variety of mobile genetic elements (MGEs), such as viruses and plasmids, which infect them. Restriction-modification systems, abortive infection, and clustered regularly interspaced short palindromic repeats (CRISPR) and CRISPR-associated (Cas) genes have all evolved as a result of this selection pressure (Hampton et al. 2020b, Rostøl and Marraffini 2019).

CRISPR–Cas loci have been found in the genomes of roughly 40% of bacteria and 85% of archaea, and they are occasionally carried by a wide spectrum of MGEs, demonstrating their evolutionary and ecological significance (Acter et al. 2020, Al-Shayeb et al. 2020, Faure et al. 2019, Pinilla-Redondo et al. 2020a). Cells can recall, recognize, and thwart recurrently infecting pathogens using this mode of protection. Adaptation, processing/biogenesis, and interference are the three key phases of CRISPR–Cas immunity (Hille et al. 2018). During adaptation, bits of an invading genetic element are used as "spacers" between repeat sequences in CRISPR arrays, resulting in a heritable record of previous genetic intruders. The CRISPR array is subsequently translated into a lengthy transcript (pre-crRNA), which is then processed into single CRISPR RNAs (crRNAs), which direct

Cas nucleases to invading nucleic acids with a complementary sequence to the spacer (referred to as protospacer) (Pinilla-Redondo et al. 2020b).

Many MGEs have produced inhibitors of CRISPR–Cas function called anti-CRISPR (Acr) proteins in response of the high selective pressure exerted by CRISPR–Cas immunity (Borges et al. 2017b). The first acr genes were found in phages that suppress type I–F bacteria. Many other non-homologous Acr proteins have since been discovered for various CRISPR–Cas types (for example, types II, III, and V, as well as some on non-phage MGEs) (Athukoralage et al. 2020, Bhoobalan-Chitty et al. 2019, Marino et al. 2018, Pawluk et al. 2016, Rauch et al. 2017, Watters et al. 2018).

MGEs can persist despite regular targeting by host spacer sequences, thanks to the discovery of Acr proteins. Acrs is expected to be a major driver of CRISPR–Cas system diversification in nature, as well as the accumulation of additional defense systems in prokaryotic genomes. The study of Acrs allows a better understanding of MGE-host interactions and the horizontal transfer potential of MGE-encoded traits (e.g., antibiotic resistance) across microbiomes because MGEs facilitate host genome rearrangements and provide the foundation for vast prokaryotic gene exchange networks. Acr proteins are useful in phage-based treatments and plasmid-based delivery platforms, as well as providing a control mechanism for CRISPR–Cas-derived biotechnologies (Marino et al. 2020).

Antibacterial Defense Systems

Bacterial resistance mechanisms can be classified into several classes based on a variety of factors. Antimicrobial resistance, for example, can be divided into two categories based on the acquired ways of resistance for bacteria to antimicrobial agents, namely intrinsic and acquired modes of resistance (Olaitan et al. 2016).

The first line of defense is bacterial biofilms; the second line of defense is the cell wall, cell membrane, and encased efflux pumps; and the third line of defense is intracellular biochemistry and genetic responses, which play an important role in resistance and are considered the third line of defense when bactericides eventually get into the bacterial cells. The development of these defense line ideas will aid us in better understanding the main mechanisms by which bacteria defeat antimicrobial agents (Zhou et al. 2015).

The First Line of Defense: Bacterial Biofilms

Biofilms are defined as a thin layer of microbial communities adhering to each other on organic or inorganic surfaces and surrounded by extracellular polymeric material matrices generated by the bacteria (EPS) (Costerton et al. 2003). Microorganisms are known to reside as biofilms rather than as solitary entities for the majority of their microbial lives (Hall-Stoodley et al. 2004). Biofilms can cause major environmental issues like biofouling as well as a variety of human illnesses like cystic fibrosis and urinary catheter cystitis (Costerton et al. 1999, Flemming 2002, Hall-Stoodley et al. 2004). When bacteria adhere to a solid biotic or abiotic surface, they create hydrated EPS over time, eventually forming the typical spatial patterns of biofilms (Drenkard 2003, Flemming and Wingender 2010, Høiby et al. 2010). Biofilms typically go through numerous stages of development, from initial bacterial cell attachment to maturation and final dispersion (Costerton et al. 2003, Stoodley et al. 2003). Biofilm development not only provides bacterial cells with a protected mode of growth in harsh conditions, but it also increases bacteria's resistance to antimicrobial agents (Hoyle and Costerton 1991). Although numerous mechanisms have been postulated to explain enhanced antimicrobial resistance in bacterial biofilms, it is becoming clear that only a combination of multi-factorial mechanisms or a collective resistance mechanism may explain the observed resistances in biofilm communities (Drenkard 2003, Vega and Gore 2014).

The Second Line of Defense: Bacterial Cell Wall and Cell Membrane

Antimicrobial drugs must reach a sufficiently high concentration at intracellular target locations in order to exert their antibacterial activity. They must pass through the bacterial cell wall and membrane in order to reach their target site(s), which are critical for maintaining cell shape and transferring nutrients or signaling chemicals. At the same time, numerous antimicrobial agents, such as -lactams, glycopeptides, fosfomycin, daptomycin, polymyxin, and ionophore antibiotics, target the cell wall and membrane. Resistance may occur as a result of changes in cell wall or membrane conformations, as well as restricted antimicrobial drug penetration through these two physical barriers. Furthermore, limiting access to antimicrobial drugs or efficiently eliminating them via efflux pumps enclosed in the cell wall and membrane contributes to higher levels of resistance. As a result, the cell wall, membrane, and enclosed efflux pumps serve as bacteria's second line of defense against antimicrobial chemicals (Zhou et al. 2015).

The Third Line of Defense: Intracellular Alteration

Despite the fact that there are two layers of protection outside of the bacterial cells, as mentioned above, antimicrobial drugs can still penetrate the cells and exert their activity. Antimicrobial drugs can stop bacteria from growing or kill them by damaging their metabolic systems and gene expression patterns. As a result, bacterial cells will use a variety of techniques to compete with antimicrobial drugs, including altering target locations, producing antagonistic chemicals, and regulating gene expression. The third line of defense is comprised of all resistance tactics found within bacterial cells (Zhou et al. 2015).

To summarize, bacteria protect themselves against antimicrobial chemicals by constructing three primary lines of defense at spatially different places. Firstly, most bacteria will dwell in biofilms to raise their antimicrobial resistance levels through various resistance mechanisms such as restricted antimicrobial drug penetration, altered metabolic rates and gene regulation, and the development of persisted cells. Second, antimicrobial agents must enter cells through the cell wall and membrane in order to work. Antibiotic resistance can be conferred on cells by mutations that change the cell wall and membrane. Meanwhile, efflux pumps contained in the cell wall and membrane contribute to antimicrobial resistance by pumping antimicrobial chemicals out of bacterial cells. Third, even if antimicrobial drugs effectively enter bacterial cells, they must remain stable and accumulate to inhibitory concentrations at the target locations. The three main bacterial lines of defense against antimicrobials theory, which are classified according to their spatially distinct sites of action and distribution, will aid us in quickly grasping and conceptualizing the most common antibiotic resistance mechanisms in bacteria, as well as developing reasonable strategies to overcome these resistances (Zhou et al. 2015).

Conclusions

In conclusion, MGEs play an essential role in microbial evolution by promoting genomic rearrangements and the spread of genes coding for adaptive traits such as antibiotic resistance and virulence genes among bacterial populations. However, MGEs can also decrease the host fitness, which has led to the evolution of bacterial defense mechanisms against these elements. The ability of MGEs to transfer adaptive genetic traits among bacteria is of great public health concern. Understanding the biology of MGEs provides insight into their role in horizontal gene transfer and their interactions with their hosts, which is essential in developing effective strategies to control the emergence and spread of bacterial pathogens. For example, using bacteriophages as an alternative to antibiotics to treat infections caused by multidrug- resistant bacteria is an active research theme, and understanding the development of bacterial defense mechanisms against bacteriophages is critical to develop more effective phage therapy. Further research is needed to fully understand the biochemical mechanisms and evolution of MGEs and their potential application in biotechnology and medicine.

References

Aaron, E. Darling and M.A.R. István Miklós. 2008. Dynamics of genome rearrangement in bacterial populations. PLoS Genet. 7(4): 128. https://doi.org/10.1371/Citation.

Acman, M., L. van Dorp, J.M. Santini and F. Balloux. 2020. Large-scale network analysis captures biological features of bacterial plasmids. Nat. Commun. 11(1): 1–11. https://doi.org/10.1038/s41467-020-16282-w.

Acter, T., N. Uddin, J. Das, A. Akhter, T.R. Choudhury and S. Kim. 2020. Evolution of severe acute respiratory syndrome coronavirus 2 (SARS-CoV-2) as coronavirus disease 2019 (COVID-19) pandemic: A global health emergency. Sci. Total Environ. 15:730: 138996. doi: 10.1016/j.scitotenv.2020.138996. Epub 2020 Apr 30.

Adams, V., I.S. Lucet, D. Lyras and J.I. Rood. 2004. DNA binding properties of TnpX indicate that different synapses are formed in the excision and integration of the Tn4451 family. Mol. Microbiol. 53(4): 1195–1207. https://doi.org/10.1111/J.1365-2958.2004.04198.X.

Al-Shayeb, B., R. Sachdeva, L. X. Chen, F. Ward, P. Munk, A. Devoto et al. 2020. Clades of huge phages from across Earth's ecosystems. Nature. 578(7795): 425–431. https://doi.org/10.1038/s41586-020-2007-4.

Ammann, A., H. Neve, A. Geis, K.J. Heller. 2008. Plasmid transfer via transduction from *Streptococcus thermophilus* to *Lactococcus lactis*. J. Bacteriol. 190(8): 3083–3087. https://doi.org/10.1128/JB.01448-07/ASSET/1F1044FA-1067-48BF-A49C-1165B44C9C73/ASSETS/GRAPHIC/ZJB0080877270002.JPEG.

Athukoralage, J.S., S.A. McMahon, C. Zhang, S. Grüschow, S. Graham, M. Krupovic et al. 2020. An anti-CRISPR viral ring nuclease subverts type III CRISPR immunity. Nature 577(7791): 572–575. https://doi.org/10.1038/s41586-019-1909-5.

Auchtung, J. M., N. Aleksanyan, A. Bulku and M.B. Berkmen. 2016. Biology of ICEBs1, an integrative and conjugative element in *Bacillus subtilis*. Plasmid. 86: 14–25. https://doi.org/10.1016/J.PLASMID.2016.07.001.

Azlan, A., N. Dzaki and G. Azzam. 2016. Argonaute: The executor of small RNA function. J. Genet. Genomics 43(8): 481–494. https://doi.org/10.1016/J.JGG.2016.06.002.

Azubuike, C.C., C.B. Chikere and G.C. Okpokwasili. 2016. Bioremediation techniques–classification based on site of application: principles, advantages, limitations and prospects. World J. Microbiol. Biotechnol. 32(11): 1–18. https://doi.org/10.1007/s11274-016-2137-x.

Barrangou, R., C. Fremaux, H. Deveau, M. Richards, P. Boyaval, S. Moineau et al. 2007. CRISPR provides acquired resistance against viruses in prokaryotes. Science 315(5819): 1709–1712. https://doi.org/10.1126/SCIENCE.1138140/SUPPL_FILE/BARRANGOU.SOM.PDF.

Barrangou, R. and P. Horvath. 2011. Lactic acid bacteria defenses against phages. Stress Responses of Lactic Acid Bacteria. Food Microbiology and Food Safety. Springer, Boston, MA. 459–478. https://doi.org/10.1007/978-0-387-92771-8_19.

Bassett, C.L. and W.J. Janisiewicz. 2003. Electroporation and stable maintenance of plasmid DNAs in a biocontrol strain of *Pseudomonas syringae*. Biotechnol. Lett. 25(3): 199–203. https://doi.org/10.1023/A:1022394716305.

Baxter, J.C. and B.E. Funnell. 2014. Plasmid partition mechanisms. Microbiol. Spectr. 2(6). https://doi.org/10.1128/MICROBIOLSPEC.PLAS-0023-2014.

Bhoobalan-Chitty, Y., T.B. Johansen, N. Di Cianni and X. Peng. 2019. Inhibition of Type III CRISPR-Cas immunity by an archaeal virus-encoded anti-CRISPR protein. Cell. 179(2): 448–458.e11. https://doi.org/10.1016/J.CELL.2019.09.003.

Bickle', T.A. and D.H. Kruger.1993. Biology of DNA restriction. Microbiol Rev. 57(2): 434–450. https://doi.org/10.1128/MR.57.2.434-450.1993.

Bohmert, K., I. Camus, C. Bellini, D. Bouchez, M. Caboche and C. Banning. 1998. AGO1 defines a novel locus of Arabidopsis controlling leaf development. EMBO J. 17(1): 170–180. https://doi.org/10.1093/EMBOJ/17.1.170.

Bondy-Denomy, Joe., A. Pawluk, K.L. Maxwell and A.R. Davidson. 2012. Bacteriophage genes that inactivate the CRISPR/Cas bacterial immune system. Nature 493(7432): 429–432. https://doi.org/10.1038/nature11723.

Bondy-Denomy, Joseph., B. Garcia, S. Strum, M. Du, M.F. Rollins, Y. Hidalgo-Reyes et al. 2015. Multiple mechanisms for CRISPR–Cas inhibition by anti-CRISPR proteins. Nature 526(7571): 136–139. https://doi.org/10.1038/nature15254.

Boram, W. and J. Abelson. 1971. Bacteriophage Mu integration: On the mechanism of Mu-induced mutations. J. Mol. Biol. 62(1): 171–178. https://doi.org/10.1016/0022-2836(71)90137-9.

Borges, A.L., A.R. Davidson and J. Bondy-Denomy. 2017a. The discovery, mechanisms, and evolutionary impact of anti-CRISPRs. Annu. Rev. Virol. 4: 37–59. https://doi.org/10.1146/ANNUREV-VIROLOGY-101416-041616.

Borges, A.L., A.R. Davidson and J. Bondy-Denomy. 2017b. The discovery, mechanisms, and evolutionary impact of anti-CRISPRs. Annu. Rev. Virol. 4: 37–59. https://doi.org/10.1146/ANNUREV-VIROLOGY-101416-041616.

Broniewski, J.M., M.A.W. Chisnall, N.M. Høyland-Kroghsbo, A. Buckling and E.R. Westra. 2021. The effect of Quorum sensing inhibitors on the evolution of CRISPR-based phage immunity in *Pseudomonas aeruginosa*. ISME J. 15(8): 2465–2473. https://doi.org/10.1038/s41396-021-00946-6.

Brüssow, H., C. Canchaya, W.-D. Hardt. 2004. Phages and the Evolution of Bacterial Pathogens: from Genomic Rearrangements to Lysogenic Conversion. *Microbiology and* Mol Biol Rev. 68(3): 560–602. https://doi.org/10.1128/mmbr.68.3.560-602.2004.

Buckner, M. M. C., M. L. Ciusa, L. J. V. Piddock. 2018. Strategies to combat antimicrobial resistance: anti-plasmid and plasmid curing. FEMS Microbiol Rev. 42(6): 781–804. https://doi.org/10.1093/FEMSRE/FUY031.

Burmeister, A.R. 2015. Horizontal Gene Transfer. Evol. Med. Public Health. 2015(1): 193–194. https://doi.org/10.1093/EMPH/EOV018.

Cao, M., P. Du, X. Wang, Y. Q. Yu, Y. H. Qiu, W. Li et al. 2014. Virus infection triggers widespread silencing of host genes by a distinct class of endogenous siRNAs in Arabidopsis. Proc. Natl. Acad. Sci. U.S.A. 111(40): 14613–14618. https://doi.org/10.1073/PNAS.1407131111.

Carattoli, A., A. Bertini, L. Villa, V. Falbo, K.L. Hopkins and E.J. Threlfall. 2005. Identification of plasmids by PCR-based replicon typing. J. Microbiol. Methods 63(3): 219–228. https://doi.org/10.1016/J.MIMET.2005.03.018.

Casjens, S.R. and R.W. Hendrix. 2015. Bacteriophage lambda: Early pioneer and still relevant. Virology 479–480: 310–330. https://doi.org/10.1016/J.VIROL.2015.02.010.

Chiang, Y.N., J.R. Penadés and J. Chen. 2019. Genetic transduction by phages and chromosomal islands: The new and noncanonical. PLOS Pathog. 15(8): e1007878. https://doi.org/10.1371/JOURNAL.PPAT.1007878.

Chopin, M.C., A. Chopin and E. Bidnenko. 2005. Phage abortive infection in lactococci: Variations on a theme. Curr. Opin. Microbiol. 8(4): 473–479. https://doi.org/10.1016/J.MIB.2005.06.006.

Cieślik, M., N. Bagińska, E. Jończyk-Matysiak, A. Węgrzyn, G. Węgrzyn and A. Górski. 2021. Temperate Bacteriophages— The Powerful Indirect Modulators of Eukaryotic Cells and Immune Functions. Viruses 13(6): 1013. https://doi.org/10.3390/V13061013.

Cerutti, L., M. Mian and A. Bateman. 2000. Domains in gene silencing and cell differentiation proteins : the novel PAZ domain and redefinition of the Piwi domain. Trends Biochem. Sci. 25(10): 481–482. https://ci.nii.ac.jp/naid/10027562695/.

Clark, A.J. and G.J. Warren. 2003. CONJUGAL TRANSMISSION OF PLASMIDS. Annu Rev Genet. 13: 99–125. https://doi.org/10.1146/ANNUREV.GE.13.120179.000531.

Costerton, J.W., P.S. Stewart and E.P. Greenberg. 1999. Bacterial biofilms: A common cause of persistent infections. Science. 284(5418): 1318–1322. https://doi.org/10.1126/SCIENCE.284.5418.1318.

Costerton, J. William, Z. Lewandowski, D.E. Caldwell, D.R. Korber and H.M. Lappin-Scott. 2003. Microbial biofilms. Annu. Rev. Microbiol. 49: 711–745. https://doi.org/10.1146/ANNUREV.MI.49.100195.003431.

Crozat, E., F. Fournes, F. Cornet, B. Hallet and P. Rousseau. 2014. Resolution of multimeric forms of circular plasmids and chromosomes. Microbiol. Spectr. 2(5). https://doi.org/10.1128/MICROBIOLSPEC.PLAS-0025-2014.

Cui, L., X. Wang, D. Huang, Y. Zhao, J. Feng, Q. Lu et al. 2020. CRISPR-cas3 of *Salmonella* upregulates bacterial biofilm formation and virulence to host cells by targeting quorum-sensing systems. Pathogens. 9(1): 53. https://doi.org/10.3390/PATHOGENS9010053.

Curcio, M.J. and K.M. Derbyshire. 2003. The outs and ins of transposition: from Mu to Kangaroo. Nat. Rev. Mol. Cell Biol. 4(11): 865–877. https://doi.org/10.1038/nrm1241.

De Bolle, X., C.D. Bayliss, D. Field, T. Van De Ven, N.J. Saunders, D.W. Hood et al. 2000. The length of a tetranucleotide repeat tract in *Haemophilus influenzae* determines the phase variation rate of a gene with homology to type III DNA methyltransferases. Mol. Microbiol. 35(1): 211–222. https://doi.org/10.1046/J.1365-2958.2000.01701.X.

De La Cruz, F., L.S. Frost, R.J. Meyer and E.L. Zechner. 2010. Conjugative DNA metabolism in Gram-negative bacteria. FEMS Microbiol. Rev. 34(1): 18–40. https://doi.org/10.1111/J.1574-6976.2009.00195.X.

del Solar, G., R. Giraldo, M.J. Ruiz-Echevarría, M. Espinosa and R. Díaz-Orejas. 1998. Replication and control of circular bacterial plasmids. Microbiol. Mol. Biol. Rev. 62(2): 434–464. https://doi.org/10.1128/MMBR.62.2.434-464.1998.

Domingues, S. and K.M. Nielsen. 2017. Membrane vesicles and horizontal gene transfer in prokaryotes. Curr. Opin. Microbiol. 38: 16–21. https://doi.org/10.1016/J.MIB.2017.03.012.

Domingues, S., G.J. da. Silva and K.M. Nielsen. 2012. Integrons. Mob. Genet. Elements 2(5): 211–223. https://doi.org/10.4161/MGE.22967.

Drenkard, E. 2003. Antimicrobial resistance of *Pseudomonas aeruginosa* biofilms. Microbes Infect. 5(13): 1213–1219. https://doi.org/10.1016/J.MICINF.2003.08.009.

Dubey, G.P. and S. Ben-Yehuda. 2011. Intercellular nanotubes mediate bacterial communication. Cell. 144(4): 590–600. https://doi.org/10.1016/J.CELL.2011.01.015.

Dy, R.L., C. Richter, G.P.C. Salmond and P.C. Fineran. 2014. Remarkable Mechanisms in Microbes to Resist Phage Infections. Annu. Rev. Virol. 1(1): 307–331. https://doi.org/10.1146/ANNUREV-VIROLOGY-031413-085500.

Dybvig, K., R. Sitaraman and C.T. French. 1998. A family of phase-variable restriction enzymes with differing specificities generated by high-frequency gene rearrangements. Proc. Natl. Acad. Sci. U.S.A. 95(23): 13923–13928. https://doi.org/10.1073/PNAS.95.23.13923.

Elufisan, T.O., O.O. Oyedara and B. Oyelade. 2012. Updates on microbial resistance to drugs. Afr. J. Microbiol. Res. 6(23): 4833–4844. https://doi.org/10.5897/AJMR11.436.

Faruque, S.M., I. Bin Naser, M.J. Islam, A.S.G. Faruque, A.N. Ghosh, G.B. Nair et al. 2005. Seasonal epidemics of cholera inversely correlate with the prevalence of environmental cholera phages. Proc. Natl. Acad. Sci. U.S.A. 102(5): 1702–1707. https://doi.org/10.1073/pnas.0408992102.

Faure, G., S.A. Shmakov, W.X. Yan, D.R. Cheng, D.A. Scott, J.E. Peters et al. 2019. CRISPR–Cas in mobile genetic elements: counter-defence and beyond. Nat. Rev. Microbiol. 17(8): 513–525. https://doi.org/10.1038/s41579-019-0204-7.

Flemming, H.C. 2002. Biofouling in water systems – cases, causes and countermeasures. Appl. Microbiol. Biotechnol. 59(6): 629–640. https://doi.org/10.1007/S00253-002-1066-9.

Flemming, Hans Curt and J. Wingender. 2010. The biofilm matrix. Nat. Rev. Microbiol. 8(9): 623–633. https://doi.org/10.1038/nrmicro2415.

Franke, A.E. and D.B. Clewell. 1981. Evidence for a chromosome-borne resistance transposon (Tn916) in *Streptococcus faecalis* that is capable of "conjugal" transfer in the absence of a conjugative plasmid. J. Bacteriol. 145(1): 494–502. https://doi.org/10.1128/JB.145.1.494-502.1981.

Frost, L.S., R. Leplae, A.O. Summers and A. Toussaint. 2005. Mobile genetic elements: The agents of open source evolution. Nat. Rev. Microbiol. 3(9): 722–732. https://doi.org/10.1038/nrmicro1235.

Fukuyo, M., A. Sasaki and I. Kobayashi. 2012. Success of a suicidal defense strategy against infection in a structured habitat. Sci. Rep. 2(1): 1–8. https://doi.org/10.1038/srep00238.

Gaddum, J.H., A. Silver, A.A.B. Swan, Q.J. Exptl Physiol, D.W. Woolley, E. Shaw et al. 1955. Fine structure of a genetic region in bacteriophage. Proc. Natl. Acad. Sci. U.S.A. 41(6): 344. https://doi.org/10.1073/PNAS.41.6.344.

Garcillán-Barcia, M. Pilar., R. Pluta, F. Lorenzo-Díaz, A. Bravo and M. Espinosa. 2022. The facts and family secrets of plasmids that replicate via the rolling-circle mechanism. Microbiol. Mol. Biol. Rev. 86(1). https://doi.org/10.1128/MMBR.00222-20.

Garcillán-Barcia, Maria Pilar., A. Alvarado and F. De la Cruz. 2011. Identification of bacterial plasmids based on mobility and plasmid population biology. FEMS Microbiol. Rev. 35(5): 936–956. https://doi.org/10.1111/J.1574-6976.2011.00291.X.

Garcillán-Barcia, María Pilar., M.V. Francia and F. De La Cruz. 2009. The diversity of conjugative relaxases and its application in plasmid classification. FEMS Microbiol. Rev. 33(3): 657–687. https://doi.org/10.1111/J.1574-6976.2009.00168.X.

Ghaly, T.M., L. Chow, A.J. Asher, L.S. Waldron and M.R. Gillings. 2017. Evolution of class 1 integrons: Mobilization and dispersal via food-borne bacteria. PLoS One. 12(6): e0179169. https://doi.org/10.1371/JOURNAL.PONE.0179169.

Gholizadeh, P., Ş. Köse, S. Dao, K. Ganbarov, A. Tanomand, T. Dal et al. 2020. How CRISPR-Cas system could be used to combat antimicrobial resistance. Infect Drug Resist. 13: 1111–1121. doi: 10.2147/IDR.S247271.

Goessweiner-Mohr, N., K. Arends, W. Keller and E. Grohmann. 2014. Conjugation in gram-positive bacteria. Microbiol. Spectr. 2(4). https://doi.org/10.1128/MICROBIOLSPEC.PLAS-0004-2013.

Grüll, M.P., M.E. Mulligan and A.S. Lang. 2018. Small extracellular particles with big potential for horizontal gene transfer: membrane vesicles and gene transfer agents. FEMS Microbiol. Lett. 365(19): 192. https://doi.org/10.1093/FEMSLE/FNY192.

Guglielmini, J. and L. Van. Melderen. 2011. Bacterial toxin-antitoxin systems. Mob. Genet. Elements. 1(4): 283–306. https://doi.org/10.4161/MGE.18477.

Hacker, J. and E. Carniel. 2001. Ecological fitness, genomic islands and bacterial pathogenicity A Darwinian view of the evolution of microbes. EMBO Rep. 2(5): 376–381. http://doi: 10.1093/embo-reports/kve097.

Hall-Stoodley, L., J.W. Costerton and P. Stoodley. 2004. Bacterial biofilms: From the Natural environment to infectious diseases. Nat. Rev. Microbiol. 2(2): 95–108. https://doi.org/10.1038/nrmicro821.

Hall, J.P.J., J. Botelho, A. Cazares and D.A. Baltrus. 2022a. What makes a megaplasmid? Philos. Trans. R Soc. B. 377(1842). https://doi.org/10.1098/RSTB.2020.0472.

Hall, J.P.J., E. Harrison and D.A. Baltrus. 2022b. Introduction: The secret lives of microbial mobile genetic elements. Philos Trans. R Soc. B: Biol. Sci. 377(1842). https://doi.org/10.1098/rstb.2020.0460.

Hamilton, W.D. 1964. The genetical evolution of social behaviour. II. J. Theor. Biol. 7(1): 17–52. https://doi.org/10.1016/0022-5193(64)90039-6.

Hampton, H.G., B.N.J. Watson and P.C. Fineran. 2020a. The arms race between bacteria and their phage foes. Nature. 577(7790): 327–336. https://doi.org/10.1038/s41586-019-1894-8.

Hampton, H.G., B.N.J. Watson and P.C. Fineran. 2020b. The arms race between bacteria and their phage foes. Nature. 577(7790): 327–336. https://doi.org/10.1038/s41586-019-1894-8.

Handa, N. and Kobayashi, I. 1999. Post-segregational killing by restriction modification gene complexes: Observations of individual cell deaths. Biochimie. 81(8–9): 931–938. https://doi.org/10.1016/S0300-9084(99)00201-1.

Hernández-Arriaga, A.M., W.T. Chan, M. Espinosa and R. Díaz-Orejas. 2014. Conditional Activation of Toxin-Antitoxin Systems: Postsegregational Killing and Beyond. Microbiol. Spectr. 2(5). https://doi.org/10.1128/MICROBIOLSPEC.PLAS-0009-2013.

Hickman, A.B. and F. Dyda. 2014. CRISPR-Cas immunity and mobile DNA: A new superfamily of DNA transposons encoding a Cas1 endonuclease. Mob DNA. 5(1): 1–4. https://doi.org/10.1186/1759-8753-5-23/FIGURES/1.

Hickman, A.B. and F. Dyda. 2015. Mechanisms of DNA Transposition. Microbiol Spectr. 3(2). https://doi.org/10.1128/MICROBIOLSPEC.MDNA3-0034-2014.

Hille, F., H. Richter, S.P. Wong, M. Bratovič, S. Ressel and E. Charpentier. 2018. The Biology of CRISPR-Cas: Backward and Forward. Cell. 172(6): 1239–1259. https://doi.org/10.1016/J.CELL.2017.11.032.

Hinnebusch, J. and K. Tilly. 1993. Linear plasmids and chromosomes in bacteria. Mol. Microbiol. 10(5): 917–922. https://doi.org/10.1111/J.1365-2958.1993.TB00963.X.

Höck, J. and G. Meister. 2008. The Argonaute protein family. Genome Biol. 9(2): 1–8. https://doi.org/10.1186/GB-2008-9-2-210/FIGURES/4.

Høiby, N., T. Bjarnsholt, M. Givskov, S. Molin and O. Ciofu. 2010. Antibiotic resistance of bacterial biofilms. Int. J. Antimicrob. Agents. 35(4): 322–332. https://doi.org/10.1016/J.IJANTIMICAG.2009.12.011.

Hoyle, B.D. and J.W. Costerton. 1991. Bacterial resistance to antibiotics: The role of biofilms. Prog. Drug Res. 37: 91–105. https://doi.org/10.1007/978-3-0348-7139-6_2.

Hughes, V.M. and N. Datta. 1983. Conjugative plasmids in bacteria of the "pre-antibiotic" era. Nature 302(5910): 725–726. https://doi.org/10.1038/302725a0.

Hutvagner, G. and M.J. Simard. 2008. Argonaute proteins: key players in RNA silencing. Nat. Rev. Mol. Cell Biol. 9(1): 22–32. https://doi.org/10.1038/nrm2321.

Hyland, E.M., E.W.J. Wallace and A.W. Murraya. 2014. A model for the evolution of biological specificity: A cross-reacting DNA-binding protein causes plasmid incompatibility. J. Bacteriol. 196(16): 3002–3011. https://doi.org/10.1128/JB.01811-14/SUPPL_FILE/ZJB999093263SO1.PDF.

Iwakawa, H. oki and Y. Tomari. 2015. The Functions of MicroRNAs: mRNA decay and translational repression. Trends Cell Biol. 25(11): 651–665. https://doi.org/10.1016/J.TCB.2015.07.011.

Jain, A. and P. Srivastava. 2013. Broad host range plasmids. FEMS Microbiol. Lett. 348(2): 87–96. https://doi.org/10.1111/1574-6968.12241.

Johnson, C.M. and A.D. Grossman. 2015. Integrative and Conjugative Elements (ICEs): What they do and how they work. Annu. Rev. Genet. 49: 577–601. https://doi.org/10.1146/ANNUREV-GENET-112414-055018.

Jonas, S. and E. Izaurralde. 2015. Towards a molecular understanding of microRNA-mediated gene silencing. Nat Rev Genet. 16(7): 421–433. https://doi.org/10.1038/nrg3965.

Keen, E.C. 2015. A century of phage research: Bacteriophages and the shaping of modern biology. BioEssays. 37(1): 6–9. https://doi.org/10.1002/BIES.201400152.

Koonin, E.V., K.S. Makarova and F. Zhang. 2017. Diversity, classification and evolution of CRISPR-Cas systems. Curr. Opin. Microbiol. 37: 67–78. https://doi.org/10.1016/J.MIB.2017.05.008.

Krupovic, M., P. Béguin and E.V. Koonin. 2017. Casposons: mobile genetic elements that gave rise to the CRISPR-Cas adaptation machinery. Curr. Opin. Microbiol. 38: 36–43. https://doi.org/10.1016/J.MIB.2017.04.004.

Krupovic, M., K.S. Makarova, P. Forterre, D. Prangishvili and E.V. Koonin. 2014. Casposons: A new superfamily of self-synthesizing DNA transposons at the origin of prokaryotic CRISPR-Cas immunity. BMC Biol. 12(1): 1–12. https://doi.org/10.1186/1741-7007-12-36/FIGURES/4.

Kües, U. and U. Stahl. 1989. Replication of plasmids in gram-negative bacteria. Microbiol. Rev. 53(4): 491–516. https://doi.org/10.1128/MR.53.4.491-516.1989.

Kulakauskas, S., A. Lubys and S.D. Ehrlich. 1995. DNA restriction-modification systems mediate plasmid maintenance. J. Bacteriol. 177(12): 3451–3454. https://doi.org/10.1128/JB.177.12.3451-3454.1995.

Labrie, S.J., J.E. Samson and S. Moineau. 2010a. Bacteriophage resistance mechanisms. Nat. Rev. Microbiol. 8(5): 317–327. https://doi.org/10.1038/nrmicro2315.

Labrie, S.J., J.E. Samson and S. Moineau. 2010b. Bacteriophage resistance mechanisms. Nat Rev Microbiol. 8(5): 317–327. https://doi.org/10.1038/nrmicro2315.

Lacks, S.A. and S.S. Springhorn. 1984. Transfer of recombinant plasmids containing the gene for DpnII DNA methylase into strains of *Streptococcus pneumoniae* that produce DpnI or DpnII restriction endonucleases. J. Bacteriol. 158(3): 905–909. https://doi.org/10.1128/JB.158.3.905-909.1984.

Lacroix, B. and V. Citovsky. 2016. Transfer of DNA from bacteria to eukaryotes. MBio. 7(4). https://doi.org/10.1128/MBIO.00863-16.

Landsmann, J., M. Kroger and G. Hobom. 1982. The rex region of bacteriophage lambda: Two genes under three-way control. Gene. 20(1): 11–24. https://doi.org/10.1016/0378-1119(82)90083-X.

Lang, A.S., A.B.Westbye and J.T. Beatty. 2017. The distribution, evolution, and roles of gene transfer agents in prokaryotic genetic exchange. Annu. Rev. Virol. 4: 87–104. https://doi.org/10.1146/ANNUREV-VIROLOGY-101416-041624.

Lang, A.S., O. Zhaxybayeva and J.T. Beatty. 2012. Gene transfer agents: phage-like elements of genetic exchange. Nat. Rev. Microbiol. 10(7): 472–482. https://doi.org/10.1038/nrmicro2802.

Lanza, V.F., A.P. Tedim, J.L. Martínez, F. Baquero and T.M. Coque. 2015. The Plasmidome of Firmicutes: Impact on the Emergence and the Spread of Resistance to Antimicrobials. Microbiol. Spectr. 3(2). https://doi.org/10.1128/MICROBIOLSPEC.PLAS-0039-2014.

Lau, R.K., Q. Ye, E.A. Birkholz, K.R. Berg, L. Patel, I.T. Mathews et al. 2020. Structure and mechanism of a cyclic trinucleotide-activated bacterial endonuclease mediating bacteriophage immunity. Mol Cell. 77(4): 723–733. e6. https://doi.org/10.1016/J.MOLCEL.2019.12.010.

Levy, R.M., A. Haldane and W.F. Flynn. 2017. Potts Hamiltonian models of protein co-variation, free energy landscapes, and evolutionary fitness. Curr. Opin. Struct. Biol. 43: 55–62. https://doi.org/10.1016/J.SBI.2016.11.004.

Lilly, J. and M. Camps. 2015. Mechanisms of Theta Plasmid Replication. Microbiol. Spectr. 3(1). https://doi.org/10.1128/MICROBIOLSPEC.PLAS-0029-2014.

Lindstrom, J.E., R.C. Prince, J.C. Clark, M.J. Grossman, T.R. Yeager, J.F. Braddock et al. 1991. Microbial populations and hydrocarbon biodegradation potentials in fertilized shoreline sediments affected by the T/V Exxon Valdez oil spill. Appl. Environ. Microbiol. 57(9): 2514–2522. https://doi.org/10.1128/aem.57.9.2514-2522.1991.

Liu, Y.Y., Y. Wang, T.R. Walsh, L. X. Yi, R. Zhang, J. Spencer et al. 2016. Emergence of plasmid-mediated colistin resistance mechanism MCR-1 in animals and human beings in China: A microbiological and molecular biological study. Lancet Infect. Dis. 16(2): 161–168. https://doi.org/10.1016/S1473-3099(15)00424-7.

Loftie-Eaton, W., H. Yano, S. Burleigh, R.S. Simmons, J.M. Hughes, L.M. Rogers et al. 2016. Evolutionary Paths That Expand Plasmid Host-Range: Implications for Spread of Antibiotic Resistance. Mol. Biol. Evol. 33(4): 885–897. https://doi.org/10.1093/MOLBEV/MSV339.

López-Guerrero, M.G., E. Ormeño-Orrillo, J.L. Acosta, A. Mendoza-Vargas, M.A. Rogel, M.A. Ramírez et al. 2012. *Rhizobial extrachromosomal* replicon variability, stability and expression in natural niches. Plasmid. 68(3): 149–158. https://doi.org/10.1016/j.plasmid.2012.07.002.

Makarova, K.S., Y.I. Wolf, O.S. Alkhnbashi, F. Costa, S.A. Shah, S.J. Saunders et al. 2015. An updated evolutionary classification of CRISPR–Cas systems. Nat Rev Microbiol. 13(11): 722–736. https://doi.org/10.1038/nrmicro3569.

Makarova, K.S., Y.I. Wolf, J. van der Oost and E.V. Koonin. 2009. Prokaryotic homologs of Argonaute proteins are predicted to function as key components of a novel system of defense against mobile genetic elements. Biol. Direct. 4(1): 29. https://doi.org/10.1186/1745-6150-4-29/FIGURES/5.

Marino, N.D., R. Pinilla-Redondo, B. Csörgő and J. Bondy-Denomy. 2020. Anti-CRISPR protein applications: natural brakes for CRISPR-Cas technologies. Nat. Methods. 17(5): 471–479. https://doi.org/10.1038/s41592-020-0771-6.

Marino, N.D., J.Y. Zhang, A.L. Borges, A.A. Sousa, L.M. Leon, B.J. Rauch et al. 2018. Discovery of widespread type i and type v CRISPR-Cas inhibitors. Science 362(6411): 240–242. https://doi.org/10.1126/SCIENCE.AAU5174/SUPPL_FILE/AAU5174-MARINO-SM.PDF.

Marraffini, L.A. 2015. CRISPR-Cas immunity in prokaryotes. Nature. 526(7571): 55–61. https://doi.org/10.1038/nature15386.

Marrs, B. 1974. Genetic recombination in *Rhodopseudomonas capsulata*. Proc. Natl. Acad. Sci U.S.A. 71(3): 971–973. https://doi.org/10.1073/PNAS.71.3.971.

Mashburn-Warren, L.M. and M. Whiteley. 2006. Special delivery: vesicle trafficking in prokaryotes. Mol. Microbiol. 61(4): 839–846. https://doi.org/10.1111/J.1365-2958.2006.05272.X.

Meister, G. 2013. Argonaute proteins: Functional insights and emerging roles. Nat. Rev. Genet. 14(7): 447–459. https://doi.org/10.1038/nrg3462.

Mendiola, M.V., I. Bernales and F. De La Cruz. 1994. Differential roles of the transposon termini in IS91 transposition. Proc. Natl. Acad Sci. U.S.A. 91(5): 1922–1926. https://doi.org/10.1073/PNAS.91.5.1922.

Midonet, C. and F.-X. Barre. 2014. Xer Site-Specific Recombination: Promoting vertical and horizontal transmission of genetic information. Microbiol. Spectr. 2(6). https://doi.org/10.1128/MICROBIOLSPEC.MDNA3-0056-2014.

Millan, A.S., J.A. Escudero, D.R. Gifford, Di. Mazel and R.C. MacLean. 2016. Multicopy plasmids potentiate the evolution of antibiotic resistance in bacteria. Nat. Ecol. Evol. 1(1): 1–8. https://doi.org/10.1038/s41559-016-0010.

Millan, A.S. and R.C. MacLean. 2017. Fitness costs of plasmids: A limit to plasmid transmission. Microbiol. Spectr. 5(5). https://doi.org/10.1128/MICROBIOLSPEC.MTBP-0016-2017.

Million-Weaver, S. and M. Camps. 2014. Mechanisms of plasmid segregation: Have multicopy plasmids been overlooked? Plasmid. 75: 27–36. https://doi.org/10.1016/J.PLASMID.2014.07.002.

Millman, A., S. Melamed, G. Amitai and R. Sorek. 2020. Diversity and classification of cyclic-oligonucleotide-based anti-phage signalling systems. Nat. Microbiol. 5(12): 1608–1615. https://doi.org/10.1038/s41564-020-0777-y.

Moser, D.P., D. Zarka and T. Kallas. 1993. Characterization of a restriction barrier and electrotransformation of the cyanobacterium Nostoc. PCC 7121. Arch Microbiol. 160(3): 229–237. https://doi.org/10.1007/BF00249129.

Mullany, P., M. Pallen, M. Wilks, J.R. Stephen and S. Tabaqchali. 1996. A group II intron in a conjugative transposon from the gram-positive bacterium, *Clostridium difficile*. Gene. 174(1): 145–150. https://doi.org/10.1016/0378-1119(96)00511-2.

Murray, N.E. 2002. Immigration control of DNA in bacteria: Self versus non-self. Microbiology. 148(1): 3–20. https://doi.org/10.1099/00221287-148-1-3/CITE/REFWORKS.

Nordström, K. 2006. Plasmid R1—Replication and its control. Plasmid. 55(1): 1–26. https://doi.org/10.1016/J.PLASMID.2005.07.002.

Norman, A., L.H. Hansen and S.J. Sørensen. 2009. Conjugative plasmids: Vessels of the communal gene pool. Philos. Trans. R Soc B: Biol. Sci. 364(1527): 2275–2289. https://doi.org/10.1098/rstb.2009.0037.

Noureen, M., T. Kawashima and M. Arita. 2021. Genetic markers of genome rearrangements in *Helicobacter pylori*. Microorganisms. 9(3): 1–12. https://doi.org/10.3390/microorganisms9030621.

Novick, R.P. 1987. Plasmid Incompatibility. Microbiol. Rev. 51(4): 381–395. https://journals.asm.org/journal/mr.

Ochman, Howard, G. Lawrence Jeffrey and A. Groisman, Eduardo. 2000. Lateral gene transfer and the nature of bacterial innovation. Nature. 405(6784): 299–304.

Ogura, T. and S. Hiraga. 1983. Mini-F plasmid genes that couple host cell division to plasmid proliferation. Proc. Natl. Acad. Sci. U.S.A. 80(15): 4784–4788. https://doi.org/10.1073/PNAS.80.15.4784.

Okinaka, R.T., K. Cloud, O. Hampton, A.R. Hoffmaster, K.K. Hill, P. Keim et al. 1999. Sequence and organization of pXO1, the large *Bacillus anthracis* plasmid harboring the anthrax toxin genes. J. Bacteriol. 181(20): 6509–6515. https://doi.org/10.1128/jb.181.20.6509-6515.1999.

Olaitan, A.O., S. Morand and J.M. Rolain. 2016. Emergence of colistin-resistant bacteria in humans without colistin usage: a new worry and cause for vigilance. Int. J. Antimicro. Agents. 47(1): 1–3. https://doi.org/10.1016/J.IJANTIMICAG.2015.11.009.

Olovnikov, I., K. Chan, R. Sachidanandam, D.K. Newman and A.A. Aravin. 2013. Bacterial Argonaute Samples the Transcriptome to Identify Foreign DNA. Mol. Cell. 51(5): 594–605. https://doi.org/10.1016/J.MOLCEL.2013.08.014.

Parker, J.S. and D. Barford. 2006. Argonaute: A scaffold for the function of short regulatory RNAs. Trends Biochem. Sci. 31(11): 622–630. https://doi.org/10.1016/J.TIBS.2006.09.010.

Parker, J.S., S.M. Roe and D. Barford. 2005. Structural insights into mRNA recognition from a PIWI domain–siRNA guide complex. Nature. 434(7033): 663–666. https://doi.org/10.1038/nature03462.

Partridge, S.R., S.M. Kwong, N. Firth and S.O. Jensen. 2018. Mobile genetic elements associated with antimicrobial resistance. Clin. Microbiol. Rev. 31(4): 1–61. https://doi.org/10.1128/CMR.00088-17.

Pawluk, A., N. Amrani, Y. Zhang, B. Garcia, Y. Hidalgo-Reyes, J. Lee et al. 2016. Naturally occurring off-switches for CRISPR-Cas9. Cell. 167(7): 1829–1838.e9. https://doi.org/10.1016/J.CELL.2016.11.017.

Pawluk, A., J. Bondy-Denomy, V.H.W. Cheung, K.L. Maxwell and A.R. Davidson. 2014. A new group of phage anti-CRISPR genes inhibits the type I-E CRISPR-Cas system of *Pseudomonas aeruginosa*. MBio. 5(2). https://doi.org/10.1128/MBIO.00896-14/SUPPL_FILE/MBO002141803SF02.JPG.

Pilla, G. and C.M. Tang. 2018. Going around in circles: virulence plasmids in enteric pathogens. Nat. Rev. Microbiol. 16(8): 484–495. https://doi.org/10.1038/s41579-018-0031-2.

Pinedo, C.A. and B.F. Smets. 2005. Conjugal TOL transfer from Pseudomonas putida to *Pseudomonas aeruginosa*: Effects of restriction proficiency, toxicant exposure, cell density ratios, and conjugation detection method on observed transfer efficiencies. Appl. Environ. Microbiol. 71(1): 51–57. https://doi.org/10.1128/AEM.71.1.51-57.2005.

Pinilla-Redondo, R., D. Mayo-Muñoz, J. Russel, R.A. Garrett, L. Randau, S.J. Sørensen et al. 2020a. Type IV CRISPR–Cas systems are highly diverse and involved in competition between plasmids. Nucleic Acids Res. 48(4): 2000–2012. https://doi.org/10.1093/NAR/GKZ1197.

Pinilla-Redondo, R., S. Shehreen, N.D. Marino, R.D. Fagerlund, C.M. Brown, S.J. Sørensen et al. 2020b. Discovery of multiple anti-CRISPRs highlights anti-defense gene clustering in mobile genetic elements. Nat. Commun. 11(1): 1–11. https://doi.org/10.1038/s41467-020-19415-3.

Projan, S.J., M. Monod, C.S. Narayanan and D. Dubnau. 1987. Replication properties of pIM13, a naturally occurring plasmid found in *Bacillus subtilis*, and of its close relative pE5, a plasmid native to *Staphylococcus aureus*. J. Bacteriol. 169(11): 5131–5139. https://doi.org/10.1128/JB.169.11.5131-5139.1987.

Pupo, G.M., R. Lan and P.R. Reeves. 2000. Multiple independent origins of *Shigella* clones of *Escherichia coli* and convergent evolution of many of their characteristics. Proc. Natl. Acad Sci. U.S.A. 97(19): 10567–10572. https://doi.org/10.1073/pnas.180094797.

Purdy, D., T.A.T. O'Keeffe, M. Elmore, M. Herbert, A. McLeod, M. Bokori-Brown et al. 2002. Conjugative transfer of clostridial shuttle vectors from *Escherichia coli* to *Clostridium difficile* through circumvention of the restriction barrier. Mol. Microbiol. 46(2): 439–452. https://doi.org/10.1046/J.1365-2958.2002.03134.X.

Raeside, C., J. Gaffé, D.E. Deatherage, O. Tenaillon, A.M. Briska, R.N. Ptashkin et al. 2014. Large chromosomal rearrangements during a long-term evolution experiment with *Escherichia coli*. MBio. 5(5). https://doi.org/10.1128/mBio.01377-14.

Ramsay, J.P., S.M. Kwong, R.J.T. Murphy, K.Y. Eto, K.J. Price, Q.T. Nguyen et al. 2016. An updated view of plasmid conjugation and mobilization in *Staphylococcus*. Mob. Genet. Elements. 6(4): e1208317. https://doi.org/10.1080/2159256X.2016.1208317.

Rankin, D.J., E.P.C. Rocha and S.P. Brown. 2011. What traits are carried on mobile genetic elements, and why. Heredity. 106(1): 1–10. https://doi.org/10.1038/hdy.2010.24.

Rauch, B.J., M.R. Silvis, J.F. Hultquist, C.S. Waters, M.J. McGregor, N.J. Krogan et al. 2017. Inhibition of CRISPR-Cas9 with Bacteriophage Proteins. Cell. 168(1–2): 150–158.e10. https://doi.org/10.1016/J.CELL.2016.12.009.

Ravin, N.V. 2011. N15: The linear phage–plasmid. Plasmid. 65(2): 102–109. https://doi.org/10.1016/J.PLASMID.2010.12.004.

Raza, S., K. Matuła, S. Karoń and J. Paczesny. 2021a. Resistance and Adaptation of Bacteria to Non-Antibiotic Antibacterial Agents: Physical Stressors, Nanoparticles, and Bacteriophages. Antibiotics. 10(4): 435. https://doi.org/10.3390/ANTIBIOTICS10040435.

Raza, S., K. Matuła, S. Karoń and J. Paczesny. 2021b. Resistance and adaptation of bacteria to non-antibiotic antibacterial agents: Physical stressors, nanoparticles, and bacteriophages. Antibiotics. 10(4): 435. https://doi.org/10.3390/ANTIBIOTICS10040435.

Reeks, J., J.H. Naismith and M.F. White. 2013. CRISPR interference: a structural perspective. Biochem. 453(2): 155–166. https://doi.org/10.1042/BJ20130316.

Reis, R.S. 2017. The entangled history of animal and plant microRNAs. Funct. Integr. Genom. 17(2–3): 127–134. https://doi.org/10.1007/S10142-016-0513-0/FIGURES/1.

Reznikoff, W.S. 1993. The Tn5 transposon. Annu. Rev. Microbiol. 47(1): 945–964. https://doi.org/10.1146/ANNUREV. MI.47.100193.004501.

Roberts, R.J., M. Belfort, T. Bestor, A.S. Bhagwat, T.A. Bickle, J. Bitinaite et al. 2003. A nomenclature for restriction enzymes, DNA methyltransferases, homing endonucleases and their genes. Nucleic Acids Res. 31(7): 1805–1812. https://doi.org/10.1093/NAR/GKG274.

Rodríguez-Beltrán, J., J. DelaFuente, R. León-Sampedro, R.C. MacLean and Á. San Millán. 2021. Beyond horizontal gene transfer: The role of plasmids in bacterial evolution. Nat. Rev. Microbiol. 19(6): 347–359. https://doi.org/10.1038/s41579-020-00497-1.

Rostøl, J.T. and L. Marraffini. 2019. (Ph)ighting Phages: How bacteria resist their parasites. Cell Host Microbe. 25(2): 184–194. https://doi.org/10.1016/J.CHOM.2019.01.009.

Salje, J. 2010. Plasmid segregation:How to survive as an extra piece of DNA. Crit. Rev. Biochem. Mol. Biol. 45(4): 296–317. https://doi.org/10.3109/10409238.2010.494657.

Schouler, C., M. Gautier, S.D. Ehrlich and M.C. Chopin. 1998. Combinational variation of restriction modification specificities in *Lactococcus lactis*. Mol. Microbiol. 28(1): 169–178. https://doi.org/10.1046/J.1365-2958.1998.00787.X.

Seed, K.D. 2015. Battling Phages: How bacteria defend against viral attack. PLoS Pathog. 11(6): e1004847. https://doi.org/10.1371/JOURNAL.PPAT.1004847.

Seferbekova, Z., A. Zabelkin, Y. Yakovleva, R. Afasizhev, N.O. Dranenko, N. Alexeev et al. 2021. High rates of genome rearrangements and pathogenicity of *Shigella* spp. Front Microbiol. 12(April): 1–15. https://doi.org/10.3389/fmicb.2021.628622.

Sheng, G., H. Zhao, J. Wang, Y. Rao, W. Tian, D.C. Swarts et al. 2014. Structure-based cleavage mechanism of *Thermusthermophilus* argonaute DNA guide strand-mediated DNA target cleavage. Proc. Natl. Acad Sci. U.S.A. 111(2): 652–657. https://doi.org/10.1073/PNAS.1321032111.

Shintani, M., Z.K. Sanchez and K. Kimbara. 2015. Genomics of microbial plasmids: Classification and identification based on replication and transfer systems and host taxonomy. Front. Microbiol. 6(MAR): 242. https://doi.org/10.3389/FMICB.2015.00242/ABSTRACT.

Shmakov, S., O.O. Abudayyeh, K.S. Makarova, Y.I. Wolf, J.S. Gootenberg, E. Semenova et al. 2015. Discovery and functional characterization of diverse class 2 CRISPR-Cas Systems. Mol. Cell. 60(3): 385–397. https://doi.org/10.1016/J.MOLCEL.2015.10.008.

Siguier, P., E. Gourbeyre and M. Chandler. 2017. Known knowns, known unknowns and unknown unknowns in prokaryotic transposition. Curr. Opin. Microbiol. 38: 171–180. https://doi.org/10.1016/J.MIB.2017.06.005.

Siguier, P., E. Gourbeyre, A. Varani, B. Ton-Hoang and M. Chandler. 2015. Everyman's Guide to Bacterial Insertion Sequences. Microbiol. Spectr. 3(2):MDNA3-0030-2014. doi: 10.1128/microbiolspec.MDNA3-0030-2014.

Sing, W.D. and T.R. Klaenhammer. 1990. Plasmid-induced abortive infection in lactococci: A Review. J. Dairy Sci. 73(9): 2239–2251. https://doi.org/10.3168/JDS.S0022-0302(90)78904-7.

Sitaraman, R. and K. Dybvig. 1997. The hsd loci of *Mycoplasma pulmonis*: Organization, rearrangements and expression of genes. Mol. Microbiol. 26(1): 109–120. https://doi.org/10.1046/J.1365-2958.1997.5571938.X.

Smillie, C., M.P. Garcillán-Barcia, M.V. Francia, E.P.C. Rocha and F. de la Cruz. 2010. Mobility of Plasmids. Microbiol. Mol. Biol. rev. 74(3): 434–452. https://doi.org/10.1128/MMBR.00020-10/SUPPL_FILE/SUPPLEMENTARY_TABLES2. DOC.

Smith, C.J. and A.C. Parker. 1993. Identification of a circular intermediate in the transfer and transposition of Tn4555, a mobilizable transposon from *Bacteroides* spp. J. Bacteriol. 175(9): 2682–2691. https://doi.org/10.1128/JB.175.9.2682-2691.1993.

Snyder, L. and K. McWilliams. 1989. The rex genes of bacteriophage lambda can inhibit cell function without phage superinfection. Gene. 81(1): 17–24. https://doi.org/10.1016/0378-1119(89)90332-6.

Soler, N. and P. Forterre. 2020. Vesiduction: The fourth way of HGT. Environ. Microbiol. 22(7): 2457–2460. https://doi.org/10.1111/1462-2920.15056.

Solioz, M., H.C. Yen and B. Marrs. 1975. Release and uptake of gene transfer agent by *Rhodopseudomonas capsulata*. J. Bacteriol. 123(2): 651–657. https://doi.org/10.1128/JB.123.2.651-657.1975.

Solioz, Marc and B. Marrs. 1977. The gene transfer agent of *Rhodopseudomonas capsulata*,: Purification and characterization of its nucleic acid. Arch Biochem. Biophys. 181(1): 300–307. https://doi.org/10.1016/0003-9861(77)90508-2.

Starikova, I., M. Al-Haroni, G. Werner, A.P. Roberts, V. Sørum, K.M. Nielsen et al. 2013. Fitness costs of various mobile genetic elements in *Enterococcus faecium* and *Enterococcus faecalis*. J. Antimicrob. Chemother. 68(12): 2755–2765. https://doi.org/10.1093/jac/dkt270.

Stevenson, C., J.P.J. Hall, E. Harrison, A.J. Wood and M.A. Brockhurst. 2017. Gene mobility promotes the spread of resistance in bacterial populations. ISME J. 11(8): 1930–1932. https://doi.org/10.1038/ismej.2017.42.

Stoodley, P., K. Sauer, D.G. Davies and J.W. Costerton. 2003. Biofilms as complex differentiated communities. Annu. Rev. Microbiol. 56: 187–209. https://doi.org/10.1146/ANNUREV.MICRO.56.012302.160705.

Sullivan, J.T., J.R. Trzebiatowski, R.W. Cruickshank, J. Gouzy, S.D. Brown, R.M. Elliot et al. 2002. Comparative sequence analysis of the symbiosis island of *Mesorhizobium loti* strain R7A. J. Bacteriol. 184(11): 3086–3095. https://doi.org/10.1128/JB.184.11.3086-3095.2002.

Sumby, P. and M.C.M. Smith. 2003. Phase variation in the phage growth limitation system of *Streptomyces coelicolor* A3(2). J. Bacteriol. 185(15): 4558–4563. https://doi.org/10.1128/JB.185.15.4558-4563.2003/ASSET/5FD749A0-5423-4BA3-B1E4-14338872CD75/ASSETS/GRAPHIC/JB1530258003.JPEG.

Swarts, D.C., J.W. Hegge, I. Hinojo, M. Shiimori, M.A. Ellis, J. Dumrongkulraksa et al. 2015. Argonaute of the archaeon *Pyrococcus furiosus* is a DNA-guided nuclease that targets cognate DNA. Nucleic Acids Res. 43(10): 5120–5129. https://doi.org/10.1093/NAR/GKV415.

Swarts, D.C., M.M. Jore, E.R. Westra, Y. Zhu, J.H. Janssen, A.P. Snijders et al. 2014a. DNA-guided DNA interference by a prokaryotic Argonaute. Nature. 507(7491): 258–261. https://doi.org/10.1038/nature12971.

Swarts, D.C., K. Makarova, Y. Wang, K. Nakanishi, R.F. Ketting, E.V. Koonin et al. 2014b. The evolutionary journey of Argonaute proteins. Nat. Struct. Mol. Biol. 21(9): 743–753. https://doi.org/10.1038/nsmb.2879.

Tabara, H., M. Sarkissian, W.G. Kelly, J. Fleenor, A. Grishok, L. Timmons et al. 1999. The rde-1 gene, RNA interference, and transposon silencing in *C. elegans*. Cell. 99(2): 123–132. https://doi.org/10.1016/S0092-8674(00)81644-X.

Thoma, L. and G. Muth. 2016. Conjugative DNA-transfer in *Streptomyces*, a mycelial organism. Plasmid. 87–88: 1–9. https://doi.org/10.1016/J.PLASMID.2016.09.004.

Thomas, C.M. and K.M. Nielsen. 2005. Mechanisms of, and barriers to, horizontal gene transfer between bacteria. Nat. Rev. Microbiol. 3(9): 711–721. https://doi.org/10.1038/NRMICRO1234.

Touchon, M., C. Hoede, O. Tenaillon, V. Barbe, S. Baeriswyl, P. Bidet et al. 2009. Organised genome dynamics in the *Escherichia coli* species results in highly diverse adaptive paths. PLoS Genet. 5(1). https://doi.org/10.1371/journal.pgen.1000344.

Van Der Oost, J., E.R. Westra, R.N. Jackson and B. Wiedenheft. 2014. Unravelling the structural and mechanistic basis of CRISPR–Cas systems. Nat. Rev. Microbiol. 12(7): 479–492. https://doi.org/10.1038/nrmicro3279.

van Houte, S., A. Buckling and E.R. Westra. 2016a. Evolutionary ecology of prokaryotic immune mechanisms. Microbiol. Mol. Biol. Rev. 80(3): 745–763. https://doi.org/10.1128/MMBR.00011-16/FORMAT/EPUB.

van Houte, S., A. Buckling and E.R. Westra. 2016b. Evolutionary ecology of prokaryotic immune mechanisms. Microbiol. Mol. Biol. Rev. 80(3): 745–763. https://doi.org/10.1128/MMBR.00011-16.

Varani, A., S. He, P. Siguier, K. Ross and M. Chandler. 2021. The IS6 family, a clinically important group of insertion sequences including IS26. Mob. DNA. 12(1): 1–18. https://doi.org/10.1186/S13100-021-00239-X/FIGURES/11.

Vega, N.M. and J. Gore. 2014. Collective antibiotic resistance: Mechanisms and implications. Curr. Opin. Microbiol. 21: 28–34. https://doi.org/10.1016/J.MIB.2014.09.003.

Vieira, J. and J. Messing. 1982. The pUC plasmids, an M13mp7-derived system for insertion mutagenesis and sequencing with synthetic universal primers. Gene. 19(3): 259–268. https://doi.org/10.1016/0378-1119(82)90015-4.

Von Wintersdorff, C.J.H., J. Penders, J.M. Van Niekerk, N.D. Mills, S. Majumder, L.B. Van Alphen et al. 2016. Dissemination of antimicrobial resistance in microbial ecosystems through horizontal gene transfer. Front Microbiol. 7(FEB): 173. https://doi.org/10.3389/FMICB.2016.00173/BIBTEX.

Vrancianu, C.O., L.I. Popa, C. Bleotu and M.C. Chifiriuc. 2020. Targeting plasmids to limit acquisition and transmission of antimicrobial resistance. Front. Microbiol. 11: 761. https://doi.org/10.3389/FMICB.2020.00761/BIBTEX.

Wang, R., L. Van Dorp, L.P. Shaw, P. Bradley, Q. Wang, X. Wang et al. 2018. The global distribution and spread of the mobilized colistin resistance gene mcr-1. Nat. Comm. 9(1): 1–9. https://doi.org/10.1038/s41467-018-03205-z.

Watters, K.E., C. Fellmann, H.B. Bai, S.M. Ren and J.A. Doudna. 2018. Systematic discovery of natural CRISPR-Cas12a inhibitors. Science. 362(6411): 236–239. https://doi.org/10.1126/SCIENCE.AAU5138/SUPPL_FILE/AAU5138_WATTERS_SM.PDF.

Weil, J. and E.R. Signer. 1968. Recombination in bacteriophage λ: II. Site-specific recombination promoted by the integration system. J. Mol. Biol. 34(2): 273–279. https://doi.org/10.1016/0022-2836(68)90252-0.

West, S.A., A.S. Griffin and A. Gardner. 2007. Social semantics: Altruism, cooperation, mutualism, strong reciprocity and group selection. J. Evol. Biol. 20(2): 415–432. https://doi.org/10.1111/j.1420-9101.2006.01258.x.

Wiedenheft, B., S.H. Sternberg and J.A. Doudna. 2012. RNA-guided genetic silencing systems in bacteria and archaea. Nature. 482(7385): 331–338. https://doi.org/10.1038/nature10886.

Wilkins, B.M., P.M. Chilley, A.T. Thomas and M.J. Pocklington. 1996. Distribution of restriction enzyme recognition sequences on broad host range plasmid RP4: Molecular and evolutionary implications. J. Mol. Biol. 258(3): 447–456. https://doi.org/10.1006/JMBI.1996.0261.

Willkomm, S., A. Zander, D. Grohmann and T. Restle. 2016. Mechanistic insights into archaeal and human argonaute substrate binding and cleavage properties. PLoS ONE. 11(10): e0164695. https://doi.org/10.1371/JOURNAL.PONE.0164695.

Yen, H.C., N.T. Hu and B.L. Marrs. 1979. Characterization of the gene transfer agent made by an overproducer mutant of *Rhodopseudomonas capsulata*. J. Mol. Biol. 131(2): 157–168. https://doi.org/10.1016/0022-2836(79)90071-8.

Yuan, Y.R., Y. Pei, J.B. Ma, V. Kuryavyi, M. Zhadina, G. Meister et al. 2005. Crystal structure of *A. aeolicus* Argonaute, a site-specific DNA-guided endoribonuclease, provides insights into RISC-Mediated mRNA cleavage. Mol. Cell. 19(3): 405–419. https://doi.org/10.1016/J.MOLCEL.2005.07.011.

Zhou, G., Q.S. Shi, X.M. Huang and X.B. Xie. 2015. The Three Bacterial Lines of Defense against Antimicrobial Agents. Int. J. Mol. Sci. 16(9): 21711–21733. https://doi.org/10.3390/IJMS160921711.

Section II
Molecular Basis of Viruses and Phytoplasma

CHAPTER 8

Molecular Basis of Genetic Variation in Viruses Present in Soil

Anna Marzec-Grządziel* and Jarosław Ciepiel

Introduction

Viruses are one of the most abundant biological groups on Earth and represent the largest gap in knowledge of environmental biodiversity (Emerson 2019). Their presence in the environment can affect the entire ecosystem by transferring genes, reducing biodiversity, and disrupting the metabolism. As agents responsible for horizontal gene transfer, they shape the host population, thereby influencing their evolutionary process. This is a new, unexplored field of science with tremendous learning potential. The development of sequencing techniques allows researchers to gain knowledge about microorganisms living in a given environment. Most current biodiversity research focuses on metataxonomic analysis of bacterial and fungal composition, based on next-generation sequencing techniques using conserved gene sequences such as 16S rRNA (bacteria) and ITS (fungi) (Galazka and Grzadziel 2016, 2018, Grządziel and Gałązka 2018). In the case of viruses, there is no universal sequence that can be used to analyze their taxonomic affiliation. In addition, the diversity of viral genetic material (ssRNA, dsRNA, ssDNA, dsDNA, dsDNA) also poses a huge challenge in the study of this biological group (Simmonds et al. 2017). Analysis of the viral population in the environment therefore requires more complex and multi-stage studies. However, its result provides a lot of important information, not only about the taxonomic composition of viruses, but also about the genes present in a given environment. The constantly developing next-generation sequencing techniques allow scientists to study biodiversity in greater depth. Metataxonomic analysis is increasingly supported by metagenomics, using, for example, shotgun sequencing methods (Guo et al. 2016, Donovan et al. 2018). Metagenomic analysis of viral particles isolated from a range of samples makes it possible to detect new viral species, previously unknown genes and the proteins they encode (Schulz et al. 2018). The analysis of genes present in the environment allows the development of modern biotechnology through the detection of new enzymes of applied importance (Schoenfeld et al. 2010). Scientists looking for new viral species or genes and their products with potential biotechnological and economic significance choose extreme environments, such as deserts and polar regions (Zablocki et al. 2017). The aim of this chapter is to systematize the existing knowledge about viruses present in the environment and the molecular basis of their variability.

Department of Agricultural Microbiology, Institute of Soil Science and Plant Cultivation, Puławy, 24–100, Poland.
* Corresponding author: agrzadziel@iung.pulawy.pl

Viruses- an Unexplored Area

High-throughput sequencing methods have increased viral sequence databases to more than 15M (IMG/VR), including 12M records from environmental samples, 89% of which are aquatic and less than 1.2M are soil. Genomes from host-associated samples account for 10.5%, of which about 40% are associated with the human gastrointestinal tract (Camargo et al. 2022).

The most important factors affecting the composition of viruses in a given environment are pH, temperature (Emerson et al. 2018), salinity (Roux et al. 2016b), and water content (Kimura et al. 2008). The pH not only affects the abundance of viruses in the soil, but is also important in the isolation of viral particles (Narr et al. 2017). Evidence of the importance of soil temperature and moisture for virus abundance has been presented in analyses of desert, polar and subpolar areas (Williamson et al. 2017). Higher temperature increases the number of microorganisms and consequently increases enzymatic activity. It has been proven that the abundance of viruses in soil is related to the metabolic activity of microorganisms in this environment (Williamson et al. 2017, Kuzyakov and Mason-Jones 2018). The problem with analyzing the impact of environmental conditions on the abundance and composition of viral populations stems from the fact that there are currently no studies that combine all the major aspects and physical, chemical and biological properties of the environment. It is necessary to develop some kind of gold standard for such studies so that their results are comparable and reproducible. Many years of agronomic research have shown that there is a significant gap in knowledge about the availability of elements crucial for proper plant development in the soil, which is likely to be filled by studies on the importance of viruses in the environment and their impact on the cycles of transformation of these substances. The enrichment of databases with new viral sequences will provide a wealth of important information that is crucial to viral ecology not only at the regional level, but will also be of global significance. Access to such data by scientists from all over the world will allow analysis in different ways, which will not only increase the level of knowledge, but may also lead to answers to still-open questions about the importance of viruses in the environment. Such a comparative analysis was conducted by Delgado-Baquerizo et al. on bacterial diversity, using data from 237 locations on six continents. The goal of this experiment was to determine the relationship between various environmental factors and the presence of specific species. The study found that only 2% of bacterial species can be considered core microorganisms, which are found in about half of the samples, while the rest are considered endemic bacteria associated with specific regions (Delgado-Baquerizo et al. 2018). This type of knowledge about viruses is also essential, but still incomplete.

Viruses- a Source of Crucial Genetic Information

During the lysogenic cycle, a virus that integrates its genome with the host's genetic material can simultaneously introduce genes into the host that lead to cellular changes that manifest phenotypically. An example is bacteriophages which, by integrating genes into the host's DNA, confer resistance to other phages (resistance to over-infection) (Bondy-Denomy et al. 2016). Bacteriophages present in the oceans are responsible for the transfer of 1025–1028 bp of DNA per year (Sandaa 2009). The virus can affect host metabolism through the transfer of auxiliary metabolic genes (AMGs) (Hurwitz and U'Ren 2016). The main reason for the existence of such a phenomenon is to increase the vital capacity of the host cell in such a way that the virus can be replicated as intensively, efficiently and in as many copies as possible (Brussow et al. 2004). These genes can affect the ability of microorganisms to survive in extreme conditions or cause changes in the pathogenesis of bacterial diseases (Waldor and Mekalanos 1996). Examples of such genes include sequences encoding proteins involved in the metabolic cycles of sulfur (*dsr, sox*) and nitrogen (*P-II, amoC*) (Roux et al. 2016a). Studies have also shown the presence of AMGs, which enable control of transcription and translation processes in host cells (Sharon et al. 2011). As a major agent of horizontal gene transfer (Keen et al. 2017), viruses may also be responsible for the spread of bacterial resistance to antibiotics (Balcázar 2018). This is particularly dangerous from the point of view of epidemiology and the spread of human diseases. At the same time, the genomes of soil viruses may be the source of sequences of new endolysins, which may have practical applications in therapies for many bacterial diseases (Fernández-Ruiz et al. 2018). Studies conducted on bacteriophages present in the soil revealed the presence of the trzN gene, which encodes an enzyme involved in the degradation of atrazine, an herbicide commonly used in corn cultivation. Soil bacteriophages, which are reservoirs of such genes, can positively influence the process of biological decomposition of substances harmful to both the environment and humans as direct consumers of products from areas where these substances are commonly used (Ghosh et al. 2008).

Bacteriophages isolated from natural environments are still a rich source of enzymes used in biotechnology today. Metagenomics and metatranscriptomics data can provide insight into the functional potential of viruses, which remains largely untapped (Hayes et al. 2017). Soil viruses may provide information on new, previously unknown enzymes with biogeochemical potential involved in carbohydrate breakdown (Jin et al. 2019). Studies have shown that viruses were also important in the formation of their bacterial host's genome (Ohnishi et al. 2001, Pedulla et al. 2003). Although many AMG genes have already been discovered, most of the proteins potentially responsible for changes in host cell function have yet to be investigated (Williamson et al. 2005).

Genetic Variation of Viruses

Genetic variability underlying the functioning of organisms on our planet, being a condition for the origin and maintenance of early life forms, is also a condition for the evolution of the present world (Domingo 2020). Viruses exist wherever there is life, standing at the intersection of living and non-living matter. Unlike bacteria, they do not have a cellular structure and do not metabolize outside their host cells. Activation occurs only after the virus enters the cell, where virus replication takes place, resulting in release of its copies which are able to infect subsequent cells (Louten 2016).

However, DNA is the most stable known molecule. There are five main mechanisms that determine the genetic structure and evolution of biological populations: mutation, recombination, natural selection, genetic drift and migration (Moya et al. 2004).

Mutations can result from polymerase errors during replication, transcription, or external chemical and physical factors (Sanjuán and Domingo-Calap 2016). However, most of these changes are fixed by repair systems. We can divide mutations into: 1) substitutions, where one nucleotide is changed to another, 2) deletions and insertions, where a single or multiple nucleotides are deleted or added, respectively, and 3) transpositions, where nucleotides are moved within a given genome (Berdan et al. 2021) (Figure 1). Despite the existence of efficient repair mechanisms in many DNA polymerases, it happens that the wrong nucleotide is incorporated. It is estimated that *E. coli* DNA polymerase III (with fully functional repair activity) inserts the wrong nucleotide during elongation about 1 in 10^9 times (Jee et al. 2016). Misincorporation of a pyrimidine nucleotide opposite another pyrimidine nucleotide or a purine nucleotide opposite another purine nucleotide can occur, but with much lower frequency. These less frequent misincorporations lead to transversions (Dangerfield et al. 2022). Changing the isomeric form of the purine or pyrimidine bases in a nucleotide can also lead to mutations (Fedeles et al. 2022). The frequency of misincorporation can be increased by the presence of a large number of the same nucleoside in the sequence (Drosopoulos and Prasad 1998). Improper incorporation during replication is a major pathway for introducing transversions into DNA. DNA must undergo local structural changes to accommodate these unusual and erroneous base pairs (Yang 2008). Slippage errors during replication result in the addition or deletion of nucleotides. Sometimes the leading strand can form a loop-like secondary structure that DNA polymerase does not read. In such a case, a deletion is formed in the resulting strand (Pray 2008).

Many mutations are not the result of errors in replication. Chemical reagents can oxidize and alkylate bases in DNA, sometimes changing their pairing properties. Radiation can also damage DNA (Shrivastav et al. 2009). When the amino bases, adenine and cytosine, are oxidized, they also lose an amino group. Thus, the amine is replaced by a ketone group in the product of this oxidative deamination reaction. The oxidation of cytosine, for example, produces uracil, which base pairs with adenine. Similarly, oxidation of adenine yields hypoxanthine, which pairs with cytosine. Oxidation of guanine produces xanthine, which in DNA xanthine will pair with cytosine, just like the original guanine, so this particular change is not mutagenic (Cadet et al. 2014, 2015). Many mutagens are alkylating agents. This means that they will add an alkyl group, such as a methyl or ethyl group, to a base in the DNA. The methyl group (or other alkyl group) causes a deformation of the helix. The distorted helix can alter the base-pairing properties (Fu et al. 2012). Conversely, some compounds cause the loss of nucleotides from DNA. If these deletions occur in the protein-coding region of genomic DNA, they cause a frameshift mutation, which can lead to severe changes in the structure of the final product.

The DNA sequence can also change in large regions as a result of recombination and transposition processes. Recombination involves the production of a new DNA molecule from parent DNA molecules or different regions of the same molecule (Figure 1). Recombination occurring in living organisms is divided into four types (Alberts 2003). General, homologous recombination occurs between DNA molecules with a high degree of sequence similarity, such as homologous chromosomes. The opposite is non-homologous recombination, where no sequence similarity is seen between the parent DNA, such as translocations between different chromosomes, deletions that delete several genes along a chromosome. Site-specific recombination occurs between dissimilar parental molecules, in the boron of specific short sequences. Replicative recombination generates a new copy of a DNA segment. This mechanism is used by many transposable elements to generate a new copy of the transposable element at a new location (Carroll 2013).

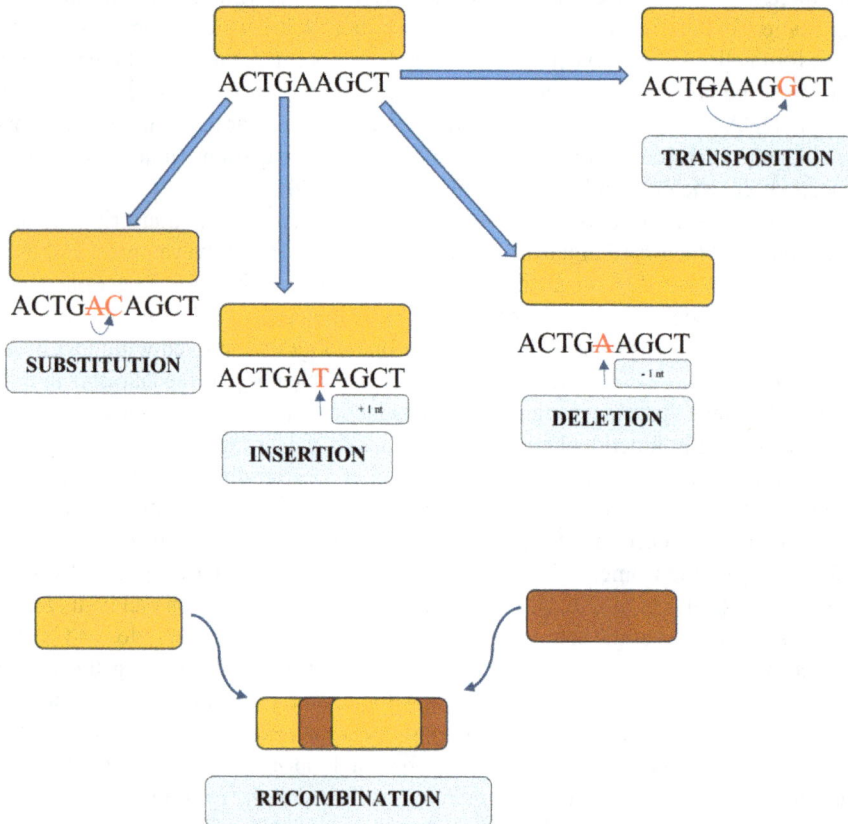

Figure 1. Genetic mechanisms that determine the structure and evolution of viral populations.

Viruses have highly heterogeneous genome structures and replication strategies (Sanjuán and Domingo-Calap 2021). Viruses contain DNA or RNA genomes that control their own replication and transmission from one host cell to the next one. They come in both linear and circular forms (Louten 2016). The mutation rate depends on the composition of the genome (RNA or DNA), its size and structure (single- or double-stranded) (Domingo 2020). The genome of viruses undergoes the same mechanisms of molecular genetic variation as in other organisms, unicellular or multicellular. During the replication process, genetic variation occurs through mutations, antigenic shift and recombination (Rubio et al. 2013).

There are the following forms of the viral genome, their types and configurations:

A - double-stranded DNA showing linear (Bacteriophages) or circular structure (Bacteriophages, Polyomavirus, Papillomavirus),

ssDNA - single-stranded DNA, showing linear or circular structure (Parvovirus B 19),

(+/-) dsRNA - double-stranded RNA, which is also non-infectious; in the presence of viral RNA polymerase it undergoes transcription, which ultimately leads to the formation of a new double-stranded viral genome (Rotavirus A, B),

(+) ssRNA - single-stranded RNA with positive polarity (sense) acts as mRNA, which enables the synthesis of viral RNA polymerase (Enterovirus, Hepatitis A virus, Human rhinovirus),

(-) ssRNA - single-stranded RNA with negative polarity (antisense- must be transcribed) (Influenza A virus, Human parainfluenza virus 1, 3, Measles virus),

The listed forms of genomes, their types and configurations determine the entire process of replication of genetic material (Louten 2016) (Figure 2).

Mutations can occur through substitutions, insertions or deletions of nucleotides in viral genomes. They are introduced at the replication stage by RNA polymerases, which lack mechanisms for proofreading (Domingo 2020). Mutations initially alter the genome sequence on a small scale. However, over time, the accumulation of mutations results in large-scale changes, new genomes are formed, visible as new strains with different characteristics (LaTourrette and Garcia-Ruiz 2022). The effects of mutations in the viral genome can be grounded in coding and non-coding regions. Single nucleotide changes can cause a so-called silent mutation, which does not manifest itself

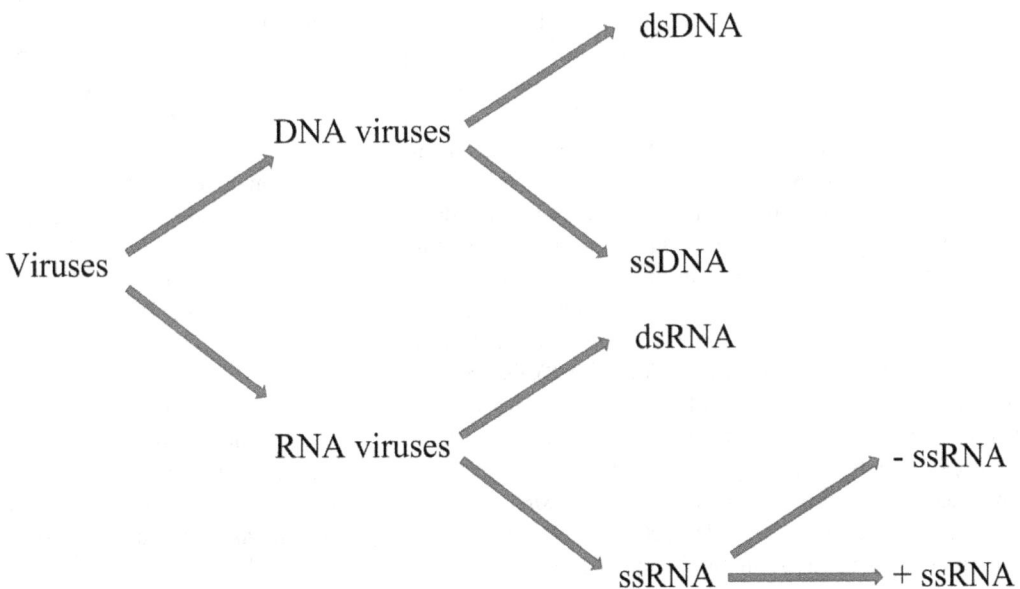

Figure 2. The genetic material diversity in viruses.

as a change in the structure of the genome, resulting in a change in the gene product, which is a protein. However, some mutations in coding genes can cause stop codons that result in a new form of protein. Neutral mutations do not cause functional effects. Non-neutral mutations, on the other hand, can lead to both positive and negative effects. A positive one is, for example, adaptation to the prevailing environmental conditions, the development of traits that determine the ability to infect a larger host group (Payne 2017, Domingo 2020).

Viruses evolve through a combination of population genetic variation, genetic drift, and selection mediated by host, vector, and environmental factors, demonstrating their continued adaptation. Because they replicate faster than their host can reproduce, viruses will continue to evolve, avoiding host immune responses while maintaining functionality across hosts and vectors (Payne 2017).

Since viruses do not have repair mechanisms, they experience frequent spontaneous mutations when replicating their genetic material. This is the reason for the high diversity of viruses within a single species. There is also frequent recombination of genetic material, such as between two viruses attacking the same cell, as well as between the virus and the host cell (Pérez-Losada et al. 2015). Segmentation of the genome positively affects recombination. The influenza virus, for example, mutates in this way (White and Lowen 2018). These processes underlie the ability of viruses to generate and maintain genetic diversity, which is key to circumventing and adapting to host and vector processes.

Within species, mutations are directly involved in creating new biological properties. For example, a single amino acid mutation enables PVC (pox virus C) and PVY (potato virus Y) viruses to change hosts (Calvo et al. 2014, Vassilakos et al. 2016). While a single mutation cannot result in a new species, it can result in a new strain if it changes the biological properties of the virus (Díaz-Pendón et al. 2019). Specific mutations have also enabled viruses to overcome resistance in important crops such as soybean (Eggenberger et al. 2008) and melon (Guiu-Aragonés et al. 2015). In addition, mutations can alter symptoms in a virus-infected host (Wieczorek and Obrepalska-Steplowska 2016), change the function of viral agents (Xu et al. 2012) or affect vector transmission (Peters et al. 2008). From a biological point of view, mutations in the viral genome are necessary to adapt to the variability present in host proteins (Brosseau et al. 2020).

Viruses - Importance and Impact on the Environment

Viral particles consist mainly of proteins and genetic material. Depending on the type of virus, the ratio of these two components can vary. For example, in the case of Caudovirales, the best-characterized phages, the ratio is 1:1, while filamentous viruses contain only 6–14% genetic material. The size of viruses is 8,000–125,000 times smaller than the size of a bacterial cell, except for giant viruses, which can exceed 500 nm. Viral genetic material can be single- or double-stranded DNA or RNA (ssDNA, dsDNA, ssRNA or dsRNA) (Kuzyakov and Mason-Jones 2018).

The life cycle of viruses has been studied in great detail. It can be lytic, leading to the death of the host cell, or lysogenic, which involves the incorporation of viral genetic material into the host genome (Clokie et al. 2011). During the lytic cycle, host cell processes target the multiplication of viral genetic material and the production of proteins that form the viral capsid. The result is cell death and the release of newly formed viral particles, the number of which is referred to as "*burst size*" (Kuzyakov and Mason-Jones 2018). During the lysogenic cycle, viral protein expression is silenced. This phenomenon allows viruses to survive under unfavorable environmental conditions (Erez et al. 2017). Under the right environmental conditions or as a result of stress factors, the virus can switch from the lysogenic to the lytic cycle. Recent studies have shown that substances produced by bacteria during the *quorum sensing* process can cause the virus to switch from the lysogenic to the lytic cycle (Liang et al. 2019). Consequently, the decline in microbial populations during changes in environmental conditions may be related not only to a lack of adaptation to the new environment, but also to the fact that viruses that are in the lysogenic to lytic cycle transition (Brum et al. 2016). The number of viruses in the soil shows a positive correlation with the number of bacteria. This may support the theory that most viruses present in soil belong to bacteriophages (Williamson et al. 2017). However, some studies have shown that the number of viruses in samples containing more microorganisms is lower. This phenomenon has been explained by the "*Piggyback-the-Winner*" (PtW) model, which promotes the lysogenic cycle observed in many environments, including soil (Knowles et al. 2016, Touchon et al. 2017). Based on previous studies, it can be concluded that the majority of bacteria found in soil have bacteriophage-derived viral material embedded in their genome (Ghosh et al. 2008). Viruses are most likely responsible for bacterial mortality in niches of the soil environment which, due to their structure, exhibit limited predatory activity belonging to *protozoa*, *collembola* and *nematodes* (Kuzyakov and Mason-Jones 2018).

Microorganisms have developed defense mechanisms against the attack of bacteriophages, preventing their binding in the results of changes in the structure of host cell membranes. Viruses, however, as the fastest-changing molecules present in the environment, can affect their own and the host's structure, allowing them to successfully penetrate the bacterial cell. The phenomenon of two-way evolution to develop features that prevent the virus from binding to the host cell, in the case of bacteria, or vice versa, in the case of viruses, has been called an "evolutionary arms race" (Koskella and Brockhurst 2014). The lytic cycle of the virus leads to the death of the host cell which, in a kill-the-winner (KtW) strategy, allows the number of bacteria dominating the environment to be reduced (Breitbart 2012).

Methods for Studying the Viral Population

Knowing both the abundance and composition of the viral population in a given environment is elementary in determining the role viruses play in it. Finding sensitive methods to learn about the viral population is crucial.

The oldest method used to determine the abundance of viruses in a given sample is the plate test. This method involves mixing a sample with a culture of the target bacteria and seeding such a mixture onto a solid medium. After a predetermined incubation time, the presence of baldness in the bacterial culture indicating the action of bacteriophage for a given microorganism is determined. This method is also used to isolate pure cultures of the viruses in question. Its greatest limitation is that it can only be used for specific bacteria capable of growth under laboratory conditions (Van Twest and Kropinski 2009, Williamson et al. 2017).

Unlike the plate test, direct counting of viral particles can provide reliable information on the total abundance of viruses in given samples. Current direct virus counting methods are based on two-step methods consisting of a viral particle isolation step and the actual determination of abundance by microscopic methods. Transmission electron microscopy (TEM), epifluorescence microscopy (EFM) and, much less frequently, flow cytometry are typically used to determine viral particle abundance (Williamson et al. 2013). In EFM microscopy, a suspension of viral particles is stained with a fluorochrome such as as SYBR Gold, SYBR Green or DAPI, which binds to nucleic acids (Williamson et al. 2003). With TEM microscopy, additional purification of the slide may be necessary to clearly visualize the virus particle. The advantage of the TEM method is that it allows visualization of the morphology of the viral particle. However, it requires significantly higher operating costs and has a lower sample throughput compared to EFM. A disadvantage of the EFM method can be the presence of background signals in the image, not coming directly from the viral particles, which can significantly overestimate the analysis results (Ferris et al. 2002).

The lack of a universal, conserved genetic sequence similar to prokaryotic 16S rRNA is a major barrier to studies of viral population phylogeny. Marker genes, highly conserved within specific viral families, can be used to assess viral populations. The most commonly used marker gene in viral population studies is *g23*, which encodes the major capsid protein of T4-type phage, which are infectious particles of *Escherichia coli* (Filée et al. 2005). Since the discovery of this sequence, this gene has been used to assess the genetic diversity of T4 phages in soils in Japan (Wang et al.

2009), and China (Liu et al. 2012, Zheng et al. 2013). However, viral populations carrying the g23 gene represent a small fraction of all viral populations in a given environment, which is a significant limitation when trying to determine the phylogenetic composition of viruses in a given sample.

The number of viral genome sequencing analyses is significantly lower than those for prokaryotic genomes. Knowing the genomes of viruses that infect representative organisms is essential to better understand the genetic and functional diversity of viruses in a given environment. In relation to soils, many sequenced viral genomes belong to crop pathogens. While these viruses are undoubtedly important because of their impact on crop production and economics, much less is known about viruses that infect microorganisms present in the soil (Klenk and Göker 2010). The key information from genomic studies is knowledge of genome structure, functional and genetic diversity, and mechanisms underlying phenotypic and evolutionary variation within populations. A prerequisite for such studies is virus isolation on host cells to obtain biological material for nucleic acid extraction for sequencing. This limits the exploratory capabilities of the method to only viruses that are phages for specific, culturable groups of microorganisms. Additionally, it makes it impossible to include such studies in a metagenomic approach. Another major problem associated with genomic analysis is the so-called bioinformatics bottleneck. Isolation and sequencing of the viral genome may be a relatively quick and inexpensive method; however, complex bioinformatics analysis is labor-intensive and time-consuming. Thus, there is an ongoing need for up-to-date and more efficient bioinformatics tools to identify genes and predict viral gene function (Williamson et al. 2017).

The next-generation sequencing technique has become a method for studying genetic diversity within entire environmental viral clusters, excluding host culture, as well as restrictions on reference genes. The number of viral genomes from different environments obtained by metagenomics methods is shown in Table 1 (Camargo et al. 2022). Completed studies of soil viriomes show that several are dominated by small ssDNA viruses. These findings challenge the dogma that soils are dominated by dsDNA phages (Reavy et al. 2015, Green et al. 2018). Viral metagenomics is a powerful approach to obtain genetic information about viral populations without being limited to first isolating the virus in tissue culture. However, the speed, and quantity, of metagenomic sequences generated, compared to the speed with which they can be rigorously analyzed, causes the bioinformatics bottleneck problem mentioned earlier (Scholz et al. 2012). An additional problem is that the resulting metagenomic data often contain a high percentage of reads unrelated to known sequences. In published studies on viral genomics, 54.5–97.3% of all reads cannot be matched to a specific taxonomic source, a specific protein function, or both, rendering these reads useless for most further analyses. It is interesting to note that the abundance of viruses in soils is much higher than the corresponding values for aquatic

Table 1. The number of viral genomes from different environments obtained by metagenomics ([1]aquaculture, floodplain, sediments, thermal springs; [2] rock – dwelling, geologic, floodplain, oil reservoir, asphalt lakes, mud volcano, volcanic fumaroles).

Environment		Number of genomes
Aquatic	Marine	5 687 030
	Freshwater	4 963 767
	Non-marine saline and alkaline	432 245
	Deep subsurface	191 015
	Other [1]	123 306
Terrestrial	Soil	1 270 132
	Deep subsurface	87 243
	Nest	40 257
	Plant litter	32 817
	Other [2]	21 522
Air	Outdoor air	639

systems. This, combined with the high percentage of unrecognized sequences in the metagenomes of soil viruses, supports the idea that soil viruses may represent the highest concentration of genetic diversity (Williamson et al. 2017).

Conclusions

Viruses represent a huge gap in the knowledge of environmental biodiversity while being one of the most numerous biological groups. They affect the surrounding environment through gene transfer, reduction of biodiversity, and metabolic disruption. They shape the host population while influencing their evolutionary process, changes manifested phenotypically, through horizontal gene transfer. This is a new, unexplored field of science with tremendous learning potential. Most current biodiversity research focuses on metataxonomic analysis of bacterial and fungal composition, based on next-generation sequencing techniques using conserved gene sequences such as 16S rRNA (bacteria) and ITS (fungi). For viruses, the lack of a universal sequence prevents the use of these methods to analyze their taxonomic affiliation. Another challenge in the study of this biological group is the diversity of viral genetic material (ssRNA, dsRNA, ssDNA, dsDNA, dsDNA). The solution has become the ever-developing next-generation sequencing techniques that allow scientists to study biodiversity in greater depth. Metataxonomic analysis is increasingly supported by metagenomics, using shotgun sequencing methods, for example. Metagenomic analysis of viral particles isolated from environmental samples makes it possible to detect new viral species, previously unknown genes and the proteins they encode, often with application significance.

This paper is an outcome of the data obtained under the project NCN 2022/45/B/NZ8/02398 (National Science Centre, Poland) "Interaction between microbiome, mycobiome and metaviriome of the rhizosphere and endorhizosphere of ruderal plants and their role in passive and active remediation of soil heavily degraded and historically contaminated with crude oil" (2023–2027).

References

Alberts, B. 2003. DNA replication and recombination. Nature 421: 431–135.

Balcázar, J.L. 2018. How do bacteriophages promote antibiotic resistance in the environment? Clin. Microbiol. Infect. 24: 447–449.

Berdan, E.L., A. Blanckaert, T. Slotte, A. Suh, A.M. Westram and I. Fragata. 2021. Unboxing mutations: Connecting mutation types with evolutionary consequences. Mol. Ecol. 30: 2710–2723.

Bondy-Denomy, J., J. Qian, E.R. Westra, A. Buckling, D.S. Guttman, A.R. Davidson et al. 2016. Prophages mediate defense against phage infection through diverse mechanisms. ISME J. 10: 2854–2866.

Breitbart, M. 2012. Marine viruses: Truth or dare. Ann. Rev. Mar. Sci. 4: 425–448.

Brosseau, C., A. Bolaji, C. Roussin-Léveillée, Z. Zhao, S. Biga and P. Moffett. 2020. Natural variation in the *Arabidopsis* AGO2 gene is associated with susceptibility to potato virus X. New Phytol. doi:10.1111/nph.16397.

Brum, J.R., B.L. Hurwitz, O. Schofield, H.W. Ducklow and M.B. Sullivan. 2016. Seasonal time bombs: Dominant temperate viruses affect Southern Ocean microbial dynamics. ISME J. 10: 437–449.

Brussow, H., C. Canchaya and W.-D. Hardt. 2004. Phages and the evolution of bacterial pathogens: From genomic rearrangements to lysogenic conversion. Microbiol. Mol. Biol. Rev. 68: 560–602.

Cadet, J., T. Douki and J.L. Ravanat. 2015. Oxidatively generated damage to cellular DNA by UVB and UVA radiation. Photochem. Photobiol. 91: 140–155.

Cadet, J., J.R. Wagner, V. Shafirovich and N.E. Geacintov. 2014. One-electron oxidation reactions of purine and pyrimidine bases in cellular DNA. Int. J. Radiat. Biol. 90: 423–432.

Calvo, M., T. Malinowski and J.A. García. 2014. Single amino acid changes in the 6K1-CI region can promote the alternative adaptation of Prunus- and Nicotiana-propagated plum pox virus C isolates to either host. Mol. Plant-Microbe Interact. doi:10.1094/MPMI-08-13-0242-R.

Camargo, A.P., S. Nayfach, I.-M.A. Chen, K. Palaniappan, A. Ratner, K. Chu et al. 2022. IMG/VR v4: An expanded database of uncultivated virus genomes within a framework of extensive functional, taxonomic, and ecological metadata. Nucleic Acids Res. 51: 733–743.

Carroll, D. 2013. Genetic Recombination. 277–280, Maloy, S., Hughes, K. (Eds.). Brenner's Encyclopedia of Genetics: Second Edition, Academic Press.

Clokie, M.R.J.J., A.D. Millard, A.V. Letarov and S. Heaphy. 2011. Phages in nature. Bacteriophage 1: 31–45.

Dangerfield, T.L., S. Kirmizialtin and K.A. Johnson. 2022. Conformational dynamics during misincorporation and mismatch extension defined using a DNA polymerase with a fluorescent artificial amino acid. J. Biol. Chem. 298: 1–23.

Delgado-Baquerizo, M., A.M. Oliverio, T.E. Brewer, A. Benavent-González, D.J. Eldridge, R.D. Bardgett et al. 2018. A global atlas of the dominant bacteria found in soil. Science (80). 359: 320–325.

Díaz-Pendón, J.A., S. Sánchez-Campos, I.M. Fortes and E. Moriones. 2019. Tomato yellow leaf curl sardinia virus, a begomovirus species evolving by mutation and recombination: A challenge for virus control. Viruses. doi:10.3390/v11010045.

Domingo, E. 2020. Molecular basis of genetic variation of viruses. Virus as Populations,. doi:10.1016/b978-0-12-816331-3.00002-7.

Donovan, P.D., G. Gonzalez, D.G. Higgins, G. Butler and K. Ito. 2018. Identification of fungi in shotgun metagenomics datasets. PLoS One 13: 1–16.

Drosopoulos, W.C. and V.R. Prasad. 1998. Increased misincorporation fidelity observed for nucleoside analog resistance mutations M184V and E89G in human immunodeficiency virus Type 1 reverse transcriptase does not correlate with the overall error rate measured *in vitro*. J. Virol. 72: 4224–4230.

Eggenberger, A.L., M.R. Hajimorad and J.H. Hill. 2008. Gain of virulence on Rsv1-genotype soybean by an avirulent Soybean mosaic virus requires concurrent mutations in both P3 and HC-Pro. Mol. Plant-Microbe Interact. doi: 10.1094/MPMI-21-7-0931.

Emerson, J.B. 2019. Soil Viruses: A New Hope. mSystems 4: 1–4.

Emerson, J.B., S. Roux, J.R. Brum, B. Bolduc, B.J. Woodcroft, H. Bin Jang et al. 2018. Host-linked soil viral ecology along a permafrost thaw gradient. Nat. Microbiol. 3: 870–880.

Erez, Z., I. Steinberger-Levy, M. Shamir, S. Doron, A. Stokar-Avihail, Y. Peleg et al. 2017. Communication between viruses guides lysis-lysogeny decisions. Nature 541: 488–505.

Fedeles, B.I., D. Li and V. Singh. 2022. Structural insights into tautomeric dynamics in nucleic acids and in antiviral nucleoside analogs. Front. Mol. Biosci. 8: 1–13.

Fernández-Ruiz, I., F.H. Coutinho and F. Rodriguez-Valera. 2018. Thousands of novel endolysins discovered in uncultured phage genomes. Front. Microbiol. 9: 1–8.

Ferris, M.M., C.L. Stoffel, T.T. Maurer and K.L. Rowlen. 2002. Quantitative intercomparison of transmission electron microscopy, flow cytometry, and epifluorescence microscopy for nanometric particle analysis. Anal. Biochem. 304: 249–256.

Filée, J., F. Tétart, C.A. Suttle and H.M. Krisch. 2005. Marine T4-type bacteriophages, a ubiquitous component of the dark matter of the biosphere. Proc. Natl. Acad. Sci. U. S. A. 102: 12471–12476.

Fu, D., J.A. Calvo and L.D. Samson. 2012. Balancing repair and tolerance of DNA damage caused by alkylating agents. Nat. Rev. Cancer 12: 104–120.

Galazka, A. and J. Grzadziel. 2016. The Molecular-based methods used for studying bacterial diversity in soils contaminated with PAHs (The Review). 85–104, Soil Contamination—Current Consequences and Further Solutions, IntechOpen, Rijeka.

Gałazka, A. and J. Grzadziel. 2018. Fungal genetics and functional diversity of microbial communities in the soil under long-term monoculture of maize using different cultivation techniques. Front. Microbiol. 9: 76–90.

Ghosh, D., K. Roy, K.E. Williamson, D.C. White, K.E. Wommack, K.L. Sublette et al. 2008. Prevalence of lysogeny among soil bacteria and presence of 16S rRNA and trzN genes in viral-community DNA. Appl. Environ. Microbiol. 74: 495–502.

Green, J.C., F. Rahman, M.A. Saxton and K.E. Williamson. 2018. Quantifying aquatic viral community change associated with stormwater runoff in a wet retention pond using metagenomic time series data. Aquat. Microb. Ecol. 81: 19–35.

Grządziel, J. and A. Gałązka. 2018. Microplot long-term experiment reveals strong soil type influence on bacteria composition and its functional diversity. Appl. Soil Ecol. 124: 117–123.

Guiu-Aragonés, C., J.A. Díaz-Pendón and A.M. Martín-Hernández. 2015. Four sequence positions of the movement protein of Cucumber mosaic virus determine the virulence against cmv1-mediated resistance in melon. Mol. Plant Pathol. doi:10.1111/mpp.12225.

Guo, J., J.R. Cole, Q. Zhang, T. Brown and J.M. Tiedje. 2016. Microbial community analysis with ribosomal gene fragments from shotgun metagenomes. Appl. Environ. Microbiol. doi:10.1128/AEM.02772-15.

Hayes, S., J. Mahony, A. Nauta and D. Van Sinderen. 2017. Metagenomic approaches to assess bacteriophages in various environmental niches. Viruses 9: 1–22.

Hurwitz, B.L. and J.M. U'Ren. 2016. Viral metabolic reprogramming in marine ecosystems. Curr. Opin. Microbiol. 31: 161–168.

Jee, J., A. Rasouly, I. Shamovsky, Y. Akivis, S.R. Steinman, B. Mishra et al. 2016. Rates and mechanisms of bacterial mutagenesis from maximum-depth sequencing. Nature 30: 693–696.

Jin, M., X. Guo, R. Zhang, W. Qu, B. Gao and R. Zeng. 2019. Diversities and potential biogeochemical impacts of mangrove soil viruses. Microbiome 7: 1–15.

Keen, E.C., V.V. Bliskovsky, F. Malagon, J.D. Baker, J.S. Prince, J.S. Klaus et al. 2017. Novel "superspreader" bacteriophages promote horizontal gene transfer by transformation. MBio. 8: 1–12.

Kimura, M., Z.J. Jia, N. Nakayama and S. Asakawa. 2008. Ecology of viruses in soils: Past, present and future perspectives. Soil Sci. Plant Nutr. 54: 1–32.

Klenk, H.P. and M. Göker. 2010. En route to a genome-based classification of Archaea and Bacteria? Syst. Appl. Microbiol. 33: 175–182.

Knowles, B., C.B. Silveira, B.A. Bailey, K. Barott, V.A. Cantu, A.G. Cobian-Güemes et al. 2016. Lytic to temperate switching of viral communities. Nature 531: 466–470.

Koskella, B. and M.A. Brockhurst. 2014. Bacteria-phage coevolution as a driver of ecological and evolutionary processes in microbial communities. FEMS Microbiol. Rev. 38: 916–931.

Kuzyakov, Y. and K. Mason-Jones. 2018. Viruses in soil: Nano-scale undead drivers of microbial life, biogeochemical turnover and ecosystem functions. Soil Biol. Biochem. 127: 305–317.

LaTourrette, K. and H. Garcia-Ruiz. 2022. Determinants of virus variation, evolution, and host adaptation. Pathog. (Basel, Switzerland) 11. doi:10.3390/pathogens11091039.

Liang, X., R.E. Wagner, B. Li, N. Zhang and M. Radosevich. 2019. Prophage induction mediated by quorum sensing signals alters soil bacterial community structure. bioRxiv. doi: 10.1101/805069.

Liu, Junjie, G. Wang, Q. Wang, Judong Liu, J. Jin and X. Liu. 2012. Phylogenetic diversity and assemblage of major capsid genes (g23) of t4-type bacteriophages in paddy field soils during rice growth season in northeast china. Soil Sci. Plant Nutr. 58: 435–444.

Louten, J. 2016. Virus Structure and Classification. Essential Human Virology, doi:10.1016/b978-0-12-800947-5.00002-8.

Moya, A., E.C. Holmes and F. González-Candelas. 2004. The population genetics and evolutionary epidemiology of RNA viruses. Nat. Rev. Microbiol. doi:10.1038/nrmicro863.

Narr, A., A. Nawaz, L.Y. Wick, H. Harms and A. Chatzinotas. 2017. Soil viral communities vary temporally and along a land use transect as revealed by virus-like particle counting and a modified community fingerprinting approach (fRAPD). Front. Microbiol. 8: 1–14.

Ohnishi, M., K. Kurokawa and T. Hayashi. 2001. Diversification of *Escherichia coli* genomes: Are bacteriophages the major contributors? Trends Microbiol. 9: 481–485.

Payne, S. 2017. Virus evolution and genetics. Viruses,. doi:10.1016/b978-0-12-803109-4.00008-8.

Pedulla, M.L., M.E. Ford, J.M. Houtz, T. Karthikeyan, C. Wadsworth, J.A. Lewis et al. 2003. Origins of highly mosaic mycobacteriophage genomes. Cell 113: 171–182.

Pérez-Losada, M., M. Arenas, J.C. Galán, F. Palero and F. González-Candelas. 2015. Recombination in viruses: Mechanisms, methods of study, and evolutionary consequences. Infect. Genet. Evol. doi:10.1016/j.meegid.2014.12.022.

Peters, K.A., D. Liang, P. Palukaitis and S.M. Gray. 2008. Small deletions in the potato leafroll virus readthrough protein affect particle morphology, aphid transmission, virus movement and accumulation. J. Gen. Virol. doi:10.1099/vir.0.83625-0.

Pray, L. 2008. DNA replication and causes of mutation. Nat. Educ. 214: 1–4.

Reavy, B., M.M. Swanson, P.J.A. Cock, L. Dawson, T.E. Freitag, B.K. Singh et al. 2015. Distinct circular single-stranded DNA viruses exist in different soil types. Appl. Environ. Microbiol. 81: 3934–3945.

Roux, S., J.R. Brum, B.E. Dutilh, S. Sunagawa, M.B. Duhaime, A. Loy et al. 2016a. Ecogenomics and potential biogeochemical impacts of globally abundant ocean viruses. Nature 537: 689–693.

Roux, S., F. Enault, V. Ravet, J. Colombet, Y. Bettarel, J.C. Auguet et al. 2016b. Analysis of metagenomic data reveals common features of halophilic viral communities across continents. Environ. Microbiol. 18: 889–903.

Rubio, L., J. Guerri and P. Moreno. 2013. Genetic variability and evolutionary dynamics of viruses of the family Closteroviridae. Front. Microbiol. doi:10.3389/fmicb.2013.00151.

Sandaa, R.-A. 2009. Viruses, environmental. Encyclopedia of Microbiology, doi:10.1016/b978-012373944-5.00366-7.

Sanjuán, R. and P. Domingo-Calap. 2016. Mechanisms of viral mutation. Cell. Mol. Life Sci. 73: 4433–4448.

Sanjuán, R. and P. Domingo-Calap. 2021. Genetic diversity and evolution of viral populations. Encyclopedia of Virology. doi:10.1016/b978-0-12-809633-8.20958-8.

Schoenfeld, T., M. Liles, K.E. Wommack, S.W. Polson, R. Godiska and D. Mead. 2010. Functional viral metagenomics and the next generation of molecular tools. Trends Microbiol. 18: 20–29.

Scholz, M.B., C.C. Lo and P.S.G. Chain. 2012. Next generation sequencing and bioinformatic bottlenecks: The current state of metagenomic data analysis. Curr. Opin. Biotechnol. 23: 9–15.

Schulz, F., L. Alteio, D. Goudeau, E.M. Ryan, F.B. Yu, R.R. Malmstrom et al. 2018. Hidden diversity of soil giant viruses. Nat. Commun. 9: 4881–4889.

Sharon, I., N. Battchikova, E.M. Aro, C. Giglione, T. Meinnel, F. Glaser et al. 2011. Comparative metagenomics of microbial traits within oceanic viral communities. ISME J. 5: 1178–1190.

Shrivastav, N., D. Li and J.M. Essigmann. 2009. Chemical biology of mutagenesis and DNA repair: Cellular responses to DNA alkylation. Carcinogenesis 31: 59–70.

Simmonds, P., M.J. Adams, M. Benk, M. Breitbart, J.R. Brister, E.B. Carstens et al. 2017. Consensus statement: Virus taxonomy in the age of metagenomics. Nat. Rev. Microbiol. 15: 161–168.

Tangherlini, M., C. Corinaldesi and A. Dell'Anno. 2012. Viral metagenomics: A new and complementary tool for environmental quality assessment. Chem. Ecol. 28: 497–501.

Touchon, M., J.A. Moura de Sousa and E.P. Rocha. 2017. Embracing the enemy: The diversification of microbial gene repertoires by phage-mediated horizontal gene transfer. Curr. Opin. Microbiol. 38: 66–73.

Van Twest, R. and A.M. Kropinski. 2009. Bacteriophage enrichment from water and soil. Methods Mol. Biol. 501: 15–21.

Vassilakos, N., V. Simon, A. Tzima, E. Johansen and B. Moury. 2016. Genetic determinism and evolutionary reconstruction of a host jump in a plant virus. Mol. Biol. Evol. doi:10.1093/molbev/msv222.

Waldor, M.K. and J.J. Mekalanos. 1996. Lysogenic Conversion by a Filamentous Phage Encoding Cholera Toxin. Science (80-.). 272: 1910–1914.

Wang, G., M. Hayashi, M. Saito, K. Tsuchiya, S. Asakawa and M. Kimura. 2009. Survey of major capsid genes (g23) of T4-type bacteriophages in Japanese paddy field soils. Soil Biol. Biochem. 41: 13–20.

White, M.C. and A.C. Lowen. 2018. Implications of segment mismatch for influenza A virus evolution. J. Gen. Virol. doi:10.1099/jgv.0.000989.

Wieczorek, P. and A. Obrepalska-Steplowska. 2016. A single amino acid substitution in movement protein of tomato torrado virus influences ToTV infectivity in *Solanum lycopersicum*. Virus Res. doi:10.1016/j.virusres.2015.11.008.

Williamson, K.E., K.A. Corzo, C.L. Drissi, J.M. Buckingham, C.P. Thompson and R.R. Helton. 2013. Estimates of viral abundance in soils are strongly influenced by extraction and enumeration methods. Biol. Fertil. Soils 49: 857–869.

Williamson, K.E., J.J. Fuhrmann, K.E. Wommack and M. Radosevich. 2017. Viruses in soil ecosystems: An unknown quantity within an unexplored territory. Annu. Rev. Virol. 4: 201–219.

Williamson, K.E., M. Radosevich and K.E. Wommack. 2005. Abundance and diversity of viruses in six Delaware soils. Appl. Environ. Microbiol. 71: 3119–3125.

Williamson, K.E., K.E. Wommack and M. Radosevich. 2003. Sampling natural viral communities from soil for culture-independent analyses. Appl. Environ. Microbiol. 69: 6628–6633.

Xu, J., X. Wang, L. Shi, Y. Zhou, D. Li, C. Han et al. 2012. Two distinct sites are essential for virulent infection and support of variant satellite RNA replication in spontaneous beet black scorch virus variants. J. Gen. Virol. doi:10.1099/vir.0.045641-0.

Yang, W. 2008. Structure and mechanism for DNA lesion recognition. Cell Res. 18: 184–197.

Zablocki, O., E.M. Adriaenssens, A. Frossard, M. Seely, J.-B. Ramond and D. Cowan. 2017. Metaviromes of extracellular soil viruses along a namib desert aridity gradient. Genome Announc. 5: e01470-16.

Zheng, C., G. Wang, J. Liu, C. Song, H. Gao and X. Liu. 2013. Characterization of the Major Capsid Genes (g23) of T4-Type Bacteriophages in the Wetlands of Northeast China. Microb. Ecol. 65: 616–625.

CHAPTER 9

How Do Phytoplasmas Manipulate Host Plants' Molecular Mechanisms?

Beata Zimowska

Introduction

Phytoplasmas (class *Mollicutes*, genus *Phytoplasma*), as the insect-transmitted wall-less bacteria, have devastating effects on yields in diverse low- and high-value crops and plants worldwide. Coconut palm lethal yellowing disease, Paulownia witches' broom disease, and many other diseases caused by phytoplasmas induce fatal damage in plants and crops all over the world (Bertaccini 2007). Phytoplasmas infect more than 700 plant species and bring about dramatic changes in plant development, including witches' broom, dwarfism, proliferation (the growth of shoots from floral organs), phyllody, virescence, sterility of flowers, bolting, purple tops (the reddening of leaves and stems), generalized yellowing, and phloem necrosis (Christensen et al. 2005, Hogenhout et al. 2008). Phytoplasmas reside within the sieve elements of plant phloem and within the cell in insects and are transmitted from plant to plant by insect vectors.

Like many plant pathogens, phytoplasmas produce virulence factors (i.e., effectors), that interfere with the host's normal life processes, changing them in favor of the pathogen. This process is based on molecular mechanisms. Phytoplasma also modulate immune response of the sap-feeding hemipteran (including plant hoppers, leafhoppers and psyllids) insect vectors in such a way that they spread these bacteria to other plants. Since phytoplasmas have no cell wall and reside inside the host cells, their membrane proteins and secreted proteins (effectors) can function in the cytoplasm of the host plant or insect cells, and are predicted to have important roles in the interplay between pathogen and host (Oshima et al. 2013, Sugio et al. 2011a,b, 2014).

Several effector molecules have been identified by various workers that have been proposed to genetically reprogram the host plants enabling the colonization, survival and spread of phytoplasma. The genes encoding functional proteins that provided interaction with the host plant (membrane transport, proteases, DNA methylases, effectors, etc.) differed from each other and from strains of other species (Bai et al. 2009).

Hence, this chapter will focus on different aspects of molecular biology and pathogenicity of phytoplasmas.

Effector Proteins that Can Reprogram Plant Development in a Number of Impressive Ways- TENGU

The first such effector protein described, the "tengu-suinducer" (TENGU), was isolated from onions (*Allium* sp.) infected with phytoplasma which caused yellowing. TENGU is a 70 amino acid

Department of Plant Protection, Subdepartment of Plant Pathology and Mycology, University of Life Sciences in Lublin, 7 K. St. Leszczyńskiego Street, 20–068 Lublin, Poland. Email: beata.zimowska@up.lublin.pl protein (4.5kDa) and is restricted to 16SrI group and shows high level of sequence conservation. It is located near ABC transporter genes in Onion Yellows strain M (OY-M) and MBSP and near PMU like genome elements in Aster Yellows strain Witches' Broom (AY-WB) genome (Sugio and Hogenhout 2012). Hoshi et al. (2009) first characterized virulence factor, tengu-su inducer (TENGU), by expressing it in *Nicotiana benthamiana* plants. They screened 30 putative secretory proteins from *Candidatus* Phytoplasma asteris (OY) strains and each protein was expressed in *N. benthamiana*. PAM765 (one of the secreted protein) gene was transformed into *N.benthamiana* using *Agrobacterium tumefaciens*. These transgenic plants expressing PAM765 developed clear symptoms of witches' broom and dwarfism. Apart from this, these plants also had malformed phyllotaxis. This was further confirmed by inoculating *Arabidopsis thaliana* with OY phytoplasma strain which exhibited the same symptoms. Therefore, PAM765 was termed as Tengu, a virulence factor that induces phytoplasma related witches' broom, dwarfism and abnormal reproductive organogenesis. This protein is transported via the phloem into other cells, including cells of the apical and axillary meristem, and causes characteristic symptoms, the witches' broom and dwarfism. TENGU also induces the sterility of male and female flowers by inhibiting the signaling pathway of jasmonic acid (JA). The reduction in endogenous JA levels is thought to contribute to attracting insect vectors (Minato et al. 2014). The N-terminus of TENGU contains an 11 amino acid signal peptide which is cleaved *in vivo* during proteolysis by plant serine protease. It is assumed that this fragment at the N-terminus of the protein directly induces the development of the observed symptoms. TENGU proteins were detected not only in phloem tissues but also in the parenchyma and meristem tissues via immunohistochemical analysis using an anti-TENGU antibody. This suggests that TENGU has the ability to be transported from phloem to non-phloem cells. Besides, TENGU proteins were strongly detected in the tip region of stems and the branching region of axillary buds, and even in the apical meristem region, suggesting that TENGU can be transported into apical buds (Hoshi et al. 2009).

TENGU may suppress the auxin signaling and biosynthesis pathways. Auxin is known to be involved in apical dominance, which is where an apical bud inhibits development of an axillary bud growth. Auxin biosynthesized in the apical bud is transported to the root, and inhibits the growth of axillary buds (Mori et al. 2005).This growth inhibition of axillary buds is released by loss of or damage to the apical bud (Cline 1997). The studies conducted by Hoshi et al. (2009) have proved that expression levels of early auxin-responsive genes, auxin efflux-related genes, and dormancy-associated genes were reduced in the *tengu*-transgenic plant, which suggests that TENGU suppresses plant auxin responses, resulting in the growth inhibition of apical buds in *tengu*-transgenic or OY-infected plants.

Secreted Aster Yellows Witches' Broom Protein 54 (SAP54)

SAP54 leads to the degradation of a specific subset of floral homeotic proteins of the MIKC-type MADS-domain family via the ubiquitin-proteasome pathway. As a consequence, the developing flowers show the homeotic transformation of floral organs into vegetative leaf-like structures. The molecular mechanism of SAP54 action involves binding to the keratin-like domain of MIKC-type proteins and to some RAD23 proteins, which translocate ubiquitylated substrates to the proteasome (Aurin et al. 2020).

For the first time, MacLean et al. (2014) elucidated the mechanism by which phytoplasma alters floral development to convert flowers into vegetative tissues by producing a novel effector protein (SAP54) that interacts with members of the MADS domain transcription factor (MTF) family, such as SEPALLATA3 and APETALA1 and occupy central positions in the regulation of floral development. SAP54 mediates degradation of MTFs by interacting with proteins of the RADIATION SENSITIVE23 (RAD23) family-eukaryotic proteins that transport substrates to the proteasome as the conversion of flowers into leaflike tissues diminished in Arabidopsis rad23 mutants in the presence of SAP54. Remarkably, plants with SAP54-induced leaf-like flowers are more attractive for colonization by phytoplasma leafhopper vectors and this colonization preference is dependent on RAD23. Why do phytoplasmas induce symptoms accompanied by peculiar morphological changes, such as witches' broom and phyllody? Both symptoms increase the prevalence of short branches and small young leaves, which are preferred by sap-feeding insects. Moreover, phyllody flowers remain green even when healthy flowers wither. These features are likely to attract insect vectors and thus the spread of phytoplasmas. Such manipulations of the morphology of host plants appear to be a common strategy for the survival of phytoplasmas. Thus, SAP54 has been proposed to act as a link between two key pathways of the host to alter development resulting in sterile flowers/ inflorescences (MacLean et al. 2011, 2014, Maejima et al. 2014a).

The main determinant of altered flower development is SAP54 (MacLean et al. 2014, Rümpler et al. 2015). A close homolog of SAP54 was characterized in parallel in another investigation and termed PHYLLOGEN1 (PHYL1). The developmental reprogramming relies on specific interactions of SAP54 with a relatively small subset of MIKC-type MADS-domain TFs. Thereby, SAP54 destine the MADS-domain TFs for degradation via the ubiquitin/26S proteasome pathway. Intriguingly, SAP54 interacts, among others, with AP1 and SEP proteins. MIKC-type MADS-domain TFs consist of a DNA-binding MADS-domain (M), an Intervening domain (I), a Keratin-like domain (K), and a C-terminal domain (C) (Kaufmann et al. 2005). It is this characteristic domain structure from which the term MIKC was derived. The phytoplasma infected 'zombie plants' and the floral homeotic mutants are thus not only morphologically similar but their phenotypes may also be brought about by convergent molecular mechanisms that all eventually lead to the depletion of floral homeotic protein complexes that determine organ identity (Maejima et al. 2014b, Rümpler et al. 2015).

It should also be highlighted that phytoplasmas are not only out of the ordinary for reprograming plant development but they also bear one of the smallest bacterial genomes known to date (Oshima et al. 2004, Rümpler et al. 2015). Despite being so small, the genome harbors a significant number of repetitive sequences that can make up some 20% of the genomic DNA (Oshima et al. 2004, Bai et al. 2006). Most of this repetitive DNA is organized in few relatively large 'potential mobile units' (PMUs) that possess features of transposons. Phytoplasmas display a high degree of genome plasticity and PMUs probably play a vital role in generating and maintaining this plasticity (Bai et al. 2006, Sugio and Hogenhout 2012). Intriguingly, the gene encoding SAP54 is part of a PMU. It is thus tempting to speculate that the evolution of SAP54 and its location in a PMU are causally linked. However, the molecular mechanisms that contributed to the origin of SAP54 remain to be determined (Rümpler et al. 2015).

Rümpler et al. (2015) claimed that the degradation of some floral homeotic proteins may not be the only mechanism by which phytoplasmas bring about aberrant floral phenotypes, even though SAP54 orthologs might be involved. A recent study provided evidence that a SAP54 ortholog inhibits the expression of a microRNA (miR396), which inhibits the translation of SHORT VEGETATIVE PHASE (SVP), yet another MADS-domain protein. Therefore, in phytoplasma-infected plants, SVP is upregulated, which may also contribute to abnormal flower development (Yang et al. 2015).

The Small Phytoplasma Virulence Effectors Contain Distinct Domains Required for Nuclear Targeting (SAP11)

The phytoplasma effectors SAP11 enhance the proliferation of axillary meristems, which is responsible for the witches' broom symptom (Figure 1) observed in phytoplasma-infected plants of *A. thaliana* TCP (TEOSINTE-BRANCHED, CYCLOIDEA, PROLIFERATION FACTOR 1 and 2) transcription factors (TFs). The TCP TFs are known to repress branching and mediate leaf development. Thus, SAP11 results in induction of axillary branches with crinkled leaves (Sugio et al. 2014). This small (10 kDa) virulence effector SAP11 of Aster Yellows phytoplasma strain Witches' Broom (AY-WB) binds and destabilizes *Arabidopsis* CIN (CINCINNATA) TCP (TEOSINTE-BRANCHED, CYCLOIDEA, PROLIFERATION FACTOR 1 AND 2) transcription factors, resulting in dramatic changes in leaf morphogenesis and increased susceptibility to

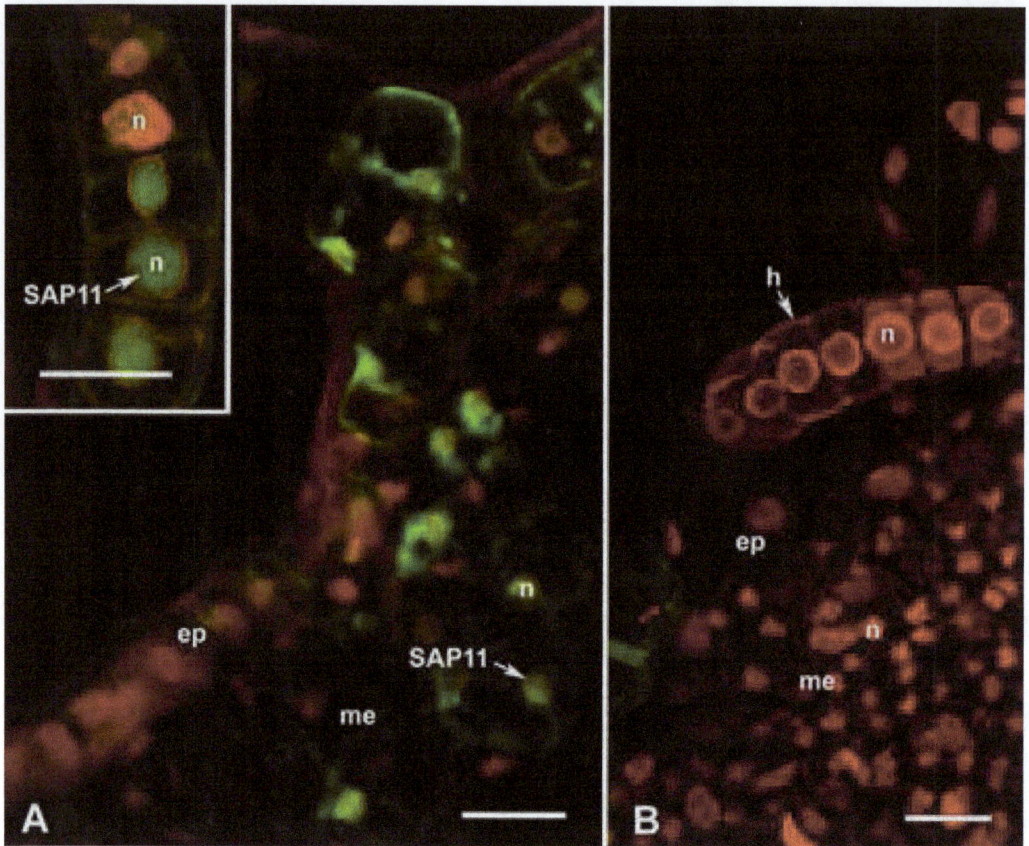

Figure 1. Aster yellows phytoplasma strain witches' broom (AY-WB) protein SAP11 located to the nuclei of plant cells beyond the phloem. A and inset: SAP11 detected in the nuclei (n), with an Alexa-fluor-conjugated antibody (green fluorescence), in sections of AY-WB-infected China aster leaves. Inset shows the detection of SAP11 in nuclei of a trichome. B: Sections from healthy China aster leaves that were processed in the same way as A (negative control). In both A and B, propidium iodide (red fluorescence) specifically stains nucleic acids. Hence, nuclei fluoresce bright red. Abbreviations: ep, epidermis; h, leaf hair; me, mesophyll. Bars = 20 μm (Bai et al. 2009).
Available at: https://apsjournals.apsnet.org/doi/pdf/10.1094/MPMI-22-1-0018

phytoplasma insect vectors. SAP11 contains a bipartite nuclear localization signal (NLS) that targets this effector to plant cell nuclei (Sugio et al. 2011a, b). In addition to their effect on the proliferation of axillary meristems, SAP11 down-regulates the synthesis of jasmonic acid (JA), resulting in flower sterility. Sugio et al. (2011a) proved that SAP11 binds to and destabilizes class II CIN-TCPs, but not class I PCF-TCPs, while Chang et al. (2018) characterized that these effectors destabilize class II TB/CYC-TCPs, the key factors controlling axillary meristem development. TB/CYC-TCPs that act as negative regulators to control axillary meristem development have been discovered in different plant species (Nicolas and Cubas 2016). Among these TB/CYC-TCPs, TB1/BRC1 is considered a central integrator of multiple endogenous and environmental pathways that modulate axillary bud outgrowth (Rameau et al. 2015, Teichmann and Muhr 2015). Up to the results published by Chang et al. (2018), only two phytoplasma-secreted virulence factors, SAP11 and TENGU, have been identified to enhance the proliferation of axillary meristems. SAP11 effectors are putatively present in a range of phylogenetically distant phytoplasmas, including the 16SrI, 16SrII, 16SrIII, 16SrIX, 16SrX, and 16SrXII groups (Figure 2); these effectors presented great ability to destabilize class II TB/CYC-TCPs (Figure 3).

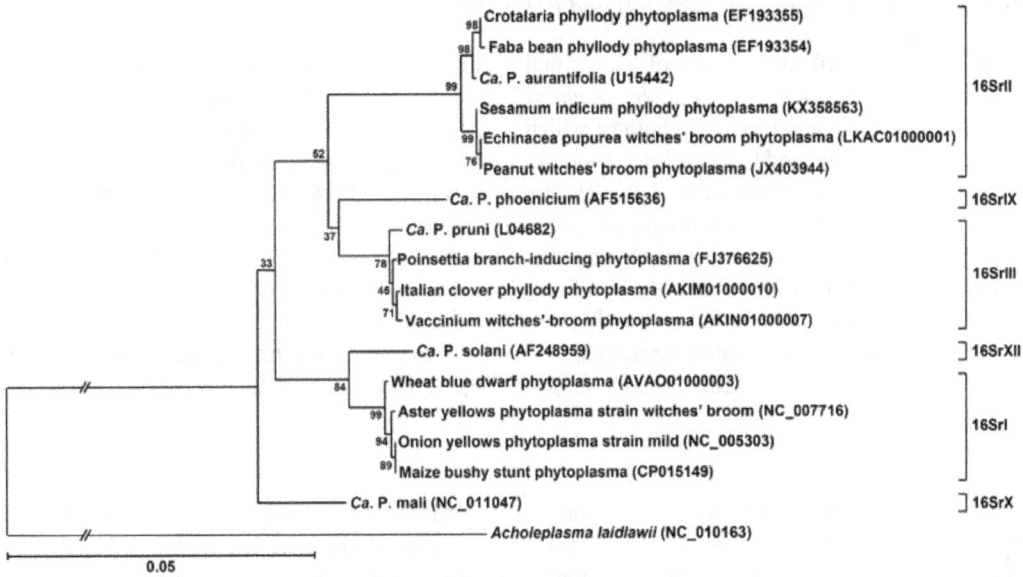

Figure 2. A phylogenetic tree was constructed based on the comparison of 16S rRNA gene sequences of phytoplasmas containing putative SAP11 homologues. The neighbor-joining method followed by bootstrap analysis was used with MEGA 7.0 software. *A. laidlawii* served as an outgroup. The numbers at the branch points are bootstrap values that represent the percentages of replicate trees based on 1000 repeats (Chang et al. (2018)).
Available at: https://academic.oup.com/jxb/article/69/22/5389/5085359.

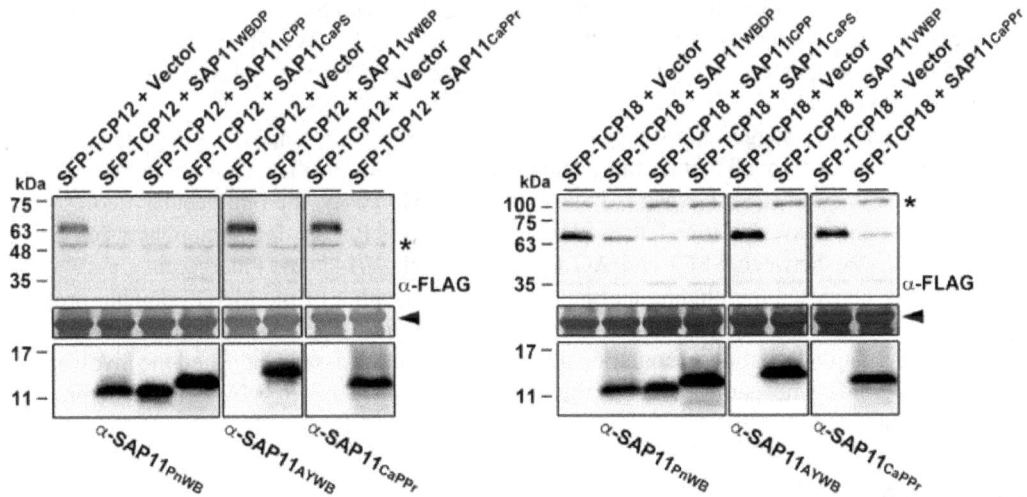

Figure 3. Co-expression assays of SAP11 effector-mediated destabilization of class II TB/CYC-TCP transcription factors. Co-expression assays were conducted in *N. benthamiana* through agroinfiltration. The relative abundance levels of Arabidopsis TCP12 and TCP18 were examined in the presence of SAP11 effectors. Western blotting was conducted to examine the expression levels of FLAG-tagged TCPs (upper panel) and SAP11 effectors (lower panel) using a monoclonal antibody against FLAG tag and polyclonal antibodies against SAP11 effectors. SAP11WBDP, SAP11ICPP, and SAP11CaPS could be recognized by a-SAP11PnWB, and SAP11VWBP could be recognized by a-SAP11AYWB. As a loading control, the large subunit of RuBisCO visualized with Coomassie Brilliant Blue staining is indicated by the arrowhead (middle panel). Non-specific bandings recognized by antibodies are indicated by asterisks (Chang et al. (2018)).
Available at: https://academic.oup.com/jxb/article/69/22/5389/5085359.

Protein in Malus Expressed 2 (PME2) and PM19_00185

This is a putative effector protein of 'Ca. P. mali' infecting *Malus* x *domestica* that was expressed in the mesophyll protoplasts of *N. benthamiana* and *N. occidentalis*. Two genetic variants were detected, PME2AT and PME2ST, both of which translocated to the nucleus of the host plant cells. PME2ST was recognized to interfere in protoplast integrity. PME2 was implied to moderate in 'Ca. P. mali' virulence (Mittelberger et al. 2019). The protein PM19_00185 of 'Ca. P. mali' strain PM19 has been reported to interact with six ubiquitin conjugating enzymes belonging to three different UBC groups of *A. thaliana*. It was suggested to have an E3 ligase activity possibly concerned in transferring the ubiquitin chain to the specific target. The E3 ligase activity of PM19_00185 was also observed when tested with E2 conjugating enzyme of *Malus domestica*. This effector also conduces suppression of basal defense response in *A. thaliana* during phytoplasma pathogenesis, thus enhancing the susceptibility to the pathogens (Strohmayer et al. 2019).

PHYL1

This effector depicts 88% amino acid similarity with SAP54. A 125 amino acid protein is encoded by PHYL1 gene which contains a 34 amino acid signal peptide at the N-terminal. The 91 amino acid mature protein (10.6 kDa) is secreted from the phytoplasma into the host cell. The expression profile of PHYL1 was detected to be threefold higher in the infected plants than in insects. Maejima et al. (2014a) pointed at the significance of the 8th amino acid at the N-terminal of the mature protein in inducing alterations in the floral phenotype in diseased plants. PHYL1 has been announced to interact with MADS domain containing proteins, namely, SEP3, AP1 and CAL. Even the truncated mutant PHYL8 interacted with the three MADS box proteins, indicating that the eight amino acids at the N-terminal of PHYL1 are not essential for its interaction with SEP3, AP1 and CAL. However, no interaction was observed between PHYL1 and WUSCHEL (WUS), a transcription factor containing a homeodomain required for maintaining the integrity of shoot and floral meristems (Laux et al. 1996).

PHYL1 has been suggested to impede the functioning of SEP3 and AP1 genes. Quantitative RT-PCR experiments proved that in transgenic plants expressing PHYL1, their ability to modulate the expression of downstream genes is highly impaired. Moreover, the bimolecular fluorescence complementation assay confirmed that PHYL1 interferes with the interaction between SEP3 and AP1 and also between SEP3 and AG (Maejima et al. 2014a, b). Fluorescence assays and Western blots suggested the binding and induction of degradation of the MADS domain proteins by PHYL1 and also highlighted the importance of N-terminus eight amino acids of PHYL1 in inducing degradation. Immunoprecipitation and immuno-blotting assays indicated the involvement of ubiquitin-proteasome pathway in PHYL1-mediated degradation of MADS domain proteins (Maejima et al. 2014a).

Conclusions

Phytoplasma effectors have a crucial role in re-programming plant cellular processes leading to the induction of specific plant phenotypes. These effectors modulate the behaviors of phytoplasma insect vectors. Current outcomes reported in many papers indicate that admittedly, the modelling approach will enable the dissection of the mechanisms involved in phytoplasma epidemics, but using experimental data of how phytoplasma effectors change plant development and plant interactions with insect vectors, it is possible to model the impacts of phytoplasma effector genes on the wider environment, including how the changes in insect vector behaviors may contribute to the spread of phytoplasma to other plants within the field, to neighboring fields, and across longer distances. This can be achieved through multi-layered mechanistic modeling. In mechanistic models, relationships between the variables in the data set are specified in terms of the biological processes that are thought to have given rise to the data, in contrast to phenomenological and statistical models, where the relationship seeks to best describe the data.

References

Aurin, M.B., M. Haupt, M. Görlach, F. Rümpler and G. Theißen. 2020. Structural requirements of the phytoplasma effector protein SAP54 for causing homeotic transformation of floral organs. Mol. Plant-Microbe In. 33(9): 1129–1141. https://doi.org/10.1094/MPMI-02-20-0028-R.

Bai, X.D., X. Bai, J. Zhang, A. Ewing, S.A. Miller, A. Jancso Radek et al. 2006. Living with genome instability: The adaptation of phytoplasmas to diverse environments of their insect and plant hosts. J. Bacteriol. 188: 3682–3696. https://doi.org/10.1128%2FJB.188.10.3682–3696.2006.

Bai, X., V.R. Correa, T.Y. Toruño, D. Ammar, S. Kamoun and S.A. Hogenhout. 2009. AY-WB phytoplasma secretes a protein that targets plant cell nuclei. Mol. Plant Microbe Int. 22: 18–30. doi:10.1094/MPMI -22-1-0018.

Bertaccini, A. 2007. Phytoplasmas: Diversity, taxonomy, and epidemiology. Front Biosci. 2: 673–689.

Christensen, N.M, K.B. Axelsen, M. Nicolaisen and A. Schulz. 2005. Phytoplasmas and their interactions with hosts. Trends Plants Sci. 10: 526–535.

Cline, M. 1997. Concepts and terminology of apical dominance. Am. J. Bot. 84: 1064–1069.

Chang, S.H., C.M. Tan, C.T. Wu, T.H. Lin, S.Y. Jiang, R.C. Liu et al. 2018. Alterations of plant architecture and phase transition by the phytoplasma virulence factor SAP11. J. Exp. Bot. 69: 5389–5401. doi:10.1093/jxb/ery318.

Hoshi, A., K. Oshima, S. Kakizawa, Y. Ishii, J. Ozeki, M. Hashimoto et al. 2009. A unique virulence factor for proliferation and dwarfism in plants identified from a phytopathogenic bacterium. P. Natl. Acad. Sci. 106(15): 6416–6421. https://doi.org/10.1073/pnas.0813038106.

Hogenhout, S.A., K. Oshima, E.D. Ammar, S. Kakizawa, H.N. Kingdom and S. Namba. 2008. Phytoplasmas: Bacteria that manipulate plants and insects. Mol. Plant Pathol. 9: 403– 423 DOI :101111J136 4–3703200800472 X.

Kaufmann, K., R. Melzer and G. Theißen. 2005. MIKC-type MADS-domain proteins: Structural modularity, protein interactions and network evolution inland plants. Gene 347: 183–198 https://doi.org/10.1016/j.gene.2004.12.014.

Laux T., K.F. Mayer J. Berger and G. Jurgens. 1996. The WUSCHEL gene is required for shoot and floral meristem integrity in Arabidopsis. Development 122(1): 87–96.

MacLean, A.M., A. Sugio, O.V. Makarova, K.C. Findlay, V.M. Grieve and R. Tóth. 2011. Phytoplasma effector SAP54 induces indeterminate leaflike flower development in Arabidopsis plants. Plant Physiol. 157: 831–841. doi: 10.1104/pp.111.181586.

MacLean, A.M., Z. Orlovskis, K. Kowitwanich, A.M. Zdziarska, G.C. Angenent, R.G. Immink et al. 2014. Phytoplasma effector SAP54 hijacks plant reproduction by degrading MADS-box proteins and promotes insect colonization in a RAD23-dependent manner. Plos Biology 12(4): e1001835. https://doi.org/10.1371/journal.pbio.1001835.

Maejima, K., R. Iwai. M. Himeno, K. Komatsu, Y. Kitazawa, N. Fujita et al. 2014a. Recognition of floral homeotic MADS domain transcription factors by a phytoplasmal effector, phyllogen, induces phyllody. Plant J. 78(4): 541–554. doi: 10.1111/tpj.12495.

Maejima, K., K. Oshima and S. Namba. 2014b. Exploring the phytoplasmas, plant pathogenic bacteria. J. Gen. Plant Pathol. 80(3): 210–221. DOI 10.1007/s10327-014-0512-8.

Maejima, K., Y. Kitazawa, T. Tomomitsu, A. Yusa, Y. Neriya, M. Himeno et al. 2015. Degradation of class E MADS-domain transcription factors in Arabidopsis by a phytoplasmal effector, phyllogen. Plant Signal. Behav. Published online July 15, 2015.http://dx.doi.org/10.1080/15592324.2015.1042635.

Minato, N., M. Himeno, A. Hoshi, K. Maejima, K. Komatsu, Y. Takebayashi et al. 2014. The phytoplasmal virulence factor TENGU causes plant sterility by down regulating of the jasmonic acid and auxin pathways. Sci. Rep. 4: 7399. DOI: 10.1038/srep07399.

Mittelberger, C., H. Stellmach, B. Hause, C. Kerschbamer, K. Schlink, T. Letschka et al. 2019. A novel effector protein of apple proliferation phytoplasma disrupts cell integrity of *Nicotiana* spp. protoplasts. Int. J. Mol. Sci. 20(18): 4613. https://doi.org/10.3390/ijms20184613.

Mori, Y., T. Nishimura and T. Koshiba. 2005. Vigorous synthesis of indole-3-acetic acid in the apical very tip leads to a constant basipetal flow of the hormone in maize coleoptiles. Plant Sci. 168: 467–473. https://doi.org/10.1016/j.plantsci.2004.09.010.

Nicolas, M. and P. Cubas. 2016. The role of TCP transcription factors in shaping flower structure, leaf morphology, and plant architecture. In: Gonzalez DH, ed. Plant transcription factors. Boston: Academic Press, 249–267.

Oshima, K., S. Kakizawa, H. Nishigawa, H.Y. Jung, W. Wei, S. Suzuki et al. 2004. Reductive evolution suggested from the complete genome sequence of a plant–pathogenic phytoplasma. Nat. Genet. 36: 27–29.

Oshima, K., K. Maejima and S. Namba. 2013. Genomic and evolutionary aspects of Phytoplasma. Front. Microbiol. 4: 230. https://doi.org/10.3389/fmicb.2013.00230.

Rameau, C., J. Bertheloot, N. Leduc, B. Andrieu, F. Foucher and S. Sakr. 2015. Multiple pathways regulate shoot branching. Front. Plant Sci. 5: 741. https://doi.org/10.3389/fpls.2014.00741.

Rümpler, F., L. Gramzow, G. Theißen and R. Melzer. 2015. Did convergent protein evolution enable phytoplasmas to generate 'zombie plants'? Trends Plant Sci. 20(12): 798–806. https://doi.org/10.1016/j.tplants.2015.08.004.

Saccardo, F., M. Martini, S. Palmano, P. Ermacora, M. Scortichini, N. Loi et al. 2012. Genome drafts of four phytoplasma strains of the ribosomal group 16SrIII. Microbiology 158(11): 2805–2814. doi: 10.1099/mic.0.061432-0.

Strohmayer, A., M. Moser, A. Si-Ammour, G. Krczal and K. Boonrod. 2019. 'Candidatus Phytoplasma mali' genome encodes a protein that functions as a E3 ubiquitin ligase and could inhibit plant basal defense. Mol. Plant Microbe In. 32(11): 1487–1495. https://doi.org/10.1094/MPMI-04-19-0107-R.

Sugio, A., A.M. MacLean, H.N. Kingdom, V.M. Grieve, R. Manimekalai and S.A. Hogenhout. 2011a. Diverse targets of phytoplasma effectors: from plant development to defense against insects. Annu. Rev. Phytopathol. 49: 175–195. doi:10.1146/annurev-phyto-072910-095323.

Sugio, A., H.N. Kingdom, A.M. MacLean, V.M. Grieve and S.A. Hogenhout. 2011b. Phytoplasma protein effector SAP11 enhances insect vector reproduction by manipulating plant development and defense hormone biosynthesis. P. Natl. Acad. Sci., USA 108: E1254–E1263. https://doi.org/10.1073/pnas.1105664108.

Sugio, A. and S.A. Hogenhout. 2012. The genome biology of phytoplasma: Modulators of plants and insects. Curr. Opin. Microbiol. 15: 247–254. https://doi.org/10.1016/j.mib.2012.04.002.

Sugio, A., A.M, MacLean and S.A. Hogenhout. 2014. The small phytoplasma virulence effector SAP 11 contains distinct domains required for nuclear targeting and CIN-TCP binding and destabilization. New Phytologist. 202(3): 838–848. doi: 10.1111/nph.12721.

Teichmann, T. and M. Muhr. 2015. Shaping plant architecture. Front. Plant Sci. 6: 233. https://doi.org/10.3389/fpls.2015.00233.

Yang, C.Y., Y.H. Huang, C.P. Lin, Y.Y. Lin, H.C. Hsu, C.N. Wang and S.S. Lin. 2015. MicroRNA396-targeted SHORT VEGETATIVE PHASE is required to repress flowering and is related to the development of abnormal flower symptoms by the phyllody symptoms1 effector. Plant Physiol. 168(4): 1702–1716. https://doi.org/10.1104%2Fpp.15.00307.

Section III
Bacterial Genetics

CHAPTER 10

Genetic Methods of Identification, Classification, and Differentiation of Bacteria

Agata Janczarek and Anna Gałązka*

Introduction

Microbiology is a biological science that studies microorganisms: bacteria, fungi, algae, parasites and viruses (Sandle 2016). These microorganisms are associated with fields such as biotechnology, food science, medicine, and genetic engineering. Their unique properties enable the production of food, antibiotics, hormones, and other therapeutic compounds. Moreover, microorganisms carry out the decomposition of various components, for example, lignocellulosic biomass into ethanol or biogas. On the other hand, some genetic traits and biochemical abilities of microorganisms can prove dangerous to human health and industry (Franco-Duarte et al. 2019).

In 1677, Antonie van Leeuwenhoek published his first work titled "Letter on protozoa" which contained the first detailed description of bacteria from various environments (Lane 2015). Except for soil, water, and the external environment, bacteria live within plants and animals, entering into symbiotic and parasitic relationships with them. They are a component of the human microbiome (Venkova et al. 2018). The number of eukaryotic cells in relation to the bacterial symbionts of the human body is only 10 percent (Dykhuizen 2005).

Bacteria play an important role in the environment and human life. They have found wide industrial applications. They are used in the synthesis of enzymes, organic acids, vitamins, antibiotics, antibodies, hormones, carotenoids, steroids, alkaloids, interferons and vaccines. They take part in wastewater treatment, biorecycling of precious metals and metalloids, biocontrol, production of prebiotics and probiotics, and environmentally friendly production of nanomaterials (Akoijam et al. 2022). In the environment, bacteria are responsible for the health and condition of plants. They carry out processes of bioremediation of soil contaminants, decomposition of organic matter and improving soil structure (Sankari et al. 2021, Sobiczewski 2021). In addition to bacteria that have positive effects on other organisms and the environment, there are also pathogenic bacteria. Avoidance of vaccinations by people and increasing resistance to antimicrobial preparations have made it important to monitor their epidemiology (Venkova et al. 2018). Growing antibiotic resistance, emerging food and water contamination, and nosocomial infections are serious threats to humans (Paczesny and Mierzejewski 2021). Due to the high variety of functions, potential applications, and the threat of bacteria, new and effective methods for their detection and identification are being sought.

Phenotypic identification of microorganisms involves comparing morphological, physiological and biochemical characteristics. Morphological features of the studied microorganisms such as cell

---

Department of Agricultural Microbiology, Institute of Soil Science and Plant Cultivation-State Research Institute (IUNG-PIB), Czartoryskich 8, 24–100 Puławy, Poland.

* Corresponding author: agalazka@iung.pulawy.pl

shape, color, dimension and colony form are learned by macroscopic and microscopic evaluation (light microscopy and electron microscopy). For phenotypic identification, information about oxygen demand, sensitivity to pH value and temperature are needed. Among biochemical characteristics, enzymatic activity, gas production and compound metabolism are important (Donelli et al. 2013). These methods are widely used in laboratories because of their low cost. However, it should be remembered that they require a lot of time and work. Additionally, the phenotype expression of microorganisms is affected by the composition of the medium and culture conditions. The small number of characterized microorganism limits the use of the Phenetic Classification Database which have a larger clinical application than industry application. It is also difficult to differentiate phenotypically similar species (Lee et al. 2012, Sandle 2014, Buszewski et al. 2017). In contrast, methods of genotypic identification of microorganisms based on differences in genomes are accurate and fast which have enriched numerous databases of microorganisms. They are increasingly becoming an alternative or complement to phenotypic methods (Tang et al. 1998, Sandle 2014, Brzozowski et al. 2017).

One of the most popular techniques that support the genetic identification of microorganisms is polymerase chain reaction (PCR). During this method, DNA or RNA fragments are exponentially amplified using a thermostable DNA polymerase (Bzducha 2007, Al-Obaidi et al. 2018). To perform PCR, there is a need of DNA templates of the typed organisms, the template DNA whose fragment is to be duplicated, two oligonucleotide primers and four nucleotides (deoxyribonucleoside triphosphates) (Al-Obaidi et al. 2018). The PCR reaction consists of thirty cycles, with each cycle consisting of three steps. The first step is denaturation. Hydrogen bonds are broken and the DNA helix is separated into single strands by applying high temperatures (92–96°C). Annealing of primers can occur at 37–72°C. The choice of temperature depends on the melting point of the primers. Elongation occurs at 72°C and begins at the 3' end of the primer (Adaszek et al. 2018).

Methods for Genotypic Identification of Bacteria

Based on Brzozowski et al. (2017), genotyping of bacteria can be made by two groups of methods. The first one does not use nucleic acid sequencing and the second one uses sequencing (Figure 1). Among the methods that do not use sequencing include REA-PFGE (Restriction Enzyme Analysis-Pulsed Field Gel Electrophoresis). This analysis begins by isolating DNA and then digesting it with a rare-cutting restriction enzyme. The DNA fragments are then placed in an agarose gel and electrophoresed in a plus electric field, which results in a periodic change of the electric field by a specific angle. The molecules in the gel change conformation and reorientation so that they move. The migration rate of DNA molecules in the gel depends on their size (Adzitey et al. 2013, Khosravi et al. 2014, Simner et al. 2015, Brzozowski et al. 2017). Longer DNA fragments need more time to reorient which leads to their slower migration. Other factors affecting the migration rate are the electric field strength, agarose gel composition, electrophoresis time, temperature, angle of electric field reversal and the composition of the buffer used for electrophoresis (Goering 2010). As a result of the analysis, electrophoretic bands are obtained, which are used to determine bacterial clonal assignment. Computer programs are used to assess bacterial relatedness. The criteria state that the presence of two to three band differences indicates close affinity, the presence of four to six band differences indicates probable affinity and seven or more band differences indicates no affinity (Neoh et al. 2019). The PFGE method has found applications in the identification of bacteria from environmental or clinical samples, identification of antibiotic-resistant strains, etiological studies, classification of bacteria, and even in determining affinity between different strains of a single species (Parizad et al. 2016). PFGE gives stable and reproducible DNA patterns. But it requires a lot of time to perform it and the DNA patterns can vary depending on the technique used (there are several types of PFGE such as FIGE, CHEF, OFAG, TAFE, PHOGE and PACE). Moreover, bands of the same size can come from different parts of the chromosome (Parizad et al. 2016, Ochoa-Díaz et al. 2017).

```
                                                    ┌─── REA-PFGE

                                                    ├─── RAPD

                                     ┌─ No using ───┼─── RFLP
                                     │  sequencing  │
                                     │              ├─── AFLP
                                     │              │
 ┌───────────────────────────┐      │              ├─── MLVA
 │ Genotypic identification   ├──────┤              │
 │ of bacteria                │      │              └─── Microarray DNA
 └───────────────────────────┘      │
                                     │              ┌─── MLSTA and MLSA
                                     └─ Using ──────┤
                                        sequencing  └─── WGS
```

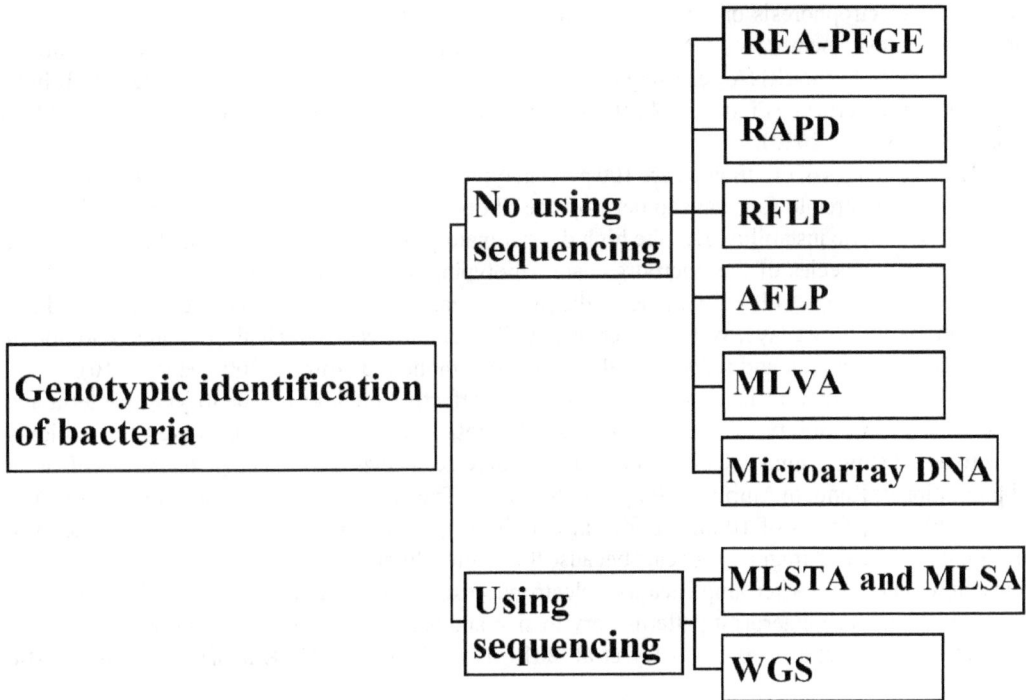

Figure 1. Genetic methods for identifying bacteria using and not using sequencing (own elaboration).

RFLP (Restriction Fragments Length Polymorphism) is an inexpensive and sensitive method to identify or differentiate bacterial strains (Adzitey et al. 2013). The method looks for small and specific differences in the sequence of double-stranded DNA. The nucleic acid is cut into smaller fragments at specific locations by restriction endonucleases. The next step is electrophoresis. A sequence-specific array of bands is formed. The arrangement of the bands depends on the length of the DNA fragments (Mittal et al. 2013, Chatterjee and Raval 2019). Then, DNA is transferred from the agarose gel to nitrocellulose or nylon membranes. The DNA is hybridized with specially prepared probes. This technique enables the distinction of closely related bacterial strains and the determination of interspecies variability (Krishna et al. 2019). That method is a modification of RFLP, namely PCR-RFLP, and can also be used to identify bacteria. This technique differs from RFLP by carrying out DNA amplification preceded by cutting with endonucleases. The amplified sequences are polymorphic; therefore, different restriction patterns are produced after electrophoretic separation. This allows RFLP-PCR to identify strains of microorganisms (Tabit 2016, Brzozowski et al. 2017, Hashim and Al-Shuhaib 2019). The results are analyzed by comparing the fingerprints for a given set of tested isolates (Säde and Björkroth 2014). The effectiveness of RFLP has been proved by Chowdhury et al. (2010), who conducted studies using this method to evaluate the diversity of *Vibrio cholerae* strains, while Qi et al. (2016) used RFLP to study the taxonomy of *Yersinia pestis*. This method is difficult to perform and time-consuming because large amounts of DNA are used and it is not possible to find single base changes (Adzitey et al. 2013).

Performing identification and characterization of bacteria is possible through the ribotyping method, which uses rRNA-based phylogenetic analysis. This is one of the most popular variations of the RFLP-PCR technique. The ribosomal RNA of bacterial cells consists of 23S, 16S and 5S molecules that vary in size. The high conservativeness of DNA encoding rRNA makes it possible to learn the evolutionary line of bacterial species, classify bacteria and establish taxonomy, conduct epidemiological studies and learn about population biology based on differences in the 16S rRNA gene (Austin and Pagotto 2003, Łyszcz and Gałązka 2017, Franco-Duarte et al. 2019). Ribotyping begins with cell lysis and extraction of genomic DNA. The DNA digested by restriction enzymes is used for electrophoresis on an agarose gel. After separation, the DNA is transferred to a nylon or nitrocellulose membrane. DNA probing is then carried out using labeled DNA fragments complementary to the rRNA gene sequence. DNA fragments containing the rRNA gene will light up to form a fingerprint (Figure 2) (Austin and Pagotto 2003, Hata 2010, Franco-Duarte et al. 2019, Qurban and Ameen 2020).

The advantages of this method are 100 percent typicality and high reproducibility. Disadvantages are that it takes three to four days to perform the ribotyping procedure, it is a complex method and sensitive to genetic instability. Despite high discriminatory power (ability to distinguish) among the species and subspecies of the microorganism, ribotyping shows low discriminatory power at the strain level (Adzitey et al. 2013). Due to the labor-intensive nature of ribotyping, the method has been automated. Nowadays, systems such as RiboPrinter® are used, which allow standardization of results, high throughput and reduced analysis time (Grimont and Grimont 2001, Hata 2010).

RAPD has shown to be a convenient and cost-effective method for identifying bacteria and estimating the variability of microorganisms. This test has more than once proved to be highly effective in distinguishing bacterial strains among diverse species (Saxena et al. 2014). RAPD is an abbreviation of Random Amplified Polymorphic DNA. The method is based on low-restriction PCR using arbitrary primers of 10 nucleotides in length. Amplification from a few to several dozen of different nucleic acid fragments occurs because the primers hybridize at low temperatures at multiple sites simultaneously. After amplification, electrophoresis is made on agarose gel and stained with ethidium bromide. Fingerprint patterns vary in size and are specific to a particular bacterial strain (Figure 2) (Baker et al. 2002, Adzitey et al. 2013, Donelli et al. 2013, Kumari and Thakur 2014,

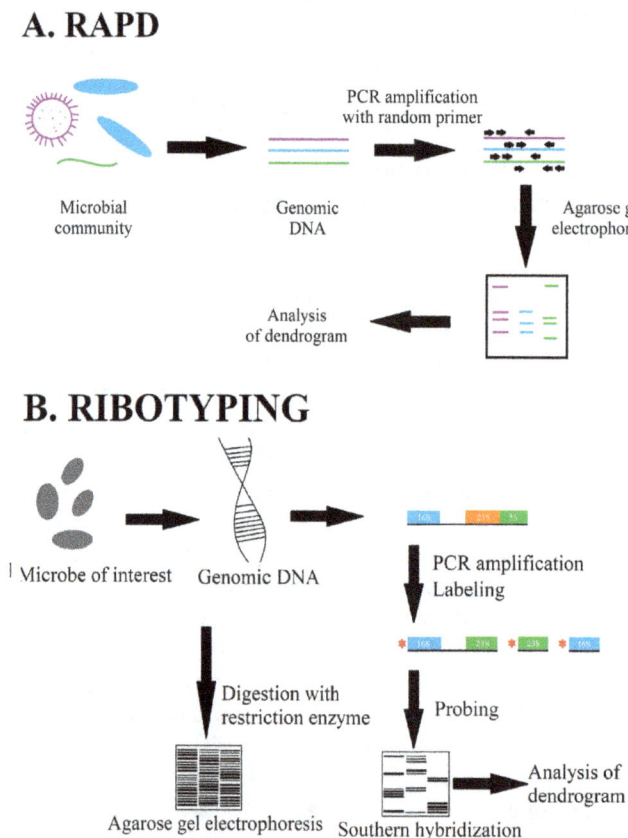

A. RAPD

B. RIBOTYPING

Figure 2. Selected methods of genetic identification of bacteria. A. Random Amplified Polymorphic DNA (RAPD) and B. Ribotyping.

Łyszcz and Gałązka 2017). Changes in the number of components and the reaction conditions affect the results of the analysis. When typing unknown strains of a given species, one should remember about the optimal selection of these factors (Stefańska et al. 2022). The quality of the genome of the analyzed sample also affects the results (Matsumoto et al. 2022).

The natural habitat of *Alicyclobacillus* bacteria is soil. These bacteria are also found in fruits and in hot springs. *Alicyclobacillus* spores are capable of surviving under standard pasteurization conditions by contaminating food in the fruit industry. They are often isolated from spoiled fruit juices (Dąbrowska and Kunicka-Styczyńska 2018). Despite their lack of negative effects on human health, they can contribute to economic losses (Sokołowska et al. 2014). Assessment of the genetic diversity of bacteria of the genus *Alicyclobacillus* can be carried out through the RAPD method. Yamazaki et al. (1997) tested a group of 42 primers from which they selected three primers for distinguishing strains belonging to the species *A. acidoterrestris* from *A. acidocaldarius, Alicyclobacillus acidoterrestris* from *Alicyclobacillus acidocaldarius* and other related bacteria. Groenewald et al. (2009) used these primers to differentiate *Alicyclobacillus* strains isolated from orchard soil and processing plant environments (Dekowska 2011).

In the Amplified Fragment Length Polymorphism (AFLP) method, fragmentation of genomic DNA is carried out using two restriction endonucleases- a rare-cutting enzyme and a frequent-cutting enzyme. After DNA digestion, oligonucleotide adaptors are ligated, then sets of restriction fragments are selectively amplified. By using two different enzymes, the number of fragments used for amplification can be decided (Vos et al. 1995, Blears et al. 1998, Wang et al. 2015). The prepared fragments are electrophoretically separated resulting in a characteristic fingerprint (Meudt and Clarke 2007). Two options can be used for visualization. Automated genome analyzers allow the characterization of several samples simultaneously using fluorescent labeling (Hata 2010, Bertani et al. 2019, Shen 2019). The AFLP technique has high reproducibility and typing ability but is time-consuming and requires large financial outlays (Adzitey et al. 2013). Applied Biosystems AFLP Microbial Fingerprinting Kit has been widely used for AFLP analysis over the past years. The adaptor digestion and ligation step were performed at the same time, the dilution of digested-bound DNA fragments used as templates in the first amplification reaction was reduced, and the number of PCR cycles was increased, which shortened the analysis time and more accurate results were obtained (Bertani et al. 2019). The method is used to assess genetic diversity within a species or between closely related species, infer phylogeny at the population level and biogeographic patterns, generate genetic maps and determine relatedness between varieties. Although AFLP shows higher reproducibility (compared to RAPD), one should remember to test reproducibility for each batch of reactions (Paun and Schönswetter 2012).

Technological advances have made it possible to use the DNA microarray technique to identify and isotype bacteria (Al-Khaldi et al. 2012). The technique is particularly applicable to the analysis of large numbers of samples of both clinical and environmental origin. Microarray work is based on the hybridization reaction of two complementary DNA sequences (Al-Khaldi et al. 2012, Ranjbar et al. 2017). DNA microarray, also called CHIP DNA, is formed by immobilizing single-stranded oligonucleotide probes or cDNA probes on a plate.There can be thousands to hundreds of thousands of probes on a single DNA microarray. A labeled sample is applied to the plates and scanned with a laser reader. A set of luminescent dots is received, for which the value of fluorescence intensity is determined. The results obtained are subjected to bioinformatics analysis (Gentry and Zhou 2006, Łyszcz and Gałązka 2016, Ranjbar et al. 2017).

The usefulness of using DNA microarrays is limited by the variability of the data and the high cost of analysis (Al-Khaldii et al. 2012). It is important to remember that according to the principles of hybridization kinetics, the measured signal is linear only over a limited concentration range of species hybridizing to the array. If the DNA sequences are similar, it is difficult to select a set of probes in which the DNA sequences will not bind to the same probe. Furthermore, the DNA array is designed to detect specific sequences, so the lack of probes complementary to the sequences found in the hybridized solution is associated with a lack of identification of some species (Bumgarner 2013).

Bacterial genes and intergenic regions can be equipped with DNA repeat loci that are different between strains in the number of repeat units or primary structure. These are called the variable number of tandem repeat regions (VNTRs). Using knowledge of the *Haemophilus influenzae* genome sequence, a method has been developed to analyze the variable number of tandem repeats - Multiple Locus VNTR analysis (MLVA) (Van Belkum 2007). The method is inexpensive, fast and not technically demanding. It was first designed for typing *H. influenzae, M. tuberculosis, B. anthracis* and *Y. pestis/Haemophilus influenzae, Mycobacterium tuberculosis, Bacillus anthracis* and *Yersinia pestis* but over time the potential for genotyping other bacterial species has been demonstrated (Haguenoer et al. 2011, Guinard et al. 2017). MLVA begins by running PCR amplification of repetitive and flanking regions. The primers used in the PCR reaction are fluorescently labeled. The next step is capillary electrophoresis to segregate amplicons by size (Nadon et al. 2013, Van Stelten and Nightingale 2014, Helldal et al. 2017). Differences in the length of different VNTR sequences are used to determine the strain of the microorganism under study. This method has found application in the identification of rapidly evolving bacteria, as VNTR regions show a high potential for storing changes (Wang et al. 2015, Brzozowski et al. 2017). Apart to strain typing, the MLVA technique provides insight into the population structure of microorganisms (Dahyot et al. 2018). Limitations of this method may include the need to design a new set of loci for each species or serotype studied, or the creation of an uninformative null allele as a result of the loss of unstable loci of some strains or lines (Zaluga et al. 2013).

Methods for Genotypic Identification of Bacteria Using Sequencing

The DNA sequencing method, also known as the chain termination method, was developed in 1977 by Frederick Sanger. It involves the addition of nucleotides to single-stranded DNA of unknown sequence according to the principle of complementarity of nitrogenous bases. The reaction is catalyzed by DNA polymerase (Totomoch-Serra et al. 2017). The Sanger method was followed by other sequencing methods called generations. Second-generation sequencing uses high-throughput sequencers to learn the sequence of millions of DNA fragments. Fragments of 35–700 bp (base pair) in length can be read at a single throw. Third-generation sequencing allows a single DNA molecule to be read without amplification. The first next-generation method developed was pyrosequencing, in which DNA synthesis is monitored in real-time. Pyrophosphate released during the addition of complemented nucleotides is converted to adenosine triphosphate (ATP), which is a cofactor for luciferase, an enzyme that oxidizes luciferin to oxyluciferin and light (Kotowska and Zakrzewska-Czerwińska 2010, Harrington et al. 2013, Al-Obaidi et al. 2018).

The sequence of the 16S rRNA gene is the most commonly used genetic marker for studying bacterial phylogeny and taxonomy because it is found in almost all bacteria, often in the form of operons. The size of the gene (1,500 bp) allows for computerized analyses. Its function does not change over time by which modifications in the nucleotide sequence demonstrate evolutionary variation. Previous studies suggest that 16S rRNA gene sequencing has 90 percent efficiency in bacterial genus identification. Efficiencies of 65 – 83 percentage are observed in species identification, with 1 to 14 percentage of isolates remaining unidentified (Janda and Abbott 2007, Jaroszewska and Misiewicz 2012). Sequencing of the 16S rRNA gene is preceded by an amplification reaction. Identification of the microorganism is carried out by comparing the obtained nucleotide sequences with available databases. The presence of variable and conserved regions means that genome-wide universal primers are used. The disadvantages of sequencing are the duration (two-three days), limited availability and cost of performing the procedure (Jaroszewska and Misiewicz 2012, Król 2012, Adzitey et al. 2013, Raina et al. 2019).

A method similar to 16S rRNA sequencing is multifocal sequencing. In this technique, fragments of multiple "housekeeping" genes are combined and the resulting long sequence is

Table 1. Characterization of genetic methods' identification of bacteria. Used "-" to indicate the absence of a feature and "+" to indicate the occurrence of a feature.

Stage Methods	PCR	Restriction enzymes (ERs)-	Electrophoresis	Transfer to nitrocellulose or nylon membranes	Visualization	Citation
Unknown sequence						
REA-PFGE	-	+	Pulsed field gel electrophoresis	-	The gel is stained by ethidium bromie and can be observed under UV	Sharma-Kuinkel et al. 2016
RFLP	-	+	Agarose gel electrophoresis	+	DNA hybridizes with radioactive probes and is detected by autoradiography on X-ray film	Varshney et al. 2004, Ben-Ari and Lavi 2012
Ribotyping	+	+	Agarose gel electrophoresis	+	Radioactively labeled probes and autoradiography (alternatively, fluorescent or chemiluminescent-labeled probes can be used)	Kashyap et al. 2014
RAPD	+	-	Agarose gel electrophoresis	-	Stained by ethidium bromide	Kumari and Thakur 2014
AFLP	+	+	Polyacrylamide gel electrophoresis or capillary electrophoresis	-	Fluorescently or radioactively labeled	Hata 2010, Shen 2019
MLVA	+	-	Capillary electrophoresis	-	Fluorescently labeled	Pruvost et al. 2014
Microarray DNA	-	+	-	-	Fluorescent labeling of the hybrid	Beier et al. 2004, Bzducha et al. 2007, Łyszcz and Gałązka 2016
Known sequence						
MLST and MLSA	+	-	-	-	Using an automated DNA sequencer to know allele profile	Nunney et al. 2012, Ruppitsch 2016
WGS	+	-	-	-	Labeled by probes DNA reading by a sequencing	Brzozowski et al. 2017, Ochoa-Díaz et al. 2018

compared with other sequences. Proteins encoded by "housekeeping" genes regulate the course of important cellular processes. There are many such proteins, an example being the rpoA and rpoB proteins, which are the alpha/α and beta/β subunits of RNA polymerase. The strategies currently in use include multilocus sequence typing (MLST) and multilocus sequence analysis (MLSA) (Emerson et al. 2008).

MLST typically consists of sequences of five or seven housekeeping genes or loci located in a chromosome (Rong and Huang 2014, Dingle and MacCannell 2015). This results in allele

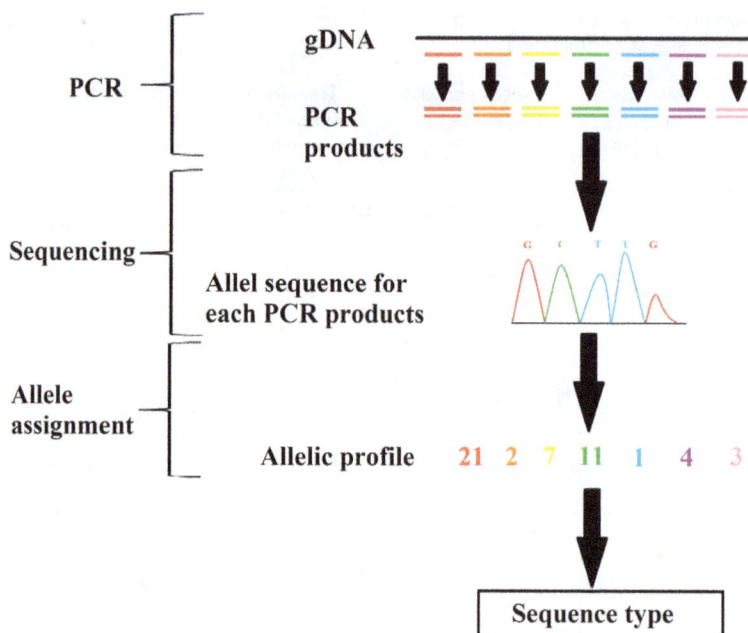

Figure 3. Diagram showing steps of MLST - amplifying seven "housekeeping" genes, comparing the number of each allele and determining the type of sequence (Ruppitsch 2016).

combinations called sequence type (ST) by which strains, isolates, samples, etc. are compared, enabling identification of the microorganism without previously sequencing the entire genome (Nunney et al. 2012, Paris et al. 2015). Each allele that is part of a ST is given a number. Strains with the same allele profiles refer to the same sequence type and strains with some common alleles and refer to sequence complexes (Figure 3) (Glaeser and Kämpfer 2015). Laboratories can compare results through a public online database (Oliver and Jones 2015). The data obtained from the analysis are used to identify bacterial species, distinguish similar bacterial species and divide the genus into species (Pérez-Losada et al. 2011). The use of the MLST technique is associated with high cost and work intensity, which according to some authors, can be offset by the use of the newest sequencers (Brzozowski et al. 2017). The MLSA technique similar to MLST is based on housekeeping sequencing of genes or loci, but the comparative analysis narrows the range in gene selection to less than six (Emerson et al. 2008).

Plants in the *Fabaceae* family inhabit many terrestrial biomes. Some of them enter into symbiosis with several types of bacteria together called rhizobia. As a result of this interaction, biological fixation of atmospheric nitrogen occurs, which improves the global nitrogen cycle and has a positive effect on agriculture. Bacteria belonging to *Bradyrhizobium* are the common ancestor of nitrogen-fixing rhizobia species (Delago et al. 2007, Martyniuk 2012, Ferguson 2017). Research on this bacterial genus may allow us to learn more about the evolution of rhizobia. However, the low variability of the 16S rRNA gene means that the sequencing method may not be effective for assessing diversity. Therefore, the MLSA technique is increasingly being used in phylogenetics and taxonomic analysis. In a study by Delamuta et al. (2012), MLSA analysis of five housekeeping genes was conducted. The results allowed the determination of phylogeny and identification of new subgroups indicative of new *Bradyrhizobium* species.

Whole genome sequencing (WGS) is a next-generation sequencing method that can serve as a widely available and inexpensive tool for the identification and characterization of bacteria (Al-Obaidi et al. 2018, Franco-Duarte et al. 2019). WGS begins by cutting DNA into smaller fragments, which are combined with DNA tags that function as barcodes read by the sequencer. Behind typing bacteria and learning about evolutionary lines, WGS makes it possible to understand antimicrobial resistance in bacteria and conduct epidemiological studies. Technological and analytical advances have made it possible to reduce the cost of analysis while increasing its efficiency and speed (Quainoo et al. 2017, Franco-Duarte et al. 2019).

The use of this technique generates large amounts of data which makes it difficult to interpret (Larsen et al. 2012). Analysis of WGS data can be carried out in a number of ways. This is due to the use of a nucleotide or allele as a unit of interest and whether the method is reference or reference-free (Franz et al. 2016). In addition, different pipelines, software and platforms are used to analyze sequencing data. Standardization and harmonization of key requirements and quality criteria are needed to compare results (Franz et al. 2016, Rantsiou et al. 2018). Assembling a genome most often involves transforming sets of contigs or scaffolds into the most likely sequence orientation. Contigs are formed by linking individual and overlapping sequencing reads to form a continuous DNA sequence. Scaffolds are formed by combining contigs into a whole. The resulting genome has regions not covered by contigs. These gaps are either folding artifacts or natural deletions (Nowak 2010, Kremer et al. 2017). The file with the assembled sequence is uploaded to a database such as the National Center for Biotechnology Information (NCBI) (Rantsiou et al. 2018).

In order to minimize data loss and more accurately map genomic features, closing the gaps created in the genome is carried out. The traditional way starts with the primer design based on the edges of adjacent contigs. The next steps are PCR amplification, Sanger sequencing and local assembly, usually with manual curation. To reduce the time of this process, new *in vitro* methods have been developed (Kremer et al. 2017). Examples include PCR-based technologies such as multiplex PCR, inverse PCR, restriction site PCR, and capture PCR (Rogers et al. 2005). Also, optical mapping (Huang et al. 2021) and hybrid sequencing and assembly (Ribeiro et al. 2012).

Conclusion

Genotyping methods are among the popular methods for identifying microorganisms including bacteria. They are increasingly replacing phenotypic methods due to the receptivity and sensitivity of identification. However, they differ in their application, ease of execution and the amount of cost involved in carrying them out. REA-PFGE is a method in which identification of bacteria is carried out on the basis of differences in the length of DNA fragments separated by pulsed-field electrophoresis. AFLP, MLVA, RAPD and 16S rRNA sequencing methods require a PCR-based amplification step to perform. To visualize the nucleic acid fragments gained after digestion, the RFLP and ribotyping techniques are used. Microarrays use the hybridization of two complementary DNA sequences. WGS is about knowing the sequence of nucleic acid fragments linked to DNA markers. MLST typically sequences five to seven-housekeeping genes or loci getting characteristic combinations to identify bacteria. The huge variety makes it possible to choose a method tailored to the skills of laboratory personnel, laboratory equipment, type of sample and expected results.

Acknowledgments

This paper is an outcome of the data obtained under the project NCN 2022/45/B/NZ8/02398 (National Science Centre, Poland) *"Interaction between microbiome, mycobiome and metaviriome of the rhizosphere and endorhizosphere of ruderal plants and their role in passive and active remediation of soil heavily degraded and historically contaminated with crude oil"* (2023–2027).

Abbreviations

REA-PFGE   –   Restriction Enzyme Analysis–Pulsed Field Gel Electrophoresis
RFLP   –   Restriction Fragments Length Polymorphism
AFLP   –   Amplified Fragment Length Polymorphism
RAPD   –   Random Amplified Polymorphic DNA
MLVA   –   Multiple-locus VNTR Analysis
MLST   –   Multilocus Sequence Typing
MLSA   –   Multilocus Sequence Analysis
WGS   –   Whole Genome Sequencing
ST   –   Sequence Type

References

Adaszek, Ł., B. Dzięgiel, Ł. Mazurek and S. Winiarczyk. 2018. Zastosowanie techniki PCR w badaniach bakteriologicznych. Życie Weterynaryjne 93(4): 234–237.

Adzitey, F., N. Huda and G.R. Ali. 2013. Molecular techniques for detecting and typing of bacteria, advantages and application to food borne pathogens isolated from ducks. 3 Biotech. 3(2): 97–107.

Akoijam, N., D. Kalita and S.R. Joshi. 2022. Bacteria and their industrial importance. pp. *In*: Verma, P. [ed.]. Industrial Microbiology and Biotechnology. Springer, Singapore.

Al-Khaldi, S.F., M.M. Mossoba, M.M. Allard, E.K. Lienau and E.D. Brown. 2012. Bacterial identification and subtyping using DNA microarray and DNA sequencing. pp. *In*: Navid, A. [ed.]. Microbial Systems Biology. Methods in Molecular Biology, Humana Press, Totowa, NJ.

Al-Obaidi, M.M.J., Z. Suhaili and M.N. Mohd Desa. 2018. Genotyping approaches for identification and characterization of *Staphylococcus aureus*. pp. 35–42. *In*: I. Abdurakhmonov [ed.]. Genotyping. IntechOpen.

Austin, J.W. and F.J. Pagotto. 2003. MICROBIOLOGY | Detection of foodborne pathogens and their toxins. pp. 3886–3892. *In*: Caballero, B. [ed.]. Encyclopedia of Food Sciences and Nutrition (Second Edition), Academic Press, Oxford.

Baker, J.C., R.E. Crumley and T.T. Eckdahl. 2002. Random amplified polymorphic DNA PCR in the microbiology teaching laboratory: Identification of bacterial unknowns. Biochem. Mol. Biol. Educ. 30(6): 394–397.

Beier, V., A. Bauer, M. Baum and J.D. Hoheisel. 2004. Fluorescent sample labeling for DNA microarray analyses. pp. 127–135. *In*: Niemeyer, C.M. [ed.]. Bioconjugation Protocols: Strategies and Methods, Humana Press, Totowa, NJ.

Ben-Ari, G. and U. Lavi. 2012. Marker-assisted selection in plant breeding. pp. 163–184. *In*: Altman, A. and P.M. Hasegawa [eds.]. Plant Biotechnology and Agriculture, Academic Press, San Diego.

Bertani, G., M.L. Savo Sardaro, E. Neviani, C. Lazzi. 2019. AFLP protocol comparison for microbial diversity fingerprinting. J App Genet. 60: 217–223.

Blears, M. J., S.A. De Grandis, H. Lee and J.T. Trevors. 1998. Amplified fragment length polymorphism (AFLP): Review of the procedure and its applications. J. In. Microbiol. Biot. 21(3): 99–114.

Brzozowski, M., P. Kwiatkowski, D. Kosik-Bogacka and J. Jursa-Kulesza. 2017. Metody genotypowe i fenotypowe wykorzystywane w typowaniu drobnoustrojów do celów epidemiologicznych. POST. MIKROBIOL. 56(3): 353–366.

Bumgarner, R. 2013. Overview of DNA microarrays: types, applications, and their future. Curr. Protocols Molecul. Biol. 22.

Buszewski, B., A. Rogowska, P. Pomastowski, M. Złoch and V. Railen-Plugaru. 2017. Identification of microorganisms by modern analytical techniques. J. AOAC Int. 100(6): 1607–1623.

Bzducha, A. 2007. Szybkie metody identyfikacji mikroorganizmów w żywności. Medycyna Weterynaryjna 63(7): 773–777.

Chatterjee, S. and I.H. Raval. 2019. Pathogenic microbial genetic diversity with reference to health. pp. 559–577. *In*: Das, S. and H.R. Dash [eds.]. Microbial Diversity in The Genomic Era. Academic Press.

Chowdhury, N., M. Asakura, S.B. Neogi, A. Hinenoya, S. Haldar, T. Ramamurthy et al. 2010. Development of simple and rapid PCR-finger printing methods for *Vibrio cholerae* on the basis of genetic diversity of the superintegron. J. Appl. Microbiol. 109: 304–312.

Dahyot, S., J. Lebeurre, X. Argemi, P. François, L. Lemée, G. Prévost et al. 2018. Multiple-Locus Variable Number Tandem Repeat Analysis (MLVA) and Tandem Repeat Sequence Typing (TRST), help fultools for subtyping *Staphylococcus lugdunensis*. Sci. Rep. 8: 11669.

Dąbrowska, J. and A. Kunicka-Styczyńska. 2018. *Alicyclobacillus* – bakterie nadal poznawane. POST. MIKROBIOL. 57(2): 117–124.

Dekowska, A. 2011. Zastosowanie metod biologii molekularnej w diagnostyce bakterii z rodzaju *Alicyclobacillus*. Postępy Nauki i Technologii Przemysłu Rolno-Spożywczego 66(2): 34–43.

Delago, M.J., S. Casella and E.J. Bedmar. 2007. Denitrification in rhizobia-legume symbiosis. pp. 83–91. *In*: Bothe, H., Ferguson, S.J. and Newton, W.E. [eds.]. Biology of the Nitrogen Cycle, Elsevier, Amsterdam.

Delamuta, J.R., R.A. Ribeiro, P. Menna, E.V. Bangel and M. Hungria. 2012. Multilocus sequence analysis (MLSA) of *Bradyrhizobium* strains: Revealing high diversity of tropical diazotrophic symbiotic bacteria. Braz. J. Microbiol. 43(2): 698–710.

Dingle, T.C. and D.R. MacCannell. 2015. Molecular strain typing and characterisation of toxigenic *Clostridium difficile*. pp. 329–357. *In*: Sails, A. and Y.-W. Tang [eds.]. Current and Emerging Technologies for the Diagnosis of Microbial Infections, Methods in Microbiology, Academic Press.

Donelli, G., C. Vuotti and P. Mastromarino. 2013. Phenotyping and genotyping are both essential to identify and classify a probiotic microorganism. Microb. Ecol. Helath D. 24: 20105.

Dykhuizen, D. 2005. Species numbers in bacteria. Proc. Calif Acad. Sci. 56(6,1): 62–71.

Emerson, D., L. Agulto, H. Liu and L. Liu. 2008. Identifying and Characterizing Bacteria in an Era of Genomics and Proteomics, BioScience 58(10): 925–936.

Ferguson, B.J. 2017. Rhizobia and Legume Nodulation Gene ☆. Reference Module in Life Sciences, Elsevier.

Franco-Duarte, R., L. Černáková, S. Kadam, K.S. Kaushik, B. Salehi, A. Bevilacqua et al. 2019. Advances in chemical and biological methods to identify microorganisms-from past to present. Microorganisms 7(5): 130.

Franz, E., L.M. Gras and T. Dallman. 2016. Significance of whole genome sequencing for surveillance, source attribution and microbial risk assessment of foodborne pathogens. Curr. Opin. Food Sci. 8: 74–79.

Gentry, T.J. and J. Zhou. 2006. Microarray-based microbial identification and characterization. pp. 276–290. Advanced Techniques in Diagnostic Microbiology. Springer, Boston.

Glaeser, S.P. and P. Kämpfer. 2015. Multilocus sequence analysis (MLSA) in prokaryotic taxonomy. Systematic and Applied Microbiology 38(4): 234–245.

Goering, R.V. 2010. Pulsed field gel electrophoresis: A review of application and interpretation in the molecular epidemiology of infectious disease. Infect. Genet. Evol. 10: 866–875.

Grimont, P.A.D. and F. Grimont. 2001. rRNA gene restriction pattern determination (Ribotyping) and computer interpretation. pp. 107–133. *In*: Dijkshoorn, L., K.J. Towner and M. Struelens [eds.]. New Approaches for the Generation and Analysis of Microbial Typing Data, Elsevier Science B.V., Amsterdam.

Groenewald, W.H., P.A. Gouws and R.C. Witthuhn. 2009. Isolation, identification and typification of *Alicyclobacillus acidoterrestris* and *Alicyclobacillus acidocaldarius* strains from orchardsoil and the fruit processing environment in South Africa. Food Microbiol. 26: 71–76.

Guinard, J., A. Latreille, F. Guérin, S. Poussier and E. Wicker. 2017. New Multilocus Variable-Number Tandem-Repeat Analysis (MLVA) scheme for fine-scale monitoring and microevolution-related study of *Ralstonia pseudosolanacearum* Phylotype I Populations. AEM 83(5): e03095–16.

Haguenoer, E., G. Baty, C. Pourcel, M.-F. Lartigue, A.-S. Domelier, A. Rosenau et al. 2011. A multilocus variable number of tandem repeat analysis (MLVA) scheme for *Streptococcus agalactiae* genotyping. BMC Microbiol. 11: 171.

Harrington, C.T., E.I. Lin, M.T. Olson and J.R. Eshleman. 2013. Fundamentals of pyrosequencing. Arch. of Pathol. Lab. Med. 137(9): 1296–1303.

Hashim, H.O. and M.B. Al-Shuhaib. 2019. Exploring the potential and limitations of PCR-RFLP and PCR-SSCP for SNP Detection: A review. JABR 6(4): 137–144.

Hata, D.J. 2010. Molecular methods for identification and characterization of *Acinetobacter* spp. pp. 313–326. *In*: Grody, W.W., R.M. Nakamura, C.M. Strom and F.L. Kiechle. Molecular Diagnostics Techniques, Academic Press, San Diego.

Helldal, L., N. Karami, C. Welinder-Olsson, E.R.B. Moore and C. Åhren. 2017. Evaluation of MLVA for epidemiological typing and outbreak detection of ESBL-producing *Escherichia coli* in Sweden. BMC Microbiol. 17: 8.

Huang, B., G. Wei, B. Wang, F. Ju, Y. Zhong, Z. Shi et al. 2021. Filling gaps of genome scaffolds via probabilistic searching optical maps against assembly graph. BMC Bioinformatics 22: 533.

Janda, J.M. and S.L. Abbott. 2007. 16S rRNA gene sequencing for bacterial identification in the diagnostic laboratory: pluses, perils, and pitfalls. J. Clin. Microbiol. 45(9): 2761–2764.

Jaroszewska, E. and A. Misiewicz. 2012. Wybrane molekularne metody identyfikacji mikroorganizmów w kolekcjach kultur drobnoustrojów. Postępy Nauki i Technologii Przemysłu Rolno-Spożywczego 67(4): 67–74.

Kashyap, S.K., S. Maherchandani and N. Kumar. 2014. Ribotyping: A Tool for molecular taxonomy. pp. 327–344. *In*: Verma, A.S. and A. Singh [eds.]. Animal Biotechnology, Academic Press, San Diego.

Khosravi, A.D., S. Vatani, M.M. Feizabadi, E.A. Montazeri and A. Jolodar. 2014. Application of pulsed field gel electrophoresis for study of genetic diversity in *Mycobacterium tuberculosis* strains isolated from tuberculosis patients. Jundishapur. J. Microbiol. 7(5): e9963.

Kotowska, M. and J. Zakrzewska-Czerwińska. 2010. Kurs szybkiego czytania DNA - nowoczesne techniki sekwencjonowania. Biotechnologia 4(91): 24–38.

Kremer, F.S., A.J.A. McBride and L.S. Pinto. 2017. Approaches for *in silico* finishing of microbial genome sequences. Genet. Mol. Biol. 40(3): 553–576.

Krishna, R., W.A. Ansari, J.P. Verma and M. Diaz. 2019. Modern molecular and omicstools for understanding the plant growth-promoting rhizobacteria. pp. 39–53. *In*: Kumar, A., A.K. Singh and K.K. Choudhary [eds.]. Role of Plant Growth Promoting Microorganisms in Sustainable Agriculture and Nanotechnology, Woodhead Publishing.

Król, J. 2012. Identyfikacja fenotypowa oraz genotypowa bakterii z rodziny *Pasteurellaceae* izolowanych od psów i kotów. Wydawnictwo Uniwersytetu Przyrodniczego we Wrocławiu, Współczesne Pro. Med. Weter. 6: 1–80.

Kumari, N. and S.K. Thakur. 2014. Randomly amplified polymorphic DNA - a brief review. Am. J. Anim. Vet. 9(1): 6–13.

Lane, N. 2015. The unseenworld: Reflections on Leeuwenhoek (1677) 'Concerning little animals'. Philos. Trans. R Soc. Lond. B Biol. Sci. 370(1666): 20140344.

Larsen, M.V., S. Cosentino, S. Rasmussen, C. Friis, H. Hasman, R.L. Marvig et al. 2012. Multilocus sequence typing of total-genome-sequenced bacteria. J. Clin. Microbiol. 50(4): 1355–1361.

Lee, C.C., W.C. Lo, S.M. Lai, YP.P. Chen, C.Y. Tang, P.C. Lyu. 2012. Metabolic classification of microbial genomes using functional probes. BMC Genomics 13: 157.

Łyszcz, M. and A. Gałązka. 2016. Wybrane metody molekularne wykorzystywane w ocenie bioróżnorodności mikroorganizmów glebowych. Post. Mikrobiol. 55(3): 309–319.

Łyszcz, M. and A. Gałązka. 2017. Metody oparte o amplifikację DNA techniką PCR wykorzystywane w ocenie bioróżnorodności mikroorganizmów glebowych. Kosmos – Problemy Nauk Biologicznych 66(2): 193–206.

Martyniuk S. 2012. Naukowe i praktyczne aspekty symbiozy roślin strączkowych z bakteriami brodawkowymi (Scientific and practical aspects of legumes symbiosis with root-nodule bacteria). PJA 9: 17–22.

Matsumoto, S., K. Watanabe, H. Kiyota, M. Tachibana, T. Shimizu and M. Watarai. 2022. Distinction of *Paramecium* strains by a combination method of RAPD analysis and multiplex PCR. PLOS ONE 17(3): e0265139.

Meudt, H.M. and A.C. Clarke. 2007. Almost forgotten or latest practice? AFLP applications, analyses and advances. Trends Plant Sci. 12(3): 106–117.

Mittal B., P. Chaturvedi and S. Tulsyan. 2013. Restriction fragment length polymorphism. pp. 190–193. *In*: Maloy, S. and K. Hughes [eds.]. Brenner's Encyclopedia of Genetics (Second Edition), Academic Press, San Diego.

Nadon, C.A., E. Trees, L.K. Ng, E. Møller Nielsen, A. Reimer, N. Maxwell et al. 2013. Development and application of MLVA methods as a tool for inter-laboratory surveillance. Euro Surveill. 18(35): 20565.

Neoh, H.M., X.E. Tan, H.F. Sapri and T.L. Tan. 2019. Pulsed-field gel electrophoresis (PFGE): A review of the "gold standard" for bacteria typing and current alternatives. Infect. Genet. Evol. 74: 103935.

Nowak, J.K. 2010. Sekwencjonowanie największego chromosomu *Paramecium tetraurelia* – przykład opracowania metody. Biotechnologia 4(91): 76–90.

Nunney, L., S. Elfekih and R. Stouthamer. 2012. The importance of multilocus sequence typing: Cautionary tales from the Bacterium *Xylella fastidiosa*. Phytopathology 102(5): 456–460.

Ochoa-Díaz, M.M., S. Daza-Giovannetty and D. Gómez-Camargo. 2018. Bacterial genotyping methods: From the basics to modern. pp. 13–20. *In*: Medina,C. and F. López-Baena [eds.]. Host-Pathogen Interactions. Methods in Molecular Biology. Humana Press, New York, NY.

Oliver, J.D. and J.L. Jones. 2015. *Vibrio parahaemolyticus* and *Vibrio vulnificus*. pp. 1169–1186. *In*: Tang, Y.-W., M. Sussam, D. Liu, I. Poxton and J. Schwartzman [eds.]. Molecular Medical Microbiology (second Edition). Academic Press, Boston.

Paczesny, J. and P.A. Mierzejewski. 2021. The use of probes and bacteriophages for the detection of bacteria. pp. 49–93. *In*: Gurtler, V. [ed.]. Fluorescent Probes, Methods in Microbiology. Academic Press.

Paris, D.H., A.L. Richards and N.P.J. Day. 2015. Orientia. pp. 2057–2096. *In*: Tang, Y-W., D. Liu, I. Poxton and J. Schwartzman [eds.]. Molecular Medical Microbiology (Second Edition), Academic Press, Boston.

Parizad, E.G., E.G. Parizad and A. Valizadeh. 2016. The application of pulsed field gel electrophoresis in clinical studies. J. Clin. Diagn. 10(1): DE01-4.

Paun, O. and P. Schönswetter. 2012. Amplified fragment length polymorphism: An invaluable fingerprinting technique for genomic, transcriptomic, and epigenetic studies. Methods Mol. Biol. 862: 75–87.

Pérez-Losada, M., M.L. Porter, R.P. Viscidi and K.A. Crandall. 2011. Multilocus sequence typing of pathogens. pp. 503–521. *In*: Tibayrenc, M. [ed.] Genetics and Evolution of Infectious Disease, Elsevier, London.

Pruvost, O., M. Magne, K. Boyer, A. Leduc, C. Tourterel, C. Drevet et al. 2014. A MLVA genotyping scheme for global surveillance of the citrus pathogen *Xanthomonascitri pv. citri* suggests a worldwide geographical expansion of a single genetic lineage. PLOS ONE 9(6): e98129.

Qi, Z., Y. Cui, Q. Zhang and R. Yang. 2016. Taxonomy of *Yersinia pestis*. pp. 35–78. *In*: Yang, R. and A. Anisimov [eds.]. *Yersinia pestis*: Retrospective and Perspective. Advances in Experimental Medicine and Biology, Springer, Dordrecht.

Quainoo, S., J.P.M. Coolen, S.A.F.T. van Hijum, M.A. Huynen, W.J.G. Melchers, W. van Schaik et al. 2017. Whole-genome sequencing of bacterial pathogens: The future of nosocomial outbreak analysis. Clin. Microbiol. Rev. 3(4): 1015–1063.

Qurban, A. and A. Ameen. 2020. Bacterial Identification by 16S Ribotyping: A Review. Biotechnol. Ind. J. 16(2): 204.

Raina, V., T. Nayak, L. Ray, K. Kumari and M. Suar. 2019. A polyphasic taxonomic approach for designation and description of novel microbial species. pp. 137–152. *In*: Das, S. and H.R. Dash [eds.]. Microbial Diversity in the Genomic Era, Academic Press.

Ranjbar, R., P. Behzadi, A. Najafi and R. Roudi. 2017. DNA microarray for rapid detection and identification of food and water borne bacteria: From dry to wet lab. Open J. Med. Microbiol. 11: 330–338.

Rantsiou, K., S. Kathariou, A. Winkler, P. Skandamis, M.J. Saint-Cyr, K. Rouzeau-Szynalski et al. 2018. Next generation microbiological risk assessment: Opportunities of whole genome sequencing (WGS) for foodborne pathogen surveillance, source tracking and risk assessment. Int. J. Food Microbiol. 287: 3–9.

Rogers, Y.C., A.C. Munk, L.J. Meincke and C.S. Han. 2005. Closing bacterial genomic sequence gaps with adaptor-PCR. BioTechniques 39(1): 31–34.

Rong, X. and Y. Huang. 2014. Multi-locus sequence analysis: taking prokaryotic systematics to the next level. pp. 221–251. *In*: Goodfellow, M., I. Sutcliffe and J. Chun [eds.]. New Approaches to Prokaryotic Systematics Methods in Microbiology, Academic Press.

Ruppitsch, W. 2016. Molecular typing of bacteria for epidemiological surveillance and outbreak investigation. Die Bodenkultur: Journal of Land Management, Food and Environment 67(4): 199–224.

Säde, E. and J. Björkroth. 2014. Identification methods | DNA fingerprinting: Restriction fragment-length polymorphism. pp. 274–281. *In*: Batt, C.A. and M.L. Tortorello [eds.]. Encyclopedia of Food Microbiology (Second Edition), Academic Press, Oxford.

Sandle, T. 2014. Biochemical and modern identification techniques | Food-Poisoning Microorganisms. pp. 238–243. *In*: C.A. Batt and M.L. Tortorello [eds.]. Encyclopedia of Food Microbiology (Second Edition), Academic Press, Oxford.

Sandle, T. 2016. Introduction to pharmaceutical microbiology. pp. 1–14. *In*: Sandle, T. [ed.]. Pharmaceutical Microbiology. Woodhead Publishing, Oxford.

Sankari Meena, K., S.P. Deepa, P. Manju and S. Vijayakumar. 2021. Importance of Bacteria in Agriculture. Argiculture and Food: E-newsletter 3(12): 272–274.

Saxena, S., J. Verma and D.R. Shikha, Modi. 2014. RAPD-PCR and 16S rDNA phylogenetic analysis of alkaline protease producing bacteria isolated from soil of India: Identification and detection of genetic variability. JGEB 12(1): 27–35.

Sharma, A., S. Lee and Y.S. Park. 2020. Molecular typing tools for identifying and characterizing lactic acid bacteria: A review. Food Sci. Biotechnol. 29(10): 1301–1318.

Sharma-Kuinkel, B.K., T.H. Rude, V.G. Fowler. Jr. 2016. Pulse field gel electrophoresis. pp. 117–130. *In*: Bose, J. [eds.]. Methods in Molecular Biology, Humana Press, New York, NY.

Shen, C.-H. 2019. Characterization of nucleic acids and proteins. pp. 249–276. *In*: Shen, C.-H. [ed.]. Diagnostic Molecular Biology, Academic Press.

Simner, P.J., R. Khare and N.L. Wengenack. 2015. Rapidly growing mycobacteria. Molecular medical microbiology (Second Edition). pp. 1679–1690. *In*: Tang, Yi-W., M. Sussman, D. Liu, I. Poxton and J. Schwartzman [eds.] Academic Press, Boston.

Sobiczewski, P. 2021. Bakterie w środowisku roślin – wrogowie i sprzymierzeńcy. Kosmos – Problemy Nauk Biologicznych 70(4:333): 685–696.

Sokołowska, B. 2014. *Alicyclobacillus* – termofilne kwasolubne bakterie przetrwalnikujące – charakterystyka i występowanie. ŻYWNOŚĆ. Nauka. Technologia. Jakość 4(95): 5–17.

Stefańska, I., E. Kwiecień, M. Górzyńska, A. Sałamaszyńska-Guz and M. Rzewuska. 2022. RAPD-PCR-based fingerprinting method as a tool for epidemiological analysis of *Trueperella pyogenes* Infections. Pathogens 11(5): 562.

Tabit, F.T. 2016. Advantages and limitations of potential methods for the analysis of bacteria in milk: A review. JFST 53(1): 42–49.

Tang, Y.W., N.M. Ellis, M.K. Hopkins, D.H. Smith, D.E. Dodge and D.H. Persing. 1998. Comparison of phenotypic and genotypic techniques for identification of unusual aerobic pathogenic gram-negative bacilli. J. Clin. Microbiol. 36(12): 3674–9.

Totomoch-Serra, A., M.F. Marquez and D.E. Cervantes-Barragán. 2017. Sanger sequencing as a first-line approach for molecular diagnosis of Andersen-Tawil syndrome. F1000Research 6: 1016.

Van Belkum, A. 2007. Tracing isolates of bacterial species by multilocus variable number of tandem repeat analysis (MLVA). FEMS Immunology and Medical Microbiology 49(1): 22–27.

Van Stelten, A. and K.K. Nightingale. 2014. Microbiological analysis | DNA Methods. pp. 294–300. *In*: Dikeman, M. and D. Carrick [eds.]. Encyclopedia of Meat Sciences (Second Edition), Academic Press, Oxford.

Varshney, A., T. Mohapatra and R.P. Sharma. 2004. Molecular mapping and marker assisted selection of traits for crop improvement. pp. 289–330. *In*: Srivastava, A. and S. Narula [eds.]. Plant Biotechnology and Molecular Markers, Anamaya, New Delhi, India.

Venkova, T., C.C. Yeo and M. Espinosa. 2018. Editorial: The good, the bad, and the ugly: multiple roles of bacteria in human life. Front. Microbiol. 9: 1702.

Vos, P., R. Hogers, M. Bleeker, M. Reijans, T. Lee, M. Hornes et al. 1995. AFLP: a new technique for DNA fingerprinting. Nucleic Acids Res. 23(21): 4407–4414.

Wang, X., I.K. Jordan and L.W. Mayer. 2015. A phylogenetic perspective on molecular epidemiology. pp. 517–536. *In*: Tang, Y.-W., M. Sussman, D. Liu, I. Poxton and J. Schwartzman [eds.]. Molecular Medical Microbiology (Second Edition), Academic Press, Boston.

Yamazaki, K., T. Okubo, N. Inoue and H. Shinano. 1997. Randomly amplified polymorphic DNA (RAPD) for rapid identification of spoilage bacterium *Alicyclobacillus acidoterrestris*. Biosci. Biotechnol. Biochem. 61: 1016–1018.

Zaluga J., P. Stragier, J. Van Vaerenbergh, M. Maes and P. De Vos .2013. Multilocus Variable-Number-Tandem-Repeats Analysis (MLVA) distinguishes a clonal complex of *Clavibacter michiganensis* subsp. michiganensis strains isolated from recent outbreaks of bacterial wilt and canker in Belgium. BMC Microbiol. 13: 126.

CHAPTER 11

The Secretome Landscape of *Ralstonia*

Juan Carlos Ariute Oliveira,[1,2] *Lucas Gabriel Rodrigues Gomes,*[3]
Arun Kumar Jaiswal,[3] *Sandeep Tiwari,*[4,5] *Vasco Azevedo,*[3]
Ana Maria Benko-Iseppon[2,†,*] and *Flávia Figueira Aburjaile*[1,†,*]

Introduction

The *Ralstonia* genus comprises rod-shaped, aerobic, non-fermenting, Gram-negative pathogen bacteria capable of surviving in environments such as soil and water (Yabuuchi et al. 1995, Ryan and Adley 2014). The genus was finally settled when former species of the *Burkholderia* and *Alcaligenes* genera, such as *R. eutropha*, *R. solanacearum*, and the type species, *R. picketti*, were reclassified in 1995, based on phylogenetic analyses of 16S rRNA nucleotide sequences and rRNA-DNA hybridization, besides other phenotypic characteristics (Yabuuchi et al. 1995, Zhang and Qiu 2016). Nowadays, the genus comprises 13 representative species with the *Cupriavidus* genus (Parte 2014). However, not all receive the same attention in the research field. For example, *Ralstonia basilensis*, *Ralstonia gilardii*, *Ralstonia oxalatica,* and *Ralstonia paucula* have no genomes available on public databases. Moreover, not long ago, researchers classified *R. solanacearum* as a species complex with four phylotypes according to their center of diversification and origin. However, new phylogenetic analyses separated them into two other species: *R. pseudosolanacearum* and *R. syzygii* (Safni et al. 2014, Santiago et al. 2016). Currently, over 400 genomes of varied *Ralstonia* species are available in public databases, with *R. solanacearum* genomes as the most abundant (Table 1).

In terms of pathogenesis, *Ralstonia* species exhibit diverse ecological niches: some of them are emergent opportunistic pathogens of humans, such as *R. picketti*, *R. mannitolilytica*, and *R. insidiosa,* while others are well-known plant pathogens, such as *R. pseudosolanacearum, R. syzygii,* and *R. solanacearum* (Ryan and Adley 2014, Santiago et al. 2016). Since they share a broad range of hosts, the latter species infect important crops all around the globe, causing substantial economic losses every year in banana, potato, and tomato plantations, for instance (Lopes and Rossato 2018).

Many genomic studies have been carried out on the *R. solanacearum* genome to elucidate how the pathogenesis occurs on the host, emphasizing the development of sequencing technologies throughout the last decades. It's been announced that *R. solanacearum* has a bipartite genome

---

[1] Laboratory of Integrative Bioinformatics, Preventive Veterinary Medicine Department, Veterinary School, Federal University of Minas Gerais, Belo Horizonte 31270-901, Minas Gerais, Brazil.

[2] Laboratory of Plants Genetics and Biotechnology, Genetics Department, Biosciences Center, Federal University of Pernambuco, Recife, 50740–600, Pernambuco, Brazil.

[3] Laboratory of Cellular and Molecular Genetics, Department of Genetics, Ecology and Evolution, Institute of Biological Sciences, Federal University of Minas Gerais, Belo Horizonte 31270–901, Minas Gerais, Brazil.

[4] Post-Graduate Program in Microbiology, Institute of Biology, Federal University of Bahia, Salvador, Brazil.

[5] Post-Graduate Program in Immunology, Department of Biochemistry and Biophysics, Institute of Health Sciences, Federal University of Bahia, Salvador, 40231–300, Bahia, Brazil.

[†] These authors equally contributed to this work.

[*] Corresponding authors: ana.iseppon@gmail.com; faburjaile@gmail.com

Table 1. Several genome assemblies for each species of *Ralstonia* are available in the NCBI and PATRIC databases.

Species' Scientific Name	Number of Genomes Assembled	Host/Source	Country
Ralstonia eutropha (Davis 1969) (Yabuuchi et al. 1996)	2	-	-
Ralstonia insidiosa (Coenye et al. 2003)	20	Medical sample, food processing center	USA, China, Russia
Ralstonia mannitolilytica corrig (De Baere et al. 2001)	13	Medical sample and *Cupriavidus*	USA, China, Turkey,
Ralstonia pickettii (Ralston et al. 1973) (Yabuuchi et al. 1996)	87	Rice, medical samples	USA, China, Japan, New Zealand
Ralstonia pseudosolanacearum (Safni et al. 2014)	34	Soil-borne, tomato, tobacco, rose, *Cupriavidus*	Brazil, China, and the Netherlands
Ralstonia solanacearum (Smith 1896, Yabuuchi et al. 1996)	297	Soil-borne, banana, platano, tobacco, eggplant, potato, eucalyptol, bell pepper, etc.	North America, South America, Asia, and Africa
Ralstonia syzygii (Roberts et al. 1990) (Vaneechoutte et al. 2004)	6	Banana, tobacco, and *Cupriavidus.*	China and Sumatra

consisting of one chromosome and one megaplasmid of 3.7 and 2.1 megabases, respectively, with a wide array of virulence factors, especially those secreted by types II, III, and IV secretion systems (Salanoubat et al. 2002, Genin and Boucher 2004). These virulence factors have also been severely studied in secretomics, i.e., proteomics focusing on secreted proteins by either eukaryotic or prokaryotic cells. This approach aims to elucidate the cellular context at a given time and condition (Tjalsma et al. 2000, Hathout 2007), accessing essential molecular features for virulence and host defense. The secretome can be estimated with tools combining comprehensive genome analyses, genes annotation, and proteomics approaches such as mass spectrometry (Hathout 2007). The secretome of *R. solanacearum* sheds light on type III effectors (T3Es), confirming that T3Es from other Gram-negative pathogenic bacteria play a critical role in host-pathogen relationships to cause pathogenesis, including *Yersinia, Shigella,* and *Pseudomonas*, for instance (Cornelis and Van Gijsegem 2000).

Ralstonia's type III effectors are often called "Rips" – *Ralstonia* injected proteins, primarily dependent on *hrp* genes (Mukaihara et al. 2010). Besides them, proteins secreted through the type II secretion system have been associated with success for colonization and multiplication inside the host, achieving adaptation success, as in the case of *R. solanacearum*, which occurs in different environments (Kang 1994, Genin and Boucher 2004). This chapter will focus on how *Ralstonia*'s secretome influences its pathogenicity through the effectome, focusing mainly on T3SS and T2SS, their peculiarities, bringing insights on host specificity, challenges, and perspectives of this field.

Ralstonia's Secretion Systems Make-Up a Powerful Secretome

To fulfill many functions in the bacteria cell, proteins are usually translocated to cellular compartments other than the ones they are synthesized in, as soon as the cell machinery recognizes their signal peptides (Tjalsma et al. 2000). Secretion systems are usually responsible for protein translocations in and outside the cell. Secretome studies value secreted proteins because they are intimately related to many roles in the cell, such as homeostatic maintenance in normal physiologic and pathogenic conditions (Hathout 2007). In this context, secretion systems of different types play a significant role in bacterial interaction with the environment or other competitor bacteria, host recognition, and colonization. All these aspects receive equal attention in secretome analysis.

Bacterial cells display three basic types of secretion systems: the Tat (twin-arginine translocation) and Sec (general secretion) pathways and ABC (ATP-binding cassette) transporters, which are the most used machinery for protein translocation between plasmatic membranes (Natale et al. 2008, Green and Mecsas 2016). Secretion systems may cross from one to three phospholipidic membranes: the inner plasmatic membrane, both plasmatic membranes or both the pathogen and the host's membranes (Green and Mecsas 2016). However, for the effective transport of essential proteins, such as virulence factors, more complex secretion systems are commonly associated with the basic ones, especially in Gram-negative bacteria. The main difference between Gram-positive and Gram-negative secretion systems is that the latter can secrete proteins between and out of the cell plasmatic membrane, while the first may also secrete to the cell wall (Tjalsma et al. 2000).

To date, seven different types of specialized secretion systems have been identified in bacteria, and even though not every bacteria displays all of them, Gram-negative bacteria, like *Ralstonia*, usually have more types of secretion systems than Gram-positive bacteria (Thanassi and Hultgren 2000, Abdallah et al. 2007, Gama et al. 2016, Green and Mecsas 2016). Most genomes of the plant pathogen *R. solanacearum* count with secretion system type 1 (T1SS), T2SS, T3SS, T4SS, T5SS, and the recently discovered T6SS (Genin and Denny 2012, Gama et al. 2016) (Figure 1).

Multiple essential techniques used in genomics, transcriptomics, and proteomics may be optimized to investigate one organism's secretome. By far, the best method for secretomic approaches is mass spectrometry, which can be executed with a range of combined techniques, like liquid chromatography, 2D gel electrophoresis, and SILAC (stable-isotope labeling by amino acids in cell culture) (Mann 2006, Hathout 2007, Lonjon et al. 2016). Genome-wide analysis has been crucial to refining secretome studies in the past years since they can identify sequences inside the genome, including predicted targets on a metabolic pathway especially when lacking *in vivo* and *in vitro* data (Gagic et al. 2016). Tools for *in silico* protein prediction might work based on Bayesian inferences associated with Markov-chains, network-based inferences, and machine-learning algorithms, amongst other bioinformatics and statistic models (Zhou et al. 2010, Caccia et al. 2013, Gagic et al. 2016). Transcriptomes can also be helpful for secretome insights. Still, it must be noted that not always the level of expression of one transcript corresponds to the

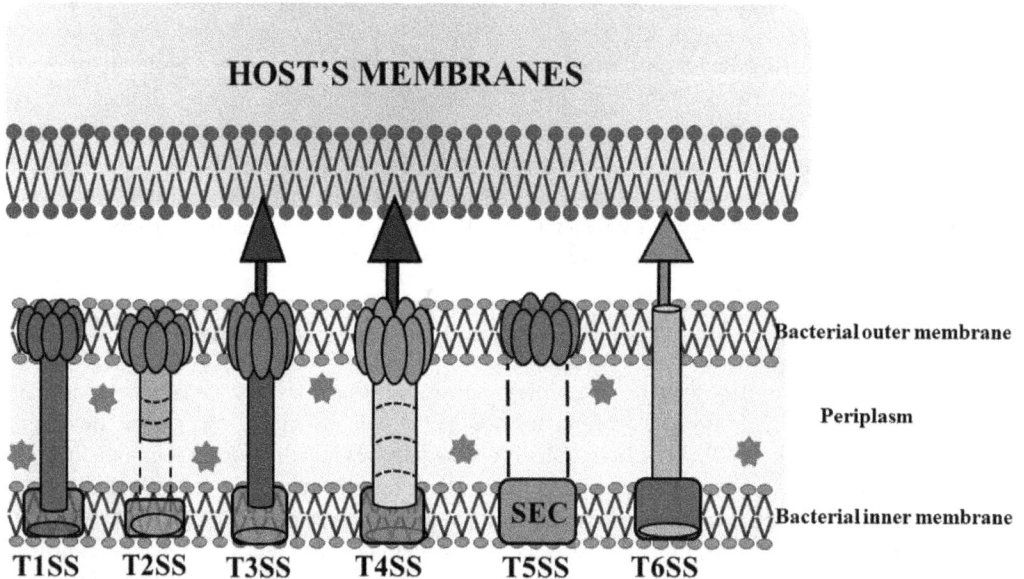

Figure 1. Hypothetical Gram-negative bacterial cell with the six mentioned types of secretion systems. Syringe-like secretion systems (T3SS, T4SS, and T6SS) can inject effectors (septagon-shape) directly into the host cell, while the reminiscent (T1SS, T2SS, and T5SS) only transport effectors from the cytoplasm and periplasm and out of the cell.

exact amount of the respective protein, demanding validation using other approaches under the same condition, such as mass spectrometry (Hathout 2007).

In *Ralstonia,* a special attention is given to their type III effectors, with many different approaches applied to enhance their prediction and characterization (De Ryck et al. 2023). Genome-based prediction involve identification of motifs/domains specific to these effectors (Cunnac et al. 2004), but transcriptomics are still essential to understand the participation of the effector in pathogenesis context (Busset et al. 2021). Further, proteome-based approaches have been used to understand interactions between the effector and host proteins (De Meyer et al. 2020).

As Gram-negative bacteria first need to transport proteins into the periplasm before translocation out of the cell or into the host, Tat and Sec pathways were the first targets in secretome studies involving *R. solanacearum*. Bioinformatics predictions of proteins containing Tat-associated motifs have indicated that approximately 70 *R. solanacearum* proteins are exported to the periplasm through the Tat pathway before potential secretion through T2SS (González et al. 2007, Poueymiro and Genin 2009). Subsequent studies focused on T2SS and T3SS mutants of this same species revealed many insights into how hundreds of effectors secreted by them were crucial for plant cell wall degradation, colonization, and multiplication inside the host, besides triggering host resistance mechanism, for instance (Genin and Boucher 2004, Liu et al. 2005, Jones and Dangl 2006, Poueymiro and Genin 2009). Most recent data has revealed that the secretome of *R. solanacearum* comprises around 228 proteins involved in many cellular activities, and over one-third of them belong to Rips (Lonjon et al. 2016).

Those cellular types of machinery collectively contribute to bacterial success in the environment and inside hosts. The following pages will go through the most relevant secretion systems for *Ralstonia* species: T2SS and T3SS. We will go over their structure, main functions, and why they are essential for these bacteria. Finally, the role of the reminiscent secretion systems will be first discussed.

Ralstonia's Arsenal of Secretion Systems

Type I secretion system is the most spread one across diverse Gram-negative bacteria, including animal and plant pathogens, like *Ralstonia* species (Gama et al. 2016, Green and Mecsas 2016). It is an ABC transporter dependent system that secretes proteins across the cell membranes composed of three essential components: an ABC transporter, which captures the proteins on the inner membrane; an MFP (membrane fusion protein), which transports molecules across the periplasm into the outer membrane; and an OMF (outer membrane factor), a porin-like protein that finally secretes the molecules (Thomas et al. 2014, Morgan et al. 2017). One example of proteins secreted by T1SS is RTX-like toxin genes, important for host cell rupture, found in the *R. solanacearum* genome (Genin and Boucher 2004, Green and Mecsas 2016).

Genes related to the type IV secretion system are also found in *R. solanacearum* genomes. This system is generally responsible for bacteria-bacteria interaction via conjugation but can also secrete protein-protein and protein-DNA complexes into a broad range of target cells, including eukaryotic host cells (Cascales and Christie 2003, Green and Mecsas 2016). Usually, 12 proteins belonging to the VirB/D family assemble this system, comprising an ATP-power system on the inner membrane, a secretion channel through the periplasm and its associated proteins, besides an extracellular T pilus that recognizes and delivers molecules into the target cells (Fronzes et al. 2009, Green and Mecsas 2016). This system serves different purposes depending on the species where they are found. Still, they are believed to uptake hosts' DNA sequences related to resistance and adaptation and deliver virulence effectors that numb the host immune responses, providing a better environment for pathogen colonization (Isberg et al. 2009, Gama et al. 2016). Recently, the T4SS has been related to the acquisition of arsenite resistance genes through horizontal transference in gentamicin resistant strains of *R. pickettii* (Ferro et al. 2021).

The type V secretion system is unique as it is composed of autotransporter proteins that cross the outer membrane and secrete molecules themselves (Gama et al. 2016, Green and Mecsas 2016). The superfamily of autotransporter proteins found in T5SS usually exhibits one peptide with a C-terminal pore-forming β-barrel domain for substrate secretion and another peptide with an N-terminal exodomain, allowing this system to channel into the membrane and transport molecules which were delivered to the periplasm via the Sec pathway (Leyton et al. 2012, Green and Mecsas 2016). The T5SS is divided into three categories according to the number of proteins that participate in the secretion process: (i) type Va, the classical model; (ii) type Vb, the two-partner-secretion model, composed of two proteins (one with the exodomain and the pore-forming β-barrel domain) who are translated and translocated into the membranes separately; and (iii) type Vc, the trimeric AT adhesin, which is composed of polypeptides that are translated together and act together as a trimeric since they have short C-terminal β-barrel chains (Jacob-Dubuisson et al. 2004, Meng et al. 2006, Leyton et al. 2012). The genome of *R. solanacearum* exhibits genes for at least two hemagglutinin-related proteins with autotransporter features, which may aid in the secretion of diverse virulence factors such as adhesins and toxins (Genin and Boucher 2004).

At last, the type VI secretion system is also a syringe-like apparatus responsible for the secretion of toxin substrates, which is crucial for bacterial interspecies competition and useful for pathogenesis success (Russell et al. 2014). Despite the late identification and characterization of T6SS coding genes in some *R. solanacearum* genomes Zhang et al. 2012, 2014), their role in virulence contribution to host infection has only been recently proved in mutants essays infecting eggplants (Asolkar and Ramesh 2020).

The Type II Secretion System: Wall-Breaker of Host's Cell

The translocation through T2SS is a two-step process requiring the protein first to be transported into the periplasmic environment through the Sec or Tat secretion pathways (Gama et al. 2016, Green and Mecsas 2016), and all of them are required for proper molecules delivery. It comprises a versatile system conserved in several Gram-negative bacteria, and has been shown to have a significant function in their pathogenesis. For example, *Vibrio cholerae* and *Vibrio vulnificus* mutants lacking PilD (which plays an essential role in pseudopili formation) are attenuated in animal models (Fullner and Mekalanos 1999, Paranjpye and Strom 2005). Furthermore, the T2SS was one of the first described for *R. solanacearum* (Kang 1994). While not considered as vital to pathogenesis as the type III secretion system, T2SS is essential for *R. solanacearum*'s virulence, responsible for the secretion of various cell wall degrading enzymes (Kang 1994, Liu et al. 2005, Poueymiro and Genin 2009). Lastly, the T2SS counts with highly conserved 15 protein complexes that transport folded proteins from the periplasm to the extracellular environment.

The T2SS comprises four subunits: an ATPase, an inner-membrane platform, an outer-membrane complex, and a pseudopilus. The ATPase, located in the cytoplasm, provides energy to the system, allowing protein transport. The inner-membrane platform extends into the periplasm, facilitating signaling communication between the pilus, the secretin, and ATPase. The outer-membrane complex serves as a transmembrane channel composed of multimeric secretin. The pseudopilus, which is structurally similar to type IV pili found on the bacterial cell surface, theoretically pushes the substrate through the outer membrane complex (Gama et al. 2016, Green and Mecsas 2016). An analysis of the *Ralstonia solanacearum* genome shows that it contains the complete set of core T2SS genes located in the RSc3105–RSc3116 gene cluster.

The general secretion (Sec) pathway is responsible for transporting unfolded proteins from the cytoplasm to the periplasmic environment. Because the T2SS only secretes folded substrates, proteins transported through the Sec pathway must be folded in the periplasm before secretion. The Sec system is composed of three distinct subunits: SecB, a protein targeting component that recognizes and binds to pre-secretory proteins, SecA, a motor protein and ATPase providing energy for transport, and SecYEG, a membrane-integrated translocase (Papanikou et al. 2007, Green and Mecsas 2016).

The twin-arginine protein translocation (Tat) pathway is vital to *R. solanacearum* virulence. Unlike the Sec pathway, the Tat pathway transports proteins that have already been folded. It is mainly related to the secretion of proteins that go through post-translational modifications or cannot be folded within the periplasm (Natale et al. 2008, Green and Mecsas 2016). The Tat pathway for Gram-negative bacteria is composed of three proteins: TatA and TatB, which bind to the signal peptide of Tat-secreted proteins, and TatC, which is then recruited to form a transmembrane channel. The signal peptide in Tat-secreted proteins is recognized by the "twin" arginines at the N-terminus of the folded protein (González et al. 2007, Natale et al. 2008, Green and Mecsas 2016). As for its importance in pathogenesis, a *R. solanacearum* strain with a mutated TatC gene lost much of its capability to cause bacterial wilt, proving that the Tat pathway participates in the secretion of vital virulence factors (González et al. 2007).

The most relevant studies on *R. solanacearum* revealed that the T2SS main contribution to pathogenesis is to secrete a variety of extracellular proteins with degrading features. Those proteins are often called cell-wall-degrading enzymes (CWDEs), important in damaging plant cells for nutrient uptake (Gama et al. 2016). Among these, there are pectic enzymes (PehA, PehB, PehC, and Pme), and cellulolytic enzymes (Egl, CbhA), as well as extracellular nucleases (NucA, NucB), which serve to degrade and avoid host defenses. Mutant strains of *R. solanacearum* lacking these CWDEs have been shown to have diminished virulence (Kang 1994, Liu et al. 2005, González et al. 2007, Poueymiro and Genin 2009). Furthermore, another *R. solanacearum* mutant strain with an inactivated T2SS secretin gene has been shown to significantly attenuate virulence, even with the mutant strains lacking all known cell-wall-degrading enzymes. This diminished virulence points to further unidentified virulence factors associated with T2SS in *Ralstonia solanacearum*.

The Type III Secretion System: An Effectome Powerplant

Of all secretion systems, the type III must be the critical player for host invasion and colonization success, as it secretes proteins capable of numbing the host basal immune responses (Staskawicz et al. 2001, Gama et al. 2016). Like T4SS and T6SS, the T3SS structure is described as a needle or syringe that allows it to secrete protein substrates directly into the target cell's cytoplasm (Green and Mecsas 2016). This system comprises nine core proteins, but up to 20 accessory proteins can be associated with them, depending on the target species (Gama et al. 2016, Green and Mecsas 2016). The basic structure called the "injectisome" comprises two basal body rings, one in the inner and another on the outer membrane, making a hollow channel for protein translocation; one inner rod that guides the protein from inside the periplasm; and an export pilus-like apparatus, which carries a needle for protein secretion inside the host cell's cytoplasm (Burkinshaw and Strynadka 2014). Since this system is essential for pathogenesis, it requires a complex temporal and spatial regulation of the genes involved for its correct assembly.

The coding genes responsible for the T3SS are the *hrp/hrc* (hypersensitive response and pathogenicity) gene clusters. As they're located in few operons inside the genome, they are commonly horizontally transferred (i.e., via pathogenicity islands and plasmids) amongst pathogenic bacteria, so one bacterium may present different types of T3SS, even when compared to close related strains of the same species (Troisfontaines and Cornelis 2005, Green and Mecsas 2016). The nine core Hrp proteins are homologous and highly conserved in both animal and plant Gram-negative pathogen bacteria, so their respective genes were renamed *hrc* (hypersensitive response conserved) (Bogdanove et al. 1996, Van Gijsegem et al. 1998, Gama et al. 2016, de Pedro-Jové et al. 2021).

In *R. solanacearum*, this cluster is located in an operon region and organized on at least seven transcriptional units regulated by *hrr*G and *hrp*B genes. Interestingly, this cluster is located on the megaplasmid (instead of the main chromosome) and lacks pathogenicity islands (PAI) transference characteristics, such as the absence of nearby mobile elements and GC content similar to the core genome (Tampakaki et al. 2010). Moreover, many studies on *R. solanacearum* and other phytopathogenic bacteria have revealed that HrpB is a crucial regulator to many effector proteins, including the ones secreted by different types of secretion systems, like T2SS (Furutani et al. 2004, Kang et al. 2008, Mukaihara et al. 2010, Jeong et al. 2011).

The first studies on T3SS involving *Ralstonia* species tried to understand the importance of this system in plant interactions, as the host signal during infection induces the *hrp/hrc* cluster expression and T3SS assembly. They proved that mutants with *hrp*-depleted genomes had significantly reduced capacity for both host colonization and ability to cause disease since they could not assemble T3SS and other genomic regions regulated by the *hrp/hrc* cluster were prejudiced (Kanda et al. 2003, Mukaihara et al. 2010). In *R. solanacearum*, the *hrp*Y gene is responsible for the T3SS pilus assembly, involved in either protein and harpins secretion, and host signal receptor (Staskawicz et al. 2001, Kvitko et al. 2007). Finally, later studies have shown that the T3SS in *R. solanacearum* not only impacts the ability to cause pathogenicity but also influences how the host interacts with other microbes, such as mutualistic nodulating bacteria and the ability for nodulation itself (Guan et al. 2013, Benezech et al. 2021). The following section will go deeper into effectors secreted by the T3SS and their impact on *R. solanacearum* pathogenicity and regulation.

Hrp Regulators' Refined Roles in the Pathogenicity Scenario

Gene regulation analyses were crucial to explain how *R. solanacearum* sharply tunes its pathogenicity. The transcription of the *hrp* family genes - which encode the T3SS - and its related effectors were found to be managed by the HrpB transcriptional activator (de Pedro-Jové et al. 2021). The first studies regarding this gene revealed that the *hrp*B gene is located downstream of a regulatory cascade induced by the contact of the bacterium with the plant cell wall (Brito et al. 1999). In *R. solanacearum,* HrpB (an AraC-type regulator), and HrpG (an upstream OmpR-like response regulator), are responsible for controlling *hrp/hrc* family gene expression (Peeters et al. 2013b). Moreover, HrpB is a direct trigger for the transcription of T3SS genes, likely by binding to the *hrpII* box found in their promoter regions, and its expression is governed by HrpG (Peeters et al. 2013b). Both genes are genetically and functionally conserved/preserved in *Xanthomonas* spp. Still, a unique feature of the *R. solanacearum* genome is the presence of upstream regulators, which can specifically induce *hrp*G expression when the bacterium identifies a plant cell-wall component (Aldon et al. 2000). PrhA perceives the signal from detecting cell-wall components, an outer membrane receptor, which transmits the signal to PrhI and PrhR, membrane-associated proteins that trigger *hrp/hrc* expression through the transcriptional regulators PrhJ, HrpG, and HrpB (Brito et al. 2002, Peeters et al. 2013b).

Gene expression studies have shown that *R. solanacearum hrp* genes and T3SS effectors were transcribed *in planta* at late infection stages (Monteiro et al. 2012). Other studies of bacteria infecting tomato, banana, and potato roots have confirmed these findings (Jacobs et al. 2012, Ailloud et al. 2016, Puigvert et al. 2017). Further, transcriptomics has indicated that the *hrp* regulators may control additional functions other than the T3SS and its effectors. For instance, the HrpB regulator has previously been involved in chemotaxis and the biosynthesis of certain bacterial compounds, such as the Hrp-dependent factor (HDF), which possibly stimulates a cell density-dependent LuxR system (Delaspre et al. 2007). In addition, the HrpG regulator has also been shown to be involved in a variety of genetic regulatory processes unrelated to bacterial secretion (Valls et al. 2006, Peeters et al. 2013b). Plener and collaborators (2012) have shown that HrpG is involved in methionine synthesis regulation. It has been proposed that it may be a promoter of MetE, which synthesizes methionine without vitamin B12. This process is vital for methionine biosynthesis in vitamin-poor environments in the plant. Thus, it can be said that HrpG is central to the regulation of *R. solanacearum* pathogenicity through the T3SS and controls a wide array of unrelated virulence genes (Plener et al. 2012).

Ralstonia Injected Proteins (Rips): Diversity, Function, and Importance

As we just mentioned, the type III secretion system is vital to *Ralstonia solanacearum* pathogenesis and virulence due to its ability to inject effector proteins directly into the host cell (Landry et al. 2020). These type 3 effectors (T3Es) are thus referred to as *Ralstonia* injected proteins (Rips), a highly diverse repertoire of virulence factors that enable *R. solanacearum* to infect a wide array of host organisms (Lonjon et al. 2016, Landry et al. 2020, Bocsanczy et al. 2022). The bacterium's vast array of T3Es contributes to virulence and pathogen fitness *in planta*, with many roles in pathogenicity (Poueymiro and Genin 2009, Genin 2010, Lonjon et al. 2016).

The first Rips were cloned and identified in the 1990s, discovered in 1990, and initially named AvrA, now referred to as RipAA (Carney and Denny 1990, Landry et al. 2020). Since then, various technologies and techniques have allowed for the identification and functional characterization of additional Rips. The advent of cheaper, more accessible sequencing technologies has allowed the development of sequence-based methods to identify Rips in the genome by searching for homology to known Rips and known molecular patterns in the sequence (Landry et al. 2020). In addition, bioinformatics-based tools capable of predicting and identifying Rip sequences have been developed, further improving our capacity to map virulence-associated T3Es (Peeters et al. 2013a, Sabbagh et al. 2019). Additionally, the ongoing evolution of genetic engineering techniques has allowed the advent of regulation-based methods to identify Rips, as the hrp genes regulate the T3SS expression, and their manipulation can be used in assays aimed at finding T3Es and, consequently, Rips (Furutani et al. 2004, Kang et al. 2008, Mukaihara et al. 2010, Jeong et al. 2011).

Rips have many functions in *Ralstonia*'s life cycle and pathogenicity, some identified and described over the last few decades. Over 50 have already been characterized in some level of detail (Landry et al. 2020). Several Rips, such as RipP2, RipAY, and RipX (the first Rip characterized), are understood to interfere with plant immunity and trigger hypersensitive responses on hosts (Arlat et al. 1994, Le Roux et al. 2015, Fujiwara et al. 2016, 2020). Others, such as RipTAL, act as a transcriptional activator-like effector, targeting and manipulating plant metabolism to boost the production of specific antimicrobial molecules, possibly inhibiting the growth of *R. solanacearum* competitors (Wu et al. 2019). At last, some Rips suppress host recognition and response to other effectors, such as RipAY, RipAC, and RipAK (Sun et al. 2017, Sang et al. 2020, Yu et al. 2020). A micro review published by Landry and colleagues (2020) has synthesized the many functions Rips fulfill for a successful infection.

Recent studies on the multitude of available *R. solanacearum* strains have sought to identify this bacterium and its subspecies' pan-effectome or the common effectors found within the species and its variants. The pan-effectome of *Ralstonia solanacearum* was reported to include 102 type III effector genes and 16 hypothetical T3E genes, with each strain having, on average, 64 T3E genes (Sabbagh et al. 2019). There is also a significant amount of variation between strains, as the core effectome (effectors that are shared by over 95% of the strains) of the bacterium had only 16 proteins, including RipG5, RipB, RipW, RipAC, RipAB, RipR, RipE1, RipAM, RipAN, RipAY, RipAJ, RipF1, and RipAI (Ailloud et al. 2015, Landry et al. 2020). Both the core-effectome and the pan-effectome of *R. solanacearum* are very large and diverse compared with other phytopathogens, such as *Pseudomonas syringae* and *Xanthomonas campestris* (Roux et al. 2015, Dillon et al. 2019, Landry et al. 2020). A couple of Rips families found for *R. solanacearum* to date is available at Table 2. This relatively large and diverse repertoire of Rips has been theorized to be related to this bacterium's ubiquity, although few host-specificity determinants have been found so far (Sabbagh et al. 2019, Bocsanczy et al. 2022).

The pan-effectome of *R. solanacearum* is highly redundant, which may, in part, be related to its size (Angot et al. 2006, Solé et al. 2012, Chen et al. 2014). These redundant effectors point toward a robust virulence strategy: having many genes means having an additive effect on their functions, making the bacterium less susceptible to the deleterious impacts that may arise from mutations in those genes (Landry et al. 2020). A remarkable feature of strains from the former *R. solanacearum*

Table 2. List of type III effectors (T3Es) genes currently identified in the *R. solanacearum* species.

Type III effectors (T3Es)	Representative gene member	Other Name	References
RipA1	RSc2139	AWR1	Solé et al. 2012
RipA2	RSp0099	RipA, Rip29, Hpx31, AWR2	Cunnac et al. 2004, Mukaihara et al. 2010
RipA3	RSp0846	Rip44, Hpx32, AWR3	Mukaihara et al. 2010
RipA4	RSp0847	Rip45, Hpx4, AWR4	Mukaihara et al. 2010
RipA5	RSp1024	Rip56, Hpx10, AWR5	Mukaihara et al. 2010
RipAA	RSc0608	AvrA, Rip5, Brg46	Mukaihara et al. 2010
RipAB	RSp0876	PopB, Rip48	Gueneron et al. 2000, Mukaihara et al. 2010
RipAC	RSp0875	PopC, Rip47	Gueneron et al. 2000; Mukaihara et al. 2010
RipAD	RSp1601	Rip72	Mukaihara et al. 2010
RipAE	RSc0321	Rip4	Mukaihara et al. 2010
RipAF1	RSp0822	Rip40	Mukaihara et al. 2010
RipAF2	R. syzygii RALSY_20037	-	Peeters et al. 2013a
RipAG	RSc0824	Rip6	Mukaihara et al. 2010
RipAH	RSc0895	Rip11	Mukaihara et al. 2010
RipAI	RSp0838	Rip41	Mukaihara et al. 2010
RipAJ	RSc2101	Rip21, Hpx18	Mukaihara et al. 2010
RipAK	RSc2359	Rip23, Hpx28, Brg36	Mukaihara et al. 2010
RipAL	UW551 RRSL_02221	Rip38	Mukaihara et al. 2010
RipAM	RSc3272	Brg40	Peeters et al. 2013a
RipAN	RSp0845	Rip43, Hpx33, Brg33	Mukaihara et al. 2010
RipAO	RSp0879	Rip50, Hpx2, Brg34	Mukaihara et al. 2010
RipAP	UW551 RRSL_04655	Rip60	Mukaihara et al. 2010
RipAQ	RSp0885	Rip51, Brg35	Mukaihara et al. 2010
RipAR	RSp1236	Rip61	Mukaihara et al. 2010
RipAS	RSp1384	Rip66, Hpx9, Brg43	Mukaihara et al. 2010
RipAT	RSp1388	Rip67, Brg48	Mukaihara et al. 2010
RipAU	RSp1460	Rip68, Hpx8, Brg45	Mukaihara et al. 2010
RipAV	RSp0732	Rip39, Hpx27, Brg39	Mukaihara et al. 2010
RipAW	RSp1475	Rip69	Mukaihara et al. 2010, Peeters et al. 2013a
RipAX2	RSp0572	Rip36, Brg14	Mukaihara et al. 2010
RipAY	RSp1022	Rip55, Hpx21, Brg37	Mukaihara et al. 2010
RipAZ1	RSp1582	Rip71	Mukaihara et al. 2010
RipAZ2	R. syzygii RALSY_20407	-	Peeters et al. 2013a
RipB	Rsc0245	RipB, Rip2, Hpx11	Mukaihara et al. 2010
RipBA	RSc0227, RSp0228 [pseudogene]	-	Peeters et al. 2013a
RipBB	Psi07 RPSI07_mp0573	-	Peeters et al. 2013a
RipBC	CFBP2957 RCFBP_mp30170	-	Peeters et al. 2013a
RipBD	R. syzygii RALSY_20184	-	Peeters et al. 2013a

Table 2 contd. ...

...Table 2 contd.

Type III effectors (T3Es)	Representative gene member	Other Name	References
RipBE	RS1000	Rip10	Mukaihara et al. 2010
RipBF	Psi07 RPSI07_2863	-	Peeters et al. 2013a
RipBG	Molk2 RSMK00763	-	Peeters et al. 2013a
RipBH	Psi07 RPSI07_mp1715	-	Peeters et al. 2013a
RipBI	CFBP2957 RCFBP_mp30113	-	Peeters et al. 2013a
RipBJ	GMI1000 RSp0213	-	Lonjon et al. 2016
RipBK	YC45_c025370	HopAM1	Chang et al. 2005
RipBL	YC45_m001910	HopAO1	Chang et al. 2005
RipBM	Psi07 RSPsi07_1860	-	Sabbagh et al. 2019
RipBN	CMR15v4_30917	AvrRpt2	Eschen-Lippold et al. 2016
RipBO	GM1000 RSc3174	-	Sabbagh et al. 2019
RipBP	OE1-1_24290	HopW1 (homologs in *Xanthomonas*)	Zumaquero et al. 2010
RipBQ	KACC10722_38580	HopK1 XopAK	Chang et al. 2005
RipC1	RSp1239	Rip62	Cunnac et al. 2004, Mukaihara et al. 2010
RipC2	CFBP2957 RCFBP_mp20032	-	Mukaihara et al. 2010
RipD	RSp0304	Rip34, Hpx25, Brg8	Mukaihara et al. 2010
RipE1	RSc3369	Rip26, Brg9	Mukaihara et al. 2010
RipE2	CFBP2957 RCFBP_mp10565	-	Mukaihara et al. 2010
RipF1	RSp1555	PopF1, PopF2, Rip70	Meyer et al. 2006, Mukaihara et al. 2010
RipF2	CFBP2957 RCFBP_mp30453	-	Meyer et al. 2006, Mukaihara et al. 2010
RipG1	RSp0914	Gala1, Rip53	Mukaihara et al. 2010
RipG2	RSp0672	Gala2, Rip37, Hpx20	Mukaihara et al. 2010
RipG3	RSp0023	Gala3, Rip28	Mukaihara et al. 2010
RipG4	RSc1800	Gala4, Rip17, Hpx15	Mukaihara et al. 2010
RipG5	RSc1801	Gala5, Rip18, Hpx16	Mukaihara et al. 2010
RipG6	RSc1356	RipG, Gala6, Rip13, Hpx13	Baltrus et al. 2011, Cunnac et al. 2004
RipG7	RSc1357	Gala7, Rip14, Hpx14	Mukaihara et al. 2010
RipG8	CMR15 CMR15v4_10224	Gala8	Mukaihara et al. 2010
RipH1	RSc1386	HLK1, Rip15, Brg19	Mukaihara et al. 2010
RipH2	RSp0215	HLK2, Rip32	Mukaihara et al. 2010
RipH3	RSp0160	HLK3, Rip30, Brg18	Mukaihara et al. 2010
RipH4	Psi07 RPSI07_mp0161	HLK4	Mukaihara et al. 2010
RipI	RSc0041	Rip1	Mukaihara et al. 2010
RipJ	RSc2132	Rip22	Mukaihara et al. 2010
RipK	CFBP2957 RCFBP_mp10024	-	Mukaihara et al. 2010
RipL	RSp0193	Rip31, Brg22	Mukaihara et al. 2010
RipM	RSc1475	Rip16, Brg42	Mukaihara et al. 2010
RipN	RSp1130	Rip58, Hpx26, Brg44	Mukaihara et al. 2010

Table 2 contd. ...

...Table 2 contd.

Type III effectors (T3Es)	Representative gene member	Other Name	References
RipO1	RSp0323	Rip35, Brg12	Mukaihara et al. 2010
RipO2	R. syzygii RALSY_mp30159	-	Peeters et al. 2013a
RipP1	RSc0826	PopP1, Rip7	Lavie et al. 2002, Mukaihara et al. 2010
RipP2	RSc0868	PopP2, Rip8	Cunnac et al. 2004, Mukaihara et al. 2010
RipP3	UW163 [GenBank accession : CAF32358.1]	PopP3	Peeters et al. 2013a
RipQ	RSp1277 Rip63	Hpx23	Mukaihara et al. 2010
RipR	RSp1281	Rip64, Hpx24, Brg15, PopS	Mukaihara et al. 2010
RipS1	RSc3401	SKWP1, Rip27, Hpx37	Mukaihara et al. 2010
RipS2	RSp1374	SKWP2, Rip65, Hpx36	Mukaihara et al. 2010
RipS3	RSp0930	SKWP3, Rip54	Mukaihara et al. 2010
RipS4	RSc1839	SKWP4, Rip20, Hpx30	Mukaihara et al. 2010
RipS5	RSp0296	SKWP5, Rip33, Hpx34	Mukaihara et al. 2010
RipS6	RSc2130	SKWP6	Peeters et al. 2013a
RipS7	Molk2 RSMK02658	SKWP7	Peeters et al. 2013a
RipS8	Psi07 RSPsi07_1850	SKWP8	Peeters et al. 2013a
RipT	RSc3212	RipT, Rip25	Cunnac et al. 2004, Mukaihara et al. 2010
RipTAL1	Rsc1815	Rip19, Hpx17, Brg11	Mukaihara et al. 2010
RipTPS	RSp0731	-	Peeters et al. 2013a
RipU	RSp1212	Rip59	Mukaihara et al. 2010
RipV1	RSc1349	Rip12, Hpx29, Brg17	Mukaihara et al. 2010
RipV2	Psi07 RSPsi07_1895	-	Mukaihara et al. 2010
RipW	RSc2775	PopW, Rip24	Li et al. 2010, Mukaihara et al. 2010
RipX	RSp0877	PopA, Rip49	Mukaihara et al. 2010, Yang et al. 2000
RipY	RSc0257	Rip3, Brg23	Mukaihara et al. 2010
RipZ	RSp1031	Rip57, Brg38	Mukaihara et al. 2010
Hyp17	RS244_m000380	-	Sabbagh et al. 2019
Hyp18	CMR15v4_mp10535	-	Sabbagh et al. 2019

species complex is the presence of genes for at least one of these Rip families: RipG (GALA), RipS (SKWP), RipA (AWR), RipH (HLK), or RipP (PopP) (Landry et al. 2020), as discussed below.

The RipG (GALA) family is composed of seven genes with leucine-rich repeats on their sequences, initially identified by their homology with plant proteins with F-box domains, which participate in the ubiquitination of a broad range of proteins (Kirkpatrick et al. 2005, Angot et al. 2006). These effectors are believed to associate with SKP-1-like proteins and promote the ubiquitination of chloroplast proteins, leading to plant cell death (Wang et al. 2016, Landry et al. 2020). The RipG family can be subdivided into two clades based on neofunctionalization: RipG1, 3, 4, 5, and RipG2, 6, 7, in which RipG1, 2, and RipG5 are ancestral genes, and other acquired new functions through diversification (Remigi et al. 2011). In addition, RipG4 is responsible for callose deposition inhibition in *A. thaliana*, and RipG1 and 3 inhibit host immune responses dependent on salicylic acid (Remigi et al. 2011, Medina-Puche et al. 2020). Besides, the core effector RipG7 is one of the most studied in this family and is attributed to promoting full virulence in tomato and *A. thaliana* infections and compatible interaction with other hosts (Angot et al. 2006, Wang et al. 2016).

When the RipS (SKWP) family was discovered, it was described as the family with the largest effectors of plant and animal pathogenic bacteria. It comprises six effectors (RipS1-6) characterized by a 42-long novel motif in 12–18 tandem repeats, called the SKWP repeats (Mukaihara and Tamura 2009, Sabbagh et al. 2019). *R. solanacearum* mutants with RipS deleted genes did not lose their capacity to infect tobacco roots and leaves; however, they were necessary to achieve full virulence on the host (Peeters et al. 2013a). Effectors that portray long domains, like SKWP, usually interact with plant factors to fulfill virulence functions (Mukaihara and Tamura 2009). Encoded by initially five genes (*awr*1–5), the RipA (AWR) family proteins have triad-domains of alanine-tryptophan-arginine, and the core effectors RipA2, RipA3, and RipA5 are crucial for *R. solanacearum* virulence, multiplication inside the host, and cell death (Solé et al. 2012).

The RipH (HLK) family comprises three effectors (RipH1-3) that are 700-800 amino acids long and have no particular known motifs or domains, being named after the histidine-leucine-lysine triad conserved on the sequence C-terminal region (Poueymiro and Genin 2009, Chen et al. 2014). Two of the coding genes (*hlk*2 and *hlk*3) are located on the megaplasmid, while the other remains on the main chromosome (Chen et al. 2014). Collectively, these proteins have been associated with promoting full virulence in tomatoes, but mutant strains with *rip*H deleted genomes were as virulent as wild strains in tobacco and eggplant (Chen et al. 2014). The RipP (PopP) family comprises two effectors, RipP1 and Rip2, that belong to the YopJ/AvrRxv superfamily of proteins related to the avirulence/immune responses on both animal and plant hosts. Once the bacteria are inside the host, RipP2 plays a crucial role in Avr hypersensitive reactions. It promotes virulence on hosts numbing the immune responses via acetylation of WRKY transcriptions factors while RipP1 is somehow related to avirulence responses in some petunia lines (Deslandes et al. 2003, Macho et al. 2010).

Ralstonia's T3Es and Host Specificity

When the former RSSC was grouped in phylotypes, the clades were also characterized based on the host range of the strains and the varied diseases they would cause, often referred to as ecotypes. Common ecotypes of RSSC are the classic bacterial wilt, Moko disease, brown rot of potato, banana blood disease, and the emergent NPB (non-pathogenic to banana) (Genin and Denny 2012, Santiago et al. 2016, 2017). This variability of hosts directly reflects the genetic variability of strains, which portrays different molecular tools to infect and cause disease on a given host, which in turn also responds in a specific manner. For example, when a pathogen invades a plant cell, a standard, robust defense mechanism is to induce the release of many chemical and biochemical compounds and alter ions effluxes, leading to cell death and thus stopping the pathogen from spreading across the host's systems (Figure 2). This mechanism is called a hypersensitive response, or HR.

The simplest model explains that an HR occurs due to an incompatible interaction between the products of a pathogen *avr* gene and a plant resistance gene (*R*), with a compatible interaction leading to further disease on the host (Bent 1996, Morel and Dangl 1997). The *hrp* gene cluster on *Ralstonia's* genome is crucial for inducing and suppressing HRs since they also regulate *avr/rip* genes (Lam et al. 2001, He 2006, Gama et al. 2016). The pathogen's virulence effectors will subvert metabolic responses and diminish the host's defense molecules on a susceptible host. In this context, *Ralstonia*'s T3Es injected inside the host cell may be recognized (or not) by plant R proteins, undergoing a selective pressure that occasionally causes loss of recognized effectors to infect a broader range of hosts. That explains why some *Ralstonia* strains can infect some hosts while others cannot (Kraepiel and Barny 2016).

Phylogenetic and comparative assays are indispensable to understanding which groups of strains are able to infect a given host. One of the first studies focused on Japanese strains of RSSC revealed that despite most strains belonging to phylotype I, a specific genetically related group

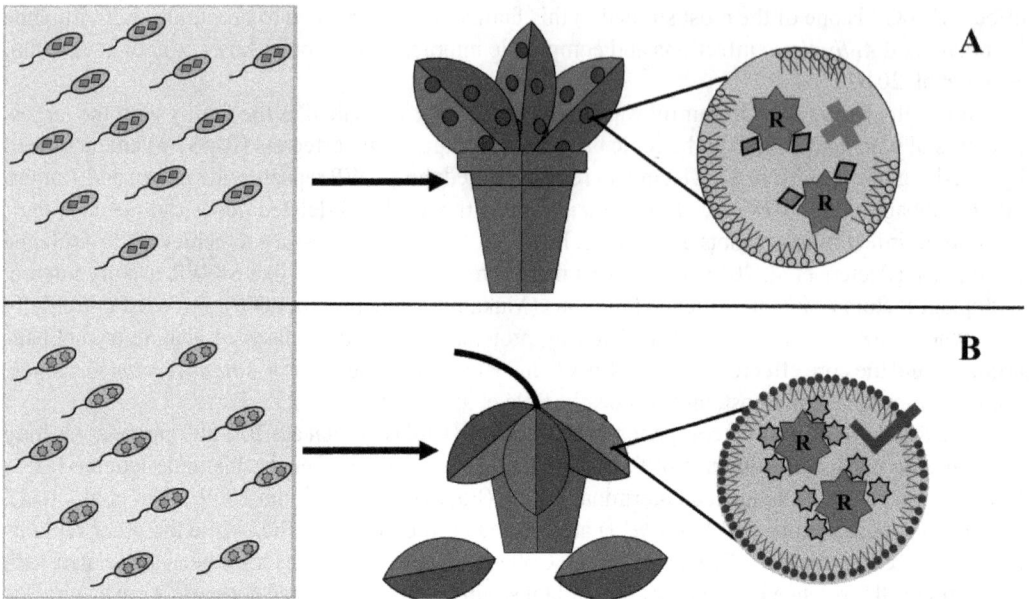

Figure 2. Scheme for pathogen adaptation for new hosts via a change in the Rips repertoire. One given Rip triggers a robust hypersensitive response protecting the host from infection (A), while the other has a compatible interaction with host defense R protein (B), leading to disease.

of strains would initiate HR on tobacco, and others would cause bacterial wilt (Liu et al. 2009). When looking deeper into the effectome, more correlations between Rips and host specificity are found. Comparative genomic studies on *R. solanacearum* strains of different ecotypes showed that the emergent NPB and brown rot ecotypes probably changed their host range due to losing a few (but essential) conserved genes in Moko disease strains including *ripAA* and *ripAU*, respectively (Ailloud et al. 2015). Contributing to this correlation, comparative analysis of emergent strains of *R. pseudosolanacearum* pathogenic to new different hosts, like blueberry, tea rose, mandevilla, and osteospermum, has revealed that despite strains being phylogenetically close, their ability to infect specific hosts are related to absence and presence of specific assets of T3Es (Bocsanczy et al. 2022)

Conclusion

This chapter aimed to update the findings of the *Ralstonia* secretome, addressing the main secretion systems and the characterization of new discoveries about *Ralstonia* Injected Proteins.

Different omics data revealed the existence of a significant genetic diversity of these elements resulting from the coevolutionary host-pathogen response. With an in-depth understanding of this topic, we can encompass new perspectives of biotechnological applications, including developing diagnostic kits to detect and control emerging pathogens and the prospection of targets to combat *Ralstonia´*s phytopathogens of worldwide relevance.

References

Abdallah, A.M., N.C. Gey van Pittius, P.A.D. Champion, J. Cox, J. Luirink, C.M.J.E. Vandenbroucke-Grauls et al. 2007. Type VII secretion--mycobacteria show the way. Nat. Rev. Microbiol. 5: 883–891.

Ailloud, F., T. Lowe, G. Cellier, D. Roche, C. Allen and P. Prior. 2015. Comparative genomic analysis of *Ralstonia solanacearum* reveals candidate genes for host specificity. BMC Genomics 16: 270.

Ailloud, F., T.M. Lowe, I. Robène, S. Cruveiller, C. Allen and P. Prior. 2016. *In planta* comparative transcriptomics of host-adapted strains of *Ralstonia solanacearum*. PeerJ 4: e1549.

Aldon, D., B. Brito, C. Boucher and S. Genin. 2000. A bacterial sensor of plant cell contact controls the transcriptional induction of *Ralstonia solanacearum* pathogenicity genes. EMBO J. 19: 2304–2314.

Angot, A., N. Peeters, E. Lechner, F. Vailleau, C. Baud, L. Gentzbittel et al. 2006. *Ralstonia solanacearum* requires F-box-like domain-containing type III effectors to promote disease on several host plants. Proc. Natl. Acad. Sci. U. S. A. 103: 14620–14625.

Arlat, M., F. Van Gijsegem, J.C. Huet, J.C. Pernollet and C.A. Boucher. 1994. PopA1, a protein which induces a hypersensitivity-like response on specific Petunia genotypes, is secreted via the Hrp pathway of *Pseudomonas solanacearum*. EMBO J. 13: 543–553.

Asolkar, T. and R. Ramesh. 2020. The involvement of the Type Six Secretion System (T6SS) in the virulence of *Ralstonia solanacearum* on brinjal. 3 Biotech 10: 324.

Benezech, C., A. Le Scornet and B. Gourion. 2021. *Medicago-Sinorhizobium-Ralstonia*: A model system to investigate pathogen-triggered inhibition of Nodulation. Mol. Plant-Microbe Interactions® 34: 499–503.

Bent, A. 1996. Plant disease resistance genes: Function meets structure. Plant Cell 8: 1757–1771.

Bocsanczy, A.M., P. Bonants, J. van der Wolf, M. Bergsma-Vlami and D.J. Norman. 2022. Identification of candidate type 3 effectors that determine host specificity associated with emerging *Ralstonia pseudosolanacearum* strains. Eur. J. Plant Pathol. 163: 35–50.

Bogdanove, A.J., S.V. Beer, U. Bonas, C.A. Boucher, A. Collmer, D.L. Coplin et al. 1996. Unified nomenclature for broadly conserved hrp genes of phytopathogenic bacteria. Mol. Microbiol. 20: 681–683.

Brito, B., D. Aldon, P. Barberis, C. Boucher and S. Genin. 2002. A signal transfer system through three compartments transduces the plant cell contact-dependent signal controlling *Ralstonia solanacearum hrp* Genes. Mol. Plant-Microbe Interactions® 15: 109–119.

Brito, B., M. Marenda, P. Barberis, C. Boucher and S. Genin. 1999. prhJ and hrpG, two new components of the plant signal-dependent regulatory cascade controlled by PrhA in *Ralstonia solanacearum*. Mol. Microbiol. 31: 237–251.

Burkinshaw, B.J. and N.C.J. Strynadka. 2014. Assembly and structure of the T3SS. Biochim. Biophys. Acta BBA - Mol. Cell Res., Protein trafficking and secretion in bacteria 1843: 1649–1663.

Caccia, D., M. Dugo, M. Callari and I. Bongarzone. 2013. Bioinformatics tools for secretome analysis. Biochim. Biophys. Acta 1834: 2442–2453.

Carney, B.F. and T.P. Denny. 1990. A cloned avirulence gene from *Pseudomonas solanacearum* determines incompatibility on *Nicotiana tabacum* at the host species level. J. Bacteriol. 172: 4836–4843.

Cascales, E. and P.J. Christie. 2003. The versatile bacterial type IV secretion systems. Nat. Rev. Microbiol. 1: 137–149.

Chang, J.H., J.M, Urbach, T.F. Law, L.W. Arnold, A. Hu, S. Gombar. et al. 2005. A high-throughput, near-saturating screen for type III effector genes from *Pseudomonas syringae*. Proc. Natl. Acad. Sci. 102: 2549–2554.

Chen, L., M. Shirota, Z. Yong, A. Kiba, Y. Hikichi and K. Ohnishi. 2014. Involvement of HLK effectors in *Ralstonia solanacearum* disease development in tomato. J. Gen. Plant Pathol. 80.

Cornelis, G.R. and F. Van Gijsegem. 2000. Assembly and function of type III secretory systems. Annu. Rev. Microbiol. 54: 735–774.

Cunnac, S., C. Boucher and S. Genin. 2004. Characterization of the cis-acting regulatory element controlling HrpB-mediated activation of the type III secretion system and effector genes in *Ralstonia solanacearum*. Journal of Bacteriology 186(8): 2309–2318.

de Pedro-Jové, R., M. Puigvert, P. Sebastià, A.P. Macho, J.S. Monteiro, N.S. Coll et al. 2021. Dynamic expression of *Ralstonia solanacearum* virulence factors and metabolism-controlling genes during plant infection. BMC Genomics 22: 170.

De Ryck, J., P. Van Damme and S. Goormachtig. 2023. From prediction to function: Current practices and challenges towards the functional characterization of type III effectors. Frontiers in Microbiology, 14.

Delaspre, F., C.G. Nieto Peñalver, O. Saurel, P. Kiefer, E. Gras, A. Milon et al. 2007. The *Ralstonia solanacearum* pathogenicity regulator HrpB induces 3-hydroxy-oxindole synthesis. Proc. Natl. Acad. Sci. 104: 15870–15875.

Deslandes, L., J. Olivier, N. Peeters, D.X. Feng, M. Khounlotham, C. Boucher et al. 2003. Physical interaction between RRS1-R, a protein conferring resistance to bacterial wilt, and PopP2, a type III effector targeted to the plant nucleus. Proc. Natl. Acad. Sci. 100: 8024–8029.

Dillon, M.M., R.N.D. Almeida, B. Laflamme, A. Martel, B.S. Weir, D. Desveaux et al. 2019. Molecular evolution of *Pseudomonas syringae* Type III Secreted Effector Proteins. Front. Plant Sci. 10.

Eschen-Lippold, L., X. Jiang, J.M. Elmore, D. Mackey, L. Shan, G. Coaker et al. 2016. Bacterial AvrRpt2-like cysteine proteases block activation of the *Arabidopsis* Mitogen-activated protein Kinases, MPK4 and MPK11. Plant Physiol. 171: 2223–2238.

Ferro, P., I. Vaz-Moreira and C.M. Manaia. 2021. Evolution of gentamicin and arsenite resistance acquisition in *Ralstonia pickettii* water isolates. Res. Microbiol. 172: 103790.

Fronzes, R., P.J. Christie and G. Waksman. 2009. The structural biology of type IV secretion systems. Nat. Rev. Microbiol. 7: 703–714.

Fujiwara, S., A. Ikejiri, N. Tanaka and M. Tabuchi. 2020. Characterization of the mechanism of thioredoxin-dependent activation of γ-glutamylcyclotransferase, RipAY, from *Ralstonia solanacearum*. Biochem. Biophys. Res. Commun. 523: 759–765.

Fujiwara, S., T. Kawazoe, K. Ohnishi, T. Kitagawa, C. Popa, M. Valls et al. 2016. RipAY, a plant pathogen effector protein, exhibits robust γ-Glutamyl Cyclotransferase Activity When Stimulated by Eukaryotic Thioredoxins*. J. Biol. Chem. 291: 6813–6830.

Fullner, K.J. and J.J. Mekalanos. 1999. Genetic characterization of a new type IV-A Pilus Gene cluster found in both classical and El Tor Biotypes of Vibrio cholerae. Infect. Immun. 67: 1393–1404.

Furutani, A., S.Tsuge, K. Ohnishi, Y. Hikichi, T. Oku, K. Tsuno et al. 2004. Evidence for HrpXo-dependent expression of type II secretory proteins in *Xanthomonas oryzae* pv. *oryzae*. J. Bacteriol. 186: 1374–1380.

Gagic, D., M. Ciric, W.X. Wen, F. Ng and J. Rakonjac. 2016. Exploring the secretomes of microbes and microbial communities using filamentous phage display. Front. Microbiol. 7.

Gama, M.A.S., A. Nicoli, W.J. De Oliveira and M.H.J. de Lima-Carvalho. 2016. Genética da interação bactéria-planta. In: Estado da Arte em Fitobacterioses Tropicais. EDFRPE - Editora Universitária da UFRPE, Recife, Pernambuco, Brasil, pp. 43–64.

Genin, S. 2010. Molecular traits controlling host range and adaptation to plants in *Ralstonia solanacearum*. New Phytol. 187: 920–928.

Genin, S. and C. Boucher. 2004. Lessons learned from the genome analysis of *Ralstonia solanacearum*. Annu. Rev. Phytopathol. 42: 107–134.

Genin, S. and T.P. Denny. 2012. Pathogenomics of the *Ralstonia solanacearum* species complex. Annu. Rev. Phytopathol. 50: 67–89.

González, E.T., D.G. Brown, J.K. Swanson and C. Allen. 2007. Using the *Ralstonia solanacearum* tat secretome to identify bacterial wilt virulence factors. Appl. Environ. Microbiol. 73: 3779–3786.

Green, E.R. and J. Mecsas. 2016. Bacterial secretion systems: An Overview. Microbiol. Spectr. 4.

Guan, S.H., C. Gris, S. Cruveiller, C. Pouzet, L. Tasse, A. Leru et al. 2013. Experimental evolution of nodule intracellular infection in legume symbionts. ISME J. 7: 1367–1377.

Hathout, Y. 2007. Approaches to the study of the cell secretome. Expert Rev. Proteomics 4: 239–248.

He, S.Y. 2006. Role of the type III protein secretion system in bacterial infection of plants. In: Virulence Mechanisms of Bacterial Pathogens. ASM Press, Washington, pp. 209–220.

Isberg, R.R., T.J. O'Connor and M. Heidtman. 2009. The *Legionella pneumophila* replication vacuole: Making a cosy niche inside host cells. Nat. Rev. Microbiol. 7: 13–24.

Jacob-Dubuisson, F., R. Fernandez and L. Coutte. 2004. Protein secretion through autotransporter and two-partner pathways. Biochim. Biophys. Acta 1694: 235–257.

Jacobs, J.M., L. Babujee, F. Meng, A. Milling and C. Allen. 2012. The *In Planta* Transcriptome of *Ralstonia solanacearum*: Conserved Physiological and Virulence Strategies during Bacterial Wilt of Tomato. mBio 3: e00114–12.

Jeong, Y., H. Cheong, O. Choi, J.K. Kim, Y. Kang, J. Kim et al. 2011. An HrpB-dependent but type III-independent extracellular aspartic protease is a virulence factor of *Ralstonia solanacearum*. Mol. Plant Pathol. 12: 373–380.

Jones, J.D.G. and J.L. Dangl. 2006. The plant immune system. Nature 444: 323–329.

Kanda, A., S. Ohnishi, H. Tomiyama, H. Hasegawa, M. Yasukohchi, A. Kiba et al. 2003. Type III secretion machinery-deficient mutants of *Ralstonia solanacearum* lose their ability to colonize resulting in loss of pathogenicity. J. Gen. Plant Pathol. 69: 250–257.

Kang, Y. 1994. Dramatically reduced virulence of mutants of *Pseudomonas solanacearum* defective in export of extracellular proteins across the outer membrane. Mol. Plant. Microbe Interact. 7: 370.

Kang, Y., J. Kim, S. Kim, H. Kim, J.Y. Lim, M. Kim et al. 2008. Proteomic analysis of the proteins regulated by HrpB from the plant pathogenic bacterium *Burkholderia glumae*. PROTEOMICS 8: 106–121.

Kirkpatrick, D.S., C. Denison and S.P. Gygi. 2005. Weighing in on Ubiquitin: The expanding role of mass spectrometry-based Proteomics. Nat. Cell Biol. 7: 750–757.

Kraepiel, Y. and M.-A. Barny. 2016. Gram-negative phytopathogenic bacteria, all hemibiotrophs after all? Mol. Plant Pathol. 17: 313–316.

Kvitko, B.H., A.R. Ramos, J.E. Morello, H.-S. Oh and A. Collmer. 2007. Identification of harpins in *Pseudomonas syringae* pv. *tomato* DC3000, which are functionally similar to HrpK1 in promoting translocation of type III secretion system effectors. J. Bacteriol. 189: 8059–8072.

Lam, E., N. Kato and M. Lawton. 2001. Programmed cell death, mitochondria and the plant hypersensitive response. Nature 411: 848–853.

Landry, D., M. González-Fuente, L. Deslandes and N. Peeters. 2020. The large, diverse, and robust arsenal of *Ralstonia solanacearum* type III effectors and their in planta functions. Mol. Plant Pathol. 21: 1377–1388.

Le Roux, C., G. Huet, A. Jauneau, L. Camborde, D. Trémousaygue, A. Kraut et al. 2015. A Receptor Pair with an Integrated Decoy Converts Pathogen Disabling of Transcription Factors to Immunity. Cell 161: 1074–1088.

Leyton, D.L., A.E. Rossiter and I.R. Henderson. 2012. From self sufficiency to dependence: mechanisms and factors important for autotransporter biogenesis. Nat. Rev. Microbiol. 10: 213–225.

Liu, H., S. Zhang, M.A. Schell and T.P. Denny. 2005. Pyramiding unmarked deletions in *Ralstonia solanacearum* shows that secreted proteins in addition to plant cell-wall-degrading enzymes contribute to virulence. Mol. Plant-Microbe Interact. MPMI 18: 1296–1305.

Liu, Y., A. Kanda, K. Yano, A. Kiba, Y. Hikichi, M. Aino et al. 2009. Molecular typing of Japanese strains of *Ralstonia solanacearum* in relation to the ability to induce a hypersensitive reaction in tobacco. J. Gen. Plant Pathol. 75: 369–380.

Lonjon, F., M. Turner, C. Henry, D. Rengel, D. Lohou, Q. van de Kerkhove et al. 2016. Comparative secretome analysis of *Ralstonia solanacearum* Type 3 secretion-associated mutants reveals a fine control of effector delivery, essential for bacterial pathogenicity. Mol. Cell. Proteomics 15: 598–613.

Lopes, C.A. and M. Rossato. 2018. History and status of selected hosts of the *Ralstonia solanacearum* species complex causing bacterial wilt in Brazil. Front. Microbiol. 9: 1228.

Macho, A.P., A. Guidot, P. Barberis, C.R. Beuzón and S. Genin. 2010. A competitive index assay identifies several *Ralstonia solanacearum* Type III effector mutant strains with reduced fitness in host plants. Mol. Plant-Microbe Interactions® 23: 1197–1205.

Mann, M. 2006. Functional and quantitative proteomics using SILAC. Nat. Rev. Mol. Cell Biol. 7: 952–958.

Medina-Puche, L., H. Tan, V. Dogra, M. Wu, T. Rosas-Diaz, L. Wang et al. 2020. A defense pathway linking plasma membrane and chloroplasts and co-opted by pathogens. Cell 182: 1109–1124.e25.

Meng, G., N.K. Surana, J.W. St Geme and G. Waksman. 2006. Structure of the outer membrane translocator domain of the *Haemophilus influenzae* Hia trimeric autotransporter. EMBO J. 25: 2297–2304.

Meyer, M.D., J.D. Ryck, S. Goormachtig and P. Van Damme. 2020. Keeping in Touch with Type-III secretion system effectors: mass spectrometry-based proteomics to study effector–host protein–protein interactions. International Journal of Molecular Sciences. 21(18): 6891

Monteiro, F., S. Genin, I. van Dijk and M. Valls. 2012. A luminescent reporter evidences active expression of *Ralstonia solanacearum* type III secretion system genes throughout plant infection. Microbiology 15: 2107–2116.

Morel, J.-B. and J.L. Dangl. 1997. The hypersensitive response and the induction of cell death in plants. Cell Death Differ. 4: 671–683.

Morgan, J.L.W., J.F. Acheson and J. Zimmer. 2017. Structure of a Type-1 secretion system ABC transporter. Structure 25: 522–529.

Mukaihara, T. and N. Tamura. 2009. Identification of novel *Ralstonia solanacearum* type III effector proteins through translocation analysis of hrpB-regulated gene products. Microbiology 155: 2235–2244.

Mukaihara, T., N. Tamura and M. Iwabuchi. 2010. Genome-wide identification of a large repertoire of *Ralstonia solanacearum* Type III effector proteins by a new functional screen. Mol. Plant-Microbe Interactions® 23: 251–262.

Natale, P., T. Brüser and A.J.M. Driessen. 2008. Sec- and Tat-mediated protein secretion across the bacterial cytoplasmic membrane--distinct translocases and mechanisms. Biochim. Biophys. Acta 1778: 1735–1756.

Papanikou, E., S. Karamanou and A. Economou. 2007. Bacterial protein secretion through the translocase nanomachine. Nat. Rev. Microbiol. 5: 839–851.

Paranjpye, R.N., M.S. Strom. 2005. A Vibrio vulnificus Type IV Pilin Contributes to Biofilm Formation, Adherence to Epithelial Cells, and Virulence. Infect. Immun. 73: 1411–1422.

Parte, A.C. 2014. LPSN—list of prokaryotic names with standing in nomenclature. Nucleic Acids Res. 42: D613–D616.

Peeters, N., S. Carrère, M. Anisimova, L. Plener, A.-C. Cazalé and S. Genin. 2013a. Repertoire, unified nomenclature and evolution of the Type III effector gene set in the *Ralstonia solanacearum* species complex. BMC Genomics 14: 859.

Peeters, N., A. Guidot, F. Vailleau and M. Valls. 2013b. *Ralstonia solanacearum* , a widespread bacterial plant pathogen in the post-genomic era: *Ralstonia solanacearum* and bacterial wilt disease. Mol. Plant Pathol. 14: 651–662.

Plener, L., P. Boistard, A. González, C. Boucher and S. Genin. 2012. Metabolic Adaptation of *Ralstonia solanacearum* during Plant Infection: A Methionine Biosynthesis Case Study. PLoS ONE 7: e36877.

Poueymiro, M. and S. Genin. 2009. Secreted proteins from *Ralstonia solanacearum*: a hundred tricks to kill a plant. Curr. Opin. Microbiol., Host–Microbe Interactions: Bacteria 12: 44–52.

Puigvert, M., R. Guarischi-Sousa, P. Zuluaga, N.S. Coll, A.P. Macho, J.C. Setubal et al. 2017. Transcriptomes of *Ralstonia solanacearum* during root colonization of *Solanum commersonii*. Front. Plant Sci. 8.

Remigi, P., M. Anisimova, A. Guidot, S. Genin and N. Peeters. 2011. Functional diversification of the GALA type III effector family contributes to *Ralstonia solanacearum* adaptation on different plant hosts. New Phytol. 192: 976–987.

Roux, B., S. Bolot, E. Guy, N. Denancé, M. Lautier, M.-F. Jardinaud et al. 2015. Genomics and transcriptomics of *Xanthomonas campestris* species challenge the concept of core type III effectome. BMC Genomics 16: 975.

Russell, A.B., S.B. Peterson and J.D. Mougous. 2014. Type VI secretion system effectors: Poisons with a purpose. Nat. Rev. Microbiol. 12: 137–148.

Ryan, M.P. and C.C. Adley. 2014. *Ralstonia* spp.: Emerging global opportunistic pathogens. Eur. J. Clin. Microbiol. Infect. Dis. 33: 291–304.

Sabbagh, C.R.R., S. Carrere, F. Lonjon, F. Vailleau, A.P. Macho, S. Genin et al. 2019. Pangenomic type III effector database of the plant pathogenic *Ralstonia* spp. PeerJ 7: e7346.

Safni, I., I. Cleenwerck, P. De Vos, M. Fegan, L. Sly and U. Kappler. 2014. Polyphasic taxonomic revision of the *Ralstonia solanacearum* species complex: proposal to emend the descriptions of *Ralstonia solanacearum* and *Ralstonia syzygii* and reclassify current *R. syzygii* strains as *Ralstonia syzygii* subsp. *syzygii* subsp. nov., *R. solanacearum* phylotype IV strains as *Ralstonia syzygii* subsp. indonesiensis subsp. nov., banana blood disease bacterium strains as *Ralstonia syzygii* subsp. *celebesensis* subsp. nov. and *R. solanacearum* phylotype I and III strains as *Ralstonia pseudosolanacearum* sp. nov. Int. J. Syst. Evol. Microbiol. 64: 3087–3103.

Salanoubat, M., S. Genin, F. Artiguenave, J. Gouzy, S. Mangenot, M. Arlat et al. 2002. Genome sequence of the plant pathogen *Ralstonia solanacearum*. Nature 415: 497–502.

Sang, Y., W. Yu, H. Zhuang, Y. Wei, L. Derevnina, G. Yu et al. 2020. Intra-strain elicitation and suppression of plant immunity by *Ralstonia solanacearum* Type-III effectors in *Nicotiana benthamiana*. Plant Commun., Special Issue on Plant-Pathogen Interactions (Organizing Editors: Paul Birch, Savithramma Dinesh-Kumar, Hui-Shan Guo, Ping He, Xin Li, Frank Takken, Yuanchao Wang) 1: 100025.

Santiago, T.R., C.A. Lopes, G. Caetano-Anollés and E.S.G. Mizubuti. 2017. Phylotype and sequevar variability of *Ralstonia solanacearum* in Brazil, an ancient centre of diversity of the pathogen. Plant Pathol. 66: 383–392.

Santiago, T.R., C.A. Lopes and E.S. Mizubuti. 2016. Diversidade e Variabilidade de *Ralstonia* spp. In: Estado da Arte em Fitobacterioses Tropicais. EDFRPE - Editora Universitária da UFRPE, Recife, Pernambuco, Brasil, pp. 243–256.

Solé, M., C. Popa, O. Mith, K.H. Sohn, J.D.G. Jones, L. Deslandes et al. 2012. The awr gene family encodes a novel class of *Ralstonia solanacearum* Type III Effectors Displaying Virulence and Avirulence Activities. Mol. Plant-Microbe Interactions® 25: 941–953.

Staskawicz, B.J., M.B. Mudgett, J.L. Dangl and J.E. Galan. 2001. Common and contrasting themes of plant and animal diseases. Science 292: 2285–2289.

Sun, Y., P. Li, M. Deng, D. Shen, G. Dai, N. Yao et al. 2017. The *Ralstonia solanacearum* effector RipAK suppresses plant hypersensitive response by inhibiting the activity of host catalases. Cell. Microbiol. 19: e12736.

Tampakaki, A.P., N. Skandalis, A.D. Gazi, M.N. Bastaki, F. S. Panagiotis, S.N. Charova et al. 2010. Playing the "Harp": evolution of our understanding of hrp/hrc Genes. Annu. Rev. Phytopathol. 48: 347–370.

Thanassi, D.G. and S.J. Hultgren. 2000. Multiple pathways allow protein secretion across the bacterial outer membrane. Curr. Opin. Cell Biol. 12: 420–430.

Thomas, S., I.B. Holland and L. Schmitt. 2014. The Type 1 secretion pathway—The hemolysin system and beyond. Biochim. Biophys. Acta BBA - Mol. Cell Res. Protein Trafficking and Secretion in Bacteria 1843: 1629–1641.

Tjalsma, H., A. Bolhuis, J.D.H. Jongbloed, S. Bron and J.M. van Dijl. 2000. Signal peptide-dependent protein transport in *Bacillus subtilis*: a genome-based survey of the secretome. Microbiol. Mol. Biol. Rev. 64: 515–547.

Troisfontaines, P. and G.R. Cornelis. 2005. Type III Secretion: More Systems Than You Think. Physiology 20: 326–339.

Valls, M., S. Genin and C. Boucher. 2006. Integrated Regulation of the Type III Secretion System and Other Virulence Determinants in *Ralstonia solanacearum*. PLoS Pathog. 2: e82.

Van Gijsegem, F., M. Marenda, B. Brito, J. Vasse, C. Zischek, S. Genin et al. 1998. The *Ralstonia solanacearum* hrp Gene Region: Role of the Encoded Proteins in Interactions with Plants and Regulation of Gene Expression. pp. 178–183. *In*: Prior, P., C. Allen and J. Elphinstone [Eds.]. Bacterial Wilt Disease: Molecular and Ecological Aspects. Springer, Berlin, Heidelberg.

Wang, K., P. Remigi, M. Anisimova, F. Lonjon, I. Kars, A. Kajava et al. 2016. Functional assignment to positively selected sites in the core type III effector RipG7 from *Ralstonia solanacearum*. Mol. Plant Pathol. 17: 553–564.

Wu, D., E. von Roepenack-Lahaye, M. Buntru, O. de Lange, N. Schandry, A.L. Pérez-Quintero et al. 2019. A plant pathogen type III effector protein subverts translational regulation to boost host polyamine levels. Cell Host Microbe. 26: 638–649.e5.

Yabuuchi, E., Y. Kosako, I. Yano, H. Hotta and Y. Nishiuchi. 1995. Transfer of two *Burkholderia* and an *Alcaligenes* species to *Ralstonia* gen. Nov.: Proposal of *Ralstonia pickettii* (Ralston, Palleroni and Doudoroff 1973) comb. Nov., *Ralstonia solanacearum* (Smith 1896) comb. Nov. and *Ralstonia eutropha* (Davis 1969) comb. Nov. Microbiol. Immunol. 39: 897–904.

Yu, G., L. Xian, H. Xue, W. Yu, J.S. Rufian, Y. Sang et al. 2020. A bacterial effector protein prevents MAPK-mediated phosphorylation of SGT1 to suppress plant immunity. PLOS Pathog. 16: e1008933.

Zhang, L., Jingsheng Xu, Jin Xu, K. Chen, L. He and J. Feng. 2012. TssM is essential for virulence and required for type VI secretion in *Ralstonia solanacearum*. J. Plant Dis. Prot. 119: 125–134.

Zhang, L., Jingsheng Xu, Jin Xu, H. Zhang, L. He and J. Feng. 2014. TssB is essential for virulence and required for Type VI secretion system in *Ralstonia solanacearum*. Microb. Pathog. 74: 1–7.

Zhang, Y. and S. Qiu. 2016. Phylogenomic analysis of the genus *Ralstonia* based on 686 single-copy genes. Antonie Van Leeuwenhoek 109: 71–82.

Zhou, M., D. Theunissen, M. Wels and R.J. Siezen. 2010. LAB-Secretome: a genome-scale comparative analysis of the predicted extracellular and surface-associated proteins of Lactic Acid Bacteria. BMC Genomics 11: 651.

Zumaquero, A., A.P. Macho, J.S. Rufián and C.R. Beuzón. 2010. Analysis of the Role of the Type III Effector Inventory of *Pseudomonas syringae* pv. *phaseolicola* 1448a in Interaction with the Plant. J. Bacteriol. 192: 4474–4488.

CHAPTER 12

Use of *Salmonella* Typhimurium as Tester Strains (*Salmonella*/Microsome) for Evaluating Mutagenicity of Compounds

Débora Antunes Neto Moreno,[1] Natália Tribuiani,[1] Edson Hideaki Yoshida,[1] Jocimar de Souza,[1] Ederson Constantino,[1] Carolina Alves dos Santos,[1] Marco Vinicius Chaud,[1] Eduardo Matheus Ricciardi Suzuki,[2] Bruno Cabral de Oliveira,[3] Hellen Cristine Boschilha Lastra,[3] Marcela Pelegrini Peçanha,[2] Flavia Aparecida Resende Nogueira,[4] José Carlos Cogo,[5] Silvio Roberto Consonni,[6] Mahendra Rai[7] and Yoko Oshima-Franco[1,]*

Introduction

According to National Cancer Institute, "cancer is a genetic disease—that is, cancer is caused by certain changes to genes that control the way our cells function, especially how they grow and divide" (National Cancer Institute 2017). A gene is the basic unit of heredity in humans and almost all other organisms made up of deoxyribonucleic acid, DNA. Genes either act as instructions to make– or to not - molecules called proteins. Humans have between 20,000 and 25,000 genes, which vary in size from a few hundred DNA bases to more than 2 million bases. Figure 1 shows a gene along the length of a chromosome, which in turn is found in the nucleus of each cell. Each chromosome is made up of DNA tightly coiled many times around proteins called histones that support its structure (MedlinePlus [Internet] 2020).

Chromosomes have a p (from "petit" allusive to small) arm, a q (next alphabetical word) arm, and a centromere. They are made up of DNA wrapped around histone proteins (Figure 2).

The genetic changes when present in germ cells inherited from parents are found in every cell of the offspring, but cancer-causing genetic changes can also be acquired during the lifetime, after conception, and are called genetic changes in somatic cells (National Cancer Institute 2017).

It is known that DNA damage can affect just one unit of DNA, called a nucleotide, or involve larger stretches of DNA, or sometimes the changes are not in the actual sequence of DNA. It means

[1] Post-Graduate Program in Pharmaceutical Sciences, University of Sorocaba, 18023–000 Sorocaba, SP, Brazil.
[2] Pharmacy Course. University of Sorocaba, 18023-000, Sorocaba, SP, Brazil.
[3] Bioprocess and Biotechnology Engineer Course, University of Sorocaba, 18023–000, Sorocaba, SP, Brazil.
[4] Department of Biological Sciences and Health, University of Araraquara, 14801–340, Araraquara, SP, Brazil.
[5] Post-Graduate Program in Bioengineering, Technological and Scientific Institute, Brazil University, 08230–030, São Paulo, SP, Brazil.
[6] Department of Biochemistry and Tissue Biology, Institute of Biology, State University of Campinas, Campinas, SP, Brazil.
[7] Department of Biotechnology, SGB Amravati University, Amravati-444 602, Maharashtra, India.
* Corresponding author: yoko.franco@prof.uniso.br

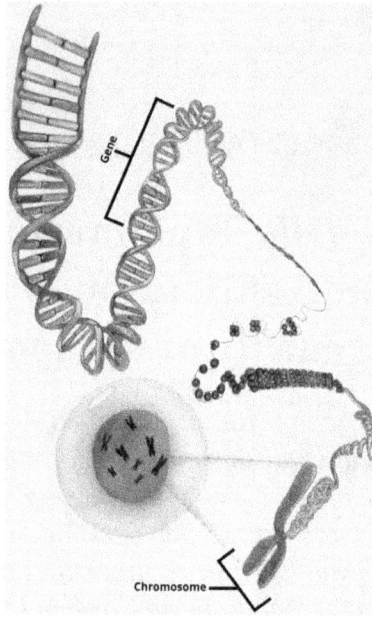

Figure 1. A gene is shown along the length of a chromosome (modified from Medline Plus [Internet] 2020).

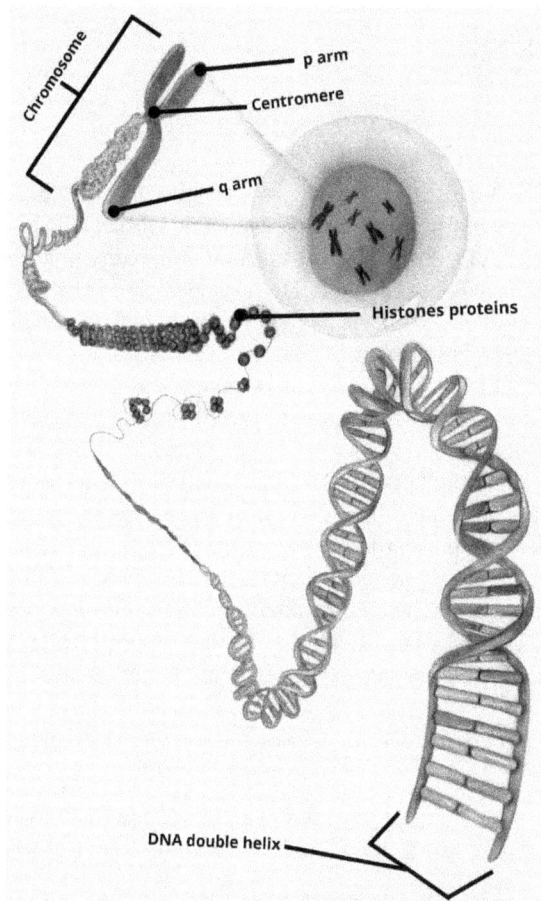

Figure 2. Chromosomes' tridimensional structure and DNA double helix (modified from MedlinePlus [Internet] 2020).

that changes can occur not only on genes, but also during their expression, a condition called epigenetic modifications, that is, whether and how much messenger ribonucleic acid, RNA, is produced, which in turn is translated to produce the proteins encoded by the DNA (National Cancer Institute 2017).

According to Nature portfolio (2022), "cancer is a complex disease in which cells in a specific tissue are no longer fully responsive to the signals within the tissue that regulate cellular differentiation, survival, proliferation and death. As a result, these cells accumulate within the tissue, causing local damage and inflammation. There are over 200 different types of cancer".

Cells are the basic building blocks of all living things, which contain the body's hereditary material and can make copies of themselves. Cells have many parts (Figure 3), each with a different function (MedlinePlus [Internet] 2020): cytoplasm that is made up of a jelly-like fluid (called the cytosol) and other structures that surround the nucleus; cytoskeleton is a network of long fibers that make up the cell's structural framework; endoplasmic reticulum (ER) helps process molecules created by the cell and transports these molecules to their specific destinations either inside or outside the cell, as well; Golgi apparatus packages molecules processed by the endoplasmic reticulum to be transported out of the cell; lysosomes and peroxisomes are the recycling center of the cell; mitochondria are complex organelles that convert energy from food into a form that the cell can use, besides having their own genetic material, separate from the DNA in the nucleus, and can make copies of themselves; nucleus serves as the cell's command center, sending directions to the cell to grow, mature, divide, or die. It also houses DNA, the cell's hereditary material. The nucleus is surrounded by a membrane called the nuclear envelope, which protects the DNA and separates the nucleus from the rest of the cell; plasma membrane is the outer lining of the cell; ribosomes are organelles that process the cell's genetic instructions to create proteins.

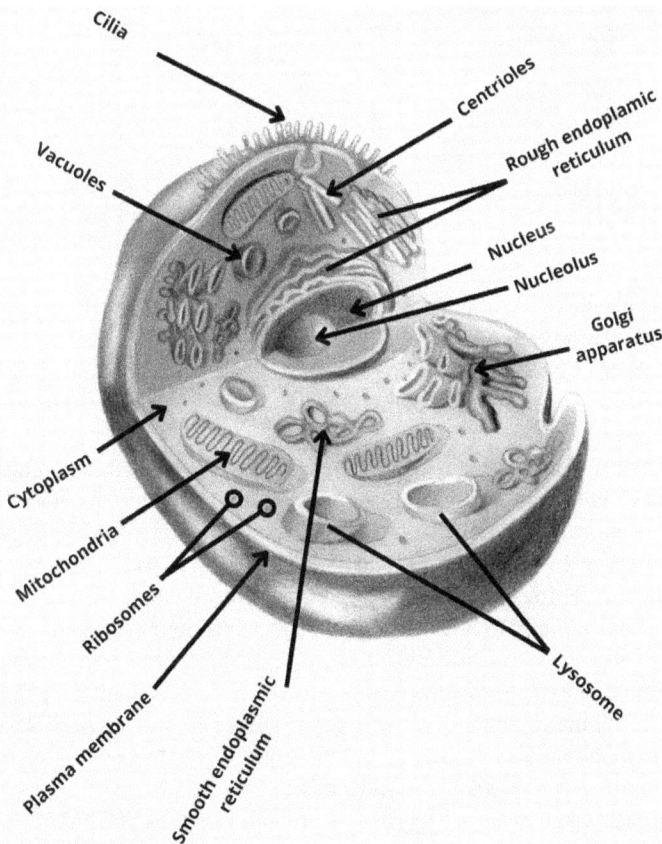

Figure 3. Human cells composition (modified from MedlinePlus [Internet] 2020).

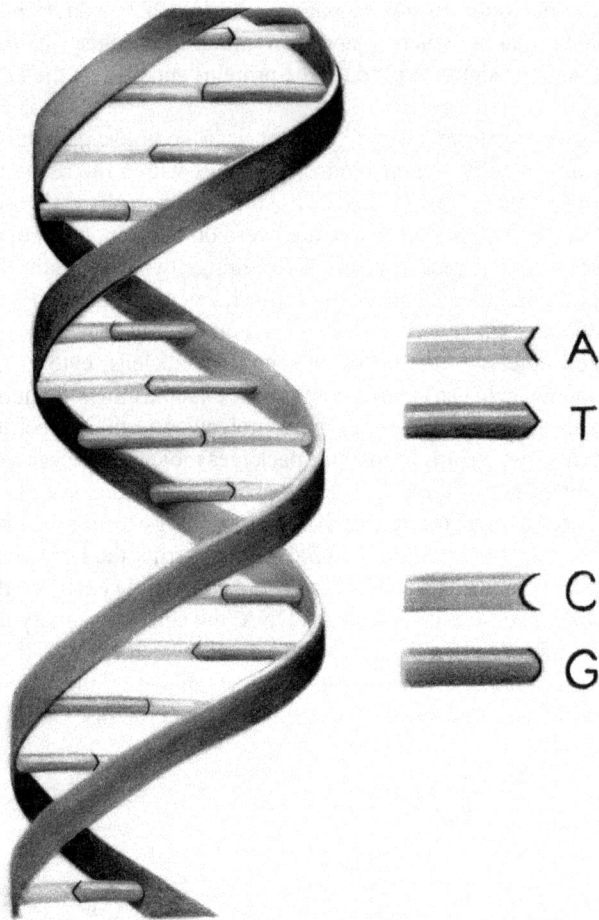

Figure 4. Base-pairs of DNA (modified from Medline Plus [Internet] 2020)

DNA is made up of base pairs (adenine-thymine and guanine-cytosine) and a sugar phosphate backbone (Figure 4).

Although the DNA, the basic unit of inheritance is an intrinsically reactive molecule and is highly susceptible to chemical modifications by endogenous and exogenous agents, cells are equipped with intricate and sophisticated systems—DNA repair, damage tolerance, cell cycle checkpoints and cell death pathways—that collectively function to reduce the deleterious consequences of DNA damage (Chatterjee and Walker 2017).

When the pathways of DNA repair, DNA damage tolerance, and DNA damage response (DDR) are disrupted or deregulated, an increase of mutagenesis and genomic instability is noticed, which in turn can promote the cancer progression (Bouwman and Jonkers 2012, Ghosal and Chen 2013, Wolters and Schumacher 2013, Chatterjee and Walker 2017), as well by aging due to attrition of chromosomal ends and failing capacities of a combination of these pathways. The study of the importance of mechanisms of DNA damage and repair and their implications for human health rendered the 2015 Nobel Prize in Chemistry to Drs. Lindahl, Modrich and Sancar.

Despite the importance of genetic changes in germ cells, in this chapter we direct our concern to acquired changes during the lifetime on somatic cells. Therefore, mutagenesis assays using bacteria can contribute to screen candidate-compounds in causing DNA damage, since mutagenesis is the first event from a multi-step process of carcinogenesis (Basu 2018).

The main goal of this study was to highlight the importance of bacteria to evaluate the mutagenic activity of cancer-inducing agents and their applicability using doxycycline, silver nanoparticles

(AgNPs), *Bothrops jararacussu* snake venom and the influence of AgNPs in the simultaneous treatment with venom as unprecedented examples by the Ames test to investigate their safety potential to be used in the health and medical fields.

Mutagenicity Assays Using Bacterium

Two basic premises guided Bruce N. Ames, according to current knowledge, in 1970s in the development of Ames test: first, a carcinogen is a mutagen; second, a microbe is a suitable model organism for assaying mutagenicity as it occurred in human cells (Creager 2014). Bacteria such as *Salmonella* Typhimurium and *Escherichia coli* are widely used for detecting chemically induced gene mutations as an endpoint (Gatehouse 2012, Föllmann et al. 2013). In this chapter, we focused on the *Salmonella* enterica serotype Typhimurium (*S.* Typhimurium).

The wild *Salmonella* enterica serotype Typhimurium, a Gram-negative enteric pathogen and a facultative anaerobe, causes the severe food-borne illness, gastroenteritis (Wall et al. 2010). Bruce N. Ames, in 1973, used four mutant strains of *Salmonella*, all deficient in their ability to synthesize the amino acid histidine (auxotrophic for histidine, His(-)), which must be supplemented in the growth media. The original strains TA1531, TA1532, and TA1534 detected frameshift (insertion or deletion of one or more nucleotide pairs) mutagens, whereas TA1530, containing a base-pair change, detected mutations involving base-pair substitutions. Besides, all strains also included a mutation in the *uvrB* gene that disabled DNA excision repair. Concerning mutagens, a specific type of mutation is induced in the DNA sequence of each strain, named as revertants or reverse mutations due the ability of mutagen in inducing a new mutation on strain to wild *Salmonella* (Creager 2014). Only those bacteria that revert to amino-acid independence (His(+)) will grow to form visible colonies (Goodson-Gregg and De Stasio 2021), via different mechanisms. The number of spontaneously induced revertant colonies per plate is relatively constant. However, in presence of a mutagen, the number of revertant colonies per plate is increased, usually in a dose-related manner (Gatehouse 2012).

The Rationale of Ames Test

Since DNA is chemically the same in all organisms, bacteria can be primarily used to test potential mutagens, taking advantage of mammals to develop cancer. Thus, mutant strains of the bacteria *Salmonella* Typhimurium were used to design the rationale of mutagenicity assay.

These haploid bacteria already contain particular mutations in the gene, encoding an enzyme used to synthesize the amino acid histidine; their genotype is given as His(-). Since the bacteria require histidine to make many of their proteins, these mutant bacteria will die unless the media in which they are grown contains histidine. It is known that secondary mutations occur at a low spontaneous rate; these mutants are called revertants because they have reverted to the His(+) genotype and phenotype and can now grow just fine in media lacking histidine.

The assay then involves plating *S.* Thyphimurium onto media with trace amounts histidine and adding chemicals to be tested for mutagenicity. The number of colonies growing on the plate indicates the number of revertants. In a true testing situation, a variety of concentrations of each chemical would be tested to generate a dose-response curve (De Stasio 2012).

The Genotypic Characteristics of Tester Strains

The test uses several strains of *S.* Typhimurium (Table 1) which carry different mutations in various genes of the histidine operon, and *E. coli* which carry the same AT base pair at the critical mutation site within the trpE gene, component I of anthranilate synthase (*E. coli K12*) (Levin et al. 1982b, 1984, Jurado et al. 1993, Gatehouse 2012, Hamel et al. 2016, Sugiyama et al. 2016, Vijay et al. 2018).

Table 1. Genotypic characterization of the Salmonella strains grouped by similarities.

Strain designation	TA100	TA1535	TA97	TA97a	TA1537	E.coli WP2 uvrA	TA102	TA98	TA1538
DNA target	*hisG46*	*hisG46*	*hisO1242*	*hisD6610*	*hisC3076*	*trpE56*	*hisG428*	*hisD3052*	*hisD3052*
Repair deficiency	uvrB	uvrB	uvrB	uvrB	uvrB	uvrA	-	uvrB	uvrB
LPS defect	rfa (*gal chl bio*)	rfa (*gal chl bio*)	rfa (*gal chl bio*)	rfa (*gal chl bio*)	rfa (*gal chl bio*)	-	rfa (*gal*)	rfa (*gal chl bio*)	rfa (*gal chl bio*)
Plasmids	pKM101	No plasmid	pKM101	pKM101	No plasmid	pKM101	pKM101 pAQ1	pKM101	No plasmid
Mutagen-susceptible sequence (hotspot)	G.C base pairs	G.C base pairs	-C-C-C-C-C- (+1 cytosine at run of C's)	-C-C-C-C- (+1 cytosine at run of C's)	+1 frameshift (near –CC–C– run)	Ochre mutation	A.T base pairs	CGCGCGCG	CGCGCGCG
Main mechanism, Sensitivity	Most base pair substitution	Most base pair substitution	Frameshift, intercalation	Frameshift, intercalation	Frameshift, intercalation	Base substitution	Base substitution, small deletions, cross linking, and oxidizing agents	Frameshift	Frameshift

Gal chl bio are genes which means galactose, chlorate and biotin, respectively. Plasmid pKM101 carries an ampicillin resistance gene and *mucAB* genes encoding analogs of UmuD/C proteins of *E. coli*, which are involved in error-prone DNA repair (Maron and Ames 1983, Mortelmans 2006). The pAQ1 is a derivative of pBR322 and carries the target DNA sequence for reversion, *hisG428*, a part of the histidine biosynthetic operon originated from *S. Typhimurium*. Thus, *hisG428* is a self-cloned gene. The vector pBR322 consists of the following DNA segments assembled *in vitro*: the tetracycline resistance gene, ampicillin resistance gene, and the replicator regions derived from colicin plasmid, pMB1 (Bettlach et al. 1976, Bolivar et al. 1977). Base pair substitution mutagens are agents that cause a base change in DNA. In a reversion test, this change may occur at the site of the original mutation, or at a second site in the bacterial genome (OECD 1997). Frameshift mutagens are agents that cause the addition or deletion of one or more base pairs in the DNA, thus changing the reading frame in the RNA (OECD 1997).

The Plate Incorporation Using Preincubation Method

The plate incorporation assay (Figure 5) using various *Salmonella* Typhimurium LT2 and *E. coli* WP2 strains (showed as strain) is a short-term bacterial reverse mutation assay specifically designed to detect a wide range of chemical substances capable of causing DNA damage leading to gene mutations (Gatehouse 2012).

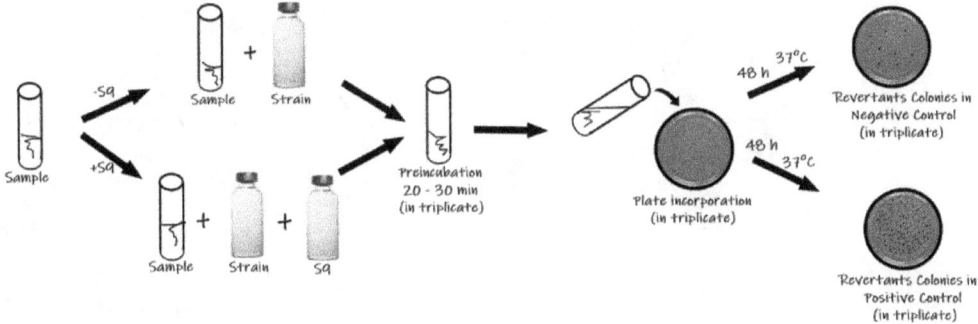

Figure 5. The plate incorporation using preincubation method.

Table 2. Validated positive controls used in mutagenicity assay.

	Bacterial strain	Without S9 mix (-S9) (µg/plate)	With S9 mix (+S9) (µg/plate)
S. Typhimurium	TA97/TA97a	9-Aminoacridine, 9AC (50)	2-Aminoanthracene, 2AA (1–5)
	TA98	2-Nitrofluorene, 2NF (1.0) or 4-Nitro-o-phenylenediamine, NOPD (2.5)	2-Aminoanthracene, 2AA (1–5) or Benz[a]pyrene, BaP (5)
	TA100	Sodium azide, NaAz (5)	2-Aminoanthracene, 2AA (1–5) or Benz[a]pyrene, BaP (5)
	TA102	Mitomycin C, MC (0.5)	2-Aminoanthracene (5–10) or Danthron, DAN (25)
	TA1535	Sodium azide, NaAz (5)	2-Aminoanthracene, 2AA (2–10)
	TA1537	Neutral red (10)	2-Aminoanthracene, 2AA (2–10) or Benz[a]pyrene, BaP (5)
	TA1538	4-Nitro-o-phenylenediamine, NOPD (2.5)	2-Aminoanthracene, 2AA (2–10)
E. Coli	WP2 uvrA	Nifuroxime (5–15) or 4-Nitroquinoline-1-oxide, NQO (0.5)	2-Aminoanthracene, 2AA (1–10) or (20)
	WP2uvrA (pKM101)	Cumene hydroperoxide (75–200) or 4-Nitroquinoline-1-oxide, NQO (0.2)	2-Aminoanthracene, 2AA (1–10) or (20)

The Positive and Negative Controls

The mutagenicity test is carried out under validated positive controls selected on the basis of the type of bacteria strains used, with (+S9) and without (-S9) metabolic activation. Table 2 shows suggested chemicals able to induce bacterial reverse mutation from His- to His+ (Gatehouse et al. 1994, Gatehouse 2012, Hamel et al. 2016). Negative controls are usually carried out using solvent, as dimethyl sulfoxide (DMSO), or vehicle alone, without a test substance (OECD 2020). Figure 6 shows negative (A) and positive (B) controls.

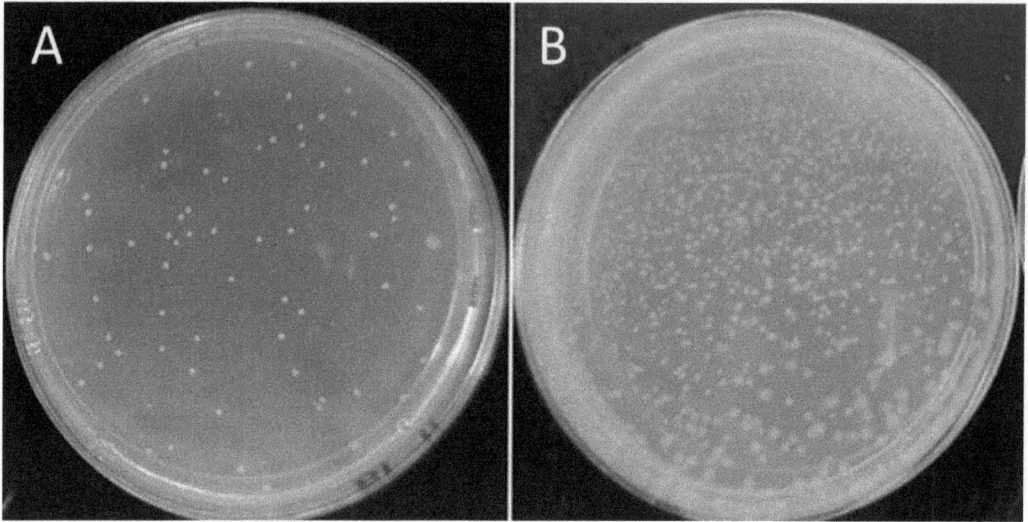

Figure 6. Revertant colonies of Salmonella Typhimurium in negative (A, DMSO) and positive (B, Sodium azide) controls from TA100 strain.

The Metabolic Activation

Nowadays, commercial lyophilized preformulated S9 mix is available to carry out mutagenicity assay with advantages of quality control certificate supplied by manufacturer. Thus, the recommendations of manufacturer must be followed.

Requirements (False-positives and False-negatives)

The test is used worldwide as an initial screen to determine the mutagenic potential of new chemicals and drugs (Gatehouse 2012). Since Ames test only shows positive results with about 60% of genotoxic agents, other *in vitro* tests such as the mouse lymphoma test and the micronucleus test have also been recommended. Hence, to ensure that the whole spectrum of genetic damage is detected, a battery of genotoxicity tests including both Ames and *in vitro* mammalian assays such as in mice or rats is required by many regulatory agencies before being labeled as a human carcinogen (Tchounwou 2013, De Stasio 2022). To avoid false-negatives, preliminary toxicity using the maximum concentration of test substance is also required, but the highest achievable level that avoids the death of bacteria.

Mutagenic Index (MI)

The mutagenic index (MI) is calculated for each concentration tested, as the average number of revertants per plate of each compound divided by the average number of revertants per plate with the negative (solvent) control. A sample is considered mutagenic when a dose-response relationship is detected and a two-fold increase in the number of mutants (MI $\geq$ 2) is observed with at least one concentration (Varella et al. 2004, Yoshida et al. 2016).

Statistical Analysis

The unprecedented results of the mutagenicity tests shown in this chapter were analyzed with the Salanal statistical software package (U.S. Environmental Protection Agency, Monitoring Systems Laboratory, Las Vegas, NV, version 1.0, from Research Triangle Institute, RTP, North Carolina,

USA) adopting the Bersntein et al.'s (1982) model. The data (revertants/plate) were also assessed by analysis of variance (One-way ANOVA) and Tukey's test to compare data sets.

Applicability: Some Unprecedented Examples Carried Out in Our Laboratory

Safety Assessment of Doxycycline by Mutagenicity Assay in Salmonella Typhimurium Microsomes

Doxycycline has shown promise for the treatment of chronic neurodegenerative diseases, such as Parkinson's disease (Blum et al. 2004, Smilack 1999), but the effects of chronic oral administration are not known. The prolonged half-life of doxycycline allows once daily administration (Cunha et al. 2012), which is advantageous in repurposing concerning doxycycline (Tribuiani et al. 2022). Although the *Salmonella*/microsome mutagenicity assay can provide information on the mutagenic potential of numerous compounds and is one of the first screenings for registration or acceptance of chemicals, such as drugs and biocides, this study faced a challenge since doxycycline is an antibiotic, which in theory could kill bacteria. To avoid it, a preliminary test is always performed to define the toxic dosage (Mortelmans and Zeiger 2000).

Here, we carried out the preliminary toxicity assay using TA100 strain. The initial concentration of 5,000 µg/mL doxycycline-HCl (Sigma-Aldrich, St. Louis, MO) killed all bacteria. Some colonies of bacteria have grown up with 3,125 µg/mL. Doxycycline was solubilized in water (Table 3).

Table 3. Preliminary toxicity of Doxycycline. Revertants/plate and standard deviation for the strain TA100 of *S.* Typhimurium after treatment with various doses of doxycycline, without metabolic activation (-S9).

Treatments (mg/plate)		Number of revertants M ± SD/plates
		TA 100
		- S9
Control -		117 ± 35
Doxycycline	0.5000	0 (0)
	0.3125	58 ± 9 (0.5)
	0.1562	105 ± 12 (0.9)
	0.0781	117 ± 8 (1.0)
	0.039	128 ± 14 (1.0)
	0.0195	140 ± 21 (1.1)
Control +		1310 ± 197

Representation of the number of reversing colonies (± SD), starting from the doxycycline solution at a concentration of 5,000 µg/mL, which kills bacteria. Control ˙ - Negative control: water. Control + - Positive control: sodium azide (1.25 µg/plate). Values in parentheses express MI, mutagenic index.

When observing the number of revertants per plate, except the 0.5000 mg/plate which kills all bacteria, 0.3125 mg/plate showed to be toxic, whereas all the subsequent ones were non-toxic compared to the vehicle control as negative control. After, descending concentrations from the highest toxic were submitted to Ames assays itself, in the absence and presence of metabolic activation.

The methodology for the mutagenic activity is described as follows using the *S.* Typhimurium tester strains TA 97a, TA98, TA100 and TA102 (OECD 1997) kindly provided by B.N. Ames (Berkeley, CA, USA), with and without metabolization by the preincubation method (Maron and Ames 1983). The strains from frozen cultures were grown overnight for 12–14 h in Oxoid Nutrient Broth No.2 (United Kingdom – UK). The metabolic activation system named here simply as S9, prepared from livers of Sprague-Dawley rats treated with the polychlorinated biphenyl mixture Aroclor 1254 (500 mg/kg), was purchased from Moltox (Molecular Toxicology Inc) or Sigma (San Luis, Missouri, EUA) and freshly prepared before each test (Maron and Ames 1983).

After confirmation that Doxycycline at a high concentration as 5,000 µg/mL did cause the death of bacteria, a concentration of 3,125 µg/mL was preincubated with 0.5 mL of 0.2M sodium phosphate buffer (pH 7.4), or 0.5 mL de 4% S9 mixture, with 0.1 mL of bacterial culture and then incubated at 37°C for 20 min. Next, 2 mL of top agar (0.6% agar, histidine, and biotin 0.5 mM each, and 0.5% NaCl) was added and the mixture was poured onto a plate containing minimal glucose agar (1.5% Bacto-Difco agar and 2% glucose in Vogel Bonner medium E). The plates were incubated at 37°C for 48 h and the His(+) revertant colonies were counted manually. All experiments were carried out in triplicate. The standard mutagens used as positive controls in experiments without S9 mix were 4-nitro-O-phenylenediamine (NPD) (10 µg/plate) for TA98 and TA97a, sodium azide (1.25 µg/plate) for TA100 and mitomycin (0.5 µg/plate) for TA102, whereas 2-anthramine (1.25 µg/plate) was used for all strains in the experiments with metabolic activation. Water served as the negative control.

From the chosen concentration, the Ames test was performed with and without metabolic activation; the results are shown in Table 4.

Antibiotics are generally believed to be mutagenic since some antibacterials can attack genetic material (Çelik and Dilek 2011). Different antibiotics, often of the same chemical class, have different mechanisms of action and, therefore, caution is needed when extrapolating data from a "typical" member even though they are in the antibiotic class.

In this study, doxycycline, belonging to the tetracycline family, did not show mutagenic changes under the *S.* Typhimurium strains parameter without metabolic activation. When in the presence of the microsomal fraction S9, doxycycline also did not demonstrate mutagenicity in any concentration and strains studied, indicating that even if biotransformation occurs, the molecule formed is not mutagenic. It is believed to be N-monodemethyldoxycycline, the metabolite of doxycycline, according to studies with HPLC (Böcker 1983).

Neurodegeneration, a pathological process, leads to the destruction of neuronal tissue (Brezovakova et al. 2018) and diseases such as Alzheimer's, Parkinson's, transmissible spongiform encephalopathies (or prion diseases), and multiple sclerosis, among others, are related to this neuronal

Table 4. Revertants/plate, standard deviation, and mutagenicity index (in brackets) for the strains TA97a, TA98, TA100 and TA102 of S. Typhimurium after treatment with various doses of doxycycline, with (+S9) and without (−S9) metabolic activation.

Treatments (mg/plate)		Number of revertants (M ± SD)/plates and (MI)							
		TA 97 a		TA 98		TA 100		TA 102	
		- S9	+ S9 [a]	- S9	+ S9 [a]	- S9	+ S9 [a]	- S9	+ S9 [b]
Control -		114 ± 29	127 ± 5	20 ± 5.5	43 ± 1.3	117 ± 35	174 ± 6.5	350 ± 57	366 ± 26
Doxycycline	0.3125	136 ± 0,7* (1.2)	76 ± 15* (0.6)	10 ± 21* (0.5)	38 ± 16.6* (0.9)	58 ± 9* (0.5)	34 ± 3.6* (0.2)	315 ± 28* (0.9)	476 ± 17.4 (1.3)
	0.1562	102 ± 114* (0.9)	101 ± 4.0* (0.8)	16 ± 15* (0.8)	60 ± 30* (1.4)	105 ± 12* (0.9)	69 ± 4.2* (0.4)	280 ± 6* (0.8)	402 ± 8.0 (1.1)
	0.0781	57 ± 18* (0.5)	127 ± 58.5* (1.0)	20 ± 21* (1.0)	43 ± 12.7* (1.0)	117 ± 8* (1.0)	104 ± 18* (0.6)	280 ± 32* (0.8)	403 ± 21 (1.1)
	0.039	137 ± 88* (1.2)	101 ± 7.6* (0.8)	14 ± 6.5* (0.7)	38 ± 6.0* (0.9)	128 ± 14* (1.1)	139 ± 60* (0.8)	245 ± 47* (0.7)	366 ± 21.3 (1.0)
	0.0195	148 ± 26* (1.3)	102 ± 1.5* (0.8)	28 ± 29* (1.4)	43 ± 12.5* (1.0)	140 ± 21* (1.2)	261 ± 16* (1.5)	244 ± 36* (0.7)	403 ± 10.5 (1.1)
Control +		767.6 ± 146.9[1]	2189 ± 85[2]	1060 ±113[1]	1932 ± 52[2]	1310 ± 197[3]	1291 ± 90[2]	1203 ± 79[4]	399 ± 6 [5]

Representation of the number of reversing colonies (± SD), starting from the doxycycline solution at a concentration of 3,125 µg/mL. [a] – S9 from Moltox, [b] – S9 from Sigma.
Control - Negative control: water. Positive control (Control +) in the absence of S9 (-S9) for: TA97a and TA98 strains, [1], 4 -nitro-o-phenylenediamine (NOPD – 10.0 µg/plate); TA100, [3], sodium azide (1.25 µg/plate); TA102, [4], mitomycin (0.5 µg/plate). Positive control (Control +) in the presence of S9 (+S9) for: TA97a, TA98, TA100, [2], 2-anthramine (1.25 µg/plate); and TA102, [5], 2-aminofluorene (10.0 µg/plate). *, p<0.05. Values in parentheses express MI, mutagenic index.

dysfunction. Therapeutic agents that confer neuroprotection, as doxycycline, by counteracting these shared characteristics can therefore have beneficial effects in a wide range of neurodegenerative diseases (Noble et al. 2009).

In conclusion, when considering a direction for chronic drug treatment, doxycycline has been shown to be safe based on mutagenicity testing (Ames test), which can stimulate the long-term use in neurodegenerative diseases. More studies with experimental models for brain analysis are suggested.

Silver Nanoparticles

AgNPs have potential action against viruses, fungi, and bacteria with wide application in several areas, but the toxicity assays, including mutagenicity of all obtained AgNPs with different sizes, are recommended (Warheit and Donner 2010, Kim et al. 2013, Proença-Assunção et al. 2021a). Besides, the Ames test of various kinds of nanoparticles has been predominantly negative for several reasons as suggested by numerous studies (Kisin et al. 2007, Landsiedel et al. 2009).

We briefly described the AgNPs' obtention according to Oliveira et al. (2019) and characterized by Santos et al. (2017). A silver nitrate ($AgNO_3$, 45 mg) in an aqueous solution was the agent of the chemical reaction, using sodium citrate at the final proportion of 1:0.34 molar. Pyrrolidone polyvinyl and alcohol polyvinyl stabilizers were added to a solution resulting in a concentration of 135 mg. All the process was carried out under agitation and controlled temperature to obtain silver nanoparticles at 50 nm (0.081 mg/mL). Silver nanoparticles were characterized by Transmission Electron Microscopy (TEM) using an LEO 906-Zeiss transmission electron microscope (Carl

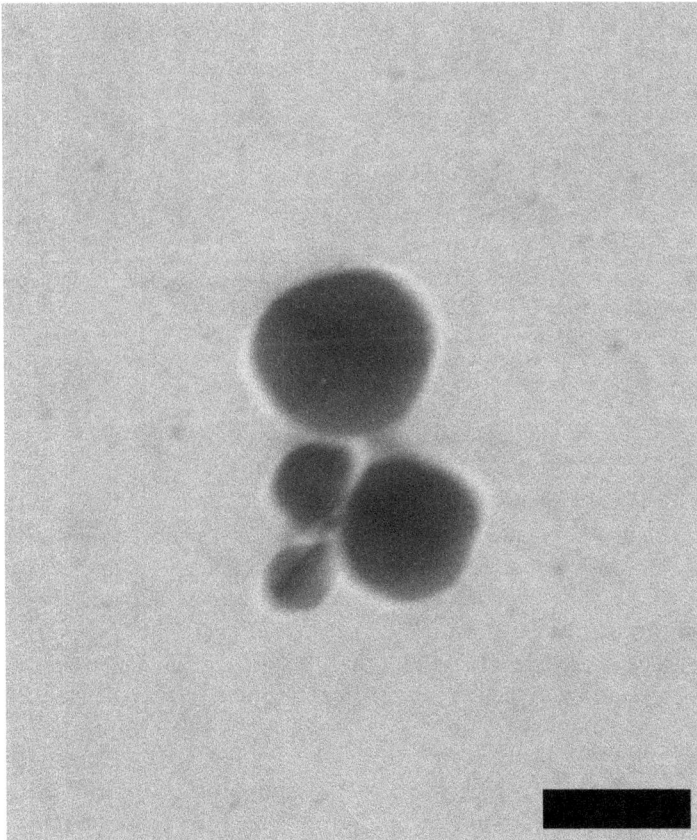

Figure 7. Image of AgNPs (0.081 mg mL^{-1}) by TEM. Magnification of 100.000X.

Table 5. Preliminary toxicity of AgNPs 0.081 mg/mL. Revertant per plate and standard deviation (M±SD) for the strains TA98 and TA100, of *S*. Typhimurium after treatment with various doses of AgNPs, without metabolic activation (−S9).

Treatment (mg/plate)		Number of revertants per plate (M ± SD) and Mutagenic Index (MI)	
		TA 98	TA 100
		- S9	- S9
	Control -	20 ± 5.5	117 ± 35
AgNPs	0.002	10 ± 4.5 (0.3)	117 ± 1.5 (1.0)
	0.004	2 ± 0.5 (0.1)	58.5 ± 3.7 (0.5)
	0.0081	0.2 ± 0.0 (0.01)	3.5 ± 1.0 (0.03)
	Control +	1060 ± 113[1]	1310 ± 197[2]

Control - Negative control: dimethylsulfoxide (DMSO - 50 µL/plate). [1]Positive control (Control +) for TA98, in absence of S9 (-S9): 4 -nitro-*o*-phenylenediamine (NOPD – 10.0 µg/ plate). [2]Positive control (Control +) for TA100, in absence of S9 (-S9): sodium azide (1.25 µg/ plate). Values in parentheses express MI, mutagenic index.

Table 6. Revertant per plate, standard deviation, and mutagenicity index (in parentheses) for the strains TA98, TA100, TA102 and TA97a of S. Typhimurium after treatment with various doses of AgNPs 0.081 mg/mL with (+S9) and without (−S9) metabolic activation.

Treatment (mg/plate)		Number of revertant/plate (M ± SD) and MI ( )							
		TA 97a		TA 98		TA 100		TA 102	
		-S9	+S9	-S9	+S9	-S9	+S9	-S9	+S9
Control -		114 ± 29	127 ± 5	20 ± 5.5	43 ± 1.3	117 ± 35	174 ± 6.5	350 ± 57	366 ± 26
AgNPs	0.00050	114 ± 6.5 (1.0)	139 ± 20 (1.1)	20 ± 1.5 (1.0)	73 ± 2** (1.7)	118 ± 38 (1.0)	191 ± 2.1 (1.1)	525 ± 18 (1.5)	256 ± 63 (0.7)
	0.00101	182 ± 51 (1.6)	114 ± 32 (0.9)	18 ± 1.7 (0.9)	81 ± 8.8** (1.9)	128 ± 69 (1.1)	261 ± 1.0** (1.5)	560 ± 11 (1.6)	329 ± 44 (0.9)
	0.00202	160 ± 20 (1.4)	139 ± 28 (1.1)	18 ± 1.7 (0.9)	137 ± 14.1** (3.2)	105 ± 46 (0.9)	348 ± 26.4** (2.0)	561 ± 14 (1.6)	328 ± 109 (0.9)
	0.00405	34 ± 19 (0.3)	127 ± 1 (1.0)	10 ± 1 (0.5)	154 ± 16.6* (3.6)	81 ± 6.2 (0.7)	452 ± 26.4** (2.6)	315 ± 22 (0.9)	257 ± 63 (0.7)
Control +		767.6 ± 146.9[1]	2189 ± 85[2]	1060 ±113[1]	1932 ± 52[2]	1310 ± 197[3]	1291 ± 90[2]	1203 ± 79[4]	1929 ± 231[5]

Representation of the number of reversing colonies (± SD). Control- Negative control: dimethylsulfoxide (DMSO- 50 µL/plate). Positive control (Control +) in the absence of S9 (-S9) for: TA97a and TA98 strains, [1], 4-nitro-*o*-phenylene-diamine (NOPD– 10.0 µg/plate); TA100, [3], sodium azide (1.25 µg/plate); TA102, [4], mitomycin (0.5 µg/plate). Positive control (Control +) in the presence of S9 (+S9) for: TA97a, TA98, TA100, [2], 2-anthramine (1.25 µg/plate); and TA102, [5], 2-aminofluorene (10.0 µg/plate). *, p < 0.05. **, p < 0.001. Values in parentheses express MI, mutagenic index.

Zeiss Microscopy GmbH, Germany) at an accelerating voltage of 60 kV (Figure 7) as described by Proença-Assunção et al. (2021b).

As recommend (OECD 1997), before applying the *Salmonella*/Microsome test, preliminary toxicity was carried out to select the highest amount of test substance to cause cytotoxicity, which may be detected by a reduction in the number of revertant colonies or by clearing or diminution of the background lawn. The recommended maximum test concentration for soluble cytotoxic substances is 5 mg/plate or 5,000 µL/plate. Thus, a concentration of 0.081 mg/mL of AgNPs (taking 100 µL) was assayed using TA98 and TA100 strains (Table 5), without metabolic activation (-S9).

Notice that the major tested concentration (0.0081 mg/plate) was completely toxic to TA98 and TA100 strains.

For subsequent *Salmonella*/microsome assay, we applied the same methodology described for doxycycline (tester strains, positive and negative controls as described in each table's legend, metabolic activation system, and statistical analysis). Decreasing concentrations of AgNPs from 0.00405 mg/plate were carried out (Table 6).

Interestingly, all concentrations of AgNPs studied here were not directly mutagenic in absence of metabolic activation; but in presence of S9, they were mutagenic at 0.00202 and 0.00405 mg/plate when exposed to TA98 and TA100 strains. Concerning hisD3052 from TA98, as all Salmonella tester strains, it also has a mutation in the histidine operon (His-), uvrB deletion mutation, rfa mutation, and introduction of plasmid pKM101 (Mortelmans and Zeiger 2000). The type of reversion event caused by TA98 is frameshift, whereas TA100 is the base-pair substitution reversion. The latter is originated from *S.* Typhimurium LT2, with the following genotype: *hisG46*, which carries a *rfa* (deep rough) mutation for permeation of test chemicals. The uvrB gene deletion enhances the sensitivity of Ames test; as well as, *gal*, *chl*, and *bio* genes; and plasmid pKM101 which carries an ampicillin resistance gene and mucAB genes, which are involved in error-prone DNA repair (Ames et al. 1973, Maron and Ames 1983, Mortelmans 2006, Sugiyama et al. 2016). Even in the presence of S9, AgNPs did not interfere with either TA97a or with TA102 strains.

Bothrops Jararacussu Snake Venom

Natural products, such as snake venoms, are a source of bioactive molecules that can inspire the development of new drugs for a range of conditions such as cardiovascular diseases, cancer, and neurological disorders (Shibata et al. 2018).

On the other hand, snakebites are a relevant public health problem affecting mainly the rural population. Taking into consideration that metalloproteinases, lectins, and phospholipases play important roles in cytotoxic damage in snake envenomation, paradoxically they can constitute the future arsenal against life-threatening diseases (Chakrabarty and Sarkar 2017). For example, snake cytotoxins could speed the destruction of blood vessels that supply nutrients to tumors (Bateman et al. 2013).

In opposition to the advances against cancer using isolated toxins, a few have been made in this area using crude venom. After all, snake venoms are a complex multifunctional cocktail that can act synergistically to rapidly immobilize prey and deter predators (Ferraz et al. 2019), including human beings. Cancer associated with snakebites remains largely unknown due to the infrequent and isolated nature of these occurrences.

It is known that genotoxicity is one of the major causes of cancer. Genotoxins are agents that can cause damage to DNA or chromosomal structure thereby causing mutations. This damage in the somatic cells will lead to various diseases ranging from cancer to death, whereas the damage to the germ cell will lead to heritable diseases (Mohamed et al. 2017). Only three studies were found in the literature about the genotoxicity of crude snake venoms, which were applied to human lymphocyte DNA, Micronucleus test, Comet assay (Marcussi et al. 2011, 2013), and MTT, Comet assay, and *in vivo* Micronucleus assay (Zobiole et al. 2015).

Firstly, Marcussi et al. (2011) studied the mutagenic potential and genotoxic effects of *Crotalus durissus terrificus* snake venom and its isolated toxins on human lymphocytes, using the micronucleus and comet assays. The main findings were: crotoxin and crotapotin caused damage to DNA; the crude venom induced the formation of micronuclei; and in the Comet assay, all the toxins were tested (crotamine, crotoxin, crotapotin and basic phospholipase A_2) and the crude venom showed genotoxic activity. In another study, Marcussi et al. (2013) studied the genotoxic effect on human lymphocyte DNA using *Bothrops* snake venoms (*B. jararacussu, B. atrox, B. moojeni, B. alternatus* and *B. brazili*) and isolated toxins (MjTX-I, BthTX-I and II myotoxins, BjussuMP-II metalloprotease, and BatxLAAO L-amino acid oxidase). Significant DNA damages were observed, indicating genotoxic potential after exposure of the lymphocytes to the toxins BthTX-I, II, and BatxLAAO compared to untreated and Cisplatin-treated controls, which were able to induce the greater formation of micronuclei. *B. brazili, B. jararacussu,* and *B. atrox* crude venoms also showed genotoxic potential and the latter two induced DNA breakage 5 times more often than in normal environmental conditions (control without treatment). *B. jararacussu* venom and its isolated toxins, as well as an LAAO from *B. atrox,* were able to cause lymphocyte DNA breakage in the Comet

Figure 8. *Bothrops jararacussu snake.* Photo by: Giuseppe Puorto (Instituto Butantan, SP, Brazil).

Table 7. Preliminary toxicity of Bothrops jararacussu venom. Revertants/plate expressed in mean ± standard deviation (M±SD) for the strains TA98 and TA100 of *S.* Typhimurium after treatment with various doses of *B. jararacussu* venom, without metabolic activation (−S9).

Treatment (mg/plate)		Number of revertant/plate (M ± SD) and Mutagenic Index (MI)	
		TA 98	**TA 100**
		-S9	**-S9**
Control -		20 ± 5.5	117 ± 35
Snake Venom	**0.31**	12 ± 3.7 (0.6)	128 ± 5.1 (1.1)
	0.63	10 ± 2.0 (0.5)	46.8 ± 2.8 (0.4)
	1.25	12 ± 2.8 (0.6)	58.5 ± 5.9 (0.5)
	2.5	14 ± 3.8 (0.7)	2.3 ± 0.5 (0.02)
	5.00	16 ± 1.0 (0.8)	0
Control +		1060 ±113[1]	1310 ± 197[2]

Control - Negative control: dimethylsulfoxide (DMSO - 50 µL/plate). [1]Positive control (Control +) for TA98, in absence of S9 (-S9): 4 -nitro-*o*-phenylenediamine (NOPD – 10.0 µg/ plate). [2]Positive control (Control +) for TA100, in absence of S9 (-S9): sodium azide (1.25 µg/ plate). Values in parentheses express MI, mutagenic index.

test with more than 85% damage levels. Zobiole et al. (2015) evaluated *in vitro* and *in vivo* the genotoxicity of *Bothrops moojeni* snake venom. The authors concluded that *B. moojeni* presented cyto- and genotoxicity in a dose-dependent way.

In this study, the Ames test as routinely standardized was applied to evaluate the mutagenicity of *Bothrops jararacussu* (Figure 8), since the Ames test is one of the most accurate and commonly used procedures to detect genotoxic carcinogens which cause two classes of gene mutation,

Table 8. Revertant per plate, standard deviation and mutagenicity index (in parentheses) for the TA98, TA100, TA102 and TA97a of S. Typhimurium strains after treatment with various doses of B. jararacussu venom 50 mg/mL, with (+S9) and without (−S9) metabolic activation.

Treatment (mg/plate)		Number of revertant/plate (M ± SD) and MI ( )							
		TA 97a		TA 98		TA 100		TA 102	
		-S9	+S9	-S9	+S9	-S9	+S9	-S9	+S9
Control -		114 ± 29	127 ± 5	20 ± 5.5	43 ± 1.3	117 ± 35	174 ± 6.5	350 ± 57.5	366 ± 26.4
Venom	0.31	176 ± 1.7 (1.6)	228 ± 5.3** (1.8)	12 ± 3.7 (0.6)	44 ± 32.9 (1.0)	128 ± 5.1 (1.1)	330 ± 16** (1.9)	299 ± 29 (0.8)	187 ± 23 (0.5)
	0.63	175 ± 23 (1.5)	259 ± 13** (2.0)	10 ± 2.0 (0.5)	98 ± 49.3* (2.3)	46.8 ± 2.8 (0.4)	382 ± 5.3** (2.2)	353 ± 17 (1.0)	395 ± 13.9 (1.0)
	1.25	174 ± 1.5 (1.5)	450 ± 16** (3.5)	12 ± 2.8 (0.6)	163 ± 38.9** (3.8)	58.5 ± 5.9 (0.5)	469 ± 13** (2.7)	259 ± 6.2 (0.7)	501 ± 6.0** (1.3)
	2.50	177 ± 2.6 (1.5)	605 ± 15** (4.8)	14 ± 3.8 (0.7)	190 ± 83.5** (4.4)	2.3 ± 0.5 (0.02)	556 ± 23.5** (3.2)	205 ± 14 (0.6)	732 ± 15** (2.0)
Control +		767.6 ± 146.9[1]	2189 ± 85[2]	1060 ±113[1]	1932 ± 52[2]	1310 ± 197[3]	1291 ± 90[2]	1203 ± 79[4]	1929 ± 231[5]

Representation of the number of reversing colonies (± SD). Control - Negative control: dimethylsulfoxide (DMSO - 50 µL/plate). Positive control (Control +) in the absence of S9 (-S9) for: TA97a and TA98 strains, [1], 4-nitro-*o*-phenylenediamine (NOPD – 10.0 µg/plate); TA100, [3], sodium azide (1.25 µg/plate); TA102, [4], mitomycin (0.5 µg/plate). Positive control (Control +) in the presence of S9 (+S9) for: TA97a, TA98, TA100, [2], 2-anthramine (1.25 µg/plate); and TA102, [5], 2-aminofluorene (10.0 µg/plate). *, p < 0.05. **, p < 0.001. Values in parentheses express MI, mutagenic index.

base-pair substitution, and small frameshift (Ames et al. 1973). The venom was collected from male and female adult specimens kept in the serpentarium of Centro de Estudos da Natureza – CEN (SMA 15.380/2012) from the University of Vale do Paraiba, Sao Jose dos Campos, SP, Brazil), lyophilized, and kept refrigerated at 4°C until its use.

As a routine, the preliminary toxicity assay was before carried out using 50 mg/mL *B. jararacussu* venom (Table 7), dissolved in saline, to give the recommended concentration of 5 mg/plate.

It can be noticed in Table 7 that with a high concentration of snake venom, 5 mg/plate, there was total inhibition of the bacteria growth in TA100, but viable colonies for TA98 were kept. In general, mutagenicity index values < 0.5 are toxic to bacteria.

The *B. jararacussu* venom was donated by prof. Dr. José Carlos Cogo. He is as an author (Table 8).

The concentration of the snake venom was assayed as recommended by regulatory agencies and according to the Organisation for Economic Cooperation and Development (OECD) test guideline 471 (OECD 1997).

B. jararacussu venom alone in absence of metabolic activation was not mutagenic. It is known that it is possible for the snake to milk up to 1000 mg (dry weight) from its venom glands on a single occasion (Milani Júnior et al. 1997), a great amount of the venom in a bite. There is no report about cancer development at the local site in humans bitten by snakes. Snakebites are typically singular events that occur infrequently, but there have been reports of a snake delivering venomous bites to two individuals during a single incident (Mamum et al. 2000, Amim et al. 2008). Thus, the clinical importance of this finding is irrelevant in an accident, but not when it is thought of as medicine.

According to Levy et al. (2019), the test substance is a mutagen if a positive response is seen in any one strain/activation combination, whereas a test is non-mutagenic if showed to be negative without S9 or with S9.

In the presence of S9 metabolic activation, *B. jararacussu* venom was mutagenic for all strains (TA98, TA100, and TA97a: from 0.63 to 2.5 mg/plate; TA102: 2.5 mg/plate), even using toxic concentrations as shown in absence of microsomal enzymes, showing that the components can suffer biotransformation acting indirectly on the organism. More complementary studies are needed to attribute carcinogenicity to the venom since it is known that a carcinogen is termed genotoxic if it can bind covalently to cellular DNA, and if left unrepaired, this could potentially initiate mutations by causing the incorrect incorporation of bases during the DNA replication process (Moschel 2001). These mutagenic findings are corroborated elsewhere (Marcussi et al. 2013) that previously showed the genotoxic potential of *B. jararacussu* in other experimental assays (lymphocyte DNA breakage in the comet test). According to Rojas et al. (1999), the Comet assay is a sensitive, reliable, and rapid method for the DNA double- and single-strand breaks' detections, alkali-labile sites, and delayed repair-site detection in eukaryotic individual cells. Although the Ames test only detects DNA damage of prokaryotic cells (Chen et al. 2003), we showed here predictable data and certain parallelism between both assays.

Ability of Silver Nanoparticles Against the Mutagenicity of Bothrops Jararacussu Venom

In this topic, we speculate if the obtained AgNPs (0.081 mg/mL, 50 nm) could interfere with *Bothrops jararacussu* venom (5 mg/plate) mutagenicity, using the same version of the *Salmonella/* microsome assay. A suggested procedure involved preparing a solution with *B. jararacussu* venom dissolved in saline to achieve a concentration of 5 mg per plate. Additionally, silver nanoparticles (AgNPs) were prepared at a concentration of 0.081 mg/mL. To create a mixture, both the venom and AgNPs were combined at equal concentrations, with a final volume of 50, 25, 12.5 and 6.25 µL of each of them. The same procedure described before was carried out in this set of experiments. The plates were incubated at 37°C for 48 h and the revertant colonies were counted manually (Table 9). All experiments were analyzed in triplicate.

The genotype characteristics of TA100 were already discussed before. Concerning TA102 genotype characteristics: insertion of the mutation *hisG428* on the multicopy plasmid pAQ1 was made, aiming to amplify the number of target sites, in addition to plasmid pKM101; there is

Table 9. Revertant per plate, standard deviation and mutagenicity index (in parentheses) for the strains TA98, TA100, TA102 and TA97a of S. Typhimurium after treatment with various doses of *B. jararacussu* venom + AgNPs mixture, with (+S9) and without (−S9) metabolic activation.

Treatment (mg/plate)		Number of revertant/plate (M ± SD) and MI ( )							
		TA 97a		TA 98		TA 100		TA 102	
		-S9	+S9	-S9	+S9	-S9	+S9	-S9	+S9
Control -		114 ± 29	127 ± 5	20 ± 5.5	43 ± 1.3	117 ± 35	174 ± 6.5	350 ± 57	366 ± 26
Mixture	(0.31 + 0.00050)	250 ± 46* (2.2)	279± 19** (2.2)	82 ± 3.7** (4.1)	25 ± 4.3 (0.6)	81 ± 5.1 (0.7)	139 ± 22 (0.8)	105 ± 74 (0.3)	219 ± 17 (0.6)
	(0.63 + 0.00101)	114 ± 11 (1.0)	317 ± 83 (2.5)	74 ± 20* (3.7)	60 ± 1.7** (1.4)	105 ± 28 (0.9)	191 ± 21 (1.1)	140 ± 18 (0.4)	549 ± 1 (1.5)
	(1.25 + 0.00202)	102 ± 16 (0.9)	342 ± 13* (2.7)	74 ± 28 (3.7)	73 ± 6.2** (1.7)	93 ± 59 (0.8)	174 ± 17 (1.0)	175 ± 31 (0.5)	439 ± 40 (1.2)
	(2.5 + 0.00405)	80 ± 15* (0.7)	342 ± 19** (2.7)	108 ± 68 (5.4)	116 ± 28** (2.7)	82± 51 (0.7)	87 ± 23 (0.5)	280 ± 14 (0.8)	685 ± 21 (1.6)
Control +		767.6 ± 146.9[1]	2189 ± 85[2]	1060 ±113[1]	1932 ± 52[2]	1310 ± 197[3]	1291 ± 90[2]	1203 ± 79[4]	1929 ± 231[5]

Representation of the number of reversing colonies (± SD). Control - Negative control: dimethylsulfoxide (DMSO - 50 µL/plate). Positive control (Control +) in the absence of S9 (-S9) for: TA97a and TA98 strains, [1], 4 -nitro-*o*-phenylen-ediamine (NOPD – 10.0 µg/plate); TA100, [3], sodium azide (1.25 µg/plate); TA102, [4], mitomycin (0.5 µg/plate). Positive control (Control +) in the presence of S9 (+S9) for: TA97a, TA98, TA100, [2], 2-anthramine (1.25 µg/plate); and TA102, [5], 2-aminofluorene (10.0 µg/plate). *, p<0.05. **, p<0.001.AgNPs were able to counteract the mutagenic response of *B. jararacussu* venom exposed to TA100 and TA102, in presence of metabolic activation, but not on TA97a and TA98. Values in parentheses express MI, mutagenic index.

Δ(*gal-bio-uvrB*) genes mutation, that renders the strains' excision-repair deficient (ΔuvrB) (Hamel et al. 2016). However, on metabolic activation, the mixture, even at the minor association venom + AgNPs, was mutagenic when exposed to TA97a strain, *hisD6610*, containing *uvrB* repair deficiency, *rfa* mutation and plasmid pKM101. There are multiple modes of reversion in all strains where each strain has a particular mutagen-susceptible sequence, i.e., hotspot, as in the case of *hisD6610* producing frameshift mutation (Barnes 1982).

The DNA target for TA100 is –G-G-G- causing a reversion event of base-pair substitution (Levin et al. 1982a); for TA98, it is –C-G-C-G-C-G-CG- causing a reversion event of frameshift (Isono and Yourno 1974); for TA97, it is -C-C-C-C-C-C- reversing by frameshift (Levin et al. 1982a); and for TA102, TAA (ochre) which reversion event is transitions/transversions (Levin et al. 1982b). On the other hand, *Bothrops* venoms are a complex mixture of components (phospholipases A_2, metalloproteases, serine proteases, L-amino acid oxidases, and C-type lectin) (Teixeira et al. 2003, 2009, Correa-Netto et al. 2010, Klein et al. 2015). The use of crude venom is interesting to know the mutagenic potential, but inadequate to predict which substance can activate the reversion event of each *S*. Typhimurium TA tester strain. It would of great interest to submit isolated fractions of *B. jararacussu* venom to clear this question and to verify the toxic potential each of them.

We can conclude that the crude *B. jararacussu* venom and the silver nanoparticles (0.081 mg/mL, 50 nm) were mutagenic through *Salmonella*/microsome parameter. But, looking closely, AgNPs were not mutagenic in absence of metabolic activation and the direct use on certain substances as the venom can be useful, as shown by interaction in the mixture submitted to TA100 and TA102. The chemical mechanism between silver nanoparticles and crude venom interaction remains to be clear.

Final Considerations and Conclusion

Using Ames test, a validated test for assessing mutagenicity, we assayed a variety of compounds such as doxycycline, AgNPs, *Bothrops jararacussu* snake venom and the resulting interaction between AgNPs and snake venom. Each test substance was judged to be a mutagen if treatment with the test chemical caused an increase in the number of revertant colonies in any one of the tester strain-S9 combinations. Our results showed that Doxycycline is not mutagenic, AgNPs have no direct mutagenicity (in absence of S9), but indirect mutagenicity (in presence of S9), *B. jararacussu* venom has no direct mutagenicity (in absence of S9), but has indirect mutagenicity (in presence of S9), and AgNPs is able to avoid mutations on TA100 and TA102 (either in absence or in presence of S9) of snake venom. The Ames test provides useful data about the compounds' safety and the purpose of use of any substance must be carefully taken into account.

References

Ames, B.N., F.D. Lee and W.E. Durston. 1973. An improved bacterial test system for the detection and classification of mutagens and carcinogens. Proc. Natl. Acad Sci. U.S.A. 70: 782–786.

Amin, M.R., S.M.H. Mamun, N.H. Chowdhury, M.I. Rahman, A.V. Ghose, A.V. Al Hasan et al. 2008. Consecutive bites on two persons by the same cobra: a case report. J. Venom Anim. Toxins Incl. Trop Dis. 14: 725–737.

Barnes, W. 1982. Base-sequence analysis of his+ revertants of the hisG46 missense mutation in *Salmonella* Typhimurium. Environ Mutagen. 4: 297.

Basu, A.K. 2018. DNA Damage, mutagenesis and cancer. Int. J. Mol. Sci. 2018 19: 970. Doi: 10.3390/ijms19040970.

Bateman, E., M. Venning, P. Mirtschin and A. Woods. 2013. The effects of selected Australian snake venoms on tumor-associated microvascular endothelial cells (TAMECs) *in vitro*. J. Venom Res. 4: 21–30.

Bernstein, L., J. Kaldor, J. Mccann and M.C. Pike. 1982. An empirical approach to the statistical analysis of mutagenesis data from the *Salmonella* Test. Mutat Res/Environ Mutagen Relat Subj. 97: 267–281. Doi: 10.1016/0165-1161(82)90026-7.

Betlach, M.C., V. Hershfield, L. Chow, W. Brown, H.M. Goodman, H.W. Boyer et al. 1976. A restriction endonuclease analysis of the bacterial plasmid controlling the *Eco*RI restriction and modification of DNA. Fed Proc. 35: 2037–2043.

Bolivar, F., R.L. Rodriguez, P.J. Greene, M.C. Betlach, H.L. Heyneker, H.W. Boyer et al. 1977. Construction and characterization of new cloning vehicles. II. A multipurpose cloning system. Gene. 2: 95–113. Doi: 10.1016/0378-1119(77)90000-2.

Bouwman, P. and J. Jonkers. 2012. The effects of deregulated DNA damage signalling on cancer chemotherapy response and resistance. Nat. Rev. Cancer. 12: 587–598.

Blum, D., A. Chtarto, L. Tenenbaum, J. Brotchi and M. Levivier. 2004. Clinical potential of minocycline for neurodegenerative disorders. Neurobiol. Dis. 17: 359–366. Doi: 10.1016/j.nbd.2004.07.012.

Böcker, R. 1983. Analysis and quantitation of a metabolite of doxycycline in mice, rats, and humans by high-performance liquid chromatography. J. Chromatogr B: Biomed. Sci. Appl. 274: 255–262. Doi: 10.1016/S0378-4347(00)84428-X.

Brezovakova, V., B. Valachova, J. Hanes, M. Novak and S. Jadhav. 2018. Dendritic cells as an alternate approach for treatment of neurodegenerative disorders. Cell Mol. Neurobiol. 38: 1207–1214. Doi: 10.1007/s10571-018-0598-1.

Çelik, A. and D. Dilek. 2011. The assessment of cytotoxicity and genotoxicity of tetracycline antibiotic in human blood lymphocytes using CBMN and SCE analysis, *in vitro*. Int. J. Hum. Genet. 11: 23–29. Doi: 10.1080/09723757.2011.11886119.

Chakrabarty, D. and A. Sarkar. 2017. Cytotoxic effects of snake venoms. pp. 327–342. *In*: Gopalakrishnakone, P., H. Inagaki, A. Mukherjee, T. Rahmy and C.W. Vogel. [Eds.]. Snake Venoms. Toxinol., Springer, Dordrecht.

Chatterjee, N. and G.C. Walker. 2017. Mechanisms of DNA damage, repair, and mutagenesis. Environ. Mol. Mutagen. 58: 235–263. Doi: 10.1002/em.22087.

Chen, S.C., C.M. Kao, M.H. Huang, M.K. Shih, Y.L. Chen, S-P. Huang et al. 2003. Assessment of genotoxicity of benzidine and its structural analogues to human lymphocytes using Comet assay. Toxicol. Sci.72: 283–288.

Correa-Netto, C., R. Teixeira-Araujo, A.S. Aguiar, A.R. Melgarejo, S.G. De-Simone, M.R. Soares et al. 2010. Immunome and venome of *Bothrops jararacussu*: A proteomic approach to study the molecular immunology of snake toxins. Toxicon. 55: 1222–1235.

Creager, A.N.H. 2014. The political life of mutagens: A history of the Ames test. pp. 46–64. *In*: Boudia, S. and N. Jas [Eds.]. Powerless science: science and politics in a toxic world. New York [u.a.]: Berghahn Books.

Cunha, B.A., C.M. Sibley and A.M. Ristuccia. 1982. Doxycycline. Ther Drug Monit. 4: 115–35. Doi: 10.1097/00007691-198206000-00001.

De Stasio, E. The Ames test. Available at: <http://legacy.genetics-gsa.org/education/pdf/GSA_DeStasio_Ames_Student_Resources.pdf>. Accessed on 07.01.2022.

Ferraz, C.R., A. Arrahman, C. Xie, N.R. Casewell, R.J. Lewis, H.W. Boyer et al. 2019. Multifunctional toxins in snake venoms and therapeutic implications: from pain to hemorrhage and necrosis. Front Ecol. Evol. 7: 1–19.

Föllmann, W., G. Degen and J.G. Hengstler. 2013. Ames test. pp. 104–107. *In*: Brenner's Encyclopedia of Genetics (Second Edition) (Stanley Maloy and Kelly Hughes), Academic Press: Elsevier 2013.

Gatehouse, D. 2012. Bacterial mutagenicity assays: Test methods. Methods Mol. Biol. 817: 21–34. Doi: 10.1007/978-1-61779-421-6 2.

Gatehouse, D., S. Haworth, T. Cebula, E. Gocke, L. Kier, J. Kool et al. 1994. Recommendations for the performance of bacterial mutation assays. Mutat Res. 312: 217–233.

Ghosal, G. and J. Chen. 2019. DNA damage tolerance: a double-edged sword guarding the genome. Transl Cancer Res. 2: 107–129.

Goodson-Gregg, N. and E.A. de Stasio. 2009. Reinventing the Ames test as a quantitative lab that connects classical and molecular genetics. Genetics 181: 23–21. Doi: 10.1534/genetics.108.095588.

Goodson-Gregg, N. and E.A. de Stasio. 2021. The Ames Test. Network for Integrating Bioinformatics into Life Sciences Education, (Version 2.0). QUBES Educational Resources. Doi:10.25334/P18R-F263.

Hamel, A., M. Roy and R. Proudlock. 2016. The bacterial reverse mutation test. pp. 79–138. *In*: Proudlock R. [Ed.]. Genetic Toxicology Testing. Elsevier Inc.

Isono, K. and J. Yourno. 1974. Chemical carcinogens as frameshift mutagens: *Salmonella* DNA sequence sensitive to mutagenesis by polycyclic carcinogens. Proc. Natl. Acad. Sci. U.S.A. 71: 1612–1617.

Jurado, J., E. Alejandre-Durán ans C. Pueyo. 1993. Genetic differences between the standard Ames tester strains TA100 and TA98. Mutagenesis. 8: 527–32. Doi: 10.1093/mutage/8.6.527.

Kim, H.R., Y.J. Park, D.Y. Shin, S.M. Oh and K.H. Chung. 2013. Appropriate in vitro methods for genotoxicity testing of silver nanoparticles. Environ Anal Health Toxicol. 28: e2013003. Doi: 10.5620/eht.2013.28.e2013003.

Kisin, E.R., A.R. Murray, M.J. Keane, X.C. Shi, D. Schwegler-Berry, O. Gorelik et al. 2007. Single-walled carbon nanotubes: Geno- and cytotoxic effects in lung fibroblast V79 cells. J. Toxicol. Environ. Health A. 70: 2071–2079.

Klein, R.C., M.H. Fabres-Klein, L.L, de Oliveira, R.N. Feio, F. Malouin and O. Ribon Ade. 2015. A C-type lectin from *Bothrops jararacussu* venom disrupts Staphylococcal biofilms. PLoS One. 10: e0120514. Doi: 10.1371/journal.pone.0120514.

Landsiedel, R., M.D. Kapp, M. Schulz, K. Wiench and F. Oesch. 2009. Genotoxicity investigations on nanomaterials: methods, preparation and characterization of test material, potential artifacts and limitations - many questions, some answers. Mutat. Res. 681: 241–258.

Levin, D.E., E. Yamasaki and B.N. Ames. 1982a. A new Salmonella tester strain, TA97, for the detection of frameshift mutagens: A run of cytosines as a mutational hot-spot. Mutat Res. 94: 315–330.

Levin, D.E., M. Hollstein, M. F. Christman, E.A. Schwiers and B.N. Ames. 1982b. A new *Salmonella* tester strain (TA102) with A X T base pairs at the site of mutation detects oxidative mutagens. Proc. Natl. Acad. Sci. U.S.A. 72: 7445–7449.

Levin, D.E., L.J. Marnett and B.N. Ames. 1984. Spontaneous and mutagen-induced deletions: mechanistic studies in *Salmonella* tester strain TA102. Proc. Natl. Acad Sci. U S A. 81: 4457–61. Doi: 10.1073/pnas.81.14.4457.

Levy, D.D., E. Zeiger, P.A. Escobar, A. Hakura, B.J.M. van der Leede, M. Kato et al. 2019. Recommended criteria for the evaluation of bacterial mutagenicity data (Ames test). Mutat Res Gen Tox En. 841: 403074.

Mamun, S.M.H., M.A. Faiz, R. Rahman, F.E. Chowdhury, M.A. Wahab and Q.S. Ataher. 2000. Two consecutive venomous bites by the same snake: a case report. Society of Medicine 1st Conference. Doi: 10.1590/S1678-91992008000400014.

Marcussi, S., P.R. Santos, D.L. Menaldo, L.B. Silveira, N.A. Santos-Filho, M.V. Mazzi et al. 2011. Evaluation of the genotoxicity of *Crotalus durissus terrificus* snake venom and its isolated toxins on human lymphocytes. Mutat. Res. 724: 59–63.

Marcussi, S., R.G. Stábeli, N. Santos-Filho, D.L. Menaldo, L.L. Silva-Pereira, J.P. Zuliani et al. 2013. Genotoxic effect of *Bothrops* snake venoms and isolated toxins on human lymphocyte DNA. Toxicon. 65: 9–14.

Maron, D.M. and B.N. Ames. 1983. Revised methods for the *Salmonella* mutagenicity test. Mutat. Res./Environ. Mutagen. Relat. Subj. 113: 3–4, 173–215. Doi:10.1016/0165-1161(83)90010-9.

MedlinePlus [Internet]. Bethesda (MD): National Library of Medicine (US); [updated Jun 24; cited 2020 Jul 1]. Available from: https://medlineplus.gov/download/genetics/understanding/basics.pdf. Accessed on 11/01/2022.

Milani Júnior, R., M.T. Jorge, F.P. de Campos, F.P. Martins, A. Bousso, J.L. Cardoso et al. 1997. Snake bites by the jararacuçu (*Bothrops jararacussu*): clinicopathological studies of 29 proven cases in São Paulo State, Brazil. QJM. 90: 323–334.

Mohamed, S., U. Sabita, S. Rajendra and D. Raman. 2017. Genotoxicity: mechanisms, testing guidelines and methods. Glob J. Pharmaceu. Sci.1: 1–6.

Mortelmans, K. 2006. Isolation of plasmid pKM101 in the Stocker laboratory. Mutat. Res. 612: 151–164.

Mortelmans, K. and E. Zeiger. 2000. The Ames Salmonella/microsome mutagenicity assay. Mutat Res-Fund Mol. M. 455: 1–2, 29–60. Doi: 10.1016/S0027-5107(00)00064-6.

Moschel, R.C. 2001. Carcinogens, Encyclopedia of Genetics. pp. 271–272.

National Cancer Institute. 2017. The Genetics of Cancer (*The Genetics of Cancer was originally published by the National Cancer Institute*). https://www.cancer.gov/about-cancer/causes-prevention/genetics. Accessed on 12/05/2022.

Nature portfolio. 2022. (https://www.nature.com/subjects/cancer/). Accessed on 10.01.2022.

Noble, W., C.J. Garwood and D.P. Hanger. 2009. Minocycline as a potential therapeutic agent in neurodegenerative disorders characterized by protein misfolding. Prion, 3, 2: 78–83. Doi: 10.4161/pri.3.2.8820.

OECD - Organization For Economic Cooperation And Development. 1997. Guideline for testing of chemicals-bacterial reverse mutation test.

OECD. 2020. Test No. 471: Bacterial Reverse Mutation Test, OECD Guidelines for the Testing of Chemicals, Section 4, OECD Publishing, Paris, https://doi.org/10.1787/9789264071247-en.

Oliveira, I.C.F., M.O. de Paula, H.C.B. Lastra, B.B. Alves, D.A.N. Moreno, E.H. Yoshida et al. 2019. Activity of silver nanoparticles on prokaryotic cells and *Bothrops jararacussu* snake venom. Drug Chem Toxicol. 42: 60–64. Doi: 10.1080/01480545.2018.1478850.

Proença-Assunção, J.C., E. Constantino, A.P. Farias-de-França, F.A.R. Nogueira, S.R. Consonni, M.V. Chaud et al. 2021a. Mutagenicity of silver nanoparticles synthesized with curcumin (Cur-AgNPs). J. Saudi Chem. Soc. 25: 101321. Doi: 10.1016/j.jscs.2021.101321.

Proença-Assunção, J.C., A.P. Farias-de-França, N. Tribuiani, J.C. Cogo, R. de C. Collaço, P. Randazzo-Moura et al. 2021b. The influence of silver nanoparticles against toxic effects of *Philodryas olfersii* venom. Int. J. Nanomed. 16: 3555–3564.

Redbook 2000: IV.C.1.a. Bacterial Reverse Mutation Test. Toxicological Principles for the Safety Assessment of Food Ingredients. 2018. Available at: <https://www.fda.gov/regulatory-information/search-fda-guidance-documents/redbook-2000-ivc1a-bacterial-reverse-mutation-test>. Accessed in 16/09/2019.

Rojas, E., M.C. Lopez and M. Valverde. 1999. Single cell gel electrophoresis assay: methodology and applications. J. Chromatogr B Biomed Sci. Appl. 722: 225–254.

Santos, C.A., V.M. Balcão, M.V. Chaud, M.M. Seckler, M. Rai, M.M.D.C. Vila et al. 2017. Production, stabilization and characterization of silver nanoparticles coated with bioactive polymers pluronic F68, PVP and PVA. IET Nanobiotech. 11: 552–556.

Shibata, H., T. Chijiwa, N. Oda-Ueda, H. Nakamura, K. Yamaguchi, S. Hattori et al. 2018. The habu genome reveals accelerated evolution of venom protein genes. Sci. Rep. 8: 11300.

Smilack, J.D. 1999. The tetracyclines. *In*: Mayo Clinic Proceedings. Elsevier, 727–729. Doi: 10.4065/74.7.727.

Sugiyama, K., M. Yamada, T. Awogi and A. Hakura. 2016. The strains recommended for use in the bacterial reverse mutation test (OECD guideline 471) can be certified as non-genetically modified organisms. Genes Environ. 38: 2. Doi: 10.1186/s41021-016-0030-3.

Tchunwou, P.B. 2013. Genotoxic stress. *In*: Brenner's Encyclopedia of Genetics (Second Edition) (Stanley Maloy and Kelly Hughes), Academic Press: Elsevier 2013, pp. 313–317.

Teixeira, C., Y. Cury, V. Moreira, G. Picolo and F. Chaves. 2009. Inflammation induced by *Bothrops asper* venom. Toxicon. 54: 67–76.

Teixeira, C.F., E.C. Landucci, E. Antunes, M. Chacur and Y. Cury. 2003. Inflammatory effects of snake venom myotoxic phospholipases A2. Toxicon. 42: 947–962.

Tribuiani N., J. de Souza, M.A. de Queiroz Junior, D.A. Baldo, V. de C. Orsi, Y. Oshima-Franco et al. 2022. *In situ* effects of Doxycycline on neuromuscular junction in mice. Curr. Mol. Med. 22: 349–353. Doi: 10.2174/1566524021666210 521125553.

Varella, S.D., G.L. Pozetti, W. Vilegas and E.A. Varanda. 2004. Mutagenic Activity of Sweepings and Pigments from a Household-Wax Factory Assayed with *Salmonella typhimurium*. Food Chem. Toxicol. 42: 2029–2035. Doi: 10.1016/j. fct.2004.07.019.

Vijay, U., S. Gupta, P. Mathur, P. Suravajhala and P. Bhatnagar. 2018. Microbial Mutagenicity Assay: Ames Test. Bio-protocol, 8: e2763. DOI:10.21769/BioProtoc.2763.

Yoshida, E.H., N. Tribuiani, G. Sabadim, D.A. Neto Moreno, E.A. Varanda and Y. Oshima-Franco. 2016. Evaluation of betulin mutagenicity by *Salmonella*/Microsome Test. Adv. Pharm. Bull. 6: 443–447. Doi: 10.15171/apb.2016.057.

Wall, D.M., C.V. Srikanth and B.A. McCormick. 2010. Targeting tumors with *Salmonella typhimurium* - potential for therapy. Oncotarget. 1: 721–728. Doi: 10.18632/oncotarget.101208.

Warheit, D.B. and E.M. Donner. 2010. Rationale of genotoxicity testing of nanomaterials: regulatory requirements and appropriateness of available OECD test guidelines. Nanotoxicol. 4: 409–413. Doi: 10.3109/17435390.2010.485704.

Zeiger, E. 2019. The test that changed the world: The Ames test and the regulation of chemicals. Mutat. Res. Genet. Toxicol. Environ. Mutagen. 841: 43–48. Doi: 10.1016/j.mrgentox.2019.05.007.

Zobiole, N.N., T. Caon, J.W. Bertol, C.A.S. Pereira, B.M. Okubo et al. 2015. *In vitro* and *in vivo* genotoxic evaluation of *Bothrops moojeni* snake venom. Pharm. Biol. 53: 930–934.

Section IV
Fungal and Protozoan Genetics

CHAPTER 13

Fungal Genetics, Transcriptomics, Proteomics, and Metabolomics

Monika Bielecka

Introduction

Fungi are the largest group of eukaryotic organisms that ranges from yeast and slime molds to mushrooms. With an estimated 1.5 to 5.1 million species worldwide, fungi represent one of the largest branches of the Tree of Life (Hawksworth 1991, Blackwell 2011, Hawksworth and Lücking 2017). They are also the most successful Eucaryotes on Earth - due to their small size and their cryptic lifestyle in soil, dead and decomposing matter. Fungi have evolved strategies to survive in the most diverse and extreme environments and to build mutualistic associations with organisms from other kingdoms, such as bacteria, algae, other fungi species, fungi, plants and animals (Onofri et al. 2007, Chávez et al. 2015, Mohanta and Bae 2015). These abilities enabled fungal organisms for Earth domination from tropical to temperate and polar habitats as well as for inhabiting the extreme marine environment (Ma et al. 1999, Arnold et al. 2000, Pointing et al. 2009, Raghukumar et al. 2014, Murgia et al. 2019).

Fungi act as major recyclers and decomposers in nature, but they are also major pathogens for plants and animals, causing economic losses to agriculture and affecting animal and human health (Yuste et al. 2011, Dean et al. 2012, Köhler et al. 2015, Kumar and Chandra 2020). On the other hand, humans have made use of fungi since early history for their various applications. Selected fungi are directly used as human food and food additives and yeasts are used in the bread industry and alcohol fermentation. Fungi also produce valuable enzymes and a variety of specialized metabolites, which can act as antibiotics, pesticides or mycotoxins. Therefore, fungi are utilized in a remarkable variety of biotechnological applications, from the food and beverage industry to the production of antibodies, enzymes, biofuels, and pharmaceutical compounds (de Vries 2003, Lange et al. 2012, Demain 2014, Kazemi Shariat Panahi et al. 2022, Chatterjee and Venkata Mohan 2022).

Because of their characteristic intrinsic features and importance to mankind, fungi are subjected to intense physiological, ecological, phylogenetic, and molecular studies. For decades, fungi are used extensively in both fundamental research and industrial applications. To meet fundamental and industrial research needs, gaining knowledge at different levels within the cellular hierarchy of the information flow in fungal organisms is highly desirable. Fungus, like all other cellular forms of life, share the same fundamental scheme of genome replication and expression that has been formulated by Francis Crick and called the Central Dogma of Molecular Biology (Guo 2014). Over the past two decades, the advancement of so-called 'omics' approaches, including genomics, phylogenomics, transcriptomics, proteomics and metabolomics, has enormously accelerated the way to understand fungal physiology and diversity at different taxonomic levels. The relationship

Department of Pharmaceutical Biology and Biotechnology, Wroclaw Medical University, 50–556 Wroclaw, Poland.
Email: monika.bielecka@umw.edu.pl

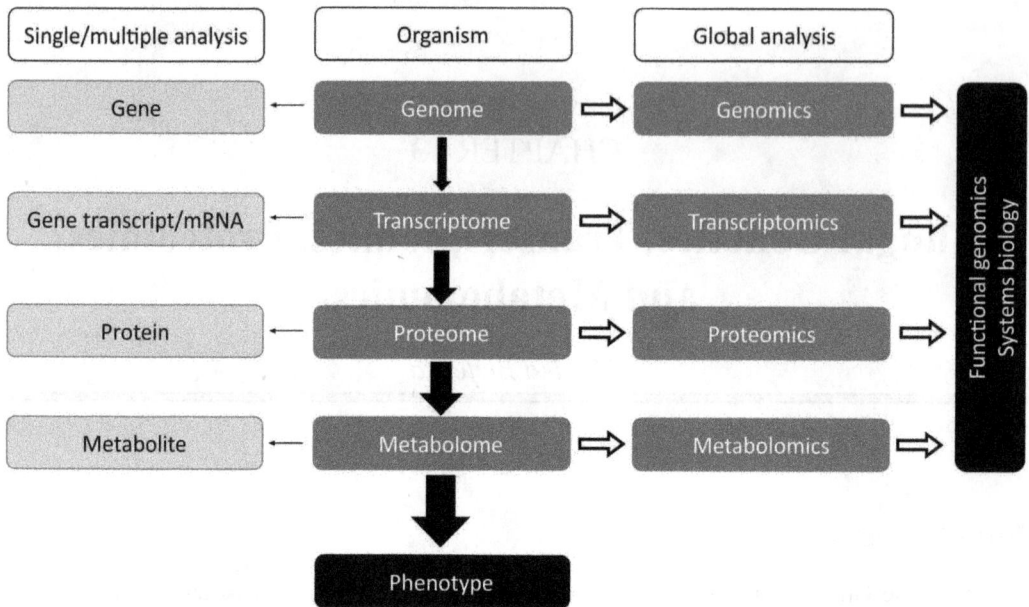

Figure 1. General schematic of 'omics' organization with respect to the central dogma of molecular biology.

between the 'omics' with respect to the general dogma of molecular biology is illustrated in Figure 1. Used individually, each of these applications can generate a plethora of data that could be mined by researchers for years. When combined, the 'omics' techniques can comprehensively dissect a system at the genomic, transcriptional, translational and metabolic levels. They have also paved a new way for the advancement in the functional genomics, whose ultimate goal is the assignment of function to every gene within a genome (Hofmann et al. 2003). The traditional method of assigning functions to genes was screening of mutant collections (Forsburg 2001, Casselton and Zolan 2002). Nowadays, functional genomics can feed information from all fields of 'omics' technology and thus connect genes to morphological and physiological characteristics. The abundance of 'omics' data facilitates the discovery of fungal metabolites, proteins, enzymes and pathways. Moreover, from analysis of the 'omics' it has become possible to map the interactions between all the components of the cell and describe them also at the quantitative level, which is referred to as systems biology (Ideker et al. 2001, Nielsen and Olsson 2002, Hofmann et al. 2003) (Figure 1).

Fungal Genetics

Early History of Fungi Genome Sequencing

Genetics Versus Genomics

The study of individual genes, genetic variation, and their roles in inheritance is called genetics. Genomics is an interdisciplinary field of biology focusing on the structure, function, evolution, mapping, and editing of genomes. In contrast to genetics, genomics aims at the collective characterization and quantification of all of an organism's genes, their actions, interrelations and influence on the organism. Genomics involves DNA sequencing, the identification of genes and the development of methods used to achieve this. Nowadays, advanced bioinformatics tools are also needed to assemble and analyze the function and structure of entire genomes (Hofmann et al. 2003).

Under this definition, the first highlight of fungal genomics was the sequencing of the whole genome of *Saccharomyces cerevisiae* (Goffeau et al. 1996). After this first breakthrough, there was a long gap before such advancements could be matched in other organisms. In 2001, the Whitehead Institute Center for Genome Research made its first release of *Neurospora crassa* genome, the

Aspergillus nidulans genome was sequenced by Monsanto and DSM company announced the completion of the *Aspergillus niger* genome. These commercially produced data, however, had restricted or no access to the public (Hofmann et al. 2003). The beginning of the new era was marked by the publishing of the genome sequence of the fission yeast *Schizosaccharomyces pombe*, and since then the number of sequenced fungi genomes increased exponentially (Wood et al. 2002). In 2005, over 40 complete fungal genomes have been publicly released with an equal number currently being sequenced, representing then the widest sampling of genomes from any eukaryotic kingdom (Galagan et al. 2005). Advancements in sequencing strategies and technologies, which took place over recent 45 years, facilitated the transition from rather simple genetics to genomics studies, also in the field of fungal research.

Sanger Sequencing

Developed in 1977, Sanger sequencing is based on the random incorporation of chain-terminating dideoxynucleotides (ddNTPs) by DNA polymerase during *in vitro* DNA replication via polymerase chain reaction (PCR) (Sanger et al. 1977). The ddNTPs may be radioactively or fluorescently labelled for detection in automated sequencing machines that allow for electrophoretic separation of labelled products, thus enabling the readout of the DNA sequence of the matrix DNA.

Although Sanger sequencing was initially used for multiple whole-genome sequencing projects, including the human genome (Craig Venter et al. 2001) and above-mentioned sequencing of fungal genomes, it was a monumental, time-consuming, costly task. From a technical point of view, fungal genomes were considered the perfect target for a genome sequencing program, being relatively small and having high gene density and few repetitive regions (Mohanta and Bae 2015). However, the sequencing of the relatively small and compact yeast's genome (about 12 Mb) required a huge international effort of about 600 scientists and several million dollars in funding. This is because Sanger sequencing suffers from many limitations, which makes it suboptimal for whole-genome projects. The limitations include non-specific binding of the primer to the DNA, affecting accurate read-out of the DNA sequence, and DNA secondary structures affecting the fidelity of the sequence. The method can directly sequence only relatively short (300-1000 nucleotides long) DNA fragments in a single reaction. Also, a small sample handling capacity makes this technique hardly reach the hight-troughput goal. Apart from the limitations coming from the sequencing technique, a major challenge was the necessity for the construction of clone libraries in high-capacity vectors and the tedious filling-in of the gaps in clone coverage. A strategy developed to overcome the limitations of clone-by-clone sequencing was the whole-genome shotgun sequencing and a hybrid strategy that involved both clone-by-clone and shotgun-sequencing components that have been adopted in many whole-genome projects (Green 2001). The success of this strategy was determined by the reduced complexity of library preparation methods, thus shotgun sequencing became in fact one of the precursor technologies that was responsible for enabling whole-genome sequencing with next-generation sequencing (NGS) technologies.

Next-generation Sequencing (NGS)

NGS methods enable faster and cheaper nucleotide sequencing than the traditional Sanger method, opening a new era in genomics. Compared to the Sanger sequencing method, NGS technologies provide higher throughput data at lower costs and better support genome research (Park and Kim 2016). These new sequencing methods share three major advantages. First, NGS libraries are prepared in a cell-free system instead of requiring bacterial cloning of DNA fragments. Second, they produce thousands to millions of sequencing reactions in parallel. Finally, base interrogation is performed cyclically and in parallel, so that the sequencing output is detected without electrophoresis. As a result, the enormous number of reads generated by NGS enables genome sequencing at an unprecedented speed (Dijk et al. 2014).

Next-generation sequencing technologies which use the principle of sequencing by synthesis (SBS) are now referred as second-generation sequencing technology. Despite some technical differences, in SBS technologies the shotgun sequencing of randomly fragmented genomic DNA is performed and adapter sequences and primers are ligated to the fragmented DNA for the construction of template libraries. Library amplification is performed on a solid surface (Solexa-Ilumina) or on beads while isolated within miniature emulsion droplets (SOLiD, Roche, Ion Torrent). Nucleotide incorporation is monitored by luminescence detection (Solexa-Ilumina) or by changes in electrical charge (IonTorrent) during the sequencing procedure (Aguilar-Pontes et al. 2014).

During the past decade, NGS technologies have evolved through the incorporation of revolutionary innovations to tackle the complexities of genomes (Goodwin et al. 2016). Short-read sequencing approaches, such as sequencing by synthesis, ion semiconductor, and nano ball sequencing, maximize the number of bases sequenced in a short time, generating a wealth of data that can help understand complex phenotypes.

Although the enormous numbers of reads generated by NGS have enabled genome sequencing at an unprecedented speed, this technology is not free from limitations either. One of them is the higher sequencing error rate, compared to Sanger sequencing, which is compensated by the possibility of extremely high genome coverage. Another limitation of the NGS approaches utilizing a PCR-based library preparation protocols is the PCR biases which can lead to low coverage of the GC-rich DNA regions. This was, however, mitigated by implementing PCR-free library protocols. The contiguity of the assemblies produced with short reads benefited from the use of longer insert size libraries coupled with paired reads technology, which is one of the reasons that made Illumina the dominant second-generation sequencing technology (Kozarewa et al. 2009). Another drawback of NGS sequencing consists of relatively short reads, making genome assembly more difficult, with a need for the development of novel alignment algorithms (Dijk et al. 2014). Genomes of many organisms (especially of plants and fungi) are highly complex with many repetitive elements, copy number, and structural variations which are relevant for evolution and adaptation. Many of these elements are so long that they cannot be resolved through short-read paired-end technologies (Goodwin et al. 2016). Therefore, other NGS methods have been developed, out of which the so-called third-generation methods could be especially promising for the field of fungal genomics.

Third-generation Sequencing

Long-read sequencing techniques aim to sequence longer contiguous DNA sectors, which is required to resolve structurally complex regions. Third-generation long-read sequencing has the potential to overcome many limitations of the short-read approaches, e.g., the resolution of repeated sequences and large genomic rearrangements (Goodwin et al. 2016, Kumar et al. 2019).

Third-generation methods allow for the detection of single molecules, but sequencing occurs in real-time as an additional common feature (Schadt et al. 2010). Reads in excess of several kilobases are delivered by such long-read sequencing, which allows for the resolution of these large structural features. These long reads can span repetitive regions with a single continuous read, which eliminates ambiguity in the position or size of genomic elements.

Out of the third-generation sequencing technologies, two main types of long-read technologies, in particular, seem to be most suitable for fungal genomics, that is the single molecule real-time sequencing approaches delivered by Pacific Biosciences (PacBio SMRT) (Eid et al. 2009) and nanopore sequencing implemented by Oxford Nanopore Technologies (Clarke et al. 2009). PacBio SMRT technology is based on sequencing by synthesis of circularized DNA created by ligating hairpin adapters to both ends of the target DNA molecule. It provides reads up to 80 kilobases, which is limited by the DNA template length and the efficiency of polymerase. All four nucleotides are added simultaneously and measured in real-time by fluorescence; circularization of the template allows to sequence it several times providing a high accuracy (>99.9%; single pass error rate is instead ~15%) (Eid et al. 2009, Pushkarev et al. 2009). The principle of nanopore sequencing (Oxford Nanopore) is completely different. Here, each nucleotide of a template DNA strain is detected by a nanopore constituted by a mutated form of the *Mycobacterium smegmatis* porin A (MspA) which serves as a biosensor (Derrington et al. 2010). A voltage bias generated at the two sides of a membrane produces current that drives ssDNA (or RNA) through the nanopore. An enzyme (polymerase or helicase) is bound to the nucleic acid and works as a ratchet, controlling the step-wise passage of the nucleotides, allowing the detection of the different current perturbation generated by each nucleotide when passing through the narrowest part of the nanopore. Reads length limit is mostly determined by the length of the template DNA generating ultra-long reads, up to almost 1 Mb. The initially low accuracy (85%) was increased to 97% when sequencing both strands (Branton et al. 2008).

The advantages of nanopore sequencing for studying plant genomes, especially in the context of addressing challenges for producing high-quality, highly contiguous plant genome assemblies was recently demonstrated by Oxford Nanopore Technology (Oxford Nanopore Technologies 2020). There are more and more examples of using the advantages of the third-generation technology in the field of fungal genomics. The accurate long-read sequences obtained through the PacBio Sequel Systems were used to assemble the draft genomes of fungal endophytes isolated from *G. biloba* (Zou et al. 2021). The genome of an endophytic strain of *Amphirosellinia nigrospora* (JS-1675), isolated from *Pteris cretica*, was recently sequenced with the PacBio RS II technology (Jeon et al. 2019).

Third-generation sequencing technologies' performances were tested alone or in combination with second-generation technologies on *Saccharomyces* (Jenjaroenpun et al. 2018) and *Leptosphaeria* species (Dutreux et al. 2018). The genome sequence of an endophytic strain of *Gaeumannomyces* sp. (JS-464) was generated using both the HiSeq 2000 (Illumina) and PacBio RSII platform (Kim et al. 2017). The hybrid assembly approach - which on one side, exploits long reads to obtain high contiguity and on the other, short reads for high accuracy and error correction - is now becoming the gold standard.

Bioinformatic Data Analysis

The advancement in genome sequencing technologies has boosted genomics research by enabling many small laboratories to achieve sequencing of entire genomes. A huge amount of genomic data was produced in the last two decades, which brought both – a great potential and challenges. This, in turn, boosted the rapid development of a relatively new discipline of bioinformatics, which become vital for managing the large amounts of data produced by genomic projects.

The first step, after the reads have been generated and the base calling is done, is a genome assembly. This process refers to assembling the sequencing reads in the correct order to reconstruct the whole genome sequence. The assembling process maps DNA reads by finding the overlapping regions between reads with or without a reference genome. If the reference genome is not available, which is often the case for non-model organisms, the reads have to be assembled *de novo* (without knowing the genomic sequence of the target species).

To efficiently manage sequences from various sequencing technologies, and to assemble genomes of different sizes and complexity, various assembly strategies have been developed. The assembly algorithms are constantly updated to follow the development in the sequencing technologies and to solve issues that prevent to get higher quality genome assemblies, such as sequencing errors, repetitive regions and uneven reads depth. The most used assemblers in fungi genomics and ascomycetes are described in the literature (Aguilar-Pontes et al. 2014, Muggia et al. 2020).

After the genomic sequence is assembled, the process of identifying structures and functions of genes within a genome (called annotation) must be performed to make the genome available for the scientific community exploitation. Two strategies can be used to delimit genes and gene structures in the genome: intrinsic (ab initio) gene prediction, relying on the genomic sequence itself, and extrinsic – utilizing other similar sequences such as transcripts. These strategies are usually combined in the genome annotation pipelines and the examples of tools which have been used for fungal genome annotation are summarized in the literature (Aguilar-Pontes et al. 2014, Muggia et al. 2020).

Rapid improvements in sequencing technology have greatly increased the number of fungal genomes. In the past few years, the number of available fungal genomes grew exponentially and the next is in the pipeline. There are several databases dedicated to fungal research, the analysis of which, however, shows that despite the astonishing number of fungi species, we do not know a lot about the genomics of this diverse kingdom.

Up to date (the 9th of January 2023), the Joint Genome Institute (JGI) fungi-dedicated database, Mycocosm (https://mycocosm.jgi.doe.gov/mycocosm/home), holds 2,411 fungi genomes (Grigoriev et al. 2014). The number of sequenced genomes in each clade of the fungal tree of life is shown in Figure 2. The ENSEMBLFungi database (https://fungi.ensembl.org/index.html) currently holds 1,506 complete genomes. The current release of the e-Fungi database (http://www.cs.man.ac.uk/~cornell/eFungi/index.html) provides genomic and functional data for the 34 fungal genomes. The FungiDB (https://fungidb.org/fungidb/app), a part of the Eukaryotic Pathogen Genomics Database Resource (EuPathDB, eupathdb.org), which is a free online resource for data mining and functional genomics analysis for fungal and oomycete species, currently holds genomic sequences for 286 fungal organisms (Basenko et al. 2018).

Richer resources are found in the GenBank, a part of the International Nucleotide Sequence Database Collaboration, which comprises the DNA DataBank of Japan (DDBJ), the European Nucleotide Archive (ENA), and GenBank at the National Center for Biotechnology Information (NCBI) (Benson et al. 2013). Up to date (the 9th of January 2023), 13,175 fungi genomes at various assembly levels are stored at GeneBank/NCBI, out of which 3,804 have been annotated and 3,986 are regarded as the reference genomes (https://www.ncbi.nlm.nih.gov/data-hub/genome/?taxon=4751). However, if looked only for complete genomes (with the highest assembly level), the number

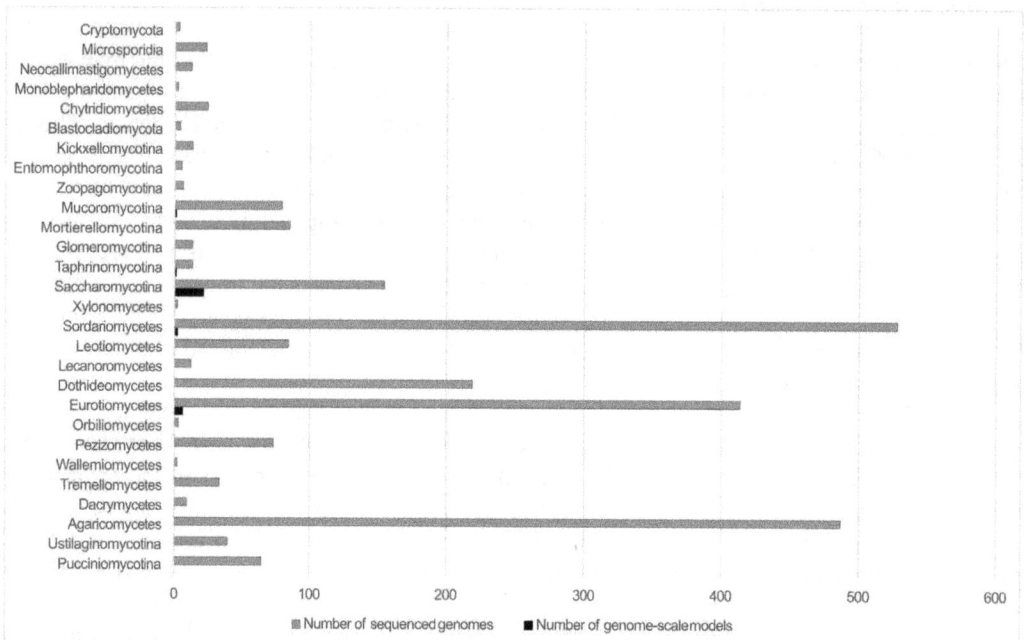

Figure 2. Representation of sequenced fungal genomes available for each clade in the fungal tree of life. JGI data retrieved from https://mycocosm.jgi.doe.gov/mycocosm/home, GSM data retrieved from https://systemsbiology.ucsd.edu/InSilicoOrganisms/OtherOrganisms (accessed 09.01.2023).

of available genomes changes to 248, out of which only 91 are annotated and 94 are reported as reference genomes.

Non-model Fungi vs GSM

As seen from the above, the current decade witnesses enormous proliferation of sequenced, yet understudied (non-model), fungal genomes. This is due to increasingly affordable next-generation sequencing technology and investments by federally funded agencies, such as the Department of Energy's JGI. (Wilken et al. 2019). Despite many sequenced fungal genomes being available in various databases and many more being in progress, very few genome-scale models (GSMs) have been reconstructed from fungal 'omics' data per clade. A GSM is a mathematical representation of the metabolic network of a microorganism, a detailed mathematical model of cellular metabolism and physiology (Feist et al. 2008). The number of GSMs available for each clade in the kingdom of fungi is depicted in Figure 2. The generation of sequenced genomes (currently 2411) far outpaces the generation of GSMs (31 in total). The significant difference between the number of sequenced genomes and available GSMs highlights the long-standing knowledge gap in our systems-level understanding of fungi. To effectively utilize the enzymes and pathways of the fungal world, a multifaceted approach that combines sequence informatics, systems modeling, and traditional metabolic engineering approaches is required (Wilken et al. 2019).

Functional and Comparative Genomics

The ultimate goal of functional genomics is the assignment of function to every gene within a genome. In the pre-NGS era, reductive approaches used genetic screens to reveal phenotypes from collections of mutants. However, reductive approaches suffer numerous limitations. Firstly, filamentous fungi typically feature about 15,000 genes; knocking them all out and characterizing them would be a monumental task. Secondly, genes may have multiple phenotypes, including fatal ones; several genes may have redundant functions for certain phenotypes or in certain environments; and lastly, the development of complexity and its regulation is widely believed to involve interactions of many genes through gene-networks that are sparsely revealed by associations detected between single genes and single developmental phenotypes (Wang et al. 2018a).

The availability of whole genome sequences has also fundamentally changed the methods used for the identification and mapping of genes. Over the years, numerous comparative genomics projects have been published. To name just a few, comparative genomics revealed high biological diversity and specific adaptations in the industrially and medically important fungal genus *Aspergillus* (de Vries et al. 2017), the dynamic genome evolution was studied through comparative genomics for the most commonly occurring *Trichoderma* species (Kubicek et al. 2019), and in host specialist ectomycorrhizal fungi (Lofgren et al. 2021), comparative genomics of xylose-fermenting fungi was performed for enhanced biofuel production (Wohlbach et al. 2011).

Advances in technologies for genome sequencing have generated a huge quantity of high-throughput and high-dimensional biological data (Marx 2013). The increasing availability of fungal genomes, coupled with complementary 'omics' data sets, (transcriptomics, proteomics, and metabolomics) holds great promise for future discoveries.

Fungal Transcriptomics

According to the central dogma of molecular biology (Figure 1), the information flow within a cell is transmitted from genes by their transcription into the messenger RNA (mRNA) and the translation of the latter into peptides (Guo 2014). While the DNA sequence of a gene is considered relatively stable, its transcription level is usually highly variable. Gene expression varies in time and between different cell types, it also depends on the organism's developmental status and is influenced by the environment. The cell's transcriptome represents its complete set of transcripts, at a certain moment in time. Transcriptomics is the quantification of the transcriptome, for a specific developmental stage or physiological condition. It provides the ability to identify quantitative and qualitative differences in gene expression when comparing multiple mRNA populations (Tan et al. 2009, Finotello and Di Camillo 2015). Hence, the field of transcriptomics bridges the gap between genes and proteins and helps connect genes to a phenotype expressed as morphological and physiological characteristics.

Gene Expression Analysis in the Pre-NGS Era

One of the first methods developed to study gene expression was a DNA/RNA probe-based technique called Northern blot (Krumlauf 1994). Despite its obvious limitations, it was successively used to study the expression of some fungal genes (Appleyard et al. 1995, Qi and Yang 2002). Later, various PCR-based methods were implemented to monitor the transcript abundance, such as semi-quantitative RT-PCR (Marone et al. 2001) and real-time quantitative RT-PCR (real-time qRT-PCR), which were also successfully implemented into fungal gene expression studies (Choquer et al. 2003, Rantakokko-Jalava et al. 2003, Aslam et al. 2017a). Although considered fast and effective, real-time qRT-PCR is rather suitable for small- to mid-size projects due to its mid-throughput feature. A further challenge in employing qRT-PCR in expression analysis of fungal genes is the choice of a suitable reference gene, which expression would be stable and independent from the influence of environmental parameters and growth conditions. In the experiment with the lignin-degrading basidiomycete *Pleurotus ostreatus*, it was demonstrated that the use of non-validated reference genes as internal controls lead to biased results and misinterpretations of the biological responses underlying expression changes (Castanera et al. 2015). This has motivated the introduction of novel, high-throughput technologies in order to ameliorate these limitations. Hybridization techniques such as microarrays were able to partially address these issues by increasing the amounts of genes to be analysed per run (Ijoma et al. 2021).

In the pre-NGS era, hybridization-based approaches, such as microarrays, were the most used solutions for gene expression profiling and differential expression analysis, thanks to their high-throughput and relatively low costs. These technologies consisted in an array of probes, whose sequences represented particular regions of the genes to be monitored. The sample under investigation was washed over the array, and RNAs were free to hybridize to the probes with a complementary sequence. A fluorescent dye was used to label the RNAs, so that image acquisition of the whole array enabled the quantification of the expressed genes (Xiang and Chen 2000, Finotello and Di Camillo 2015).

The first fungal microarray studies were performed on *S. cerevisiae* (DeRisi et al. 1997). DNA microarrays containing virtually every gene of *S. cerevisiae* were used to carry out a comprehensive investigation of the temporal program of gene expression accompanying the metabolic shift from fermentation to respiration.

Since then, thousands of fungal microarray studies have been performed on different aspects of fungal transcriptomics. Microarrays were used for investigation of pathogenicity genes in *Fusarium graminearum* during barley infection (Güldener et al. 2006), analysis of molecular basis of pathogenicity in *Magnaporthe grisea* causing the rice blast (Gowda et al. 2006) and for the reconstruction of the complete metabolic network of *Aspergillus nidulans* – a model organism for aspergilli (David et al. 2006). A microarray-based technique was used to perform comparative transcript profiling in the grass fungal endophytes *Neotyphodium coenophialum*, *Neotyphodium lolii*, and *Epichloë festucae*, which allowed to investigate genome-specific gene expression, novel endophyte genes, and general plant–endophyte interactions through revealing factors supporting the establishment and maintenance of the mutualistic association (Felitti et al. 2006). The first 50 microarray studies performed in over 20 species of filamentous fungi, which encompassed a wide variety of research areas, were reviewed by Breakspear and Momany (2007). The studies have addressed aspects of fungal metabolism (i.e., identification of various fungal biosynthesis genes or gene clusters), development (i.e., fruiting body development genes, light-induced genes), symbiosis (i.e., mycorrhizal development genes, host-specific symbiosis-regulated genes), pathogenicity (i.e., host-specific pathogenic genes) and industrial applications of *Aspergillus niger*, *Aspergillus nidulans* and *Trichoderma reesei*.

The main microarray platforms used worldwide for fungi were: Affymetrix (currently acquired by Thermo Fisher Scientific), Agilent (https://www.agilent.com), Illumina (https://www.illumina.com) and Nimblegen (currently a member of the Roche Group) (Aguilar-Pontes et al. 2014).

Although widely used in quantitative transcriptomics, microarray-based techniques have several limitations: high background levels due to cross-hybridization between quasi-complementary sequences, limited dynamic range due to background noise and signal saturation and the need for normalization to compare data from different arrays. The biggest obstacle, however, is the necessity to rely on a prior knowledge about the genome, which is required to develop complementary probes. As a consequence, microarrays give the possibility to only monitor the changes in expression of previously known genes and no information about the sequence of analysed transcripts is generated (Finotello and Di Camillo 2015). Despite that, the microarray-based technique allowed for transcriptome profiling in a high-throughput manner, thus becoming an important step toward global transcriptomic analysis.

Next-generation Sequencing of Trancriptomes

RNA sequencing (RNA-seq), a methodology for RNA profiling based on next-generation sequencing, has efficiently replaced microarrays for the study of gene expression (Shendure 2008). RNA-seq allows for simultaneous analysis of thousands of known and unknown genes without the need for the reference gene. Moreover, RNA-seq enables the reconstruction of both known and novel transcripts at single-base level; it has a broad dynamic range (not limited by signal saturation) and high levels of reproducibility. These features have revolutionized transcriptomics and quickly established RNA-seq as the preferred methodology for the study of gene expression (Wang et al. 2009, Finotello and Di Camillo 2015).

The sequencing framework of RNA-seq enables to investigate at high resolution all the RNAs present in a sample, characterizing their sequences and quantifying their abundances at the same time. In practice, millions of reads are sequenced from random positions of the input RNAs. These reads can then be computationally mapped on a reference genome to reveal a 'transcriptional map', where the number of reads aligned to each gene gives a measure of its level of expression. If the reference genome is not available, which is often the case for non-model organisms, the transcriptome assembly can be performed *de novo*. In a traditional RNA-seq study aiming at the detection of differentially expressed genes, a basic data processing pipeline consists of the following steps: quality control of the raw reads, reads mapping and alignment, counts computation, counts normalization and detection of differentially expressed genes (Finotello and Di Camillo 2015, Ijoma et al. 2021).

To make the counts comparable across samples or experiments, they have to be normalized to remove the influence of all possible biases. Normalized measures such as RPKM (reads per kilobase of exon model per million reads), FPKM (fragments per kilobase of exon model per million mapped reads), and TPM (transcripts per million) are then used to enable transcript quantification and report expression values (Conesa et al. 2016). Differential expression analysis can then be carried out by comparing values from different samples.

RNA-seq studies produce large and complex data sets, for the managing and interpretation of which several computational tools have been developed. Computational tools themselves are chasing the development of sequencing technologies to accommodate changes in the data features, and the assessment and comparison of state-of-the-art methods must be constantly performed to implement an updated RNA-seq computational pipeline. Computational tools designed for differential gene expression studies performed with NGS methods have been reviewed by Finotello and di Camillo

(2015) and an update in recent bioinformatic methods applied in fungi research was done by Ijoma and co-workers (2021). Various tools used for *de novo* transcriptome assembly haev been described by Geniza and Jaiswal (2017).

RNA-seq has been widely used in fungal research. In 2008, the RNA-seq technology (Illumina) was applied to sequence the transcriptome of *S. cerevisiae* (Nagalakshmi et al. 2008). The SOLiD platform was used to generate antisense transcript reads of *Aspergillus niger* grown on different carbon sources (Delmas et al. 2012) and to compare the transcriptional response of *Trichoderma reesei* and *A. niger* to lignocellulose (Ries et al. 2013). The RNA-seq on Illumina platform was used to obtain insight into the degradative ability of a white-rot *Lentinus squarrosulus* based on the expression of genes encoding for degrading enzymes (Ravichandran et al. 2020). RNA-seq followed by *de novo* assembly of *Pycnoporus cinnabarinus* transcriptome was used to investigate the differential expression of its ligninolytic enzymes, which turned out to be dependent on the nature of the substrate (Henske et al. 2018). *De novo* RNA-seq of endophytic fungi transcriptomes has allowed exploring their genetic pathways and resulted in the discovery of fungal structural genes potentially involved in the biosynthesis of plant bioactive metabolites, i.e., paclitaxel in *Cladosporium cladosporioides* MD2 (Miao et al. 2018), resveratrol in *Alternaria* sp. MG1 (Che et al. 2016), swainsonine in *Embellisia oxytropis* (Li and Lu 2019) and huperzine A in *Colletotrichum gloeosporioides* Cg01 (Zhang et al. 2015), which was recently reviewed by Bielecka and co-workers (2022). Similarly, with the genomic studies, the third-generation sequencing technology generating long reads can also be used in fungal transcriptomic research, considering that they can span entire mRNA transcripts, thus allowing to identify the precise connectivity of exons and discern gene isoforms (Dijk et al. 2014).

The powerful features of RNA-seq, such as high resolution and broad dynamic range, have boosted unprecedented progress of transcriptomics research, producing an impressive amount of data worldwide. With the development of transcriptomic technologies, it has become common practice to submit transcriptome data to public repositories. Nowadays, the main gene expression database is the Gene Expression Omnibus (GEO) (Edgar et al. 2002). GEO is an international public repository that archives and freely distributes microarray, next-generation sequencing, and other forms of high-throughput functional genomics data submitted by the research community.

Fungal Proteomics

Translation of mRNA into functional proteins is the second step of the central dogma; it is considered a precursor to metabolomics studies and most cellular biochemical activities (Figure1). A research field that attempts to understand the expression, function and regulation of the entire set of proteins, called 'proteome', in a cell, tissue or organism, is refered as proteomics. Proteomics aims at the identification and quantification of all proteins present in a given cell, tissue, or organism (Graves and Haystead 2002, Aslam et al. 2017b). Proteomics provides a comprehensive insight into the protein profile of an organism and is therefore often used as a complementary technique to transcriptomics. Often, however, the abundance of a protein detected by proteomics does not necessarily correlate to the expression of the corresponding mRNA in a quantitative manner. This difference in the abundance of a protein and the gene transcript levels measured by transcriptomics could be caused by post-translational regulation of gene products, protein consumption and accumulation, or other mechanisms. Thus, proteomics, describing proteins as effectors of biological function, would be considered as the most relevant data set to characterize a biological system (Aguilar-Pontes et al. 2014).

The Development of Methodology

Proteomics also involves the applications of technologies for the identification and quantification of overall proteins present in content of a cell, tissue or organism. The technological and applicative aspects of proteomics have been comprehensively reviewed by Aslam and co-workers (2017b). According to them, proteomics technologies can be divided into conventional, advanced, quantitative and high-throughput technologies.

The conventional techniques for the purification of proteins are chromatography based such as ion exchange chromatography (IEC), size exclusion chromatography (SEC) and affinity chromatography (Hage et al. 2012). For analysis of selective proteins, enzyme-linked immunosorbent assay (ELISA) and western blotting can be used. These techniques, however, enable analysis of a few individual proteins at once, they are also incapable of defining protein expression levels. Sodium dodecyl sulfate-polyacrylamide gel electrophoresis (SDS-PAGE), two-dimensional gel electrophoresis (2-DE) and two-dimensional differential gel electrophoresis (2D-DIGE) techniques are used for the separation of complex protein samples (Marouga et al. 2005, Issaq and Veenstra 2008).

Protein microarrays or chips have been established for high-throughput and rapid expression analysis (Zhu et al. 2001, Hall et al. 2007, Sutandy et al. 2013). Diverse proteomics approaches such as mass spectrometry (MS) have been developed to analyze complex protein mixtures with higher sensitivity (Yates 2011). Additionally, Edman degradation has been developed to determine the amino acid sequence of a particular protein (Smith 2001). Isotope-coded affinity tag (ICAT) labeling, stable isotope labeling with amino acids in cell culture (SILAC) and isobaric tag for relative and absolute quantitation (iTRAQ) techniques have been developed for quantitative proteomic (Ong et al. 2006, Shiio and Aebersold 2006, Wiese et al. 2007). X-ray crystallography and nuclear magnetic resonance (NMR) spectroscopy are two major high-throughput techniques that provide three-dimensional (3D) structure of protein that might be helpful to understand its biological function (Smyth and Martin 2000, Gerothanassis et al. 2002).

Bioinformatics Tools for Fungal Protein Analysis

High-throughput proteomics technologies generate a huge volume of proteomics data. Bioinformatics databases have been established to handle the enormous quantity of data and its storage (Perez-Riverol et al. 2015). Various bioinformatics tools are developed for 3D structure prediction, protein domain and motif analysis, rapid analysis of protein–protein interaction and data analysis of MS. The alignment tools are helpful for sequence and structure alignment to discover the evolutionary relationship (Vihinen 2001).

Apart from the classical integration of proteome data with such annotation databases as Gene Ontology (GO) and pathway databases (KEGG), functional insights into the fungal proteome data can be additionally obtained resorting to readily available free online resources for data mining like the comprehensive database of protein sequence and functional information UniProt (http://www.uniprot.org) – that includes manual annotation of fungi-specific proteins and protein families (Fungal Protein Annotation Project; https://www.uniprot.org/program/Fungi/) (Muggia et al. 2020). More specific bioinformatic tools, which were designed for prediction of various protein's characteristics, are solely based on the amino acidic sequence. Tools for the localization of intra- and extracellular proteins in fungi have been developed and include: WoLF PSORT, MultiLoc2, SherLoc2, MSLoc-DT, SCLpred and BUSCA. Advanced platforms, e.g., SignalP 5.0 (http://www.cbs.dtu.dk/services/SignalP/) and Phobius (http://phobius.sbc.su.se/cgi-bin/predict.pl), enable analyzing proteins for sequence features like signal peptides and transmembrane regions. The Fungal Secretome Knowledge Base (FunSecKB and FunSecKB2; http://bioinformatics.ysu. edu/secretomes/fungi.php) and the Fungal Secretome Database (FSD; http://fsd.snu.ac.kr/) have been established to collect extracellular proteins from all available fungal protein data in the NCBI RefSeq database (Muggia et al. 2020).

Proteomics of S. cerevisiae

Shortly after the publication of *S. cerevisiae* genome (Goffeau et al. 1996) and transcriptome analysis by microarray (DeRisi et al. 1997), the intense research on the proteome of this model species had begun. In 1999, by using isotope-coded affinity tags, protein expression in the *S. cerevisiae* grown on ethanol or galactose as a carbon source was analysed quantitatively (Gygi et al. 1999). The year 2000 brought a comprehensive analysis of protein-protein interaction in *S. cerevisiae* (Uetz et al. 2000). Two large-scale yeast two-hybrid screens were undertaken to identify protein–protein interactions between full-length open reading frames predicted from the *S. cerevisiae* genome sequence. These approaches resulted in the detection of 957 putative interactions involving 1,004 *S. cerevisiae* proteins, which attached many functionally unclassified proteins to a biological function and linked biological functions together into larger cellular processes. In 2001, Zhu and co-workers published the first proteome chip for *S. cerevisiae* (2001). In this study, yeast proteome microarray containing 5800 proteins was designed and screened for their ability to interact with proteins and phospholipids. This proved the concept that microarrays of an entire eukaryotic proteome can be prepared and screened for diverse biochemical activities. In 2002, two-dimentional polyacrylamide gel electrophoresis (2-D PAGE) coupled with mass spectrometry analysis of yeast proteome was published (Bader and Hogue 2002). In the same year, a more comprehensive, whole-cell approach was undertaken, where changes in the *S. cerevisiae* proteome were related to changes in the transcriptome of cultures grown on galactose versus those grown on ethanol (Griffin et al. 2002).

This shows that proteomics research on *S. cerevisiae* was again ahead of that on all other fungi; however, fungi that are pathogenic to humans, such as *Candida albicans* (Niimi et al. 1999) and *Aspergillus fumigatus* (Bruneau et al. 2001), have been also early targeted in terms of proteome investigation. 2D-PAGE with mass spectrometry approach was used to analyse the proteome of *Trichoderma reesei*, a major producer of cellulases; proteins associated with the cell wall were investigated on cellulase-inducing and cellulase-repressing media, respectively (Lim et al. 2001). The same analytical approach was used to analyse the effect of an antibiotic produced by *Streptomyces*, concanamycin A, on protein levels in the filamentous fungus *Aspergillus nidulans* (Melin et al. 2002).

Different areas of proteomics studies have been applied to fungi: descriptive proteomics (intracellular and subcellular) (Neto et al. 2014), differential production of proteins (Jun et al. 2013), post-translational modifications, protein–protein interaction (Schwikowski et al. 2000), and the extracellular proteome studies referred as secretomics (Tsang et al. 2009).

The study of protein expression, modification, structure, and function by means of proteomics has brought progress in various fields of science and technology because proteins are usually the end-products of gene expression and the sought-after bioproducts (Ijoma et al. 2021). The most extensive analyses of fungal cell wall proteome were performed on *Candida albicans* (Hernáez et al. 2010, Vialás et al. 2012). Proteomic analysis of pathogenic fungi revealed highly expressed conserved, fungi-typical cell wall proteins, which are of interest to the development of vaccines or drug targets (Champer et al. 2012, 2016, Voltersen et al. 2018). Investigations of fungi's proteome enabled detection of biomarkers and virulence determinants of the human pathogenic fungi (Nogueira et al. 2010). Global secretome characteristation enabled detection of pathogenesis-related proteins in human opportunistic fungal pathogen *Candida glabrata* (Rasheed et al. 2020) and *Aspergillus fumigatus* (Wang et al. 2018b). Proteome studies have helped in elucidating the dynamics of important proteins during plant–pathogen interaction and plant disease development (Mehta et al. 2008, Kniemeyer et al. 2011).

Fungal proteomics investigations also involve the routine identification of yeasts, dimorphic and filamentous fungi in culture, by using MALDI-ToF-MS on intact cells (Paul et al. 2017). Proteome phenotype profiling has been performed on many fungal genera such as *Aspergillus*, *Fusarium*, *Penicillium* or *Trichoderma*, including common and uncommon clinically relevant isolates (Putignani et al. 2011, Del Chierico et al. 2012, Ranque et al. 2014, Chalupová et al. 2014). Therfore, MALDI–ToF-MS has a great potential to become an alternative to identification techniques using ribosomal DNA genes sequencing.

Several comprehensive reviews of the research performed to date on fungal proteome has been published, i.e., reviews dedicated to proteomic work on filamentous (Kim et al. 2007), pathogenic (González-Fernández et al. 2010, Gonzalez-Fernandez and Jorrin-Novo 2012, Karkowska-Kuleta and Kozik 2015), industrial fungi (De Oliveira and De Graaff 2011) or edible mushrooms (Al-Obaidi 2016). Methodologies and perspectives of fungi proteomics have been reviewed by Bianco and Perrotta (2015). The adoption and application of advanced proteomic technologies in understanding plant–pathogen interactions has been reviewed by Ashwin and co-workers (2017). Ball and co-workers reviewed advancements in MS-based proteomics as it relates to fungal pathogenesis and interactions between these fungi and the host (2019). Proteomic researches on lignocellulolytic enzymes originating from various microorganisms including fungi have been recently reviewed twice (Guo et al. 2018, 2022). A very recent overview of proteomics works for 53 ascomycota species has been published by Muggia and co-workers (2020).

Fungal Metabolomics

Usually, any chemical compound of the cell that is not genetically encoded and is a substrate, intermediate or product of metabolism, is referred as metabolite. The metabolome, in turn, is defined as the quantitative complement of low-molecular weight metabolites present in a cell under a given set of physiological conditions (Kell et al. 2005). Metabolomics is the term used to describe metabolome analysis aiming to identify and quantify the entire collection of intracellular and extracellular metabolites. Metabolomics uses high-throughput methods to analyze as many metabolites as possible with a single analysis method.

Analysis of the metabolome is viewed as a complementary technique to other 'omics' techniques. However, compared to transcriptomics and proteomics, which unite the linear predictive potential of the genome, it offers distinct advantages. Firstly, the metabolome is directly related to the phenotype, which means that it more directly influences the phenotype than either transcripts or proteins (Figure 1). Secondly, changes in the metabolome are often amplified relative to changes in the transcriptome or proteome. Thus, metabolomic studies signify the non-linear, final bioactive products of the genome, arising from the complex systems that regulate the expression of the genome (Tan et al. 2009, Keshri et al. 2021).

Moreover, what also differentiates metabolomics from other 'omics' techniques, is that the technology involved in metabolomics is generic. For example, glucose extracted from *S. cerevisiae* is the same glucose molecule as that extracted from humans or plants. The same cannot be said for the DNA, RNA and subsequent protein sequences of components involved in the metabolism of glucose from these different organisms (Tan et al. 2009).

Metabolomics Methodology and its Development for Fungal Research

Conceptually, there are two basic approaches used in metabolomics methodologies. Untargeted metabolomics is an intended comprehensive analysis of all measurable analytes in a sample including chemical unknowns (Vinayavekhin and Saghatelian 2010, Patti et al. 2012). Targeted metabolomics focuses on the measurement of defined groups of chemically characterized and biochemically annotated metabolites (Roberts et al. 2012). Untargeted and semi-targeted methods are typically applied in hypothesis-generating investigations, while targeted approaches analyze a relatively smaller subset of biochemically important and relevant metabolites.

The general technical route of metabolomics includes following steps: sample preparation, data acquisition, data processing and data analysis. The choice of sample-preparation method is extremely important in metabolomic studies because it affects both the observed metabolite content and the biological interpretation of the data. To ensure the most accurate, efficient and reproducible results, an appropriate sample collection and preparation should be performed for a given metabolic analysis. An ideal sample preparation method for global metabolomics should be as non-selective as possible to ensure adequate depth of metabolite coverage; be simple and fast to prevent metabolite loss and/ or degradation during the preparation procedure and enable high-throughput; be reproducible; and incorporate a metabolism-quenching step to represent true metabolome composition at the time of sampling (Vuckovic 2012). General sample preparation steps include rapid sampling, quenching, and sample extraction.

Several analytical methods have been developed for qualitative and quantitative analysis of metabolic extracts, often coupling one of the separation techniques with a detection method. The commonly used analytical methods for fungal metabolomics include gas chromatography/ mass spectrometry (GC–MS), liquid chromatography/mass spectrometry (LC–MS), nuclear magnetic resonance spectroscopy/mass spectrometry (NMR–MS), capillary electrophoresis/ mass spectrometry (CE–MS), and matrix-assisted laser desorption ionization mass spectrometry (MALDI–MS). To analyze substances of high thermal stability and volatility, such as sugars, amino acids, organic acids, lipids, amines and phosphorylated metabolites, the GC–MS method might be of choice. It exhibits extraordinary robustness, excellent separation ability, high selectivity, effective sensitivity, and reproducibility. It is not suitable, however, for analysis of nonvolatile compounds, which require prior derivatization, such as silanization, acylation, alkylation and esterification (Beale et al. 2018). Generally, the high-performance liquid chromatography (HPLC) methods are suitable for thermally labile compounds, nonvolatile compounds, polar compounds, and compounds that are macromolecules. HPLC offers several advantages, such as simple preparation, high sensitivity, signal reproducibility, minimal noise, and high qualitative and quantitative ability. An ultra-performance liquid chromatography (UHPLC) offers an improved resolution of peaks and faster analysis (Xiao et al. 2012, Di Guida et al. 2016). NMR techniques have the advantages of high reproducibility, accurate quantification, simple sample preparation, measurable analytes in various solvents, clear identification of unknown metabolites, and complete metabolite detection. The disadvantage is low sensitivity, which severely limits the use of NMR in metabolomics (Markley et al. 2017). Capillary electrophoresis techniques are suitable for the analysis of highly polar, charged metabolites, such as nucleic acids, amino acids, carboxylic acids, and sugar phosphates (Ramautar et al. 2006, Zhang et al. 2017).

The next step is the processing procedure of the original data obtained by the analytical instrument, which mainly includes noise reduction and baseline correction, peak detection and deconvolution, normalization, and data standardization (Goodacre et al. 2007, Di Guida et al. 2016). As the univariate analysis method cannot accurately distinguish groups of a small difference, multivariate analysis is used to analyze the changes in single metabolites between different groups and the dependent structure of individual molecules (Bartel et al. 2013). Out of several multivariate analysis, an unsupervised learning method – the principal component analysis (PCA) - and two supervised learning methods, such as partial least squares discriminant analysis (PLS-DA) and partial least squares discriminant analysis based on orthogonal signal correction (OPLS-DA) are the most frequently used in the field of metabolomics (Westerhuis et al. 2010).

A single analytical platform that would directly and precisely characterize or quantify thousands of small-molecule metabolites involved in fungal metabolic processes, does not exist. As each analytical tool has advantages and disadvantages, the right combination of tools is often needed depending on the experimental situation to better analyze the target fungi. A comprehensive review of analytical tools and methodology applied in fungal metabolomics was recently published by Li and co-workers (2022).

Early Metabolomics in S. cerevisiae

The concept of metabolomics was introduced into the field of fungal research in 2001, by Raamsdonk and co-workers (2001). The approach to functional analysis, called FANCY (functional analysis by co-responses in yeast) (Oliver et al. 1998), was implemented, in which comparative metabolomics of several *S. cerevisiae* mutants was used to reveal functions of unstudied genes. Later, another functional genomics approach, based on a combination of metabolome analysis combined with *in silico* pathway analysis, was implemented to elucidate the function of orphan genes using metabolome data (Förster et al. 2002). A high-performance anion exchange chromatography (HPAEC) coupled to electrospray ionization tandem mass spectrometry was used for the selective and quantitative analysis of several glycolytic intermediates in cells of *S. cerevisiae* (Van Dam et al. 2002). LC–MS has been exploited to map metabolic activity and flexibility through dynamic analysis of intracellular metabolites during the yeast cell cycle (Wittmann et al. 2005) and to determine the effect of culture age on metabolite pools (Mashego et al. 2005). GC–MS method coupled to a statistical data-mining strategy was applied for the integrated analysis of clearly identified and quantified intracellular and extracellular metabolites in *S. cerevisiae* (Villas-Bôas et al. 2005). This work has demonstrated that differential metabolite level data provides insight into specific metabolic pathways and lays the groundwork for integrated transcription–metabolism studies of yeasts. The power of metabolomics was demonstrated later for an applied research, where metabolite engineering was used to show the effect of the genetic engineering of *S. cerevisiae* on metabolites that are important to the flavor of wine (Eglinton et al. 2002).

Application of Metabolomics in Fungal Research

Similarly to other 'omics' approaches, metabolomics has been applied to all different research fields of fungi, including taxonomic identification, the discovery of fungal bioactive natural products, mycotoxin research and the area of host–fungal interactions.

The huge metabolic chemodiversity within the Mycota kingdom has become an indispensable tool for the classification and identification of fungi. The secondary metabolism of fungi stands second only to that of plants, and secondary metabolites, which are generally secreted, represent a highly interesting subset of the fungal exometabolome, which has been exploited for fungal taxonomy (Jewett et al. 2006). Several successful attempts have been made to prove the suitability of metabolomics for the classification and identification of fungi. A chemical taxonomy-based approach on secondary metabolite profiling of seven species of *Trichoderma* was found to be advantageous over other classification methods such as morphological and even rDNA-based identification (Kang et al. 2011). A very interesting study employing NMR metabolomics has been proven to allow for successful differentiation between Chinese and Korean *Ganoderma lucidum* and for the differentiation of the cultivated origin of Chinese *Ganoderma lucidum* (Wen et al. 2010). A comprehensive data review of metabolomics studies utilized for taxonomic identification of fungi has been published by Li and co-workers (2022).

Mycotoxins are toxic secondary metabolites of low molecular weight produced by filamentous fungi (Bennett and Klich 2003). The ability to detect mycotoxins is critical for maintaining a safe supply of food and feed for human and animal consumption. Detection and quantification of mycotoxins by advanced analytical tools and chemometrics has been of interest for years (Krska et al. 2006). Nielsen and Smedsgaard have collected high-performance liquid chromatography (HPLC)–UV data of 474 mycotoxins and published them to facilitate detection and identification when standards are not available (Nielsen and Smedsgaard 2003). Later, a multi-target method for the determination of 191 mycotoxins and other fungal metabolites in almonds, hazelnuts, peanuts and pistachios was developed (Varga et al. 2013). More recent and advanced studies have shown the usefulness of 'omics' approaches, including metabolomics, for mycotoxin research (Bhatnagar et al. 2008, Garcia-Cela et al. 2018). Metabolomics analysis applications and detection performance criteria for targeted and untargeted mycotoxin detection have been recently reviewed by Esheli and co-workers (2018).

Fungi can provide diverse and unique secondary metabolites of various bioactivities, making them potential drug sources. With the advancement of analytical technology platforms, MS-based metabolomics workflows are mainly suitable for screening hundreds of natural products simultaneously for dereplication studies and extractions of bioactive compounds, which is of great benefit for the comprehensive exploration of potentially useful secondary metabolites (Keller 2019, Li et al. 2022, Shankar and Sharma 2022). Using metabolomics approach, several novel metabolites have been discovered in fungi, i.e. cordycepin (Joshi et al. 2019), orsellinic acid (Yao et al. 2016), aspergillide B1 (El-Hawary et al. 2021), diorcinol, botryoisocoumarin A, and mullein (Tawfike et al. 2019), to name just a few.

A large number of potentially high-value bioactive compounds with pharmaceutical importance were discovered from cultivable endophytic fungi as reviewed by (Kaul et al. 2012, Nicoletti and Fiorentino 2015, Venugopalan and Srivastava 2015, Zimowska et al. 2020, Keshri et al. 2021). With the aid of metabolomics, it was possible to detect the production of plant-like bioactive metabolites in endophytic fungi, which makes the endophytic fungal communities an alternative source of bioactive compounds to be used for industrial applications (Nagarajan et al. 2022). Moreover, Sagita and co-workers recently described several successful attempts to discover novel secondary metabolites in endophytic fungi by using an approach of metabolic phenotyping (Sagita et al. 2021).

During a long period of co-existence, endophytic fungi have developed mutually beneficial relationships with the hosts, while other fungi can be harmful and cause various plant diseases. Understanding the interactions between fungi and plants is essential for preventing and controlling plant diseases. In the last decade, metabolomics technologies have been widely applied in various research fields on fungal plant interactions, such as determining the mechanism of infection and detecting the interaction between fungi and the host. The applications of metabolomics can help us to understand the pathogenesis and plant defense mechanisms of pathogenic fungi and develop effective prevention and treatment strategies for fungal diseases (Chen et al. 2019, Castro-Moretti et al. 2020, Gupta et al. 2022).

The ability of fungi to colonize and persist within the human host is accompanied by an adaptation of fungal metabolism that allows them to withstand the conditions within the human body. Thus, investigations of the unique metabolic fingerprints of pathogenic microorganisms and the infection-associated changes to the host's metabolism can provide a more complete impression of the infection process. The metabolome analyses in human pathogenic fungi have been extensively reviewed (Brandt et al. 2020, Oyedeji et al. 2021).

Research methods of metabolomics and the application of metabolomics in the abovementioned and other fungal research areas have been reviewed by Li and co-workers (2022).

Conclusions

The past decade has seen the launch of uncountable 'omics' projects to uncover the different aspects of fungal diversity, spanning from fungal cell metabolism to interactions with the environment and evolution of Mycota. Existing examples of 'omics' research have already proven that using a variety of advanced technological platforms for in-depth profiling enables one to utterly mine the potential of fungi.

Each of the modern technologies involved in genomics, transcriptomics, proteomics and metabolomics can generate enormous amount of information in a unprecedented high-throughput manner. However, it is their use in a combination that is the most exciting. The simultaneous measuring of thousands of molecules in the same biological sample, such as DNA, RNA, proteins and metabolites, will promote future systems biology studies. The need for a coordinated multi-omics approach to studying organisms has already been expressed by leading scientists from various fields of biology. Based on the *S. cerevisiae* example, demand for generating information in a multi-omics manner in Mycota was expressed by Hoffman and co-workers (2003). Since then, more and more researchers pointed at multi-omics approaches as those having the potential to revolutionize fungal research (Tan et al. 2009, Hautbergue et al. 2018, Ijoma et al. 2021, Keshri et al. 2021). Recently, a comprehensive review expressing the need for an integrated multi-omic approach in microbiome science was published (Ferrocino et al. 2023).

Although the multi-omics approach holds an utterly new potential to biology studies, it also brings new challenges. Strategies that should be implemented for a successful outcome of a multi-omics study have been proposed (Ferrocino et al. 2023). Firstly, they include a proper experimental design and sampling procedures to ensure accurate analysis with complementary techniques. Predefined standards should be implemented regulating the experimental design and workflow, sample collection and preservation as well as data storage at repositories to enable its reuse and metadata collection for multi-omics studies by a scientific community (Hu et al. 2022). Secondly, they should involve optimal combinations of 'omics' technologies for each studied ecosystem to ensure validation of the hypothesis. Lastly, as the development of multi-omics strategies produces big data, new approaches and bioinformatics tools must be implemented to address the integration of various multi-omics data (Santiago-Rodriguez and Hollister 2021). Data integration can be achieved by either network analysis (Jiang et al. 2019) or overlying the multi-omics data onto existing metabolic pathways.

Apart from the 'omics' discussed in this chapter, there are more and more in-depth technologies available that could be used simultaneously in fungal research, such as metagenomics, culturomics, lipidomics, mirnaomics, epigenomics, secretomics, phenomics, to name just a few. Combining data from several 'omics' studies may find correlations or other associations that are unable to be revealed by individual studies alone. However, efforts must be focused on the accessibility and comparability of fungal 'omics' datasets to enable data mining by the scientific community. Existing examples have already proven that further accumulation and exploration of fungi multi-omics data is inevitable and will certainly provide further insights into the biology of Mycota in the near future.

References

Aguilar-Pontes, M.V., R.P. de Vries and M. Zhou. 2014. (Post-)Genomics approaches in fungal research. Brief. Funct. Genomics 13: 424–439. doi: 10.1093/bfgp/elu028.

Al-Obaidi, J.R. 2016. Proteomics of edible mushrooms: A mini-review. Electrophoresis 37: 1257–1263. doi:10.1002/elps.201600031.

Appleyard, V.C.L., S.E. Unkles, M. Legg and J.R. Kinghorn. 1995. Secondary metabolite production in filamentous fungi displayed. MGG Mol. Gen. Genet. 247: 338–342. doi:10.1007/BF00293201.

Arnold, A.E., Z. Maynard, G.S. Gilbert, P.D. Coley and T.A. Kursar. 2000. Are tropical fungal endophytes hyperdiverse? Ecol. Lett. 3: 267–274. doi:10.1046/J.1461-0248.2000.00159.X.

Ashwin, N.M.R., L. Barnabas, A. Ramesh Sundar, P. Malathi, R.Viswanathan, A. Masi et al. 2017. Advances in proteomic technologies and their scope of application in understanding plant–pathogen interactions. J. Plant Biochem. Biotechnol. 26: 371–386. doi:10.1007/S13562-017-0402-1/FIGURES/4.

Aslam, S., A.Tahir, M.F. Aslam, M.W. Alam, A.A. Shedayi and S. Sadia. 2017a. Recent advances in molecular techniques for the identification of phytopathogenic fungi – a mini review. J. Plant Interact. 12: 493–504. doi:10.1080/17429145.2017.1397205.

Aslam, B., M. Basit, M.A. Nisar, M. Khurshid and M.H. Rasool. 2017b. Proteomics: Technologies and Their Applications. J. Chromatogr. Sci. 55: 182–196. doi:10.1093/CHROMSCI/BMW167.

Bader, G.D. and C.W.V. Hogue. 2002. Analyzing yeast protein–protein interaction data obtained from different sources. Nat. Biotechnol. 2002 2010 20: 991–997. doi:10.1038/nbt1002-991.

Ball, B., A. Bermas, D. Carruthers-Lay, J. Geddes-McAlister. 2019. Mass Spectrometry-Based Proteomics of Fungal Pathogenesis, Host–Fungal Interactions, and Antifungal Development. J. Fungi 2019, Vol. 5, Page 52 5: 52. doi:10.3390/JOF5020052.

Bartel, J., J. Krumsiek, F.J. Theis. 2013. Statistical methods for the analysis of high-throughput metabolomics data. Comput. Struct. Biotechnol. J. 4: e201301009. doi:10.5936/CSBJ.201301009.

Basenko, E.Y., J.A. Pulman, A. Shanmugasundram, O.S. Harb, K. Crouch, D. Starns et al. 2018. FungiDB: An integrated bioinformatic resource for fungi and oomycetes. J. Fungi 4: 1–28. doi:10.3390/jof4010039.

Beale, D.J., F.R. Pinu, K.A. Kouremenos, M.M. Poojary, V.K. Narayana, B.A. Boughton et al. 2018. Review of recent developments in GC–MS approaches to metabolomics-based research. Metabolomics 14: 1–31. doi:10.1007/S11306-018-1449-2/FIGURES/4.

Bennett, J.W. and M. Klich. 2003. Mycotoxins. Clin. Microbiol. Rev. 16: 497–516. doi:10.1128/CMR.16.3.497-516.2003/ASSET/249F0296-4B2E-4460-BCFD-A33A80A6C166/ASSETS/GRAPHIC/CM0330050010.JPEG.

Benson, D.A., M. Cavanaugh, K. Clark, I. Karsch-Mizrachi, D.J. Lipman, J. Ostell et al. 2013. GenBank. Nucleic Acids Res. 41. doi:10.1093/NAR/GKS1195.

Bhatnagar, D., K. Rajasekaran, G. Payne, R. Brown, J. Yu and T. Cleveland. 2008. The "omics" tools: genomics, proteomics, metabolomics and their potential for solving the aflatoxin contamination problem. World Mycotoxin J. 1: 3–12. doi:10.3920/wmj2008.x001.

Bianco, L. and G. Perrotta. 2015. Methodologies and Perspectives of Proteomics Applied to Filamentous Fungi: From Sample Preparation to Secretome Analysis. Int. J. Mol. Sci. 2015, 16: 5803–5829 16: 5803–5829. doi:10.3390/IJMS16035803.

Bielecka, M., B. Pencakowski and R. Nicoletti. 2022. Using Next-Generation Sequencing Technology to Explore Genetic Pathways in Endophytic Fungi in the Syntheses of Plant Bioactive Metabolites. Agric. 12. doi:10.3390/agriculture12020187.

Blackwell, M. 2011. The fungi: 1, 2, 3 ... 5.1 million species? Am. J. Bot. 98: 426–438. doi:10.3732/ajb.1000298.

Brandt, P., E. Garbe and S. Vylkova. 2020. Catch the wave: metabolomic analyses in human pathogenic fungi. PLoS Pathog. 16: 1–9. doi:10.1371/journal.ppat.1008757.

Branton, D., D.W. Deamer, A. Marziali, H. Bayley, S.A. Benner, T. Butler et al. 2008. The potential and challenges of nanopore sequencing. Nat. Biotechnol. 2008 2610 26: 1146–1153. doi:10.1038/nbt.1495.

Breakspear, A. and M. Momany. 2007. The first fifty microarray studies in filamentous fungi. Microbiology 153: 7–15. doi:10.1099/mic.0.2006/002592-0.

Bruneau, J.M., T. Magnin, E. Tagat, R. Legrand, M. Bernard, M. Diaquin et al. 2001. Proteome analysis of Aspergillus fumigatus identifies glycosylphosphatidylinositol-anchored proteins associated to the cell wall biosynthesis. Electrophoresis 22: 2812–2823. doi:10.1002/1522-2683(200108)22:13<2812::AID-ELPS2812>3.0.CO;2-Q.

Casselton, L. and M. Zolan. 2002. The art and design of genetic screens: filamentous fungi. Nat. Rev. Genet. 2002 39 3: 683–697. doi:10.1038/nrg889.

Castanera, R., L. López-Varas, A.G. Pisabarro and L. Ramíre. 2015. Validation of reference genes for transcriptional analyses in Pleurotus ostreatus by using reverse transcription-quantitative PCR. Appl. Environ. Microbiol. 81: 4120–4129. doi:10.1128/AEM.00402-15.

Castro-Moretti, F.R., I.N. Gentzel, D. Mackey and A.P. Alonso. 2020. Metabolomics as an emerging tool for the study of plant–pathogen interactions. Metabolites 10: 1–23. doi:10.3390/metabo10020052.

Chalupová, J., M. Raus, M. Sedláŕová and M. Šebela. 2014. Identification of fungal microorganisms by MALDI-TOF mass spectrometry. Biotechnol. Adv. 32: 230–241. doi:10.1016/J.BIOTECHADV.2013.11.002.

Champer, J., D. Diaz-Arevalo, M. Champer, T.B. Hong, M. Wong, M. Shannahoff et al. 2012. Protein targets for broad-spectrum mycosis vaccines: quantitative proteomic analysis of Aspergillus and Coccidioides and comparisons with other fungal pathogens. Ann. N. Y. Acad. Sci. 1273: 44–51. doi:10.1111/J.1749-6632.2012.06761.X.

Champer, J., J.I. Ito, K. V. Clemons, D.A. Stevens and M. Kalkum. 2016. Proteomic Analysis of Pathogenic Fungi Reveals Highly Expressed Conserved Cell Wall Proteins. J. Fungi 2016, 2: 6 2, 6. doi:10.3390/JOF2010006.

Chatterjee, S. and S. Venkata Mohan. 2022. Fungal biorefinery for sustainable resource recovery from waste. Bioresour. Technol. 345: 126443. doi:10.1016/J.BIORTECH.2021.126443.

Chávez, R., F. Fierro, R.O. García-Rico and I. Vaca. 2015. Filamentous fungi from extreme environments as a promising source of novel bioactive secondary metabolites. Front. Microbiol. 6: 903. doi:10.3389/FMICB.2015.00903/BIBTEX.

Che, J., J. Shi, Z. Gao and Y. Zhang. 2016. Transcriptome analysis reveals the genetic basis of the resveratrol biosynthesis pathway in an endophytic fungus (*Alternaria* sp. MG1) isolated from Vitis vinifera. Front. Microbiol. 7: 1–12. doi:10.3389/fmicb.2016.01257.

Chen, F., R. Ma and X.L. Chen. 2019. Advances of Metabolomics in Fungal Pathogen–Plant Interactions. Metab. 2019, Vol. 9, Page 169 9: 169. doi:10.3390/METABO9080169.

Choquer, M., M. Boccara and A. Vidal-Cros. 2003. A semi-quantitative RT-PCR method to readily compare expression levels within Botrytis cinerea multigenic families in vitro and in planta. Curr. Genet. 43: 303–309. doi:10.1007/s00294-003-0397-0.

Clarke, J., H.C. Wu, L. Jayasinghe, A. Patel, S. Reid and H. Bayley. 2009. Continuous base identification for single-molecule nanopore DNA sequencing. Nat. Nanotechnol. 4: 265–270. doi:10.1038/nnano.2009.12.

Conesa, A., P. Madrigal, S. Tarazona, D. Gomez-Cabrero, A. Cervera, A. McPherson et al. 2016. A survey of best practices for RNA-seq data analysis. Genome Biol. 2016 171 17: 1–19. doi:10.1186/S13059-016-0881-8.

Craig Venter, J., M.D. Adams, E.W. Myers, P.W. Li, R.J. Mural, G.G. Sutton et al. 2001. The sequence of the human genome. Science (80-.). 291: 1304–1351. doi:10.1126/SCIENCE.1058040/SUPPL_FILE/1058040S3-4_MED.GIF.

David, H., G. Hofmann, A.P. Oliveira, H. Jarmer and J. Nielsen. 2006. Metabolic network driven analysis of genome-wide transcription data from Aspergillus nidulans. Genome Biol. 7. doi:10.1186/gb-2006-7-11-r108.

De Oliveira, J.M.P.F. and L.H. De Graaff. 2011. Proteomics of industrial fungi: Trends and insights for biotechnology. Appl. Microbiol. Biotechnol. 89: 225–237. doi:10.1007/S00253-010-2900-0/FIGURES/2.

de Vries, R.P. 2003. Regulation of Aspergillus genes encoding plant cell wall polysaccharide-degrading enzymes; relevance for industrial production. Appl. Microbiol. Biotechnol. 61: 10–20. doi:10.1007/S00253-002-1171-9/METRICS.

de Vries, R.P., R. Riley, A. Wiebenga, G. Aguilar-Osorio, S. Amillis, C.A. Uchima et al. 2017. Comparative genomics reveals high biological diversity and specific adaptations in the industrially and medically important fungal genus *Aspergillus*. Genome Biol. 2017 181 18, 1–45. doi:10.1186/S13059-017-1151-0.

Dean, R., J.A.L. Van Kan, Z.A. Pretorius, K.E. Hammond-Kosack, A. Di Pietro, P.D. Spanu et al. 2012. The Top 10 fungal pathogens in molecular plant pathology. Mol. Plant Pathol. 13: 414–430. doi:10.1111/J.1364-3703.2011.00783.X.

Del Chierico, F., A. Masotti, M. Onori, E. Fiscarelli, L. Mancinelli, G. Ricciotti et al. 2012. MALDI-TOF MS proteomic phenotyping of filamentous and other fungi from clinical origin. J. Proteomics 75: 3314–3330. doi:10.1016/J. JPROT.2012.03.048

Delmas, S., S.T. Pullan, S. Gaddipati, M. Kokolski, S. Malla, M.J. Blythe et al. 2012. Uncovering the Genome-Wide Transcriptional Responses of the Filamentous Fungus *Aspergillus niger* to Lignocellulose Using RNA Sequencing. PLOS Genet. 8, e1002875. doi:10.1371/JOURNAL.PGEN.1002875.

Demain, A.L. 2014. Valuable Secondary Metabolites from Fungi 1–15. doi:10.1007/978-1-4939-1191-2_1.

DeRisi, J.L., V.R. Iyer and P.O. Brown. 1997. Exploring the metabolic and genetic control of gene expression on a genomic scale. Science (80-.). 278: 680–686. doi:10.1126/science.278.5338.680.

Derrington, I.M., T.Z. Butler, M.D. Collins, E. Manrao, M. Pavlenok, M. Niederweis et al. 2010. Nanopore DNA sequencing with MspA. Proc. Natl. Acad. Sci. U. S. A. 107: 16060–16065. doi:10.1073/PNAS.1001831107/SUPPL_FILE/ PNAS.1001831107_SI.PDF.

Di Guida, R., J. Engel, J.W. Allwood, R.J.M. Weber, M.R. Jones, U. Sommer et al. 2016. Non-targeted UHPLC-MS metabolomic data processing methods: a comparative investigation of normalisation, missing value imputation, transformation and scaling. Metabolomics 12: 1–14. doi:10.1007/S11306-016-1030-9/TABLES/5.

Dijk, E.L. van, H. Auger, Y. Jaszczyszyn and C. Thermes. 2014. Ten years of next-generation sequencing technology. Trends Genet. 30: 418–426. doi:10.1016/j.tig.2014.07.001.

Dutreux, F., C. Da Silva, L. D'agata, A. Couloux, E.J. Gay, B. Istace et al. 2018. *De novo* assembly and annotation of three *Leptosphaeria* genomes using Oxford Nanopore MinION sequencing. Sci. Data 2018 51 5: 1–11. doi:10.1038/ sdata.2018.235.

Edgar, R., M. Domrachev and A.E. Lash. 2002. Gene Expression Omnibus: NCBI gene expression and hybridization array data repository. Nucleic Acids Res. 30: 207–210. doi:10.1093/NAR/30.1.207.

Eglinton, J.M., A.J. Heinrich, A.P. Pollnitz, P. Langridge, P.A. Henschke and E.M. De Barros Lopes. 2002. Decreasing acetic acid accumulation by a glycerol overproducing strain of *Saccharomyces cerevisiae* by deleting the ALD6 aldehyde dehydrogenase gene. Yeast 19: 295–301. doi:10.1002/YEA.834.

Eid, J., A. Fehr, J. Gray, K. Luong, J. Lyle, G. Otto et al. 2009. Real-time DNA sequencing from single polymerase molecules. Science (80-.). 323: 133–138. doi:10.1126/SCIENCE.1162986/SUPPL_FILE/PAP.PDF.

El-Hawary, S.S., R. Mohammed, H.S. Bahr, E.Z. Attia, M.H. El-Katatny, N. Abelyan et al. 2021. Soybean-associated endophytic fungi as potential source for anti-COVID-19 metabolites supported by docking analysis. J. Appl. Microbiol. 131: 1193–1211. doi:10.1111/JAM.15031.

Eshelli, M., M.M. Qader, E.J. Jambi, A.S. Hursthouse and M.E. Rateb. 2018. Current status and future opportunities of omics tools in mycotoxin research. Toxins (Basel). 10: 1–26. doi:10.3390/toxins10110433.

Feist, A.M., M.J. Herrgård, I. Thiele, J.L. Reed and B. Palsson. 2008. Reconstruction of biochemical networks in microorganisms. Nat. Rev. Microbiol. 2009 72 7: 129–143. doi:10.1038/nrmicro1949.

Felitti, S., K. Shields, M. Ramsperger, P. Tian, T. Sawbridge, T. Webster et al. 2006. Transcriptome analysis of Neotyphodium and Epichloë grass endophytes. Fungal Genet. Biol. 43: 465–475. doi:10.1016/j.fgb.2006.01.013.

Ferrocino, I., K. Rantsiou, R. Mcclure, T. Kostic, R. Soares Correa De Souza, L. Lange et al. 2023. The need for an integrated multi-OMICs approach in microbiome science in the food system. doi:10.1111/1541-4337.13103.

Finotello, F. and B. Di Camillo. 2015. Measuring differential gene expression with RNA-seq: Challenges and strategies for data analysis. Brief. Funct. Genomics 14: 130–142. doi:10.1093/bfgp/elu035.

Forsburg, S.L. 2001. The art and design of genetic screens: yeast. Nat. Rev. Genet. 2001 29 2, 659–668. doi:10.1038/35088500.

Förster, J., A.K. Gombert and J. Nielsen. 2002. A functional genomics approach using metabolomics and *in silico* pathway analysis. Biotechnol. Bioeng. 79: 703–712. doi:10.1002/BIT.10378.

Galagan, J.E., M.R. Henn, L.J. Ma, C.A. Cuomoand B. Birren. 2005. Genomics of the fungal kingdom: Insights into eukaryotic biology. Genome Res. 15: 1620–1631. doi:10.1101/GR.3767105.

Garcia-Cela, E., C. Verheecke-Vaessen, N. Magan and A. Medina 2018. The ``-omics'' contributions to the understanding of mycotoxin production under diverse environmental conditions. Curr. Opin. Food Sci. 23: 97–104. doi:10.1016/J. COFS.2018.08.005.

Geniza, M. and P. Jaiswal. 2017. Tools for building *de novo* transcriptome assembly. Curr. Plant Biol. 11–12: 41–45. doi:10.1016/J.CPB.2017.12.004.

Gerothanassis, I.P., A. Troganis, V. Exarchou and K. Barbarossou. 2002. Nuclear magnetic resonance (NMR) spectroscopy: basic principles and phenomena, and their applications to chemistry, biology and medicine. Chem. Educ. Res. Pr. 3: 229–252. doi:10.1039/b2rp90018a.

Goffeau, A., G. Barrell, H. Bussey, R.W. Davis, B. Dujon, H. Feldmann et al. 1996. Life with 6000 Genes. Science (80-.). 274: 546–567. doi:10.1126/SCIENCE.274.5287.546.

Gonzalez-Fernandez, R. and J.V. Jorrin-Novo. 2012. Contribution of proteomics to the study of plant pathogenic fungi. J. Proteome Res. 11: 3–16. doi:10.1021/pr200873p.

González-Fernández, R., E. Prats and J.V. Jorrín-Novo. 2010. Proteomics of plant pathogenic fungi. J. Biomed. Biotechnol. 2010. doi:10.1155/2010/932527.

Goodacre, R., D. Broadhurst, A.K. Smilde, B.S. Kristal, J.D. Baker, R. Beger et al. 2007. Proposed minimum reporting standards for data analysis in metabolomics. Metabolomics 3, 231–241. doi:10.1007/S11306-007-0081-3/FIGURES/2.

Goodwin, S., J.D. McPherson and W.R. McCombie. 2016. Coming of age: ten years of next-generation sequencing technologies. Nat. Rev. Genet. 17: 333–351. doi:10.1038/nrg.2016.49.

Gowda, M., R.C. Venu, M.B. Raghupathy, K. Nobuta, H. Li, R. Wing et al. 2006. Deep and comparative analysis of the mycelium and appressorium transcriptomes of Magnaporthe grisea using MPSS, RL-SAGE, and oligoarray methods. BMC Genomics 7: 1–15. doi:10.1186/1471-2164-7-310.

Graves, P.R. and T.A.J. Haystead. 2002. Molecular Biologist's Guide to Proteomics. Microbiol. Mol. Biol. Rev. 66: 39–63. doi:10.1128/MMBR.66.1.39-63.2002/ASSET/AFF762F9-2F7E-4B7A-B184-214282F5BCAA/ASSETS/GRAPHIC/MR0120003015.JPEG.

Green, E.D. 2001. Strategies for the systematic sequencing of complex genomes. Nat. Rev. Genet. 2001 28 2: 573–583. doi:10.1038/35084503.

Griffin, T.J., S.P. Gygi, T. Ideker, B. Rist, J. Eng, L. Hood et al. 2002. Complementary profiling of gene expression at the transcriptome and proteome levels in Saccharomyces cerevisiae. Mol. Cell. Proteomics 1: 323–333. doi:10.1074/mcp.M200001-MCP200.

Grigoriev, I. V., R. Nikitin, S. Haridas, A. Kuo, R. Ohm, R. Otillar et al. 2014. MycoCosm portal: Gearing up for 1000 fungal genomes. Nucleic Acids Res. 42: 699–704. doi:10.1093/nar/gkt1183.

Güldener, U., K.Y. Seong, J. Boddu, S. Cho, F. Trail, J.R. Xu et al. 2006. Development of a Fusarium graminearum Affymetrix GeneChip for profiling fungal gene expression in vitro and in planta. Fungal Genet. Biol. 43: 316–325. doi:10.1016/j.fgb.2006.01.005.

Guo, H., T. He and D.J. Lee. 2022. Contemporary proteomic research on lignocellulosic enzymes and enzymolysis: A review. Bioresour. Technol. 344: 126263. doi:10.1016/J.BIORTECH.2021.126263.

Guo, H., X.D. Wang, D.J. Lee. 2018. Proteomic researches for lignocellulose-degrading enzymes: A mini-review. Bioresour. Technol. 265: 532–541. doi:10.1016/J.BIORTECH.2018.05.101.

Guo, J. 2014. Editorial: Transcription: the epicenter of gene expression. J. Zhejiang Univ-Sci B (Biomed Biotechnol) 15: 409–411. doi:10.1631/jzus.B1400113.

Gupta, S., M. Schillaci and U. Roessner. 2022. Metabolomics as an emerging tool to study plant-microbe interactions. Emerg. Top. Life Sci. 6: 175–183. doi:10.1042/ETLS20210262.

Gygi, S.P., B. Rist, S.A. Gerber, F. Turecek, M.H. Gelb and R. Aebersold. 1999. Quantitative analysis of complex protein mixtures using isotope-coded affinity tags. Nat. Biotechnol. 1999 1710 17: 994–999. doi:10.1038/13690.

Hage, D.S., J.A. Anguizola, C. Bi, R. Li, R. Matsuda, E. Papastavros et al. 2012. Pharmaceutical and biomedical applications of affinity chromatography: Recent trends and developments. J. Pharm. Biomed. Anal. 69: 93–105. doi:10.1016/J.JPBA.2012.01.004.

Hall, D.A., J. Ptacek and M. Snyder. 2007. Protein microarray technology. Mech. Ageing Dev. 128: 161–167. doi:10.1016/J.MAD.2006.11.021.

Haridas, S., C. Breuill, J. Bohlmann and T. Hsiang. 2011. A biologist's guide to de novo genome assembly using next-generation sequence data: A test with fungal genomes. J. Microbiol. Methods 86: 368–375. doi:10.1016/J.MIMET.2011.06.019.

Hautbergue, T., E.L. Jamin, L. Debrauwer, O. Puel and I.P. Oswald. 2018. From genomics to metabolomics, moving toward an integrated strategy for the discovery of fungal secondary metabolites. Nat. Prod. Rep. 35: 147–173. doi:10.1039/c7np00032d.

Hawksworth, D.L. 1991. The fungal dimension of biodiversity: magnitude, significance, and conservation. Mycol. Res. 95: 641–655. doi:10.1016/S0953-7562(09)80810-1.

Hawksworth, D.L. and R. Lücking. 2017. Fungal diversity revisited: 2.2 to 3.8 million species. The Fungal Kingdom 79–95. doi:10.1128/9781555819583.ch4.

Henske, J.K., S.D. Springer, M.A. O'Malley and A. Butler. 2018. Substrate-based differential expression analysis reveals control of biomass degrading enzymes in *Pycnoporus cinnabarinus*. Biochem. Eng. J. 130: 83–89. doi:10.1016/J.BEJ.2017.11.015.

Hernáez, M.L., P. Ximénez-Embún, M. Martínez-Gomariz, M.D. Gutiérrez-Blázquez, C. Nombela and C. Gil. 2010. Identification of Candida albicans exposed surface proteins in vivo by a rapid proteomic approach. J. Proteomics 73: 1404–1409. doi:10.1016/J.JPROT.2010.02.008.

Hofmann, G., M. McIntyre and J. Nielsen. 2003. Fungal genomics beyond Saccharomyces cerevisiae? Curr. Opin. Biotechnol. 14: 226–231. doi:10.1016/S0958-1669(03)00020-X.

Hu, B., S. Canon, E.A. Eloe-Fadrosh, M. Anubhav, Babinski, Y. Corilo, K. Davenport et al. 2022. Challenges in Bioinformatics Workflows for Processing Microbiome Omics Data at Scale. Front. Bioinforma. 1: 89. doi:10.3389/FBINF.2021.826370.

Ideker, T., T. Galitski and L. Hood. 2001. A NEW APPROACH TO DECODING LIFE: Systems Biology. Annu. Rev. Genomics Hum. Genet. 5: 343–372.

Ijoma, G.N., S.M. Heri, T.S. Matambo and M. Tekere. 2021. Trends and applications of omics technologies to functional characterisation of enzymes and protein metabolites produced by fungi. J. Fungi 7. doi:10.3390/jof7090700.

Issaq, H.J. and T.D. Veenstra. 2008. Two-dimensional polyacrylamide gel electrophoresis (2D-PAGE): Advances and perspectives. Biotechniques 44: 697–700. doi:10.2144/000112823/ASSET/IMAGES/LARGE/FIGURE1.JPEG.

Jenjaroenpun, P., T. Wongsurawat, R. Pereira, P. Patumcharoenpol, D.W. Ussery, J. Nielsen et al. 2018. Complete genomic and transcriptional landscape analysis using third-generation sequencing: a case study of *Saccharomyces cerevisiae* CEN.PK113-7D. Nucleic Acids Res. 46: e38–e38. doi:10.1093/NAR/GKY014.

Jeon, J., S.-Y. Park, J.A. Kim, N.H. Yu, A.R. Park, J.-C. Kim et al. 2019. Draft Genome Sequence of Amphirosellinia nigrospora JS-1675, an Endophytic Fungus from Pteris cretica. Microbiol. Resour. Announc. 1–2.

Jewett, M.C., G. Hofmann, J. Nielsen. 2006. Fungal metabolite analysis in genomics and phenomics. Curr. Opin. Biotechnol. 17: 191–197. doi:10.1016/j.copbio.2006.02.001.

Jiang, D., C.R. Armour, C. Hu, M. Mei, C. Tian, T.J. Sharpton et al. 2019. Microbiome Multi-Omics Network Analysis: Statistical Considerations, Limitations, and Opportunities. Front. Genet. 10: 995. doi:10.3389/FGENE.2019.00995/BIBTEX.

Joshi, R., A. Sharma, K. Thakur, D. Kumar, G. Nadda. 2019. Metabolite analysis and nucleoside determination using reproducible UHPLC-Q-ToF-IMS in Ophiocordyceps sinensis. https://doi.org/10.1080/10826076.2018.1541804 41: 927–936. doi:10.1080/10826076.2018.1541804.

Jun, H., H. Guangye, C. Daiwen. 2013. Insights into enzyme secretion by filamentous fungi: Comparative proteome analysis of Trichoderma reesei grown on different carbon sources. J. Proteomics 89: 191–201. doi:10.1016/J.JPROT.2013.06.014

Kang, D., J. Kim, J.N. Choi, K.H. Liu, C.H. Lee. 2011. Chemotaxonomy of Trichoderma spp. Using Mass Spectrometry-Based Metabolite Profiling. J. Microbiol. Biotechnol. 21: 5–13. doi:10.4014/JMB.1008.08018.

Karkowska-Kuleta, J., A. Kozik. 2015. Cell wall proteome of pathogenic fungi. Acta Biochim. Pol. 62: 339–351. doi:10.18388/abp.2015_1032.

Kaul, S., S. Gupta, M. Ahmed, M.K. Dhar. 2012. Endophytic fungi from medicinal plants: A treasure hunt for bioactive metabolites. Phytochem. Rev. 11: 487–505. doi:10.1007/s11101-012-9260-6.

Kazemi Shariat Panahi, H., M. Dehhaghi, G.J. Guillemin, V.K. Gupta, S.S. Lam, M. Aghbashlo et al. 2022. A comprehensive review on anaerobic fungi applications in biofuels production. Sci. Total Environ. 829: 154521. doi:10.1016/J.SCITOTENV.2022.154521.

Kell, D.B., M. Brown, H.M. Davey, W.B. Dunn, I. Spasic, S.G. Oliver. 2005. Metabolic footprinting and systems biology: the medium is the message. Nat. Rev. Microbiol. 2005 37 3: 557–565. doi:10.1038/nrmicro1177.

Keller, N.P. 2019. Fungal secondary metabolism: regulation, function and drug discovery. Nat. Rev. Microbiol. 17: 167–180. doi:10.1038/s41579-018-0121-1.

Keshri, P.K., N. Rai, A. Verma, S.C. Kamble, S. Barik, P. Mishra et al. 2021. Biological potential of bioactive metabolites derived from fungal endophytes associated with medicinal plants. Mycol. Prog. 20: 577–594. doi:10.1007/s11557-021-01695-8.

Kim, J.A., J. Jeon, K.-T. Kim, G. Choi, S.-Y. Park, H.-J. Lee et al. 2017. Draft Genome Sequence of an Endophytic Fungus, Gaeumannomyces sp. Strain JS-464, Isolated from a Reed Plant, Phragmites communis. Microbiol. Resour. Announc. 5: 1–2.

Kim, Y., M.P. Nandakumar, M.R. Marten. 2007. Proteomics of filamentous fungi. Trends Biotechnol. 25: 395–400. doi:10.1016/J.TIBTECH.2007.07.008.

Kniemeyer, O., A.D. Schmidt, M. Vödisch, D. Wartenberg, A.A. Brakhage. 2011. Identification of virulence determinants of the human pathogenic fungi Aspergillus fumigatus and Candida albicans by proteomics. Int. J. Med. Microbiol. 301: 368–377. doi:10.1016/J.IJMM.2011.04.001.

Köhler, J.R., A. Casadevall, J. Perfect. 2015. The Spectrum of Fungi That Infects Humans. Cold Spring Harb. Perspect. Med. 5, a019273. doi:10.1101/CSHPERSPECT.A019273.

Kozarewa, I., Z. Ning, M.A. Quail, M.J. Sanders, M. Berriman, D.J. Turner. 2009. Amplification-free Illumina sequencing-library preparation facilitates improved mapping and assembly of (G+C)-biased genomes. Nat. Methods 2009 64 6: 291–295. doi:10.1038/nmeth.1311.

Krska, R., E. Welzig, F. Berthiller, A. Molinelli, B. Mizaikoff. 2006. Advances in the analysis of mycotoxins and its quality assurance. http://dx.doi.org/10.1080/02652030500070192 22: 345–353. doi:10.1080/02652030500070192.

Krumlauf, R. 1994. Analysis of gene expression by Northern blot. Mol. Biotechnol. 2: 227–242. doi:10.1007/BF02745879.

Kubicek, C.P., A.S. Steindorff, K. Chenthamara, G. Manganiello, B. Henrissat, J. Zhang et al. 2019. Evolution and comparative genomics of the most common Trichoderma species. BMC Genomics 20: 1–24. doi:10.1186/S12864-019-5680-7/FIGURES/6.

Kumar, A., R. Chandra. 2020. Ligninolytic enzymes and its mechanisms for degradation of lignocellulosic waste in environment. Heliyon 6, e03170. doi:10.1016/J.HELIYON.2020.E03170.

Kumar, K.R., M.J. Cowley, R.L. Davis. 2019. Next-Generation Sequencing and Emerging Technologies. Semin. Thromb. Hemost. 45: 661–673. doi:10.1055/s-0039-1688446.

Lange, L., L. Bech, P.K. Busk, M.N. Grell, Y. Huang, M. Lange et al. 2012. The importance of fungi and of mycology for a global development of the bioeconomy. IMA Fungus 3: 87–92. doi:10.5598/imafungus.2012.03.01.09.

Li, G., T. Jian, X. Liu, Q. Lv, G. Zhang, J. Ling. 2022. Application of Metabolomics in Fungal Research. Molecules 27: 7365. doi:10.3390/molecules27217365.

Li, X., P. Lu. 2019. Transcriptome Profiles of Alternaria oxytropis Provides Insights into Swainsonine Biosynthesis. Sci. Rep. 9: 2–9. doi:10.1038/s41598-019-42173-2.

Lim, D., P. Hains, B.Walsh, P. Bergquist, H. Nevalainen. 2001. Proteins associated with the cell envelope of Trichoderma reesei: A proteomic approach. Proteomics 1: 899–910. doi:10.1002/1615-9861(200107)1:7<899::aid-prot899>3.3.co;2-r.

Lofgren, L.A., N.H. Nguyen, R. Vilgalys, J. Ruytinx, H.L. Liao, S. Branco et al. 2021. Comparative genomics reveals dynamic genome evolution in host specialist ectomycorrhizal fungi. New Phytol. 230: 774–792. doi:10.1111/NPH.17160.

Ma, L., C.M. Catranis, W.T. Starmer, S.O. Rogers. 1999. Revival and characterization of fungi from ancient polar ice. Mycologist 13: 70–73. doi:10.1016/S0269-915X(99)80012-3.

Markley, J.L., R. Brüschweiler, A.S. Edison, H.R. Eghbalnia, R. Powers, D. Raftery et al. 2017. The future of NMR-based metabolomics. Curr. Opin. Biotechnol. 43: 34–40. doi:10.1016/J.COPBIO.2016.08.001.

Marone, M., S. Mozzetti, D. De Ritis, L. Pierelli, G. Scambia. 2001. Semiquantitative RT-PCR analysis to assess the expression levels of multiple transcripts from the same sample. Biol. Proced. Online • 3: 19–25.

Marouga, R., S. David, E. Hawkins. 2005. The development of the DIGE system: 2D fluorescence difference gel analysis technology. Anal. Bioanal. Chem. 382: 669–678. doi:10.1007/S00216-005-3126-3/FIGURES/6.

Marx, V. 2013. The big challenges of big data. Nature 498, 255–260. doi:10.1038/498255a.

Mashego, M.R., M.L.A. Jansen, J.L. Vinke, W.M. Van Gulik, J.J. Heijnen. 2005. Changes in the metabolome of Saccharomyces cerevisiae associated with evolution in aerobic glucose-limited chemostats. FEMS Yeast Res. 5: 419–430. doi:10.1016/J.FEMSYR.2004.11.008.

Mehta, A., A.C.M. Brasileiro, D.S.L. Souza, E. Romano, M.A. Campos, M.F. Grossi-De-Sá et al. 2008. Plant–pathogen interactions: what is proteomics telling us? FEBS J. 275: 3731–3746. doi:10.1111/J.1742-4658.2008.06528.X.

Melin, P., J. Schnürer, E. Wagner. 2002. Proteome analysis of Aspergillus nidulans reveals proteins associated with the response to the antibiotic concanamycin A, produced by Streptomyces species. Mol. Genet. Genomics 267: 695–702. doi:10.1007/s00438-002-0695-0.

Miao, L.Y., X.C. Mo, X.Y. Xi, L. Zhou, G. De, Y.S. Ke et al. 2018. Transcriptome analysis of a taxol-producing endophytic fungus Cladosporium cladosporioides MD2. AMB Express 8. doi:10.1186/s13568-018-0567-6.

Mohanta, T.K., H. Bae, 2015. The diversity of fungal genome. Biol. Proced. Online 17: 1–9. doi:10.1186/s12575-015-0020-z

Muggia, L., C.G. Ametrano, K. Sterflinger, D. Tesei. 2020. An overview of genomics, phylogenomics and proteomics approaches in ascomycota. Life 10: 1–75. doi:10.3390/life10120356.

Murgia, M., M. Fiamma, A. Barac, M. Deligios, V. Mazzarello, B. Paglietti et al. 2019. Biodiversity of fungi in hot desert sands. Microbiologyopen 8, e00595. doi:10.1002/MBO3.595.

Nagalakshmi, U., Z. Wang, K. Waern, C. Shou, D. Raha, M. Gerstein et al. 2008. The transcriptional landscape of the yeast genome defined by RNA sequencing. Science (80-.). 320: 1344–1349. doi:10.1126/SCIENCE.1158441/SUPPL_FILE/1158441_TABLES_S2_TO_S6.ZIP.

Nagarajan, K., B. Ibrahim, A.A. Bawadikji, J.W. Lim, W.Y. Tong, C.R. Leong et al. 2022. Recent Developments in Metabolomics Studies of Endophytic Fungi. J. Fungi 8. doi:10.3390/jof8010028

Neto, A.G.B., M.C. Pestana-Calsa, M.A. de Morais, T. Calsa. 2014. Proteome responses to nitrate in bioethanol production contaminant Dekkera bruxellensis. J. Proteomics 104: 104–111. doi:10.1016/J.JPROT.2014.03.014.

Nicoletti, R., A. Fiorentino. 2015. Plant Bioactive Metabolites and Drugs Produced by Endophytic Fungi of Spermatophyta. Agric. 5: 918–970. doi:10.3390/agriculture5040918.

Nielsen, J., L. Olsson. 2002. An expanded role for microbial physiology in metabolic engineering and functional genomics: moving towards systems biology. FEMS Yeast Res. 2: 175–181. doi:10.1111/J.1567-1364.2002.TB00083.X.

Nielsen, K.F., J. Smedsgaard. 2003. Fungal metabolite screening: database of 474 mycotoxins and fungal metabolites for dereplication by standardised liquid chromatography–UV–mass spectrometry methodology. J. Chromatogr. A 1002: 111–136. doi:10.1016/S0021-9673(03)00490-4.

Niimi, M., R.D. Cannon, B.C. Monk. 1999. Candida albicans pathogenicity: A proteomic perspective. doi:10.1002/(SICI)1522-2683(19990801)20:11.

Nogueira, S.V., F.L. Fonseca, M.L. Rodrigues, V. Mundodi, E.A. Abi-Chacra, M.S.Winters et al. 2010. Paracoccidioides brasiliensis enolase is a surface protein that binds plasminogen and mediates interaction of yeast forms with host cells. Infect. Immun. 78: 4040–4050. doi:10.1128/IAI.00221-10/ASSET/C5493CB0-88BD-4EED-B926-58FE402465E0/ASSETS/GRAPHIC/ZII9990987690006.JPEG.

Oliver, S.G., M.K. Winson, D.B. Kell, F. Baganz. 1998. Systematic functional analysis of the yeast genome. Trends Biotechnol. 16: 373–378. doi:10.1016/S0167-7799(98)01214-1.

Ong, S.E., B. Blagoev, I. Kratchmarova, L.J. Foster, J.S. Andersen, M. Mann. 2006. Stable Isotope Labeling by Amino Acids in Cell Culture for Quantitative Proteomics. Cell Biol. A Lab. Handb. 427–436. doi:10.1016/B978-012164730-8/50239-2.

Onofri, S., L. Selbmann, G.S. de Hoog, M. Grube, D. Barreca, S. Ruisi et al. 2007. Evolution and adaptation of fungi at boundaries of life. Adv. Sp. Res. 40: 1657–1664. doi:10.1016/J.ASR.2007.06.004.

Oxford Nanopore Technologies. 2020. Closing the gap in plant genomes. White Pap.

Oyedeji, A.B., E. Green, J.A. Adebiyi, O.M. Ogundele, S. Gbashi, M.A. Adefisoye et al. 2021. Metabolomic approaches for the determination of metabolites from pathogenic microorganisms: A review. Food Res. Int. 140: 110042. doi:10.1016/j.foodres.2020.110042.

Park, S.T., J. Kim. 2016. Trends in next-generation sequencing and a new era for whole genome sequencing. Int. Neurourol. J. 20: 76–83. doi:10.5213/inj.1632742.371.

Patti, G.J., O. Yanes, G. Siuzdak. 2012. Metabolomics: the apogee of the omics trilogy. Nat. Rev. Mol. Cell Biol. 2012 134 13: 263–269. doi:10.1038/nrm3314.

Paul, S., P. Singh, S.M. Rudramurthy, A. Chakrabarti, A.K. Ghosh. 2017. Matrix-assisted laser desorption/ionization–time of flight mass spectrometry: protocol standardization and database expansion for rapid identification of clinically important molds. https://doi.org/10.2217/fmb-2017-0105 12: 1457–1466. doi:10.2217/FMB-2017-0105.

Perez-Riverol, Y., E. Alpi, R. Wang, H. Hermjakob, J.A. Vizcaíno. 2015. Making proteomics data accessible and reusable: Current state of proteomics databases and repositories. Proteomics 15: 930–950. doi:10.1002/PMIC.201400302.

Pointing, S.B., Y. Chan, D.C. Lacap, M.C.Y. Lau, J.A. Jurgens, R.L. Farrell. 2009. Highly specialized microbial diversity in hyper-arid polar desert. Proc. Natl. Acad. Sci. U. S. A. 106: 19964–19969. doi:10.1073/PNAS.0908274106/SUPPL_FILE/0908274106SI.PDF.

Pushkarev, D., N.F. Neff, S.R. Quake. 2009. Single-molecule sequencing of an individual human genome. Nat. Biotechnol. 2009 279 27: 847–850. doi:10.1038/nbt.1561.

Putignani, L., F. Del Chierico, M. Onori, L. Mancinelli, M. Argentieri, P. Bernaschi et al. 2011. MALDI-TOF mass spectrometry proteomic phenotyping of clinically relevant fungi. Mol. Biosyst. 7: 620–629. doi:10.1039/C0MB00138D.

Qi, M., Y. Yang. 2002. Quantification of Magnaporthe grisea during infection of rice plants using real-time polymerase chain reaction and northern blot/phosphoimaging analyses. Phytopathology 92: 870–876. doi:10.1094/PHYTO.2002.92.8.870.

Raamsdonk, L.M., B. Teusink, D. Broadhurst, N. Zhang, A. Hayes, M.C. Walsh et al. 2001. A functional genomics strategy that uses metabolome data to reveal the phenotype of silent mutations. Nat. Biotechnol. 2001 191 19: 45–50. doi:10.1038/83496.

Raghukumar, S., C. Raghukumar, C.S. Manohar. 2014. Fungi living in diverse extreme habitats of the marine environment. Mycol. Society of India. https://drs.nio.org/xmlui/handle/2264/4639.

Ramautar, R., A. Demirci, G.J. d. Jong. 2006. Capillary electrophoresis in metabolomics. TrAC Trends Anal. Chem. 25: 455–466. doi:10.1016/J.TRAC.2006.02.004.

Ranque, S., A.C. Normand, C. Cassagne, J.B. Murat, N. Bourgeois, F. Dalle et al. 2014. MALDI-TOF mass spectrometry identification of filamentous fungi in the clinical laboratory. Mycoses 57: 135–140. doi:10.1111/MYC.12115.

Rantakokko-Jalava, K., S. Laaksonen, J. Issakainen, J. Vauras, J. Nikoskelainen, M.K. Viljanen et al. 2003. Semiquantitative detection by real-time PCR of Aspergillus fumigatus in bronchoalveolar lavage fluids and tissue biopsy specimens from patients with invasive aspergillosis. J. Clin. Microbiol. 41: 4304–4311. doi:10.1128/JCM.41.9.4304-4311.2003.

Rasheed, M., N. Kumar, R. Kaur. 2020. Global Secretome Characterization of the Pathogenic Yeast Candida glabrata. J. Proteome Res. 19: 49–63. doi:10.1021/ACS.JPROTEOME.9B00299/SUPPL_FILE/PR9B00299_SI_002.XLSX.

Ravichandran, A., A.P. Kolte, A. Dhali, S.M. Gopinath, M. Sridhar. 2020. Transcriptomic analysis of Lentinus squarrosulus provide insights into its biodegradation ability. bioRxiv 2020.09.28.316471. doi:10.1101/2020.09.28.316471.

Ries, L., S.T. Pullan, S. Delmas, S. Malla, M.J. Blythe, D.B. Archer. 2013. Genome-wide transcriptional response of Trichoderma reesei to lignocellulose using RNA sequencing and comparison with Aspergillus niger. BMC Genomics 14: 1–12. doi:10.1186/1471-2164-14-541/FIGURES/3.

Roberts, L.D., A.L. Souza, R.E. Gerszten, C.B. Clish. 2012. Targeted Metabolomics. Curr. Protoc. Mol. Biol. 98: 30.2.1-30.2.24. doi:10.1002/0471142727.MB3002S98.

Sagita, R., W.J. Quax, K. Haslinger. 2021. Current State and Future Directions of Genetics and Genomics of Endophytic Fungi for Bioprospecting Efforts. Front. Bioeng. Biotechnol. 9. doi:10.3389/fbioe.2021.649906.

Sanger, F., S. Nicklen, A.R. Coulson. 1977. DNA sequencing with chain-terminating inhibitors. Proc. Natl. Acad. Sci. 74: 5463–5467. doi:10.1073/PNAS.74.12.5463.

Santiago-Rodriguez, T.M., E.B. Hollister. 2021. Multi 'omic data integration: A review of concepts, considerations, and approaches. Semin. Perinatol. 45: 151456. doi:10.1016/J.SEMPERI.2021.151456.

Schadt, E.E., S. Turner, A. Kasarskis. 2010. A window into third-generation sequencing. Hum. Mol. Genet. 19: 227–240. doi:10.1093/hmg/ddq416.

Schwikowski, B., P. Uetz, S. Fields. 2000. A network of protein–protein interactions in yeast. Nat. Biotechnol. 2000 1812 18: 1257–1261. doi:10.1038/82360.

Shankar, A., K.K. Sharma. 2022. Fungal secondary metabolites in food and pharmaceuticals in the era of multi-omics. Appl. Microbiol. Biotechnol. 106: 3465–3488. doi:10.1007/s00253-022-11945-8.

Shendure, J. 2008. The beginning of the end for microarrays? Nat. Methods 5: 585–587. doi:10.1038/nmeth0708-585.

Shiio, Y., R. Aebersold. 2006. Quantitative proteome analysis using isotope-coded affinity tags and mass spectrometry. Nat. Protoc. 2006 11 1: 139–145. doi:10.1038/nprot.2006.22.

Smith, J.B. 2001. Peptide Sequencing by Edman Degradation.

Smyth, M.S., J.H.J. Martin. 2000. x Ray crystallography. Mol. Pathol. 53: 8. doi:10.1136/MP.53.1.8.

Sutandy, F.X.R., J. Qian, C.S. Chen, H. Zhu. 2013. Overview of Protein Microarrays. Curr. Protoc. Protein Sci. 72: 27.1.1–27.1.16. doi:10.1002/0471140864.PS2701S72.

Tan, K.C., S.V.S. Ipcho, R.D. Trengove, R.P. Oliver, P.S. Solomon. 2009. Assessing the impact of transcriptomics, proteomics and metabolomics on fungal phytopathology. Mol. Plant Pathol. 10: 703–715. doi:10.1111/j.1364-3703.2009.00565.x.

Tawfike, A.F., M. Romli, C. Clements, G. Abbott, L. Young, M. Schumacher et al. 2019. Isolation of anticancer and anti-trypanosome secondary metabolites from the endophytic fungus Aspergillus flocculus via bioactivity guided isolation and MS based metabolomics. J. Chromatogr. B 1106–1107: 71–83. doi:10.1016/J.JCHROMB.2018.12.032.

Tsang, A., G. Butler, J. Powlowski, E.A. Panisko, S.E. Baker. 2009. Analytical and computational approaches to define the Aspergillus niger secretome. Fungal Genet. Biol. 46: S153–S160. doi:10.1016/J.FGB.2008.07.014.

Uetz, P., L. Glot, G. Cagney, T.A. Mansfield, R.S. Judson, J.R. Knight et al. 2000. A comprehensive analysis of protein–protein interactions in Saccharomyces cerevisiae. Nat. 2000 4036770 403: 623–627. doi:10.1038/35001009.

Van Dam, J.C., M.R. Eman, J. Frank, H.C. Lange, G.W.K. Van Dedem, S.J. Heijnen. 2002. Analysis of glycolytic intermediates in Saccharomyces cerevisiae using anion exchange chromatography and electrospray ionization with tandem mass spectrometric detection. Anal. Chim. Acta 460: 209–218.

Varga, E., T. Glauner, F. Berthiller, R. Krska, R. Schuhmacher, M. Sulyok. 2013. Development and validation of a (semi-) quantitative UHPLC-MS/MS method for the determination of 191 mycotoxins and other fungal metabolites in almonds, hazelnuts, peanuts and pistachios. Anal. Bioanal. Chem. 405: 5087–5104. doi:10.1007/S00216-013-6831-3/TABLES/7.

Venugopalan, A., S. Srivastava. 2015. Endophytes as in vitro production platforms of high value plant secondary metabolites. Biotechnol. Adv. 33: 873–887. doi:10.1016/j.biotechadv.2015.07.004.

Vialás, V., P. Perumal, D. Gutierrez, P. Ximénez-Embún, C. Nombela, C. Gil et al. 2012. Cell surface shaving of Candida albicans biofilms, hyphae, and yeast form cells. Proteomics 12: 2331–2339. doi:10.1002/PMIC.201100588.

Vihinen, M. 2001. Bioinformatics in proteomics. Biomol. Eng. 18: 241–248. doi:10.1016/S1389-0344(01)00099-5.

Villas-Bôas, S.G., J.F. Moxley, M. Åkesson, G. Stephanopoulos, J. Nielsen. 2005. High-throughput metabolic state analysis: the missing link in integrated functional genomics of yeasts. Biochem. J. 388: 669–677. doi:10.1042/BJ20041162.

Vinayavekhin, N., A. Saghatelian. 2010. Untargeted Metabolomics. Curr. Protoc. Mol. Biol. 90: 30.1.1-30.1.24. doi:10.1002/0471142727.MB3001S90.

Voltersen, V., M.G. Blango, S. Herrmann, F. Schmidt, T. Heinekamp, M. Strassburger et al. 2018. Proteome analysis reveals the conidial surface protein CcpA essential for virulence of the pathogenic fungus Aspergillus fumigatus. MBio 9. doi:10.1128/MBIO.01557-18/SUPPL_FILE/MBO004184034ST2.PDF.

Vuckovic, D. 2012. Current trends and challenges in sample preparation for global metabolomics using liquid chromatography-mass spectrometry. Anal. Bioanal. Chem. doi:10.1007/s00216-012-6039-y.

Wang, Z., M. Gerstein, M. Snyder. 2009. RNA-Seq: a revolutionary tool for transcriptomics. Nat. Rev. Genet. 10: 57–63.

Wang, Z., A. Gudibanda, U. Ugwuowo, F. Trail, J.P. Townsend. 2018a. Using evolutionary genomics, transcriptomics, and systems biology to reveal gene networks underlying fungal development. Fungal Biol. Rev. 32: 249–264. doi:10.1016/j.fbr.2018.02.001.

Wang, D., L. Zhang, H. Zou, L. Wang. 2018b. Secretome profiling reveals temperature-dependent growth of Aspergillus fumigatus. Sci. China Life Sci. 61: 578–592. doi:10.1007/S11427-017-9168-4/METRICS. Wang, D., L. Zhang, H. Zou, L. Wang. 2018. Secretome profiling reveals temperature-dependent growth of Aspergillus fumigatus. Sci. China Life Sci. 61: 578–592. doi:10.1007/S11427-017-9168-4/METRICS.

Wen, H., S. Kang, Youngmin, Song, Yonghyun, Song, S.H. Sung, S. Park. 2010. Differentiation of cultivation sources of Ganoderma lucidum by NMR-based metabolomics approach. Phytochem. Anal. 21: 73–79. doi:10.1002/PCA.1166.

Westerhuis, J.A., E.J.J. van Velzen, H.C.J. Hoefsloot, A.K. Smilde. 2010. Multivariate paired data analysis: Multilevel PLSDA versus OPLSDA. Metabolomics 6: 119–128. doi:10.1007/S11306-009-0185-Z/FIGURES/7.

Wiese, S., K.A. Reidegeld, H.E. Meyer, B. Warscheid. 2007. Protein labeling by iTRAQ: A new tool for quantitative mass spectrometry in proteome research. Proteomics 7: 340–350. doi:10.1002/PMIC.200600422.

Wilken, S.E., C.L. Swift, I.A. Podolsky, T.S. Lankiewicz, S. Seppälä, M.A. O'Malley. 2019. Linking 'omics' to function unlocks the biotech potential of non-model fungi. Curr. Opin. Syst. Biol. 14: 9–17. doi:10.1016/j.coisb.2019.02.001.

Wittmann, C., M. Hans, W.A. Van Winden, C. Ras, J.J. Heijne. 2005. Dynamics of intracellular metabolites of glycolysis and TCA cycle during cell-cycle-related oscillation in Saccharomyces cerevisiae. Biotechnol. Bioeng. 89: 839–847. doi:10.1002/BIT.20408.

Wohlbach, D.J., A. Kuo, T.K. Sato, K.M. Potts, A.A. Salamov, K.M. LaButti et al. 2011. Comparative genomics of xylose-fermenting fungi for enhanced biofuel production. Proc. Natl. Acad. Sci. U. S. A. 108: 13212–13217. doi:10.1073/PNAS.1103039108/SUPPL_FILE/SD01.TXT.

Wood, V., R. Gwilliam, M.A. Rajandream, M. Lyne, R. Lyne, A. Stewart et al. 2002. The genome sequence of Schizosaccharomyces pombe. Nat. 2002 4156874 415, 871–880. doi:10.1038/nature724.

Xiang, C.C., Y. Chen. 2000. cDNA microarray technology and its applications. Biotechnol. Adv. 18: 35–46. doi:10.1016/S0734-9750(99)00035-X.

Xiao, J.F., B. Zhou, H.W. Ressom. 2012. Metabolite identification and quantitation in LC-MS/MS-based metabolomics. TrAC Trends Anal. Chem. 32: 1–14. doi:10.1016/J.TRAC.2011.08.009.

Yao, L., L.P. Zhu, X.Y. Xu, L.L. Tan, M. Sadilek, H. Fan et al. 2016. Discovery of novel xylosides in co-culture of basidiomycetes Trametes versicolor and Ganoderma applanatum by integrated metabolomics and bioinformatics. Sci. Reports 2016 61 6: 1–13. doi:10.1038/srep33237.

Yates, J.R. 2011. A century of mass spectrometry: from atoms to proteomes. Nat. Methods 2011 88 8: 633–637. doi:10.1038/nmeth.1659.

Yuste, J.C., J. Peñuelas, M. Estiarte, J. Garcia-Mas, S. Mattana, R. Ogaya et al. 2011. Drought-resistant fungi control soil organic matter decomposition and its response to temperature. Glob. Chang. Biol. 17: 1475–1486. doi:10.1111/J.1365-2486.2010.02300.X.

Zhang, G., W. Wang, X. Zhang, Q. Xia, X. Zhao, Y. Ahn et al. 2015. De Novo RNA sequencing and transcriptome analysis of Colletotrichum gloeosporioides ES026 reveal genes related to biosynthesis of huperzine A. PLoS One 10: 1–16. doi:10.1371/journal.pone.0120809.

Zhang, W., T. Hankemeier, R. Ramautar. 2017. Next-generation capillary electrophoresis–mass spectrometry approaches in metabolomics. Curr. Opin. Biotechnol. 43: 1–7. doi:10.1016/J.COPBIO.2016.07.002.

Zhu, H., M. Bilgin, R. Bangham, D. Hall, A. Casamayor, P. Bertone et al. 2001. Global analysis of protein activities using proteome chips. Science (80-.). 293:2101–2105. doi:10.1126/science.1062191.

Zimowska, B., M. Bielecka, B. Abramczyk, R. Nicoletti. 2020. Bioactive products from endophytic fungi of sages (Salvia spp.). Agric. 10: 1–16. doi:10.3390/agriculture10110543.

Zou, K., X. Liu, Q. Hu, D. Zhang, S. Fu, S. Zhang et al. 2021. Root Endophytes and Ginkgo biloba Are Likely to Share and Compensate Secondary Metabolic Processes, and Potentially Exchange Genetic Information by LTR-RTs. Front. Plant Sci. 12. doi:10.3389/fpls.2021.704985.

CHAPTER 14

Practical Aspects of Sequence-Based Species Delimitation in Fungi

Andrea Becchimanzi

Introduction

Importance of Taxonomical Assignations in Fungi – Why Do We Need to Discriminate Between Species?

Taxonomic classification provides a standardized method for naming and classifying organisms that allow scientists and even the general public to communicate effectively about specific species and their characteristics. In essence, it provides a framework to organize our understanding of the diversity of life on Earth.

Fungi, whose known species comprise only ~7% of the predicted species number (Mora et al. 2011), play a critical role in many ecosystems. Their correct taxonomic assignment can help us better understand the relationships between different fungal species, how they interact with other organisms, and how they have evolved over time. This, in turn, can provide insight into ecosystem functioning and lead to better conservation and management decisions. In addition, this can be important for a variety of purposes, including agricultural and medical research. For example, some species of fungi are important decomposers that break down organic matter and return nutrients to the soil. This is essential for maintaining ecosystem health and promoting plant growth. By understanding the taxonomic classification of these fungi, we can better understand their role in ecosystems and develop strategies to support their growth and function. In the medical and agricultural fields, understanding the taxonomic assignments of fungi can also be important for identifying fungal pathogens. Different species of fungi can cause different types of infections, and knowledge of their specific characteristics can help effectively diagnose and treat fungal diseases. Correct identification provides access to information about the biology, distribution, ecology, host range, and control of fungal species, as well as the risks associated with them. On the other hand, incorrect identification can lead to unnecessary or incorrect control measures and restrictions (Maharachchikumbura et al. 2021).

Species Concepts

The species concept in fungi is a contentious topic, with different approaches leading to different conclusions regarding boundaries and the number of species. Currently, biologists use over 10 different species concepts in fungal taxonomy, but none has been adopted as a unified standard,

Department of Agricultural Sciences, University of Naples Federico II, 80055 Portici, Italy.
Email: andrea.becchimanzi@unina.it

making the field subjective. The most commonly used species concepts are the morphological, biological, and evolutionary/phylogenetic concepts (Stengel et al. 2022). However, each of these approaches has its own advantages and disadvantages, and the lack of a clear distinction between theoretical species concepts and working criteria has led to confusion and debate in the field (Matute and Sepúlveda 2019).

The morphological species concept (MSC) is a way of defining species in fungi based on their physical and cultural characteristics, such as their size and shape. This concept is widely used in fungal taxonomy because it is simple and straightforward, but it has been criticized for its subjectivity and lack of biological significance. Some fungi are pleomorphic, meaning they have multiple forms, which can make it difficult to apply the MSC and accurately classify them. Additionally, the MSC can obscure the evolutionary origins of certain features and does not account for the developmental plasticity of fungi. Despite these limitations, the MSC is still commonly used in fungal taxonomy because it is practical for identification and cataloguing (Xu 2020).

The biological species concept (BSC) is a theory in evolutionary biology that defines a species as a group of populations that can interbreed and produce viable offspring. This concept cannot be applied to asexual fungi or fungi that can self-fertilize. Recent studies have suggested that some fungi known only from their asexual forms may also reproduce sexually, but the BSC cannot be applied to these taxa because the sexual stages remain unknown (Taylor et al. 2000, Dettman et al. 2003, Cai et al. 2011). BSC is also not applicable to interspecific hybridization in fungi, as it does not account for partial interfertility or differences in interbreeding barriers among closely related fungal species. Despite these limitations, the BSC has been widely used by organismal biologists to recognize and delimit new fungal species. It has been applied to a range of fungi and has been used to resolve cryptic species within morphological species.

The phylogenetic species concept (PSC) is a theory in evolutionary biology that defines a species as a monophyletic group sharing molecular characteristics derived from a common ancestor. This concept relies on genetic data as a signature of speciation and is the most widely applied for recognizing species in fungi, as it allows for the identification of species before changes in mating behavior or morphology have occurred (Taylor et al. 2000, Bobay and Ochman 2017). DNA sequences can be viewed as genetic 'barcodes' that are embedded in every cell. DNA barcoding, through sequencing of short DNA sequences that are present in a wide range of species, is currently the most popular molecular tool used to identify, classify and analyze species diversity. Barcoding is optimal when interspecific variation of the target sequence exceed intraspecific variation (barcode gap) (Hebert et al. 2003). Species delimitation approaches based on barcode gap, such as ABGD, have proven to be fast and repeatable, providing a valuable tool for establishing a basic understanding of species boundaries that can be validated with subsequent studies (Kekkonen and Hebert 2014) OTUs.

The phylogenetic and biological species concepts have a common ground. Indeed, according to the BSC, speciation occurs in instances where reproductive isolation exists (Matute and Sepúlveda 2019). The reproductive isolation, underlying speciation, lead to genetic differentiation between the putative species which should leave a signature of genetic discontinuity across the whole genome (Bobay and Ochman 2017). Under the PSC, species are diagnosed as a cluster of individuals which is sufficiently differentiated from other clusters as revealed by DNA sequences. The PSC has several advantages over other species recognition methods, including the ability to be applied to asexual organisms. It has been successfully used to resolve cryptic speciation within plant and human pathogenic fungi.

Species Delimitation Approaches

The PSC in fungi has two broadly-defined forms, the strict genealogic concordance (SGC) and the coalescent-based species delimitation (CBD) approaches (revised in Taylor et al. 2000). The SGC approach is a method used to assess the extent of genetic concordance across different loci in order to determine whether reproductive isolation has occurred and differentiate between structured populations and true species. The SGC approach involves two main rules: the concordance rule, which states that if all gene genealogies from unlinked loci in the genome are congruent with each other, speciation may have occurred; and the support rule, which states that if a group of putative species forms reciprocally monophyletic groups with high Bayesian support and bootstrap values (at least 90% and 70%, respectively), they are likely to be reproductively isolated (Dettman et al. 2003). Notably, support rule is more commonly used than the concordance rule likely due to technical reasons. Indeed, once a topology is inferred, assessing the level of concordance requires different approaches, while assessing the level of support of the branches is almost automatic (Matute and Sepúlveda 2019).

Coalescent-based delimitation (CBD) is a set of approaches that use coalescent theory to understand how multiple species are related by modeling the genealogical history of individuals back to a common ancestor. These approaches involve estimating the likelihood of obtaining a given set of gene genealogies given a species tree, which allows for the estimation of the probability of a species tree and the probability of a particular species delimitation given multilocus data. There are three main families of CBD methods: methods that sequentially collapse nodes to identify potential species using a pre-specified topology (e.g., BP&P (Yang 2015)); methods that explore the full space of possible species' tree topologies to delimit species and infer the species tree simultaneously (e.g., STEM (Kubatko et al. 2009), SpedeSTEM (Ence and Carstens 2011), DISSECT (Jones et al. 2015), STACEY (Jones 2017)); and methods that use distinct branching patterns between divergence and intraspecific diversification to distinguish between species and populations. The main assumption of these methods is that within-species branching, events will be more frequent than between-species events. These methods use either Yule, such as GMYC (Generalized Mixed Yule Coalescent; (Pons et al. 2006), such as PTP (Poisson Tree Processes (Zhang et al. 2013)).

The goal of this chapter is to provide a brief guide to conducting sequence-based species delimitation analysis in fungi, from data collection to discussion of results. The methods presented here are used in several taxonomy studies on fungi, although the main protocol steps can be broadly applied to other organisms such as animals and plants.

Materials

In general, sequence-based analyses for species delimitation do not require large computational resources and can be performed on an ordinary laptop computer (8 Gb RAM), especially when the sequence number is less than 1000 and the sequence length is less than 5000. However, in some cases (e.g., as sequence length and number increase), CPU time and memory requirements can increase exponentially, both for alignment (Kumar et al. 2016), as well as phylogenetic reconstruction (Tang et al. 2014), which is required for most sequence-based species delimitation methods. Although most of the tools used are available online and thus easily accessible, a basic knowledge of the Unix command language is strongly recommended to better control each step of the analysis.

A Simple Integrative Approach for Sequence-Based Species Delimitation in Fungi

DNA-based species delimitation methods are increasingly employed for resolving fungal taxonomy (Parnmen et al. 2012, Haelewaters et al. 2018, Bustamante et al. 2019), although these tools are recommended as part of an integrative approach to establish well-supported boundaries among species (Carstens et al. 2013, Maharachchikumbura et al. 2021). In this brief tutorial, we will provide the basic information for molecular species identification of fungi from data production and analysis, to comparison and integration of results obtained from different methods (Figure 1).

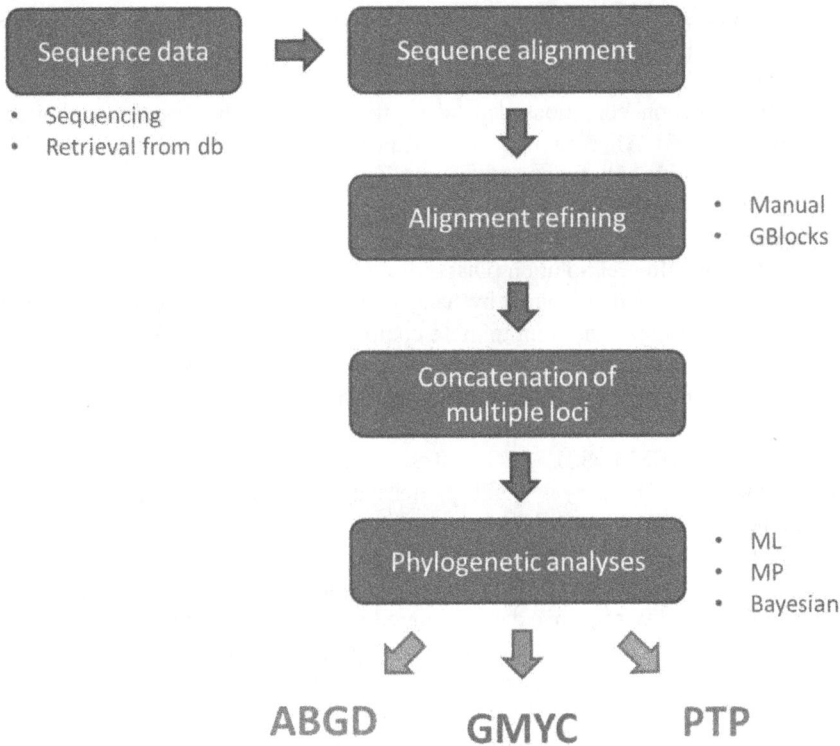

Figure 1. Overview of the sequence-based species delimitation analysis pipeline.

Sequence Data

Sequence data to be used in species delimitation analysis can be obtained by sequencing (original data) or retrieved from databases such as GenBank (NCBI). Sequence data production steps include DNA isolation, PCR amplification and/or DNA sequencing. These steps won't be described in detail in the present chapter; however, accurate protocols are available following the cited literature (e.g., Cubero et al. 1999).

Generating Sequence Data

DNA isolation can be performed either from isolated strains or from environmental sources, such as host plants or animals. Most commonly used methods of DNA isolation are based on CTAB (cetyl trimethyl ammonium bromide) extraction buffers (Doyle 1991, Cubero et al. 1999, Huang et al. 2018) or on Genomic DNA Extraction Kits (e.g., QIAGEN GmbH, QIAGEN Strasse 1, BioFlux®). PCR amplification can target different genomic sequences, depending on the genus, family or order of interest. The most widely used region is ITS with more than 170,000 full-length fungal sequences deposited in GenBank, the most popular sequence database. Moreover, the advantages of using ITS are the high success rate in PCR using universal primers and the high capacity of discrimination between species, with a more clearly defined barcode gap (Schoch et al. 2012). For these reasons, ITS represents the primary fungal barcode marker for the Consortium for the Barcode of Life and the curated database BOLD (Barcode of Life Data), although supplementary barcodes may be developed for particular narrowly circumscribed taxonomic groups (Schoch et al. 2012). The appropriate gene regions (with proper universal primers) to be sequenced for the taxonomical group of interest can be identified through relevant scientific literature. The most commonly used markers are nuclear loci (e.g., ITS, TEF1-α, β-TUB, RPB2), chosen for being single copy genes, harboring polymorphism, thus not being completely constrained by purifying selection (Matute and

Sepúlveda 2019). A reliable phylogenetic reconstruction and species delimitation analysis should be based on the use of multiple informative markers, evaluated individually or combined by sequence concatenation.

The PCR amplification conditions depend on the length of the targeted region and on the particular primer pair used. After running the PCR products on an agarose gel comprised of DNA intercalating dyes, quantity and quality of the amplicons can be visualized under UV light. The commonly used, but mutagenic, intercalating agent is ethidium bromide, but it is being replaced by a range of safer but more expensive alternatives, such as GelRed™, GelGreen™, SYBR™ safe, SafeView and EZ-Vision®In-Gel Solution (Dissanayake 2020).

DNA sequences are usually obtained by Sanger sequencing of PCR amplicons, performed by sequencing companies. Samples are sent to these company along with the PCR primers, which will be used for the sequencing in separate reactions: forward primer and reverse primer reactions. The output of each sequencing reaction is a chromatogram, which contains information about nucleotide calling and quality of such calling. This can be assessed by analyzing the sharpness of the peaks using tools, such as abiviewer (EMBOSS), MEGA's Trace File Editor and BioEdit v.5, which recognize ABI sequencer trace files (Dissanayake 2020). After quality check, trimming of low quality 3' and 5' ends can be performed using a common text editor. Thus, forward and reverse sequences are aligned and a consensus sequence generated, by considering only nucleotides in common between the two sequences. This approach based on consensus sequence is the most conservative and allows the generation of high quality sequence data. Trimmed consensus sequences, in fasta format, can be subjected to BLAST searches against nucleotide collection (nr/nt) database (via https://blast.ncbi. nlm.nih.gov), which allow a preliminary taxa assignation. This should be done checking at least the top ten hits of BLAST results in NCBI database, in order to have a proper idea about the isolates assignation. This step is important for determining the taxa range to be included in the study and for choosing the correct outgroup. After taxa range determination, a literature check should be conducted in order to assess if other particular loci are needed to be sequenced.

Sequence Data from Databases

Usually, sequence-based species delimitation analyses include both in-house generated data and sequence data derived from already published studies and retrieved from several external databases (e.g., GenBank). The database-retrieved taxa (reference sequences) are important in order to give the right context to newly generated data, and depend upon the extent of the study (e.g., introducing a new species/genus, revising a family, etc.). For example, when introducing a new species, it is recommended to include all type species and at least 2–3 strains from each species of the genus. Indeed, these choices should always avoid taxon sampling problems (Hillis et al. 2003, Talavera et al. 2013), related to uneven sampling of the selected taxa (e.g., high variability in the number of strains per each species). After determining the taxa range of the ingroup (i.e., the group of taxa which is being studied), another important choice is the outgroup: a cluster of taxa which falls outside the clade being studied but is closely related to that clade and acts as the comparison point of the ingroup, allowing the phylogeny to be rooted (Wilberg 2015, Dissanayake 2020). When studying a particular genus (ingroup), a good candidate for the outgroup is one or more species belonging to the most closely related genus. Outgroup sampling is of primary importance in phylogenetic analyses, because it affects ingroup relationships and characters' polarization (Lyons-Weiler et al. 1998).

Fasta sequences can be searched and downloaded in several ways for phylogenetic analysis:

- anual searches on GenBank db (e.g., using "*Cladosporium cladosporioides* [organism] AND ITS" as keywords).

- By using the accession numbers listed in a particular scientific publication (there's often a table reporting all the used sequences) through the Batch Entrez download tool (via https://www.

ncbi.nlm.nih.gov/sites/batchentrez), which allows the batch download of the corresponding fasta files.

At this stage, a good practice is to keep track of the sequences of interest using an excel spreadsheet with columns reporting species name, isolate name and GenBank accession number.

Tutorial 1 - Dataset Retrieval from GenBank

As a case study, I provide sequence data of *Cladosporium* species (Becchimanzi et al. 2021) which can be downloaded by entering the accession number list in Batch Entrez:

1) Go to https://doi.org/10.3390/pathogens10091167. Select the whole content of Table 1 and 2, and copy-paste it in a new Excel spreadsheet. Select only *ITS* column and copy-paste it into a new text file named "its.txt" (Figure 2A).

2) Navigate to https://www.ncbi.nlm.nih.gov/sites/batchentrez, select "Nucleotide" and click on "Select file". Browse to your "its_list.txt" and click on "Retrieve" (Figure 2B).

3) Download your multi-fasta file as "its.fa" (Figure 2C). Do the same for translation elongation factor (*tef1*) and actin (*act*) loci.

   INPUT: accession list
   OUTPUT: its.fa, tef1.fa, act.fa

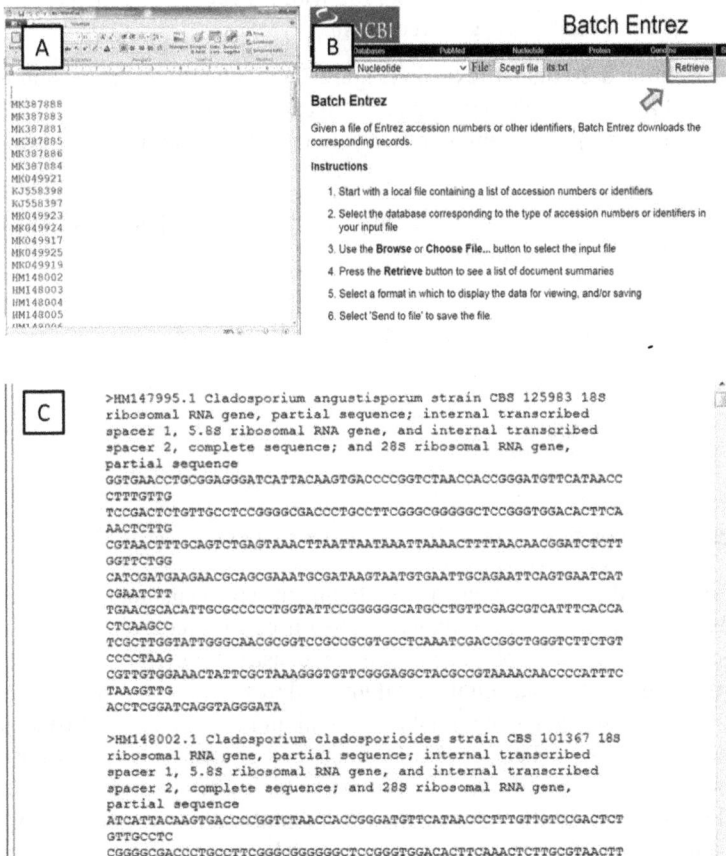

Figure 2. Dataset retrieval from GenBank. A list of accession number of ITS sequences (A) is selected as input for Batch Entrez (B), allowing the download of multi fasta files (C).

Now each file contains a single locus dataset which will be aligned, combined and then processed for two classes of phylogenetic analysis (Maximum Likelihood and Bayesian Inference) by providing details in each step for Linux/Mac OSX users (and for Windows users where possible).

Sequence Alignment

In order to be compared and subjected to phylogenetic analyses, sequences need to be aligned. Practical aspects of alignment and phylogenetic reconstruction steps are described by several authors and accurately reviewed by Dissanayake et al. (2020). The most popular alignment software is currently T-Coffee (Notredame et al. 2000), MAFFT (Katoh et al. 2002) and Muscle (Edgar 2004). Each one of these can be accessed via web interface or can be easily installed on a local machine by Linux/Mac OSX users. Windows users can access these and other alignment tools by using JALVIEW (Waterhouse et al. 2009), a powerful free software for alignment visualization and editing.

Tutorial 2 – Alignment

Before aligning the sequences, we will change the fasta header of each sequence of the three files with the corresponding strain ID. This is important because after the alignment, we will carry out a concatenation of our dataset based on the strain ID.

To perform this step:

1) Change the long headers of your multi-fasta file with the accession number
   awk '{print $1}' its.fa | sed 's/\.1//g' > 1col.its.fa

2) Create a tab-delimited 2-column text file (Figure 3A) with Old ID (accession number) in the first column and New ID (strain ID) in the second one (e.g., use awk 'BEGIN {FS=OFS= "\t"} {print $7, $2}' to print specific tab-delimited columns from a tab-delimited file)

3) Change the headers
   awk 'FNR==NR { a[">"$1]=$2; next } $1 in a { sub(/>/,">"a[$1]"\t",$1)}1' names.txt 1col.its.fa | awk '{print $1}' > renamed.its.fa

4) Align
   muscle3.8.31_i86linux64 -in renamed.its.fa -out its.ali

5) Repeat 1–4 for each of the selected loci

   INPUT: its.fa, tef1.fa, act.fa
   OUTPUT: its.ali, tef1.ali, act.ali

Refining the Alignment

Each position (also called site or column) in an alignment represents homologous character states in the sequences. These homologous columns are then used to infer a phylogeny through various inference procedures which lead to phylogenetic tree reconstruction. Thus, the quality of a multiple sequence alignment has great impact on the final inferred tree (Talavera and Castresana 2007).

In order to check overall alignment quality, each alignment file can be visualized using tools such as JALVIEW or AliView (Larsson 2014). Indeed, alignment software is not error-free and sequences are manually adjusted, if necessary, to create high-quality alignments. AliView handles alignments of limitless size in the formats most commonly used, i.e., fasta, phylip, nexus and clustal. Manual trimming of poorly aligned, and so not evolutionary informative, 3' and 5' regions is frequently performed at this stage (Figure 3B). Another important feature to check are indel events (insertion and/or deletion), which are indicated by gaps in the alignment (-). Gaps in alignments

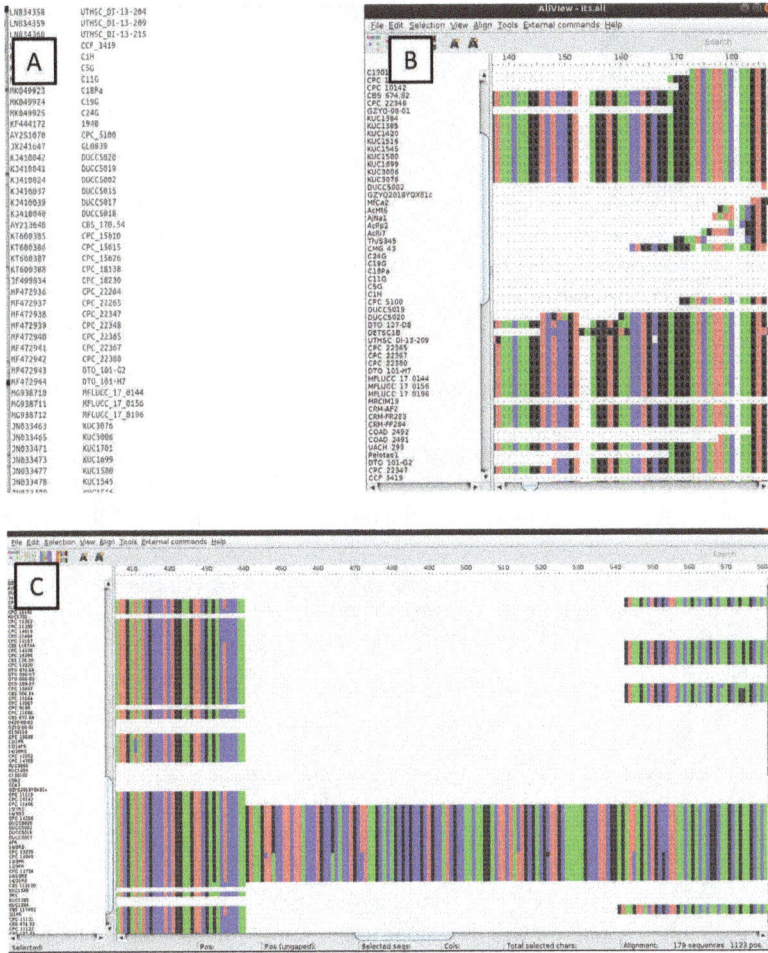

Figure 3. Sequence alignment visualized through Aliview. A tab-delimited 2-column text file (A) with Old ID (accession number) in the first column and New ID (strain ID) in the second one is used to change the multi-fasta header (step 2 of Tutorial 2). Alignment regions at 5' (B) and introns/gaps (C) are visually checked and eventually trimmed out to avoid errors and/or uninformative columns in the subsequent analyses.

suggest alterations in sequences due to genetic rearrangements, and are treated in different ways by the different phylogenetic methods (Rodriguez-Murillo and Salem 2013). However, the most common way to treat them is to consider gaps as missing data (- = N). Moreover, the aligned sequences should be manually checked in order to identify introns, which are frequent, for example, in *tef1* (Matheny et al. 2007) and are characterized by the presence of GT-AG nucleotides (5'–3'). Checking gaps and introns is important because the major part of alignment errors and ambiguities occurs around these regions (Figure 3C). However, identifying misaligned regions by eye in large alignments can be tricky and several tools have been developed for the automatic detection of conflicting signals in the data and the removal of uninformative columns.

Although many authors consider such removal, called *alignment masking*, to be beneficial (Castresana 2000, Löytynoja and Milinkovitch 2001, Rajan 2013), others think that there is loss of information upon removing any part of the sequence (Lee 2001, Aagesen 2004). Whether masking is beneficial may depend on the data being analysed, and our suggestion is to apply different strategies (manual, automatic, manual + automatic, etc.) and compare the results. One of the most popular tools for alignment masking is GBLOCKS (Castresana 2000, Talavera and Castresana 2007), designed to identify conserved blocks in an alignment and exclude sections that are variable beyond a threshold.

Phylogenetic Analyses of Concatenated Loci

The 'final alignment' can be a single locus or a combination of multiple loci, according to the purpose of the study. A commonly used approach is based on the comparison of single and multiple loci analyses. However, combined data analysis generally outperforms other methods of phylogenetic inference such as trees derived from individual gene fragments (Thomson and Shaffer 2010, Kupczok et al. 2010, Chesters and Vogler 2013). The separately aligned single locus can be concatenated using tools such as BioEdit (Hall 1999), MegaX (Kumar et al. 2018), SequenceMatrix (Vaidya et al. 2011) or Concatenator (Vences et al. 2022).

Phylogenetic analyses required to perform the most used sequence-based species delimitation methods are of two categories: Maximum Likelihood (ML) and Bayesian analyses. Both approaches aim to estimate the most likely tree given a set of observed sequence data, but they differ in how they approach this estimation. The basic idea behind ML is to compare the likelihood of different tree topologies (the arrangement of branches on the tree) given a particular model of evolution. The tree topology that maximizes the likelihood is considered the most probable, or the maximum likelihood, tree (Felsenstein 1981). Bayesian phylogenetics, on the other hand, is a probabilistic approach that uses Bayesian statistical techniques to estimate the posterior probability of different tree topologies. This approach involves specifying a prior probability distribution over tree topologies and then using the observed data to update this distribution and estimate the posterior probability of each tree topology. The tree topology with the highest posterior probability is considered the most probable tree (Huelsenbeck and Ronquist 2001). One key difference between ML and Bayesian phylogenetics is that ML estimates a single most likely tree, whereas Bayesian phylogenetics estimates the posterior probability distribution over all possible trees. This means that Bayesian phylogenetics can provide a measure of uncertainty about the estimated tree, which can be useful in some situations. However, Bayesian phylogenetics can be computationally more intensive than ML, so it may not always be practical to use this approach.

Tutorial 3 – Phylogenetic Analyses of Concatenated Loci

To concatenate the 3 aligned loci, we will download a useful script from https://github.com/nylander/catfasta2phyml. After concatenation, we will perform a ML and a Bayesian analysis using RAxML (Stamatakis 2006) and BEAST (Bouckaert et al. 2014), respectively. To complete this step:

1) Concatenate the aligned loci. Make sure you changed the fasta header as described in Practice 2
   catfasta2phyml.pl --intersect -f its.ali tef1.ali act.ali > combined.ali.

2) Open the alignment using AliView to check alignment quality and make the proper manual adjustments. You can download the manual refined version of the combined alignments from https://doi.org/10.5281/zenodo.5152222 ("reali.fasta.fasta"). Save the final alignment both in fasta and in nexus format.

3) Make an ML tree using the rapid Bootstrap analysis and search for the best-scoring ML tree in one single RAxML run, with 1000 bootstrap replications.
   raxmlHPC-SSE3 -f a -m GTRGAMMA -p 12345 -x 12345 -# 1000 -s combined.ali -n combined.ali.fastboot1000.

4) Open Beauti, the input compiler tool of BEAST (see https://beast.community/first_tutorial for a complete tutorial). Import alignments, set GTR, relaxed clock log normal, coalescent constant population, 50mln chain length, store every -1. Save the output file.

5) Launch BEAST using the beauti output file. When BEAST ends, open the tracer software to check convergence of the chain and sufficient effective sampling size (ESS; ESS > 200).

6) Open Logcombiner and select subsample at 1000 to obtain between 1000 and 10000 trees.

7) Launch TreeAnnotator with 25% burn-in to infer the maximum clade credibility tree, with the highest product of individual clade posterior probabilities.
treeannotator -b 25 logcombiner.trees tree.beast.nex.

8) Use FigureTree to open the newick and nexus files containing the trees generated from RAxML and BEAST, respectively. Look at the file *RAxML_bipartitions*. You can open it in FigTree - it will ask you what the labels represent. They represent bootstrap proportions. In the view menu on the left in FigTree, set the node labels to display bootstrap proportions. Select the outgroup (*C. hillianum*) and set the root of the tree. Tip and node labels can be simply modified. FigTree can export trees in nexus, newick, and graphic formats, including pdf, png, etc. (http://tree.bio.ed.ac.uk/software/figtree/).

INPUT: its.ali, tef1.ali, act.ali
OUTPUT: RaxML_bipartitions.combined.new, tree.beast.nex

Species Delimitation

In this section, we will provide the basic information to perform one sequence-based and three tree-based methods of species delimitation, which are frequently used in studies on fungal diversity (Parnmen et al. 2012, Haelewaters et al. 2018, Bustamante et al. 2019, Zimowska et al. 2021): the automatic barcode gap discovery (ABGD), the general mixed Yule-coalescent (GMYC) model and the Poisson Tree Processes (PTP).

ABGD

ABGD is a sequence-based method that sorts the sequences into hypothetical species based on the barcode gap, which can be observed whenever intra-specific is smaller than inter-specific divergence (Puillandre et al. 2012).

The input file for this analysis is the alignment of the nucleotide sequences you want to analyze in fasta format, making this method the fastest in splitting a sequence alignment dataset into candidate species, because it does not require a phylogenetic tree. However, its output should be interpreted by complementation with other methods (Puillandre et al. 2012).

Tutorial 4 – ABGD

Before performing the analysis with the program, you have to set some priors: the nucleotide substitution model for pairwise nucleotide distance estimation; a value that is a proxy for minimum width of the gap to look for; the range of nucleotide distances within which the gap is supposed to be found; the number of steps to perform while looking for the gap within the range of distances.
To perform this step:

1) Navigate to ABGD web server (https://bioinfo.mnhn.fr/abi/public/abgd/abgdweb.html).

2) Choose your aligned fasta file as input.

3) Set the model criteria as follows: variability (P) between 0.001 (Pmin) and 0.1 (Pmax), gap width ($\times$) of 0.1, Kimura (K80) parameters and 50 screening steps (Figure 4A).

4) Click on the partition corresponding to prior intraspecific divergence (P) = 0.0115 (Figure 4B) and save the hypothetical species list (Figure 4C).

INPUT: combined.ali
OUTPUT: hypothetical species list

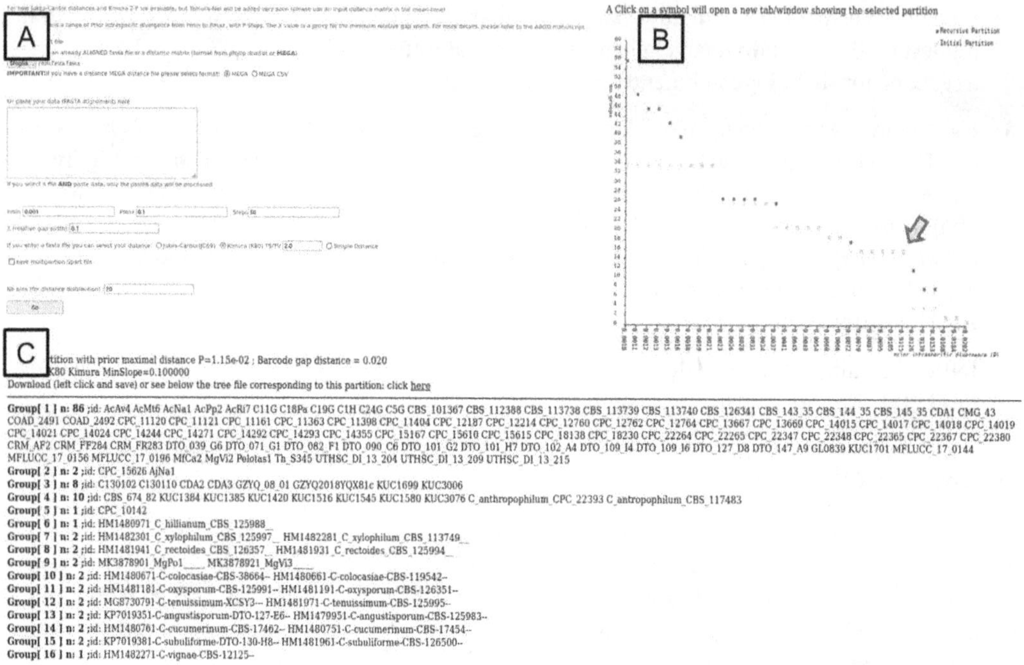

Figure 4. ABGD is a sequence-based species delimitation method which is accessible via web interface (A). By clicking on the partition corresponding to a certain prior intraspecific divergence (B), one can save the hypothetical species list (C).

The program produces as output a histogram of the distribution of the nucleotide distances between the sequences of the dataset you are analyzing; the same result is represented also as ranked ordered values, and the same result is represented also as ranked ordered values, which allow the identification of the barcode gap. This latter is located in the vicinity of a sudden slope increase of the curve (Figure 4B). Moreover, the program gives as output the list of the detected partitions, each one associated with the prior intraspecific divergence value (Figure 4C). In the partitions, the groups (hypothetical species) in which your sequences can be split are listed. The best partition (the best delimitation for the analyzed data) has to be chosen in accordance with the delimitation obtained with other molecular delimitation methods or different approaches.

A tutorial for the command line version of the program is available at https://naturalis.github.io/mebioda/doc/week1/w1d3/lecture3.html.

ASAP is a new tool published by the same authors as ABGD, eight years after ABGD (Puillandre et al. 2021). This is a single locus species delimitation tool, based on the same idea of ABGD and was developed to be more user-friendly than the latter, improving the choice of priors and the optimal delimitation, which in ABGD are left to the user. ASAP can be run either through a graphical web-interface (https://bioinfo.mnhn.fr/abi/public/asap/) or downloaded and used locally (same url).

As ABGD, the input file for this analysis is the alignment of the nucleotide sequences of interest, in fasta format. The only prior you have to set for performing ASAP analyses is the substitution model for computing the pairwise nucleotide distances. The main output of this program is the list of the partitions detected, including the number of delimited groups (hypothetical species) and the groups composition (the sequences composing each group) (Puillandre et al. 2021).

PTP

PTP is a species delimitation tool that looks for the point of transition between intra- and interspecific processes assuming a two-parameter model that accounts for speciation and coalescent process based on the Poisson distribution of branch lengths (Zhang et al. 2013). After the first publication of PTP, some implemented versions of the tool were published: the Bayesian version of PTP, namely bPTP, and the multirate PTP, mPTP. bPTP is an updated version of PTP, that includes Bayesian support (BS) values to delimited species on the input tree. mPTP, the further update of PTP, was created for resolving the shortcomings of PTP (Kapli et al. 2017). With mPTP, the branching events of each delimited species are fit to a distinct exponential distribution in order to suit different sampling depth and population characteristics. Both PTP and bPTP can be run on the web server http:// mPTP.h-its.org and https://species.h-its.org/, respectively. Alternatively, they can be used through dedicated python scripts (available at https://github.com/zhangjiajie/PTP). PTP method is useful when speed is a priority because it does not require the use of ultrametric trees (e.g., generated by BEAST), which can be time-consuming to generate. However, the PTP method assumes that branching events (divergences in the evolutionary lineages) are related to the number of substitutions, rather than time, which may not be accurate if substitution rates are not constant across the tree (Tang et al. 2014).

Tutorial 5 – PTP

For PTP/bPTP, we will use as input the tree produced with RAxML, as described above, and default settings.

To perform this step:

1) Navigate to PTP/bPTP web server (https://species.h-its.org/).

2) Choose your *RAxML_bestTree* file as input (the server will only accept Newick format or NEXUS format with no annotations on the tree, thus *RAxML_bipartitions* and *RAxML_bootstrap* don't work) (Figure 5A).

3) Set the root by typing the correct name displayed in the tree file (e.g., HM1480971_C_ hillianum_CBS_125988__).

4) Enter your e-mail and click on submit.

5) Check the output figure "Maximum likelihood solution" (PTP) and download the delimitation results (Figure 5B). In this tutorial, we won't download "Highest Bayesian supported solution" (bPTP) data, because this method is oversplitting our dataset, identifying 93 different species.

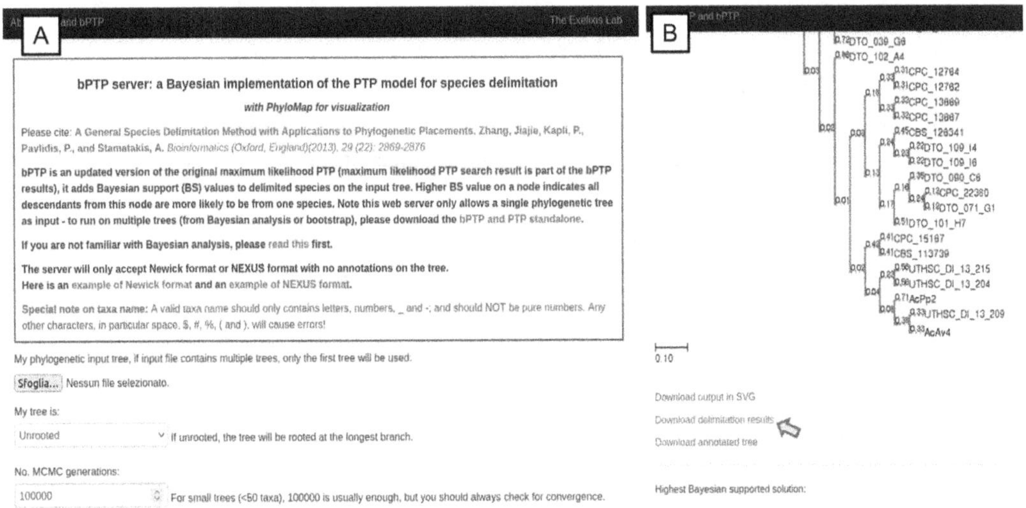

Figure 5. PTP is a tree-based species delimitation method which is accessible via web server (A), allowing the download of delimitation results in SVG format and as a list (B).

INPUT: RAxML_bestTree (Newick format or NEXUS format with no annotations on the tree)
OUTPUT: hypothetical species list

GMYC

The GMYC model uses maximum likelihood and an ultrametric gene tree to model the transition between inter- and intraspecific branching patterns (Pons et al. 2006). Indeed, this method is based on the prediction that independent evolution leads to the appearance of distinct genetic clusters, separated by longer internal branches in a gene tree (Fujisawa and Barraclough 2013). Likewise, PTP tries to determine the transition point from a between- to a within-species process using a two-parameter model, one for the speciation and one for the coalescent process (Zhang et al. 2013). Using the GMYC method with an ultrametric BEAST tree can provide accurate estimates of diversity, but the process of obtaining a BEAST tree is computationally expensive. However, if there is a high level of rate heterogeneity (differences in the rates at which evolutionary changes occur) and this can be accounted for in the tree estimation process, then more sophisticated dating and diversity estimation methods might be beneficial.

GMYC analysis can be performed both using the dedicated R package (named *splits*) or the on-line web server at the following link https://species.h-its.org/gmyc , which currently doesn't work (last access: 27 December 2022). The input file for the analysis is an ultrametric fully dichotomous tree. It can be a single marker tree, or a tree inferred from concatenated alignments. If the tree is inferred using BEAST, it is already ultrametric, whereas when other tools such as RAxML or IQtree are used, the inferred tree should be made ultrametric. For obtaining an ultrametric tree from a non ultrametric one, different tools can be used (one of them is the *ape* R library, but in *gmyc* R function help they suggest also other methods). For a more detailed tutorial, please refer to Tomochika Fujisawa's site (https://tmfujis.wordpress.com/2013/04/23/how-to-run-gmyc/).

Tutorial 6 – GMYC

To perform the GMYC delimitation method, an ultrametric tree was constructed in BEAST 2, as described above. After removing 25% of the trees as burn-in, the remaining trees were used to generate a single summarized tree in TreeAnnotator v.2.0.2 (part of the BEAST v.2.0.2 package) as an input file for GMYC analyses. To perform this step in R:

1) Launch R at the terminal and set the working directory (e.g., where the ultrametric tree is saved) /usr/bin/R.
   setwd("~/Desktop/cladosporioides/")

2) Install and load the packages
   install.packages(c("MASS", "ape", "paran", "rncl"))
   install.packages("splits", repos = "http://R-Forge.R-project.org")
   library(splits)

3) Load the ultrametric tree in nexus format (e.g., the tree obtained by BEAST)
   tree <- read.nexus("tree.beast.nex")

   If the tree is in newick format, use
   tree2 <- read.tree("tree.newick")

4) Launch gmyc and summarize the result, which includes the number of delimited species and the significance of the GMYC model
   result <- gmyc(tree)
   summary(result)

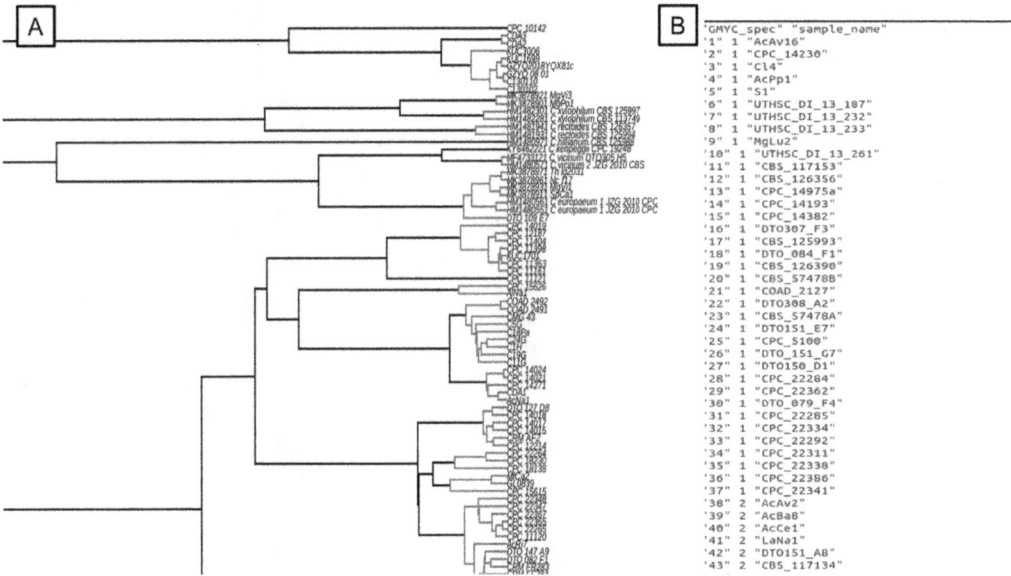

Figure 6. Delimited species are indicated by gray branches on the gene tree (A). A detailed association of samples to species (B) can be generated using the *spec.list* command in R (Tutorial 6).

5) By using the *plot* function, you can plot a lineage through time plot, a likelihood surface and a tree with delimited species plot (result)
The red vertical line indicates the threshold time between inter – intraspecific branching. The red branches on the gene tree are delimited species (Figure 6A). For a better plot function, refer to https://tmfujis.wordpress.com/2016/01/26/a-better-plot-function-for-gmyc-result.

6) If you want detailed associations of samples to species (Figure 6B), use *spec.list*.
spec.list (result)

This function returns a dataframe showing the sample-species associations. You may save this result of delimitation by calling the *write.table* function and use it for other analyses.
spec_list <- spec.list(result) ; df<- data.frame(spec_list) ; write.table(df, file = "speclist.tab")
INPUT: tree.beast.nex (NEXUS format ultrametric tree)
OUTPUT: hypothetical species list

Interpretation of Results and Evaluation of Methods

For an easier result comparison, hypothetical species recognized by each of the 3 tested methods can be reported on a bar graph near the ultrametric tree (Figure 7), as observed in several studies (Parnmen et al. 2012, Haelewaters et al. 2018, Zimowska et al. 2021, Becchimanzi et al. 2021) making species determination difficult and resulting in different classifications accepting between one to eight species. Multi-locus DNA sequence data provide an avenue to test species delimitation scenarios using genealogical and coalescent methods, employing gene and species trees. Here we tested species delimitation in the complex using molecular data of four loci (nuITS and IGS rDNA, protein-coding GAPDH and Mcm-7. Concordance among our results (Figure 7) suggests that several strains of *C. cladosporioides* (group G) represent a new putative species that requires morphological characterization prior to formal taxonomic changes.

A certain discordance among methods is reported for groups A to D. For these groups, the most conservative method (ABGD) indicate one species, while GMYC and PTP indicate seven species. These results are in line with many studies reporting the tendency of ABGD to collapse multiple taxa into one (Song et al. 2018, Huang et al. 2020), as well as the tendency of GMYC

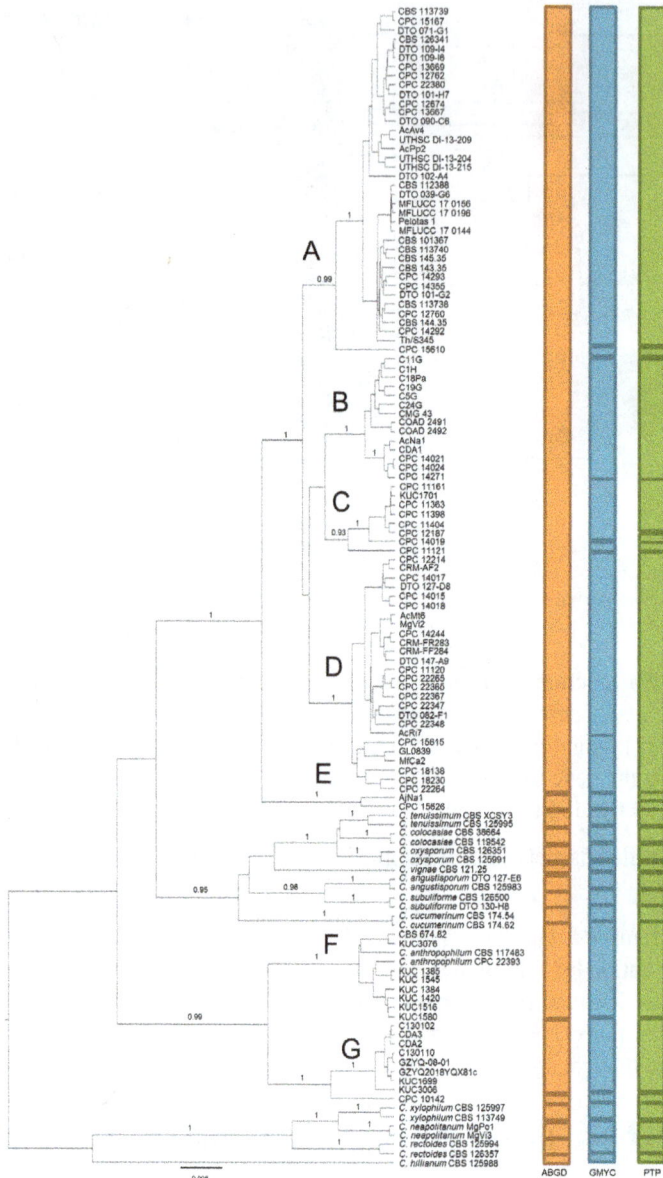

Figure 7. Ultrametric tree phylogeny of *C. cladosporioides* showing the results of the sequence-based species delimitation methods. The tree is the result of a Bayesian analysis performed in BEAST on the concatenated ITS, tef1, act dataset. For each node, posterior probabilities (if >0.90) are presented above the branch leading to that node. Results of species delimitation analyses are represented by colored boxes to the right. Main groups identified by phylogenetic reconstruction are indicated by letters A, B, C, D, E, F and G. Different colors indicate the different methods used. (Source: adapted from (Becchimanzi et al. 2021) under the terms of Creative Commons Attribution Licence CC-BY 4.0 via *MDPI Pathogens*).

and PTP to split (Talavera et al. 2013, Kapli et al. 2017). Such discordance can be managed using a conservative approach, to minimize the risk of oversplitting (i.e., the inclusion of false positives) and, thus, delimiting entities that do not represent actual evolutionary lineages (Jörger et al. 2012, Carstens et al. 2013, Weigand et al. 2013). This promising strategy represents a precious tool for elucidating diversity in directly collected specimens, as well as in public repositories of molecular data, such as GenBank. Public repositories represent a fundamental resource for studying cryptic diversity using molecular data, although data they contain are not error-free and require careful examination before being used for taxonomic purposes (Becchimanzi et al. 2021).

Combining in a single analysis several criteria of species delimitation likely brings out discordance among methods, which can be efficiently treated using conservative approaches (i.e., only species recognized by all the methods used are considered *true species*).

Conclusions

Sequence-based species delimitation methods are becoming more and more popular for resolving cryptic diversity in fungi and nowadays should be part of the mycologist's toolkit for taxonomic assignation. Here we presented some of these methods from a practical perspective, providing a mini-guide that can be useful, even for researchers who are not familiar with phylogenetics and molecular tools for species characterization. Although species discovery using the DNA-based methods here described should be integrated with other approaches (e.g., morphological species concept), the use of such molecular methods for species delimitation represents a first step of key importance for delineating new taxonomic entities in the highly diverse kingdom of Fungi.

Acknowledgments

Thanks are due to Giulia Magoga, University of Milan (Italy), for a critical reading of an early draft of the manuscript.

References

Aagesen, L. 2004. The information content of an ambiguously alignable region, a case study of the trnL intron from the Rhamnaceae. Org. Divers Evol. 4(1): 35–49. https://doi.org/10.1016/j.ode.2003.11.003.

Becchimanzi, A., B. Zimowska and R. Nicoletti. 2021. Cryptic diversity in cladosporium cladosporioides resulting from sequence-based species delimitation analyses. Pathogens. 10(9): 1167. https://doi.org/10.3390/pathogens10091167.

Bobay, L-M. and H. Ochman. 2017. Biological Species Are Universal across Life's Domains. Genome Biol Evol. 9(3): 491–501. https://doi.org/10.1093/gbe/evx026.

Bouckaert, R., J. Heled, D. Kühnert, T. Vaughan, C-H. Wu, D. Xie et al. 2014. BEAST 2: A Software Platform for Bayesian Evolutionary Analysis. PLOS Comput Biol. 10(4): e1003537. https://doi.org/10.1371/journal.pcbi.1003537.

Bustamante, D.E., M. Oliva, S. Leiva, J.E. Mendoza, L. Bobadilla, G. Angulo et al. 2019. Phylogeny and species delimitations in the entomopathogenic genus Beauveria (Hypocreales, Ascomycota), including the description of *B. peruviensis* sp. nov. MycoKeys. 58: 47–68. https://doi.org/10.3897/mycokeys.58.35764.

Cai, L., T. Giraud, N. Zhang, D. Begerow, G. Cai and R.G. Shivas. 2011. The evolution of species concepts and species recognition criteria in plant pathogenic fungi. Fungal Divers. 50(1): 121. https://doi.org/10.1007/s13225-011-0127-8.

Carstens, B.C., T.A. Pelletier, N.M. Reid and J.D. Satler. 2013. How to fail at species delimitation. Mol. Ecol. 22(17): 4369–4383. https://doi.org/10.1111/mec.12413.

Castresana, J. 2000. Selection of Conserved Blocks from Multiple Alignments for Their Use in Phylogenetic Analysis. Mol. Biol. Evol. 17(4): 540–552. https://doi.org/10.1093/oxfordjournals.molbev.a026334.

Chesters, D. and A.P. Vogler. 2013. Resolving Ambiguity of Species Limits and Concatenation in Multilocus Sequence Data for the Construction of Phylogenetic Supermatrices. Syst Biol. 62(3): 456–466.

Cubero, O.F., A. Crespo, J. Fatehi and P.D. Bridge. 1999. DNA extraction and PCR amplification method suitable for fresh, herbarium-stored, lichenized, and other fungi. Pl Syst Evol. 216(3): 243–249. https://doi.org/10.1007/BF01084401.

Dettman, J.R., D.J. Jacobson and J.W. Taylor. 2003. A Multilocus Genealogical Approach to Phylogenetic Species Recognition in the Model Eukaryote Neurospora. Evolution. 57(12): 2703–2720. https://doi.org/10.1111/j.0014-3820.2003.tb01514.x.

Dissanayake, A. 2020. Applied aspects of methods to infer phylogenetic relationships amongst fungi. Mycosphere. 11(1): 2652–2676. https://doi.org/10.5943/mycosphere/11/1/18.

Doyle, J. 1991. DNA Protocols for Plants. In: Hewitt GM, Johnston AWB, Young JPW, editors. Molecular Techniques in Taxonomy [Internet]. Berlin, Heidelberg: Springer; [accessed 2022 Dec 14]; p. 283–293. https://doi.org/10.1007/978-3-642-83962-7_18.

Edgar, R.C. 2004. MUSCLE: multiple sequence alignment with high accuracy and high throughput. Nucleic Acids Res. 32(5): 1792–1797. https://doi.org/10.1093/nar/gkh340.

Ence, D.D. and B.C. Carstens. 2011. SpedeSTEM: a rapid and accurate method for species delimitation. Mol. Ecol. Resour. 11(3): 473–480. https://doi.org/10.1111/j.1755-0998.2010.02947.x.

Felsenstein, J. 1981. Evolutionary trees from DNA sequences: A maximum likelihood approach. J Mol Evol. 17(6): 368–376. https://doi.org/10.1007/BF01734359.

Fujisawa, T. and T.G. Barraclough. 2013. Delimiting Species Using Single-Locus Data and the Generalized Mixed Yule Coalescent Approach: A Revised Method and Evaluation on Simulated Data Sets. Syst Biol. 62(5): 707–724. https://doi.org/10.1093/sysbio/syt033.

Haelewaters, D., A. De Kesel and D.H. Pfister. 2018. Integrative taxonomy reveals hidden species within a common fungal parasite of ladybirds. Sci Rep. 8(1): 15966. https://doi.org/10.1038/s41598-018-34319-5.

Hall, T.A. 1999. BioEdit: a user-friendly biological sequence alignment editor and analysis program for Windows 95/98/NT. In: Nucleic acids symposium series. Vol. 41. [place unknown]: [London]: Information Retrieval Ltd., c1979-c2000.; p. 95–98.

Hebert, P.D.N., A. Cywinska, S.L. Ball and J.R. deWaard. 2003. Biological identifications through DNA barcodes. Proc Biol Sci. 270(1512): 313–321. https://doi.org/10.1098/rspb.2002.2218.

Hillis, D.M., D.D. Pollock, J.A. McGuire and D.J. Zwickl. 2003. Is Sparse Taxon Sampling a Problem for Phylogenetic Inference? Systematic Biology. 52(1): 124–126. https://doi.org/10.1080/10635150390132911.

Huang, W., X. Xie, L. Huo, X. Liang, X. Wang and X. Chen. 2020. An integrative DNA barcoding framework of ladybird beetles (Coleoptera: Coccinellidae). Sci Rep. 10(1): 10063. https://doi.org/10.1038/s41598-020-66874-1.

Huang, X., N. Duan, H. Xu, T.N. Xie, Y.-R. Xue and C.-H. Liu. 2018. CTAB-PEG DNA Extraction from Fungi with High Contents of Polysaccharides. Mol. Biol. 52(4): 621–628. https://doi.org/10.1134/S0026893318040088.

Huelsenbeck, J.P. and F. Ronquist. 2001. MRBAYES: Bayesian inference of phylogenetic trees. Bioinformatics. 17(8): 754–755. https://doi.org/10.1093/bioinformatics/17.8.754.

Jones, G. 2017. Algorithmic improvements to species delimitation and phylogeny estimation under the multispecies coalescent. J. Math Biol. 74(1): 447–467. https://doi.org/10.1007/s00285-016-1034-0.

Jones, G., Z. Aydin and B. Oxelman. 2015. DISSECT: an assignment-free Bayesian discovery method for species delimitation under the multispecies coalescent. Bioinformatics. 31(7): 991–998. https://doi.org/10.1093/bioinformatics/btu770.

Jörger K.M., J.L. Norenburg, N.G Wilson and M. Schrödl. 2012. Barcoding against a paradox? Combined molecular species delineations reveal multiple cryptic lineages in elusive meiofaunal sea slugs. BMC Evol Biol. 12: 245. https://doi.org/10.1186/1471-2148-12-245.

Kapli, P., S. Lutteropp, J. Zhang, K. Kobert, P. Pavlidis, A. Stamatakis et al. 2017. Multi-rate Poisson tree processes for single-locus species delimitation under maximum likelihood and Markov chain Monte Carlo. Bioinformatics. 33(11): 1630–1638. https://doi.org/10.1093/bioinformatics/btx025.

Katoh, K., K. Misawa, K. Kuma and T. Miyata. 2002. MAFFT: a novel method for rapid multiple sequence alignment based on fast Fourier transform. Nucleic Acids Res. 30(14): 3059–3066. https://doi.org/10.1093/nar/gkf436.

Kekkonen, M. and P.D.N. Hebert. 2014. DNA barcode-based delineation of putative species: efficient start for taxonomic workflows. Mol. Ecol. Resour. 14(4): 706–715. https://doi.org/10.1111/1755-0998.12233.

Kubatko, L.S., B.C. Carstens and L.L. Knowles. 2009. STEM: species tree estimation using maximum likelihood for gene trees under coalescence. Bioinformatics. 25(7): 971–973. https://doi.org/10.1093/bioinformatics/btp079.

Kumar, S., G. Stecher, M. Li, C. Knyaz and K. Tamura. 2018. MEGA X: Molecular Evolutionary Genetics Analysis across Computing Platforms. Mol. Biol. Evol. 35(6): 1547–1549. https://doi.org/10.1093/molbev/msy096.

Kumar S., G. Stecher and K. Tamura. 2016. MEGA7: Molecular Evolutionary Genetics Analysis Version 7.0 for Bigger Datasets. Mol. Biol. Evol. 33(7): 1870–1874. https://doi.org/10.1093/molbev/msw054.

Kupczok, A., H.A. Schmidt and A. von Haeseler. 2010. Accuracy of phylogeny reconstruction methods combining overlapping gene data sets. Algorithm Mol Biol. 5(1): 37. https://doi.org/10.1186/1748-7188-5-37.

Larsson, A. 2014. AliView: a fast and lightweight alignment viewer and editor for large datasets. Bioinformatics. 30(22): 3276–3278. https://doi.org/10.1093/bioinformatics/btu531.

Lee, M.S.Y. 2001. Unalignable sequences and molecular evolution. Trends Ecol. Evol. 16(12): 681–685. https://doi.org/10.1016/S0169-5347(01)02313-8.

Löytynoja, A. and M.C. Milinkovitch. 2001. SOAP, cleaning multiple alignments from unstable blocks. Bioinformatics. 17(6): 573–574. https://doi.org/10.1093/bioinformatics/17.6.573.

Lyons-Weiler, J., G.A. Hoelzer and R.J. Tausch. 1998. Optimal outgroup analysis. Biol. J. Linn. Soc. 64(4): 493–511. https://doi.org/10.1111/j.1095-8312.1998.tb00346.x.

Maharachchikumbura, S.S.N., Y. Chen, H.A. Ariyawansa, K.D. Hyde, D. Haelewaters, R.H. Perera et al. 2021. Integrative approaches for species delimitation in Ascomycota. Fungal Divers. https://doi.org/10.1007/s13225-021-00486-6.

Matheny, P.B., Z. Wang, M. Binder, J.M. Curtis, Y.W. Lim, R.H. Nilsson et al. 2007. Contributions of rpb2 and tef1 to the phylogeny of mushrooms and allies (Basidiomycota, Fungi). Mol. Phylogenet. Evol. 43(2): 430–451. https://doi.org/10.1016/j.ympev.2006.08.024.

Matute, D.R. and V.E. Sepúlveda. 2019. Fungal species boundaries in the genomics era. Fungal Genet Biol. 131: 103249. https://doi.org/10.1016/j.fgb.2019.103249.

Mora, C., D.P. Tittensor, S. Adl, A.G.B. Simpson and B. Worm. 2011. How Many Species Are There on Earth and in the Ocean? PLoS Biol. 9(8): e1001127. https://doi.org/10.1371/journal.pbio.1001127.

Notredame, C., D.G. Higgins and J. Heringa. 2000. T-Coffee: A novel method for fast and accurate multiple sequence alignment. J. Mol. Biol. 302(1): 205–217. https://doi.org/10.1006/jmbi.2000.4042.

Parnmen, S., A. Rangsiruji, P. Mongkolsuk, K. Boonpragob, A. Nutakki and H.T. Lumbsch. 2012. Using phylogenetic and coalescent methods to understand the species diversity in the *Cladia aggregata* Complex (Ascomycota, Lecanorales). PLoS ONE 7(12): e52245. https://doi.org/10.1371/journal.pone.0052245.

Pons, J., T.G. Barraclough, J. Gomez-Zurita, A. Cardoso, D.P. Duran, S. Hazell et al. 2006. Sequence-Based Species Delimitation for the DNA Taxonomy of Undescribed Insects. Syst Biol. 55(4): 595–609. https://doi.org/10.1080/10635150600852011.

Puillandre, N., S. Brouillet and G. Achaz. 2021. ASAP: assemble species by automatic partitioning. Mol Ecol Resour. 21(2): 609–620. https://doi.org/10.1111/1755-0998.13281.

Puillandre, N., A. Lambert, S. Brouillet and G. Achaz. 2012. ABGD, Automatic Barcode Gap Discovery for primary species delimitation. Mol. Ecol. 21(8): 1864–1877. https://doi.org/10.1111/j.1365-294X.2011.05239.x.

Rajan, V. 2013. A Method of Alignment Masking for Refining the Phylogenetic Signal of Multiple Sequence Alignments. Mol Biol Evol. 30(3): 689–712. https://doi.org/10.1093/molbev/mss264.

Schoch, C.L., K.A. Seifert, S. Huhndorf, V. Robert, J.L. Spouge, C.A. Levesque et al. 2012. Nuclear ribosomal internal transcribed spacer (ITS) region as a universal DNA barcode marker for Fungi. P. Natl. Acad Sci. USA. 109(16): 6241–6246. https://doi.org/10.1073/pnas.1117018109.

Song, C., X-L. Lin, Q. Wang and X-H. Wang. 2018. DNA barcodes successfully delimit morphospecies in a superdiverse insect genus. Zool. Scr. 47(3): 311–324. https://doi.org/10.1111/zsc.12284.

Stamatakis, A. 2006. RAxML-VI-HPC: maximum likelihood-based phylogenetic analyses with thousands of taxa and mixed models. Bioinformatics. 22(21): 2688–2690. https://doi.org/10.1093/bioinformatics/btl446.

Stengel, A., K.M. Stanke, A.C. Quattrone and J.R. Herr. 2022. Improving Taxonomic Delimitation of Fungal Species in the Age of Genomics and Phenomics. Front Microbiol. 13. https://www.frontiersin.org/articles/10.3389/fmicb.2022.847067.

Talavera, G. and J. Castresana. 2007. Improvement of Phylogenies after Removing Divergent and Ambiguously Aligned Blocks from Protein Sequence Alignments. Syst. Biol. 56(4): 564–577.

Talavera, G., V. Dincă and R. Vila 2013. Factors affecting species delimitations with the GMYC model: insights from a butterfly survey. Methods Ecol. Evol. 4(12): 1101–1110. https://doi.org/10.1111/2041-210X.12107.

Tang, C.Q., A.M. Humphreys, D. Fontaneto and T.G. Barraclough. 2014. Effects of phylogenetic reconstruction method on the robustness of species delimitation using single-locus data. Methods Ecol. Evol. 5(10): 1086–1094. https://doi.org/10.1111/2041-210X.12246.

Taylor, J.W., D.J. Jacobson, S. Kroken, T. Kasuga, D.M. Geiser, D.S. Hibbett et al. 2000. Phylogenetic Species Recognition and Species Concepts in Fungi. Fungal Genet Biol. 31(1): 21–32. https://doi.org/10.1006/fgbi.2000.1228.

Thomson, R.C. and H.B. Shaffer. 2010. Sparse supermatrices for phylogenetic inference: taxonomy, alignment, rogue taxa, and the phylogeny of living turtles. Syst Biol. 59(1): 42–58. https://doi.org/10.1093/sysbio/syp075.

Vaidya, G., D.J. Lohman and R. Meier. 2011. SequenceMatrix: concatenation software for the fast assembly of multi-gene datasets with character set and codon information. Cladistics. 27(2): 171–180. https://doi.org/10.1111/j.1096-0031.2010.00329.x.

Vences, M., S. Patmanidis, V. Kharchev, S.S. Renner. 2022. Concatenator, a user-friendly program to concatenate DNA sequences, implementing graphical user interfaces for MAFFT and FastTree. Bioinform Adv. 2(1): vbac050. https://doi.org/10.1093/bioadv/vbac050.

Waterhouse, A.M., J.B. Procter, D.M.A. Martin, M. Clamp, G.J. Barton. 2009. Jalview Version 2—a multiple sequence alignment editor and analysis workbench. Bioinformatics. 25(9): 1189–1191. https://doi.org/10.1093/bioinformatics/btp033.

Weigand, A.M., A. Jochum, R. Slapnik, J. Schnitzler, E. Zarza and A. Klussmann-Kolb. 2013. Evolution of microgastropods (Ellobioidea, Carychiidae): integrating taxonomic, phylogenetic and evolutionary hypotheses. BMC Evol. Biol. 13:18. https://doi.org/10.1186/1471-2148-13-18.

Wilberg, E.W. 2015. What's in an Outgroup? The Impact of Outgroup Choice on the Phylogenetic Position of Thalattosuchia (Crocodylomorpha) and the Origin of Crocodyliformes. Syst. Biol. 64(4): 621–637. https://doi.org/10.1093/sysbio/syv020.

Xu, J. 2020. Fungal species concepts in the genomics era. Genome. 63(9): 459–468. https://doi.org/10.1139/gen-2020-0022

Yang, Z. 2015. The BPP program for species tree estimation and species delimitation. Curr. Zool. 61(5): 854–865. https://doi.org/10.1093/czoolo/61.5.854.

Zhang, J., P. Kapli, P. Pavlidis and A. Stamatakis. 2013. A general species delimitation method with applications to phylogenetic placements. Bioinformatics. 29(22): 2869–2876. https://doi.org/10.1093/bioinformatics/btt499.

Zimowska, B., A. Becchimanzi, E.D. Krol, A. Furmanczyk, K. Bensch and R. Nicoletti 2021. New *Cladosporium* species from normal and galled flowers of Lamiaceae. Pathogens. 10(3): 369. https://doi.org/10.3390/pathogens10030369.

CHAPTER 15

Pathogenesis and Virulence of *Phakopsora pachyrhizi*
An Insight into the Genetic and Molecular Features

Md. Motaher Hossain

Introduction

Soybean (*Glycine max*) is among the most important crops. It is widely utilized for human, animal, and industrial purposes. Soybean is cultivated in substantial quantities in numerous regions in the Americas, Asia, Europe, and Africa. Since 1964, the global production and demand for soybeans have increased annually (FAO Stat 2020). To satisfy the growing demand for soybeans, production must increase. However, a number of serious foliar diseases, such as Asian soybean rust, pose a substantial threat to enhanced soybean yield. Asian rust of soybean is caused by the obligate Basidiomycete fungus *Phakopsora pachyrhizi*. This is one of the most destructive plant pathogens of the twentieth century. The disease is common in all soybean-growing areas in Asia, Africa, and the Americas, significantly threatening food security worldwide (Hossain et al. 2022). The fungus can infect soybean plants at any age and impede their growth and development. Even a small amount of infection (0.05%) can upset yields, and if it is not appropriately controlled, there have been reports of yield losses of up to 80% (Scherm et al. 2009, Yorinori et al. 2005). The disease is gradually requiring more fungicides, which is raising production costs. Chemical treatments were first used in Brazil to fight the disease during the growing season of 2002/2003 (Yorinori et al. 2005). During the next growing season, fungicides were sprayed on about 20 million hectares of soybeans to cease the spread of this disease (Yorinori et al. 2005, Melo Reis et al. 2015). It is estimated that managing *P. pachyrhizi* costs more than $2 billion per season in Brazil alone (Melo Reis et al. 2015).

The pathogen has a high adaptive capacity and quickly evolves into new variants. Many of the newly emerged races or pathotypes are highly virulent and can defeat host resistance (Paul et al. 2013). Moreover, the fungus is becoming more resistant to the main types of site-specific fungicides, which makes chemical control less effective (Godoy et al. 2016, Barro et al. 2021). Another unique characteristic of this obligate biotrophic pathogen is that it can live on various hosts. As of today, it can infect 153 different species of legumes from 54 genera (Bonde et al. 2008, Slaminko et al. 2008). A broad host range is crucial from an epidemiological perspective because it enables the pathogen to survive in environments and seasons where soybeans are not produced.

The interaction of *P. pachyrhizi* with soybean plants elicits a series of cellular and molecular events in both the pathogen and the host plants (Schneider et al. 2011). This aggressive mode of infection prevents the breeding efforts for durable soybean rust resistance from being successful. The destructive capability of *P. pachyrhizi*, combined with its widespread prevalence, has

Department of Plant Pathology, Bangabandhu Sheikh Mujibur Rahman Agricultural University, Gazipur 1706.
Email: hossainmm@bsmrau.edu.bd

sparked interest in understanding the molecular interactions of *P. pachyrhizi* with its host plants (Loehrer et al. 2014). Acquiring insight into the infection courses of *P. pachyrhizi* on soybean plants is crucial to explore novel approaches to counter the pathogen attacks. Over the past decades, significant progress has been made in understanding the molecular mechanisms of pathogenesis and virulence of *P. pachyrhizi* in soybean. This review aims to summarize current knowledge of the infection mechanisms of *P. pachyrhizi* in soybean plants at the genomic and molecular levels.

Causal Organisms of Soybean Rust

The first identification of the causative agent of soybean rust was made on yam beans (*Pachyrhizus ahipa*) in 1902 in Japan (Hennings 1903). However, the first name *P. pachyrhizi* Syd. & P. Syd. was given to an isolate obtained from the leguminous host crop *Pachyrhizus erosus* (L.) Urb. (= *Pachyrhizus angulatus*) in Taiwan (Sydow and Sydow 1914). On the contrary, *Phakopsora* isolates from Puerto Rico were shown to be significantly less potent and to produce fewer urediniospores on soybean than isolates from the Eastern Hemisphere (Vakili and Bromfield 1976). At that time, although it was thought that *P. pachyrhizi* was the cause of the soybean rust disease in Puerto Rico, isozyme analysis indicates the existence of two different species of *Phakopsora* causing rust on soybeans: one in Asia and the Eastern Hemisphere and the other one in the "New World" (South and Central America and the Caribbean) (Bonde et al. 1988). The links between rust pathogens on legumes were further established by thorough research on the morphology of Phakopsora species, which made it clear that at least two species are responsible for soybean rust (Ono et al. 1992). In the "New World," where *P. meibomiae* (Arthur) Arthur has been identified as the pathogenic agent of non-soybean legumes since 1899, *P. pachyrhizi* was shown to be the causal agent of soybean rust (Ono et al. 1992). These species have recently undergone DNA-level differentiation. According to an examination of the internal transcribed spacer (ITS) region of the ribosomal DNA, *P. pachyrhizi* and *P. meibomiae* have 80% identical nucleotide sequences (Frederick et al. 2002).

Small brown or brick-red patches on leaves are the earliest symptoms of soybean rust caused by *P. pachyrhizi* (Figure 1A). *P. meibomiae* can also cause symptoms similar to those of *P. pachyrhizi*. Although seedlings are susceptible to infection, these spots often appear in the lower canopy either

Figure 1. Symptoms and signs of *Phakopsora pachyrhizi* on soybean leaves. (A) Numerous chlorotic lesions develop on the upper surface of the leaves. (B) Severely infected leaves. (C) Uredinia capable of producing urediniospores are prevalent on the underside of the leaf. (D) Urediniospores are produced in clusters in the uredinia. (E) Uredinia are cone-shaped with an opening at the top. (F) *P. pachyrhizi* urediniospores form a long germ tube after germination.

during or after flowering. Typically, lesions first appear at the leaflet base, adjacent to the petiole and the leaf veins. Because this area of the leaflet holds dew for an unusually long time, making it presumably an ideal environment for infection. Although the average size of the lesions remains unchanged, their prevalence increases as the disease progresses (Figure 1B). These lesions, which are most prevalent on the underside of the leaf, develop into pustules (Figure 1C) called uredinia and can produce numerous urediniospores. The raised pustules, especially when sporulating, are visible to the naked eye and do not require a microscope. Masses of urediniospores can completely cover lesions when pustules mature (Figure 1D). When studied under a microscope, uredinia appear as cone-shaped structures with an open top (Figure 1E). Soybean rust disease is distinctive from other rusts because the urediniospores of soybean rust range in color from a pale yellow-brown to colorless and are ornamented with echinulate (short) spines, while those of most others are a reddish brown (rust-colored). When *P. pachyrhizi* urediniospores germinate, they form a germ tube that terminates into an appressorium, which the fungus uses for direct or indirect host penetration (Figure 1F).

Spread of *Phakopsora pachyrhizi* Throughout the World

P. pachyrhizi was initially reported in Japan in 1903. Within a few decades, it had been found across Asia and even as far south as Australia. In later years, its distribution grew further, affecting the entire soybean-growing region. A study claims that the disease initially appeared in Africa in the 1970s (Javaid and Ashraf 1978). Between 1996 and 2003, it rapidly spread to Zimbabwe, Uganda, Kenya, Rwanda, Zambia, Nigeria, Mozambique, South Africa, and Cameroon (Jarvie 2009). However, Ghana, the Democratic Republic of the Congo, Tanzania, and Malawi have all recorded soybean rust cases in recent years (Murithi et al. 2015). Airborne urediniospores are thought to have travelled from western India to the East African coast on the wet northeast monsoon winds (Levy 2005). However, South America remained rust-free until the 2000/2001 campaign. From 2001 to 2004, the disease began to spread and became established in many South American countries (Yorinori et al. 2005). A study using ribosomal DNA nucleotide sequences found that Brazilian rust was more closely related to African samples than those from Asia (Freire et al. 2008). The Asian soybean rust probably made its way to South America through natural means, with the support of air currents that carried spores across the Atlantic. In the fall of 2004, the United States saw its first case of soybean rust. Some *P. pachyrhizi* isolates have been found in Asia, Australia, and the United States, but not in Africa or South America (Zhang et al. 2012), suggesting a different migration route was taken by the United States than the one in South America. Some *P. pachyrhizi* isolates were probably brought into the United States from Asia or Australia via the Hawaiian Islands.

Polycyclic Disease Cycle

Phakopsora pachyrhizi is a highly destructive polycyclic pathogen (Scherm et al. 2009). The fungus life cycle is quite simple, producing only urediniospores and teliospores. However, there is a lack of solid evidence that *P. pachyrhizi* is often capable of sexual reproduction and producing teliospores. Although teliospores have been seen in old lesions in Asian countries on various hosts, including soybean, their germination in nature has never been recorded (Bromfield 1984). The uredinial stage accounts for most of its life cycle (Anderson et al. 2008). After contact with a leaf surface, urediniospores germinate to begin the rapidly progressing infection process. The development of lesions on the leaf surface starts with the formation of reproductive uredinia. Within five to eight days, new urediniospores are formed within uredinia (Zambolin 2006, Morales et al. 2012). Each uredinium has the potential to produce hundreds of spores. In general, uredinia develop more prominently on the lower leaf surface than on the upper leaf surface. Once the spores have been expelled from the uredinia through the ostiole, the wind can carry them to other hosts, starting a fresh cycle of infection. Urediniospores can re-infect the soybean plants whenever they land on a wet susceptible host surface. Since *P. pachyrhizi* is an obligate biotrophic organism, it cannot subsist independently of its hosts or on debris. The fungus has a wide host range and can sporulate on 31 species in 17 genera (Ono et al. 1992).

Virulence Diversity of *Phakopsora pachyrhizi*

Understanding the virulence diversity of local *P. pachyrhizi* populations is essential for successfully incorporating resistant genes (*Rpp* genes) into soybean-breeding programs. It has been observed that *P. pachyrhizi* isolates collected from different countries exhibited varied levels of virulence when tested on differential hosts (Table 1). The first demonstration of the differential virulence of *P. pachyrhizi* was conducted in Taiwan, where six pathotypes were described (Lin 1966). In Japan, where the disease has been prevalent in the past century, researchers isolated 45 samples of rust from soybeans, *Pueraria lobata*, and *Glycine soja* and identified 18 different races of rust (Yamaoka et al. 2002). Another study found six races among 26 single lesion isolates from Tsukuba, Japan (Yamaoka et al. 2014). The same races were found on both soybean and kudzu, and two or more races existed on the same soybean cultivar. Comparing Japanese and Brazilian populations of P. *pachyrhizi* revealed that Japanese populations were less virulent than their Brazilian counterparts (Yamanaka et al. 2010). According to a comparison of pathogenicity in isolates reflecting various geographic and temporal origins, older *P. pachyrhizi* isolates from Asia and Australia in the 1970s were less virulent than more recent ones from Africa and South America in 2001 (Bonde et al. 2006). In India, 25 isolates from Northern Karnataka during Kharif 2010–2012 were classified into one of three pathotypes based on their responses to 13 differentials (Devaraj et al. 2016). In Bangladesh, 13 isolates sampled from three different soybean growing regions in 2016 were differentiated into eight pathotypes when inoculated onto 12 soybean differentials (Hossain and Yamanaka 2019). In a recent study, 34 isolates collected from five soybean-growing regions of Bangladesh were separated into 21 pathotypes when tested on the same differentials (Hossain et al. 2022). According to these findings, the Bangladeshi soybean rust population was more virulent and diverse in 2018-2019 than in 2016.

A few pieces of research have been done in Africa to determine the virulence variability of *P. pachyrhizi*. Twizeyimana et al. (2009) grouped 116 *P. pachyrhizi* isolates from Nigeria, a nation with comparatively little experience with the pandemic, into seven pathotypes that showed significant variations in the pathotype composition of *P. pachyrhizi*. However, only three were found in Uganda (Tukamuhabwa and Maphosa 2012). A total of 17 isolates, including those from Malawi, Kenya, Tanzania, and South Africa, were found to belong to four distinct pathotypes. The virulence diversity of *P. pachyrhizi* isolates from eight countries, including those 17 isolates from Africa, was evaluated using 11 soybean differentials. The 25 isolates were divided into ten distinct pathotypes, with the South African isolate being the most virulent of the *P. pachyrhizi* strains and those from Kenya, Malawi, and several of the Tanzanian isolates having the lowest virulence (Murithi et al. 2017). Another study identified 12 pathotypes among 65 isolates obtained from four East African nations over several years (Murithi et al. 2021). In this study, 54% of the isolates belonged to pathotype 1000, widespread across all countries. All pathotypes were highly virulent on soybean genotypes, expressing the *P. pachyrhizi* resistance gene *Rpp1*; however, they were avirulent on cultivars with *Rpp1b, Rpp2,* or *Rpp3* genes, as well as on cultivar No6-12-1, which carries the *Rpp2, Rpp4,* and *Rpp5* genes. Furthermore, the virulence spectrum of the isolates gathered from various nations over time varied.

The highly pathogenic fungus *P. pachyrhizi* is one of the most significant economic risks to South American soybean production. Several extensive studies using differential soybean cultivars harbouring various known *Rpp* genes documented high virulence and virulence diversity of soybean rust pathogen populations in South America, with no noticeable pattern in regional or yearly characteristics. Three pathotypes and six aggression groups were reported in a collection of American isolates (Twizeyimana and Hartman 2012). Over three growing seasons (2007/2008, 2008/2009,

Table 1. Virulence diversity of *Phakopsora pachyrhizi* populations collected from different geographical locations and time periods.

Country of origin	Number of isolates	Line used	Races or Pathotypes	References
Taiwan	9	11 legume accessions, 6 accessions of soybean, and 5 *Phaseolus* spp.	6	Lin (1966)
Australia	Not known	'Wills' and PI 200492	2	McLean and Byth (1980)
Australia, India, and Taiwan	4	PI 200492 (*Rpp1*), PI 230970 (*Rpp2*) and PI 462312 (*Rpp3*)	4	Bromfield et al. (1980)
Taiwan	50	PI 200492 (*Rpp1*), PI 462312 (*Rpp3*),	3	Yeh (1983)
Australia	8	257 accessions of four *Glycine* spp.: *Glycine canescens* (60), *G. clandestine* (63), *G. tabacina* (100), and *G. tomentella* (47)	6	Burdon and Speer (1984)
China	7	PI 200492 (*Rpp1*), PI 462312 (*Rpp3*), PI 459025 (*Rpp4*), and 5 other accessions	4	Tan and Sun (1989)
Taiwan	42	AVRDC differential lines: PI 200492 (*Rpp1*), PI 230970 (*Rpp2*), PI 462312 (*Rpp3*), PI 230971, PI 239871A, PI 239871B, PI 459024 and PI 459025, TK-5, TN-4, and Wayne	9	Hartman et al. (2011)
Japan	45	AVRDC differential lines	18	Yamaoka et al. (2002)
International	12	PI 200492 (*Rpp1*), PI 230970 (*Rpp2*), PI 462312 (*Rpp3*), and PI 459025B (*Rpp4*)	6	Bonde et al. (2006)
United States	6	PI 200492 (Rpp1), PI 594538A (*Rpp1b*), PI 462312 (*Rpp3*), 459025B (*Rpp4*), and 23 other accessions	2	Paul and Hartman (2009)
International	10	PI 200492 (*Rpp1*), PI 230970 (*Rpp2*), PI 462312 (*Rpp3*), PI 459025B (*Rpp4*), and 16 others	8	Pham et al. (2009)
Nigeria	116	PI 200492 (*Rpp1*), PI 230970 (*Rpp2*), PI 462312 (*Rpp3*), PI 459025B (*Rpp4*), PI 594538A, UG-5, TGx 1485-1D and TGx 1844-4F	7	Twizeyimana et al. (2009)
International	8	PI 200492 (*Rpp1*), PI 594538A (*Rpp1b*), PI 587866, and PI 587880A	3	
Vietnam	1 composite field population	PI 200492 (*Rpp1*), PI 594538A (*Rpp1b*), PI 462312 (*Rpp3*), 459025B (*Rpp4*), and 85 other accessions	7	Pham et al. (2010)
United States	Field populations	PI 200492 (*Rpp1*), PI 230970 (*Rpp2*), PI 462312 (*Rpp3*), PI 459025B (*Rpp4*), and over 500 other accessions	2	Walker et al. (2011)
United States	72	PI 200492 (*Rpp1*), PI 230970 (*Rpp2*), PI 462312 (*Rpp3*), PI 459025B (*Rpp4*), PI 200526 (*Rpp5*), and three others	3	Twizeyimana and Hartman (2012)
India	25	PI 459024B, PI 459025F, EC-241778, EC-241780, EC-391160, PI 230970, PI 200492, JS-335, PI 230971, PI 459025B, EC-462312, TK-5, and Wayne	3	Devaraj et al. (2016)

Table 1 contd. ...

...Table 1 contd.

Country of origin	Number of isolates	Line used	Races or Pathotypes	References
Japan	26	PI 200492 (*Rpp1*), Tainung#4, PI 230970 (*Rpp2*), PI 417125 (*Rpp2*), PI 462312 (*Rpp3*), PI 459025 (*Rpp4*), PI 200526 (*Rpp5*), PI 416764 (*Rpp3*), PI 587880A (*Rpp1*), PI 587886 (*Rpp1*), PI 587905 (*Rpp1*), TK-5, and Wayne	6	Yamaoka et al. (2014)
Bangladesh	13	PI 200492 (*Rpp1*), PI 587886 (*Rpp1*), PI 230970 (*Rpp2*), PI 462312 (*Rpp3*), PI 416764 (*Rpp3*), PI 459025 (*Rpp4*), PI 200526 (*Rpp5*), PI 567102B (*Rpp6*), PI 587880A (*Rpp1-b*), PI 594767A (*Rpp1-b*), BRS154, and No6-12-1 (*Rpp2+Rpp4+Rpp5*)	8	Hossain and Yamanaka (2019)
Bangladesh	34	PI 200492 (*Rpp1*), PI 587886 (*Rpp1*), PI 230970 (*Rpp2*), PI 462312 (*Rpp3*), PI 416764 (*Rpp3*), PI 459025 (*Rpp4*), PI 200526 (*Rpp5*), PI 567102B (*Rpp6*), PI 587880A (*Rpp1-b*), PI 594767A (*Rpp1-b*), BRS154, and No6-12-1 (*Rpp2+Rpp4+Rpp5*)	21	Hossain et al. (2022)
Argentina, Brazil, and Paraguay	59	PI 200492 (*Rpp1*), PI 368039 (*Rpp1*), PI 230970 (*Rpp2*), PI 417125 (*Rpp2*), PI 462312 (*Rpp3*) PI 459025 (*Rpp4*), PI 200526	37	Akamatsu et al. (2013)
East African Countries	65	PI 200492 (*Rpp1*), PI 587855 (*Rpp1b*), PI 230970 (*Rpp2*), PI 462312 (*Rpp3*), PI 459025B (*Rpp4*), PI 200526 (*Rpp5*), PI 567102B (*Rpp6*), Hyuuga (*Rpp3* and *Rpp5*), UG 5 (*Rpp1* and *Rpp3*), and No6-12-1 (*Rpp2*, *Rpp4*, and *Rpp5*).	12	Murithi et al. (2021)
East African Countries	25	LD0916057 (*Rpp1*), LD103005 (*Rpp* lb), PI 417125 (*Rpp2*), PI 462312 (*Rpp3*), PI 459025B (*Rpp4*), PI 200526 (*Rpp5a*), LD1014274 (*Rpp5b*), PI 567102B (Rpp6), and Hyuuga (*Rpp3* and *Rpp5*)	10	Murithi et al. (2017)

and 2009/2010), the pathogenicity of 59 rust populations from Argentina, Brazil, and Paraguay was assessed using 16 soybean differentials (Akamatsu et al. 2013). Pathogenicity profiles were identical only in two pairs of *P. pachyrhizi* populations, demonstrating high pathogenic heterogeneity among the rust populations in these countries. Pathogenic differences were found within South American *P. pachyrhizi* populations and between South American and Japanese *P. pachyrhizi* populations during a comparative analysis of 59 South American and five Japanese samples. A further study comprising 83 isolates from Argentina, Brazil, and Paraguay evaluated 16 host differentials between 2010 and 2015 and found that no two virulence profiles were the same (Akamatsu et al. 2017, Stewart et al. 2019). This indicates that *P. pachyrhizi* isolates from these countries exhibit significant virulence diversity. The Mexican rust populations (MRPs) sampled in 2015 were found to be highly pathogenic and demonstrated wide virulence variation (Garca-Rodrguez et al. 2017). A recent study examined the spatial variance in the virulence of 19 MRPs obtained in two states of Mexico (Tamaulipas and Chiapas). It revealed slight variation in the pathogenicity of MRPs from the two states toward differentials harboring *Rpp4* or *Rpp5* genes, while *Rpp1*, *Rpp1*-b, *Rpp2*, and *Rpp3* indicated a divergent phenotype (Garca-Rodrguez et al. 2022). Such knowledge of the virulence variation of *P. pachyrhizi* is vital for identifying relevant genetic sources to minimize disease and generate rust-resistant soybean cultivars for particular locales.

Genetic Diversity of *Phakopsora pachyrhizi*

Numerous reports show that R genes are ineffective when infected with fungal isolates from different continents or isolates lacking the corresponding virulence gene (Yamaoka et al. 2002, 2014, Akamatsu et al. 2013, 2017), indicating the presence of genetic variability in fungus populations. After introducing rust-resistant cultivars, the diversity of *P. pachyrhizi* populations has frequently led to the selection of particularly virulent pathotypes and the breakdown of resistant cultivars. Therefore, a better understanding of the pathogen evolution and rapid adaptation necessitates knowledge of the pathogenicity and variability of the *P. pachyrhizi* species. Such knowledge can be helpful for genetic breeding programs as well as other areas of research. The application of molecular tools is successfully used in studying the genetic variability of *P. pachyrhizi*. The transcribed internal spacer regions (ITS) have been widely used to indicate the variability at species or population levels. This is because the fungus evolves rapidly and, as a result, exhibits high intraspecific variation in its nucleotide sequences and length. Additionally, amplification through PCR (Polymerase Chain Reaction) has been commonly used (Martin 2007). AFLPs, also known as amplified fragment length polymorphisms, is a type of molecular marker that has firmly established itself as a reliable method for determining the extent of intraspecific genetic diversity (Meudt and Clarke 2007).

An investigation into the molecular genetic variations present in *P. pachyrhizi* isolates from Asia was carried out using a sequence-based methodology that involved two gene sections of the nuclear genome: ITS and the ADP-ribosylation factor (ARF) region (Zhang et al. 2012). The phylogeny based on ITS and ARF sequences showed at least six and two separate clads, respectively. The vast majority of clads comprise isolates of mixed origin, indicating that the genetic variation at the national level (country) is just as diverse as at the regional level (Asia). Rocha et al. (2015) utilized the AFLP technique to amplify *P. pachyrhizi* DNA extracted from naturally rust-infected plants in 23 production fields. The results revealed that 77% of markers were polymorphic, indicating a high pathogen diversity. The molecular variance analysis revealed greater genetic diversity within countries than between them. A temporal analysis revealed that within-year genetic variation was greater than between years. Darben et al. (2020) evaluated the genetic variability among *P. pachyrhizi* mono-uredinial isolates obtained from field samples collected in Brazil between 2007 and 2012 by comparing the sequences of ITS regions and AFLP markers. The polymorphisms among the AFLP markers (in the local population) and the ITS sequences (in the global population) demonstrated the high genetic diversity of this pathogen. The results obtained in this work show that analysis of the sequences of ITS regions and AFLP markers effectively identifies intraspecific variation among mono-uredinial isolates of *P. pachyrhizi*.

Various molecular studies have established that *P. pachyrhizi* possesses high genetic diversity and weak genetic structure across large geographic areas. Such a phenomenon is not only detected in the populations of Asia but also Africa and South America. *P. pachyrhizi* displayed approximately 90% of its genetic and pathogenicity variation inside individual soybean fields in Nigeria, with only approximately 6% of its variability spread across multiple fields (Twizeyimana et al. 2011). According to Lárzábal et al. (2022), the virulence variation occurring within a *P. pachyrhizi* in Uruguay is more remarkable than between fields. The majority (14 out of 16) of the ribotypes found in other parts of the world were discovered in samples from three different experimental stations in Brazil (Jorge et al. 2015). Compared to the levels of genetic variation observed at greater regional scales (Argentina, 'Americas+Africa,' and 'Asia+Australia'), the levels of genetic variation observed at the local scale (three experimental stations in Brazil) did not diminish. The absence of a definitive genetic structure and differentiation across a large geographic scale indicates the availability of efficient mechanisms for long-distance dispersal and significant gene flow between soybean fields.

Genetics of Resistance to *Phakopsora pachyrhizi*

It is common practice to evaluate soybeans for susceptibility or resistance to *P. pachyrhizi* based on the phenotypes of host reactions to the pathogen. Three phenotypes are usually observed against infection by *P. pachyrhizi*; susceptible, resistant, and immune (Figure 2). Tan-colored "TAN" lesions characterize the susceptible response with abundant sporulation on the abaxial sides of infected leaves (Figure 2A); reddish-brown-colored "RB" lesions typify the resistant reaction with few or no sporulation (Figure 2B); and a lack of macroscopically visible lesion distinguishes the immune response from other two phenotypes (Figure 2C). Molecular mapping analyses have identified seven loci (*Rpp1* to *Rpp7*) that provide varying degrees of resistance to *P. pachyrhizi*. Based on the soybean reference genome, all the known *Rpp* loci, except for *Rpp5* and *Rpp6*, contain clusters of genes that code for NLR (nucleotide-binding domain, leucine-rich) proteins (*Rpp1/Rpp1b*–NLR, Rpp2-TIR-NLR, *Rpp3*–TIR-NLR, *Rpp4*–CC-NLR, and *Rpp7*–NLR) (Whitham et al. 2016, Childs et al. 2018). On the other hand, only *Rpp1* and *Rpp4* have been demonstrated beyond a reasonable doubt to be encoded by a component of the linked NLR cluster (Meyer et al. 2009). Through a combination of DNA sequencing, gene expression analysis, and virus inducing gene silencing, Meyer et al. (2009) identified the first candidate gene driving rust resistance in soybean from the *P. pachyrhizi*-resistant accession PI 459025B (*Rpp4*). At the *Rpp4* locus in the susceptible cultivar Williams 82, three genes belonging to the *R* gene family known as CC-NBS-LRR (coiled-coil, nucleotide binding site, leucine-rich repeat) were detected. Further investigation revealed that these *R* genes shared similarities with the *R* gene family in lettuce, which is known as *Rgc2* (Resistant Gene Candidate 2). However, these genes had distinct alleles in the resistant accession PI 459025B, and one of those alleles (*Rpp4C4*) was primarily expressed in resistant plants both before and after they were challenged with the *P. pachyrhizi* strain LA04-1, which generated RB lesions exclusively. The loss of resistance caused by silencing the CC-NBS-LRR R gene provides evidence that this gene is the most likely candidate for the causative role in the *Rpp4* gene cluster. Pedley et al. (2018) functionally defined the *Rpp1* locus-mediated rust resistance. The investigators concluded that three genes are similar to NBS-LRR; each candidate gene is anticipated to encode the N-terminal ubiquitin-like protease 1 (ULP1) domain. The co-silencing of these potential genes rendered the *Rpp1*-resistant soybean accession PI 200492 incapable of defending itself, evidencing that *Rpp1* is a ULP1-NBS-LRR protein that plays an integral part in the resistant response.

Although a single *Rpp* gene confers strong resistance, breeders and pathologists typically do not recommend deploying these genes to manage plant diseases (Mundt 2018). These single genes are race-specific and frequently have a short lifespan. It is desirable and feasible to develop soybean cultivars carrying broad-spectrum resistance to soybean rust, given the high pathogenic variability and the emergence of new pathotypes in field populations of *P. pachyrhizi*. However, despite the well-known short-lived effectiveness of host resistance to pathotype-specific genes, deploying a single major gene may make sense for soybean-growing regions where the environment is less

Figure 2. Various disease phenotypes produced by *Phakopsora pachyrhizi* on soybean leaves. (A) Lesions characterize a susceptible phenotype with abundant production of uredinia and urediniospores. (B) A resistant response is observed as lesions with few or no formations of uredinia and urediniospores. (C) An imperfect appearance of lesions without uredinia and urediniospores specifies an immune reaction.

conducive to disease. The best example from the past is the soybean cultivars carrying the *Rpp1* and *Rpp3* genes, which were still effective in Brazil against rust for several years after being released in the field until a highly virulent Brazilian *P. pachyrhizi* population emerged in the state of Mato Grosso in 2003 (Yorinori 2008). However, deploying several genes into a single genotype may confer effective and (non)race-specific resistance, leading to a gene pyramid for long-lasting resistance (Yamanaka et al. 2013). Notably, more than one *Rpp* gene has also been found in several soybean accessions. Hyuuga (PI 506764) possesses two resistance genes, one at the *Rpp3* locus and the other at the *Rpp5* locus (Kendrick et al. 2011). Like the previous example, soybean genotypes EC 241780 and DT 2000 (PI 635999) carry *Rpp1b* and *Rpp2*, and *Rpp3* and *Rpp4*, respectively (Bhor et al. 2015, Vuong et al. 2016). In soybean, it has been demonstrated that combining *Rpp2*, *Rpp4*, and *Rpp5* into a single genotype increases rust resistance (Lemos et al. 2011, Yamanaka et al. 2013). Similarly, paired gene pyramiding of *Rpp2*, *Rpp3*, and *Rpp4* decreased sporulation and the severity of rust (Maphosa et al. 2012). Yamanaka et al. (2015) also found that the *Rpp* pyramided lines had significantly higher resistance to soybean rust than the original resistance sources carrying single genes, PI 230970 (*Rpp2*), Hyuuga (*Rpp3*) (*Rpp5*). While none of the differentials containing a single *Rpp* gene demonstrated resistance to all four *P. pachyrhizi* populations from Mexico and Bangladesh, it is interesting to note that a number of pyramided lines were resistant to all of them (García-Rodríguez et al. 2017, Yamanaka and Hossain 2019). These data demonstrate the potential of pyramided lines bearing *Rpp* gene combinations to avoid the breakdown of single-gene resistance and provide broad-spectrum resistance against soybean rust.

Common Molecular Markers for Soybean Rust Resistance

The threat of Asian soybean rust could be prevented with the help of resistant genes. An essential feature of a resistant genotype is the identification and utilization of fungus-inhibiting R genes. Molecular or genetic markers linked with resistant traits are commonly used to identify gene locations in the plant genome and select plants with resistant characteristics. Marker-assisted selection (MAS) using DNA markers has become an essential tool in a breeding program that drastically shortens the time to identify genotypes with desirable traits. Many polymorphic DNA markers have been used to map the *Rpp1* to *Rpp7* genes (Table 2). Consequently, these molecular markers can be used

Table 2. Genetic/molecular markers reported to be linked with soybean resistance against *Phakopsora pachyrhizi*.

Rust resistant gene	Soybean cultivar	Linkage group/ Chromosome	Flanking markers used in the linkage map	References
Rpp1Rpp1-b	PI 200492 PI 594767A PI 587886 PI 587880A Xiao Jin Huang Himeshirazu	LG-G/18	Sct_187 Sat_064 Sat_117 Sat_372 Satt191 SSR66 Sct_199	Hyten et al. (2007) Ray et al. (2009) Hossain et al. (2015) Yamanka et al. (2015)
Rpp2	PI 230970	LG-C2/16	Sat_255 Satt620 Satt529	Monterous et al. (2007) Yamanaka et al. (2015)
Rpp3	PI 462312	LG-J/16	Satt134 Satt460 Sat_263 Sat_251 Sat_238 Satt307	Hyten et al. (2007) Hossain et al. (2015)
Rpp4	PI 459025B PI 459025	LG-C4/18 LG-G/18	Satt288 Satt503 Satt612 Satt288 Satt517 Sat_143 Sct_199 Satt472 Satt191 Sc21–3360	Silva et al. (2008) Garcia et al. (2008) Mayer et al. (2009)
Rpp5	PI 200456 PI 200487 (Kinoshita) PI 471904 PI 200526 (Shiranui) Hyuuga	LG-N/3	Sat_275 Sat_280 Satt080 Satt125 Satt485 Satt393	Garcia et al. (2008) Lemos et al. (2011) Kendrick et al. (2011)
Rpp6	PI 567102B	LG-G/18	Satt324 Satt394	Li et al. (2012)
Rpp7	PI 605823	LG-L/19	GSM0547 GSM0548	Childs et al. (2018)

in breeding program to select breeding lines expressing *Rpp* genes-mediated rust resistance. For instance, SSR molecular marker BARC-Sct_187 on linkage group G has been used in investigations on three different soybean lines (Hyten et al. 2007). The marker has revealed a genomic area where the L85–2378 line shares an allele with the PI 200492 line and is polymorphic with the Williams 82 cultivar. Sct_187 was found to be closely linked with *Rpp1*, as evidenced by their highly significant independent assortment (Hyten et al. 2007). The location of the *Rpp1* locus has been identified as a ~1cM interval between Sct_187 and Sat_064 on chromosome 18 (LG-G) (Hyten et al. 2007). Genotype Tainung 4 demonstrated resistance due to several resistance gene loci (Hyten et al. 2007). With variation estimates of 0.46 and 0.84, respectively, the adjacent markers Sct_187 and Sat_064 are strongly linked with the *Rpp1* locus in Tainung 4, indicating high polymorphism of the SSR markers in a wide range of crossings (Cregan et al. 1999). Likewise, the *Rpp1* gene was mapped using different mapping populations between SSR markers Satt191 and Sat_064 in PI 594767A, PI 587886, PI 587880A and Xiao Jin Huang, between Sat_064 and SSR66 in PI 587905, and between Sct_187 and Sat_064 in Himeshirazu (Ray et al. 2009, Hossain et al. 2015, Yamanka et al. 2015). The *Rpp-b*, an allele of *Rpp1*, was mapped between the marker Sat_064 and Sat_372 in PI 594538A (Chakraborty et al. 2009). In PI 200492 and Himeshirazu, Sct_187 is the nearest marker to *Rpp1*, while in PI 594767A, PI 587880A, and PI 587905, Sat_064 is the closest marker to *Rpp1*, and in PI 594538A, Sat_372 is the nearest marker linked to *Rpp1*. The other markers found in the intermediate region include Sat117 and Sct_199. On the other hand, an inversion at Sat_372 and Sat_064 was reported to be one of the ways that the *Rpp1* linkage map diverges from the soybean consensus map of Song et al. (2004). Yamanaka et al. (2015) argued that the *Rpp1* of PI 594538A, PI 594767A, and PI 587905 mapped between two SSR markers Sat_064 and Sat_372 loci are, in fact, *Rpp1-b* loci (Chakraborty et al. 2009, Hossain et al. 2014). According to them, the soybean genotypes PI 200492, PI 368039, PI 587886, Xiao Jing Huang, and Himeshirazu possibly carry the original *Rpp1*, as reported by Hartwing and Bromfield (1983). The two alleles also show differences in resistance to South American ASR pathogens (Hossain et al. 2015). These findings suggest overlap and variation between *Rpp1* linkage maps derived from different genotypes.

The rust resistance in the cultivar FT2 is associated with SSR molecular markers located in the linkage group (LG) C2 (Cregan et al. 1999). Mapping a resistance gene, *Rpp2,* in the cultivar Hyuuga identified the gene location on LG-C2 at a ~3cM gap between Satt134 and Satt460 (Monterous et al. 2007). The *Rpp2* gene in a Japanese cultivar, Iyodaizu B, was identified as a QTL, and the LOD peaks of the QTLs for rust-resistant traits were found at the position between the markers Satt620 and Sat_366 (Yamanaka et al. 2015). However, the gene was identified at 8.4 cM-intervals between Sat_255 and Satt620 in PI 230970 (Silva et al. 2008). A similar ~3cM interval on LG-J between Satt134 and Satt460 has been found during the mapping of the *Rpp3* (Hyten et al. 2007). The resistant gene *Rpp3,* in soybean cultivar PI 416764 was identified as QTL using markers such as Sat_263, Sat_251, Sat_238 and Satt307 (Hossain et al. 2015). At the same time, within 1.9 cM (Silva et al. 2008) and 2.8 cM (Garcia et al. 2008) of SSR marker Satt288, the *Rpp4* locus was mapped on chromosome 18 on linkage group G. *Rpp4* of PI 459025 was mapped to 4.29 cM interval between the SSR markers Sc21-3360 and Satt288 on chromosome 18 (Meyer et al. 2009). Satt288 was the closest marker linked to *Rpp4*. Rpp5 in PI 200487 (Kinoshita) is located approximately 4 cM between the markers Sat_275 and Sat_280 on chromosome 3 (Lemos et al. 2011). The identical position of *Rpp5* was reported in PI 200456, PI 471904, and PI 200526 (Shiranui), as reported by Garcia et al. (2008), and *Rpp5* in Hyuuga, as reported by Kendrick et al. (2011). The rust resistance locus on PI 567102B was reported as Rpp6 and located between the markers Satt324 and Satt394 on chromosome 18 (Li et al. 2012). The newly discovered *Rpp7* was mapped to a 154-kb interval on chromosome 19 between GSM0547 and GSM0548 in PI 605823 (Childs et al. 2018).

Soybean rust resistance in a popular soybean variety was improved by molecular techniques (Khanh et al. 2013). Using molecular markers, the *Rpp5* gene of soybean rust resistance was successfully incorporated into HL203, a top Vietnamese soybean variety, from two contributor lines. These lines were DT2000 and Stuart 99084B-28. Markers Sat_275 and Sat_280 were found to be associated with the *Rpp5* locus in this investigation. Microsatellite DNA markers are also employed in gene pyramiding to resist soybean rust (Yamanaka et al. 2015, Viganó et al. 2018, Yamanaka and Hossain 2019). These show that DNA markers are widely seen as effective candidates for use in crop improvement.

Biphasic Infection Process of *Phakopsora pachyrhizi*

In the past ten years, researchers have focused on the infection strategies utilized by this biotrophic organism. Urediniospores are the primary and secondary infecting agents for *P. pachyrhizi*. In general, urediniospores start adhering to soybean leaves as soon as 30 minutes after landing on the host surface (Vélez-Climent and Dufault 2007). After three hours, urediniospores germinate, producing an extended germ tube that develops into a terminal appressorium. Inside the appressorium, a cone-like structure known as an appressorial cone or a penetration peg forms (Chang et al. 2014). Hyaline appressoria are found in *P. pachyrhizi*, and melanin does not appear necessary to form turgor pressure in *P. pachyrhizi* (Chang et al. 2014). After 12 hours, they penetrate the leaves (Vélez-Climent and Dufault 2007). Although most rusts enter the host tissues through the stomatal hole, *P. pachyrhizi* can penetrate the host cells directly. High osmolytes building up inside the appressorium leads to an internal turgor pressure of up to 5 or 6 MPa (Chang et al. 2014). The infection peg penetrates the host epidermis due to a combination of factors, including the mechanical force created by increased turgor pressure and the enzymatic destruction of the plant cell wall. Infection with *P. pachyrhizi* produces a specialized structure known as a haustorium and intercellular colonization for exploiting the host resources. As a biotrophic pathogen, these events occur without cell death. According to Micali et al. (2011), the primary hypha generated from the penetration peg invades the mesophyll tissues and develops into a haustorium that remains separated from the cytoplasm by the plasma membrane of the host. Although the fundamental function of haustorium is to take nutrients from the host cells, it is also responsible for creating and transporting effectors (Garnica et al. 2014). As the intracellular parasite growth continues to advance and haustorium forms within the newly acquired host cells, the development of lesions starts with the formation of reproductive pustules known as uredinia. It takes several days to finish all the infection cycle stages. It is now established that *P. pachyrhizi* infection of soybean plants causes a biphasic response, with a surge of host responses occurring within the first 12 hours (van de Mortel et al. 2007). The initial phase is followed by a quiescent period characterized by weak host responses, which last 24 to 48 hours after inoculation. During the interval time, the fungus grows undetected and unchallenged. The second phase occurs with intense host reactions. The distinct biphasic response implies that each stage of fungal development is associated with unique plant responses.

P. pachyrhizi Transcriptomes During Pathogenesis

The majority of the molecular knowledge regarding *P. pachyrhizi* has been assembled via research involving transcriptome analysis. Most of these studies have concentrated on characterizing particular molecular mechanisms active during infection. More specifically, they have paid close attention to the analyses of specific structures of the pathogen, such as the urediniospores and the haustorium, as well as the infected leaf as a whole (Posada-Buitrago and Frederick 2005, Tremblay et al. 2009, 2012, 2013, Stone et al. 2012, Link et al. 2014, Carvalho et al. 2016). As a result, candidate genes differentially expressed in soybean during infection in compatible and incompatible interactions have been identified. Using whole genome Affymetrix microarrays, the transcriptome of *P. pachyrhizi*-exposed young soybean plants (V2 development stage) was profiled 72 h after inoculation (Panthee et al. 2007). *P. pachyrhizi* infection led to the differential expression of 112 genes, 46 of which were upregulated and 66 of which were downregulated. The majority of differentially expressed genes were connected to general defense and stress. These results suggest that a modest and non-specific innate response to the pathogen may account for the failure of the tested soybean variety to develop rust resistance. Over a 7-day *P. pachyrhizi* infection time course, abundance variations of soybean mRNAs were observed in soybean accession (PI230970) harboring the *Rpp2* resistance gene and a susceptible genotype (Embrapa-48) (van de Mortel et al. 2007). The expression patterns of differentially expressed genes demonstrated a biphasic response to soybean rust in each genotype. In the first 12 hours after inoculation, both genotypes exhibited distinct gene expression alterations corresponding to fungal germination and epidermal cell penetration. By 24 hours, mRNA expression of these genes had largely reverted to levels observed in mock-inoculated plants. In the susceptible genotype, rust infection had little effect on gene expression until 96 hours after inoculation, when fast fungal growth occurred. In contrast, gene expression in the resistant genotype diverged from that of the mock-inoculated control before 72 hours, indicating that *Rpp2*-mediated defenses were activated earlier (before this period). These findings further show that soybean rust initially elicits a non-specific response that is either temporary or inhibited in both soybean genotypes once early colonization processes are accomplished (van de Mortel et al. 2007). Another study indicated that the initial surge in gene expression is associated with appressorium development and epidermal cell penetration, while the second surge follows the commencement of haustoria formation in both compatible and incompatible interactions (Schneider et al. 2011). The proliferation of haustoria coincides with the suppression of *P. pachyrhizi* growth in incompatible interactions or the onset of accelerated growth in compatible interactions. Unexpectedly, a gene expression study in *P. pachyrhizi*-infected leaves of resistant (PI459025B, *Rpp4*) and susceptible (Williams 82) cultivars across a 12-day time course identified two biphasic responses (Morales et al. 2013). In the incompatible reaction, genes induced at 12 h after infection were not differentially expressed at 24 hours after infection but were induced at 72 hours after infection. In contrast, genes repressed 12 hours after infection were not differentially expressed from 24 to 144 hours after infection but were repressed at 216 hours after infection and later. Fourteen transcription factors common to all resistant and susceptible responses and fourteen transcription factors unique to R-gene-mediated resistance responses were identified (Morales et al. 2013). The transcriptomic study of susceptible BRS184 and *Rpp3* with soybean rust isolate T1-2 identified a total of 4518 differentially expressed genes (Hossain et al. 2018). There was an increase in the expression of 52% of the genes associated with the phenylpropanoid pathway. The RT-qPCR experiment showed that the *Rpp* lines utilized these genes in a rate-limiting manner as a defensive mechanism. All the genes displayed the highest expression 12 hours after inoculation, except for glycinol 4-dimethylallyl transferase (G4DT) and chalcone reductase (CHR). However, the gene expressions between 24 and 96 hours after inoculation set these *Rpp* lines apart from their respective soybean rust isolates. Additionally, the functional coordination of arogenate dehydratase 6 (*ADT6*) and 4-hydroxy-3-methyl but-2-enyl diphosphate synthase (*ispG*), chalcone synthase (*CHS*) and *CHR*, and *G4DT* and phytyltransferase 3 (PT3) may have a significant influence on the resistance of soybeans to soybean rust (Hossain et al. 2018). Elmore et al. (2020) conducted a transcriptome study to identify *P. pachyrhizi* candidate-secreted effector proteins. It has demonstrated that effector-like actions are present in the small subset of potential effectors predicted from these transcriptome analyses.

The analysis of the transcriptomes of *P. pachyrhizi* demonstrated that nucleic acid metabolism (primarily DNA synthesis), transcription processes, cell cycle control, and metabolic signaling pathway processes function during the germination of urediniospore (Posada-Buitrago and Frederick 2005, Tremblay et al. 2013, Link et al. 2014). In addition, it is believed that *P. pachyrhizi* does not have access to host nutrients in the early stages of infection. Because of this, glycerol, necessary for penetration, is synthesized from the nutrients obtained from lipids, glycogen, and sugar present in urediniospores. This indicates the activity of these metabolic pathways in the degradation of these compounds during germination until penetration of the host tissue (Thomas et al. 2002, Both et al. 2005). Transcripts primarily associated with processes such as the uptake of sugars and amino acids (membrane transporters and carbohydrate metabolism), as well as lipid metabolism and more active biosynthesis and transcription processes, are mainly detected during haustorium formation (Both et al. 2005, Tremblay et al. 2013, Link et al. 2014). In contrast, the synthesis of carbohydrates and lipids, in addition to protein synthesis and the metabolism of amino acids, are all engaged in the later phases of fungus development, specifically during the production of uredinia and later on during sporulation (Both et al. 2005, Tremblay et al. 2013). Nucleic acid metabolism in *P. pachyrhizi* is represented chiefly by transcripts associated with DNA replication and repair as well as RNA metabolism. Both of these processes reflect the proliferation of the fungus through the synthesis of proteins and cell division, which are primarily involved in penetration and the production of urediniospores. These processes occur while the fungus penetrates its host and produces urediniospores (Bindschedler et al. 2009). After the formation of the haustorium, there is a possibility that transcripts related to nitrogen metabolism may operate during the process of absorption of compounds from the host. A putative gene for nitrate reductase was found in the genome and transcriptome of *M. lini* during the establishment of haustorium six days after the first inoculation (Nemri et al. 2014). However, this metabolic pathway may not be functional for rust species (Spanu et al. 2010, Cantu et al. 2011). Nemri et al. (2014) identified homologues of the ammonia assimilation pathway, suggesting that most of the nitrogen received from the host is assimilated in the form of ammonia. The findings of Rincão et al. (2018) also support the conclusion that the sequence related to the nitrate reductase gene had low expression levels in both the RNA-Seq and RT-qPCR analyses.

In addition, the most abundant transcripts, which contain sequences similar to those of other fungal virulence and signaling proteins, have been identified. Rincão et al. (2018) reported that nearly half of the 58 transcripts with the highest expression levels in the transcriptome study shared similarities with virulence proteins GAS1 and GAS2 of *Magneporthe grisea*. These proteins have already been found in *the P. pachyrhizi* secretome and are virulence factors that primarily function in the early phases of the infection process during the penetration of host tissues (Xue et al. 2002, Carvalho et al. 2016). Although its function is still unknown, the DUF3129 domain has been specifically identified among sequences secreted by *P. pachyrhizi* (Stone et al. 2012, Carvalho et al. 2016, Rincão et al. 2018). It also appears to be conserved among some species of rust (Saunders et al. 2012). Despite being uncommon in fungi, the phosphotyrosine signaling factor with the SH2 domain has been described to play a crucial role in numerous cell-to-cell communication pathways, including those that control proliferation, differentiation, adhesion, hormone responses, and immune defense (Hunter 2009). Furthermore, Carvalho et al. (2016) investigated three of the most highly expressed transcripts as candidate *P. pachyrhizi* effectors. These transcripts, *de novo* 2238, *de novo* 5381, and de novo 5849, demonstrated the ability of these sequences to suppress ETI responses.

Effector Proteins for Pathogenicity or Virulence of *Phakopsora pachyrhizi*

Exocytosis, also known as cellular secretion, is a critical activity during the infection of host tissues for several reasons, the most important of which relates to the secretion of proteins known as effectors during haustorium development (Catanzariti et al. 2010). These effector proteins are responsible for causing changes in the structure and function of the host cell, which ultimately results in cellular, molecular, and physiological shifts that make it easier for the pathogen to infect the host and absorb nutrients (Voegele and Mendgen 2011). The tiny proteins are parts of the *P. pachyrhizi* arsenals that are used to overcome host defense systems. Until now, a number of candidate effectors have been identified in the *P. pachyrhizi* secretome, which suppress immune responses against various pathogens (Figure 3). The first *P. pachyrhizi* secretome was investigated based on the results of an RNA-seq study using samples from the haustoria (Link et al. 2014). They identified 156 *P. pachyrhizi* contigs with open reading frames (ORFs) encoding putative secretion. These potentially secreted proteins include homologues to haustoria-expressed secreted proteins (HESP379, HESP767, HESP-C49) from flax rust (Catanzariti et al. 2006), as well as homologues to rust-transferred proteins (RTPs) in *Uromyces species* (Pretsch et al. 2013). Later, the number of potential proteins in this secretome was increased to 851 by combining laser capture microdissection of lesions and RNA-seq (Carvalho et al. 2017). Recently, a *de novo* transcriptome

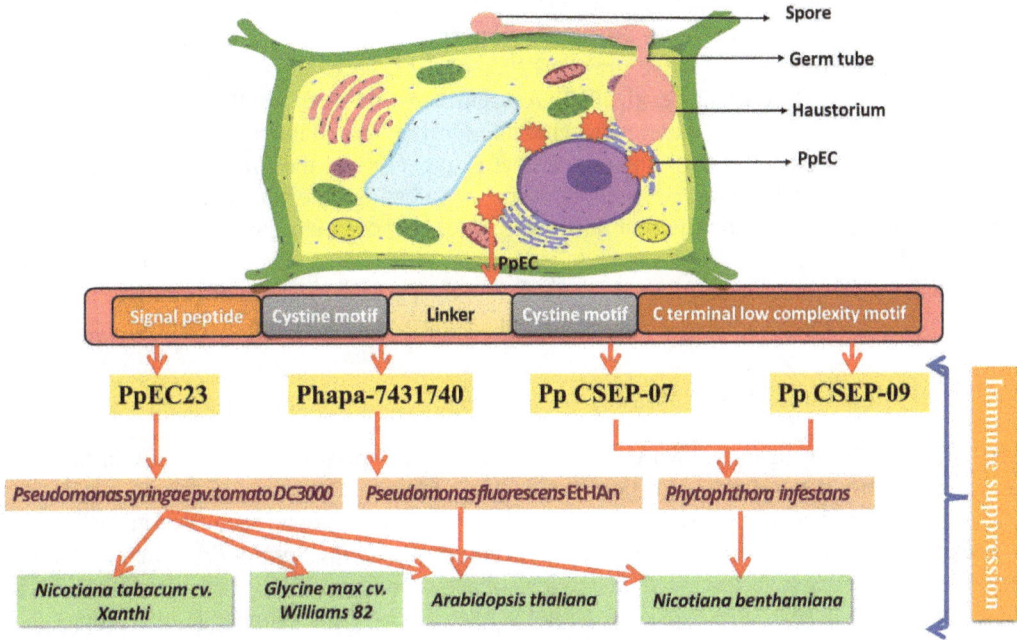

Figure 3. *Phakopsora pachyrhizi* effector candidates (PpECs) with virulence related activity in different pathosystems. Haustoria of rusts secrete effector proteins into their hosts to facilitate infection. Effector motifs depicted in the diagram are composed of signal peptide, C-terminal low complexity motif, ten-cysteine motif 1, linker and ten-cysteine motif 2. Candidate effectors Pp CSEP-07 and Pp CSEP-09 promoted the susceptibility of *Nicotiana benthamiana* to *Phytophthora infestans* significantly (Kunjeti et al. 2016). One putative effector, PpEC23, inhibited effector-triggered immunity (ETI) responses in tobacco and soybean leaves caused by *Pseudomonas syringae* pv. *tomato* DC3000 (Qi et al. 2018). Phapa-7431740 inhibits pathogen-associated molecular pattern-triggered immunity (PTI) and interferes with an antifungal PR protein glucan endo-1,3-glucosidase (*GmGLU*) to promote *P. pachyrhizi* infection of soybean.

of *P. pachyrhizi* was constructed by combining short-read (Illumina RNA-Seq) and long-read (PacBio Iso-Seq) sequencing data. This assembly was also used to identify a small subset of candidate effector proteins during infection (Elmore et al. 2020). A majority of the effectors were expressed during the later stages of infection, and many could be bioinformatically associated with haustoria. Thirty-five candidate effector genes from a cDNA library of *P. pachyrhizi* were identified (Kunjeti et al. 2016). Two (Pp CSEP-07 and Pp CSEP-09) of the candidate effectors were able to promote the susceptibility of *Nicotiana benthamiana* to *Phytophthora infestans* significantly. This suggests that these effectors play a vital role in the pathogenicity of *P. pachyrhizi* on its hosts (Kunjeti et al. 2016). A study using various heterologous expression systems found that 17 of the 82 *P. pachyrhizi* effector candidates (PpECs) were able to suppress the basal defense responses associated with pathogen-associated molecular pattern-triggered immunity (PTI) (Qi et al. 2018). The molecular level identification and characterization of only one putative effector, PpEC23, and its host interactor, have been completed. In yeast and plant cells, PpEC23 was found to interact with the soybean transcription factor known as *GmSPL12l*. This interaction might cause PpEC23 to interfere directly with the activity of *GmSPL12l* or its post-translational regulation (Qi et al. 2016). Several *P. pachyrhizi* candidate effectors were functionally verified by transient overexpression in tobacco, soybean and Arabidopsis leaves (Carvalho et al. 2016). These effector candidates inhibited effector-triggered immunity (ETI) responses in tobacco and soybean leaves caused by *Pseudomonas syringae* pv. *tomato* DC3000 (Qi et al. 2018). Hierarchical cluster analysis of families of candidate secreted proteins encoded by rust fungi placed candidate effector, *de novo* 3939, in a family (family 3), which contains members of secreted, small, cysteine-rich proteins that also possess potentially conserved motifs, [AFY] xC and [FY] xC (Carvalho et al. 2017). It has been suggested that these motifs play a significant role in the translocation of effector proteins across the fungal extrahaustorial membrane. Candidates expressed by rust fungi are being considered for this role (Godfrey et al. 2010, Saunders et al. 2012). The *de novo* 3939 corresponded to the gene model Phapa-7431740 based on the reference genome of the strain MT2006 of *P. pachyrhizi* recently released by the Joint Genome Institute (JGI) (Carvalho et al. 2017). A recent study indicates that Phapa-7431740 inhibits pathogen-associated molecular pattern-triggered immunity (PTI) and interacts with soybean glucan endo-1,3-glucosidase (*GmGLU*), a PR protein that is a member of the *PR-2* family (Bueno et al. 2022). According to transcriptional profiling during an infection time course, the *GmGLU* mRNA is strongly upregulated during the first hours after infection, coinciding with the peak expression of Phapa-7431740. The effector was able to reduce *GmGLU* activity in a dose-dependent manner. These findings are consistent with the hypothesis that Phapa-7431740 is an effector that interferes with an antifungal PR protein (*GmGLU*) to promote *P. pachyrhizi* infection of soybean.

Transcription Factors for *Phakopsora pachyrhizi* Resistance

Several studies have suggested that transcription factors (TF) play an important part in the plant responses against *P. pachyrhizi* infection. van de Mortel et al. (2007) and Schneider et al. (2011) found that TF genes are overrepresented among genes whose expression is activated in the biphasic transcriptional response in rust-resistant soybean genotypes harboring *Rpp2* or *Rpp3*. In incompatible interactions between soybean and *P. pachyrhizi*, genes encoding WRKY, bHLH, and MYB TFs were activated. However, these genes were not activated during the compatible interactions. The *Rpp2*-mediated rust resistance was eliminated when the genes *GmWRKY36*, *GmWRKY40*, *GmWRKY45*, and *GmMYB84* were individually silenced (Pandey et al. 2011). On the contrary, silencing two soybean TFs (Glyma14g11400, which belongs to the PHD superfamily, and Glyma12g30600, a zinc finger TF) partially hampered the *Rpp1*-conferred soybean rust resistance (Cooper et al. 2013). Likewise, Bencke-Malato et al. (2014) revealed that the increase of mRNA transcripts for many WRKY TFs occurred more quickly and robustly in a resistant soybean accession than in a susceptible soybean accession. The simultaneous silencing of four discovered WRKY genes in soybean plants resulted in an increase in plant susceptibility to rust infection. Other investigations also found varied patterns of TF expression in incompatible or compatible interactions between soybean and *P. pachyrhizi* (Panthee et al. 2007, Morales et al. 2013, Aoyagi et al. 2014). A significant amount of TF activity appears to be shared throughout *Rpp2*, *Rpp3*, and *Rpp4*-mediated soybean rust resistance (Morales et al. 2013). As a result, these TFs would be fantastic candidates for the engineering of soybean rust resistance. However, TFs regulate a wide variety of loci; thus, the alteration of TF balance may affect agronomic traits.

Highly Complex Genome of *Phakopsora pachyrhizi*

A high-quality genome is essential to promote innovation, create novel traits, and identify fresh ways to combat *P. pachyrhizi*. However, the remarkable size and complexity of its genome have prevented its sequencing for a long time. Several genome sequencing studies for *P. pachyrhizi* have been attempted. The genome of *P. pachyrhizi* (isolate Taiwan 72–1) was sequenced using a fosmid shotgun sequencing technique as part of the Department of Energy's Joint Genome Institute Community Sequencing Program in 2004. The 50 Mb estimate for the genome size at the time was grossly inadequate. The JGI now considers the sequencing project at the "permanent draft" stage (http://genome.jgi.doe.gov/programs/fungi/1000fungalgenomes.jsf). In contrast to the previously revealed mitochondrial genome sequence (Stone et al. 2010), no reports of nuclear genome assembly efforts have been known. During the 2005 National Soybean Rust Symposium in Nashville, attendees heard the latest research on this issue (TN, USA). The genome size was estimated to range from 300 Mb to 950 Mb (Posada-Buitrago 2005). Igor Grigoriev, head of the JGI fungal program, made a similar claim, indicating a genome size greater than 850 Mb (Duplessis et al. 2012) and deciphering the genomes of rust fungi has revealed several common traits that present significant challenges for accurate genome assembly. These include extended multigene families and a substantial proportion of transposable elements (>45%) (Loehrer et al. 2014). Loehrer and associates (2014) initiated efforts toward uncovering the genome size of *P. pachyrhizi* using their lab isolate (Brazil 05–1) and followed a strategy based on k-mer analysis. They obtained an N50 value of 569 bp after assembly and scaffolding with SOAPdenovo, allowing no gene number or length prediction.

Eventually, a unique international collaboration of 12 governmental and commercial organizations was formed to address this pressing issue. The 2Blades foundation, Bayer, the Brazilian Company of Agricultural Research (Embrapa), the German University of Hohenheim, the Institut National de la Recherche Agronomique (INRA), and the Université de Lorraine, as well as the US Department of Energy's Joint Genome Institute, KeyGene, the German University of RWTH Aachen, The Sainsbury Laboratory, Syngenta, and the Federal University of Viçosa, have participated in this collaborative effort. This consortium has finished sequencing and assembling the genomes of three different *P. pachyrhizi* strains. Small (Illumina) or lengthy reads were produced using various premium next-generation sequencing methods (PacBio). The JGI provided the consortium with knowledge regarding one isolate through their Community Science Program. For the second isolates, KeyGene assisted in enhancing the assembly of one of the genomes by making their most recent long-read sequencing platform, Promethean from Oxford Nanopore, freely accessible. As a result, the number of scaffolds needed to assemble the chromosomes for this strain was reduced by ten times. Syngenta provided significant support for the bioinformatics inputs required for a third isolate. The JGI also offers a platform for hosting and making these genomic resources accessible

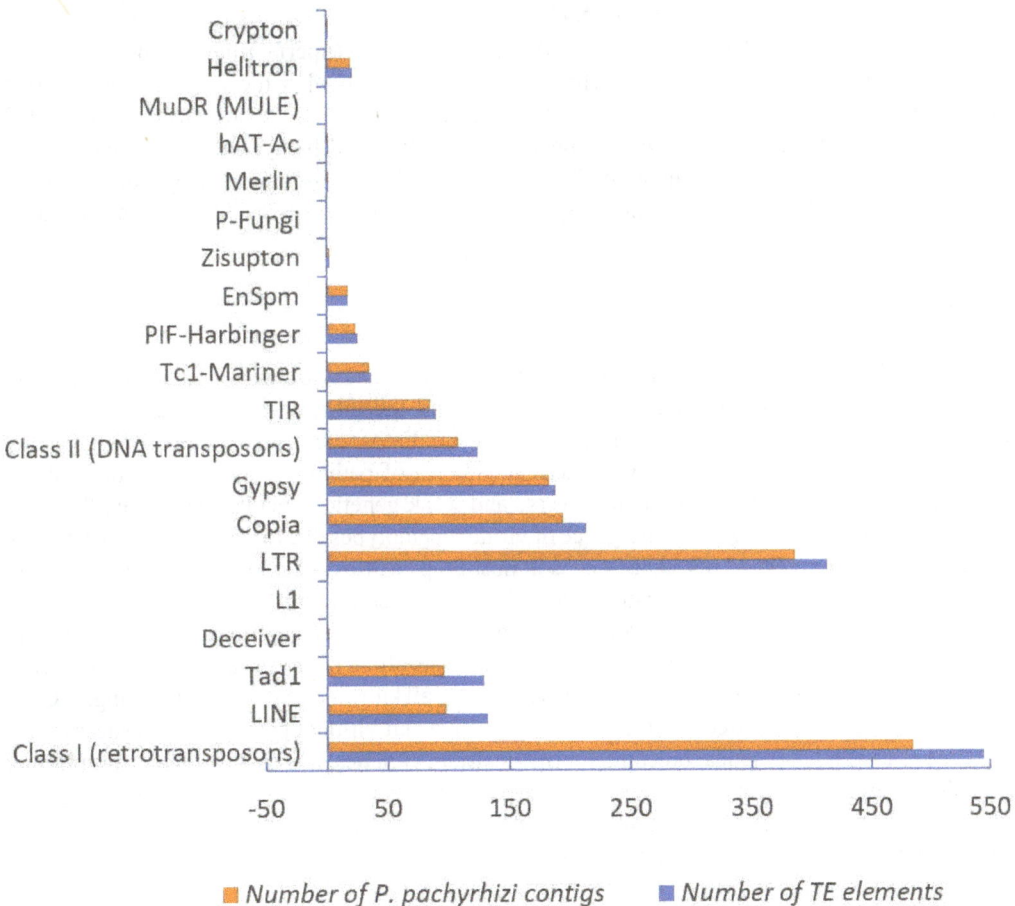

Figure 4. Transcriptionally active transposable elements (TEs) representing the classes, orders and superfamilies in *Phakopsora pachyrhizi* transcriptome. The figure is constructed using the data of Rincão et al. (2018).

to the general public to benefit the scientific community. It is possible to access all information at https://mycocosm.jgi.doe.gov/Phapa1. The completed *P. pachyrhizi* genome is 1.057 Gb, making it one of the largest plant pathogen genomes among fungi. Sequencing indicated that non-coding DNA and sequence repeats comprise nearly 90% of all sequences and are extraordinarily abundant in the genome. A number of families of transposons are also present in the genome, which may help the fungus quickly adapt to new surroundings. *P. pachyrhizi* has a vast genome; however, it only has about 20,700 genes, equivalent to the number of genes found in other rust fungi. The consortium has produced a transcriptome map of all fungal structures and infection stages to elucidate which genome regions are active at various phases of the pathogen life cycle. This information would improve annotation and further our understanding of the molecular pathways used by *P. pachyrhizi* to infect soybeans.

Recently, Gupta et al. (2023) published the genomes of three *P. pachyrhizi* strains that showed a 1.25 Gb genome with two haplotypes and a TE content of 93%. This organism has one of the highest TE contents yet documented. Class 1 retrotransposons are the most common type of TE, accounting for 54.0% of the genome. Over 80% of *P. pachyrhizi* genome comprises only two types of TEs: long terminal repeats (LTR) and terminal inverted repeats (TIR). Gypsy retrotransposons are the biggest single family of TE, accounting for 43% of the whole genome. Previously, Rinco et al. (2018) described high TE content in the *P. pachyrhizi* genome in their transcriptome investigation. The work offered a comprehensive categorization of the active TEs in *P. pachyrhizi's* transcriptome, identifying superfamilies within each class (Figure 4). Retrotransposons accounted for the vast majority of the TEs found (81.76%). The TE classes include many transposon superfamilies: Tc1-Mariner, PIF-Harbinger, EnSpm, hAT, MuDR, P, and Helitron, and retrotransposon superfamilies: Tad1, Copia, and Gypsy. The last two were the most prevalent superfamilies comprising 61.7% of all transposable elements (Figure 4). The TEs can significantly alter the structure and function of a genome. Transposon or retrotransposon translocation into or near genes can result in partial or total gene inactivation, altering gene expression regulation and likely leading to a wide range of genotypes and phenotypes. This high TE content could be an essential strategy for increasing genetic variation in *P. pachyrhizi,* which may be important in processes like host range adaptation and stress responses.

Conclusion

Soybean rust is the most notorious disease ever to affect soybeans worldwide. The causal organism *P. pachyrhizi*, a pathogen endemic to Asia-Australia and the more virulent of the two species, produces enormous quantities of wind-disseminated urediniospores and sustains through many infection cycles within the same growing season. Studies of pathotypic and genetic diversity in *P. pachyrhizi* populations reveal a highly diverse and weak genetic structure both in the field and across large geographic areas. The emergence of virulent pathotypes due to severe selection pressure caused by the deployment of resistance genes in host crops causes severe host infection. Different genetic and genomic assays demonstrate that biphasic transcriptional and molecular responses exist in both compatible and incompatible soybean-*P. pachyrhizi* interactions. Utilizing resistant cultivars of soybeans is the most effective method for controlling soybean rust over the long term. However, none of the *Rpp* genes is solely effective against all known pathotypes of this fungus. Gene pyramiding appears to be effective against highly diverse and virulent pathotypes. To successfully use gene pyramiding in practical breeding for rust-resistant varieties, it is essential to perform further research on better combinations of various *Rpp* genes. Furthermore, biotechnological approaches aimed at introducing novel resistance and increasing disease resistance in soybean cultivars have the potential to provide accurate and long-term management of soybean rust.

Acknowledgments

The research is supported by the research grant (2020-2023) from the Research Management Committee (RMC) of Bangabandhu Sheikh Mujibur Rahman Agricultural University.

References

Akamatsu, H., N.Yamanaka, R.M. Soares, A.J.G, Ivancovich, M.A. Lavilla, A.N. Bogado et al. 2017. Pathogenic variation of South American *Phakopsora pachyrhizi* populations isolated from soybeans from 2010 to 2015. Jpn. Agric. Res. Q. 51: 221–232. https://doi.org/10.6090/jarq.51.221.

Akamatsu, H., N. Yamanaka, Y. Yamaoka, R.M. Soares, W. Morel, A.J.G. Ivancovich, Bogado et al. 2013. Pathogenic diversity of soybean rust in Argentina, Brazil, and Paraguay. J. Gen. Plant Pathol. 79: 28–40. https://doi.org/10.1007/s10327-012-0421-7.

Anderson, S.J., C.L. Stone, M.L. Posada-Buitrago, J.L. Boore, B.A. Neelam, R.M. Stephens et al. 2008. Development of simple sequence repeat markers for the soybean rust fungus, *Phakopsora pachyrhizi*. Mol. Ecol. Res. 8: 1310–1312.

Aoyagi, L.N., V.S. Lopes-Caitar, M.C.C.G. de Carvalho, L.M. Darben, A. Polizel-Podanosqui, M.K., Kuwahara et al. 2014. Genomic and transcriptomic characterization of the transcription factor family R2R3-MYB in soybean and its involvement in the resistance responses to *Phakopsora pachyrhizi*. Plant Sci. 229: 32–42. doi: 10.1016/j.plantsci.2014.08.005.

Barro, J.P., K.S. Alves, C.V. Godoy, A.R. Dias, C.A. Forcelini, C.M. Utiamada et al. 2021. Performance of dual and triple fungicide premixes for managing soybean rust across years and regions in Brazil: A meta-analysis. Plant Pathol. 70: 1920–1935. https://doi.org/10.1111/ppa.13418.

Bencke-Malato, M., C. Cabreira, B. Wiebke-Strohm, L. Bücker-Neto, E. Mancini, M.B. Osorio et al. 2014. Genome-wide annotation of the soybean WRKY family and functional characterization of genes involved in response to *Phakopsora pachyrhizi* infection. BMC Plant Biol. 14: 236. doi: 10.1186/s12870-014-0236-0.

Bhor, T.J., V.P. Chimote and M.P. Deshmukh 2015. Molecular tagging of Asiatic soybean rust resistance in exotic genotype EC 241780 reveals complementation of two genes. Plant Breed. 134: 70–77.

Bindschedler, L.V., T.A. Burgis, D.J. Mills, J.T. Ho, R. Cramer and P.D. Spanu. 2009. *In planta* proteomics and proteogenomics of the biotrophic barley fungal pathogen *Blumeria graminis* f. sp. *hordei*. Mol Cell Proteomics. 8: 2368–2381.

Bonde, M.R., S.E. Nester, D.K. Berner, R.D. Frederick, W.F. Moore and S. Little. 2008. Comparative susceptibilities of legume species to infection by *Phakopsora pachyrhizi*. Plant Dis. 92(1): 30–36. https://doi.org/10.1094/PDIS-92-1-0030.

Bonde, M.R., S.E. Nester, C.N. Austin, C.L. Stone, R.D. Frederick, G.L. Hartman et al. 2006. Evaluation of virulence of *Phakopsora pachyrhizi* and *P. meibomiae* isolates. Plant Dis. 90: 708–716. https://doi.org/10.1094/PD-90-0708.

Bonde, M.R., G.L. Peterson and W.M. Dowler 1988. A comparison of isozymes of *Phakopsora pachyrhizi* from the Eastern Hemisphere and the New World. Phytopathology. 78: 1491–1494.

Both, M., M. Csukai, M.P.H. Stumpf and P.D. Spanu. 2005. Gene expression profiles of *Blumeria graminis* indicate dynamic changes to primary metabolism during development of an obligate biotrophic pathogen. Plant Cell. 17: 2107–2122.

Bromfield, K.R. and E.E. Hartwig. 1980. Resistance to soybean rust (*Phakopsora pachyrhizi*) and mode of inheritance. Crop Sci. 20: 254–255. doi: 10.1071/AR9800951.

Bromfield, K.R. 1984. Soybean Rust; American Phytopathological Society: St. Paul, MN, USA.

Bueno, T.V., P.P. Fontes, V.Y. Abe, A.S. Utiyama, R.L. Senra, L.S. Oliveira et al. 2022. A *Phakopsora pachyrhizi* Effector Suppresses PAMP-Triggered Immunity and Interacts with a Soybean Glucan Endo-1,3-β-Glucosidase to Promote virulence. Mol Plant Microbe Interact. 35(9): 779–790. doi: 10.1094/MPMI-12-21-0301-R.

Burdon, J.J. and S.S. Speer. 1984. A set of differential Glycine hosts for the identification of races of *Phakopsora pachyrhizi* Syd. Euphytica. 33: 891–896. doi: 10.1007/BF00021917.

Cantu, D., M. Govindarajulu, A. Kozik, M. Wang, X. Chen, K.K. Kojima et al. 2011. Next generation sequencing provides rapid access to the genome of *Puccinia striiformis* f. sp. *tritici*, the causal agent of wheat stripe rust. PLoS One 6: e24230.

de Carvalho, M.C., L. Costa Nascimento, L.M. Darben, A.M. Polizel-Podanosqui, V.S. Lopes-Caitar, M. Qi et al. 2017. Prediction of the in planta *P. pachyrhizi* secretome and potential effector families. Mol Plant Pathol.18:363–377.

Catanzariti A.M., P.N.V.E.T. Dodds, B. Kobe, J.G. Ellis and B.J. 2010. Staskawicz The AvrM effector from flax rust has a structured C-terminal domain and interacts directly with the M resistance protein. Mol. Plant Microbe. Interact. 23: 49–57.

Catanzariti, A.M., P.N. Dodds, G.J. Lawrence, M.A. Ayliffe and J.G. Ellis. 2006. Haustorially expressed secreted proteins from flax rust are highly enriched for avirulence elicitors. Plant Cell. 18: 243–256.

Chakraborty, N., J. Curley, R.D. Frederick, D.L. Hyten, R.L. Nelson, G.L. Hartman et al. 2009. Mapping and confirmation of a new allele at *Rpp1* from soybean PI 594538A conferring RB lesion–type resistance to soybean rust. Crop Sci. 49: 783. https://doi.org/10.2135/cropsci2008.06.0335.

Chang, H.-X., L.A. Miller and G.L. Hartman. 2014. Melanin-independent accumulation of turgor pressure in appressoria of *Phakopsora pachyrhizi*. Phytopathology. 104: 977–984.

Childs, S.P., Z.R. King, D.R. Walker, D.K. Harris, K.F. Pedley, J.W. Buck et al. 2018. Discovery of a seventh *Rpp* soybean rust resistance locus in soybean accession PI 605823. Theor. Appl. Genet. 131: 27–41. doi:10.1007/s00122-017-2983-4.

Cooper, B., K.B. Campbell, M.B. McMahon and D.G. Luster. 2013. Disruption of *Rpp1*-mediated soybean rust immunity by virus-induced gene silencing. Plant Signal. Behav. 8: e27543. doi: 10.4161/psb.27543.

Cregan, P.B., T. Jarvik, A.L. Bush, R.C. Shoemaker, K.G. Lark, A.L. Kahler et al. 1999. An integrated genetic linkage map of the soybean genome. Crop Sci. 39: 1464–1490. doi: 10.2135/cropsci1999.3951464x.

Darben, L.M., A.Yokoyama, F.M. Castanho, V.S. Lopes-Caitar, M.C. da Cruz Gallo de Carvalho, C.V. Godoy et al. 2020. Characterization of genetic diversity and pathogenicity of *Phakopsora pachyrhizi* mono-uredinial isolates collected in Brazil. Eur. J. Plant Pathol. 156: 355–372. https://doi.org/10.1007/s10658-019-01872-2.

Devaraj, L., S. Jahagirdar and G.T. Basavaraja. 2016. Prevalence of pathotypes of *Phakopsora pachyrhizi* Syd. causing Asian soybean rust in India. Sri Lanka Journal of Food and Agriculture. 2: 19–28.

Duplessis, S., D.J. Joly and P.N. Dodds. 2012. "Rust Effectors," in Effectors in Plant-Microbe Interactions, 1st Edn, eds F. Martin and S. Kamoun (Chichester: John Wiley and Sons, Ltd), 155–193.

Elmore, M. G., S. Banerjee, K.F. Pedley, A. Ruck and S.A. Whitham. 2020. *De novo* transcriptome of *Phakopsora pachyrhizi* uncovers putative effector repertoire during infection. Physiol. Mol. Plant Pathol. 110:101464. https://doi.org/10.1016/j.pmpp.2020.101464.

Frederick, R.D., C.L. Snyder, G.L. Peterson and M.R. Bonde. 2002. Polymerase chain reaction assays for the detection and discrimination of the soybean rust pathogens *Phakopsora pachyrhizi* and *P. meibomiae*. Phytopathology. 92: 217–227.

Freire, M.C.M., L.O. de Oliveira, A.M.R. de Almeida, I. Schuster, M.A. Moreira, M.M. Liebenberg et al. 2008. Evolutionary history of *Phakopsora pachyrhizi* (the Asian soybean rust) in Brazil based on nucleotide sequences of the internal transcribed spacer region of the nuclear ribosomal DNA. Genet Mol. Biol. 31: 920–931.

Garcia A., E.S. Calvo, R.A. de Souza Kiihl, A. Harada, D.M. Hiromoto and L.G. Vieira. 2008. Molecular mapping of Soybean rust (*Phakopsora pachyrhizi*) resistance genes: Discovery of a novel locus and alleles. Theor. Appl. Genet. 117: 545–553. doi: 10.1007/s00122-008-0798-z.

García-Rodríguez, J.C., Z. Vicente-Hernández, M. Grajales-Solís and N. Yamanaka. 2022. Virulence Diversity of *Phakopsora pachyrhizi* in Mexico. PhytoFrontiers. 2:1: 52–59.

García-Rodríguez, J.C., M. Morishita, M. Kato and N. Yamanaka. 2017. Características patogénicas de la roya asiática de la soya (*Phakopsora pachyrhizi*) en México. Rev. Mex. Fitopatol. 35: 338–349.

Garnica, D.P., A. Nemri, N.M. Upadhyaya, J.P. Rathjen and P.N. Dodds. 2014. The ins and outs of rust haustoria. PLoS Pathog. 10(9): e1004329. https://doi.org/10.1371/journal.ppat.1004329.

Godfrey, D., H. Böhlenius, C. Pedersen, Z. Zhang, J. Emmersen and H. Thordal-Christensen. 2010. Powdery mildew fungal effector candidates share N-terminal Y/F/WxC-motif. BMC Genomics. 11: 317. https://doi.org/10.1186/1471-2164-11-317.

Godoy, C.V., C.D.S. Seixas, R.M. Soares, F.C. Marcelino-Guimarães, M.C. Meyer, L.M. Costamilan. 2016. Asian soybean rust in Brazil: past, present, and future. Pesquisa Agropecuária Brasileira. 51: 407–421.

Gupta, Y.K., F.C. Marcelino-Guimarães, C. Lorrain, A. Farmer, S. Haridas, E.G.C. Ferreira et al. 2023. Major proliferation of transposable elements shaped the genome of the soybean rust pathogen *Phakopsora pachyrhizi*. Nat Commun. 2023 Apr 1;14(1): 1835. doi: 10.1038/s41467-023-37551-4.

Hartman, G.L., C. Hill, M. Twizeyimana, M. MR and R. Bandyopadhyay. 2011. Interaction of soybean and *Phakopsora pachyrhizi*, the cause of soybean rust. CAB Rev. Perspect. Agric. Vet. Sci. Nutr. Nat. Resour. 6: 25. doi: 10.1079/PAVSNNR20116025.

Hennings, P. 1903. Einige neue japanische Uredinales (in German). Hedwigia. IV, 107–108.

Hossain, M.Z., Y. Ishiga, N. Yamanaka, E. Ogiso-Tanaka and Y. Yamaoka. 2018. Soybean leaves transcriptomic data dissects the phenylpropanoid pathway genes as a defence response against *Phakopsora pachyrhizi*. Plant Physiol. Biochem. 132: 424–433. https://doi.org/10.1016/j.plaphy.2018.09.020.

Hossain, M.M., H. Akamatsu, M. Morishita, T. Mori, Y. Yamaoka, K. Suenaga et al. 2015. Molecular mapping of Asian soybean rust resistance in soybean landraces PI 594767A, PI 587905 and PI 416764. Plant Pathol. 64: 147–156. https://doi.org/10.1111/ppa.12226.

Hossain, M.M. and N. Yamanaka. 2019. Pathogenic variation of Asian soybean rust pathogen in Bangladesh. Journal of General Plant Pathology. 85: 90–100.

Hossain, M.M., L.Yasmin, M.T. Rubayet, H. Akamatsu and N. Yamanaka. 2022. A major variation in the virulence of the Asian soybean rust pathogen (*Phakopsora pachyrhizi*) in Bangladesh. Plant Pathology. 71: 1355–1368. https://doi.org/10.1111/ppa.13568.

Hunter, T. 2009. Tyrosine phosphorylation: Thirty years and counting. Curr. Opin. Cell Biol. 21: 140–146.

Hyten, D.L., G.L. Hartman, R.L. Nelson, R.D. Frederick, V.C. Concibido and P.B. Cregan. 2007. Map location of the *Rpp1* locus that confers resistance to Soybean rust in Soybean. Crop Sci. 47: 837–838. doi: 10.2135/cropsci2006.07.0484.

Jarvie, J.A. 2009. A review of soybean rust from a South African perspective. S A J Sci. 105: 103 -108.

Javaid, I. and M. Ashraf. 1978. Some observations on soybean diseases in Zambia and occurrence of *Pyrenochpta glycines* on certain varieties. Plant Dis. Rep. 62(1): 46–47.

Jorge, V.R., M.R. Silva, E.A. Guillin, M.C.M. Freire, I. Schuster, A.M.R. Almeida et al. 2015. The origin and genetic diversity of the causal agent of Asian soybean rust, *Phakopsora pachyrhizi*, in South America. Plant Pathol. 64: 729–737. https://doi.org/10.1111/ppa.12300.

Kendrick, M.D., D.K. Harris, B.K. Ha, D.L. Hyten, P.B. Cregan, R.D. Frederick et al. 2011. Identification of a second Asian soybean rust resistance gene in Hyuuga soybean. Phytopathology. 101: 535–543. https://doi.org/10.1094/PHYTO-09-10-0257.

Khanh, T.D., T.Q. Anh, B.C. Buu and T.D. Xuan. 2013. Applying molecular breeding to improve soybean rust resistance in Vietnamese elite soybean. American J. Plant Sci. DOI:10.4236/ajps.2013.41001.

Kunjeti, S.G., G. Iyer, E. Johnson, E. Li, K.E. Broglie, G. Rauscher et al. 2016. Identification of *Phakopsora pachyrhizi* candidate effectors with virulence activity in a distantly related pathosystem. Front. Plant Sci. 7: 269. doi: 10.3389/fpls.2016.00269.

Larzábal, J., M. Rodríguez and N. Yamanaka. 2022. Stewart S. Pathogenic variability of Asian soybean rust fungus within fields in Uruguay. Trop Plant Pathol. 26: 1–9.

Lemos, N.G., A.D.L. e Braccini, R.V. Abdelnoor, M.C.N. de Oliveira, K. Suenaga and N. Yamanaka. 2011. Characterization of genes *Rpp2*, *Rpp4*, and *Rpp5* for resistance to soybean rust. Euphytica. 182: 53–64. https://doi.org/10.1007/s10681-011-0465-3.

Levy, C. 2005. Epidemiology and chemical control of soybean rust in southern Africa. Plant Dis. 89: 669–674.

Li, S., J.R. Smith, J.D. Ray, R.D. Frederick. 2012. Identification of a new soybean rust resistance gene in PI 567102B. Theor. Appl. Genet. 125: 133–142. https://doi.org/10.1007/s00122-012-1821-y.

Lin, S.Y. 1966. Studies on the physiologic races of soybean rust fungus, *Phakopsora pachyrhizi* Syd. J. Taiwan Agric. Res. 51: 24–28.

Link, T.I., P. Lang, B.E. Scheffler, M.V. Duke, M.A. Graham, B. Cooper, M.L. Tucker et al. 2014. The haustorial transcriptomes of Uromyces appendiculatus and *Phakopsora pachyrhizi* and their candidate effector families. Mol. Plant Pathol. 15(4): 379–393. https://doi.org/10.1111/mpp.12099.

Loehrer, M., A. Vogel, B. Huettel, R. Reinhardt, V. Benes, S. Duplessis et al. 2014. On the current status of *Phakopsora pachyrhizi* genome sequencing. Front. Plant Sci. 5: 377. doi: 10.3389/fpls.2014.00377.

Maphosa, M., H. Talwana and P. Tukamuhabwa. 2012. Enhancing soybean rust resistance through *Rpp2*, *Rpp3* and *Rpp4* pairwise gene pyramiding. Afr. J. Agric. Res. 7: 4271–4277.

Martin, K.J. 2007. Introduction to molecular analysis of ectomycorrhizal communities. Soil Sci. Soc. Am. J. 71: 601–610.

McLean, R.J. and D.E. Byth. 1980. Inheritance of resistance to rust (*Phakopsora pachyrhizi*) in soybeans. Aust. J. Agric. Res. 31: 951–956. doi: 10.1071/AR9800951.

Melo Reis E., E. Deuner and M. Zanatta. 2015. *In vivo* sensitivity of *Phakopsora pachyrhizi* to DMI and QoI fungicides. Summa Phytopathologica. 41: 21–24.

Meudt, H.M. and A.C. Clarke. 2007. Almost forgotten or latest practice? AFLP applications, analysis and advances. Trends Plant Science. 12: 106–117.

Meyer, J.D.F., D.C.G. Silva, C. Yang, K.F. Pedley, C. Zhang, M. Van de Mortel et al. 2009. Identification and analyses of candidate genes for *Rpp4*-mediated resistance to Asian oybean rust in soybean. Plant Physiol. 150: 295–307. https://doi.org/10.1104/pp.108.134551.

Micali, C.O., U. Neumann, D. Grunewald, R. Panstruga and R. O'Connell. 2011. Biogenesis of a specialized plant–fungal interface during host cell internalization of *Golovinomyces orontii* haustoria. Cellular Microbiol. 13: 210–226.

Monterous M.J., A.M. Missaoui, D.V. Phillips, D.R. Walker and H.R. Boerma. 2007. Mapping and confirmation of the 'Hyuuga' red–brown lesion resistance gene for Asian Soybean rust. Crop Sci. 47: 829–836. doi: 10.2135/cropsci06.07.0462.

Morales, A.M.A.P., A. Borém, M.A. Graham and R.V. Abdelnoor. 2012. Advances on molecular studies of the interaction soybean - Asian rust. Crop Breed Appl Biotechnol. 12: 1–7.

Morales, A.M.A.P., J.A. O'Rourke, M.van de Mortel, K.T. Scheider, T.J. Bancroft, A. Borém et al. 2013. Transcriptome analyses and virus induced gene silencing identify genes in the *Rpp4*-mediated Asian soybean rust resistance pathway. Funct Plant Biol. 40: 1029–1047. https://doi.org/10.1071/FP12296.

Mundt, C.C. 2018 Pyramiding for Resistance Durability: Theory and Practice. Phytopathology. 108: 792–802. https://doi.org/10.1094/PHYTO-12-17-0426-RVW.

Murithi, H.M., F.D. Beed, M.M. Soko, J.S. Haudenshield and G.L. Hartman. 2015. First report of *Phakopsora pachyrhizi* causing rust on soybean in Malawi. Plant Dis. 99(3): 420.

Murithi, H.M., J.S. Haudenshield, F. Beed, G. Mahuku, M.H.A.J. Joosten and G.L. Hartman. 2017. Virulence diversity of *Phakopsora pachyrhizi* isolates from East Africa compared to a geographically diverse collection. Plant Dis. 101: 1194–1200. https://doi.org/10.1094/PDIS-10-16-1470-RE.

Murithi, H.M., R.M. Soares, G. Mahuku, H.P. van Esse and M.H.A.J. Joosten. 2021. Diversity and distribution of pathotypes of the soybean rust fungus *Phakopsora pachyrhizi* in East Africa. Plant Pathol. 70: 655–666. https://doi.org/10.1111/ppa.13324.

Nemri, A., D.G. Saunders, C. Anderson, N.M. Upadhyaya, J. Win, G.J. Lawrence et al. 2014. The genome sequence and effector complement of the flax rust pathogen *Melampsora lini*. Front Plant Sci. 2014; 5: 1–14.

240 *Microbial Genetics*

Ono, Y., P. Buriticá and J.F. Hennen 1992. Delimitation of *Phakopsora*, *Physopella* and *Cerotelium* and their species on Leguminosae. Mycological Research. 96: 825–850.

Pandey, A.K., C. Yang, C. Zhang, M.A. Graham, H.D. Horstman, Y. Lee et al. 2011. Functional analysis of the Asian soybean rust resistance pathway mediated by *Rpp2*. Mol. Plant Microbe Interact. 24: 194–206. doi: 10.1094/MPMI-08-10-0187.

Panthee, D.R., J.S. Yuan, D.L. Wright, J.J. Marois, D. Mailhot and C.N. Stewart. 2007. Gene expression analysis in soybean in response to the causal agent of Asian soybean rust (*Phakopsora pachyrhizi* Sydow) in an early growth stage. Funct. Integr. Genomics. 7: 291–301. https://doi.org/10.1007/s10142-007-0045-8.

Paul, C. and G.L. Hartman 2009. Sources of soybean rust resistance challenged with single-spored isolates of *Phakopsora pachyrhizi*. Crop Sci. 49: 1781–1785. https://doi.org/10.2135/cropsci2008.12.0710.

Paul C., G.L. Hartman J.J. Marois D.L. Wright and D.R. Walker. 2013. First report of *Phakopsora pachyrhizi* adapting to soybean genotypes with *Rpp1* or *Rpp6* rust resistance genes in field plots in the United States. Plant Dis. 97: 1379–1379.

Pedley, K.F., A.K. Pandey, A. Ruck, L.M. Lincoln, S.A. Whitham and M.A. Graham. 2018. *Rpp1* encodes a ULP1-NBS-LRR protein that controls immunity to *Phakopsora pachyrhizi* in soybean. Mol. Plant Microbe Interact. 32: 120–133.

Pham, A.T., J.D. Lee, J.G. Shannon and K.D. Bilyeu. 2010. Mutant alleles of FAD2-1A and FAD2-1B combine to produce soybeans with the high oleic acid seed oil trait. BMC Plant Biol. 10: 195. doi: 10.1186/1471-2229-10-195.

Pham, T.A., M.R. Miles, R.D. Frederick, C.B. Hill and G.L. Hartman. 2009. Differential responses of resistant soybean entries to isolates of *Phakopsora pachyrhizi*. Plant Dis. 93(3): 224–228. doi: 10.1094/PDIS-93-3-0224.

Posada-Buitrago, M.L. 2005. Frederick RD. Expressed sequence tag analysis of the soybean rust pathogen *Phakopsora pachyrhizi* . Fungal Genet Biol. 42: 949–962.

Posada-Buitrago, M.L., J.L. Boore and R.D. Frederick. 2005. "Soybean Rust Genome Sequencing Project," in Proceedings of the National Soybean Rust Symposium, Nashville, TN. Available at: http://www.plantmanagementnetwork.org/infocenter/topic/soybeanrust/symposium/posters/3.pdf.

Pretsch, K., A.C. Kemen, E. Kemen, M. Geiger, K. Mendgen and R.T. Voegele, 2013. The rust transferred proteins – a new family of effector proteins exhibiting protease inhibitor function. Mol. Plant Pathol. 14: 96–107.

Qi, Z., M. Liu, Y. Dong, Q. Zhu, L. Li, B. Li et al. 2016. The syntaxin protein (MoSyn8) mediates intracellular trafficking to regulate conidiogenesis and pathogenicity of rice blast fungus. New Phytologist. 209: 1655–1667.

Qi, M., J.P. Grayczyk, J.M. Seitz, Y. Lee, T.I. Link, D. Choi et al. 2018. Suppression or activation of immune responses by predicted secreted proteins of the soybean rust pathogen *Phakopsora pachyrhizi*. Mol. Plant-Microbe Interact. 31: 163–174. https://doi.org/10.1094/MPMI-07-17-0173-F.

Ray, J.D., W. Morel, J.R. Smith, R.D. Frederick and M.R. Miles. 2009. Genetics and mapping of adult plant rust resistance in soybean PI 587886 and PI 587880A. Theor. Appl. Genet. 119: 271–280. https://doi.org/10.1007/s00122-009-1036-z.

Rincão, M.P., M.C.D.C.G. Carvalho, L.C. Nascimento, V.S. Lopes-Caitar, K. Carvalho, L.M. Darben et al. 2018. New insights into *Phakopsora pachyrhizi* infection based on transcriptome analysis in planta. Genet Mol. Biol. 41(3): 671–691. doi: 10.1590/1678-4685-GMB-2017-0161.

Rocha, C.M.L., G.R. Vellicce, M.G. García, E.M. Pardo, J. Racedo, M.F. Perera et al. 2015. Use of AFLP markers to estimate molecular diversity of *Phakopsora pachyrhizi*. Electron. J. Biotechnol. 18: 439–444. https://doi.org/10.1016/j.ejbt.2015.06.007.

Saunders, D.G.O., J. Win, L.M. Cano, L.J. Szabo, S. Kamoun and S. Raffaele. 2012. Using hierarchical clustering of secreted protein families to classify and rank candidate effectors of rust fungi. PLoS One. 7: e29847. https://doi.org/10.1371/journal.pone.0029847.

Scherm H., R.S.C. Christiano, P.D. Esker, E.M. Del Ponte and C.V. Godoy. 2009. Quantitative review of fungicide efficacy trials for managing soybean rust in Brazil. Crop Protect. 28: 774–782.

Schneider, K.T., M. van de Mortel, T.J. Bancroft, E. Braun, D. Nettleton, R.T. Nelson et al. 2011. Biphasic gene expression changes elicited by *Phakopsora pachyrhizi* in soybean correlate with fungal penetration and haustoria formation. Plant Physiol. 157: 355–371. https://doi.org/10.1104/pp.111.181149.

Silva, D.C., N. Yamanaka, R.L. Brogin, C.A. Arias, A.L. Nepomuceno, A.O. Di Mauro et al. 2008. Molecular mapping of two loci that confer resistance to Asian rust in Soybean. Theor. Appl. Genet. 117: 57–63. Doi: 10.1007/s00122-008-0752-0.

Slaminko, T.L., M.R. Miles, R.D. Frederick, M.R. Bonde and G.L. Hartman. 2008. New legume hosts of *Phakopsora pachyrhizi* based on greenhouse evaluations. Plant Dis. 92: 767–771.

Song, Q., L. Marek, R.C. Shoemaker, K.G. Lark, V.C. Concibido, X. Delannay et al. 2004. A new integrated genetic linkage map of the soybean. Theor. Appl. Genet. 109(1): 122–8. doi: 10.1007/s00122-004-1602-3.

Spanu, P.D., J.C. Abbott, J. Amselem, T A. Burgis, D.M. Soanes, K. Stüber et al. 2010. Genome expansion and gene loss in powdery mildew fungi reveal tradeoffs in extreme parasitism. Science 330: 1543–1546.

Stewart, S., M. Rodríguez and N. Yamanaka. 2019. Pathotypic variation of *Phakopsora pachyrhizi* isolates from Uruguay. Trop. Plant Pathol. 44: 309–317. https://doi.org/10.1007/s40858-018-0269-2.

Stone, C.L., M.B. McMahon, L.L. Fortis, A. Nuñez, G.W. Smythers, D.G. Luster et al. 2012. Gene expression and proteomic analysis of the formation of *Phakopsora pachyrhizi* appressoria. BMC Genom. 13: 269. https://doi.org/10.1186/1471-2164-13-269.

Stone, C.L., M.L. Posada-Buitrago, J.L. Boore and R.D. Frederick. 2010. Analysis of the complete mitochondrial genome sequences of the soybean rust pathogens *Phakopsora pachyrhizi* and *P. meibomiae*. Mycologia. 102: 887–897. doi: 10.3852/09-198.

Sydow, H. and P. Sydow. 1914. A contribution to knowledge of the parasitic fungi on the island of Formosa. Ann. Mycol. 12: 105–112.

Tan, Y.J. and Y.L. Sun. 1989. Preliminary studies on physiological races of soybean rust. Soybean Sci. 8: 71–74.

Thomas, S.W., M.A. Glaring, S.W. Rasmussen, J.T. Kinane and R.P. Oliver. 2002. Transcript profiling in the barley mildew pathogen *Blumeria graminis* by serial analysis of gene expression (SAGE). Mol Plant Microbe Interact. 15(8): 847–856. https://doi.org/10.1094/MPMI.2002.15.8.847.

Tremblay, A., P. Hosseini, S. Li, N.W. Alkharouf and B.F. Matthews. 2013. Analysis of *Phakopsora pachyrhizi* transcript abundance in critical pathways at four time-points during infection of a susceptible soybean cultivar using deep sequencing. BMC Genom. 14: 614.

Tremblay, A., P. Hosseini, S. Li, N.W. Alkharouf and B.F. Matthews. 2012. Identification of genes expressed by *Phakopsora pachyrhizi*, the pathogen causing soybean rust, at a late stage of infection of susceptible soybean leaves. Plant Pathol. 61: 773–786.

Tremblay, A., S. Li, B.E. Scheffle and B.F. Matthews. 2009. Laser capture microdissection and expressed sequence tag analysis of uredinia formed by *Phakopsora pachyrhizi*, the causal agent of Asian soybean rust. Physiol. Mol. Plant Pathol. 73: 163–174.

Tukamuhabwa, P. and M. Maphosa. 2012. State of knowledge in breeding for resistance to soybean rust in the developing world. FAO Plant Production and Protection Paper. 204. FAO, Rome, Italy.

Twizeyimana, M., P.S. Ojiambo, J.S. Haudenshield, G. Caetano-Anollés, K.F. Pedley, R. Bandyopadhyay et al. 2011. Genetic structure and diversity of Phakopsora pachyrhizi isolates from soyabean. Plant Pathol. 60: 719–729. https://doi.org/10.1111/j.1365-3059.2011.02428.x.

Twizeyimana, M. and G.L. Hartman. 2012. Pathogenic variation of *Phakopsora pachyrhizi* isolates on soybean in the United States from 2006 to 2009. Plant Dis. 96: 75–81. https://doi.org/10.1094/PDIS-05-11-0379.

Twizeyimana, M., P.S. Ojiambo, K. Sonder, T. Ikotun, G.L. Hartman and R. Bandyopadhyay. 2009. Pathogenic variation of *Phakopsora pachyrhizi* infecting soybean in Nigeria. Phytopathology. 99: 353–361. https://doi.org/10.1094/PHYTO-99-4-0353.

Vakili, N.G. and K.R. Bromfield. 1976. *Phakopsora* rust on soybean and other legumes in Puerto-Rico. Plant Dis. Rep. 60: 995–999.

van de Mortel, M., J.C. Recknor, M.A. Graham, D. Nettleton, J.D. Dittman, R.T. Nelson et al. 2007. Distinct biphasic mRNA changes in response to Asian soybean rust infection. Mol. Plant-Microbe Interact. 20: 887–899. https://doi.org/10.1094/MPMI-20-8-0887.

Vélez-Climent, M.C. and N. Dufault. 2007. The adhesion and rainfallwashoff of urediniospores on soybean leaves. In: National Soybean Rust Symposium, Louisville, Proceedings. Available at: http://www.plantmanagementnetwork.org/infocenter/topic/soybeanrust/2007/presentations/Dufault.pdf.Accessed 10 Jan. 2008.

Viganó, J., A.L. Braccini, I. Schuster and V.M.P.S. Menezes. 2018. Microsatellite molecular marker-assisted gene pyramiding for resistance to Asian soybean rust (ASR). Acta Scientiarum. Agronomy. 40(1): 39619. doi:10.4025/actasciagron. v40i1.39619.

Voegele, R.T. and K.W. Mendgen. 2011. Nutrient uptake in rust fungi: How sweet is parasitic life? Euphytica. 179: 41–55.

Vuong, T.D., D.R. Walker, B.T. Nguyen, T.T. Nguyen, H.X., Dinh, D.L. Hyten et al. 2016. Molecular characterization of resistance to soybean rust (*Phakopsora pachyrhizi* Syd. & Syd.) in soybean cultivar DT 2000 (PI 635999). PLoS ONE. 11:e0164493.

Walker, D.R., H.R. Boerma, D.V. Philips, R.W. Schneider, J.B. Buckley, E.R. Shipe et al. 2011. Evaluation of USDA soybean germplasm accessions for resistance to soybean rust in the southern United States. Crop Sci. 51: 678– 693. https://doi. org/10.2135/cropsci2010.06.0340.

Whitham, S.A., M. Qi, R.W. Innes, W. Ma, V. Lopes-Caitar and T. Hewezi. 2016. Molecular soybean-pathogen interactions. Annu. Rev. Phytopathol. 54: 443–468.

Xue, C., G. Park, W. Choi, L. Zheng, R.A. Dean and J. R. Xu. 2002. Two novel fungal virulence genes specifically expressed in appressoria of the rice blast fungus. The Plant Cell. 14(9): 2107–2119. https://doi.org/10.1105/tpc.003426.

Yamanaka, N. and M.M. Hossain. 2019. Pyramiding three rust-resistance genes confers a high level of resistance in soybean (*Glycine max*). Plant Breed. 138: 686–695. https://doi.org/10.1111/pbr.12720.

Yamanaka, N., N.G. Lemos, H. Akamatsu, Y. Yamaoka, M. Kato, D.C.G. Silva et al. 2010. Development of classification criteria for resistance to soybean rust and differences in virulence among Japanese and Brazilian rust populations. Trop. Plant Pathol. 35: 153–162. https://doi.org/10.1590/S1982-56762010000300003.

Yamanaka, N., N.G. Lemos, M. Uno, H. Akamatsu, Y. Yamaoka, R.V. Abdelnoor et al. 2013. Resistance to Asian soybean rust in soybean lines with the pyramided three Rpp genes. Crop Breed. Appl. Biotechnol. 13: 75–82.

Yamanaka, N., M. Morishita, T. Mori, N.G. Lemos, M.M. Hossain, H. Akamatsu et al. 2015. Multiple *Rpp*-gene pyramiding confers resistance to Asian soybean rust isolates that are virulent on each of the pyramided genes. Trop. Plant Pathol. 40: 283–290. https://doi.org/10.1007/s40858-015-0038-4.

Yamaoka, Y., Y. Fujiwara, M. Kakishima, K. Katsuya, K. Yamada and H. Hagiwara. 2002. Pathogenic races of *Phakopsora pachyrhizi* on soybean and wild host plants collected in Japan. J. Gen. Plant Pathol. 68: 52–56.

Yamaoka, Y., N. Yamanaka, H. Akamatsu and K. Suenaga. 2014. Pathogenic races of soybean rust *Phakopsora pachyrhizi* collected in Tsukuba and vicinity in Ibaraki, Japan. J. Gen. Plant Pathol. 80: 184–188.

Yeh, C.C. 1983. Physiological Races of *Phakopsora pachyrhizi* in Taiwan. J Agric Res. 32(1): 69–74.

Yorinori J.T., W.M. Paiva, R.D. Frederick, L.M. Costamilan, P.F. Bertagnolli, G.E. Hartman et al. 2005. Epidemics of Soybean Rust (*Phakopsora pachyrhizi*) in Brazil and Paraguay from 2001 to 2003. Plant Dis. 89: 675–677.

Yorinori, J.T. 2008. Soybean germplasms with resistance and tolerance to Asian rust and screening methods. In JIRCAS Working Report No. 58: Facing the challenge of soybean rust in South America; H. Kudo, K. Suenaga, R.M. Soares, A. Toledo Eds.; JIRCAS: Tsukuba, Japan, pp. 70–87.

Zambolin, L. 2006. *In*: Zambolin, L. [ed.]. Ferrugem asiática da soja. UFV; Viçosa: Manejo integrado da ferrugem asiática da soja; p. 139.

Zhang, X.C., M.C.M. Freire, M.H. Le, L.O. De Oliveira, J.W. Pitkin, G. Segers et al. 2012. Genetic diversity and origins of *Phakopsora pachyrhizi* isolates in the United States. Asian J. Plant Pathol. 6: 52–65.

CHAPTER 16

Genetics of *Trichoderma*-plant-pathogen Interactions

Md. Motaher Hossain[1],* and *Farjana Sultana*[2]

Introduction

Microorganisms are indispensable parts of world ecosystems. They interact with all kinds of other living organisms and perform various functions. Many of the microorganisms are ecocentric to plants. Based on their interaction with plants, they can be either beneficial, neutral, or pathogenic. Beneficial microorganisms are those that form symbiotic or intimate relationships with plants, take part in nutrient mineralization and mobilization, produce various phytohormones, act as adversaries to plant pests, and/or confer resistance to environmental stresses (Hossain and Sultana 2020). These beneficial microbes are abundant in plant and soil-associated microenvironments, most notably the rhizosphere and rhizoplane. They primarily include rhizobia, plant growth-promoting fungi (PGPF), mycorrhizal fungi, actinomycetes, and diazotrophic bacteria. As dynamic microbial agents, PGPF are particularly effective at colonizing the rhizosphere and supporting plant growth following soil inoculation (Hossain et al. 2017). PGPF are considered highly valuable in agricultural fields and exploited to meet intended goals of sustainability due to their low environmental impact, extended lifespan, and multiple benefits to the host plants. Hence, numerous studies are ongoing across the world to explore the potential of PGPF for improving agricultural output by restoring soil and plant health.

Within PGPF, the highest and most diverse group of known species is found in the genus *Trichoderma*. *Trichoderma* is one of those fungi that are frequently used to stimulate plant growth. Many inhibit other microorganisms and are utilized as biological control agents (Figure 1). Some are known to induce resistance and protect distal plant parts from a wide array of pathogens and herbivores (Hossain et al. 2017). Apart from these beneficial capabilities, *Trichoderma* can also remediate toxic organic and inorganic contaminants, modulate abiotic stress tolerance, and supplement soil for biodegradation, decomposition, and nutrient enhancement (Woo et al. 2014) (Figure 1). *Trichoderma* is, therefore, one of the fungi currently marketed as an active ingredient in biofertilizers, biopesticides, and phytostimulants.

With the advancement of various "omics" technologies, the basis of *Trichoderma*-mediated beneficial effects on plants is currently genetically accessible. Genomic and genetic analyses of these species and their interactions with plants and pathogens have provided a wealth of information regarding the biosynthetic gene clusters, antimicrobial compounds, cell wall degrading enzymes, and intracellular signal transduction pathways associated with *Trichoderma*-mediated biocontrol.

[1] Department of Plant Pathology, Bangabandhu Sheikh Mujibur Rahman Agricultural University, Gazipur 1706, Bangladesh.
[2] College of Agricultural Sciences, International University of Business Agriculture and Technology, Dhaka 1230, Bangladesh.
* Corresponding author: hossainmm@bsmrau.edu.bd

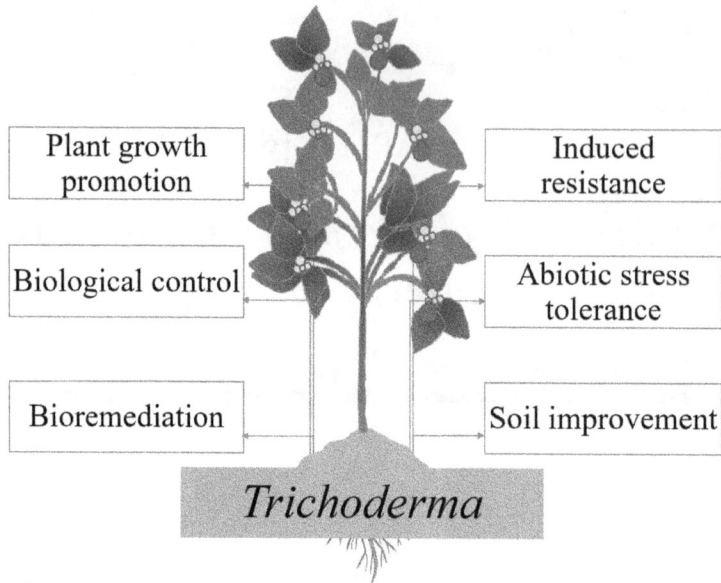

Figure 1. Schematic presentation of multiple roles and effects of *Trichoderma* on plants and soil.

Moreover, in recent years, the genomes of several *Trichoderma* species have been sequenced to identify important genes involved in critical biological processes. The present chapter aims to update recent development in the genetics and genomic features of *Trichoderma*-mediated biological control of plant diseases. This chapter would significantly advance our understanding of the biocontrol properties and identify the beneficial qualities of the genus *Trichoderma* for plants.

Trichoderma, a Worthy Genus of Numerous Bioagents

Trichoderma is an asexual, spore-producing, fungicolous ascomycete fungal genus. Taxonomically, its teleomorphs belong to the phylum Ascomycota, class-Sordariomycetes, order Hypocreales, and family Hypocreaceae. Taxonomically, its teleomorphs belong to the phylum Ascomycota, class-Sordariomycetes, order Hypocreales, and family Hypocreaceae. Fungi of the genus *Trichoderma* are quite a vast group of microorganisms. There were 375 *Trichoderma* species with recognized names as of July 2020 (Cai and Druzhinina 2021). Until now, Index Fungorum (http://www.indexfungorum.org) included 453 records of *Trichoderma* species, while the National Center for Biotechnology Information (NCBI) recorded 337 taxonomic taxa with sequences after eliminating confer (Cf.) names and uncharacterized isolates (Rush et al. 2021). Fungi of the genus *Trichoderma* are common in various habitats, primarily in soil, root environments, and decomposing organic matter. The fungi are well known for their remarkable diversities of lifestyles and interactions with various living species, including plants, animals, and other microbes. Many are omnivores, thriving on resources supplied by animals, plants, and other fungi (Kubicek et al. 2019). They can be opportunistic, free-living, and parasites on different groups of fungi. Some act as facultative endophytes (Druzhinina et al. 2011), while others have adapted to be biotrophic partners that boost plant vitality and development (Lorito et al. 2010). *Trichoderma* fungi are hardly reported as parasites on plants; instead, they can colonize most plants as a symptomless endosymbiont. Due to their affirmative effects on plants, a number of *Trichoderma* species have been the subjects of considerable interest in agricultural and biotechnological solutions (Harman et al. 2004, Harman 2006, Vinale et al. 2008, Lorito et al. 2010, Mendoza-Mendoza et al. 2018, Thambugala et al. 2020). Currently, 31 fungal genera are commonly employed as active bioagents for beneficial effects on plants, of which *Trichoderma* has by far the most species, with a total of thirteen (Rush et al. 2021). According to a corresponding study, ascomycete fungi show the maximum bioactivities against plant diseases, with *Trichoderma* being the source of the most employed bioactive ingredients (Thambugala et al. 2020). Although relatively fewer *Trichoderma* species are exploited for their growth promotion and biocontrol abilities compared to the vast number of species in the environment, this number vastly exceeds that of any other genus of fungi.

Trichoderma for Phytostimulant Applications

The PGPF *Trichoderma* helps plants grow sustainably by establishing mutualistic interactions with plants. Numerous researchers have shown that plant growth is enhanced when *Trichoderma* colonizes the rhizosphere (Yedidia et al. 2001, Mastouri et al. 2010, Nayaka et al. 2010, Kaveh et al. 2011, Delgado-Sanchez et al. 2011, Nagaraju et al. 2012, Jogaiah et al. 2013, Doni et al. 2014, Rahman et al. 2015, Lee et al. 2016, Jogaiah et al. 2018). *Trichoderma* species, such as *T. harzianum* (Nagaraju et al. 2012, Jogaiah et al. 2013), *T. virens* (Contreras-Cornejo et al. 2009, Jogaiah et al. 2018), *T. longibrachiatum* (Zhang et al. 2016), *T. asperellum* (Viterbo et al. 2010), *T. ghanense* (Martínez-Medina et al. 2014), *T. hamatum* (Martínez-Medina et al. 2014), *T. atroviride* (Chirino-Valle et al. 2016), *T. viride* (Hung et al. 2014) and *T. koningii* (Benítez et al. 1998) are extensively recognized for their growth promotion effect on diverse plant species in various agroecosystems (Table 1). The ability of the PGPF *T. harzianum* strain T-3 to stimulate a growth response in tomato plants was investigated in pots. When *T. harzianum* propagules were added into the soil, a 24% rise in seedling emergence was recorded up to 7 days after sowing. After four weeks, these plants displayed an 87% increase in cumulative root length, 68% increase in dry weight, 37% increase in shoot length and 63% increase in leaf number. Crop yield can be boosted by up to 300% when *T. hamatum* or *T. koningii* is added to fields (Chet et al. 1997, Benítez et al. 1998). Application of *Trichoderma* may result in augmented seed germination and seedling growth in rice (Rahman et al. 2015), maize (Contreras-Cornejo et al. 2009), wheat (Zhang et al. 2016), tomato (Vinale et al. 2008), pea (Akhter et al. 2015), chickpea (Hossain et al. 2013), cucumber (Yedidia et al. 2001), melon (Martínez-Medina et al. 2014), sugarcane (Yadav et al. 2009), pepper, lettuce (Vinale et al. 2008), sunflower (Lamba et al. 2008) and brinjal (Ghosh 2017). When chickpea was treated with *T. asperellum* in a mini pot trial, a 25% increase in seed germination and shoot length, a 27% rise in root length, and a 45% increase in vigour index were achieved (Ghosh and Pal 2017). The root endophytic *Trichoderma* isolates significantly increased the growth of the second-generation energy crop *Miscanthus giganteus* (Chirino-Valle et al. 2016). Italian ryegrass forage treated with *T. harzianum* IS005-12 showed a significant increase in germination and seedling growth (Banjac et al. 2021). When several fungi were given to rice seedlings in the experimental field, they all enhanced seed germination, shoot and root growth, with *T. virens* being the most effective (Ghosh and Panja 2021). It has recently been reported that combinations of *Trichoderma* strains or species stimulate plant growth more effectively than a single strain or species (Chirino-Valle et al. 2016, Chen et al. 2021). Similarly, the collective use of *Trichoderma* strains and other fungi has been described in several studies stating, for instance, that the combined use of *Trichoderma* and AM fungi can promote higher cucumber growth than their sole application (Chandanie et al. 2009). Co-inoculation of *Bacillus* OSU-142, *Glomus intraradices, Pseudomonas* BA-8, and *T. atroviride* can also augment plant yield, growth, and nutrition content (Colla et al. 2015). The same species or strain may not possess all plant growth-promoting traits or be able to promote the growth of particular crops and/or varieties. The presence of multiple species or strains in crop rhizosphere may complement each other and synergistically employ their beneficial effects on plants.

Table 1. Effect of plant growth-promoting fungi (PGPF) *Trichoderma* spp. on seed germination, plant growth and yield of various plant species

PGPF strain	Test Crop	Specific Effects	References
Trichoderma asperellum	Chickpea *(Cicer arietinum)*	~25% increase in seed germination and shoot length, a ~27% rise in root length, and a ~45% increase in vigour index	Ghosh and Pal (2017)
Trichoderma harzianum	Cucumber (*Cucumis sativus*)	~ 30% increase in seedling emergence	Yedidia et al. (2001)
T. harzianum	Maize (*Zea mays*)	Reduced *F. verticillioides* and fumonisin incidence and increased field emergence	Nayaka et al. (2010)
T. harzianum Bi	Muskmelon (*C. melo*)	Augmented seed germination	Kaveh et al. (2011)
T. koningii	*Opuntia streptacantha*	Broke seed dormancy	Delgado-Sanchez et al. (2011)
T. harzianum	Rice (*Oryzae sativa*)	Improved plant stand establishment	Rahman et al. (2015)
Trichodrma spp. SL2	Rice (*Oryzae sativa*)	Increased seed germination and seedling vigour	Doni et al. (2014)
T. harzianum	Sunflower (*Helianthus annuus*)	Increased seed germination and seedling vigour	Nagaraju et al. (2012)
T. harzianum T-22	Tomato (*Lycopersicon lycopersicum*)	Under stress, treated seed germinated consistently faster and more uniformly	Mastouri et al. (2010)
T. harzianum TriH_JSB27, TriH_JSB36	Tomato (*Lycopersicon lycopersicum*)	Enhanced early seedling emergence, seedling vigour and early flowering	Jogaiah et al. (2013)
T. viride (BBA 70239)	*A. thaliana*, Tomato (*L. lycopersicum*)	Increased shoot weight, lateral roots, chlorophyll contents, early flower and fruit development and yield	Lee et al. (2016)
T. virens Gv. 29-8	*A. thaliana*	Increased shoot biomass and lateral root production	Contreras-Cornejo et al. (2009)
T. harzinum	Chickpea (*Cicer arietinum*)	Enhanced shoot length and shoot biomass	Yadav et al. (2011)
T. harzianum GT3-2	Cucumber (*C. sativus*)	Enhanced the plant shoot dry weight	Chandanie et al. (2009)
T. harzianum T-22	GiSeLa6® (*Prunus cerasus × P. canescens*)	Improved shoot growth and ~76% increase in mean root length	Sofo et al. (2012)
T. harzianum T22	Maize (*Z. mays*)	Produced larger shoots and deeper and robust roots with a greater surface area	Harman et al. (2004)
T. harzianum, T. ghanense T. hamatum	Melon (*C. melo*)	Increased shoot and root fresh weight	Martínez-Medina et al. (2014)
T. atroviride	*Miscanthus × giganteus*	Enhanced plant height	Chirino-Valle et al. (2016)
T. virens Gv. 29-8	*Arabidopsis thaliana*	Induced production of lateral root (LR)	Contreras-Cornejo et al. (2009)
T. harzianum T-22	Cherry rootstocks (*Prunus cerasus* x *P. canescens*)	Produced larger root	Sofo et al. (2011)
T. hamatum DIS 219b	Maize (*Z. mays*)	Increased chlorophyll contents under drought	Bae et al. (2009)
T. virens	Maize (*Z. mays*)	Increased photosynthetic rate	Vargas et al. (2009)
T. atroviride TaID20G	Maize (*Z. mays*)	Improved the chlorophyll under drought stress	Guler et al. (2016)

Table 1 contd. ...

...Table 1 contd.

PGPF strain	Test Crop	Specific Effects	References
T. viride	*A. thaliana, C. forskohlii*	Showed robust and early flowering phenotype	Hung et al. (2014)
T. harzianum T-75	Chickpea (*Cicer arietinum*)	Enhanced grain yield in the field	Hossain et al. (2013)
T. harzianum	Mustard (*B. nigra*), Tomato (*L. lycopersicum*)	Enhanced yield	Haque et al. (2012)
T. harzianum T-3	Pea (*Pisum sativum*)	Enhanced seedling emergence and grain yield in the field	Akhter et al. (2015)
T. viride	Sugarcane (*Saccharum officinarum*)	Increased cane yield	Yadav et al. (2009)
T. harzianum T. viride	Sugarcane (*Saccharum officinarum*)	Increased millable canes, yield and commercial cane sugar (CCS t/ha)	Srivastava et al. (2006)
T. viride	*Coleus forskohlii*	Enhanced forskolin yield in roots	Boby and Bagyaraj (2003)
T. longibrachiatum FNBR-6	Rice (*O. sativa* L. var. IR-36) pea (*Pisum sativum* L. var. PG-3)	Enhanced carotenoid and protein content	Srivastava et al. (2012)
T. atroviride D16	*Salvia miltiorrhiz*	Higher tanshinone I (T-I) and tanshinone IIA (T-IIA) content	Ming et al. (2013)
T. harzianum	Sunflower (*H. annuus*)	Higher starch, total soluble sugars, reducing sugar, phenol, lipid and linoleic acid content	Lamba et al. (2008)
T. longibrachiatum T6	Wheat (*T. aestivum*)	Higher soluble sugar and protein content	Zhang et al. (2016)

Mechanisms of *Trichoderma*-mediated Growth Promotion

Mechanisms underlying the growth promotion effects of *Trichoderma* strains have been investigated. The course of plant growth promotion by *Trichoderma* occurs through enhanced one or more mechanisms of phytohormone production, nutrient mobilization, extenuation of plant stresses and suppression of deleterious microorganisms (Figure 2). In many *Trichoderma*-plant interactions, the significant increase in root and shoot growth is correlated with a substantial rise in IAA (Indole Acetic Acid) production by *Trichoderma* (Prasad et al. 2012, Lee et al. 2016, Jogaiah et al. 2018). *Trichoderma* may also produce abscisic acid, gibberellin, and other phytohormones that appear to play a role in promoting plant growth (Hossain et al. 2017). *Trichoderma* culture filtrates contain gibberellic acids, indole-3-acetic acid (IAA), and high phosphate concentrations that are believed to contribute to plant growth (Kang et al. 2015). Mineralizing nutrients from organic matter plays a dynamic role in the growth promotion of *Trichoderma*-colonized plants (Hossain and Sultana 2020). Another key mechanism of *Trichoderma*-mediated plant growth promotion is its ability to produce siderophores (Chen et al. 2021). Siderophore is a micromolecular iron-chelating ligand that improves Fe III solubilization and stimulates plant development. The use of multiple *T. harzianum* strains stimulates higher IAA and siderophore production, both of which are important in the process of growth promotion (Chen et al. 2021).

Some *Trichoderma* does not produce plant hormones or dissolve fixed phosphate, but they do promote plant growth. For instance, a number of *Trichoderma* isolates lacked growth-promoting traits such as P solubilization, IAA, and siderophore production but significantly improved the growth of bean seedlings (Hoyos-Carvajal et al. 2009). It is supposed that these *Trichoderma* isolates promote plant growth by accelerating soil mineralization. During the ratoon initiation stage of sugarcane, the application of *T. harzianum* strain Th 37 improved the availability of both macro- and micronutrients, as well as organic carbonate (Singh et al. 2010). The colonization of cucumber roots by *T. harzianum* increased the availability and plant uptake of essential nutrients

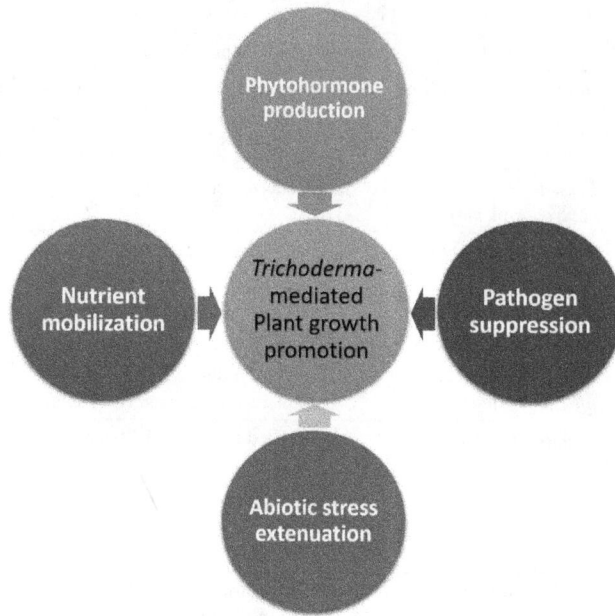

Figure 2. Schematic presentation of mechanisms potentially involved in the *Trichoderma*-mediated growth promotion of plants.

(Yedidia et al. 2001). *Trichoderma* assimilate complex polyaromatic compounds such as lignin and humic or phenolic acids more efficiently than many other microorganisms (Ruess 2003). The interaction between plants and *Trichoderma* has a significant impact on plant adaptation to biotic and abiotic stresses and subsequent plant growth. The colonization of maize roots by *T. atroviride* ID20G reduced drought stress and increased root fresh and dry weight (Guler et al. 2016). Adding *T. harzianum* to NaCl-treated mustard seedlings increased shoot, root, and plant dry weight by 13.8, 11.8 and 16.7%, respectively, compared to plants treated with NaCl (200 mM) alone (Ahmad et al. 2015). Under stress, however, *T. harzianum* T22 has little effect on tomato seedling performance, although treated seeds germinate earlier and more consistently than untreated seeds (Mastouri et al. 2010). As discussed in the latter portion of the chapter, *Trichoderma* protects and enables plants to resist harmful pathogens and promotes their improved growth. Moreover, volatile organic compounds (VOCs) generated by various *Trichoderma* species and strains increased Arabidopsis plant biomass and growth (Lee et al. 2016). VOCs are believed to primarily promote plant growth through CO_2 enrichment during co-cultivation (Kai and Piechulla 2009).

Molecular and Genetic Basis of *Trichoderma*-induced Plant Growth

The association of *Trichoderma* with plants induces a series of cellular changes in the plants leading to altered gene expression, biochemical processes, metabolic adjustments, and physiological adaptation. Recent breakthroughs in cell and molecular biology have enabled the elucidation of the genetic and molecular basis of *Trichoderma*-induced plant development. Arabidopsis plants inoculated with *T. virens* exhibited enhanced growth and the expression of auxin-regulated genes (Contreras-Cornejo et al. 2009). Mutations in genes associated with auxin transport or signalling, *AUX1, EIR1, BIG,* and *AXR1*, reduced the beneficial effects of *T. virens* on growth and root development in Arabidopsis. These data demonstrate that *T. virens* promotes plant growth via the classical auxin response pathway (Contreras-Cornejo et al. 2009). Shoresh and Harman (2008) observed that *Trichoderma* colonization of maize roots resulted in a significant increase in fructokinase (*FRK*), fructose bisphosphate aldolase (FBA), malate dehydrogenase (MDH), glyceraldehyde-3-P dehydrogenase (GAPDH), 3-phosphoglycerate kinase, β-glucosidases, and oxalate oxidases. The high expression of the *FRK2* gene in tomato leaves is required for stem growth and vascular development (Odanaka et al. 2002, Damari-Weissler et al. 2009). Suppression or decreased expression of this gene led to lower xylem and phloem cell sizes and significantly shorter plants (Odanaka et al. 2002, Damari-Weissler et al. 2009). The high level of expression of *FRK2* in stems approves a similar function. Cotton plants transformed with the tomato fructokinase gene (*LeFRK1*) have significantly increased leaf area and stem diameter (Mukherjee et al. 2015). Increased FBA content in plastids promotes tobacco plant growth (Uematsu et al. 2012). MDH is a component of the tricarboxylic acid cycle involved in the photosynthetic fixation of CO_2 (Nunes-Nesi et al. 2005). In Arabidopsis, single- and double-knockout mutants of the mitochondrial MDH isoforms lacked detectable MDH activity, resulting in tiny, slow-growing plants. These results demonstrate that *Trichoderma*-induced stimulation of glucose metabolism in plants contributes to increased shoot growth.

Utilization of *Trichoderma* for the Biocontrol of Plant Diseases

Trichoderma spp. are the most commonly utilized fungal biocontrol agents against plant diseases. Several *Trichoderma* species, such as *T. harzianumm* (Hossain et al. 2013, Akhter et al. 2015), *T. hamatum* (Kandula et al. 2015), *T. koningii* (Tsahouridou and Thanassoulopoulos 2002), *T. viride* (Khan and Gupta 1998), *T. virens* (Christopher et al. 2010, Izquierdo-García et al. 2020), *T. atroviride* (Ghazalibiglar et al. 2016), and *T. asperellum* (Thangavelu and Gopi 2015) are successfully used as biocontrol agents of plant diseases caused by various pathogens. They show great usefulness in controlling destructive soilborne diseases caused by fungi, oomycetes, protists, bacteria, nematodes, and viruses such as damping off (Akhter et al. 2015), Fusarium wilt (Prasad et al. 2016), bakanae (Watanabe et al. 2007), take-all (Zafari et al. 2008), dry root rot (Manjunatha et al. 2013), bacterial wilt (Kariuki et al. 2020), common scab (Gharate et al. 2016), soft rot (Hu et al. 2009), root-knot (Medeiros et al. 2017), cyst nematodes (Contina et al. 2017), clubroot (Cheah et al. 2000), and rhizomania (Naraghi et al. 2014). *Trichoderma* has also been used to successfully manage economically important foliar diseases, including anthracnose (Koike et al. 2001), gray mold (Mathys et al. 2012), angular leaf spot (Shoresh et al. 2005), bacterial leaf spot (Alfano et al. 2007), Phytophthora blight (Ahmed et al. 2000) and cucumber mosaic (Elsharkawy et al. 2013) in a wide range of host plants grown in the greenhouse and field (Table 2). The observed protective effects are highly variable among the *Trichoderma* strains. Studies have shown that despite promising results in lab settings, some *Trichoderma* biocontrol agents performed poorly or not at all when introduced to commercial field conditions (Hossain et al. 2017). The varied efficacy of biocontrol agents can be ascribed to their rhizosphere ineptitude (survival, colonization capacity), fluctuating production of essential metabolites or enzymes, and/or unstable product formulation (Ruocco et al. 2009). To improve the consistency and durability of the performance, combinations of *Trichoderma* species or strains are now becoming increasingly popular. As such, recently commercially available products, including Bioten® WP, BioWorks® Rootshield® Plus+ WP; Kiwivax®, VinevaxTM Bio-Dowel, and VinevaxTM Pruning Wound Dressing contain mixed strains or species of *Trichoderma* for better biocontrol (Rush et al. 2021). Different biocontrol agent consortia, composed of two or more *Trichoderma* strains or other microbial strains, are constructed to enhance the stability and efficacy of disease suppression. In many circumstances, combining multiple biocontrol agents has reported superior disease suppression. For example, co-application of *T. asperellum* GDFS1009 and *Bacillus amyloliquefaciens* ACCC1111060 was found to be more effective against infection by *Botrytis cinerea*, the causative agent of gray mold disease than either strain alone (Wu et al. 2018). The empirical mixture of *Trichoderma virens* GI006 and *Bacillus velezensis* Bs006 successfully controlled Fusarium wilt caused by *Fusarium oxysporum* f. sp. *phaseoli* in cape gooseberry and showed higher efficacy than the application of GI006 and Bs006 alone (Izquierdo-García et al. 2020). The successful colonization of the biological control agents is dependent on the germination of fungal conidiospores and *Bacillus* endospores. The interaction of

Table 2. Use of *Trichoderma* spp. for biological control of different plant pathogens in various host plants.

Disease	Pathogen name	Host plant	PGPF species	Effect on disease	Reference
Damping off	*Rhizoctonia solani*	*Pisum sativum*	*Trichoderma harzianum* isolate T-3	Reduced seedling mortality	Akhter et al. (2015)
	Rhizoctonia solani	*Lolium perenne*	*T. atroviride, T. virens*	Increased seedling emergence by 60–150% and 35–212%, respectively	Kandula et al. (2015)
	Sclerotinia trifoliorum	*Trifolium pratense*	*T. atroviride, T. hamatum*	Increased seedling emergence	Kandula et al. (2015)
	Sclerotium rolfsii	*Solanum lycopersicum*	*T. koningii* Rifai	Reduced disease incidence and increased seedling counts	Tsahouridou and Thanas-soulopoulos (2002)
	Macrophomina phaseolina	*Solanum melongena*	*T. hamatum, T. harzianum, T. viride*	Checked negative effect of *M. phaseolina* seedling mortality	Khan and Gupta (1998)
Fusarium wilt	*Fusarium udum*	*Cajanus cajan*	*T. harzianum* T-75	Reduced wilt and wet root rot incidence	Prasad et al. (2002)
	Fusarium oxysporum f. sp. *ciceris*	*Cicer arietinum*	*T. harzianum*	Reduces wilt incidence	Hossain et al. (2013) Dubey et al. (2007)
	Fusarium oxysporum f. sp. *lycopersici*	*Solanum lycopersicum*	*T. harzianum*	Reduced wilt incidence	Basco et al. (2017)
	Fusarium oxysporum f. sp. *lycopersici*	*Solanum lycopersicum*	*T. virens*	Reduced wilt incidence	Christopher et al. (2010)
	Fusarium oxysporum f. sp. *lycopersici*	*Solanum lycopersicum*	*T. atroviride*	Reduced wilt incidence	Ghazalibiglar et al. (2016)
	Fusarium oxysporum f. sp. *cubense* (Foc) race 1	Banana	*T. asperellum* prr2, *T.* sp. NRCB3	A 47% reduction of Fusarium wilt incidence	Thangavelu and Gopi (2015)
	Fusarium oxysporum f. sp. *physali* (Foph)	Cape gooseberry	*T. virens* GI006	Reduced wilt severity	Izquierdo-Garcia et al. (2020)
	Fusarium oxysporum f. sp. *radices cucumerinum*	*Cucumis sativus*	*T. harzianum* Tr6	Reduction in disease incidence index	Alizadeh et al. (2013)
	Fusarium oxysporum f. sp. *cucumerinum*	*Cucumis sativus*	*Trichoderma* sp. GT3-2	Reduction in disease index and severity	Koike et al. (2001)
Seedling blight	*Rhizopus* sp.	*Oryza sativa*	*T. harzianum* GT3-2	Reduction in disease severity	Elsharkawy et al. (2014)
Bakanae	*Fusarium semitectum Fusarium moniliformae*	*Oryza sativa*	*T. harzianum* GT3-2	Reduction in disease severity	Elsharkawy et al. (2014)
	Gibberella fujikuroi	*Oryza sativa*	*T. asperellum* SKT-1	Reduced bakanae symptoms	Watanabe et al. (2007)

Table 2. contd. ...

Disease	Pathogen	Host	Trichoderma	Effect	Reference
Take-all	*Gaeumannomyces graminis var tritici*	*Triticum aestivum*	*T. virens* T65, T90, T96, T122, *T. koningii* T77	Reduced 25 to 55% disease severity	Zafari et al. (2008)
Dry root rot	*Macrophomina Phaseolina*	*Cicer arietinum*	*T. viride*	Reduced root rot incidence	Manjunatha et al. (2013)
Bacterial wilt	*Ralstonia solanacearum*	*Solanum lycopersicum*	*Trichoderma* spp. isolate T1	Reduced incidence and severity of wilt	Kariuki et al. (2020)
			Trichoderma spp. AA2	Prevented 92-97% of the bacterial wilt infection in the field	Yendyo et al. (2017)
		Solanum tuberosum	*T. asperellum* T34	Reduced severity of the disease under greenhouse and field conditions	Mohamed et al. (2020)
			T. asperellum (T4 and T8)	Delayed wilt development and decreased disease incidence	Konappa et al. (2018)
Common scab	*Streptomyces scabies*	*Solanum tuberosum*	*T. asperellium*	Reduced disease incidence and index with minimum diseased tuber yield	Gharate et al. (2016)
			T. viride	Reduced percent tuber infection by common scab	Shiwangi and Pathak (2019)
Soft rot	*Pectobacterium carotovorum* subsp. *carotovorum*	*Brassica rapa*	*T. pseudokoningii* SMF2	Reduced bacterial soft rot infection in the field, giving protection of up to 82.08%	Hu et al. (2009), Li et al. (2014)
		Solanum tuberosum	*T. viride, T. virens, T. harzianum*	Reduced soft rot symptoms on inoculated potato tuber slices	Abd-El-Khair et al. (2021)
Root-knot	*Meloidogyne javanica*	*Solanum lycopersicum*	*T. atroviride*	Reduced number of galls, nematodes, and eggs	Medeiros et al. (2017)
	Meloidogyne incognita	*Solanum lycopersicum*	*T. asperellum* (T34), *T. harzianum* (T22)	Reduced the number of eggs per plant	Pocurull et al. (2020)
	Meloidogyne incognita	*Solanum lycopersicum*	*T. harzianum* T-78	Reduced the number of eggs per plant	Martinez-Medina et al. (2017)
Cyst nematodes	*Globodera pallida*	*Solanum tuberosum*	*T. harzianum* ThzID1-M3	Reduced infection and reproduction in soil	Contina et al. (2017)
Club root	*Plasmodiophora brassicae*	*Brassica oleracea*	*Trichoderma* spp. isolate TC32 TC45 and TC63	Reduced clubroot severity in the glasshouse and field	Cheah et al. (2000)
	Plasmodiophora brassicae	*Brassica rapa*	*T. harzianum* LTR-2	Reduced disease incidence by 45.4% and lowered the abundance of pathogen	Li et al. (2020)
	Plasmodiophora brassicae	*Brassica oleracea*	*T. hamatum, T. harzianum*	Decreased the clubroot incidence	Suada (2017)
	Plasmodiophora brassicae	*Brassica rapa*	*T. harzianum* T4	Reduced clubroot incidence	Yu et al. (2015)

Table 2. contd....

...Table 2. contd.

Disease	Pathogen name	Host plant	PGPF species	Effect on disease	Reference
Rhizomania	*Polymyxa betae*	*Beta vulgaris*	*T. harzianum*	Reduced cystosorus population and increased root weight	Naraghi et al. (2014)
Anthracnose	*Colletotrichum orbiculare*	*Cucumis sativus*	*Trichoderma* sp. GT3-2	Reduction in total lesion area, and lesion diameter	Koike et al. (2001)
	Colletotrichum graminicola	*Oryza sativa*	*T. virens* Gv29-8	Reduction in lesion area	Djonović et al. (2006)
	Cochliobolus graminicola	*Zea mays*	*T. virens* Gv29-8	Reduction in lesion area	Djonović et al. (2007)
	Colletotrichum sp.	*Gossypium* spp.	*T. virens* strain Gv29-8	Reduction in lesion area	Djonović et al. (2006)
Gray mold	*Botrytis cinerea* B05-10	*Arabidopsis thaliana*	*T. hamatum* T382	Reduction in average lesion diameter	Mathys et al. (2012)
	Botrytis. cinerea	*Arabidopsis thaliana*	*T. harzianum* Tr6	Reduction in disease incidence index	Alizadeh et al. (2013)
	Botrytis cinerea Strain B4	*Arabidopsis thaliana*	*T. harzianum* Rifai T39	Reduction in percentage of mean lesion area	Korolev et al. (2008)
	Botrytis cinerea	*Solanum lycopersicum*	*T. harzianum* T-78	Reduction in lesion size	Martinez-Medina et al. (2013)
Angular leaf spot	*Pseusomonas syringae* pv. *lachrymans*	*Cucumis sativus*	*T. asperellum* T203	Inhibited multiplication of bacterium	Shoresh et al. (2005)
	Pseusomonas syringae pv. *lachrymans*	*Cucumis sativus*	*Trichoderma* sp. GT3-2	Reduction in disease index and severity	Koike et al. (2001)
Bacterial leaf spot	*Xanthomonas euvesicatoria* 110c	*Solanum lycopersicum*	*T. hamatum* 382	Reduction in disease severity	Alfano et al. (2007)
Bacterial leaf spot	*Xanthomonas campestris* pv. *vesicatoria*	*Solanum lycopersicum*	*T. harzianum*	Reduction in spot numbers, disease severity and pathogen proliferation	Saksirirat et al. (2009)
Phytophthora blight	*Phytophthora capsici*	*Piper nigrum*	*T. harzianum* isolate 2413	Reduction in length of necrosis in the stem	Ahmed et al. (2000)
Cucumber mosaic	*Cucumber mosaic virus (CMV-Y)*	*Arabidopsis thaliana*	*T. asperellum* SKT-1	Reduction in CMV symptoms	Elsharkawy et al. (2013)
	Cucumber mosaic virus (CMV)	*Solanum lycopersicum*	*T. harzianum* T-22	Reduction in CMV infection severity and accumulation	Vitti et al. (2016)

Trichoderma conidia with *Bacillus* biofilms did not inhibit the germination and establishment of the fungus in soil (Izquierdo-Garca et al. 2020). *Trichoderma* mycelia can stimulate the production of bacterial biofilms and facilitate the migration of bacteria in the soil. Nutrients in the fungus exudates facilitated bacterial growth (Triveni et al. 2012).

Mechanisms of *Trichoderma*-mediated Biological Control of Plant Diseases

Various mechanisms have been anticipated to explain disease control by *Trichoderma* spp. However, clear proof of the involvement of a particular mechanism is challenging to establish. It is believed that *Trichoderma* controls plant diseases by engaging a coordinated network of different mechanisms. There is vital evidence for the role of direct and indirect mechanisms in the *Trichoderma*-mediated biological control of plant diseases. Direct protection is primarily achieved through the direct inhibition of pathogen activity by *Trichoderma* isolates employing one or more antagonistic mechanisms, including antibiosis, mycoparasitism, and competition for resources like nutrients or space. However, the indirect pathogen inhibition by *Trichoderma* is mainly thrived by plant defense activities that do not involve direct antagonistic interaction with the pathogen.

Mycoparasitism

Mycoparasites are fungi that establish themselves as parasites on the active hypha or dormant structure of another fungus. The parasitized fungus is frequently referred to as the host or prey. Mycoparasitism is considered as the ancestral trait in *Trichoderma*/Hypocrea and a key mechanism by which *Trichoderma* exhibits direct anti-pathogen activity. The capacity of *Trichoderma* to parasitize and kill other fungi has been a significant factor in its economic success as biofungicides, particularly against soilborne pathogens. Weindling (1932) was the first to report the mycoparasitic behaviour of *T. virens* against *R. solani*: "coiling of hyphae, growth in straight or wavy lines, coagulation of protoplasts, and loss of vacuolated structures." However, the antagonistic action of the *Trichoderma* strain was eventually ascribed to a "lethal principle," which was later recognized as gliotoxin (Weindling and Emerson 1936). The strain efficacy in suppressing *R. solani* was subsequently confirmed in pot tests (Weindling and Fawcett 1936). When *T. hamatum* was used as a seed treatment, mycoparasitism was proposed as the mechanism of biocontrol of *Pythium* spp. and *R. solani* in radish (Harman et al. 1980). White beans (*Phaseolus vulgaris*) infected with *R. solani* were protected from the disease by a mycoparasite strain of *T. virens* (Tu and Vaartaja 1981). By acting as a parasite on the sclerotia, the application of *T. koningii* was able to diminish the inoculum levels of *Sclerotinia sclerotiorum* (Helotiales) in the soil (Trutmann and Keane 1990). The biocontrol agent *T. harzianum* CECT 2413 demonstrated mycoparasitic activity against various pathogens, including *R. meloni* and *Phytophthora citrophthora* (Moreno-Mateos et al. 2007). This strain effectively controlled *Verticillium dahliae* Kleb in olive (Ruano-Rosa et al. 2015). *In vitro*, microscopic assays indicated that mycoparasitism is the potential mechanism of disease suppression by *T. harzianum*, as evidenced by coiling around *Verticillium* hypha by *Trichoderma* (Ruano-Rosa et al. 2015). Another biocontrol agent *T. asperellum* expressed chitinases and β-1,3-glucanases during parasitism of the cotton root rot pathogen *Phymatotrichopsis omnivore* (Guigón-López et al. 2015). Production of lytic enzymes such as chitinases, β-1,3-glucanases, proteases and cellulases by hyperparasite play a vital function in digesting host-pathogen cells (Harman 2000). Chitinases, glucanases and proteases are most important to degrade fungal cell walls, while cellulases are useful for colonizing oomycete hyphae (Hossain and Sultana 2020, Hossain 2022).

Competitive Exclusion

Competition for vital nutrients and colonization niche is crucial in the biological control of plant diseases by *Trichoderma* spp. Both pathogens and *Trichoderma* compete for food and space to establish themselves successfully and proliferate subsequently in the host environment. Like many other biocontrol agents, *Trichoderma* spp. are highly rhizosphere competent and specialized in nutrient uptake. They often colonize the host surface and establish themselves in the niche more quickly than pathogens. For example, *T. harzianum* T35 was able to control the fusarium wilt pathogen *Fusarium oxysporum* due to its superior ability to compete for nutrients and rhizosphere colonization (Tjamos et al. 1992). Besides, *Trichoderma* demonstrates a competitive ability to absorb scarce nutrients (Hossain et al. 2017). In oxidized and aerated soils where iron is limited, *Trichoderma* primarily synthesizes iron-chelating siderophores to combat iron scarcity caused by other pathogenic fungi (Benítez et al. 2004). Due to these two characteristics, *Trichoderma* competitively excludes soilborne pathogens from infection sites, making the pathogen less likely to establish and cause the disease.

Antibiosis

Antibiosis is an antagonistic relationship between two or more microorganisms due to the toxicity of secondary metabolites produced by one microorganism for other microorganisms. Antibiosis is a widespread phenomenon accountable for the biological control activity of many biocontrol agents. *Trichoderma* species produce a large amount of secondary metabolites, both volatile and non-volatile, that exhibit biological activities against plant pathogens, including *Sclerotium rolfsii* (Figure 3). Over the years, a vast array of *Trichoderma* molecules has been identified, and their role in suppressing plant pathogens has been studied. Based on the structures, two distinct classes of *Trichoderma* antibiotic molecules have been identified: (1) low-molecular-weight nonpolar volatile metabolites, which include polyketides such as pyrones and butenolides, simple aromatic compounds, isocyanate metabolites, and volatile terpenes, and (2) high-molecular-weight polar metabolites, which include peptaibols and diketopiperazinelike gliotoxin and gliovirin compounds (Vinale et al. 2006, Xiao-Yan et al. 2006, Vinale et al. 2009). A single *Trichoderma* strain can produce several secondary metabolites with distinct functions and antimicrobial activity against different pathogen species. For example, the antifungal effect of *T. virens* could be attributed to the production of gliotoxin, a generally nonselective antibiotic (Whipps and Lumsden 2001). The fungus produces another antibiotic gliovirin, inhibitory to *Pythium ultimum* (Howell et al. 1993, Howel 2003). Strain "Q" of *T. virens* produces gliotoxin but not gliovirin in inhibiting *Rhizoctonia solani*, whereas strain "P" produces gliovirin but not gliotoxin and is unable to inhibit *R. solani* (Howell et al. 1993). The primary bioactive elements of *T. harzianum* and *T. viride* were palmitic and acetic acids, respectively (Yassin et al. 2021). *Trichoderma harzianum* extract contains harzianic acid, which acts as an anti-mycotic agent against *Sclerotinia sclerotiorum*, *Rhizoctonia solani*, and *Pythium irregulare* (Vinale et al. 2009). Harzianic acid stimulates plant development and is considered a novel siderophore due to its affinity for Fe^{3+} (Vinale et al. 2012). 6-Pentyl pyrone (6-PP), an unsaturated

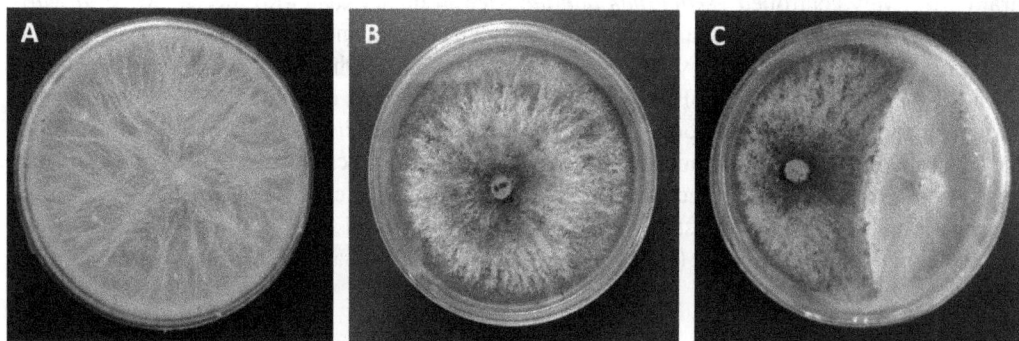

Figure 3. Antagonism of *Trichoderma harzianum* Th-2 against tomato southern blight pathogen *Sclerotium rolfsii*. (A) *Sclerotium rolfsii* culture without *Trichoderma* (B) *Trichoderma harzianum* Pb-22 (TPb-22) culture without *S. rolfsii* (C) *S. rolfsii* culture with *Trichoderma harzianum* Pb-22 (Th-2).

lactone produced by *T. atroviride* and *T. harzianum,* shows antifungal and growth-promoting activities (Vinale et al. 2008). 6-PP production was reported in strains of *T. koningii*, which inhibited mycelial growth and altered the morphology of filamentous fungal hyphae (Ismaiel and Ali 2017). Cerinolactone, a novel metabolite isolated from *T. cerinum* cultures, was active against *R. solani, B. cinerea,* and *P. ultimum* (Vinale et al. 2012).

Peptaibols have been demonstrated to limit the growth of fungi pathogens cooperatively with cell wall-degrading enzymes. On isolated plasma membranes of *B. cinerea*, the peptaibols trichorzianin TA and TB reduced β-glucan synthase activity (Lorito et al. 1996). The potent bioactivity of *T. stromaticum* strains against the cacao disease pathogen *Moniliophthora perniciosa* is likely to be caused by trichostromaticins A–E (Aime and Phillips-Mora 2005). It has been demonstrated that a combination of 20-residue peptaibols isolated from *T. citrinoviride* has antifungal efficacy against pathogens that affect forest trees (Maddau et al. 2009). Trichokonin VI, found in *T. pseudokoningii* cultures, induces cytoplasmic vacuole formation and metacaspase-independent apoptosis in Fusarium oxysporum (Shi et al. 2012). Cell-wall disintegrating enzymes and secondary metabolites sometimes collaborate, especially when *Trichoderma* suppresses plant pathogens via its highly antagonistic and mycoparasitic capabilities (Viterbo and Horwitz 2010). The metabolites produced by *Trichoderma* may have a role in the competition with other microorganisms and are also directly engaged in plant benefits such as growth, nitrogen uptake, fertilizer usage efficiency, seed germination, and tolerance to biotic and abiotic stress damages (Viterbo et al. 2007). Even some *Trichoderma* produces secondary metabolites that are antimicrobial at high concentrations but act as auxin-like molecules and microbe-associated molecular patterns (MAMPs) at low concentrations (Vinale et al. 2008).

Induced Defense Responses

In 1997, it was first revealed that root colonization by *Trichoderma* could alleviate foliar disease, a phenomenon which is popularly dubbed as induced systemic resistance (ISR) (Bigirimana et al. 1997). For the next two decades, ISR as a mechanism of plant disease management led to *Trichoderma* research (Yedidia et al. 1999, Koike et al. 2001, Djonović et al. 2006, Alfano et al. 2007, Djonović et al. 2007, Korolev et al. 2008, Contreras-Cornejo et al. 2011, Alizadeh et al. 2013, Martínez-Medina et al. 2014, Hossain et al. 2017). These studies have exhibited that *Trichoderma* spp. can help plants combat plant diseases by making them more resistant (local or systemic). Typically, *Trichoderma* colonization of the roots confers a broad spectrum of ISR in host plants and protects them from various diseases. *Trichoderma*-mediated ISR triggers physiological changes in plants, activating multiple defense mechanisms and an enhanced state of resistance to the invading pathogen (Alfano et al. 2007). Cucumber plants co-cultured with *T. harzianum* produced more peroxidases and chitinases than nontreated plants (Yedidia et al. 1999). Resistance to *Pseudomonas syringe* pv. *lachrymans* was imparted by *T. asperellum* T34 via modulation of proteins involved in stress tolerance, isoprenoid and ethylene production, photorespiration, and glucose metabolism (Segarra et al. 2009). Root colonization by *T. atroviride* and *T. virens* triggered the accumulation of hydrogen peroxide and camalexin and induction of SA (salicylic acid)-inducible pathogenesis-related reporter marker gene *pPr1a: uidA* and JA (jasmonic acid)-inducible marker gene *pLox2: uidA*, resulting in a significant reduction in disease severity caused by *B. cinerea* in *Arabidopsis thaliana* (Contreras-Cornejo et al. 2011). It was observed that treating tomato plants with *T. asperellum* and *T. atroviride* elicited the expression of plant defense- and mycoparasitic-related genes against the oomycete *Phytophthora nicotiana* (La Spada et al. 2020). The findings show that *Trichoderma* employs both induced defence and direct antagonistic activity against *P. nicotianae*. Similarly, *T. asperelloides* PSU-P1 induced pathogenesis-related protein gene expression in muskmelon (*Cucumis melo*) against gummy stem blight in the field (Intana et al. 2022). Many other findings exist that demonstrate that *Trichoderma* spp. can stimulate lignin deposition, total phenolic content, β-1, 3-glucanase, chitinase, phenyl alanine ammonia lyase, peroxidase, and phenylpropanoid activities, enabling plants to defend themselves against the invading pathogens (Koike et al. 2001, Shoresh et al. 2005, Hossain et al. 2017, Bisen et al. 2019).

Using single *Trichoderma* species is a widespread practice for inducing defense responses against pathogens. However, utilizing a *Trichoderma* consortium consisting of strains with distinct characteristics may open new avenues for investigating its agricultural applications. Brinjal seeds treated with two compatible *Trichoderma* isolates (BHU51 and BHU105) singly and in combination were evaluated for defense induction against *Sclerotium rolfsii* (Bisen et al. 2019). *Trichoderma* consortium-treated seeds had the lowest damping-off rate. At the same time, *S. rolfsii* infection of *Trichoderma* consortia-treated plants resulted in an overall increase in defense-related enzymes, antioxidants, and phenolic content, particularly shikimic acid, gallic acid, t-chlorogenic acid, and syringic acid. Likewise, another recent study revealed that biopriming of brinjal with the consortium of *Trichoderma* spp. provided robust protection against *S. sclerotiorum* and demonstrated superior activity in inducing defense-related enzymes such as phenyl alanine ammonia lyase, peroxidase, polyphenol oxidase and various phenolics with a more significant amount of each than with an individual treatment (Singh et al. 2021). A consortium of two or more microbes may produce a more robust response than a single microorganism formulation, improving disease management by eliciting multiple mechanisms and maybe being more stable across a wide range of environmental circumstances (Jain et al. 2012, Singh and Singh 2015, Keswani et al. 2019).

Peptaibols produced by *Trichoderma* have also been demonstrated to induce resistance to plants against diseases. Alamethicin, a 20-residue peptaibol from *T. viride*, triggers jasmonic acid and salicylic acid biosynthesis in lima beans (Engelberth et al. 2001). In Brussels sprout plants, alamethicin was 20-fold more potent in inducing plant defense than JA (Bruinsma et al. 2009). Peptaibols containing 18 residues from *T. virens* induce systemic defenses in cucumber against the foliar bacterial pathogen *Pseudomonas syringae* pv. *lachrymans* (Viterbo et al. 2007). *Trichoderma* peptaibols have been discovered to exert their antiviral activity via various defensive signalling mechanisms (Luo et al. 2010). Interestingly, it has been demonstrated that 6-PP activates plant defense systems and regulates plant growth in tomatoes, peas, and canola at low concentrations (1 ppm) (Vinale et al. 2008), implying that plant defense and developmental responses to *Trichoderma* have comparable components.

Biocontrol Genes in *Trichoderma*

Numerous soilborne diseases are effectively controlled through the widespread use of *Trichoderma* spp., which can efficiently synthesize cell wall-destroying enzymes and/or produce antifungal compounds. In most cases, the synergy between cell degradation and antimicrobial activity plays a vital role in the biocontrol of plant pathogens by *Trichoderma* spp. The genetic basis of mycoparasitic and antimicrobial activities of *Trichoderma* spp. have been examined comprehensively using functional genomics techniques (Marra et al. 2006). Several genes have successfully been isolated and cloned as essential biocontrol genes from *Trichoderma* spp. (Massart and Jijakli 2007). Many of these are crucial for degrading cell walls, and others are required for antimicrobial activities. Recent reports have illuminated the microbial genes and intercellular signalling pathways underlying *Trichoderma*-mediated biocontrol activities against plant pathogens (Keller 2019).

Genes for Cell Wall Degradation

The cell wall of fungal pathogen comprises polysaccharides (80%) and proteins (3–20%), with trace amounts of lipids, pigments, and inorganic ions. ß-glucan, chitin, and mannoproteins are the major macromolecular components of the cell walls of higher fungi (Ascomycetes, Basidiomycetes, and Deuteromycetes) (glycoproteins), while cellulose predominates over chitin in the lower fungi (Myxomycetes, Phycomycetes). Depending on the fungal species, genes encoding cell wall

Table 3. Microbial genes involved in the biocontrol activity of *Trichoderma* against plant pathogens.

Gene	*Trichoderma* strain	Function	References
ech-42	*Trichoderma atroviride*	Encoding chitinases against *Rhizoctonia solani*	Cortés et al. (1998)
	T. hamatum	Encoding chitinases against *Sclerotinia sclerotiorum*	Steyaert et al. (2003)
prb1	*T. atroviride*	Encoding proteinase against *Rhizoctonia solani*	Cortés et al. (1998)
	T. hamatum	Encoding proteinase against *Sclerotinia sclerotiorum*	Steyaert et al. (2003)
exc1, exc2, chit42, chit33 prb1, bgn 13.1	*T. harzianum*	Encoding for N-acetyl-β-D-glucosaminidase, protease, and β-glucanase showed mycoparasitic activity against pathogens, especially *Fusarium*	Lopez Mondejar et al. (2011)
gluc78	*T. atroviride*	Encoding ß-glucanase against *Pythium ultimum* but not against *Rhizoctonia solani*	Donzelli et al. (2001)
Th-Chit	*T. harzianum*	Encoding endochitinase against *Alternaria alternata*	Saiprasad et al. (2009)
Tga1	*T. atroviride*	Encoding chitinase against *Rhizoctonia solani*, *Botrytis cinerea*, and *Sclerotinia. sclerotiorum*	Reithner et al. (2005)
ech42	*T. harzianum*	Encoding endochitinase against pathogens *B. cinerea*, and *Rhizoctonia solani*	Woo et al. (1999)
tvsp1	*T. virens*	Encoding a serine protease against *Rhizoctonia solani*	Pozo et al. (2004)
tag83	*T. asperellum*	Encoding Exo-1,3-glucanase against *Rhizoctonia solani* cell walls	Marcello et al. (2010)
TvBgn2, TvBgn3	*T. virens*	Encoding β-1,3 and β-1,6 glucanases against *Rhizoctonia solani, Pythium ultimum*, and *Rhizopus oryzae*	Djonovic et al. (2007)
gluc78	*T. atroviride*	Encoding an antifungal glucan 1,3-β-glucosidase, which causes cell wall degradation of *Pythium ultimum*	Donzelli et al. (2001)
cre1	*T. harzianum*	Repression of genes encoding cellulase and xylanas	Saadia et al. (2008)
ThPG1	*T. harzianum*	Encoding endopolygalacturonase involved in the degradation of the cell walls of *Rhizoctonia solani* and *Pythium ultimum*	Morán-Diez et al. (2009)
Tv6Gal	*T. viride*	Encoding endo-β-(1-6)-galactanase that catalyzes the hydrolysis of β-1,3-linked galactosyl oligosaccharides	Kotake et al. (2004)

degrading enzymes such as chitinases, ß-l,3-glucanases, proteases, or cellulases play a crucial role in the mycoparasitic activity of *Trichoderma* species (Table 3). For example, a ß-glucanase gene, *gluc78* was expressed in *T. atroviride* during its confrontation with *P. ultimum* but not with *R. solani* (Donzelli et al. 2001). A gene named *tvsp1* encoding a serine protease was successfully cloned from *T. virens*, which was found to play a critical role in degrading the fungal cell wall and biocontrol of *R. solani*, a pathogen that affects cotton seedlings (Pozo et al. 2004). The expression of the *tag83* gene, encoding the cell wall degrading enzyme Exo-1,3-glucanase, was isolated and characterized from *T. asperellum*. The expression of the *tag83* gene indicated that the glucanase enzyme is required for the parasitic action of *T. asperellum* against *R. solani* (Marcello et al. 2010). *T. virens* transformant strains expressing two distinct types of β-1,3 and β-1,6 glucanase genes, namely *TvBgn2* and *TvBgn3,* produce an enzyme that degrades the fungal cell walls and aids in biocontrol activity against pathogens such as *Rhizopus oryzae, P. ultimum*, and *R. solani* (Djonovic et al. 2007). A gene, *gluc78*, from *T. atroviride* encodes an antifungal glucan 1,3-β-glucosidase, which plays a role in the cell wall degradation of *Pythium ultimum* (Donzelli et al. 2001). A glucose repressor gene, *creI*, was isolated and characterized from *T. harzianum*. This gene is responsible for the repression of genes encoding cellulase and xylanase (Saadia et al. 2008). When the *Trichoderma* strain SY was grown exclusively on cellulose as a carbon source, a gene, *Xyl* was highly expressed (Min et al. 2002). The gene *ThPG1* encoding endopolygalacturonase, isolated and characterized from *T. harzianum,* is involved in the degradation of the cell walls of *P. ultimum* and *R. solani* (MoranDiez et al. 2009). A T. *viride* gene, *Tv6Gal*, encoding endo-β-(1-6)-galactanase, was isolated, cloned, and expressed in *Escherichia coli* (Kotake et al. 2004). Galactanase enzymes are members of the arabinogalactan protein family that catalyzes the β-1,3-linked galactosyl oligosaccharides and polysaccharides (Ichinose et al. 2008).

Serine proteases are essential cell-degrading enzymes for biocontrol activity by many *Trichoderma* strains. Flores et al. (1997) observed a fivefold increase in biocontrol activity when several copies of the protease gene *prb1* were integrated into the genome of *T. atroviride*. Chitinase (*ech-42*) and protease (*prb1*) genes were both highly expressed in *T. atroviride* during its confrontation with *R. solani* (Cortés et al. 1998) and in *T. hamatum* during a confrontation with *Sclerotinia sclerotiorum* (Steyaert et al. 2003). A novel serine protease gene named *SL41* was successfully cloned and expressed in *Saccharomyces cerevisiae* from *T. harzianum* (Liu et al. 2009). From five isolates of *T. harzianum*, genes encoding for N-acetyl-β-D-glucosaminidase (*exc1* and *exc2*), chitinase (*chit42* and *chit33*), protease (*prb1*), and β-glucanase (*bgn* 13.1) were cloned and expressed, which played a major role in the mycoparasitic activity against fungal pathogens, especially *Fusarium oxysporum* (López-Mondéjar et al. 2011).

Genes for Antimicrobial Activity

Trichoderma exhibits antimicrobial activity against a wide variety of bacteria, yeasts, and filamentous fungi (Vizcano et al. 2005), due to its ability to produce diverse secondary metabolites such as peptaibols, gliotoxin, gliovirin, polyketides, pyrones, and terpenes (Vinale et al. 2008). The genomic analysis of the *Trichoderma* has identified a number of genes responsible for various antimicrobial metabolites (Table 4). A trichodiene synthase gene *tri5* identified in *T. harzianum* encodes trichothecene that inhibits protein and DNA synthesis in pathogen cells conferring toxicity to *Fusarium* species (Gallo et al. 2004). Overexpression of the *tri5* gene in *T. brevicompactum* contributes to forming an antifungal compound trichodermin, which is active against *S. cerevisiae, Kluyveromyces marxianus, Candida albicans, Candida glabrata, Candida tropicalis*, and *Aspergillus fumigatus* (Tijerino et al. 2011). The *thga1* gene knock mutant of *T. harzianum* Th-33 had a reduced growth rate, conidial yield, cAMP level, hydrophobicity, and antagonism against *R. solani* than the wild type, indicating that the *thga1* gene positively regulates the growth, conidial production, hydrophobicity, and antagonism of *T. harzianum* against *R. solani* (Sun et al. 2016). The deletion mutants of *T. atroviride* showed that *Taabc2* gene is essential for antagonism and biocontrol against *P. ultimum* and *R. solani* in tomato plants (Ruocco et al. 2009). A transcription factor gene *Thctf1,* identified in *T. harzianum,* is involved in synthesizing 6-pentyl-2H-pyran-2-one (6-PP) and exhibiting antifungal action against *R. solani, B. cinerea*, and *S. rolfsii* (Rubio et al. 2009). The *erg1* gene from *T. harzianum* encodes an enzyme named squalene epoxidase that helps synthesize an antifungal compound active against *Saccharomyces cerevisiae* (Cardoza et al. 2006). From *T. hamatum,* the monooxygenase gene was isolated, which helps in the antifungal activity against some pathogens like *Sclerotium sclerotiorum, Sclerotinia minor*, and *Sclerotium cepivorum* (Carpenter et al. 2008). A gene, *Sm1*, isolated from *T. virens* encodes a small cysteine-rich protein that induces anti-disease activity in dicot and monocot plants (Buensanteai et al. 2010). These

Table 4. Major antifungal genes isolated from various *Trichoderma* strains having roles in the antifungal activity against plant pathogens.

Gene	*Trichoderma* strain	Function	References
tri5	*T. harzianum*	Encoding the enzyme trichothecene, which inhibits protein, DNA synthesis and growth of *Fusarium* species	Gallo et al. (2004)
tri5	*T. brevicompactum*	Encoding an antifungal compound trichodermin against *S. cerevisiae, Kluyveromyces marxianus, Candida albicans, C. glabrata, C. tropicalis,* and *Aspergillus fumigatus*	Tijerino et al. (2011)
thga1	*T. harzianum*	Positively influences growth rate, conidial yield, cAMP level, hydrophobicity and antagonism against *R. solani*	Sun et al. (2016)
Taabc2	*T. atroviride*	Protects tomato plants from *Pythium ultimum* and *Rhizoctonia solani* attack	Ruocco et al. (2009)
Thctf1	*T. harzianum*	Involved in the production of 6-pentyl-2H-pyran-2-one (6-PP) and shows antifungal activity against *R. solani, B. cinerea,* and *S. rolfsii*	Rubio et al. (2009)
erg1	*T. harzianum*	Encodes an antifungal compound active against *Saccharomyces cerevisiae*	Cardoza et al. (2006)
Sm1	*T. virens*	Encoding a small cysteine-rich protein that induces anti-disease activity in dicot and monocot plants	Buensanteai et al. (2010)
TmkA	*T. virens*	Encoding mitogen-activated protein kinase that influences biocontrol properties against *R. solani* and *S. rolfsii*	Mukherjee et al. (2003)

preceding reports imply that the antifungal genes isolated from various *Trichoderma* strains have critical roles in the antifungal activity against plant pathogens.

G protein Signalling in the Biocontrol Activity of *Trichoderma*

Biocontrol by *Trichoderma* is described as a synergism of distinct processes that operate in concert to produce disease control. A number of genes related to the detection, attraction, attachment, coiling, penetration, lysis of the host fungus and secretion of metabolites are induced, which appear to be regulated by highly conserved intracellular signal transduction pathways. The G protein-mediated signal transduction system is essential for cells to communicate across the membrane. It is critical for cell regulation and extracellular signal transmission to the cells. Classical heterotrimeric G protein signalling consists of three components: a G protein-coupled receptor (GPCR), a heterotrimeric G protein (Gα, Gß, and Gγ subunits), and an effector (Neer 1995). The Gα or G-protein α-subunits are the most numerous among the G-protein subunits and have been proven to regulate vegetative growth, conidiation, and mycoparasitic responses in fungi (Omann and Zeilinger 2010). Fungal Gα subunits can be categorized into three distinct subgroups, of which Gα subgroup I is homologues of the mammalian Gi subunits. However, only a few members of Gα subgroup II are associated with a biological function or a unique phenotype in fungi (Li et al. 2007). Members of Gα subgroup III, which are homologues of the mammalian Gs family, have been shown to positively alter the internal cAMP level in fungi (Bölker 1998).

The search for G-protein signalling chemicals in various *Trichoderma* species found that they contain members of the fungal Gα subgroups I, II, and III. Biochemical evidence indicates that Gα is involved in coiling in *Trichoderma*. The application of exogenous activators of G-protein-mediated signalling, the peptide toxin mastoparan and fluoroaluminate (AlF4), results in a more than twofold increase in coiling around nylon fibres compared to the results obtained with controls (Omero et al. 1999). The addition of exogenous dibutyryl-cyclic AMP (cAMP) also promotes coiling. cAMP impacts another developmental phase in *Trichoderma* as it skips the requirement for light for sporulation. In contrast, atropine, a known inhibitor of cAMP generation in fungi, blocks sporulation even after photoinduction (Rocha-Ramirez et al. 2002). The *T. atroviride* Gα gene, *Tga1,* was assigned to the fungal subfamily with the highest similarity to the Gαi class (Rocha-Ramirez et al. 2002). *Tga1* is involved in mycoparasitic coiling and conidiation. Further characterization of the Δ*tga1* mutant revealed that this G-protein component is involved in vegetative development, antifungal metabolite secretion, and chitinase production (Reithner et al. 2005), which are involved in *Trichoderma* biocontrol of plant pathogens. In liquid culture, the Δ*tga1* mutant exhibited significantly reduced chitinase activity and decreased transcription of the *nag1* (N-acetyl-glucosaminidase-encoding) and *ech42* (endochitinase-encoding) genes (Reithner et al. 2005). Interestingly, the Δ*tga1* mutant produced less 6-pentyl—pyrone and sesquiterpene metabolites (Reithner et al. 2005), but increased peptaibols belonging to the trichorzianine family (Stoppacher et al. 2007), implying that *Tga1* has a variety of roles in controlling the production of antifungal compounds. In contrast to *Tga1* effect on growth and conidiation in *T. atroviride*, its homologue *TgaA* did not affect these traits in *T. virens*. Δ*tgaA* mutants grew and sporulated usually but were less capable of colonizing *S. rolfsii* sclerotia, although they were fully pathogenic to *R. solani* (Mukherjee et al. 2004). *Trichoderma virens* mutants lacking the *TgaB* protein (belonging to subgroup II Gα subunits) developed and sporulated normally, and its mycoparasitic activity against *R. solani* and *S. sclerotiorum* sclerotia was unaffected (Mukherjee et al. 2004).

Characterization of *Tga3* of *T. atroviride* showed that the subgroup III of the Gα subunit regulates vegetative development, conidiation, and conidial germination of the fungus (Zeilinger et al. 2005). Δ*tga3* mutants displayed decreased levels of intracellular cAMP, extracellular chitinase (despite intensely increased transcription of the chitinase-encoding genes *ech42* and *nag1*) and pathogenicity to *R. solani* or *B. cinerea* in confrontation assays. Additionally, no infection structures associated with mycoparasitism were formed, strongly implying a loss of host recognition. Apart from regulating the formation of infection structures and the synthesis of cell wall-degrading enzymes, *T. atroviride Tga3* was necessary for producing antifungal metabolites (Zeilinger et al. 2005). While there is a significant association between sporulation and the secretion of antifungal metabolites in the *T. atroviride* parental strain, Δ*tga3* mutants were utterly deficient in peptaibol synthesis despite exhibiting a hypersporulating phenotype (Komon-Zelazowska et al. 2007). Besides, mutants with a silent *Gpr1* gene demonstrated a complete lack of mycoparasitism and an inability to bind to and coil around host hyphae.

Mitogen-activated Protein Kinases in *Trichoderma* Biocontrol

It is well-known that mitogen-activated protein kinase (MAPK) cascades transduce a range of signals through the sequential activation of serine/threonine protein kinases by phosphorylation, succeeding the control of gene expression required by some biological processes in eukaryotes (Schaeffer and Weber 1999). The well-studied MAPKs in *Trichoderma* are members of the YERK1, a family of extracellular-related kinases, including MAPKs from other organisms such as Pmk1 from *Magnaporthe grisea*, Fmk1 from *Fusarium oxysporum*, Bmp1 from *B. cinerea*, and Ubc3/ Kpp2 from *Ustilago maydis*.

MAPK homolog belonging to the YERK1 class of two different strains of *T. virens* was described (Mendoza-Mendoza et al. 2003, Mukherjee et al. 2003). The genes *tmkA* and *tvk1* encode the same protein, but contradictory results have been reported regarding the role of this MAP kinase in the production of mycoparasitism-related enzymes in different strains. The enzyme activities of chitinases and proteases were increased in Δ*tvk1* mutants, but Δ*tmkA* mutants showed a delay and reduced ability to clear a chitin-rich media. Furthermore, Mukherjee et al. (2003) hypothesized that Δ*tmkA* mutants lost their biocontrol potential in a host-specific manner because they exhibited mycoparasitic coiling and lysis of *R. solani* hyphae similar to the wild-type *T. virens* IMI304061 but reduced antagonistic properties against *S. rolfsii*. Further research into the function of Tvk1 found that this MAPK is involved in the regulation of conidiation, hydrophobicity, and the production of genes encoding cell wall proteins during the development of *T. virens* (Mendoza-Mendoza et al. 2007).

Tmk1, the equivalent MAPK from *T. atroviride*, has been described and shown to be 98% identical to *T. virens* TmkA/Tvk1 (Reithner et al. 2007). Direct plate confrontation assays against *R. solani* and *B. cinerea* as hosts revealed that Δtmk1 mutants could still parasitize *R. solani* (though less effectively than the parental strain) but not *B. cinerea*, indicating that *T. atroviride* Tmk1, similar to *T. virens* TmkA, controls the mycoparasitic activity of *Trichoderma* in a host-specific manner. Microscopic examination of Δ*tmk1* mutant hyphae during interaction with *R. solani* indicated that they could still bind and coil around the host. Under chitinase-inducing conditions, *nag1* and *ech42* transcript levels, extracellular chitinase activities, and production of 6-pentyl-pyrone and peptaibols were significantly enhanced in Δ*tmk1* mutants. These findings demonstrate that mycoparasitism-related processes, such as infection structure formation (coiling), chitinase and antifungal metabolite production, and resistance to *R. solani* infection were improved by *tmk1* gene deletion in Δ*tmk1* mutants (Reithner et al. 2007).

It is now established that the *Trichoderma*-mediated plant protection is not only due to the direct mycoparasitic interaction with the pathogen but also because of a systemic defense response induced by *Trichoderma* colonization of plant roots (Harman et al. 2004). Using *T. virens* Δ*tmkA* mutants, researchers recently have investigated the significance of MAPK signalling in establishing plant systemic resistance against pathogens. When *Trichoderma* spores germinated near cucumber roots, Δ*tmkA* mutants colonized the plant roots just as the wild-type strain, but they failed to induce complete systemic resistance against the bacterial leaf pathogen *Pseudomonas syringae* pv. *lacrymans*, indicating that *T. virens* requires MAPK signaling to induce a full systemic resistance in the plant (Viterbo et al. 2005). The *T. harzianum* ThHog1, which showed a high analogy to Hog1p controlling the osmotic stress response in *S. cerevisiae* (Brewster et al. 1993), was recently identified (Delgado-Jarana et al. 2006). It was postulated that ThHog1 could neutralize stress molecules produced by the parasitized fungus during mycoparasitic activity against *R. solani*, *B. cinerea*, and *S. sclerotiorum*, such as reactive oxygen species (Delgado-Jarana et al. 2006).

cAMP Signalling During *Trichoderma* Biocontrol

cAMP signaling is involved in a wide range of processes in fungi. The cAMP pathway affects transcription and cell cycle progression (Kronstad et al. 1998). The α subunit of heterotrimeric G-proteins controls the enzymatic activity synthesizing the intracellular messenger cAMP. In eukaryotes, the majority of the effects of cAMP come from the activation of cAMP-dependent protein kinases A (PKA), consisting of two regulatory and two catalytic subunits (Dickman and Yarden 1999). In pathogenic fungi, the expression of growth, morphogenesis, and virulence traits are regulated by functional PKA (Durrenberger et al. 1998). A gene (*pkr-1*) encoding the regulatory subunit of PKA from *T. atroviride* was demonstrated to play a critical role in the regulation of light responses of the fungus (Casas-Flores et al. 2006). Finally, the cAMP and MAPK pathways appear to work in concert to regulate the response of fungi to environmental stress (Kronstad et al. 1998). There are indications that linkages between these pathways may be prevalent in regulating various fungal activities.

The cAMP can act as a positive effector of endoglucanase induction by increasing the efficiency of the induction process in *T. reesei*, a fungus capable of antagonizing and outgrowing *P. ultimum* and protecting zucchini plants in planta from *P. ultimum* blight (Seidl et al. 2006). Treatment with exogenous cAMP promoted *T. harzianum* coiling around nylon fibres in a biomimetic system (Omero et al. 1999). Media with increased levels of intracellular cAMP (e.g., caffeine, dinitrophenol, aluminium tetrafluoride) suppressed the synthesis of N-acetyl-D-glucosaminidase (Silva et al. 2004). cAMP impacts the developmental phase in *Trichoderma* as it skips the requirement for light for sporulation, whereas atropine, a known inhibitor of cAMP generation in fungi, blocks sporulation even after photoinduction (Rocha-Ramirez et al. 2002). However, exogenous cAMP did not restore mycoparasitic overgrowth or decrease the altered phenotype, but it did restore the ability to form infection structures (Zeilinger et al. 2005). Interestingly, the deficiency in host recognition and infection structure development by *T. atroviride* was repaired by exogenous cAMP addition (Omann et al. 2008), as it was also previously discovered for *T. atroviride* Δ*tga3* mutants (Zeilinger et al. 2005). This seven-transmembrane protein of the cAMP-receptor-like family of fungal G-protein-coupled receptors plays a prominent role in the antagonistic interaction with the host fungus and regulation of mycoparasitism-related processes, as revealed by the functional characterization of *T. atroviride* Gpr1 (Omann et al. 2012). The silencing of *gpr1* resulted in an avirulent phenotype and an inability to attach to host hyphae. Exogenous cAMP was capable of restoring host attachment in *gpr1*-silenced transformants, but not mycoparasitic overgrowth (Omann et al. 2012). These findings imply that Gpr1 regulates the production of infection structures via the cAMP pathway.

The role of cAMP signaling in *T. viride* and *T. atroviride* conidiation was examined (Nemcovic and Farkas 1998, Casas-Flores et al. 2006). Conidiation is the primary mechanism by which *Trichoderma* survives and spreads in the environment. Environmental conditions such as blue light and nutrient stress promote the conidiation in these mycoparasites. In *T. viride*, photoinduction of conidiation results in a rapid temporal increase in the intracellular level of cAMP (Gresik et al. 1988), and exogenous cAMP enhanced conidia production in both lit and dark colonies (Nemcovic and Farkas 1998). Deletion of a *T. virens* adenylate cyclase-encoding gene (*tac1*) reduced intracellular cAMP levels to values below the detection limit. The Δ*tac1* mutants grew at a significantly lower rate on agar, did not sporulate in darkness, showed an inability to outgrow host fungi such as *S. rolfsii*, *R. solani*, and *Pythium* sp., and produced significantly fewer secondary metabolites (Mukherjee et al. 2007).

Genetics of Root Colonization by *Trichoderma*

Root colonization refers to the ability of a fungus to live and proliferate along with growing roots for an extended length of time in the presence of the indigenous microflora (Hossain et al. 2014). Root colonization is an important trait for *Trichoderma* to establish itself in an intimate relationship with host plants. Competence in the rhizosphere is a fundamental prerequisite for *Trichoderma* to colonize host roots effectively and become a successful biocontrol agent. *Trichoderma* spp. can invade both the exterior and interior parts of plant roots. For instance, *T. harzianum* CECT 2413 adheres to tomato roots profusely, colonizes the epidermis and cortex, develops intercellular hyphae, and produces papilla-like hyphal tips (Chacón et al. 2007). *T. koningi* hyphae pierces the epidermis and makes its way into the intercellular inner cortical tissues (Masunaka et al. 2011). In biological interactions between plants and *Trichoderma*, the attraction of *Trichoderma* to plant roots is presumably the result of chemical signals exchanged between the two partners. This initial stage of the *Trichoderma*–plant interaction is relatively poorly understood than the subsequent stages, including attachment, penetration, and internal colonization of plant roots. *Trichoderma* synthesizes and modifies hormonal signals to aid in root colonization. The fungus releases auxins, which modulate root growth, increasing the available surface area for colonization (Contreras-Cornejo et al. 2009). The role of *accd*, which encodes ACC deaminase, has been established in *T. asperellum*-mediated regulation of canola root growth (Viterbo et al. 2010). To enhance anchoring/ attachment, *Trichoderma* secretes small cysteine-rich hydrophobin-like proteins. Two such proteins have been identified as capable of root attachment: *TasHyd1* from *T. asperellum* and *Qid74* from *T. harzianum* (Viterbo et al. 2006, Samolski et al. 2012). To enhance root penetration, *Trichoderma* secretes expansin-like proteins with cellulose-binding modules and endopolygalacturonase (Brotman et al. 2008, Morán-Diez et al. 2009). Once inside the roots, these fungi are capable of making intercellular growth, however only in the epidermal layer and outer cortex. Suppression of the plant's initial defenses may promote root invasion. *T. koningi*, for instance, induces a transient and decreased level of defense gene expression and inhibits phytoalexin synthesis in *Lotus japonicus* roots during its entry (Masunaka et al. 2011). Similarly, microarrays analysis of Arabidopsis roots inoculated with *T. asperelloides* T203 revealed that four of the 28 transcripts down-regulated 24 hours after *Trichoderma* inoculation were plant cytochrome P450 monooxygenases (*CYP712A2, CYP712A1, CYP93D1,* and *CYP76G1*) (Brotman et al. 2013). These genes facilitate the synthesis and metabolism of many physiologically significant primary and secondary metabolites involved in plant defense against harmful microorganisms and insects (Morant et al. 2003). Their downregulation may reflect a strategy for suppressing local defense responses to facilitate colonization. This method is also supported by elevated expressions of the transcription factor ANAC081, which has been demonstrated to be a repressor of genes encoding pathogenesis-related proteins, and overexpression of this transcription factor confers susceptibility to the soilborne fungal disease (Delessert et al. 2005).

Genome Sequences of *Trichoderma*

Molecular approaches based on ITS analysis provide a wealth of information on the taxa and species of *Trichoderma* that exist in a given habitat. However, these data typically contain little information regarding the functional roles of the microorganisms in the ecosystem. The relationship between species, origin, and function in the *Trichoderma*–plant-pathogen interaction must be investigated using whole genome sequencing. The genome sequences help understand the variety, distribution, and origin of functional genes engaged in important biogeochemical activities carried out by the fungus. In many recent studies, deep sequencing has been used to examine the diversity and function of microbial communities (Kubicek et al. 2011, 2019). Although the number of reports for *Trichoderma* is still relatively small, the genomes of the most commonly occurring and cosmopolitan *Trichoderma* species have been sequenced (Table 5). Sequences of *T. reesei, T. longibrachiatum, T. citrinoviride, T. parareesei, T. harzianum, T. guizouense, T. afroharzianum, T. virens, T. atroviride, T. gamsii, T. asperellum,* and *T. hamatum* have been made accessible to identify essential genes involved in the critical biological processes. *Trichoderma* genome size ranges from 33 to 41 Mb. The anticipated number of genes in *Trichoderma* varies between 9292 and 14,095. *T. longibrachiatum* has the shortest genome (31.4 Mb), while *T. reesei* has the lowest gene number (9,129 gene models) among the mycoparasitic species (Kubicek et al. 2019), most likely due to the deletion of mycoparasitism-specific genes. Mycoparasite genomes range in size from 36.1 Mb (*T. atroviride*, 11,863 gene models), 37.4 Mb (*T. asperellum*, 12,586 gene models), 38.8 Mb (*T. virens*, 12,427 gene models), and 40.98 Mb (*T. harzianum*, 14,095 gene models) (Table 5). *T. atroviride* and *T. asperellum* are potent antifungal antagonists (necrotrophic mycoparasites) and phylogenetically ancestral species (Kubicek et al. 2011). On the contrary, both

Table 5. Comparative genomics of most common *Trichoderma* spp.

Species	Genome size (Mbp)	Total genes	References
Trichoderma reesei	32.7–34.2	9129–10877	Kubicek et al. (2011), Li et al. (2017), Kubicek et al. (2019)
T. longibrachiatum	31.74	10938	Kubicek et al. (2019)
T. citrinoviride	33.20	9737	Kubicek et al. (2019)
T. parareesei	32.07	9292	Kubicek et al. (2019)
T. harzianum	39.40–40.90	13932–14095	Li et al. (2017), Kubicek et al. (2019)
T. afroharzianum	39.70	11297	Kubicek et al. (2019)
T. guizhouense	38.80	11297	Kubicek et al. (2019)
T. virens	40.52	12427	Kubicek et al. (2019)
T. atroviride	36.40	11863	Kubicek et al. (2019)
T. gamsii	37.90	10709	Kubicek et al. (2019)
T. asperellum	37.66	12586	Kubicek et al. (2019)
T. hamatum	38.43	10520	Kubicek et al. (2019)

T. virens and *T. harzianum* are powerful phytopathogenic fungal parasites and are highly effective in stimulating plant defensive responses (Druzhinina et al. 2011).

Comparative genome research of *T. atroviride, T. virens*, and *T. reesei* indicated that the mycoparasites had expanded multiple gene families compared to *T. reesei*. These expansions include mycoparasitism-specific genes like chitinases and certain glucanases as well as genes involved in secondary metabolite biosynthesis (Kubiceck et al. 2011). Several members of these gene families are expressed before and during interaction with the fungus that serves as the host/prey (Seidl et al. 2009a). A secretome study indicated that *Trichoderma* might have one of the most diverse collections of proteases among all fungi (Druzhinina et al. 2012). Expansion of subtilisin-like proteases of the S8 family, dipeptidyl and tripeptidyl peptidases, occurs in mycoparasites (Druzhinina et al. 2012). These findings demonstrate the critical role of these genes in attacking and killing fungal prey and support the adaptability of mycoparasitic *Trichoderma* species to their antagonistic lifestyle. Previous research on *T. reesei, T. virens*, and *T. atroviride*, established that the genus *Trichoderma* might have a limited arsenal of secondary metabolite synthases (Kubicek et al. 2011). However, a recent study demonstrates that *T. harzianum* possesses a higher number of non-ribosomal polypeptide synthetases than *Aspergillus* spp., which is thought to be particularly rich in secondary metabolites (de Vries et al. 2017). Compared to *Aspergilli, T. virens* has an average number of polyketide synthases (Schmoll et al. 2016). In the case of terpenoid synthases, *Trichoderma* has 12 to 17 genes, far outnumbering *Aspergillus* spp., which has only 2 to 10. However, the majority of these secondary metabolite synthases, particularly those producing terpenoids, have yet to be identified. The relationship between the numerous secondary metabolites described in *Trichoderma* and the genes responsible for their synthesis is thus unknown, defining yet another exciting subject for future research.

Genetic Manipulation for Saprophytic Competitiveness and Biocontrol Fitness

The ability to manipulate the plant-growth-promoting and biocontrol properties of *Trichoderma* strains embraces immense promise for developing sustainable alternatives to agrochemicals for plant disease control and crop growth stimulants. Reports have shown that genetic modification can substantially improve the plant-growth-promoting and biocontrol abilities of *Trichoderma* strains. Several transformation strategies have been utilized effectively to manipulate various *Trichoderma* species genetically. The most extensively used and optimized methods are based on protoplasts (Li et al. 2017) and transformation using *Agrobacterium tumefaciens* (Zhong et al. 2007). Other less prevalent genetic techniques, such as biolistics (gene gun) (Te'o et al. 2002), and electroporation (Wanka 2021), have been reported to utilize in the genetical engineering of *Trichoderma*. An earlier study has reported the genetic stability and ecological persistence of a genetically modified *T. virens* with a hygromycin resistance gene and a gene encoding an organophosphohydrolase (Weaver et al. 2005). The N-acetyl-d-glucosaminidase-deficient mutant of *T. hamatum*, developed by insertional mutagenesis of the corresponding gene, has shown an enhanced ability to promote plant growth without disease pressure (Ryder et al. 2012). However, its saprotrophic competitiveness during antagonistic interactions with *R. solani* in soil and its fitness as a biocontrol agent against the pre-emergence damping-off pathogen *Sclerotinia sclerotiorum* is greatly diminished. Ultimately, its potential to promote plant growth is hampered by the presence of both pathogens. These imply that while significant gains in *T. hamatum*-mediated plant growth promotion can be achieved by genetically modifying a single beneficial trait, such modification may have detrimental effects on other aspects of its biology and ecology, affecting its success as a saprotrophic competitor and antagonist of soilborne pathogens.

Gamma irradiation-induced mutagenesis of *T. atroviride* significantly improves their ability to produce antibiotics (Brunner et al. 2005). Recently, a mutant (G2) of *T. virens* was isolated using gamma-ray-induced mutagenesis that produced more secondary metabolites and showed an upregulation of genes involved in secondary metabolism, mycoparasitism, and plant interactions

(Mukherjee et al. 2019). In greenhouse experiments, this mutant outperformed the wild-type strain in terms of the biocontrol of *S. rolfsii*. Over a five-year period, superior field control of collar rot (*S. rolfsii*) was demonstrated in chickpea (*Cicer arietinum*) and lentil (*Lens culinaris*), both in "on-farm" trials and in farmers' fields. Transgenic strain *T. atroviride* SJ3-4 expressing the *Aspergillus niger* glucose oxidase-encoding gene, *goxA*, had significantly less N-acetylglucosaminidase and endochitinase activities than its nontransformed parent. But it quickly overgrew and lysed the plant pathogens *P. ultimum* and *R. solani* and the culture filtrates of the transgenic strain exhibited three-fold greater inhibition of germination of spores of *Botrytis cinerea* than those of the wild type. This work demonstrates that heterologous genes driven by pathogen-inducible promoters can increase biocontrol and systemic resistance to pathogens (Brunner et al. 2005). The improved *Chit42-9* transformant can be used to biocontrol *Sclerotiorum sclerotiorum*, the pathogen responsible for canola stem rot disease. *T. harzianum* transformants overexpressing *Chit42* displayed higher chitinase activity levels than the wild type to inhibit *S. sclerotiorum* growth in dual culture assay (Kowsari et al. 2014) and might be useful for producing a new enhanced biological control system of *Sclerotiorum sclerotiorum*.

Trichoderma reesei is one of the fungi frequently genetically modified to increase the expression of cellulase genes and then employed in the biotechnology and pharmaceutical industries (Hinterdobler et al. 2021). Compared to the parental strain (PTr1), the isolated protoplast from *T. reesei* strain PTr2 significantly increased enzyme activities in two fusants SFTr2 and SFTr3 (Prabhavathi et al. 2006). To produce muramidase enzyme for utilizing it to fatten chickens and other small fowl, *T. reesei* strain DSM 32338 was genetically engineered (Rychen et al. 2018). Since *T. reesei* has been reported to reproduce sexually, it serves as a tool for rapid and enhanced strain generation for industrial usage (Seidl et al. 2009b). However, there is a risk associated with releasing these species into the environment because they can sexually recombine with a native population. Therefore, sympatric species like *T. harzianum*, *T. atroviride*, *T. longibrachiatum* and *T. parareesei* have a clonal lifecycle and may be appropriate for genetic modification as biological control agents. Several pieces of research have already reported that *T. atroviride* (Calcáneo-Hernández et al. 2020), *T. harzianum* (Manczinger et al. 1997), and *T. longibrachiatum* (Sánchez-Torres et al. 1994) have been genetically transformed. Another example of biotechnological solutions is the *T. harzianum Thkel1* gene, which encodes a kelch-repeat protein that controls glucosidase activity (Hermosa et al. 2010, 2014). When the gene is expressed in Arabidopsis, it boosts seed germination and plant tolerance to salt and osmotic stressors. Currently, the CRISPR/Cas9-mediated genome editing system has been effectively used in *T. reesei* and administered by plasmid or ribonucleoprotein complexes (Liu et al. 2015, Hao and Su 2019). Finally, the transformation methods developed for *Trichoderma* species can be used to perform overexpression experiments on putative backbone genes or transcription factors to increase production and characterize specific metabolites produced in trace amounts or not produced under standard culture conditions.

Conclusion and Future Perspectives

Due to the agricultural and biotechnological significance, *Trichoderma* spp. are among the most prevalent fungal microorganisms extensively explored as PGPF and biocontrol agents. While various research has been conducted based on *Trichoderma*-plant-pathogen interaction, a thorough knowledge of the mechanisms from genetic and genomic perspectives is missing. The paucity of high throughput investigations in *Trichoderma* has partly been attributed to the absence of genomic studies. This scenario, however, is predicted to change now. Recent years have seen an increase in the number of studies on the signalling pathways and genes involved in this interaction. Moreover, whole genomes of a substantial number of *Trichoderma* species have been sequenced. Genome-wide expression analyses have already made some headway in this regard. Although substantial progress has been made over the past few years in unravelling molecular processes underlying antagonism conferred by *Trichoderma* spp., many unanswered questions remain, particularly regarding the activities and biotechnological implications of core genes in *Trichoderma* genomes. A collaborative international effort should be launched to elucidate the activities of each gene using high-throughput gene knockouts. Transcriptome analyses under biocontrol and plant root colonization conditions may aid in detecting new candidate genes involved in *Trichoderma*-plant-plant pathogen interaction. Once these are accomplished, the future production of genetically engineered *Trichoderma* for optimum biocontrol and other biotechnological applications should be rational. The elucidation of genetic control of *Trichoderma*-mediated biocontrol and its biotechnological application will undoubtedly be extremely valuable in devising techniques and substances to combat these diseases by *Trichoderma* spp.

Acknowledgement

The research is supported by the research grant from the Ministry of Science and Technology, Bangladesh.

References

Abd-El-Khair, H., W.M.A. El-Nagdi. 2021. Application of dry powders of six plant species, as soil amendments, for controlling *Fusarium solani* and *Meloidogyne incognita* on pea in pots. Bull Natl Res Cent. 45:116. https://doi.org/10.1186/s42269-021-00571-5.

Ahmad, P., A. Hashem, E.F. Abd-Allah, A.A. Alqarawi, R. John, D. Egamberdieva et al. 2015. Role of *Trichoderma harzianum* in mitigating NaCl stress in Indian mustard (*Brassica juncea* L) through antioxidative defense system. Front. Plant Sci. 6:868. https://doi.org/10.3389/fpls.2015.00868.

Ahmed, S.A., C. Pérez, M.E. Candela. 2000. Evaluation of induction of systemic resistance in pepper plants (*Capsicum annuum*) to *Phytophthora capsici* using *Trichoderma harzianum* and its relation with capsidiol accumulation. Eur J Plant Pathol. 106: 817–824.

Aime, M.C., W. Phillips-Mora. 2005. The causal agents of witches' broom and frosty pod rot of cacao (chocolate, *Theobroma cacao*) form a new lineage of Marasmiaceae. Mycologia. 97:1012–1022.

Akhter, W., M.K.A. Bhuiyan, F. Sultana, M.M. Hossain. 2015. Integrated effect of microbial antagonist, organic amendment and fungicide in controlling seedling mortality (*Rhizoctonia solani*) and improving yield in pea (*Pisum sativum* L.). C R Biologies. 338: 21–28fo.

Alfano, G., M.L. Ivey, C. Cakir, J.I. Bos, S.A. Miller, L.V. Madden et al. 2007. Systemic Modulation of Gene Expression in Tomato by *Trichoderma hamatum* 382. Phytopathology. 97(4): 429–437. https://doi.org/10.1094/PHYTO-97-4-0429.

Alizadeh, H., K. Behboudi, M. Ahmadzadeh, M. Javan-Nikkhah, C. Zamioudis, C.M.J. Pieterse et al. 2013. Induced systemic resistance in cucumber and *Arabidopsis thaliana* by the combination of *Trichoderma harzianum* Tr6 and *Pseudomonas* sp. Ps14. Biol Control. 65: 14–23. doi: 10.1016/j.biocontrol.2013.01.009.

Bae, H., R.C. Sicher, M.S. Kim, S-H. Kim, M.D. Strem, R.L. Melnick et al. 2009. The beneficial endophyte *Trichoderma hamatum* isolate DIS 219b promotes growth and delays the onset of the drought response in *Theobroma cacao*. J Exp Bot. 60(11): 3279–3295.

Banjac N., R. Stanisavljević I. Dimkić N. Velijević M. Soković A. Ćirić. 2021. *Trichoderma harzianum* IS005-12 promotes germination, seedling growth and seedborne fungi suppression in Italian ryegrass forage. Plant Soil Environ. 67: 130–136.

Basco, M.J., K. Bisen, C. Keswani, H.B. Singh. 2017. Biological management of fusarium wilt of tomato using biofortified vermicompost. Mycosphere. 8(3): 467–483.

Benítez, T., Delgado-Jarana, J., Rincón, A.M., Rey, M., Limón, M.C. 1998. Biofungicides: *Trichoderma* as a biocontrol agent against phytopathogenic fungi. In: Pandalai, S.G. (ed). Recent research developments in microbiology, vol. 2. Research Signpost, Trivandrum, pp 129–150.

Benítez, T., A.M. Rincon, M.C. Limon, A.C. Codon. 2004. Biocontrol mechanisms of *Trichoderma* strains. Int Microbiol. 7: 249–260.

Bigirimana, J., G. De Meyer, J. Poppe, Y. Elad, M. Höfte. 1997. Induction of systemic resistance on bean (*Phaseolus vulgaris*) by *Trichoderma harziamum*. Univ. Gent Fac. Landbouwwet. Meded. 62: 1001–1007.

Bisen K., S. Ray, S.P. Singh. 2019. Consortium of compatible *Trichoderma* isolates mediated elicitation of immune response in *Solanum melongena* after challenge with *Sclerotium rolfsii*. Arch. Phytopathol. Plant Prot. 52(7–8): 733–756.

Boby, V., D. Bagyaraj. 2003. Biological control of root-rot of *Coleus forskohlii* Briq. using microbial inoculants. World Journal of Microbiology and Biotechnology. 19: 175–180. https://doi.org/10.1023/A:1023238908028.

Bolker, M. 1998. Sex and crime: heterotrimeric G proteins in fungal mating and pathogenesis. Fungal Genet. Biol. 25(3): 143–56.

Brewster, J.L., T. de Valoir, N.D. Dwyer. 1993. An osmosensing signal transduction pathway in yeast. Science. 259(5102): 1760–3.

Brotman, Y., E. Briff, A. Viterbo, I. Chet. 2008. Role of swollenin, an expansin-like protein from Trichoderma, in plant root colonization. Plant Physiol. 147: 779–789. doi: 10.1104/pp.108.116293.

Brotman, Y., U. Landau, Á. Cuadros-Inostroza, T. Tohge, A.R. Fernie, I. Chet et al. 2013. *Trichoderma*-plant root colonization: escaping early plant defense responses and activation of the antioxidant machinery for saline stress tolerance. PLoS Pathog. 9(3): e1003221. https://doi.org/10.1371/journal.ppat.1003221.

Bruinsma, M., B. Pang, R. Mumm, J.J.A. van Loon, M. Dicke. 2009. Comparing induction at an early and late step in signal transduction mediating indirect defence in *Brassica oleracea*. J. Exp. Bot. 60:2589–2599.

Brunner, K., S. Zeilinger, R. Ciliento, S. L. Woo, M. Lorito, C.P. Kubicek et al. 2005. Improvement of the Fungal Biocontrol Agent *Trichoderma atroviride* To Enhance both Antagonism and Induction of Plant Systemic Disease Resistance. Applied and Environmental Microbiology. 71(7): 3959–3965. doi:10.1128/AEM.71.7.3959-3965.2005.

Buensanteai, N., P.K. Mukherjee, B.A. Horwitz, C. Cheng, L.J. Dangott, C.M. Kenerley. 2010. Expression and purification of biologically active *Trichoderma virens* proteinaceous elicitor Sm1 in *Pichia pastoris*. Protein Expr. Purif. 72: 131–38

Cai, F., I.S. Druzhinina. 2021. In honor of John Bissett: authoritative guidelines on molecular identification of *Trichoderma*. Fungal Diversity. 107:1–69. https://doi.org/10.1007/s13225-020-00464-4.

Calcáneo-Hernández, G., E. Rojas-Espinosa, F. Landeros-Jaime, J.A. Cervantes-Chávez and E.U. Esquivel-Naranjo. 2020. An efficient transformation system for *Trichoderma atroviride* using the *pyr4* gene as a selectable marker. Brazil. J. Microbiol. 51:1631–1643. doi: 10.1007/s42770-020-00329-7.

Cardoza, Y.J., K.D. Klepzig, K.F. Raffa. 2006. Bacteria in oral secretions of an endophytic insect inhibit antagonistic fungi. Ecol Entomol. 31: 636–645; http://dx.doi.org/10.1111/j.1365-2311.2006.00829.x.

Carpenter, M.A., H.J. Ridgway, A.M. Stringer, A. J. Hay, A. Stewart. 2008. Characterisation of a *Trichoderma hamatum* monooxygenase gene involved in antagonistic activity against fungal plant pathogens. Current genetics. 53(4): 193–205. https://doi.org/10.1007/s00294-007-0175-5.

Casas-Flores, S., M. Rios-Momberg, T. Rosales-Saavedra, P. Martinez-Hernandez, V. Olmedo-Monfil, A. Herrera-Estrella. 2006. Cross talk between a fungal blue-light perception system and the cyclic AMP signaling pathway. Eukaryot Cell. 5: 499–506.

Chacón, M. R., Rodríguez-Galán, O., Benítez, T., Sousa, S., Rey, M., Llobell, A., & Delgado-Jarana, J. 2007. Microscopic and transcriptome analyses of early colonization of tomato roots by *Trichoderma harzianum*. International Microbiology. 10(1):19-27.

Chandanie, W.A., M. Kubota, M. Hyakumachi. 2009. Interactions between the arbuscular mycorrhizal fungus *Glomus mosseae* and plant growth-promoting fungi and their significance for enhancing plant growth and suppressing damping-off of cucumber (*Cucumis sativus* L.). Appl Soil Ecol. 41: 336–341.

Cheah, L.H., S.Veerakone, G. Kent. 2000. Biological control of clubroot on cauliflower with *Trichoderma* and *Streptomyces* spp. New Zealand Plant Protection. 53:18–21.

Chen, D., Q. Hou, L. Jia, K. Sun. 2021. Combined use of two *Trichoderma* strains to promote growth of pakchoi (*Brassica chinensis* L.). Agronomy. 11: 726. doi: 10.3390/agronomy11040726.

Chet, I., J. Inbar, I. Hadar. 1997. Fungal antagonists and mycoparasites. In: Wicklow, D.T., Soderstrom, B. (eds.), The Mycota IV: Environmental and microbial relationships. Springer-Verlag, Berlin. pp. 165–184.

Chirino-Valle, I., D. Kandula, C. Littlejohn, R. Hill, M. Walker, M. Shields et al. 2016. Potential of the beneficial fungus *Trichoderma* to enhance ecosystem-service provision in the biofuel grass Miscanthus × giganteus in agriculture. Sci. Rep. 6: 1–8. doi: 10.1038/srep25109.

Christopher, D.J., T.S. Raj, S.U. Rani, R. Udhayakumar. 2010. Role of defense enzymes activity in tomato as induced by *Trichoderma virens* against Fusarium wilt caused by *Fusarium oxysporum* f sp. *lycopersici*. J Biopest. 3(1): 158–162.

Colla, G., Y. Rouphael, E. di Mattia, C. El-Nakhel, M. Cardarelli. 2015. Co-inoculation of *Glomus intraradices* and *Trichoderma atroviride* acts as a biostimulant to promote growth, yield and nutrient uptake of vegetable crops. J. Food Agric. 95: 1706–1715.

Contina, J.B., L.M. Dandurand, G.R. Knudsen. 2017. Use of GFP-tagged *Trichoderma harzianum* as a tool to study the biological control of the potato cyst nematode *Globodera pallida*. Appl. Soil Ecol. 115: 31–37. doi: 10.1016/j.apsoil.2017.03.010.

Contreras-Cornejo, H. A., L. Macías-Rodríguez, E. Beltrán-Peña, A. Herrera-Estrella and J. López-Bucio. 2011. *Trichoderma*-induced plant immunity likely involves both hormonal- and camalexin-dependent mechanisms in *Arabidopsis thaliana* and confers resistance against necrotrophic fungi *Botrytis cinerea*. Plant Signal Behav. 6(10): 1554–1563. doi:10.4161/psb.6.10.17443.

Contreras-Cornejo, H.A., L. Macías-Rodríguez, C. Cortés-Penagos and J. López-Bucio. 2009. *Trichoderma virens*, a plant beneficial fungus, enhances biomass production and promotes lateral root growth through an auxin-dependent mechanism in Arabidopsis. Plant Physiol. 149: 1579–1592. doi: 10.1104/pp.108.130369.

Cortés, C., A. Gutierrez, V. Olmedo, J. Inbar, I. Chet and A. Herrera-Estrella. 1998. The expression of genes involved in parasitism by *Trichoderma harzianum* is triggered by a diffusible factor. Mol Gen Genet. 260: 218–225. https://doi.org/10.1007/s004380050889.

Damari-Weissler, H., S. Rachamilevitch, R. Aloni, M.A. German, S. Cohen, M.A. Zwieniecki et al. 2009. *LeFRK2* is required for phloem and xylem differentiation and the transport of both sugar and water. Planta. 230(4): 795–805.

de Vries, R.P., R. Riley, A. Wiebenga, G. Aguilar-Osorio, S. Amillis, C.A. Uchima et al. 2005. The transcription factor ATAF2 represses the expression of pathogenesis-related genes in Arabidopsis. Plant J. 43:745–757.

Delgado-Jarana, J., S. Sousa, F. Gonzalez. 2006. ThHog1 controls the hyperosmotic stress response in *Trichoderma harzianum*. Microbiology. 152(6): 1687–700.

Delgado-Sánchez, P., M.A. Ortega-Amaro, J.F. Jiménez-Bremont, J. Flores. 2011. Are fungi important for breaking seed dormancy in desert species? Experimental evidence in *Opuntia streptacantha* (Cactaceae). Plant Biology. 13: 154–159.

Dickman, M.B., O. Yarden. 1999. Serine/threonine protein kinases and phosphatases in filamentious fungi. Fungal Genet. Biol. 26(2): 99–117.

Djonović, S., M.J. Pozo, L.J. Dangott, C.R. Howell, C.M. Kenerley. 2006. Sm1, a proteinaceous elicitor secreted by the biocontrol fungus *Trichoderma virens* induces plant defense responses and systemic resistance. Mol Plant Microbe Interact. 19: 838–853.

Djonović, S., W.A.Vargas, M.V. Kolomiets, M. Horndeski, A. Wiest, C.M. Kenerley. 2007. A proteinaceous elicitor Sm1 from the beneficial fungus *Trichoderma virens* is required for induced systemic resistance in maize. Plant Physiol. 145: 875–889.

Doni, F., I. Anizan, C.M.Z. Che Radziah, A. Hilmi Salman. 2014. Enhancement of rice seed germination and vigour by *Trichoderma spp*. Res J App Sci Eng Technol. 7: 4547–4552.

Donzelli, B.G.G., M. Lorito, F. Scala, G.E. Harman. 2001. Cloning, sequence and structure of a gene encoding an antifungal glucan 1,3-β-glucosidase from *Trichoderma atroviride* (*T. harzianum*). Gene. 277: 199–208.

Druzhinina, I.S., V. Seidl-Seiboth, A. Herrera-Estrella, B.A. Horwitz, C.M. Kenerley, E. Monte et al. 2011. *Trichoderma*: the genomics of opportunistic success. Nature Rev Microbiol. 16: 749–759.

Druzhinina, I.S., E. Shelest, C.P. Kubicek. 2012. Novel traits of *Trichoderma* predicted through the analysis of its secretome, FEMS Microbiology Letters. 337:1–9. https://doi.org/10.1111/j.1574-6968.2012.02665.x.

Dubey S.C., M. Suresh B. Singh. 2007. Evaluation of *Trichoderma* species against *Fusarium oxysporum* f. sp. *ciceris* for integrated management of chickpea wilt. Biological Control. 40:118–127.

Durrenberger, F., K.Wong, J.W. Kronstad. 1998. Identification of a cAMP-dependent protein kinase catalytic subunit required for virulence and morphogenesis in *Ustilago maydis*. Proc. Natl. Acad. Sci. U.S.A. 95(10): 5684–9.

Elsharkawy, M.M., N. Hassan, M. Ali, S.N. Mondal, M. Hyakumachi. 2014. Effect of zoysiagrass rhizosphere fungal isolates on disease suppression and growth promotion of rice seedlings. Acta Agric. Scand. Section B -Soil and Plant Science. 64: 135–140.

Elsharkawy, M.M., M. Shimizu, H. Takahashi, K. Ozaki, M. Hyakumachi. 2013. Induction of Systemic Resistance against *Cucumber mosaic virus* in *Arabidopsis thaliana* by *Trichoderma asperellum* SKT-1. Plant Pathol J. 29: 193–200.

Engelberth, J., T. Koch, G. Schüler, N. Bachmann, J. Rechtenbach, W. Boland. 2001. Ion channel-forming alamethicin is a potent elicitor of volatile biosynthesis and tendril coiling. Cross talk between jasmonate and salicylate signaling in lima bean. Plant Physiol. 125: 369–377.

Flores, A., I. Chet, A. Herrera-Estrella. 1997. Improved biocontrol activity of *Trichoderma harzianum* by over-expression of the proteinase-encoding gene *prb1*. Curr Genet. 31: 30–37. https://doi.org/10.1007/s002940050173.

Gallo, A., G. Mulè, M. Favilla, C. Altomare. 2004. Isolation and characterisation of a trichodiene synthase homologous gene in *Trichoderma harzianum*. Physiol. Mol. Plant Pathol. 65:11–20.

Gharate, R., N. Singh, S.M. Chaudhari. 2016. Management of common scab (*Streptomyces scabies*) of potato through eco-friendly approach. Indian Phytopathol. 69: 266–270.

Ghazalibiglar, H., D.R. Kandula, J.G. Hampton. 2016. Biological control of fusarium wilt of tomato by *Trichoderma* isolates. New Zealand Plant Protection. 69: 57–63.

Ghosh, S.K. 2017. Study of some antagonistic soil fungi for protection of fruit rot (*Phomopsis vexans*) and growth promotion of brinjal. Int. J. Adv. Res. 5: 485–494.

Ghosh, S.K. and S. Pal. 2017. Growth promotion and fusarium wilt disease management ecofriendly in chickpea by *Trichoderma asperellum*. Int. J. Curr. Res. Acad. Rev. 5(1): 14–26.

Ghosh, S.K., A. Panja. 2021. Signatures of signaling pathways underlying plant-growth promotion by fungi. Jogaiah, S. (Ed.), Biocontrol Agents and Secondary Metabolites, Woodhead Publishing, pp. 321–346. https://doi.org/10.1016/B978-0-12-822919-4.00013-2.

Gresik, M., N. Kolarova, V. Farkas. 1988. Membrane potential, ATP and cyclic AMP changes induced by light in *Trichoderma viride*. Exp. Mycol. 12: 295–301.

Guigón-López, C., F. Vargas-Albores, V. Guerrero-Prieto, M. Ruocco and M. Lorito. 2015. Changes in *Trichoderma asperellum* enzyme expression during parasitism of the cotton root rot pathogen *Phymatotrichopsis omnivore*. Fungal Biol. 119: 264–273. 10.1016/j.funbio.2014.12.013.

Guler, N.S., N. Pehlivan, S.A. Karaoglu, S. Guzel, A. Bozdeveci. 2016. *Trichoderma atroviride* ID20G inoculation ameliorates drought stress-induced damages by improving antioxidant defence in maize seedlings. Acta Physiol. Plant. 38: 132. DOI: 10.1007/s11738-016-2153-3.

Hao, Z. and X. Su. 2019. Fast gene disruption in *Trichoderma reesei* using *in vitro* assembled Cas9/gRNA complex. BMC Biotechnol. 19. doi: 10.1186/s12896-018-0498-y.

Haque, M., G.N.M. Ilias, A.H. Molla. 2012. Impact of *Trichoderma*-enriched bio-fertilizer on the growth and yield of mustard (*Brassica rapa* L.) and Tomato (*Solanum lycopersicon* Mill.). Agriculturists. 10: 109–119.

Harman, G.E., I. Chet and R. Baker. 1980. *Trichoderma hamatum* effects on seed and seedling disease induced in radish and pea by Pythium spp. or *Rhizoctonia solani*. Phytopathology 70: 1167–1172.

Harman, G.E. 2006. Overview of mechanisms and uses of *Trichoderma* spp. Phytopathology 96: 190–194.

Harman, G.E. 2000. Myths and dogmas of biocontrol. Changes in perceptions derived from research on *Trichoderma harzianum* T-22. Plant Dis. 84: 377–393.

Harman, G.E., C.R. Howell, A. Viterbo. 2004. *Trichoderma* species-opportunistic, avirulent plant symbionts. Nat. Rev. Microbiol. 2: 43–56.

Hermosa, R., S.L. Woo, M. Lorito, E. Monte. 2010. Proteomic approaches to understand *Trichoderma* biocontrol mechanisms and plant interactions. Curr Proteomics. 7: 298–305. doi: 10.2174/157016410793611783.

Hermosa, R., R.E. Cardoza, M.B. Rubio, S. Gutiérrez, E. Monte. 2014. Secondary Metabolism and Antimicrobial Metabolites of *Trichoderma*. Biotechnology and Biology of *Trichoderma*. 125–137. doi:10.1016/b978-0-444-59576-8.00010-2.

Hinterdobler, W., G. Li, K. Spiegel, S. Basyouni-Khamis, M. Gorfer, M. Schmoll. 2021. *Trichoderma reesei* isolated from Austrian soil with high potential for biotechnological application. Front. Microbiol. 12: 552301. doi: 10.3389/fmicb.2021.552301.

Hossain M.M., F. Sultana. 2020. Application and mechanisms of plant growth promoting fungi (PGPF) for phytostimulation. IntechOpen Organ. Agric. 2: 1–31. doi: 10.5772/intechopen.92338.

Hossain, M.M. 2022. Biological management of plant diseases by non-pathogenic *Phoma* spp.. In: M. Rai, B. Zimowska, G.J. Kövics (eds)., Phoma: Diversity, Taxonomy, Bioactivities, and Nanotechnology. Springer, Cham. https://doi.org/10.1007/978-3-030-81218-8_15.

Hossain, M.M., N. Hossain, F. Sultana, S.M.N. Islam, M.S. Islam, M.K.A. Bhuiyan. 2013. Integrated management of Fusarium wilt of chickpea (*Cicer arietinum* L.) caused by *Fusarium oxysporum* f.sp. *ciceris* with microbial antagonist, botanical extract and fungicide. Afr J Biotechnol. 12: 4699–4706.

Hossain, M.M., F. Sultana, S. Islam. 2017. Plant Growth-Promoting Fungi (PGPF): Phytostimulation and induced systemic resistance. In: D. Singh, H. Singh, R. Prabha (Eds.). Plant-Microbe Interactions in Agro-Ecological Perspectives, Volume 2: Microbial Interactions and Agro-Ecological Impacts. Singapore: Springer. pp. 135–191. doi: 10.1007/978-981-10-6593-4(2017).

Hossain, M.M., F. Sultana, M. Miyazawa, M. Hyakumachi. 2014. The plant growth promoting fungi *Penicillium* spp. GP15-1 enhances growth and confers protection against damping-off and anthracnose in the cucumber. J Oleo Sci. 63(4): 39–400.

Howell, C.R. 2003. Mechanisms employed by *Trichoderma* species in the biological control of plant diseases: The history and evolution of current concepts. Plant Dis. 87(1): 4–10.

Howell, C.R., R.D. Stipanovic, R. Lumsden. 1993. Antibiotic production by strains of *Gliocladium virens* and its relation to biocontrol of cotton seedling diseases. Biocontrol Sci. Technol. 3: 435–441.

Hoyos-Carvajal, L., S. Orduz, J. Bissett. 2009. Growth stimulation in bean (*Phaseolus vulgaris* L.) by *Trichoderma*. Biological Control. 51: 409–416.

Hu, M.J., X.S. Zhang, X. Cao, Z.Q. Miao, Y.Z. Zhang. 2009. Studies on the control effect of *Trichoderma koningii* against soft rot of Chinese cabbage. North Hortic. 6: 102–103.

Hung, R., S. Lee, C. Rodriguez-Saona, J. W. Bennett. 2014. Common gas phase molecules from fungi affect seed germination and plant health in *Arabidopsis thaliana*. AMB Express. 4:53. https://doi.org/10.1186/s13568-014-0053-8.

Ichinose, H., M. Yoshida, Z. Fujimoto and S. Kaneko. 2008. Characterization of a modular enzyme of exo-1,5-α-L-arabinofuranosidase and arabinan binding module from *Streptomyces avermitilis* NBRC14893. Appl Microbiol Biotechnol 80:399. https://doi.org/10.1007/s00253-008-1551-x.

Intana, W., P. Wonglom, N. Suwannarach and A. Sunpapao. 2022. *Trichoderma asperelloides* PSU-P1 Induced Expression of Pathogenesis-Related Protein Genes against Gummy Stem Blight of Muskmelon (*Cucumis melo*) in Field Evaluation. J. Fungi. 8: 156. https://doi.org/10.3390/jof8020156.

Ismaiel, A.A., D.M.I. Ali. 2017. Antimicrobial properties of 6-pentyl-α-pyrone produced by endophytic strains of *Trichoderma koningii* and its effect on aflatoxin B1 production. Biologia. 72: 1403–1415. doi: 10.1515/biolog-2017-0173.

Izquierdo-García, L.F., A. González-Almario, A.M. Cotes, C.A. Moreno-Velandia. 2020. *Trichoderma virens* Gl006 and *Bacillus velezensis* Bs006: a compatible interaction controlling Fusarium wilt of cape gooseberry. Sci. Rep. 10: 6857. doi: 10.1038/s41598-020-63689-y.

Jain A., S. Singh B. Kumar Sarma H. Bahadur Singh. 2012. Microbial consortium–mediated reprogramming of defence network in pea to enhance tolerance against *Sclerotinia sclerotiorum*. J. Appl. Microbiol. 112 (3): 537–550.

Jogaiah, S., M. Abdelrahman, L.S.P. Tran, I. Shin-ichi. 2013. Characterization of rhizosphere fungi that mediate resistance in tomato against bacterial wilt disease. J Exp Bot. 64: 3829–3842.

Jogaiah, S., M. Abdelrahman, L.S.P. Tran, S.I. Ito. 2018. Different mechanisms of *Trichoderma virens*-mediated resistance in tomato against Fusarium wilt involve the jasmonic and salicylic acid pathways. Mol. Plant Pathol. 19: 870–882. doi: 10.1111/mpp.12571.

Kai, M., B. Piechulla. 2009. Plant growth promotion due to rhizobacterial volatiles - An effect of CO2? FEBS Lett. 583: 3473–3477.

Kandula, D.R.W., E.E. Jones, A. Stewart, K.L. McLean, J.G. Hampton. 2015. *Trichoderma* species for biocontrol of soil-borne plant pathogens of pasture species. Biocontrol Sci Tech. 25(9): 1052–69.

Kang, J.W., N.Y. Lee, K.C. Cho, M.Y. Lee, D.Y. Choi, S.H. Park et al. 2015. Analysis of nitrated proteins in *Saccharomyces cerevisiae* involved in mating signal transduction. Proteomics. 15(2–3): 580–590.

Kariuki, C.K., E.W. Mutitu, W.M. Muiru. 2020. Effect of *Bacillus* and *Trichoderma* species in the management of the bacterial wilt of tomato (*Lycopersicum esculentum*) in the field. Egypt J Biol Pest Control. 30: 109. https://doi.org/10.1186/s41938-020-00310-4.

Kaveh, H., S.V. Jartoodeh, H. Aruee, M. Mazhabi. 2011. Would *Trichoderma* affect seed germination and seedling quality of two muskmelon cultivars, khatooni and qasri and increase their transplanting success? J Biol Environ Sci. 5: 169–175.

Keller, N.P. 2019. Fungal secondary metabolism: regulation, function and drug discovery. Nat. Rev. Microbiol. 17: 167–180. doi: 10.1038/s41579-018-0121-1.

Keswani C., H. Dilnashin, H. Birla, S. Singh. 2019. Unravelling efficient applications of agriculturally important microorganisms for alleviation of induced inter-cellular oxidative stress in crops. Acta. Agric. Sloven. 114(1): 121–130.

Khan M.R., J. Gupta, 1998. Antagonistic efficacy of *Trichoderma* species against *Macrophomina phaseolina* on eggplant. J. Plant Dis. Prot. 105: 387–393.

Khan, S. A., M. Hamayun, H. Yoon, H.Y. Kim, S.J. Suh, S.K. Hwang et al. 2008. Plant growth promotion and Penicillium citrinum. BMC Microbiol. 8:231. https://doi.org/10.1186/1471-2180-8-231.

Koike, N., M. Hyakumachi, K. Kageyama, S. Tsuyumu, N. Doke. 2001. Induction of systemic resistance in cucumber against several diseases by plant growth-promoting fungi: lignification and superoxide generation. Eur J Plant Pathol. 107: 523–533.

Komon-Zelazowska, M., T. Neuhof, R. Dieckmann, H. von Döhren, A. Herrera-Estrella, C.P. Kubicek et al. 2007. Formation of atroviridin by *Hypocrea atroviridis* is conidiation associated and positively regulated by blue light and the G protein GNA3. Eukaryotic Cell. 6(12): 2332–2342. https://doi.org/10.1128/EC.00143-07

Konappa, N., S. Krishnamurthy, N.C. Siddaiah, N.S. Ramachandrappa and S. Chowdappa. 2018. Evaluation of biological efficacy of *Trichoderma asperellum* against tomato bacterial wilt caused by *Ralstonia solanacearum*. Egypt J Biol Pest Control. 28: 63. https://doi.org/10.1186/s41938-018-0069-5.

Korolev, N., D. Rav David, Y. Elad. 2008. The role of phytohormones in basal resistance and *Trichoderma*-induced systemic resistance to *Botrytis cinerea* in *Arabidopsis thaliana*. Biocontrol. 53667–68310.1007/s10526-007-9103-3.

Kotake, T., S. Kaneko, A. Kubomoto, M.A. Haque, H. Kobayashi, Y. Tsumuraya. 2004. Molecular cloning and expression in Escherichia coli of a *Trichoderma viride* endo-beta-(1->6)-galactanase gene. Biochem. J. 377: 749–755. doi: 10.1042/BJ20031145.

Kowsari, M., M.R. Zamani, M. Motallebi. 2014. Enhancement of *Trichoderma harzianum* activity against *Sclerotinia sclerotiorum* by overexpression of *Chit42*. Iran J Biotech. 12(2): e13869.

Kronstad, J., A.D. De Maria, D. Funnell. 1998. Signalling via cAMP in fungi: interconnections with mitogen-activated protein kinase pathways. Arch. Microbiol. 170(6): 395–404.

Kubicek, C.P., A. Herrera-Estrella, V. Seidl-Seiboth, D.A. Martinez, I.S. Druzhinina, M. Thon et al. 2011. Comparative genome sequence analysis underscores mycoparasitism as the ancestral life style of *Trichoderma*. Genome Biol. 12:R40. doi: 10.1186/gb-2011-12-4-r40.

Kubicek, C.P., A.S. Steindorff, K. Chenthamara, G. Manganiello, B. Henrissat, J. Zhang et al. 2019. Evolution and comparative genomics of the most common *Trichoderma* species. BMC Genomics. 20:485. https://doi.org/10.1186/s12864-019-5680-7.

La Spada, F., C. Stracquadanio, M. Riolo, A. Pane, S.O. Cacciola. 2020. *Trichoderma* counteracts the challenge of *Phytophthora nicotianae* infections on tomato by modulating plant defense mechanisms and the expression of crinkler, necrosis-inducing *Phytophthora* protein 1, and cellulose-binding elicitor lectin pathogenic effectors. Front. Plant Sci. 11: 583539. doi: 10.3389/fpls.2020.583539.

Lamba, P., S. Sharma, G.D. Munshi, S.K. Munshi. 2008. Biochemical changes in sunflower plants due to seed treatment/spray application with biocontrol agents. Phytoparasitica. 36: 388.

Lee, S., M. Yap, G. Behringer, R. Hung and J. W. Bennett. 2016. Volatile organic compounds emitted by *Trichoderma* species mediate plant growth. Fungal Biol Biotechnol. 3: 7.

Li, H.Y., Y. Luo, X.S. Zhang, W.L. Shi, Z.T. Gong, M. Shi et al. 2014. Trichokonins from *Trichoderma pseudokoningii* SMF2 induce resistance against Gram-negative *Pectobacterium carotovorum* subsp. *carotovorum* in Chinese cabbage. FEMS Microbiology Letters. 354(1): 75–82. https://doi.org/10.1111/1574-6968.12427.

Li, J., J. Philp, J. Li, Y. Wei, H. Li, K. Yang et al. 2020. *Trichoderma harzianum* Inoculation Reduces the Incidence of Clubroot Disease in Chinese Cabbage by Regulating the Rhizosphere Microbial Community. Microorganisms. 8(9):1325. https://doi.org/10.3390/microorganisms8091325.

Li, L., S.J. Wright, S. Krystofova, G. Park, K.A. Borkovich. 2007. Heterotrimeric G protein signaling in filamentous fungi. Annual Review of Microbiology. 61: 423–452.

Li, W.C., C.H. Huang, C.L. Chen, Y.C. Chuang, S.Y. Tung and T.F. Wang. 2017. *Trichoderma reesei* complete genome sequence, repeat-induced point mutation, and partitioning of CAZyme gene clusters. Biotechnol Biofuels. 10: 170.

Liu, R., L. Chen, Y. Jiang, Z. Zhou, G. Zou. 2015. Efficient genome editing in filamentous fungus *Trichoderma reesei* using the CRISPR/Cas9 system. Cell Discov. 1: 15007. doi: 10.1038/celldisc.2015.7.

Liu, Y., Q. Yang and J. Song. 2009. A new serine protease gene from *Trichoderma harzianum* is expressed in *Saccharomyces cerevisiae*. Appl Biochem Microbiol. 45: 22–26. https://doi.org/10.1134/S0003683809010049.

López-Mondéjar, R., M. Ros, J.A. Pascual. 2011. Mycoparasitism-related genes expression of *Trichoderma harzianum* isolates to evaluate their efficacy as biological control agent. Biol. Control. 56: 59–66. doi: 10.1073/pnas.91.21.9799.

Lorito, M., S.L. Woo, G.E. Harman, E. Monte. 2010. Translational research on *Trichoderma*: from 'omics to the field. Annu Rev Phytopathol. 48: 395–417. https://doi.org/10.1146/annurev-phyto-073009-114314

Lorito, M., V. Farkas, S. Rebuffat, B. Bodo, C.P. Kubicek. 1996. Cell wall synthesis is a major target of mycoparasitic antagonism by *Trichoderma harzianum*. J. Bacteriol. 178: 6382–6385.

Luo, Y., D.D. Zhang, X.W. Dong, P.B. Zhao, L.L. Chen, X.Y. Song et al. 2010. Antimicrobial peptaibols induce defense responses and systemic resistance in tobacco against tobacco mosaic virus. FEMS Microbiol. Lett. 313: 120–126.

Maddau, L., A. Cabras, A. Franceschini, B.T. Linaldeddu, S. Crobu, T. Roggio et al. 2009. Occurrence and characterization of peptaibols from *Trichoderma citrinoviride*, an endophytic fungus of cork oak, using electrospray ionization quadrupole time-of-flight mass spectrometry. Microbiology. 155: 3371–3381.

Manczinger, L., O. Komonyi, Z. Antal and L. Ferenczy. 1997. A method for high-frequency transformation of *Trichoderma viride*. J. Microbiol. Methods. 29: 207–210. doi: 10.1016/S0167-7012(97)00026-2.

Manjunatha, S.V., M.K. Naik, M.F.R. Khan, R.S. Goswami. 2013. Evaluation of bio-control agents for management of dry root rot of chickpea caused by *Macrophomina phaseolina*. Crop Prot. 45: 147–150.

Marcello, C.M., A.S. Steindorff, S.P. da Silva, R. Silva, N. do, L.A. Mendes Bataus et al. 2010. Expression analysis of the exo- β-1,3-glucanase from the mycoparasitic fungus *Trichoderma asperellum*. Microbiol Res. 2010;165: 75–81. doi: 10.1016/j.micres.2008.08.002.

Marra, R., P. Ambrosino, V. Carbone, F. Vinale, S.L. Woo, M. Ruocco et al. 2006. Study of the three-way interaction between *Trichoderma atroviride*, plant and fungal pathogens by using a proteomic approach. Curr. Genet. 50: 307–321.

Martínez-Medina, A., M. Del Mar Alguacil, J.A. Pascual, S.C.M. Van Wees. 2014. Phytohormone profiles induced by *Trichoderma* isolates correspond with their biocontrol and plant growth-promoting activity on melon plants. J Chem Ecol. 40: 804–815.

Martínez-Medina, A., I. Fernández, M.J. Sánchez-Guzmán, S.C. Jung, J.A. Pascual and M.J. Pozo. 2013. Deciphering the hormonal signaling network behind the systemic resistance induced by *Trichoderma harzianum* in tomato. Frot Plant Sci. 4: 206.

Martínez-Medina, A., I. Fernandez, G.B. Lok, M.J. Pozo, C.M. Pieterse and S.C. Van Wees. 2017. Shifting from priming of salicylic acid- to jasmonic acid-regulated defences by *Trichoderma* protects tomato against the root knot nematode *Meloidogyne incognita*. New Phytol. 213: 1363– 1377.

Massart, S., H.M. Jijakli. 2007. Use of molecular techniques to elucidate the mechanisms of action of fungal biocontrol agents: a review. J Microbiol Met. 69: 229–241.

Mastouri, F., T. Björkman, G.E. Harman. 2010. Seed treatment with *Trichoderma harzianum* alleviates biotic, abiotic, and physiological stresses in germinating seeds and seedlings. Phytopathology. 100: 1213–1221.

Masunaka, A., M. Hyakumachi and S. Takenaka. 2011. Plant growth-promoting fungus, *Trichoderma koningi* suppresses isoflavonoid phytoalexin vestitol production for colonization on/in the roots of *Lotus japonicus*. Microbes Environ. 26: 128–134. doi: 10.1264/jsme2.ME10176.

Mathys, J., K. De Cremer, P. Timmermans, S. Van Kerckhove, B. Lievens, M. Vanhaecke et al. 2012. Genome-wide characterization of ISR induced in Arabidopsis thaliana by *Trichoderma hamatum* T382 against *Botrytis cinerea* infection. Front. Plant Sci. 3: 108.

Medeiros, H.A., J.V. Araújo Filho, L.G. Freitas, P. Castillo, M.B. Rubio, R. Hermosa et al. 2017. Tomato progeny inherit resistance to the nematode *Meloidogyne javanica* linked to plant growth induced by the biocontrol fungus *Trichoderma atroviride*. Sci Rep. 7:40216. https://doi.org/10.1038/srep40216.

Mendoza-Mendoza, A., M.J. Pozo, D. Grzegorski. 2003. Enhanced biocontrol activity of *Trichoderma* through inactivation of a mitogen-activated protein kinase. Proc. Natl. Acad. Sci. U.S.A. 100(26): 15965–70.

Mendoza-Mendoza, A., T. Rosales-Saavedra, C. Cortes. 2007. The MAP kinase TVK1 regulates conidiation, hydrophobicity and the expression of genes encoding cell wall proteins in the fungus *Trichoderma virens*. Microbiology. 153: 2137–47.

Mendoza-Mendoza, A., R. Zaid, R. Lawry, R. Hermosa, E. Monte, B.A. Horwitz et al. 2018. Molecular dialogues between *Trichoderma* and roots: role of the fungal secretome. Fungal Biol. Rev. 32: 62–85. doi: 10.1016/j.fbr.2017.12.001.

Min, S.Y., B.G. Kim, C. Lee, H-G. Hur, J.H. Ahn. 2002. Purification, characterization and cDNA cloning of xylanase from fungus *Trichoderma* strain SY. J Microbiol Biotechnol. 12: 890–4.

Ming, Q., C. Su, C. Zheng, M. Jia, Q. Zhang, H. Zhang et al. 2013. Elicitors from the endophytic fungus *Trichoderma atroviride* promote *Salvia miltiorrhiza* hairy root growth and tanshinone biosynthesis. J Exp Bot. 64: 5687–5694.

Mohamed, B.F.F., N.M.A. Sallam, S.A.M. Alamri, A.M. Kamal Y.S.M. Abo-Elyousr, M. Hashem. 2020. Approving the biocontrol method of potato wilt caused by *Ralstonia solanacearum* (Smith) using *Enterobacter cloacae* PS14 and *Trichoderma asperellum* T34. Egypt J Biol Pest Control. 30:61. https://doi.org/10.1186/s41938-020-00262-9.

Morán-Diez, E., R. Hermosa, P. Ambrosino, R.E. Cardoza, S. Gutiérrez, M. Lorito et al. 2009. The ThPG1 endopolygalacturonase is required for the trichoderma harzianum-plant beneficial interaction. Mol. Plant Microbe Interact. 22(8): 1021–1031. https://doi.org/10.1094/MPMI-22-8-1021.

Morant, M., S. Bak, B.L. Moller, D. Werck-Reichhart. 2003. Plant cytochromes P450: tools for pharmacology, plant protection and phytoremediation. Curr Opin Biotechnol. 14: 151–162.

Moreno-Mateos M.A., J. Delgado-Jarana A.C. Codón T. Benítez. 2007. pH and Pac1 control development and antifungal activity in *Trichoderma harzianum*. Fungal Genet. Biol. 44: 1355–1367. 10.1016/j.fgb.2007.07.012.

Mukherjee, P.K., S.T. Mehetre, P.D. Sherkhane, G. Muthukathan, A. Ghosh, A.S. Kotasthane et al. 2019. A Novel Seed-Dressing Formulation Based on an Improved Mutant Strain of *Trichoderma virens*, and Its Field Evaluation. Front. Microbiol. 10:1910. https://doi.org/10.3389/fmicb.2019.01910.

Mukherjee T., M. Ivanova, M. Dagda, Y. Kanayama, D. Granot, A.S. Holaday. 2015. Constitutively overexpressing a tomato fructokinase gene (LeFRK1) in cotton (*Gossypium hirsutum* L. cv. Coker 312) positively affects plant vegetative growth, boll number and seed cotton yield. Funct Plant Biol. 42(9): 899–908. doi: 10.1071/FP15035. PMID: 32480732.

Mukherjee, M., P.K. Mukherjee, S.P. Kale. 2007. cAMP signalling is involved in growth, germination, mycoparasitism and secondary metabolism in *Trichoderma virens*. Microbiology. 153: 1734–42.

Mukherjee, P.K., J. Latha, R. Hadar. 2003. TmkA, a mitogen-activated protein kinase of *Trichoderma virens*, is involved in biocontrol properties and repression of conidiation in the dark. Eukaryot Cell. 2(3): 446–55.

Mukherjee, P.K., J. Latha, R. Hadar. 2004. Role of two G-protein alpha subunits, TgaA and TgaB, in the antagonism of plant pathogens by *Trichoderma virens*. Appl. Environ. Microbiol. 70(1): 542–9.

Nagaraju, A., J. Sudisha, S. Mahadevamurthy, S-I. Ito. 2012. Seed priming with *Trichoderma harzianum* isolates enhances plant growth and induces resistance against *Plasmopara halstedii*, an incitant of sunflower downy mildew disease. Aust. J. Plant Pathol. 41: 609– 620.

Naraghi, L., A. Heydari, A. Hesan, K. Sharefi. 2014. Evaluation of *Talaromyces flavus* and *Trichoderma harzzianum* in biological control of sugar beet damping-off disease in the greenhouse and field. Int. J. Agric. Sci. 4(1): 65–74.

Chandra Nayaka, S., S.R. Niranjana, A. Uday Shankar, S. Niranjan Raj, M.S. Reddy, H.S. Prakash et al. 2010, February. Seed biopriming with novel strain of *Trichoderma harzianum* for the control of toxigenic *Fusarium verticillioides* and *fumonisins* in maize. Arch. Phytopathol. Plant Prot. 43(3): 264–282. https://doi.org/10.1080/03235400701803879

Neer, E.J. 1995. Heterotrimeric G proteins: organizers of transmembrane signals. Cell. 80(2): 249–257.

Nemcovic, M., V. Farkas. 1998. Stimulation of conidiation by derivates of cAMP in *Trichoderma viride*. Folia Microbiol. 43(4): 399–402.

Nunes-Nesi, A., F. Carrari, A. Lytovchenko, A.M. Smith, M. Ehlers Loureiro, R.G. Ratcliffe et al. 2005. Enhanced Photosynthetic Performance and Growth as a Consequence of Decreasing Mitochondrial Malate Dehydrogenase Activity in Transgenic Tomato Plants. Plant Physiol. 137(2): 611–622. https://doi.org/10.1104/pp.104.055566.

Odanaka, S., A.B. Bennett, Y. Kanayama. 2002. Distinct physiological roles of fructokinase isozymes revealed by gene-specific suppression of frk1 and frk2 expression in tomato. Plant Physiol. 129: 1119–1126.

Omann, M., S. Lehner, K. Brunner, M. Delic, N. Stoppacher, R. Schuhmacher et al. 2008. A cAMP receptor-like GPCR is involved in *Trichoderma atroviride* mycoparasitism. In Molecular Tools for Understanding and Improving Biocontrol-Abstract Book (p. 18). http://hdl.handle.net/20.500.12708/46279.

Omann, M.R., S. Lehner, C. Escobar Rodríguez, K. Brunner and S. Zeilinger. 2012. The seven-transmembrane receptor Gpr1 governs processes relevant for the antagonistic interaction of *Trichoderma atroviride* with its host. Microbiology. 158: 107–118. https://doi.org/10.1099/mic.0.052035-0.

Omann, M., S. Zeilinger. 2010. How a Mycoparasite Employs G-Protein Signaling: Using the Example of *Trichoderma*. J. Signal Transduct. 2010: 1–8. doi:10.1155/2010/123126.

Omero, C., J. Inbar, V. Rocha-Ramirez. 1999. G protein activators and cAMP promote mycoparasitic behaviour in *Trichoderma harzianum*. Mycol. Res. 103(12): 1637–42.

Pocurull, M., A.M. Fullana, M. Ferro, P. Valero, N. Escudero, E. Saus et al. 2020. Commercial Formulates of *Trichoderma* Induce Systemic Plant Resistance to Meloidogyne incognita in Tomato and the Effect Is Additive to That of the Mi-1.2 Resistance Gene. Front. Microbiol. 10: 3042. https://doi.org/10.3389/fmicb.2019.03042.

Pozo, M.J., J.M. Baek, J.M. Garcia, C.M. Kenerley. 2004. Functional analysis of tvsp1, a serine pro-tease-encoding gene in the biocontrol agent *Trichoderma virens*. Fungal Genet Biol. 41: 336–348.

Prabhavathi, V.R., N. Mathivanan, E. Sagadevan, K. Murugesan, D. Lalithakumari. 2006. Intra-strain protoplast fusion enhances carboxymethylcellulase activity in *Trichoderma reesei*. Enzyme Microb Technol. 38: 719–723.

Prasad, R.D., R. Rangeshwaran, S.V. Hegde, C.P. Anuroop. 2002. Effect of soil and seed application of Trichoderma harzianum on pigeonpea wilt caused by *Fusarium udum* under field conditions. Crop Prot. 21(4): 293–297. doi:10.1016/s0261-2194(01)00100-4.

Prasad, A., A. Mathur, M. Singh, M.M. Gupta, G.C. Uniyal, R.K. Lal et al. 2012. Growth and asiaticoside production in multiple shoot cultures of a medicinal herb, *Centella asiatica* (L.) Urban, under the influence of nutrient manipulations. J Nat Med. 66(2): 383–387. https://doi.org/10.1007/s11418-011-0588-9.

Prasad, L., S. Chaudhary, S. Sagar, A. Tomar. 2016. Mycoparasitic Capabilities of Diverse Native Strain of *Trichoderma* spp. against *Fusarium oxysporum* f. sp. *lycopersici*. J. Appl. Nat. Sci. 8: 769–776.

Rahman, M., J. Ali, M. Masood. 2015. Seed Priming and *Trichoderma* Application: A method for improving seedling establishment and yield of dry direct seeded boro (winter) rice in Bangladesh. Univers. J. Agric. Res. 3: 59–67.

Reithner, B., K. Brunner, R. Schuhmacher. 2005. The G protein alpha subunit Tga1 of *Trichoderma atroviride* is involved in chitinase formation and differential production of antifungal metabolites. Fungal Genet. Biol. 42(9): 749–60.

Reithner, B., R. Schuhmacher, N. Stoppacher. 2007. Signaling via the *Trichoderma atroviride* mitogen-activated protein kinase Tmk1 differentially affects mycoparasitism and plant protection. Fungal Genet. Biol. 44(11): 1123–1133. https://doi.org/10.1016/j.fgb.2007.04.001.

Rocha-Ramírez, V., C. Omero, I. Chet, B.A. Horwitz, A. Herrera-Estrella. 2002. *Trichoderma atroviride* G-protein α-subunit gene tga1 is involved in mycoparasitic coiling and conidiation. Eukaryot Cell. 1(4): 594–605.

Ruano-Rosa, D., P. Prieto, A.M. Rincón, M.V. Gómez-Rodríguez, R. Valderrama, J.B. Barroso et al. 2015. Fate of *Trichoderma harzianum* in the olive rhizosphere: time course of the root colonization process and interaction with the fungal pathogen *Verticillium dahliae*. BioControl. 61(3): 269–282. https://doi.org/10.1007/s10526-015-9706-z.

Rubio, M.B., R. Hermosa, J.L. Reino, I.G. Collado, E. Monte. 2009. Thctf1 transcription factor of *Trichoderma harzianum* is involved in 6-pentyl-2H-pyran-2-one production and antifungal activity. Fungal Genet. Biol. 46: 17–27.

Ruess, L. 2003. Nematode soil faunal analysis of decomposition pathways in different ecosystems. Nematology. 5(2): 179–181. https://doi.org/10.1163/156854103767139662.

Ruocco, M., S. Lanzuise, F. Vinale, R. Marra, D. Turrà, S. L.Woo et al. 2009. Identification of a new biocontrol gene in *Trichoderma atroviride*: the role of an ABC transporter membrane pump in the interaction with different plant-pathogenic fungi. Mol Plant Microbe Interact. 22(3): 291–301. doi: 10.1094/MPMI-22-3-0291. PMID: 19245323.

Rush, T.A., H.K. Shrestha, M. Gopalakrishnan Meena, M.K. Spangler, J.C. Ellis, J.L. Labbé et al. 2021. Bioprospecting *Trichoderma*: A Systematic Roadmap to Screen Genomes and Natural Products for Biocontrol Applications. Front. Fungal Biol. 2:716511. doi: 10.3389/ffunb.2021.716511.

Rychen, G., G. Aquilina, G. Azimonti, V. Bampidis, M.D.L. Bastos, G. Bories et al. 2018. Safety and efficacy of muramidase from *Trichoderma reesei* DSM 32338 as a feed additive for chickens for fattening and minor poultry species. EFSA Journal. 16:7. https://doi.org/10.2903/j.efsa.2018.5342.

Ryder, L.S., B.D. Harris, D.M. Soanes, M.J. Kershaw, N.J. Talbot, C.R. Thornton. 2012. Saprotrophic competitiveness and biocontrol fitness of a genetically modified strain of the plant-growth-promoting fungus *Trichoderma hamatum* GD12. Microbiology (Reading, England). 158(Pt 1): 84–97. https://doi.org/10.1099/mic.0.051854-0.

Saadia, M., S. Ahmed, A. Jamil. 2008. Isolation and cloning of cre1 gene from a filamentous fungus *Trichoderma harzianum*. Pak J Bot. 40: 421–426.

Saiprasad, G.V.S., J.B. Mythili, L. Anand, C. Suneetha, H.J. Rashmi, C. Naveena et al. 2009. Development of *Trichoderma harzianum* endochitinase gene construct conferring antifungal activity in transgenic tobacco. Indian J Biotechnol. 8: 199–206.

Saksirirat, W., P. Chareerak, W. Bunyatrachata. 2009. Induced systemic resistance of biocontrol fungus, *Trichoderma* spp. against bacterial and gray leaf spot in tomatoes. Asian J Food Agro-Industry. 2: S99–S104.

Samolski, I., A.M. Rincon, L.M. Pinzon, A. Viterbo, E. Monte. 2012. The *qid74* gene from *Trichoderma harzianum* has a role in root architecture and plant biofertilization. Microbiology. 158: 129–138. doi: 10.1099/mic.0.053140-0.

Sánchez-Torres, P., R. González, J.A. Pérez-González, L. González-Candelas and D. Ramón. 1994. Development of a transformation system for *Trichoderma longibrachiatum* and its use for constructing multicopy transformants for theegl1 gene. Appl. Microbiol. Biotechnol. 41: 440–446. doi: 10.1007/bf01982533.

Schaeffer, H.J., M.J. Weber. 1999. Mitogen-activated protein kinases: specific messages from ubiquitous messengers. Mol. Cell. Biol. 19(4): 2435–44.

Schmoll, M., C. Dattenböck, N. Carreras-Villaseñor, A. Mendoza-Mendoza, D. Tisch, M. I. Alemán et al. 2016. The genomes of three uneven siblings: footprints of the lifestyles of three *Trichoderma* species. Microbiol Mol Biol Rev. 80(1): 205–327.

Segarra, G., S. Van-der-Ent, I. Trillas, C.M.J. Pieterse. 2009. *MYB72*, a node of convergence in induced systemic resistance triggered by a fungal and a bacterial beneficial microbe. Plant Biol. 11: 90–96. doi: 10.1111/j.1438-8677.2008.00162.x.

Seidl, V., M. Schmoll, B. Scherm. 2006. Antagonism of Pythium blight of zucchini by *Hypocrea jecorina* does not require cellulase gene expression but is improved by carbon catabolite derepression. FEMS Microbiol. Lett. 257(1): 145–51.

Seidl, V., L. Song, E. Lindquist, S. Gruber, A. Koptchinskiy, S. Zeilinger et al. 2009a. Transcriptomic response of the mycoparasitic fungus *Trichoderma atroviride* to the presence of fungal prey. BMC Genomics. 10: 567. doi: 10.1186/1471-2164-10-567.

Seidl, V., C. Seibel, C.P. Kubicek, M. Schmoll. 2009b. Sexual development in the industrial workhorse *Trichoderma reesei*. Proc. Natl. Acad. Sci. U.S.A. 106: 13909–13914. doi: 10.1073/pnas.0904936106.

Shi, M., L. Chen, X.W. Wang, T. Zhang, P.B. Zhao, X.Y. Song et al. 2012. Antimicrobial peptaibols from *Trichoderma pseudokoningii* induce programmed cell death in plant fungal pathogens. Microbiology. 158: 166–175.

Shiwangi, A., S.P. Pathak. 2019. Effect of eco-friendly treatments on important fungal foliar and tuber borne diseases of potato (*Solanum tuberosum* L.). Int. J. Chem. Stud. 7(2): 6–09.

Shoresh, M., G.E. Harman. 2008. The molecular basis of shoot responses of maize seedlings to *Trichoderma harzianum* T22 inoculation of the root: a proteomic approach. Plant Physiol. 147: 2147–2163.

Shoresh, M., I. Yedidia, I. Chet. 2005. Involvement of jasmonic acid/ethylene signaling pathway in the systemic resistance induced in cucumber by *Trichoderma asperellum* T203. Phytopathology. 95: 76–84.

Silva, R.N., S.P. da Silva, R.L. Brandao. 2004. Regulation of N-acetyl-Bß-D-glucosaminidase produced by *Trichoderma harzianum*: evidence that cAMP controls its expression. Res. Microbiol. 155: 667–71.

Singh S., H. Singh. 2015. Effect of mixture of *Trichoderma* isolates on biochemical parameter in tomato fruits against *Sclerotinia sclerotiorum* rot of tomato plant. J. Environ. Biol. 36: 267.

Singh, V., P.N. Singh, R.L. Yadav, S.K. Awasthi, B.B. Joshi, R.K. Singh et al. 2010. Increasing the efficacy of *Trichoderma harzianum* for nutrient uptake and control of red rot in sugarcane. J. Hortic. For. 2: 66–71.

Singh, S.P., C. Keswani, S.P. Singh, E. Sansinenea, T.X. Hoat. 2021. *Trichoderma* spp. mediated induction of systemic defense response in brinjal against *Sclerotinia sclerotiorum*. Curr. Res. Microb. Sci. 2:100051.

Sofo, A., A. Scopa, M. Manfra, M. De Nisco, G. Tenore, J. Troisi et al. 2011. *Trichoderma harzianum* strain T-22 induces changes in phytohormone levels in cherry rootstocks (*Prunus cerasus* × *P. canescens*). Plant Growth Regul. 65(2): 421–425. https://doi.org/10.1007/s10725-011-9610-1.

Sofo, A., G. Tataranni, C. Xiloyannis, B. Dichio, A. Scopa. 2012. Direct effects of *Trichoderma harzianum* strain T-22 on micropropagated shoots of GiSeLa6® (*Prunus cerasus×Prunus canescens*) rootstock. Environ. Exp. Bot. 76: 33–38. https://doi.org/10.1016/j.envexpbot.2011.10.006.

Srivastava, P.K., B.D. Shenoy, M. Gupta. 2012. Stimulatory effects of arsenic-tolerant soil fungi on plant growth promotion and soil properties. Microbes Environ. 27(4): 477–482.

Srivastava, S., V. Singh, P.S. Gupta, O.K. Sinha, A. Baitha. 2006. Nested PCR assay for detection of sugarcane grassy shoot phytoplasma in the leafhopper vector *Deltocephalus vulgaris*: a first report. Plant Pathol. 55(1): 25–28. https://doi.org/10.1111/j.1365-3059.2005.01294.x.

Steyaert J.M., H.J. Ridgway, Y. Elad, A. Stewart. 2003. Genetic basis of mycoparasitism: A mechanism of biological control by species of *Trichoderma*. N. Z. J. Crop Hortic. Sci. 31: 281–-291. DOI: 10.1080/01140671.2003.9514263.

Stoppacher N., B. Reithner, M. Omann, S. Zeilinger, R. Krska and R. Schuhmacher. 2007. Profiling of trichorzianines in culture samples of *Trichoderma atroviride* by liquid chromatography/tandem mass spectrometry. Rapid Commun. Mass Spectrom. 21: 3963–3970.

Suada, I.K. 2017. The potential of various indigenous *Trichoderma* spp. to suppress *Plasmodiophora brassicae* the pathogen of clubroot disease on cabbage. Biodiversitas. 18(4): 1424–1429.

Sun, Q., X. Jiang, L. Pang, L. Wang, M. Li. 2016. Functions of *thga1* gene in *Trichoderma harzianum* based on transcriptome analysis. BioMed Res Int. 2016: 1–9. doi:10.1155/2016/8329513.

Te'o, V.S.J., P.L. Bergquist, K.M.H. Nevalainen. 2002. Biolistic transformation of *Trichoderma reesei* using the Bio-Rad seven barrels Hepta Adaptor system. J. Microbiol. Methods. 51: 393–399. doi: 10.1016/S0167-7012(02)00126-4.

Thambugala, K.M., D.A. Daranagama, A.J.L. Phillips, S.D. Kannangara, I. Promputtha. 2020. Fungi vs. fungi in biocontrol: an overview of fungal antagonists applied against fungal plant pathogens. Front. Cell. Infect. Microbiol. 10: 604923. doi: 10.3389/fcimb.2020.604923.

Thangavelu, R., M. Gopi. 2015. Combined application of native *Trichoderma* isolates possessing multiple functions for the control of Fusarium wilt disease in banana cv. Grand Naine. Biocontrol Sci. Technol. 25: 1147–1164. doi: 10.1080/09583157.2015.1036727.

Tijerino, A., R.E. Cardoza, J. Moraga, M.G. Malmierca, F. Vicente, J. Aleu et al. 2011. Overexpression of the trichodiene synthase gene *tri5* increases trichodermin production and antimicrobial activity in *Trichoderma brevicompactum*. Fungal Genet Biol. 48: 285–296.

Tjamos, E.C., G.C. Papavizas, R.J. Cook. 1992. Biological control of plant diseases, progress and challenges for the future. Plenum Press, New York.

Triveni, S., R. Prasanna, A.K. Saxena. 2012. Optimization of conditions for in vitro development of *Trichoderma* viride-based biofilms as potential inoculants. Folia Microbiol. 57: 431–437. doi: 10.1007/s12223-012-0154-1.

Trutmann P., P.J. Keane. 1990. *Trichoderma koningii* as a biological control agent for Sclerotinia sclerotiorum in Southern Australia. Soil Biol. Biochem. 22: 43–50, 10.1016/0038-0717(90)90058-8.

Tsahouridou, P.C., C.C. Thanassoulopoulos. 2002. Proliferation of *Trichoderma koningii* in the tomato rhizosphere and suppression of damping-off causing by *Sclerotium rolfsii*. Soil Biochem. 34: 767–776.

Tu J.C., O. Vaartaja. 1981. The effect of hyperparasite (*Gliocladium virens*) on *Rhizoctonia solani* and Rhizoctonia root rot of white beans. Can. J. Bot. 59: 22–27.

Uematsu, A., M. Matsui, C. Tanaka, T. Takahashi, K. Noguchi, M. Suzuki. et al. 2012. Developmental Trajectories of Amygdala and Hippocampus from Infancy to Early Adulthood in Healthy Individuals. PLoS ONE 7(10): e46970. https://doi.org/10.1371/journal.pone.0046970.

Vargas, W.A., J.C. Mandawe and C.M. Kenerley 2009. Plant-derived sucrose is a key element in the symbiotic association between *Trichoderma virens* and maize plants. Plant Physiol. 151(2): 792–808. https://doi.org/10.1104/pp.109.141291.

Vinale, F., Flematti, G., Sivasithamparam, K., Lorito, M., Marra, R., Skelton, B.W., Ghisalberti, E.L. 2009. Harzianic acid, an antifungal and plant growth promoting metabolite from *Trichoderma harzianum*. J. Nat. Prod. 72: 2032–2035.

Vinale, F., I.A. Girona, M. Nigro, P. Mazzei, A. Piccolo, M. Ruocco et al. 2012. Cerinolactone, a hydroxyl-lactone derivative from *Trichoderma cerinum*. J. Nat. Prod. 27:103–106.

Vinale, F., R. Marra, F. Scala, E.L. Ghisalberti, M. Lorito, K. Sivasitham- param. 2006. Major secondary metabolites produced by two commercial *Trichoderma* strains active against different phytopathogens. Lett. Appl. Microbiol. 43:143–148.

Vinale, F., K. Sivasithamparam, E.L. Ghisalberti, R. Marra, M.J. Barbetti, H. Li et al. 2008. A novel role for *Trichoderma* secondary metabolites in the interactions with plants. Physiol. Mol. Plant Pathol. 72: 80–86.

Viterbo, A. and I. Chet. 2006. TasHyd1, a new hydrophobin gene from the biocontrol agent *Trichoderma asperellum*, is involved in plant root colonization. Mol Plant Pathol. 7: 249–258. doi: 10.1111/j.1364-3703.2006.00335.x.

Viterbo, A., M. Harel, B.A. Horwitz, I. Chet, P.K. Mukherjee. 2005. *Trichoderma* mitogen-activated protein kinase signaling is involved in induction of plant systemic resistance. Appl Environ Microbiol. 71: 6241–6246.

Viterbo, A., U. Landau, S. Kim, L. Chernin, I. Chet. 2010. Characterization of ACC deaminase from the biocontrol and plant growth-promoting agent *Trichoderma asperellum* T203. FEMS Microbiol Lett. 2010; 305: 42–48. doi: 10.1111/j.1574-6968.2010.01910.x.

Viterbo, A., A. Wiest, Y. Brotman, I. Chet, C.M. Kenerley. 2007. The 18mer peptaibols from *Trichoderma virens* elicit plant defense responses. Mol. Plant Pathol. 8: 737–764.

Viterbo, A., B.A. Horwitz. 2010. Mycoparasitism. In: K.A. Borkovich, D.J. Ebbole (Eds.), Cellular and Molecular Biology of Filamentous Fungi, vol. 42. ASM Press, Washington, pp. 676–693.

Vitti, A., E. Pellegrini, C. Nali, S. Lovelli, A. Sofo, M. Valerio et al. 2016. *Trichoderma harzianum* T-22 Induces Systemic Resistance in Tomato Infected by Cucumber mosaic virus. Front. Plant Sci. 7: 1520. doi: 10.3389/fpls.2016.01520.

Vizcaino, J.A., L. Sanz, A. Basilio, F. Vicente, S. Gutierrez, M.R. Hermosa et al. 2005. Screening of antimicrobial activities in *Trichoderma* isolates representing three *Trichoderma* sections. Mycol Res. 109: 1397– 1406.

Wanka, F. 2021. Open the pores: electroporation for the transformation of *Trichoderma reesei*. In A.R. Mach-Aigner, R. Martzy (eds.)., Methods in Molecular Biology. New York, NY: Humana Press Inc. 73–78.

Watanabe, S., K. Kumakura, N. Izawa, K. Nagayama, T. Mitachi, M. Kanamori et al. 2007. Mode of action of *Trichoderma asperellum* SKT-1, a biocontrol agent against *Gibberella fujikuroi*. J Pestic Sci. 32: 222–228.

Weaver M., E. Vedenyapina, C.M. Kenerley.2005. Fitness, persistence, and responsiveness of a genetically engineered strain of *Trichoderma virens* in soil mesocosms. Appl Soil Ecol. 29: 125–134.

Weindling, R. 1932. *Trichoderma lignorum* as a parasite of other soil fungi. Phytopathology. 22: 837–845.

Weindling, R., H.S. Fawcett. 1936. Experiments in the control of *Rhizoctonia* damping-off of citrus seedlings. Hilgardia.10:1–16.

Weindling, R., O. Emerson. 1936. The isolation of a toxic substance from the culture filtrate of *Trichoderma*. Phytopathology. 26: 1068–1070.

Whipps, J.M., R.D. Lumsden. 2001. Commercial use of fungi as plant disease biological control agents: status and prospects, in fungi as biocontrol agents: progress, problems and potential, ed by T.M. Butt, C. Jackson, N. Magan, CABI Publishing, Wallingford, UK, pp.9–22.

Woo, S.L., B. Donzelli, F. Scala, R.L. Mach, G.E. Harman, C.P. Kubicek et al. 1999. Disruption of the ech42 (endochitinase-encoding) gene affects biocontrol activity in *Trichoderma harzianum* P1. Mol Plant Microbe Interact. 12: 419–429.

Woo, S.L., M. Ruocco, F. Vinale, M. Nigro, R. Marra, N. Lombardi et al. 2014. *Trichoderma*-based products and their widespread use in agriculture. Open Mycol. J. 8: 71–126. doi: 10.2174/1874437001408010071.

Wu, Q., M. Ni, K. Dou, J. Tang, J. Ren, C. Yu et al. 2018. Co-culture of *Bacillus amyloliquefaciens* ACCC11060 and *Trichoderma asperellum* GDFS1009 enhanced pathogen-inhibition and amino acid yield. Microb. Cell Factories. 17(1): 155. https://doi.org/10.1186/s12934-018-1004-x.

Xiao-Yan, S., S. Qing-Tao, X. Shu-Tao, C. Xiu-Lan, S. Cai-Yun, Z. Yu- Zhong. 2006. Broadspectrum antimicrobial activity and high stability of trichokonins from *Trichoderma koningii* SMF2 against plant pathogens. FEMS Microbiol. Lett. 260: 119–125.

Yadav, J., J.P. Verma, K.N. Tiwari. 2011. Plant growth promoting activities of fungi and their effect on chickpea plant growth. Asian J. Biol. Sci. 4: 291–299.

Yadav, R.L., S.K. Shukla, A. Suman, P.N. Singh. 2009. *Trichoderma* inoculation and trash management effects on soil microbial biomass, soil respiration, nutrient uptake and yield of ratoon sugarcane under subtropical conditions. Biol Fertil Soils. 45: 461–468.

Yassin, M.T., A.A.F. Mostafa, A.A. Al-Askar, S.R. Sayed, A.M. Rady. 2021. Antagonistic activity of *Trichoderma harzianum* and *Trichoderma viride* strains against some fusarial pathogens causing stalk rot disease of maize, *in vitro*. J King Saud Univ-Sci. 33: 101363.

Yedidia, I., N. Benhamou, I. Chet. 1999. Induction of defense responses in cucumber plants (*Cucumis sativus* L.) by the biocontrol agent *Trichoderma harzianum*. Appl Environ Microbiol. 65: 1061–1070.

Yedidia, I., A.K. Srivastva, Y. Kapulnik, I. Chet. 2001. Effect of *Trichoderma harzianum* on microelement concentrations and increased growth of cucumber plants. Plant Soil. 235: 235–242.

Yendyo, S., R. G.C, B.R. Pandey. 2017. Evaluation of *Trichoderma* spp., *Pseudomonas fluorescens* and *Bacillus subtilis* for biological control of Ralstonia wilt of tomato. F1000Research. 6:2028. https://doi.org/10.12688/f1000research.12448.3

Yu, X-X., Y.T. Zhao, J. Cheng, W. Wang. 2015. Biocontrol effect of *Trichoderma harzianum* T4 on brassica clubroot and analysis of rhizosphere microbial communities based on T-RFLP. Biocontrol Sci Technol. 25: 1493–1505.

Zafari, D., M.M. Koushki, E. Bazgir. 2008. Biocontrol evaluation of wheat take-all disease by *Trichoderma* screened isolates. Afr J Biotechnol. 7: 3653– 3659.

Zeilinger S., B. Reithner, V. Scala, I. Peissl, M. Lorito, R.L. Mach. 2005. Signal transduction by Tga3, a novel G protein α subunit of *Trichoderma atroviride*. Appl Environ Microbiol. 71:1591–1597.

Zhang, S., Y. Gan, B. Xu. 2016. Application of plant-growth-promoting fungi *Trichoderma longibrachiatum* T6 enhances tolerance of wheat to salt stress through improvement of antioxidative defense system and gene expression. Front Plant Sci. 7: 1405.

Zhong, Y.H., X.L. Wang, T.H. Wang, Q. Jiang. 2007. Agrobacterium-mediated transformation (AMT) of *Trichoderma reesei* as an efficient tool for random insertional mutagenesis. Appl. Microbiol. Biotechnol. 73: 1348–1354. doi: 10.1007/s00253-006-0603-3.

CHAPTER 17

Diversity and Mechanisms of Fungal–Mineral Interaction Through Molecular and Omics Studies

Shaifali Sharma,[1,5] *Jaya Sharma,*[2] *Aditi Sharma,*[3]
Nidhi Tripathi[4] and *Rohit Sharma*[5,6,]*

Introduction

Geomicrobiology studies the interaction between microbes and minerals. Geomycology, a sub-field of microbiology, focuses on studying the interaction of fungi with minerals (Burford et al. 2003b). Most of the previous research focussed on prokaryotes and lichens (Sterflinger 2000). The role of other fungi (unicellular or mycelial) remains neglected even though these fungi are relatively more efficient in bioweathering of minerals and rocks. These interactions are either mediated by biomechanical or biochemical activities viz. the release of organic acids and metabolites (Hoffland et al. 2004, Gadd 2007). In general, fungal mycelium deteriorates the rocks by penetrating cracks and cavities mediated by solubilization of minerals and rock materials (Sterflinger 2000, Daghino et al. 2006). According to Gadd (2001a, b), the low-molecular-weight organic acids acidify the substrate. The siderophores release by fungi also helps in bioweathering due to their Fe chelating properties (Winkelmann 2002).

Rock formations like sedimentary, igneous and metamorphic rocks are important sites to study the bioweathering by fungi. Although the environment is oligotrophic with low nutrition, monuments also mimic the same environment. The microhabitat can be termed as sub-aerial at the interface between minerals and atmosphere (Ortega-Morales et al. 2016). The fungi that are associated with rock and bioweathering are peculiar in the fact that they lack sexual reproductive structures. These fungi are adapted to oligotrophic habitats but are successful in colonization due to less competition and capacity to tolerate extreme environment. The special ability of these fungi help them to adapt to temperature variations, irradiation and osmotic stress preventing them from desiccation. These fungi rely on their ability to break down rocks and minerals with the help of enzymes and metabolite and also use atmospheric moisture and airborne nutrients. Such abilities

---
[1] School of Sciences, Sanjeev Agrawal Global Education University, Katara Hills, Sahara Bypass, Bhopal-462 026, Madhya Pradesh, India.
[2] School of Agriculture, Sanjeev Agrawal Global Education University, Sahara Bypass, Bhopal-462 026, Madhya Pradesh, India.
[3] Department of Geology, Vikram University, Madhav Bhawan, University Road, Near Vikram Vatika, Ujjain-456 010, Madhya Pradesh, India.
[4] Department of Biochemistry, Career College, Govindpura, Bhopal-462 023, Madhya Pradesh, India.
[5] Centre for Biodiversity Exploration and Conservation (CBEC), Tilhari, Mandla Road, Jabalpur-482 021, Madhya Pradesh, India.
[6] Department of Basic Sciences, IES University, Bhopal-462 044, Madhya Pradesh, India.
* Corresponding author: rsfungus@gmail.com

make them important candidate for bioprospecting studies on biodeterioration of monuments and nutrient deficient agriculture and forest soils (Gorbushina 2007).

There is a need to study fungal bioweathering as they have been less studied; they produce organic acids and siderophores and play an important role in the Earth's biogeochemical cycles. Even though these fungi are distributed in natural rock formations and man-made monuments, extensive research on these have been neglected. The primary reason has been slow growth, isolation problems and collaborative research between geologists and mycologists. Once these collaborations are established, it would do wonders to the research on rock inhabiting fungi and bioweathering fungal processes. Relatively recently, many new genera and species have been discovered due to the interest of mycologists, geologists and material engineers. The chapter deals with the basic rock cycle, types of rock formation and their weathering, bio-weathering by bacteria and fungi, diversity of bioweathering fungi, omics and molecular phylogeny of bio-weathering fungi, and applications of bioweathering fungi. However, the molecular phylogeny of lichens as rock- inhabiting and bioweathering fungi have been extensively covered elsewhere and is not included in the scope of the present chapter.

Basics of Rock and Weathering Process

In general, rocks are aggregation of minerals that occur in nature with defined composition and molecular structure. There are a vast number of minerals occurring in nature (combining with one another or the same mineral in different proportions) giving rise to different types of rocks. In general, when we study about rocks of Earth, we study about crust (inner and outer), mantle and core (outer and inner) wherein the crust of the Earth is composed of different types of rocks. The most common rock forming minerals are silicates, oxides, carbonates and phosphates. It is estimated that approximately 99.9 per cent of the Earth's crust consists of about 20 minerals. Among all silicates and other mineral-forming groups, feldspars are the most abundant. Further, mica group lies in between the alkali-alumina silicates and the ferromagnesian silicates. Ferromagnesian minerals have three main groups- Amphiboles, Pyroxenes and Olivine. There are other rocks forming silicates apart from what are mentioned above. If we talk about oxides, quartz, hematite, limonite, magnetite and ilmenite are important. Among carbonates, calcite and dolomite are major rock-forming minerals. Apatite of phosphates, gypsum and barytes of sulphates are also common rock-forming minerals. A brief classification of rocks is described for better understanding.

Classification of Rocks

Based on the origin of rocks which is considered as the most reliable basis of classification, there are three types of rocks: igneous, sedimentary and metamorphic.

1. **Igneous Rocks-** (*Ignum* meaning fire) Also called primary rocks (rocks of the original Earth). These rocks are predominantly crystalline, massive, unstratified and unfossiliferous. The volcanic (extrusive) igneous rocks are extremely fine-grained and may be massive, amygdaloidal or vesicular. The formation of Igneous rock involves consolidation of magma. The rocks can be crystalline or non-crystalline (glassy) depending upon the pressure. High pressure supports the formation of crystalline whereas low pressure supports the formation of non-crystalline matter. These rocks are of two types: intrusive and extrusive. Intrusive Igneous rocks are formed when magma solidifies below the surface (high pressure) and are crystalline in nature, e.g., Granite, Diorite, Syenite and others. Extrusive Igneous rocks are formed when lava solidifies on Earth's surface (low pressure) and are fine-grained, e.g., Basalt, Andesite, Rhyolite.

2. **Sedimentary Rocks-** These rocks are very common on the Earth's surface accounting for about 75% but less than 5% of the whole crust. Among sedimentary rocks, shale is the most abundant followed by sandstone and limestone whereas conglomerates, breccias and laterites

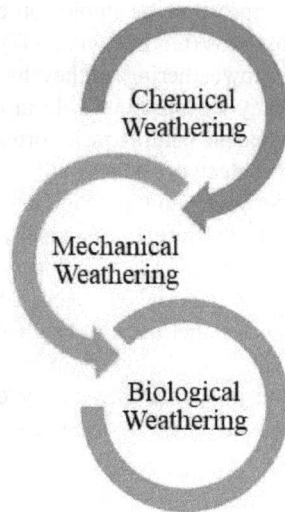

Figure 1. Types of weathering. Weathering can be split up into three types: physical, biological, and chemical weathering.

are not so common. The formation of these rocks is the result of weathering (disintegration and decomposition) of rocks which yields sediments. Further, these sediments on compaction and cementation lead to formation of sedimentary rocks (Figure 3). These are of two types: Clastic and Non-clastic. Clastic rocks are those formed by weathering, erosion and deposition, e.g., siltstone and shale. Non-clastic includes rocks of chemical and biological origin, e.g., rock salt, rock gypsum (chemical sedimentary origin) and coal (biological origin).

3. **Metamorphic Rocks-** These rocks are originated from metamorphism of Igneous (orthometamorphism) and Sedimentary rocks (parametamorphism). Metamorphism refers to mineral and textural changes in pre-existing rock. The changes include changes due to temperature, pressure or chemically active solutions. If igneous and sedimentary rocks after their formation go through above mentioned change in any of the factors, disturbing the equilibrium results in induction of metamorphism. The metamorphism can be of the following types: Thermal metamorphism- heat predominant, Cataclastic metamorphism- directed pressure predominant, Dynamic metamorphism- directed pressure predominant, Dynamo-thermal- directed pressure and heat predominant and Plutonic metamorphism- uniform pressure and heat predominant. Metamorphic rocks can be classified broadly in two groups- Foliated and Non-Foliated. Foliated rocks are those in which the mineral grains appear parallel to each other, i.e., when pressure is applied, the minerals align in a flat manner, whereas Non-Foliated rocks are those in which when pressure is applied grains are not aligned. Examples of foliated rocks are gneiss, schist, phyllite and slate. Examples of non-foliated are hornfels, quartzite and marble.

Rock Cycle

All the three types of rocks mentioned above - igneous, sedimentary and metamorphic- can transform into one another. The process of transformation takes millions of years to complete. The cycle can be summed up as follows-

1. As magma erupts onto surface, lava cools down and forms igneous rocks.

2. Weathering of the rock (Igneous and Metamorphic) will result into the formation of sediments. The sediments are transported by air, water, etc. to the place of deposition forming Sedimentary rocks.

3. When the Igneous and Sedimentary rocks gets buried deep in the crust, they are converted to metamorphic rocks due to increased temperature and pressure.

Bioweathering

Rocks and minerals present a vast reservoir of elements, many essential for life, and such elements must be released in forms that may be assimilated by the biota. These include essential metals as well as elements such as S and P (Gadd 2004, 2007, 2009, Gadd et al. 2007). Minerals influence many essential microbial processes including nutrient acquisition, cell adhesion, and biofilm formation. Microbes can remove nutrients and environmental contamination absorbed to mineral surfaces including metals and organic compounds (Vaughan et al. 2002). Other elements and nutrients may be released from minerals as a result of bio-weathering, and fungi have notable activities in this context. Bioweathering is the erosion, decay and decomposition of rocks by living organisms viz. plants, lichen, bacteria, mycelial fungi, etc. Fungi have the capacity of weathering due to diverse growth patterns, and metabolism and survive in extreme environments such as low nutrition, metal toxicity, UV radiation, and desiccation. Their mutualistic association with plants (mycorrhizas), algae and cyanobacteria (lichens) are particularly significant geo-active agents. Fungi are known to translocate water, ions and nutrients within the mycelial network thus exploiting heterogeneous environments (Sun et al. 2020).

In general terms, bioweathering is the deterioration of soil, rocks, and minerals via biological activity (Gadd 2010, 2017b, Wei et al. 2012a, b, 2013). It means that in bioweathering due to penetration of root/fungal mycelia can penetrate or bore inside the rock to a point that the rock matter gets weakened and convert/break in to small minerals with the help of cellular and biochemical process (Sterflinger 2000, Gadd 2007a, b, Cockell and Herrera 2008, Smits and Hoffland 2009). Biochemical deterioration of plant-containing soil components involves protonolysis (via secretion of protons and acidifying organic acid), redox reactions, interaction with other extracellular components (i.e., siderophores, EPS, OMVs, phenolic compounds, organic acid, peptides, and proteins), and methylation (Gadd 2013). In particular, extracellular excretion of organic acid is known to play a key role in bio-weathering, consequently modulating metal bioavailability in the soil environment. A study of lead contaminated firing range soils revealed that many fungal isolates were capable of producing an array of low molecular weight organic acids including oxalic, citric, malonic, formic, malic, and acetic acids (Sullivan et al. 2012). These isolates exhibited varying degrees of tolerance to Pb in term of hyphal growth and zones of clearing of $PbCO_3$. Organic acid production was influenced by the choice of medium, highlighting the complexity of examining metabolic interactions with metal *in vitro*. Oxalic acid solubilizes large amounts of Pb than citric, malic, and acetic acids (Debela et al. 2010). Oxalic acid has also been shown to greatly reduce Cd accumulation in *Paxillus involutus* and with other organic acids doing so to a lesser extent, including citric, formic, malic, and succinic acids (Bellion et al. 2006). Low molecular weight carboxylic acids can play a particularly important role in the chemical attack of minerals, providing protons as well as a metal-chelating anion (Gadd 2000a, b, 2008, Jacobs et al. 2002a, b, Lian et al. 2008). Some other effects include pitting and etching of mineral surfaces to complete dissolution (Adamo and Violante 2000, Adeyemi and Gadd 2005).

Rocks and minerals deteriorate through hyphal penetration and burrowing into decaying material and along crystal planes, e.g. calcitic and dolomitic rocks. Intracellular turgor pressure is considered to be one of the most significant factors in biomechanical disruption. Special exploration of the environment to locate and exploit new substrate is facilitated by a range of sensory responses that determine the direction of hyphal growth such as thigmotropism (or contact guidance). Biochemical weathering of rocks and minerals can occur through excretion of geo-active metabolites, and this is believed to be a more significant process than mechanical degradation, although a combination of mechanism is often likely. This can result in pitting and etching of surfaces to complete the dissolution of mineral grains. Bio-weathering is a highly significant process and has direct consequences not only for rock and mineral dissolution, but for the mobilization and immobilization of metals, nutrients release, and the formation of secondary minerals. Many processes of microbial interactions with metals cumulatively drive ecosystem services as the consideration of scale comes into play; landscape and plant-level processes including bioweathering, nutrient cycling, soil acidification and plant growth promotion are just a few areas of active research with regard to microbiome influence on soil meta(loid) bioavailability (Adeyemi and Gadd 2005, Bowen et al. 2007, Bonneville et al. 2009, Li et al. 2014, 2016a, b).

Fungi seem to be a component of the microbiota of all rocks, building stones, and concrete, and have been reported from a wide range of rock types, e.g. limestone, dolerite, granite, gneiss, basalt, sandstone, marble, and quartz. Rock surfaces may be subject to moisture deficit and nutrient limitation, although many species can tolerate extreme UV radiation, salinity, pH, and water potential (Burford et al. 2003a). Nutrients are obtained from atmosphere and rain using organic and inorganic residues within cracks and fissures, waste products of other microorganisms, decaying plants and insects, dust particles, aerosols, and animal feces as nutrient sources. Fungi may receive protection from environmental extremes by the presence of melanin pigments and mycosporins in their cell walls and by the production of mucilaginous exopolymeric substances that may entrap inorganic particulates, e.g. clay minerals, providing further protection. Apart from dissolution and deterioration, fungi can form patinas, films, varnishes and crusts. In soil, fungus-mineral interactions are also an integral component of environmental cycling of elements and nutrients (Burford et al. 2003b).

a. Bio-weathering by Bacteria

Rock decay is simply described as the adaptation of rocks to varying physicochemical conditions on the ground surface (Ehrlich et al. 2015). Mineralogical components of rocks are altered and modified upon exposure to Earth's surface conditions in response to different atmospheric agents and insolation that may result in disaggregation (physical weathering) or decomposition (chemical weathering) of the rock (Figure 1). Biological weathering occurs when these processes are assisted by biological processes (Frascá and Lama 2018). In biological weathering by plant roots, physical deterioration of rocks occurs due to application of pressure to form cracks. Fungi and lichens continuously degrade rock surfaces and pores. Microbial activity is considered to be one of the most important factors that controls the amount and rate of rock decay (Gorbushina and Krumbein 2005). These geological processes are accomplished by the biological activity of various kinds of microbes. Secondary minerals such as aluminosilicates and clays are formed by physical and chemical decay which is accelerated by microbes which also promotes ecological restoration (Cockell et al. 2013). However, different bacterial species have different ability of nutrient release and mineral weathering (Wu et al. 2017). Organisms found on rock surfaces form biofilms in humid environments. Biofilm is a layer of microorganisms embedded within a polysaccharide matrix and may include mixed-species populations. Rocks are broken down by microbes to obtain nutrients and minerals essential for their survival and growth (Adamo and Violante 2000). According to Gonzalez et al. (1999), *Bacillus subtilis* and *B. megaterium* were the most effective bacterial species for hematite degradation (Gonzalez et al. 1999). Moreover, bacteria could dissolve 30 times higher gypsum mineral (Sun and Friedmann 1999). The iron dissolution rate of the hornblende mineral was found to be increased approximately 20 times in the presence of bacteria in the environment. *Bacillus cereus* and *B. pumilus* cells were shown to dissolve iron, silica and aluminum elements in quartz, reduce the free iron oxide species, and dissolve the potassium from the mica mineral (Styriakov et al. 2003).

Microbes disintegrate rocks and obtain elements and minerals required for their growth and survival (Adamo and Violante 2000). Some of the bacterial species viz., *Bacillus subtilis* and *B. megaterium* are reported to be effective for their hematite degradation performance (Gonzalez et al. 1999). Bacteria are known to increase the dissolution rate of minerals in the environment.

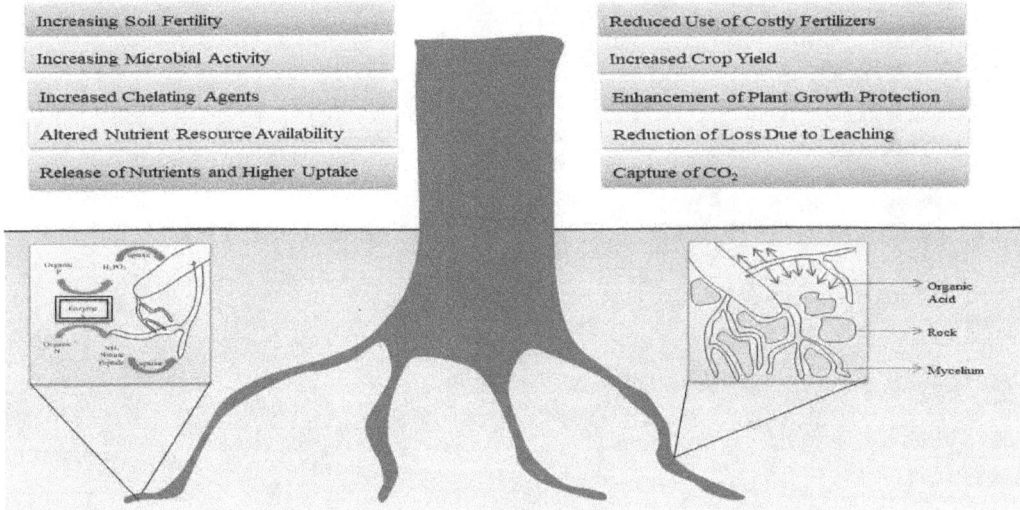

Increasing Soil Fertility		Reduced Use of Costly Fertilizers
Increasing Microbial Activity		Increased Crop Yield
Increased Chelating Agents		Enhancement of Plant Growth Protection
Altered Nutrient Resource Availability		Reduction of Loss Due to Leaching
Release of Nutrients and Higher Uptake		Capture of CO_2

Figure 2. Beneficial effects of rock-inhabiting and bioweathering fungi. These fungi have immense applications in terms of biofertilizers, and enhancement of plant growth by the release of macro- and micronutrients. It is possible by the ability of these fungi to convert rocks into minerals and rock particles.

Bacteria-containing solution degrades the pyrite mineral (Fowler et al. 2001). *Bacillus subtilis* affects granite weathering by forming pits and bacterial surfaces help in calcium precipitation and biomineralisation (Song et al. 2007). As negatively charged groups are present at a neutral pH, positively charged metal ions are bound on bacterial surfaces (Daryono et al. 2019). Finally, the amount of calcium was found to increase because of the bacterial activity performed on historical churches' frescos (Milanesi 2006). Moreover, our knowledge regarding bacterial species colonizing and reproducing in specific rock types is highly limited. Furthermore, no studies included more than one mineral, and no study was conducted directly on the rock without dusting the sample. Rock material used within buildings or structures of historical and cultural significance is not preserved effectively because of bacterial activity. Bacterial weathering of historical buildings has also not yet been thoroughly investigated.

b. Bio-weathering by Fungi

Fungi are microbes which transform metals and minerals and alter the surface structure and chemistry of rocks including their constituent minerals. "Metal-transformation" here refers to the direct and indirect role of fungi in altering metal speciation and mobility. Rock and mineral alteration processes are important in rock bio-weathering because they contribute to the formation and development of mineral soil and global biogeochemical cycles (Sterflinger 2000, Burford et al. 2003a, b, Gadd 2007) (Figure 2). Involvement of fungi in bio-weathering, soil formation, metal and mineral transformation, and element cycling is obvious as fungi are the major decomposers of organic material and can host most of the stable elements. The release of these elements can result in further interaction with environmental components, including mineral precipitation (Figure 4). Fungi utilize and metabolize organic matter thus mediating the geochemical cycle (Gadd 2007, 2010). The most significant environmental roles of fungi are as decomposers of organic materials and as animal and plant pathogens and symbionts. But there is a growing awareness among scientists that fungi are also significant geo-active agents. However, the demarcation between fungal and bacteriological research and the wider range of prokaryotic metabolic diversity has ensured that the geo-active significance of fungi has been largely unappreciated in consideration of biosphere processes. Many of the geomicrobiologist does not include fungi to be a major player in the process (Druschel and Kappler 2015). Certain physiological and morphological responses are involved in

Figure 3. Rock sample collected from Lameta formations at Jabalpur.

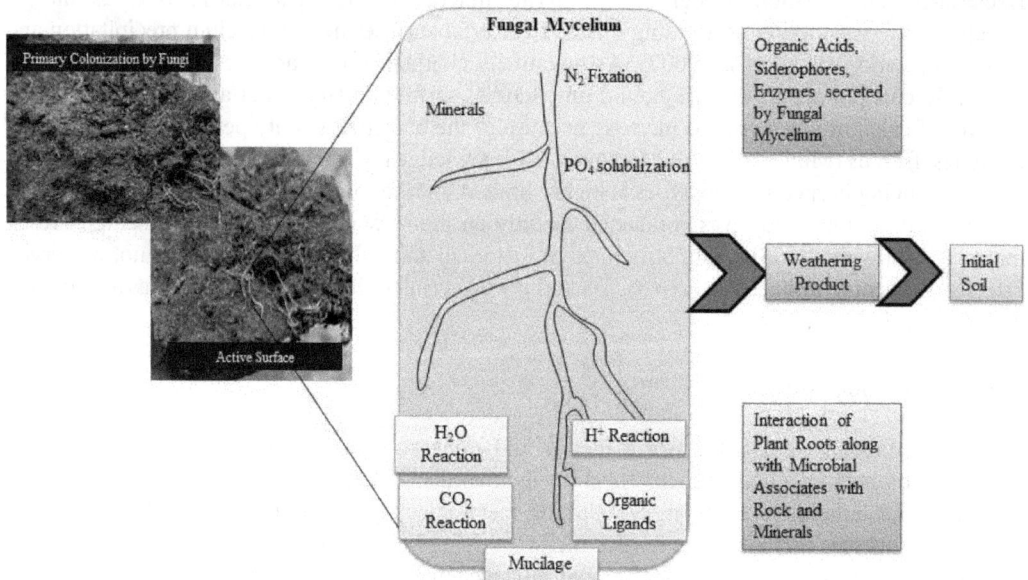

Figure 4. Fungi help in the formation of soil. Fungi help in the weathering of rocks into smaller particles and minerals. These fungi establish themselves, release CO_2 through metabolism, dissolving in water, and form carbonic acid.

the survival of these fungi viz., production of black melanized cell forms resistant to temperatures, desiccation and UV radiations (Gorbushina 2007).

Lichens are fungal growth forms, consisting of a symbiotic partnership between a fungus and a photosynthetic organism – either an alga or a cyanobacterium, but sometimes both. It is now known whether they can also contain yeast as another fungal partner. Lichens are pioneer colonizers of rocks and are the initiator of bio-weathering biofilm that are involved in the early stages of mineral soil formation. Symbiotic root-associated mycorrhizal fungi are associated with the majority of plant species and are responsible for major mineral transformations and redistributions of essential metals and phosphate in the soil (Gadd 2007). Fungi are chemoorganotrophs, which means that they use carbon-containing organic compounds for growth and energy generation, and they excrete extracellular metabolites, such as organic acids, which are key to their interactions with rocks and minerals. Fungi colonizes and transform minerals from concrete and other substances used in buildings, monuments, statues and other cultural heritage sites (Gadd 2017a). In such locations, fungi are often the most visible and destructive of the colonizing microbiota.

Fungi (both free living and symbiotic) are common inhabitants of rocks and minerals layers. Fungi may, in fact, be the most significant organism in nature that can biodeteriorate rocks and minerals (Warscheid and Braams 2000). The morphological and hysiological adaptations have ensured that fungi are found in almost all extreme environments viz., deserts, polar regions, and polluted environments - where fungi are exposed to extreme temperatures, solar irradiation and water. Most of these fungi are aerobic and even occupy unusual locations, e.g. deep aquatic sediments, hydrothermal vents, and igneous oceanic crust, where they may exist in symbiosis with prokaryotes (Iversson et al. 2016). Although rock surfaces may provide different nutrients (viz. airborne dust particles, industrial and domestic emissions and pollutants, and exudates from microbes, insects and animals), many fungi can grow in the presence of very limited amounts of nutrients.

Mechanisms of Fungal-Mineral Interaction

Diverse kinds of fungal populations inhabit rocks and their structure and diversity have mineralogical control. Rock porosity and geo-chemical diverse structures providing nutrients and essential metals affect the colonization and metabolic activity of fungi. For example, fungi may have colonized certain rock types such as granite due to the elemental and mineral constituents of the rock (Gleeson et al. 2005). Therefore, it is believed that fungi selectively do bioweathering of minerals and release elements such as K. However, this kind of effect of rock substrate may not be visible in low heterogeneous rocks like sandstone. Many mycelial fungi (which may be of soil origin) inhabit rocks that may not exhibit the hyphal mode of growth, occurring as small black melanized colonies. It results in discoloration of colonized surfaces (Marvasi et al. 2012). They are able to resist environmental stresses and are considered inhabitants of exposed rock surfaces. Filamentous mycelial hyphae developing from colonies penetrate in rock fissures, cracks and in intracrystalline penetration. This results in interaction between thin fungal hyphae and rock substrate leading to rock degradation and conversion to minerals, thus accumulating metals and minerals in cell wall and extracellular materials (Gorbushina 2007). The microscopic fungi cause pitting in rocks leading to cavities that can contain the fungal colonies. This colonization of fungi produces micropits by mechanical destruction through intracellular turgor (=hydrostatic) pressure and extracellular polysaccharide production (Marvasi et al. 2012). Micro-colonial fungi may also form casual mutualistic associations with algae in rock crevices in order to obtain carbon from their photosynthetic partner (Gorbushina 2007). The chemical and mineral structure of the rock affects the physical and biochemical aspects of fungal colonization viz., the presence of minerals such as feldspars may increase susceptibility to attack (Warscheid and Braams 2000). The transformation mechanism involves physical and biochemical processes. These processes are not mutually exclusive, rather they are generally interlinked.

The physical mechanisms of bioweathering include hyphal penetration through weak points and/ or direct boring in a weak or porous substrate (Jongmans et al. 1997, Hoppert et al. 2004). The mineral lattice gets weakened by wet and dry cycles followed by expansion or contraction of the biomass. The lichens are known to cause mechanical damage to rocks such as disaggregation and exfoliation. This can occur by penetration of their anchoring structures e.g. rhizines in foliose lichens and through expansions/contraction of the thallus on wetting/drying cycles (Chen et al. 2000). The thallus and secondary mineral products incorporate the minerals and other breakdown elements (de los Rios et al. 2002). These effects, in addition to removal of the lichen itself by animals and the weather, can lead to visible damage to a rock in just a few years (Gadd et al. 2014). Some other physical effects such as cell turgor pressure, exopolysaccharide production and secondary mineral formation are also caused by lichens. This is often considered to be a major mechanism of rock decay (Ranalli et al. 2019).

Biochemical weathering of rock and mineral substrates occurs through excretion of H^+, CO_2 organic acids, siderophores, and other metabolites, and this produces pitting, etching or dissolution (Benfield et al. 1999, Gadd 2010). Siderophores are Fe(III)-binding compounds excreted by organisms, but siderophores can also interact at lower affinity with other metals. Biogenic organic acids are very effective at mineral dissolution and are one of the most damaging agents of rock and mineral substrate, clearly underlining the importance of fungi in biochemical weathering (Gadd 2007). Oxalate is of major significance because of its ability to form metal complexes, resulting in mineral dissolution and physical damage by forming secondary metal oxalates that can expand in pores and fissures (Gadd et al. 2014). Citric and gluconic acids are other significant fungal metabolites, oxalic acid being the most significant one (Adamo and Violante 2000). In some extreme cases, and depending on the rock substrate, up to 50% of certain lichen thalli may comprise metal oxalates, which are the main secondary crystalline products of lichen bio-weathering (Purvis and Pawlik-Skowronska 2008).

Biomineralization is the biologically mediated formation of minerals, and this process forms a core component of bioweathering of rocks and minerals. The oxidation and reduction of metals and fungal metabolite excretion viz., oxalate results in fungal bioweathering. For example, soluble Mn(II) can be oxidized by many microbes, including fungi, and this results in the formation of the black Mn oxides that are a common component of rock varnish. The release of metals from rocks in mobile forms from abiotic weathering or biological mediated mechanism can, therefore, result in a variety of secondary mineral precipitates that include carbonates, phosphates, oxides and oxalates (Gadd 2010, Gadd et al. 2014). Such biomineral formation may contribute to physical disruption, staining and discoloration, and rock coating development (Gadd 2007, 2017a, b, Gorbushina 2007, Fomina et al. 2010). The dissolution and burrowing within soil minerals have been proposed to be due to fungal tunnels within soil minerals (Jongmans et al. 1997). Tunnels may also result after fungal exploration of pre-existing cracks, fissures, and pores in weatherable minerals and form the formation of a secondary minerals matrix of the same or different chemical composition as the substrate, e.g., secondary calcium carbonate or an oxalate (Fomina et al. 2010).

Fungi are involved in many environmental mineral transformations, and the role of plant-root symbiotic fungi in releasing phosphate and other nutrient from minerals is a major determinant of plant productivity. Many other transformations affect the structure and stability of rocks and minerals, leading to mineral dissolution and biodeterioration, as well as the formation of new minerals, crust, and coatings. Carbonates can be broken down by fungi, usually as a result of acid formation (Adamo and Violante 2000). Fungal attack of carbonate substrates (e.g., limestone) can result in digenesis of these substrates to secondary minerals that include dolomite [$CaMg(CO_3)$], glushinskite ($MgC_2O_4.2H_2O$), and weddellite ($CaC_2O_4.2H_2O$) (Burford et al. 2003b). Many lichens can dissolve calcite in limestone and dolomite through oxalate production, subsequent precipitation of calcium oxalate occurring at the lichen-rock interface. However, there is a possibility that many novel minerals with bioprospecting ability are present in the soil. Calcium oxalate is the most common environmental form of oxalate (Gadd et al. 2014). Calcium oxalate is produced by many free-living and symbiotic fungi, being formed by the precipitation of soluble calcium with fungal-excreted oxalate (Adamo and Violante 2000). Fungi can also produce many other metal oxalates when they interact with minerals that contain metals, including Ca, Cd, Co, Cu, Mg, Mn, Sr, Zn, Ni and Pb (Gadd 2007, Gadd et al. 2014). Lichens are associated with many secondary crystalline metal oxalates including hydrated magnesium oxalate on serpentinite and manganese ore, and hydrated copper oxalate on copper-containing rocks (Chen et al. 2000, Purvis and Pawlik-Skowronska 2008). In many arid and semi-arid regions, calcareous soils and near-surface limestones (calcretes), biomineralized fungal filaments occur secondarily cemented with calcite and calcium oxalate. The preexisting limestone is cemented by calcium carbonate formed from calcium oxalate (Bindschedler et al. 2016). Many limestone and marble monuments develop orange-brown patinas, called scialbatura, on their surfaces. Rock surfaces of limestone and marbles are protected by free-living fungi and lichens by forming calcium oxalate and calcium carbonate (Savkovic et al. 2016).

Several fungi can promote oxidation of Mn(II) to black Mn(IV)O$_2$. In many cases, fungal oxidation is affected through production of hydroxycarboxylic acid metabolites such as citrate, lactate, malate, or gluconate. Fungal MnOx material has a todorokite-like tunnel structure, which contrasts with other reported bacterial MnOx materials which have layered birnessite-type structures. In metal-bearing minerals such as siderite (FeCO$_3$) and rhodochrosite (MnCO$_3$), some fungi are able to oxidize Mn(II) and Fe(II) into oxides. Manganese and iron oxides are major components along with clay and trace elements in desert or rock varnish (Gorbushina 2007). Conversely, manganese-reducing microbes may mobilize oxidized or fixed manganese. Most fungi that reduce Mn(IV) oxides do so through the production of metabolites that can act as reductants, such as oxalate, which results in the formation of lindbergite (Mn oxalate dihydrate) (Wei et al. 2012a, b).

Fungi can liberate orthophosphate from insoluble inorganic phosphates in rocks and minerals by producing acids or chelators (e.g., gluconate, citrate, oxalate, and lactate) which complex the metal and result in dissociation. Liberated phosphate is used as a nutrient by fungi and re-precipitated with metals to form metal phosphates. Fungal solubilization of rock phosphate results in re-precipitation of calcium oxalate, while fungal dissolution of pyromorphite [Pb$_5$(PO$_4$)$_3$Cl] results in lead oxalate formation (Sayer et al. 1999). Many fungi are also capable of uranium oxide solubilization and can form secondary uranium phosphate minerals of the meta-autunite group: uramphite and/or chernikovite (Fomina et al. 2007). Fungi may also liberate orthophosphate from sources of organic phosphate in organic matter by means of phosphatase enzymes, the liberated phosphate then re-precipitates with other reactive metals. Several common soil fungi and yeasts induce lead bioprecipitation as highly insoluble pyromorphite when growing on organic phosphates (Liang et al. 2016).

Many lichens and free-living fungi play a role in silicate dissolution and therefore contribute to the genesis of clay minerals and to soil and sediment formation (Adamo and Violante 2000). In Antarctic sandstone, the reduction of cohesion between individual sandstone grains is caused by cryptoendolithic lichens resulting in surface flaking (Chen et al. 2000). Silicate dissolution may release important nutrients that otherwise would remain bound (e.g., P, K and Fe). Fungal extracellular polysaccharides can become mixed with calcium, potassium, iron, clay minerals and nanocrystalline aluminous iron oxyhydroxides in lichen weathering of silicates. Biotite-group minerals [K(Mg,Fe(II))$_3$AlSi$_3$O$_{10}$(OH,F)$_2$] can be penetrated along cleavage planes by fungal hyphae, partially converting biotite to vermiculite [(Mg,Fe(II),Al)$_3$ (Al,Si)$_4$O$_{10}$(OH)$_2$·4H$_2$O]. Clay minerals associated with lichen weathering also include allophane, goethite, halloysite, illite, kaolinite and imogolite, with halloysite, kaolinite and vermiculite being the most common, which are dependent on the lichen species involved and mineral substrate present (Chen et al. 2000).

Molecular Phylogeny of Bioweathering Fungi

There has been very few studies on the molecular phylogeny of fungi involved in bioweathering as compared to rock inhabiting fungi. Both the groups of fungi are closely related wherein the former groups of fungi have been established for their role in weathering process. Ortega-Morales et al. (2016) studied the ITS region by molecular detection methods. The phylogenetic analysis of the sequence were able to identify the isolates which belonged to 10 genera viz. *Annulohypoxylon*, *Cladosporium*, *Cochliobolus*, *Fusarium*, *Hypoxylon*, *Lasiodiplodia*, *Penicillium*, *Pestalotiopsis*, *Rosellinia*, and *Xylaria*. Calcareous and quartzitic rock surfaces depict special living conditions. The fungal colonies that grow meristimatically are *Sarcinomyces*, *Phaeosclera*, and *Botryomyces*. Sterflinger et al. (1997) compared the fungi isolated from marble in the Mediterranean region viz., *Coniosporium perforans*, *C. apollinis*, *Monodictys castaneae*, *Phaeosclera dematioides* and a *Coniosporium*-like strain by internal transcribed spacer (ITS) region sequencing and 18S rRNA gene

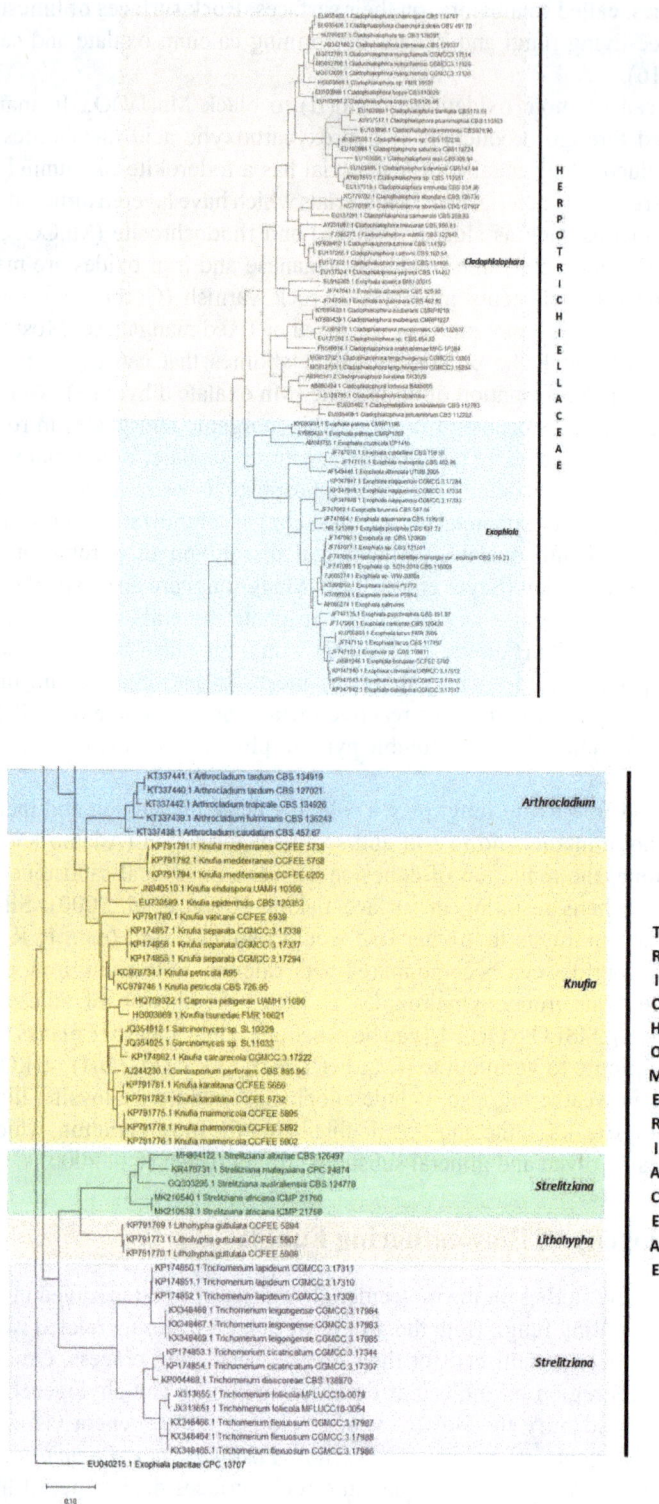

Figure 5. Phylogenetic tree generated by the maximum likelihood analysis using sequences of ITS of the class *Dothidiomycetes* (consisting of members of *Herpotrichiellaceae, Cyphellophoraceae, Trichomeriaceae*) using Tamura-Nei model substitution model. No. of Sequences: 179. No. of Sites: 743. Conserved sites are 183, Variable sites are 553, Parsimony informative sites are 478, and /singletons are 70.

phylogeny. Phylogenetic analysis showed that fungal species *Phaeosclera dematioides* is related to *Dothideales* (ascomycetous fungi) and *Monodictys castaneae* to the *Pleosporales*, whereas the three *Coniosporium* species studied are a sister group to the *Herpotrichiellaceae* (*Chaetothyriales*) (Figure 5). Many rock-inhabiting fungi are included in class *Dothideomycetes* (tolerating harsh conditions). The positions of many fungi are unclear due to a lack of phylogenetic data and inference. Based on multi-gene phylogeny, Ruibal et al. (2009) reported that members of *Myriangiales*, *Dothideales*, *Pleosporales* and *Capnodiales* belong to rock-inhabiting fungi. However, there are many lineages that are still not completely characterized but show closeness with *Dothideomycetes*, and another lineage shows affinity with *Arthoniomycetes* wherein all members are rock-inhabiting fungi. The lineages suggest that these fungi are plant pathogen and/ or saprobes but rocks and minerals are their primary substrate for growth. Studies also infer that the rock-inhabiting fungi have link with lichen habit and lichenization. The ITS region is not able to completely resolve most of the rock-inhabiting fungi. Hence, the other regions used in the phylogenetic analysis of rock-inhabiting fungi are nucILSU, nucSSU, mtSSU, RPB1 region A-D, RPB2 region 5–7, RPB2 region 7–11, etc. The phylogenetic analyses of these regions have helped in the identification as well as characterization of these fungi and some novel lineages of *Dothideomycetes* (consisting of rock-inhabiting fungi). Still, the species richness in *Dothideomycetes* is understudied.

Phylogenetic tools have resulted in providing authentic identification of these fungi associated with varied rock types viz., marble, granite, pegmatite, limestones, quartz and sandstones (Sert et al. 2007a, b, 2011, 2012, Sert and Sterflinger 2010, Martin-Sanchez et al. 2012, Egidi et al. 2014, Su et al. 2015, Isola et al. 2016). As per the updated fungal nomenclature based on MycoBank (http://www.mycobank.org), Index Fungorum (http://www.indexfungorum.org) and the phylogenetic study conducted by Abdollahzadeh et al. (2020), it is estimated that there are more than 175 rock-inhabiting fungi belonging to 64 genera, 19 families, 11 orders, 2 classes and 1 phylum. That means approximately 3.6% percent of known fungal families and 0.11% of known fungal species are known to be associated with rock and/ or involved in bioweathering. In the past, many of the taxonomic positions have been clarified through phylogenomic methods including the taxonomic status of *Lichenothelia* and *Saxomyces*. Moreover, several of the ancestral lineages belonging to Late Devonian and Middle Triassic periods have been resolved by phylogenetic analysis belonging to the orders of *Capnodiales*, *Dothideales*, and *Chaetothyriales* in a kingdom-wide phylogenetic analysis (Gostinčar et al. 2009). Thus, it is clear that the rock surfaces support diverse kinds of stress-tolerant, oligotrophic fungi which are polyphyletic in origin. These fungi have a variety of phenotypes supporting physiological adaptations (Gostinčar et al. 2012, Sterflinger et al. 2012). The phylogenetic characterization by multi-locus analyses has helped in the proper identification and classification of these fungi through phylogenetic analysis (Reblova et al. 2013, Isola et al. 2016, Sun et al. 2020).

Omics of Bioweathering Fungi

In recent years, genome sequence data has been providing significant data on the physiology and metabolism of fungi. Many of the rock-inhabiting fungi (some of them probably be involved in bioweathering) such as *Cryomyces antarcticus*, *Friedmanniomyces endolithicus*, *Friedmanniomyces simplex*, *Hortaea thailandica*, *Rachicladosporium antarcticum* and *Rachicladosporium* sp. and *Coniosporium apollinis* (Sterflinger et al. 2014, Coleine et al. 2017, 2020). Properties related to temperature tolerance, desiccation, radiation tolerance and other data related to the physiology and metabolism (stress resistance and oligotrophic nature) of the fungi is obtained through genome sequence and analyzed. Although there was no significant difference in the genome size (24 Mb) of standard fungal culture (such as *Neurispora crassa*) and any rock-inhabiting fungi and/or GC content (54%) and percent of repetitive sequences (0.33%), some fungi like *F. endolithicus* and *R. antarcticum* have double the genome size (Coleine et al. 2017, 2020). These fungi are also reported to have changes in the whole-cell proteins under extremely tolerant stress environment which

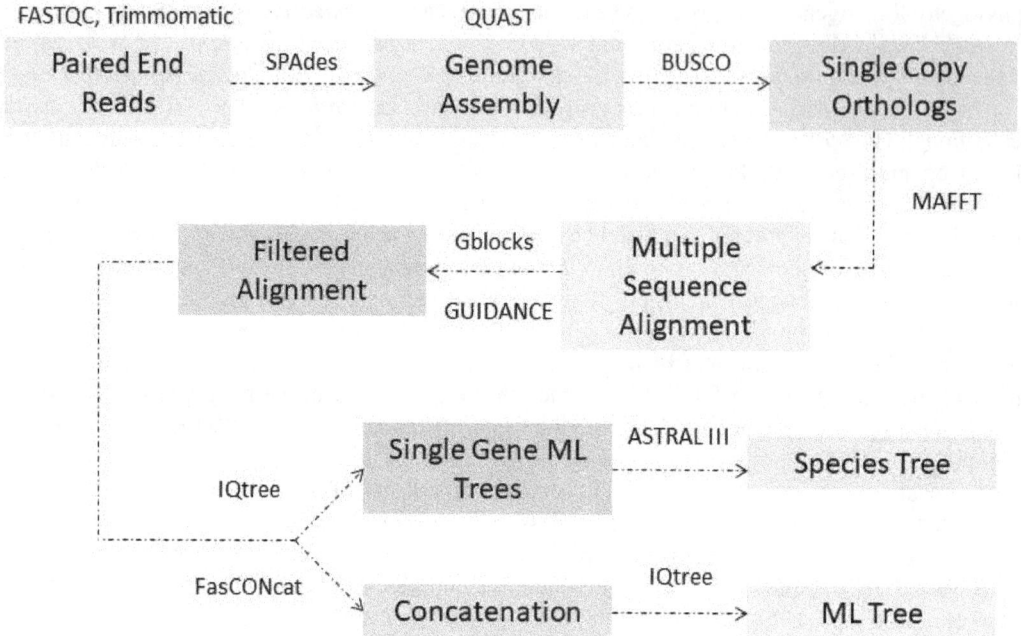

Figure 6. Flow chart of the bioinformatic pipeline used to analyze the genome sequences (Adopted from Ametrano et al. 2019).

helped in adaptation mechanisms as in *F. endolithicus, C. antarcticus, Exophiala jeanselmei* and *Knufia perforans* (=*Coniosporium perforans*) (Zakharova et al. 2013). Moreover, similar changes were observed in 2-D protein spectra of rock-inhabiting black yeasts and other mesophilic fungi wherein the mesophiles produce relatively short proteins. Moreover, it would be better for these fungi to produce specialized proteins and metabolites instead of general ones (such as heat shocking proteins) which are of high energy consumption (Naranjo-Ortiz and Gabaldón 2019). Work around the globe is focusing on several extreme environment including rock-inhabiting and bioweathering fungi viz., the "Shed Light in The Dark Lineages of the Fungal Tree of Life" (STRES) project which focuses on genome sequencing coupled with transcriptomics and metabolomics (Selbmann et al. 2020). These omics studies have helped in getting an insight in to the adaptations of these fungi while surviving in oligotrophic environments and performing their role in the geochemical cycle. The genome sequencing has shown that equal photoreceptors (similar to the plant pathogen *Botrytis cinerea*) have been detected in these fungi wherein their function is to sense and avoid sunlight stress (Schumacher and Gorbushina 2020). Some of the mechanisms such as the carbon fixation pathway of plants (CFPP) are not completely developed as enzymes like Rubisco (ribulose-1,5-bisphosphate carboxylase-oxygenase) and PRK (phosphoribulokinase) are absent (Lyu et al. 2015). However, other nutritional pathways may be developed (like oxidation of elemental sulfur, iron and manganese or light-driven rhodopsin transmembrane protons and sodium pumps (Gleason et al. 2019).

The studies on genomes' sequencing, comparison and inference is dependent on the availability of the online database. Incomplete data or data of low quality is an issue with the analysis and interpretation as far as rock-inhabiting fungi are concerned. The strategies adopted should produce consistent phylogenies even with diverse phylogenetic reconstruction (Figure 6). As more and more studies are conducted, it will allow to conduct evolutionary studies of these fungi. The subclass *Dothideomycetidae* within the class *Dothideomycetes* include orders such as *Capnodiales*, *Myriangiales* and *Dothideales* whereas the subclasses *Pleosporomycetidae* includes orders *Pleosporales*, *Mytilinidiales* and *Hysteriales* (Ruibal et al. 2009, Schoch et al. 2009, Muggia et al. 2015, Liu et al. 2017, Ametrano et al. 2019).

Application of Bioweathering Fungi

The bioweathering process has an immense impact on humans by environmental changes, biotechnological applications, biodeterioration and biocorrosion. The biodeterioration of stone and mineral artifacts represents a loss of cultural heritage (Scheerer et al. 2009, Cutler and Viles 2010). Materials which are usually used to stabilize building blocks and coating surfaces prior to painting are also susceptible to biodeterioration (Scheerer et al. 2009). Highly deteriorated stone surfaces act as soil nutrients for colonization by mosses, ferns, and higher plants (Cutler and Viles 2010, Gadd 2016). Mechanisms involved in stone deterioration are complex which includes most of the direct and indirect mechanisms previously discussed for mineral dissolution. Extracellular polymeric substances are also capable of metal complexation and weakening of mineral lattices through wetting and drying cycles, as well as the production of efflorescences, i.e., secondary minerals produced through the reaction of anions of excreted acids with cations of stone (Wright 2002). Physical damage may be caused in weak areas penetrated by hyphae, e.g., lichens cause damage due to penetration by their rhizines (composed of fungal filaments, and expansion/contraction of the thallus on wetting/drying) (Hirsch et al. 1995, Gaylarde and Morton 2002, Cockell and Herrera 2008). "Lichen acids," mainly oxalic acid, cause damage at the stone-lichen interface, and lichen thalli may accumulate up to 50% calcium oxalate, depending on the substrate (Lisci et al. 2003, Seaward 2003). Moreover in calcareous stones, calcium and magnesium carbonates are solubilized by carbonic acid produced by lichens. The fungal biodeterioration causes tunneling of hyphae on to ancient ivory viz., natural apatite, walrus tusk, etc. forming calcium oxalate and whellite in abundance (Pinzari et al. 2013). Concrete and cement materials are also biodeteriorated wherein fungi dominate the concrete-deteriorating microbiota (Gu 2009). Microbial degradation of concrete leads to biochemical and biomechanical deterioration caused by protons, inorganic and organic acids and hydrophilic slimes. Fungal species have been reported to colonize concrete samples (used as radioactive waste barriers) in Chernobyl reactor. As a result of fungal activity, the concrete samples leached iron, aluminum, silicon, and calcium and re-precipitated silicon and calcium oxalate (Fomina et al. 2007).

Mineral and metal solubilization mechanisms help in the removal of metals from industrial wastes, low-grade ores, and metal-bearing minerals. These processes find their applications in bioremediation, recycling and metal bio-recovery wherein fungi help in solubilization of metals from fly ash-contaminated soil, electronic scrap, and other waste materials. Bio-leaching of metal ores from mining sites is one of the growing area wherein microbes including fungi help in enhanced recovery of metals. It is observed that bacterial bioleaching is more efficient than fungi; fungal bioleaching may be more suited to specific bioreactor applications. The metal immobilizations viz., biosorption, bioaccumulation and bioprecipitation are a result of fungal action. Fungi are considered to be efficient agents to remove radionuclides, metals, etc. from solutions and mixtures through biosorption methods (Sun et al. 2020). Urease-positive fungi are used to precipitate metal-containing carbonates providing a means of metal biorecovery (Li et al. 2015). Fungi are also known to recover insoluble metals, metalloids, and radionuclides (e.g., oxalates, oxides, and phosphates) through the secretion of metabolites, thus helping in the production of elemental metal or metalloid forms (Gadd et al. 2012). Some biomineral and elemental products, including those of nanoscale dimensions, are relevant to the production of novel advanced biomaterials, with applications in metal and radionuclide bioremediation, antimicrobial treatments (e.g. nano-silver), solar energy and electrical battery applications, and microelectronics. In a novel approach, urease-positive *Neurospora crassa* was used to precipitate manganese carbonate. After thermal treatment at 300°C, the carbonized biomass-manganese oxide composite material was used in lithium-ion batteries and supercapacitors, where it was found to exhibit excellent electrochemical properties. In lithium-ion batteries, around 90% charge capacity was retained after 200 charge-discharge cycles (Li et al. 2016).

The ability of fungi and bacteria to transform metalloids has been successfully used for bioremediation of contaminated land and water. In San Joaquin Valley and Kesterson Reservoir, California, selenium has been obtained by a mechanism called selenium methylation which is a kind of volatization process. Mycorrhizal associations may have applications in phytoremediation - the use of plants to remove or detoxify environmental pollutants, by metal phytoextraction or by acting as a biological barrier (Adriaensen et al. 2005, Göhre and Paszkowski 2006). Metals like Cu, Cd, and Pb are sequestered by Glomalin (an insoluble glycoprotein) which is produced on hyphae of arbuscular mycorrhizal fungi. Arbuscular mycorrhizal fungi can also decrease U translocation from plant roots to shoots. For ericaceous mycorrhizas, the fungus prevents the translocation of Cu and Zn to host plant shoots. The development of stress-tolerant plant-mycorrhizal associations may be a promising strategy for phytoremediation and soil amelioration (Schützendübel and Polle 2002, Cairney and Meharg 2003). Some of the geomycological processes detailed previously may have consequences for abiotic soil treatment processes, notably the immobilization of toxic metals by phosphate formation. Apatite $[Ca_5(PO_4)_3(FClOH)]$, pyromorphite $[Pb_5(PO_4)3Cl]$, mimetite $[Pb_5(AsO_4)_3Cl]$, and vanadinite $[Pb_5(VO_4)_3Cl]$ are the most common prototypes of the apatite mineral family. Such minerals hold promise for stabilization and recycling of industrial and nuclear waste and have been explored for treatment of lead-contaminated soils and waters (Sun et al. 2020). The stability of these minerals is therefore of interest in any soil remediation strategy seeking to reduce the effects of potentially toxic elements such as Pb, V, and As. For example, Pb in nature is found in the form of insoluble lead phosphate (pyromorphite) and is unavailable for use. However, solubilization of pyromorphite and formation of lead oxalate by several free-living and symbiotic fungi demonstrate that pyromorphite may not be as effective at immobilizing lead as some previous studies have suggested. Similarly, despite the insolubility of vanadinite, fungi exerted both biochemical and biophysical effects on the mineral including etching, penetration, and the formation of new biominerals. Lead oxalate was precipitated by *Aspergillus niger* during the bioleaching of vanadinite and mimetite, which suggests a general fungal mechanism for the transformation of lead-containing apatite-group minerals (e.g., vanadinite, pyromorphite, mimetite) (Ceci et al. 2015a, b). This pattern of fungal bioweathering of lead apatites could be extended to other metal apatites, such as calcium apatite $[Ca_5(PO_4)3(OH,F,Cl)]$. Here, the formation of monohydrated (whewellite) and dihydrated (weddellite) calcium oxalate can be accomplished by many different fungal species. The ability of free-living and mycorrhizal fungi to transform toxic metal-containing minerals should therefore be taken into account in risk assessments of the long-term environmental consequences of *in situ* chemical remediation techniques, re-vegetation strategies, or natural attenuation of contaminated sites. The bioweathering potential of fungi has been suggested as a possible means for the bioremediation of asbestos-rich soils. Several fungi could extract iron from asbestos mineral fibers (e.g., 7.3% from crocidolite and 33.6% from chrysotile by a *Verticillium* sp.), thereby removing the reactive iron ions responsible for DNA damage (Fomina et al. 2005a, b, Guggiari et al. 2011).

Future Scope and Conclusion

The recently increased interest in rock-inhabiting and bioweathering fungi has helped in the expansion of the description of such fungi from around the globe. However, still many rock formations, rock surfaces and man-made structures (including monuments) remain unexplored. The bioprospecting of these fungi as biofertilizers for nutrient-poor soils, drought prone regions and other metabolites need to be explored. Moreover, detailed studies on these fungi will also help to know the damage they cause to monuments and other man-made structures. As per the new estimates of unknown fungi which are mostly based on the uncultured studies, there are exorbitantly high numbers of yet-to-be-described fungi. It needs investigation in uncommon niches like rocks, mountains, deep sea vents, endophytes, insect and animal guts, etc. which may help in knowing several new taxa. Large-scale sampling and detailed research are required to get a holistic knowledge of their diversity, distribution, habitats and environmental stress adaptation mechanisms. The molecular methods (DNA sequencing and phylogenetic classification) help in the authentic identification, characterization and placement of novel taxa isolated from rock formations, monuments and other manmade structures.

The phylogenetic analyses revealed that most rock-inhabiting fungi belong to two main lineages of *Ascomycota*, i.e., *Dothideomycetes* and *Eurotiomycetes* (Gueidan et al. 2008, 2011). However, fewer taxa have been reported from the *Eurotiomycetes*, especially in families *Trichomeriaceae* and *Herpotrihiellaceae* in order *Chaetothyriales* (Gueidan et al. 2014). The use of multi-locus phylogenetic analysis viz., ITS, SSU, *tef, tub, act* loci have been able to resolve the taxa positions in these fungi. However, ITS and nucLSU are two important loci which have helped in distinguishing novel taxa within *Trichomeriaceae* and *Herportichiellaceae* (Ruibal et al. 2009, Sert and Sterflinger 2010). A three-region phylogenetic analysis (ITS, nucSSU and mtSSU) and five-loci (nucLSU, nucSSU, mtSSU, RPB1, and RPB2) concatenated analysis have been accomplished to resolve rock-inhabiting fungi within the order *Capnodiales* and class *Dothideomycetes,* respectively (Ruibal et al. 2009). Others have also used similar analysis (four loci- nucLSU, nucSSU, mtSSU, RPB1 and six loci- ITS, nucLSU, nucSSU, mtSSU, TUB2, RPB2) in authentic identification and proper placement of taxa within common genera (viz. *Coniosporium, Rupestriomyces* and *Spissiomyces*) of rock-inhabiting fungi (Untereiner et al. 2011). Based on the previous studies, a strategy for the molecular identification and taxonomy can be put forth. Initially, ITS sequencing of the isolated fungi should be done and checked with the NCBI database (both type and non-type). Based on the percent sequence similarity of the BLASTn results of the query sequence, it should be further characterized depending on the family it belongs to (sequencing of nucLSU or nucSSU or other protein-coding genes). The species concept should be applied depending on the genera/family/ order it belonged (Sharma et al. 2015, Sun et al. 2020). The phylogenetic analysis can be done by maximum-likelihood or maximum parsimony or neighbor-joining depending on the kind of dataset and analysis required. The data can also be analyzed by the poisson tree processes (PTP) server: a Bayesian implementation.

In recent times, there are some studies available on genome sequencing of rock-inhabiting and bioweathering fungi and excellent information has been interpreted through it regarding their physiology and metabolism. However, transcriptomics and proteomics analyses are extensively required which would explain the differential gene expressions, metabolic activities and genetic composition. This would help us to understand their survival strategies under oligotrophic environment, and resumption of activities after long gaps of metabolic suspension of activities (Ametrano et al. 2019). This would also help in broadening the area of application of these fungi and thus increasing their bioprospecting prospects.

Acknowledgement

RS thanks Department of Biotechnology, New Delhi for the extramural grant no. BT/PR25368/ NER/95/1161/2017 dated 23.01.2019.

References

Adamo, P. and P. Violante. 2000. Weathering of rocks and neogenesis of minerals associated with lichen activity. Appl. Clay Sci. 165–6: 229–256.

Adeyemi, A.O. and G.M. Gadd. 2005. Fungal degradation of calcium-, lead-and silicon-bearing minerals. Biometals 183: 269.

Adriaensen, K., T. Vralstad, J.P. Noben, J. Vangronsveld and J.V. Colpaert. 2005. Copper-adapted Suillus luteus, a symbiotic solution for pines colonizing Cu mine spoils. Appl. Environ. Microbiol. 7111: 7279–7284.

Ametrano, C.G., K. Knudsen, J. Kocourková, M. Grube, L. Selbmann and L. Muggia. 2019. Phylogenetic relationships of rock-inhabiting black fungi belonging to the widespread genera Lichenothelia and Saxomyces. Mycologia 1111: 127–160.

Bellion, M., M. Courbot, C. Jacob, D. Blaudez and M. Chalot. 2006. Extracellular and cellular mechanisms sustaining metal tolerance in ectomycorrhizal fungi. FEMS Microbiol. Lett. 2542: 173–181.

Bindschedler, S., G. Cailleau and E. Verrecchia. 2016. Role of fungi in the biomineralization of calcite. Minerals 62: 41.

Bonneville, S., M.M. Smits, A. Brown, J. Harrington, J.R. Leake, R. Brydson et al. 2009. Plant-driven fungal weathering: Early stages of mineral alteration at the nanometer scale. Geology, 377: 615–618.

Bowen, A.D., F.A. Davidson, R. Keatch and G.M. Gadd. 2007. Induction of contour sensing in Aspergillus niger by stress and its relevance to fungal growth mechanics and hyphal tip structure. Fungal Genetics and Biology, 446: 484–491.

Burford, E.P., M. Fomina and G.M. Gadd. 2003a. Fungal involvement in bioweathering and biotransformation of rocks and minerals. Mineral. Mag., 676: 1127–1155.

Burford, E.P., M. Kierans G.M. Gadd. 2003b. Geomycology: fungi in mineral substrata. Mycologist, 173: 98–107.

Burford, E.P., S. Hillier and G.M. Gadd. 2006. Biomineralization of fungal hyphae with calcite CaCO3 and calcium oxalate mono-and dihydrate in carboniferous limestone microcosms. Geomicrobiol. J., 238: 599–611.

Cairney, J.W. and A.A. Meharg. 2003. Ericoid mycorrhiza: a partnership that exploits harsh edaphic conditions. Eur. J. Soil Sci., 544: 735–740.

Ceci, A., M. Kierans, S. Hillier, A.M. Persiani G.M. Gadd. 2015a. Fungal bioweathering of mimetite and a general geomycological model for lead apatite mineral biotransformations. Appl. Environ. Microbiol., 8115: 4955–4964.

Ceci, A., Y.J. Rhee, M. Kierans, S. Hillier, H. Pendlowski, N. Gray et al. 2015b. Transformation of vanadinite [Pb5 VO4 3Cl] by fungi. Environ. Microbiol., 176: 2018–2034.

Chen, J., H. P. Blume and L. Beyer. 2000. Weathering of rocks induced by lichen colonization—a review. Catena, 392, 121–146.

Cockell, C.S. and A. Herrera. 2008. Why are some microorganisms boring? Trends Microbiol., 163: 101–106.

Cockell, C.S., L.C. Kelly and V. Marteinsson. 2013. Actinobacteria—an ancient phylum active in volcanic rock weathering. Geomicrobiol. J., 308: 706–720.

Cutler, N. and H. Viles. 2010. Eukaryotic microorganisms and stone biodeterioration. Geomicrobiol. J., 276–7: 630–646.

Daghino, S., F. Turci, M. Tomatis, A. Favier, S. Perotto, T. Douki et al. 2006. Soil fungi reduce the iron content and the DNA damaging effects of asbestos fibers. Environ. Sci. Technol., 4018: 5793–5798.

Daryono, L.R., A.D. Titisari, I.W. Warmada, and S. Kawasaki. 2019. Comparative characteristics of cement materials in natural and artificial beachrocks using a petrographic method. Bull. Eng. Geol. Environ., 78: 3943–3958.

De los Ríos, A., J. Wierzchos and C. Ascaso. 2002. Microhabitats and chemical microenvironments under saxicolous lichens growing on granite. Microb. Ecol., 181–188.

Debela, F., J.M. Arocena, R.W. Thring and T. Whitcombe. 2010. Organic acid-induced release of lead from pyromorphite and its relevance to reclamation of Pb-contaminated soils. Chemosphere, 804: 450–456.

Druschel, G.K. and A. Kappler. 2015. Geomicrobiology and microbial geochemistry. Elements, 116: 389–394.

Egidi, E., G.S. De Hoog, D. Isola, S. Onofri, W. Quaedvlieg, M. De Vries et al. 2014. Phylogeny and taxonomy of meristematic rock-inhabiting black fungi in the Dothideomycetes based on multi-locus phylogenies. Fungal Diversity, 65: 127–165.

Ehrlich, H.L., D.K. Newman and A. Kappler, Eds.. 2015. Ehrlich's geomicrobiology. CRC press.

Fomina, M.A., I.J. Alexander, J.V. Colpaert and G.M. Gadd. 2005a. Solubilization of toxic metal minerals and metal tolerance of mycorrhizal fungi. Soil Biol. Biochem., 375: 851–866.

Fomina, M., S. Hillier, J.M. Charnock, K. Melville, I.J. Alexander, G.M. Gadd. 2005b. Role of oxalic acid overexcretion in transformations of toxic metal minerals by Beauveria caledonica. Appl. Environ. Microbiol., 711: 371–381.

Fomina, M., E.P. Burford, S. Hillier, M. Kierans, and G.M. Gadd. 2010. Rock-building fungi. Geomicrobiol. J., 276–7: 624–629.

Fomina, M., V.S. Podgorsky, S.V. Olishevska, V.M. Kadoshnikov, I.R. Pisanska, S. Hillier et al. 2007. Fungal deterioration of barrier concrete used in nuclear waste disposal. Geomicrobiol. J., 247–8: 643–653.

Frasca, M.H.B.D.O. and E.A. Del Lama. 2017. Biological Weathering. Encyclopedia of Engineering Geology, 1–2.

Gadd G.M., E.P. Burford, M. Fomina, K. Melville. 2007. Mineral transformations and biogeochemical cycles: a geomycological perspective, p 77–111. In G.M. Gadd, P. Dyer, S. ed. Watkinson, Fungi in the Environment. Cambridge University Press, Cambridge, United Kingdom

Gadd, G. 2008. Fungi and their role in the biosphere. In Encyclopedia of ecology pp. 1709–1717. Elsevier.

Gadd, G.M. 2000a. Bioremedial potential of microbial mechanisms of metal mobilization and immobilization. Curr. Opin. Biotechnol., 113: 271–279.

Gadd, G.M. 2000b. Heterotrophic solubilization of metal-bearing minerals by fungi.

Gadd, G.M. 2001a. Accumulation and transformation of metals by microorganisms. Biotechnology Set, 225–264.

Gadd, G.M. 2001b. Metal transformations. In British mycological society symposium series Vol. 23: pp. 359–382.

Gadd, G.M. 2004. Mycotransformation of organic and inorganic substrates. Mycologist 182: 60–70.

Gadd, G.M. 2007. Geomycology: biogeochemical transformations of rocks, minerals, metals and radionuclides by fungi, bioweathering and bioremediation. Mycol. Res. 1111: 3–49.

Gadd, G.M. 2009. Biosorption: critical review of scientific rationale, environmental importance and significance for pollution treatment. J. Chem. Technol. Biotechnol. 841: 13–28.

Gadd, G.M. 2010. Metals, minerals and microbes: geomicrobiology and bioremediation. Microbiology 1563: 609–643.

Gadd, G.M. 2013. Microbial roles in mineral transformations and metal cycling in the Earth's critical zone. Mol. Environ. Soil Sci. 115–165.

Gadd, G.M. 2017a. Geomicrobiology of the built environment. Nat. Microbiol. 24: 1–9.

Gadd, G.M. 2017b. The geomycology of elemental cycling and transformations in the environment. Microbiol. Spectrum, 51: 5–1.

Gadd, G.M., J. Bahri-Esfahani, Q. Li, Y.J. Rhee, Z. Wei, M. Fomina et al. 2014. Oxalate production by fungi: significance in geomycology, biodeterioration and bioremediation. Fungal Biol. Rev., 282–3: 36–55.

Gadd, G.M., E.P. Burford, M. Fomina and K.A.R.R.I.E. Melville. 2007b. Mineral transformations and biogeochemical cycles: a geomycological perspective. Fungi in the Environment, 77–111.

Gadd, G.M., Y.J. Rhee, K. Stephenson and Z. Wei. 2012. Geomycology: metals, actinides and biominerals. Environ. Microbiol. Rep., 43: 270–296.

Gaylarde C., G. Morton. 2002. Biodeterioration of mineral materials, p 516–528. In Bitton G ed, Environmental Microbiology, vol 1. Wiley, New York, NY

Göhre, V. and U. Paszkowski. 2006. Contribution of the arbuscular mycorrhizal symbiosis to heavy metal phytoremediation. Planta, 223: 1115–1122.

González, I., L. Laiz, B. Hermosin, B. Caballero, C. Incerti and C. Sáiz-Jiménez. 1999. Bacteria isolated from rock art paintings: the case of Atlanterra shelter south Spain. J. Microbiol. Methods, 361–2: 123–127.

Gorbushina, A.A. 2007. Life on the rocks. Environ. Microbiol., 97: 1613–1631.

Gorbushina, A.A. and W.E. Krumbein. 2005. Role of microorganisms in wear down of rocks and minerals. Microorganisms in Soils: Roles in Genesis and Functions, 59–84.

Gostinčar, C., M. Grube, S. De Hoog, P. Zalar and N. Gunde-Cimerman. 2009. Extremotolerance in fungi: evolution on the edge. FEMS Microbiol. Ecol., 711: 2–11.

Gostinčar, C., L. Muggia and M. Grube. 2012. Polyextremotolerant black fungi: oligotrophism, adaptive potential, and a link to lichen symbioses. Front. Microbiol., 3: 390.

Gu J.D. 2009. Corrosion, microbial, p 259–269. In Schaechter M ed, Encyclopedia of Microbiology. Elsevier, Amsterdam, The Netherlands

Gueidan, C., A. Aptroot, M.E. da Silva Cáceres, H. Badali and S. Stenroos. 2014. A reappraisal of orders and families within the subclass Chaetothyriomycetidae Eurotiomycetes, Ascomycota. Mycol. Prog., 13: 1027–1039.

Gueidan, C., C. Ruibal, G.S. De Hoog and H. Schneider. 2011. Rock-inhabiting fungi originated during periods of dry climate in the late Devonian and middle Triassic. Fun. Biol., 11510: 987–996.

Gueidan, C., C.R.Villaseñor, G.S. De Hoog, A.A. Gorbushina, W.A. Untereiner and F. Lutzoni. 2008. A rock-inhabiting ancestor for mutualistic and pathogen-rich fungal lineages. Stud. Mycol., 611: 111–119.

Guggiari, M., R. Bloque, M. Aragno, E. Verrecchia, D. Job and P. Junier. 2011. Experimental calcium-oxalate crystal production and dissolution by selected wood-rot fungi. Int. Biodeterior. Biodegrad., 656: 803–809.

Hirsch, P., F.E.W. Eckhardt and R.J. Palmer Jr. 1995. Methods for the study of rock-inhabiting microorganisms—a mini review. J. Microbiol. Methods, 232: 143–167.

Hoppert, M., C. Flies, W. Pohl, B. Günzl and J. Schneider. 2004. Colonization strategies of lithobiontic microorganisms on carbonate rocks. Environ. Geol., 463–4: 421–428.

Isola, D., L. Zucconi, S. Onofri, G. Caneva, G.S. De Hoog, L. Selbmann. 2016. Extremotolerant rock inhabiting black fungi from Italian monumental sites. Fun. Div., 76: 75–96.

Ivarsson, M., S. Bengtson and A. Neubeck. 2016. The igneous oceanic crust–Earth's largest fungal habitat? Fun. Ecol., 20: 249–255.

Jacobs, H., G.P. Boswell, F.A. Harper, K. Ritz, F.A. Davidson and G.M. Gadd. 2002a. Solubilization of metal phosphates by Rhizoctonia solani. Mycol. Res., 10612: 1468–1479.

Jacobs, H., G.P. Boswell, K. Ritz, F.A. Davidson and G.M. Gadd. 2002b. Solubilization of calcium phosphate as a consequence of carbon translocation by Rhizoctonia solani. FEMS Microbiol. Ecol., 401: 65–71.

Jongmans, A.G., N. Van Breemen, U. Lundström, P.A.W. Van Hees, R.D. Finlay, M. Srinivasan et al. 1997. Rock-eating fungi. Nature, 3896652: 682–683.

Li, Q., L. Csetenyi and G.M. Gadd. 2014. Biomineralization of metal carbonates by Neurospora crassa. Environ. Sci. Technol., 4824: 14409–14416.

Li, Q., L. Csetenyi, G.I. Paton and G.M. Gadd. 2015. $CaCO_3$ and $SrCO_3$ bioprecipitation by fungi isolated from calcareous soil. Environ. Microbiol., 178: 3082–3097.

Li, Q., D. Liu, Z. Jia, L. Csetenyi and G.M. Gadd. 2016a. Fungal biomineralization of manganese as a novel source of electrochemical materials. Curr. Biol., 267: 950–955.

Li, Z., L. Liu, J. Chen and H.H. Teng. 2016b. Cellular dissolution at hypha-and spore-mineral interfaces revealing unrecognized mechanisms and scales of fungal weathering. Geology, 444: 319–322.

Lian, B., B. Wang, M. Pan, C. Liu and H.H. Teng. 2008. Microbial release of potassium from K-bearing minerals by thermophilic fungus Aspergillus fumigatus. Geochim. Cosmochim. Acta, 721: 87–98.

Liang, X., M. Kierans, A. Ceci, S. Hillier and G.M. Gadd. 2016. Phosphatase-mediated bioprecipitation of lead by soil fungi. Environ. Microbiol., 181: 219–231.

Lisci, M., M. Monte and E. Pacini. 2003. Lichens and higher plants on stone: a review. Int. Biodeterior. Biodegrad., 511: 1–17.

Liu, J.K., K.D. Hyde, R. Jeewon, A.J. Phillips, S.S. Maharachchikumbura, M. Ryberg et al. 2017. Ranking higher taxa using divergence times: a case study in Dothideomycetes. Fun. Div., 84: 75–99.

Lyu, X., C. Shen, J. Xie, Y. Fu, D. Jiang, Z. Hu et al. 2015. A "footprint" of plant carbon fixation cycle functions during the development of a heterotrophic fungus. Sci. Rep., 5(1): 1–13.

Martin-Sanchez, P.M., A. Nováková, F. Bastian, C. Alabouvette and C. Saiz-Jimenez. 2012. Two new species of the genus Ochroconis, O. lascauxensis and O. anomala isolated from black stains in Lascaux Cave, France. Fun. Biol., 1165: 574–589.

Marvasi, M., F. Donnarumma, A. Frandi, G. Mastromei, K. Sterflinger, P. Tiano et al. 2012. Black microcolonial fungi as deteriogens of two famous marble statues in Florence, Italy. Int. Biodeterior. Biodegrad., 68: 36–44.

Milanesi, C., F. Baldi, R. Vignani, F. Ciampolini, C. Faleri, M. Cresti. 2006. Fungal deterioration of medieval wall fresco determined by analysing small fragments containing copper. Int. Biodeterior. Biodegrad., 571: 7–13.

Muggia, L., J. Kocourková and K. Knudsen. 2015. Disentangling the complex of Lichenothelia species from rock communities in the desert. Mycologia, 1076: 1233–1253.

Naranjo-Ortiz, M.A. and T. Gabaldón. 2019. Fungal evolution: diversity, taxonomy and phylogeny of the Fungi. Biol. Rev., 94(6): 2101–2137.

Ortega-Morales, B.O., J. Narváez-Zapata, M. Reyes-Estebanez, P. Quintana, S.D.C. De la Rosa-García, H. Bullen et al. 2016. Bioweathering potential of cultivable fungi associated with semi-arid surface microhabitats of Mayan buildings. Front. Microbiol., 7: 201.

Pinzari, F., J. Tate, M. Bicchieri, Y.J. Rhee and G.M. Gadd. 2013. Biodegradation of ivory natural apatite: possible involvement of fungal activity in biodeterioration of the Lewis C hessmen. Environ. Microbiol, 154: 1050–1062.

Pinzari, F., M. Zotti, A. De Mico and P. Calvini. 2010. Biodegradation of inorganic components in paper documents: formation of calcium oxalate crystals as a consequence of Aspergillus terreus Thom growth. Int. Biodeterior. Biodegrad., 646: 499–505.

Purvis, O.W. and B. Pawlik-Skowrońska. 2008, January. Lichens and metals. In British Mycological Society Symposia Series Vol. 27, pp. 175–200. Academic Press.

Ranalli, G., E. Zanardini and C. Sorlini. 2019. Biodeterioration–including cultural heritage.

Réblová, M., W.A. Untereiner and K. Réblová. 2013. Novel evolutionary lineages revealed in the Chaetothyriales Fungi based on multigene phylogenetic analyses and comparison of ITS secondary structure. PLoS One, 85: e63547.

Ruibal, C., C. Gueidan, L. Selbmann, A.A. Gorbushina, P.W. Crous, J. Z. Groenewald et al. 2009. Phylogeny of rock-inhabiting fungi related to Dothideomycetes. Stud. Mycol., 641: 123–133.

Savković, Ž., N. Unković, M. Stupar, M. Franković, M. Jovanović, S. Erić et al. 2016. Diversity and biodeteriorative potential of fungal dwellers on ancient stone stela. Int. Biodeterior. Biodegrad., 115: 212–223.

Sayer, J.A., J.D. Cotter-Howells, C. Watson, S. Hillier and G.M. Gadd. 1999. Lead mineral transformation by fungi. Curr. Biol., 913: 691–694.

Scheerer, S., O. Ortega-Morales and C. Gaylarde. 2009. Microbial deterioration of stone monuments—an updated overview. Adv. Appl. Microbiol., 66: 97–139.

Schoch, C.L., P.W. Crous, J.Z. Groenewald, E.W.A. Boehm, T.I. Burgess, J. De Gruyter et al. 2009. A class-wide phylogenetic assessment of Dothideomycetes. Stud. Mycol., 641: 1–15.

Schumacher, J. and A.A. Gorbushina. 2020. Light sensing in plant-and rock-associated black fungi. Fungal Biol., 124(5): 407–417.

Schutzendubel, A. and A. Polle. 2002. Plant responses to abiotic stresses: heavy metal-induced oxidative stress and protection by mycorrhization. J. Exp. Bot., 53372: 1351–1365.

Seaward, M.R. 2003. Lichens, agents of monumental destruction. Microbiology Today, 303: 110–112.

Selbmann, L., Z. Benkő, C. Coleine, S. De Hoog, C. Donati, I. Druzhinina et al. 2020. Shed light in the daRk lineagES of the fungal tree of life—STRES. Life, 10(12): 362.

Sert, H. B. and K. Sterflinger. 2010. A new *Coniosporium* species from historical marble monuments. Mycol. Prog., 9: 353–359.

Sert, H.B., H. Suembuel and K. Sterflinger. 2007a. *Sarcinomyces sideticae*, a new black yeast from historical marble monuments in Side Antalya, Turkey. Bot. J. Linn. Soc., 1543: 373–380.

Sert, H.B., H. Sümbül and K. Sterflinger. 2007b. A new species of *Capnobotryella* from monument surfaces. Mycological research, 11110: 1235–1241.

Sert, H.B., H. Sümbül and K. Sterflinger. 2011. Two new species of *Capnobotryella* from historical monuments. Mycol. Prog., 10: 333–339.

Sert, H.B., M. Wuczkowski and K. Sterflinger. 2012. *Capnobotryella isiloglui*, a new rock-inhabiting fungus from Austria. Turk. J. Bot., 364: 401–407.

Sharma R., A.V. Polkade, Y.S. Shouche. 2015 "Species Concept" in Microbial Taxonomy and Systematics. Curr. Sci., 108 10: 1804–1814.

Smits, M.M. and E. Hoffland. 2009. Possible role of ectomycorrhizal fungi in cycling of aluminium in podzols. Soil Biol. Biochem., 413: 491–497.

Song, W., N. Ogawa, C.T. Oguchi, T. Hatta and Y. Matsukura. 2007. Effect of Bacillus subtilis on granite weathering: A laboratory experiment. Catena, 703: 275–281.

Sterflinger, K. 2000. Fungi as geologic agents. Geomicrobiol. J., 172: 97–124.

Sterflinger, K., R. De Baere, G.S. De Hoog, R. De Wachter, W.E. Krumbein and G. Haase. 1997. *Coniosporium perforans* and *C. apollinis*, two new rock-inhabiting fungi isolated from marble in the Sanctuary of Delos Cyclades, Greece. Antonie van Leeuwenhoek, 72: 349–363.

Sterflinger, K., D. Tesei and K. Zakharova. 2012. Fungi in hot and cold deserts with particular reference to microcolonial fungi. Fungal Ecol., 54: 453–462.

Sterflinger, K., K. Lopandic, R.V. Pandey, B. Blasi and A. Kriegner. 2014. Nothing special in the specialist? Draft genome sequence of Cryomyces antarcticus, the most extremophilic fungus from Antarctica. PLoS One, 9(10): e109908.

Štyriaková, I., I. Štyriak, I. Kraus, D. Hradil, T. Grygar and P. Bezdička. 2003. Biodestruction and deferritization of quartz sands by Bacillus species. Miner. Eng., 168: 709–713.

Su, L., L. Guo, Y. Hao, M. Xiang, L. Cai and X. Liu. 2015. Rupestriomyces and Spissiomyces, two new genera of rock-inhabiting fungi from China. Mycologia, 1074: 831–844.

Sullivan, T.S., N.R. Gottel, N. Basta, P.M. Jardine and C.W. Schadt. 2012. Firing range soils yield a diverse array of fungal isolates capable of organic acid production and Pb mineral solubilization. Appl. Environ. Microbiol., 7817: 6078–6086.

Sun, H.J. and E.I. Friedmann. 1999. Growth on geological time scales in the Antarctic cryptoendolithic microbial community. Geomicrobiol. J., 162: 193–202.

Sun, W., L. Su, S. Yang, J. Sun, B. Liu, R. Fu et al. 2020. Unveiling the hidden diversity of rock-inhabiting fungi: Chaetothyriales from China. J. Fungi, 64: 187.

Todar, K. 2005. The Genus Bacillus. Todar's Online Textbook of Bacteriology. University of Wisconsin-Madison Department of Bacteriology.

Untereiner, W.A., C. Gueidan, M.J. Orr and P. Diederich. 2011. The phylogenetic position of the lichenicolous ascomycete Capronia peltigerae. Fungal Diversity, 49: 225–233.

Vaughan, D.J., R.A.D. Pattrick and R. A. Wogelius. 2002. Minerals, metals and molecules: ore and environmental mineralogy in the new millennium. Mineralogical Magazine, 665: 653–676.

Warscheid, T. and J. Braams. 2000. Biodeterioration of stone: a review. Int. Biodeterior. & Biodegrad., 464: 343–368.

Wei, Z., S. Hillier and G.M. Gadd. 2012a. Biotransformation of manganese oxides by fungi: solubilization and production of manganese oxalate biominerals. Environ. Microbiol., 147: 1744–1753.

Wei, Z., M. Kierans and G.M. Gadd. 2012b. A model sheet mineral system to study fungal bioweathering of mica. Geomicrobiol. J., 294: 323–331.

Wei, Z., X. Liang, H. Pendlowski, S. Hillier, K. Suntornvongsagul, P. Sihanonth et al. 2013. Fungal biotransformation of zinc silicate and sulfide mineral ores. Environ. Microbiol., 158: 2173–2186.

Wright, J.S. 2002. Geomorphology and stone conservation: sandstone decay in Stoke-on-Trent. Structural Survey, 202: 50–61.

Winkelmann, G. 2002. Microbial siderophore-mediated transport. Biochemical Society Transactions, 30(4): 691–696.

Wu, Y.W., J.C. Zhang, L.J. Wang and Y.X. Wang. 2017. A rock-weathering bacterium isolated from rock surface and its role in ecological restoration on exposed carbonate rocks. Ecol. Eng., 101: 162–169.

Zakharova, K., D. Tesei, G. Marzban, J. Dijksterhuis, T. Wyatt and K. Sterflinger. 2013. Microcolonial fungi on rocks: a life in constant drought? *Mycopathologia*, 175: 537–547.

CHAPTER 18

Application of CRISPR/Cas Technology in Plant Pathogenic Oomycetes

Fatemeh Salmaninezhad,[1] *Reza Mostowfizadeh-Ghalamfarsa*[1,*]
and *Giles E. St. J. Hardy*[2]

Introduction

With the development of recombinant DNA technology, a new era for biology has begun. It enabled molecular biologists to manipulate DNA molecules to study the genes and utilize them to develop novel medicine and biotechnology (Lamour et al. 2012). Regardless of which aspect, a new revolution in biological research has occurred with the recent advances in genome editing tools. The new genome engineering methods make it possible to directly edit or modulate the function of DNA sequences in their endogenous context, regardless of the type of organism (Lamour and Kamoun 2009). These genetic manipulation methods elucidate the genome's functional organization in any organism at different stages and identify genetic variations. Genome engineering is a process of making targeted modifications to the gene of interest, its function, or its outputs. Hence, any technology that could delete, insert, and modify an organism's genome, enables researchers to detect the function of a specific gene or protein network on larger scales (Lamour and Kamoun 2009). Also, modifying transcriptional regulation or chromatin states at specific loci would reveal the organization of genetic material and its usage within a cell, leading to the identification of relationships between the genome architecture and function.

With regard to agriculture, plant defense methods may be supported by molecular biology and biotechnology studies, which enables scientists to choose features that could temper the pathogen's aggression. Preventing a pathogen's attack can be achieved by manipulating either a host plant or the pathogen and, in rare cases, a third party serving as an antagonist. In comparison to other biotechnological methods, genetic transformation is still generally the one that is used most since it enables researchers to insert particular gene sequences into a host plant stably (Capriotti et al. 2020).

One of the key biotechnological strategies used to balance pathogen aggressiveness and resulting production losses is the overexpression of defensive genes against crop fungal infections. In host cells, overexpression of pathogenesis-related proteins, antimicrobial peptides, secondary metabolites, and particular compounds can have an immediate impact on the target level. As an alternative, it is also possible to activate the biosynthetic pathways for host defense, for instance, by overexpressing transcription factors that improve the genes involved in plant defense (Capriotti et al. 2020). In addition to these biotechnological tactics, new breeding techniques have been developed and improved over the past few decades. Examples include genome editing mediated by CRISPR/Cas technology, a highly precise tool capable of strategically introducing targeted

---

[1] Department of Plant Protection, School of Agriculture, Shiraz University, Shiraz, Iran 7144167186.
[2] Phytophthora Science and Management, Harry Butler Institute, Murdoch University, Murdoch, WA 6150, Australia.
* Corresponding author: rmostofi@shirazu.ac.ir

mutations in the host genome, or cisgenesis/intragenesis, which enables the inclusion of gene sequences from sexually compatible plants (Capriotti et al. 2020, Pirrello et al. 2023). Another effective method of preventing pathogenic attacks while the downregulation of gene expression takes place is the RNA interference (RNAi) mechanism, in which double-strand RNA (dsRNA) molecules cause mRNA destruction or translational repression (Pirrello et al. 2023). In addition to plants, pathogens can also be genetically altered in order to better understand the role that a particular gene plays in pathogenicity or, in rare instances, to increase the production of a gene of interest to increase the plant's resistance to other pathogens (Shi et al. 2017, Wang et al. 2018, Fang et al. 2020, Salmaninezhad et al. 2022). Among all plant pathogens, fungi and oomycetes are of great importance, causing huge economical losses yearly on crops or green environment. Hence, genetic manipulation of these pathogens would help researchers to have a better understanding of their genetic structure and deployment of appropriate approaches to control them. Additionally, gene engineering can enhance the production of certain structures like fruiting bodies in edible fungi, such as *Agaricus bisporus* (J.E. Lange) Imbach (Sattar et al. 2021). This chapter aimed to introduce and compare different genetic manipulation approaches in plant pathogenic fungi and oomycetes. The advantages and disadvantages of each technique will be thoroughly discussed, with emphasis on the application of the CRISPR/Cas9 system in plant pathogenic oomycetes. Due to the abundance of acronyms used in this chapter, a list of abbreviations is provided to assist the reader.

Acronyms

Cas	:	CRISPR Associated proteins
CRISPR	:	Clustered Regularly Interspaced Short Palindromic Repeats
CRISPRa	:	CRIPSR activator
CRISPRi	:	CRISPR interference
crRNA	:	CRISPR RNA
DSB	:	Double Strand Break
gRNA	:	Guide RNA
HDR	:	Homology Directed Repair
NHEJ	:	Non-Homologous End Joining
PAM	:	Protospacer Adjacent Motif
RISC	:	RNA Induced Silencing Complex
RNAi	:	RNA Interference
RNP	:	Ribonucleoprotein
TALEN	:	Transcription Activator Effector-Like Nuclease
tracrRNA	:	Trans-activating RNA
ZFN	:	Zinc-Finger Nuclease

Oomycetes, their Genetics and Genetic Modification

Oomycetes are fungus-like microorganisms from the kingdom *Stramenopila*, occupying various ecological niches, especially humid areas (Mostowfizadeh-Ghalamfarsa and Salmaninezhad 2020). Due to the similar morphological features as the filamentous fungi, this group of microorganisms has been considered true fungi for a long time. However, they are more related to diatoms and brown algae (Lamour et al. 2012). Oomycetes' features are quite distinct from true fungi because they are diploid or polyploid for most of their life cycles, whereas true fungi are mostly haploid in their life cycles (Mostowfizadeh-Ghalamfarsa et al. 2020). Moreover, the cell wall composition of oomycetes is also different from true fungi and mainly consists of 1,3-beta-glucan, while for true fungi, the cell wall is mainly chitin (Erwin and Ribeiro 1996). This group shares some common features with protists, such as having a promoter structure in which the initiator element is overrepresented. Oomycete lifestyles are diverse; they can be plant pathogens (e.g., various species of *Phytophthora*,

Achlya, Pythium, Pythiogeton, Peronospora, etc.), animal pathogens (e.g., *Saprolegnia* spp.), human pathogens (e.g., *Pythium insidiosum* De Cock, L. Mend., A.A. Padhye, Ajello and Kaufman), antagonists (e.g., *Pythium oligandrum* Drechsler and *Globisporangium nunn* (Lifsh., Stangh. and R.E.D. Baker) Uzuhashi, Tojo and Kakish.) or saprophytes (e.g., *Saprolegnia* spp.) (Mostowfizadeh-Ghalamfarsa and Salmaninezhad 2020). Among all known oomycete species, plant pathogens are the most well-studied microorganisms and cause devastating diseases in numerous crops, ornamental, and native plants. The genus *Phytophthora* is one of the most destructive plant pathogens worldwide, causing substantial economic loss annually by infecting more than 100 plant species. In general, all dicots and some monocots are infected by *Phytophthora* species (Mostowfizadeh-Ghalamfarsa et al. 2020). The most economically important *Phytophthora* species are *P. infestans* (Mont.) de Bary, *P. sojae* Kaufm. and Gerd., *P. cinnamomi* Rands, *P. ramorum* Werres, De Cock and Man in 't Veld, and *P. palmivora* (E.J. Butler) E.J. Butler (Erwin and Ribeiro 1996).

Except for the genus *Phytophthora,* genome organizations of oomycetes are not well-studied. *Phytophthora*'s genome has abundant repetitive sequences (Lamour and Kamoun 2009). Take *P. sojae,* for example: five families of tandemly repeated sequences have been identified in this pathogen. Moreover, genomic subtraction of chromosomal DNA from different isolates has been discovered. However, the copy number of the sequences between the isolates was varied, and all the copies were localized on a single chromosome of *P. sojae* (Lamour and Kamoun 2009).

Studying oomycetes at the molecular level has always been a challenge for scientists. This is mainly due to the inefficient production of stable transformants, low regeneration rates in protoplast transformation, and the requirement for large amounts of materials in oomycetes (Lamour and Kamoun 2009). Most transformation protocols that have been used for oomycetes are inefficient in producing homokaryotic stable transformants or even gene silencing (Lamour and Kamoun 2009). For example, while electroporation of *P. infestans* zoospores produced more transformants, gene silencing was much less likely to be successful (Latijnhouwers and Govers 2003). In *Pythium* spp., there are only a few reports of stable transformation, which either used protoplast electroporation or Agrobacterium-mediated transformation (Vijn and Govers 2003, Weiland 2003, Horner et al. 2012).

Pathogenic oomycetes are the subject of intense molecular studies to find their "Achilles heels" that could be used in industry (i.e., representing promising drug or pesticide targets), agriculture (i.e., promoting antagonists and plant diseases' management), and genetic studies (i.e., finding the genes involving in growth, pathogenicity, and antagonistic behaviors) (Schuster and Kahmann 2019). Previously, homologous recombination, transposon tagging, and Agrobacterium-mediated integration have been used in oomycetes to perform genome modification (Schuster and Kahmann 2019). Nevertheless, these methods have some critical limitations. Homologous recombination could not be efficient. Another important problem is the insufficient number of selection markers that necessitate incorporating extra steps of marker recycling (Wang et al. 2017). The transposon tagging method is more frequent in filamentous fungi than in oomycetes, and obtaining the targeted manipulation can be challenging using this method (Wang et al. 2017). Furthermore, detecting and isolating the stable mutant strain from many transformants is a great challenge for researchers (Wang et al. 2017).

Modern Genetic Modification Approaches

Genome modification, regardless of which type, is based on the endogenous pathways of the DNA repair system. As a result, the performance of the gene editing approach in any organism depends on the DNA repair pathway it prefers at the moment of editing (Hsu et al. 2014). DNA repair pathways depend on either homology-directed repair (HDR pathway) or non-homologous end joining (NHEJ pathway) (Hsu et al. 2014). When performing a genetic modification, nucleases make a double-strand break (DSB) in the gene of interest. In the homology-directed repair pathway, the damaged DNA during the DSB will be corrected with a homologous DNA segment as a template for repair. For the best outcome, the result is the reconstitution of the region in the location of a DSB

(Schuster and Kahmann 2019). In the NHEJ pathway, no such homologous DNA template is needed, which is why it is referred to as non-homologous. During the NHEJ pathway, both ends of the broken DNA, will be ligated by a protein complex. The result of the NHEJ pathway would be the restoration of the original sequence, substitutions of small indels, or introduction of nucleotides (Hsu et al. 2014). In the homology-directed repair pathway, a DSB fragment containing the desired modification will be introduced into the cell, and it will replace the endogenous fragment by homologous recombination.

Consequently, even though the approach is more challenging to perform, precise genome editing and gene replacement will occur (Fang et al. 2017). However, when utilizing the NHEJ pathway, the results are subjected to error-prone repair leading to the mutations of the respective gene. The NHEJ approach is the most frequently used pathway in filamentous fungi and oomycetes (Hsu et al. 2014). Few studies concentrate on the homology-directed repair pathway by inserting a template into the cell. Gene replacement using the homology-directed repair pathway has been mostly used in filamentous fungi, such as *Neurospora crassa* Shear and B.O. Dodge, *Penicillium chrysogenum* Thom, and several *Aspergillus* species, as well as a few oomycetes, including *P. sojae* and *P. capsici* (Aramayo and Selker 2013, Matsu-ura et al. 2015, Fang et al. 2017, Wang et al. 2018, Liao et al. 2022).

Double-strand breaks (DSBs) have been introduced artificially in the genome's distinct locations by targetable endonucleases, such as zinc-finger nucleases, TALENs, CRISPR/Cas systems (see the following sections), and their variants (Gaj et al. 2016). Artificially introduced DSBs would enhance the frequency of homologous recombination at the damaged site, which makes them special for genome manipulation approaches (Gaj et al. 2016). To achieve successful and effective genome manipulation via the introduction of site-specific DNA DSBs, different engineered DNA binding proteins have been introduced. Four of these DNA binding proteins are the most important ones, including meganucleases from microbial mobile genetic elements, zinc-finger nucleases based on eukaryotic transcription factors, transcription activator-like effector nucleases from plant pathogenic bacteria, *Xanthomonas* spp., and the recent RNA-guided DNA endonuclease Cas protein from the bacterial immune system CRISPR (Gaj et al. 2016). Besides, it is important to note that not all genetic modification approaches are based on the introduction of site-specific DNA DSBs and DNA binding proteins. Some modifications are conducted by other approaches, e.g., RNA-interference (RNAi), with different mechanisms.

RNA-interference

RNA-interference (RNAi), previously named post-transcriptional gene silencing or quelling, was first described in the model nematode, *Caenorhabditis elegans,* in 1998 (Fire et al. 1998). This mechanism naturally occurs in many eukaryotes, protects the organism's cells from transposons and viruses, and regulates gene expression (Alfonzo 2015, Ali et al. 2015). It also allows reduced specific gene expression and rapidly becomes a potential reverse genetic tool in organisms where gene targeting is inefficient, time-consuming, or both (Nakayashiki et al. 2005, 2006). The RNAi main mechanism is based on the introduction of double-stranded RNA (dsRNA) molecules, which activates an endonuclease called Dicer. The Dicer cuts the double-stranded RNA molecules into short 21–25 nucleotides fragments. These fragments are called small interfering RNA (siRNA), which bind to a group of proteins from the Argonaute family. Argonaute proteins are major components of the RNA-induced silencing complex (RISC) (Schmidt 2005). Then the strands of the small interfering RNA are separated. The sense strand (passenger strand) is degraded immediately. At the same time, the RISC-complex uses the antisense strand (guide strand) as a template for recognizing a complementary mRNA sequence and induces cleavage by Argonaute. The mRNA sequence is finally degraded, and its corresponding gene will be "silenced" (Fulci and Masino 2007, Armas-Tizapantzi and Montiel-González 2016).

The targeted genes will be silenced when a synthetic complementary double-stranded RNA is introduced to the protoplasts at the post-transcriptional stage (Whisson et al. 2005). In this system, the double-stranded RNA is only introduced to the protoplast at one point in the transformation process, which means the main template will be depleted over time. The maximum efficiency of the RNAi system is approximately two weeks after the introduction. After this period, the efficiency decreases significantly, and the organism returns to regular gene expression (Whisson et al. 2005). Due to the high sequence similarities shared among gene family members, the sequence-specific characteristics of RNAi are effectively used to analyze functionally redundant gene families. This leads to a simultaneously silenced gene family by a single RNAi construct (Grenville-Briggs et al. 2008).

RNA-interference (RNAi) offers a valuable gene analysis tool for oomycetes since the RNAi machinery is known to degrade cognate mRNA in the cytoplasm. Consequently, it is likely to be operative against any mRNA in multinuclear heterokaryotic mycelia. RNA-interference or transient gene silencing was first proven in *P. infestans* using the marker gene, *gfp*, and two *P. infestans* genes, *inf1* and *cdc14* (Whisson et al. 2005). The study showed that transient gene silencing can generate detectable phenotypes in *P. infestans* and could provide a high-throughput tool for *P. infestans* functional genomics. RNA-interference was successfully used to silence a complete gene family from *P. infestans* (Grenville-Briggs et al. 2008, Asman et al. 2016). The silencing of a family of four cellulose synthase genes impaired appressorium differentiation and plant infection. Using this system, researchers showed that a DEAD-box RNA helicase is required for normal zoospore development in *P. infestans* (Grenville-Briggs et al. 2008). The first transient silencing protocol for *P. sojae* was conducted in 2011 (Zhao et al. 2011). They showed that PsCdc14 is highly expressed during sporulation, zoospore, and cyst life stages. Silencing PsCdc14 resulted in low sporangial production and abnormal development of the transformants. RNA-interference was also successfully utilized to silence a gene encoding tyrosinase SpTyr in the fish pathogenic oomycete, *Saprolegnia parasitica* Coker (Saraiva et al. 2014), and it was revealed that this gene is involved in cell wall melanization, and silenced lines showed abnormal sporangia formation. Recently, an RNAi system was described in *Aphanomyces invadans* Willoughby, R.J. Roberts and Chinabut chaperone gene, AiLhs1 (Iberahim et al. 2020), which considerably reduced the number of virulence factors in the secretome and thereby the virulence of this pathogen, while proteins involved in the general production of proteins increased in abundance.

Even though RNAi causes only a knockdown of the gene, which is not a complete loss (knockout) in gene expression, incomplete gene suppression can sometimes be beneficial. For instance, partial silencing by RNAi makes it possible to investigate the effects of an essential gene. A negative aspect of sequence-specific silencing by RNAi compared to knockout mutations is that it is impossible to verify the RNAi results by genetic complementation (Ziv and Yarden 2010).

Zinc-finger Nucleases

Zinc-finger nucleases, also known as ZFNs, were the first nucleases utilized for genome editing worldwide (Gaj et al. 2016). These nucleases are fusions between an artificially designed Cys2-His2 zinc finger protein and the FokI restriction endonuclease cleavage domain (Kim et al. 1996). In other words, these nucleases are fusion proteins consisting of an array of site-specific DNA-binding domains. The nucleases function as dimers. Each monomer recognizes a specific half-site sequence (generally 10 to 18 bp DNA) by the zinc-finger DNA binding domain. These domains are adapted from zinc-finger nucleases, with their transcription factors attached to the endonuclease domain of the bacterial restriction enzyme, FokI. Each zinc-finger domain contains a 3 to 4 bp DNA sequence and tandem domains that can bind to an extended nucleotide sequence (Gaj et al. 2016). The nucleotide sequences usually have multiples of 3, 9, up to 18 bp in length, and are unique within a cell's genome. Zinc-finger nucleases are designed as a pair to cause a cleavage in a specific site. This pair recognizes two sequences flanking the site, each on the forward and reverse strand. By binding the ZFNs on either side of the target, the pair of FokI domains form a dimer and cut the DNA at the site, resulting in a double-strand break (DSB) with 5' overhangs. The cells then automatically start repairing the DSBs using either the non-homologous end joining (NHEJ) pathway or the homology-directed repair (HDR) pathway. The NHEJ can happen during any cell cycle phase but usually leads to erroneous repair (Hsu et al. 2014, Gaj et al. 2016, Fang et al. 2017). However, the HDR repair occurs during the late S phase or G2 phase of the cell cycle when a sister chromatid is available to serve as a repair template (Hse et al. 2014).

Due to the error-prone nature of the NHEJ pathway, the introduction of frameshifts into the coding sequence of a given gene is possible, which results in knocking out the targeted gene. This procedure occurs via the combination of two main mechanisms, i.e., premature truncation of the protein and nonsense-mediated decay of the mRNA transcript. The mechanism of the nonsense-mediated decay of the mRNA transcript is not always efficient (Gaj et al. 2016). However, in the HDR pathway, insertion of a specific mutation would be possible by producing a repair template containing the desired mutation flanked by homology arms (Hsu et al. 2014). When DSBs occur in the DNA, HDR uses another closely matching DNA sequence to repair it. Consequently, HDR would proceed like homologous recombination by utilizing an exogenous double-strand DNA (dsDNA) vector as a repair template (Gaj et al. 2016). Moreover, it can use an exogenous single-strand DNA (ssDNA) oligonucleotide as a template, for which the 20 nt homology arms enable the introduction of mutations in the genome. In either case, the antibiotic selection to identify the correct targeted clones is sufficient to conclude that the transformation was highly efficient. Where antibiotic selection is not used, further steps to remove the antibiotic resistance cassette from the genome would not be necessary (Gaj et al. 2016).

Although ZFNs provide researchers with a profitable tool to edit the genome, several potential disadvantages have been reported. The assembly of zinc-finger domains to bind an extended stretch of nucleotides with high affinity is not a straightforward procedure. As a result, it would be difficult for non-specialists to feasibly engineer and obtain the optimized ZFNs (Gupta and Musunuru 2014). Besides, targeting the site to select is limited while working with the ZFN genome editing tool. This is because of the nature of the open-source ZFN components, which can only be utilized to target the binding sites every 200 bp in a random DNA sequence (Gaj et al. 2016). This could not be a potential issue for those seeking to knock out the gene of interest because a frameshift introduced anywhere in the early coding sequence of the gene can produce the desired results. However, it would be challenging when a specific site is required, e.g., to knock in a specific mutation (Gupta and Musunuru 2014).

With the introduction of other platforms to engineer optimized ZFNs, some of these obstacles have been resolved, since each platform provides different degrees of speed, flexibility in site selection, and success rates (Gaj et al. 2016). Another unsolved problem regarding using ZFNs in gene editing is the use of proteins designed to create DSBs in the genome. These proteins would create DSBs at the desired site and off-target sites. Thus, using a pair of ZFNs consisting of distinct obligate heterodimer FokI domains is suggested, which may prevent a single ZFN from binding to two adjacent off-target sites while creating a DSB. Another suggestion to solve this problem is the introduction of purified ZFN proteins into the cells (Gupta and Musunuru 2014).

Even though ZFNs are increasingly used in academia and industrial research for various purposes ranging from the generation of animal models to human therapies, no reports of their utilization in oomycetes are available.

Transcription Activator-like Effectors Nucleases

Transcription activator-like effectors (TALEs) are a class of proteins that function as bacterial effectors. These proteins were first discovered as being exclusive to a group of plant pathogenic bacteria, i.e., *Xanthomonas* spp., which led to the identification of a novel DNA-binding domain called TALE repeats (Mahfouz et al. 2014). These effectors bind to the promoter of the host genes, which activate the expression to the pathogen's benefit. Transcription activator-like effectors' repeats consist of tandem arrays with 10 to 30 repeats which can successfully bind and recognize extended DNA sequences. This discovery led to the creation of custom TALENs capable of editing and modifying almost any targeted gene (Gaj et al. 2016). Transcription activator-like effectors nucleases (TALENs) have a similar modular form and function as ZFNs with an amino-terminal TALE domain that can bind with DNA. This domain is fused to a carboxy-terminal FokI cleavage domain (Christian et al. 2010, Miller et al. 2011). Each repeat in the tandem is 33 to 35 amino acids long with two adjacent amino acids. The adjacent amino acids are called the repeat-variable di-residue (RVD). Therefore, with the existence of the amino acids, a one-to-one correspondence will occur between the repeats and the base pairs in the target DNA sequence (Malzahn et al. 2017). By elucidation and defining the RVD code, researchers can engineer a site-specific nuclease, which can fuse a domain of TALE repeats to the FokI endonuclease domain, called TAL effector nucleases (TALENs) (Gaj et al. 2016).

Like ZFNs, TALENs can generate DSBs at the desired target location in the genome, making them applicable to knock out genes or knock-in mutations in the same way (Malzahn et al. 2017). In contrast to ZFNs, TALENs are much easier to design. The RVD code has been employed to design many TALEs repeat arrays. These arrays bind, with high affinity, to the desired target genomic DNA sequences up to 96% (Gaj et al. 2016). The design and construction of TALENs can take only two days, and can include large numbers, such as hundreds at a time. In other words, it is possible to construct a library with the TALENs targeting all the genes in the genome while using TALENs as a genome editing tool (Hsu et al. 2014). Transcription activator-like effectors nucleases have advantages over ZFNs in some ways. For instance, the TALE repeat array can feasibly extend to the desired length up to 18 bp or longer, whereas engineered ZFNs can usually bind 9 to 18 bp sequences.

Furthermore, having fewer constraints on the site selection is another advantage of TALENs over ZFNs. In theory, multiple possible TALEN pairs are available for each random DNA sequence base pair. However, off-target effects significantly affect the utilization of TALENs, similar to ZFNs. Moreover, both TALENs and ZFNs with obligate heterodimer FokI domains are usually used to minimize the possibility of off-target effects. The extension of the TALE repeat array can be considered both an advantage and a disadvantage. A typical size for a cDNA coding a TALEN is generally 3kb, while in ZFN, it is approximately 1 kb. Hence, delivering and expressing a pair of TALENs into the cells is more difficult (Malzahn et al. 2017). Like ZFNs, the application of TALENs in plant pathogens, especially oomycetes, is still unknown.

Advent of the CRISPR System

The CRISPR technology advent was developed in 1987 (Hsu et al. 2014). It was reported as a curious set of 29 nt repeats downstream of the *iap* gene during an investigation of the IAP enzyme involved in isozyme conversion of alkaline phosphatase in *Escherichia coli* (Migula) Castellani and Chalmers. These 29 nt repeats were interspaced by five intervening 32 nt nonrepetitive sequences. It took 10 years to sequence other microbial genomes, revealing similar and additional repeat elements from different bacterial and archaeal genomes (Hsu et al. 2014). These repeats were ultimately classified as a unique family of clustered elements present in more than 50 and 90 percent of bacteria and archaea, respectively (Gaj et al. 2016).

The acronym CRISPR was proposed to unify the description of microbial genomic loci with interspaced repeat arrays, i.e., Clustered Regularly Interspaced Short Palindromic Repeats, and several clusters of signature CRISPR-associated genes were identified as conserved sequences and typically adjacent to the repeat elements. These repeat elements served as a basis for the final classification of different types of CRISPR systems (Wang et al. 2018).

It was also found that the variable spacers between the direct repeats have high similarity with short DNA sequences of bacteriophage, virus, and foreign plasmids (Hsu et al. 2014). These sequences play an important role in the immune system to protect the prokaryotic cells against the invading genetic elements (Hsu et al. 2014). With the development of new technologies to broaden genomic data, researchers could compare the CRISPR loci in other organisms, which resulted in discovering some conserved genes adjacent to the CRISPR region. These conserved genes were later called CRISPR-associated (Cas) genes (Gaj et al. 2016). It was proposed that these Cas proteins interfere with DNA repair and recombination mechanisms as well as their involvement in CRISPR loci formation (Morange 2015).

Further investigations demonstrated that CRISPR and Cas genes work together to form an acquired prokaryotic immune system, enabling bacteria and archaea to protect themselves from foreign genetic elements such as viruses and plasmids (Gaj et al. 2016). The main components of the CRISPR/Cas system were identified with further studies, which resulted in replicating this system under *in vitro* and *in vivo* conditions using bacterial models (Jiang and Doudna 2017, Wang et al. 2018). Finally, it was proposed that this defense system could be converted into an effective genome engineering tool (Moon et al. 2019).

CRISPR/Cas Technology

What is CRISPR/Cas Technology?

Using the DNA-editing CRISPR/Cas technology, researchers can perform genetic modifications in an inexpensive, fast, and accurate way for diverse organisms. As mentioned previously, the CRISPR/Cas system requires an endonuclease protein family to cut the DNA. There are several classes of Cas proteins; however, the most recognized one is Cas9. The CRISPR nuclease, i.e., Cas9 is targeted by a short guide RNA that recognizes the target DNA by Watson-Crick base pairing. This guide RNA usually identifies the phage sequences, making a natural immune system for the CRISPR antiviral defense of prokaryotes. However, this sequence can be easily replaced with a sequence of interest to retarget the Cas9. By just introducing a short guide RNA rather than a library of large proteins, Cas9 can target multiple sites at an unprecedented scale (Moon et al. 2019). The facility and simplicity of Cas9 targeting, its high efficiency as a site-specific nuclease, and the possibility for multiple targeting and modification have made the CRISPR/Cas9 system the most widely used genome editing tool (Verma et al. 2022). This system provides a broad range of applications for biologists and biotechnologists, from identifying a single gene function to discovering new drugs and medicines (Verma et al. 2022).

This system consists of two main components, i.e., Cas9 nuclease typically derived from *Streptococcus pyogenes* and a guide RNA (gRNA). The endonuclease Cas9 has two active catalytic domains called HNH and RuvC, which target DNA strands and cleave at the non-target DNA, respectively (Hsu et al. 2021). Even though it was previously reported that the cuts generated by Cas9 at the DSB site were limited to blunt-type, recent research demonstrated that these cuts form a 50-overhang staggered end, and this is because of the post-cleavage trimmed activity of the RuvC domain (Gaj et al. 2016, Jiang and Doudna 2017). Recognition of the target DNA by Cas9 and subsequent DNA cleavage depends on the presence of a short DNA sequence called the protospacer adjacent motif (PAM). The PAM is located in the non-target strand. With the PAM recognition by Cas9, the induction of structural alterations in the Cas9 nuclease will occur, leading to the unwinding of target DNA. Consequently, an R-loop between the target DNA and gRNA molecule forms. Besides, the gRNA guides the Cas9 nuclease to its target DNA by hybridizing the DNA (Hsu et al. 2014).

Verma et al. (2022) confirmed that the referred bacterial-immune memory is stored at the CRISPR loci after an attack by a previous invader, such as a virus or plasmid. Therefore, although the CRISPR system cuts the foreign DNA to destroy it, it simultaneously stores small pieces of this foreign DNA as short DNA sequences in the CRISPR loci. The gRNAs are the outcome of the post-processing of the CRISPR loci transcripts in the *in vivo* bacterial system. Therefore, the

CRISPR loci would be transcribed and processed into approximately 20 nt short CRISPR RNAs (crRNAs) and trans-activating RNAs (tracrRNAs) after being exposed to a second attacker. The crRNAs are derived from the stored DNA sequences. The trans-activating RNAs originate from the conserved CRISPR direct-repeats and serve as a scaffold for Cas9. Thus, to form a ribonucleoprotein (RNP) complex containing Cas9, each crRNA would be annealed to a trans-activating RNA (Jiang and Doudna 2017, Verma et al. 2022). The RNP complex immediately searches for the base-pair complementarity of the crRNA through the invader DNA. After matching, Cas9 demolishes the foreign DNA precisely with DSBs at the target locus (Moon et al. 2019). According to these findings, researchers could design a user-defined gRNA by artificially fusing the conserved trans-activating RNAs sequence with a custom-designed crRNA sequence to obtain a functional and simplified chimeric gRNA. Subsequently, the Cas9 endonuclease can target any locus of interest located immediately upstream of the PAM sequence by changing the 20 nt designed sequences within the gRNA. After cutting the target, cell repairing systems would be activated, leading to the creation of mutations and gene modification (Fang et al. 2017).

Different Types of CRISPR/Cas Systems

When the first protocols of CRISPR genome manipulation were used, one of the main limitations was the constraint on genomic sequences that could be targeted. The Cas9 derived from *S. pyogenes*, i.e., SpCas9, requires the presence of a PAM sequence at the end of the 20 nt gRNA sequence. The expression of gRNA was also mostly initiated with the U6 human polymerase III promoter due to its high efficiency in transcription initiation (Khurshid et al. 2017). However, the U6 promoter starts transcription from guanine (G). These characteristics force the U6-expressed gRNAs to be chosen from genomic sequences that fit the pattern $GN_{19}NGG$, which might not frequently occur in a gene of interest. To expand the CRISPR/Cas9 sequence recognition site, one can use a different promoter (Moon et al. 2019). Take the H1 promoter as an example. This promoter initiates transcription from A or G. Thus, H1-driven gRNAs would also be able to target $AN_{19}NGG$ sequences, which might occur with a higher frequency than $GN_{19}NGG$ within a genome. This slight change in the gRNA expression cassette doubled the number of targetable sites within the genomes of different eukaryotes (Hsu et al. 2014). Another strategy would be the removal of restrictions on the PAM sequence. Since the SpCas9 needs an NGG sequence to recognize PAM, one possible choice is manipulating the Cas protein to create mutants that can recognize alternative PAM sequences (Gaj et al. 2016). Hence, recent studies focused on structural formation analyses, bacterial selection-based directed evolution, and combinatorial design that led to the development of several Cas9 variants that would be able to identify and recognize the other PAM sequences, such as NGA, NGCG, NNGRRT, and NNHRRT (Hsu et al. 2014).

Additionally, Cas9 alternatives themselves have brought other researchers' attention to the use of CRISPR-mediated gene manipulation (Khurshid et al. 2017). For instance, *Francisella* Cpf1 (FnCpf1), a type 2 nuclease lacking the HRH domain, has been explored recently. This enzyme, unlike traditional SpCas9, recognizes a different PAM sequence (5'-TTN-3'), requires a shorter crRNA, and does not need a trans-activating RNA. In addition, this Cpf1 cuts the DNA in a staggered pattern that makes it useful for primary cell editing. Moreover, the smaller size of Cpf1 makes it easier to be carried into the vectors and subsequently makes it ideal for *in vivo* genome editing applications. Cpf1 is currently known as Cas12 (Verma et al. 2022).

Although the specificity of the CRISPR/Cas9 system is determined by gRNA sequence complementation, the CRISPR/Cas9 complex is still jeopardized by several mismatches with its target (Adli 2018). Several times, it has been reported that DSBs have been observed at locations with five or more mismatched nucleotides closely related to the gRNA sequence (Hsu et al. 2014, Gaj et al. 2016, Adli 2018, Moon et al. 2019). Recently, major efforts have attempted to improve cleavage specificity due to the variable activity of the CRISPR/Cas9 system. Therefore, several online predicted tools for gRNAs have been developed to enable the user to predict the off-target activity of the gRNAs (Fang et al. 2017, Moon et al. 2019). Another approach used to improve this cleavage specificity of CRISPR/Cas9 is the direct transfection of crRNA/Cas9 ribonucleoproteins (RNPs) into the cell (Moon et al. 2019).

Furthermore, it has been shown previously that the off-target binding of Cas9 depends on the concentration of Cas9 plasmid. In other words, the persistence and expression levels of Cas9 can be limited by delivering optimized concentrations of Cas9 protein, as opposed to differentially expressing DNA plasmids (Verma et al. 2022). The combination of Cas9 with duplexed CRISPR RNA:trans-activating RNA and delivery into cells would enable the RNP complex to cleave chromosomal target DNA promptly after delivery. The complex will be rapidly removed via the cell's endogenous degradation machine, further limiting off-target cleavage (Verma et al. 2022).

Other relatively new approaches to improving CRISPR efficiency and specificity involve Cas9 endonuclease modification. As mentioned before, Cas9 has two catalytic domains, i.e., RuvC and HNH. It has been recently shown that mutations to catalytic residues in RuvC and HNH create single-strand nicks by Cas9, which is opposed to DSBs. Targeting specific knock-out mutants would be possible by using nickase mutants of Cas9, namely Cas9n, with paired gRNAs (Wang et al. 2017). When the two loci are in the vicinity of each other, but on opposite strands, and are targeted by Cas9n, this endonuclease causes a nick in both strands and creates a DSB, consequently resulting in a mutation. While each gRNA used might have off-target binding sites within the whole genome, Cas9n only catalyzes single-strands breaks (SSBs) at each of those locations. This method can significantly decrease the frequency of unwanted indel mutations, because of the preference for SSBs to be repaired through HDR rather than NHEJ (Verma et al. 2022).

Further studies concentrated on the structure of Cas9 protein. They showed a positively charged groove between each domain of this protein, i.e., HNH and RuvC, which probably stabilizes non-target DNA within the enzyme and finally increases the off-target cleavage possibility. Hence, neutralizing this interaction would reduce the attraction between Cas9 and non-target DNA. As a result, by replacing the positively charged amino acids within the groove to alanine, researchers could successfully neutralize this interaction and called the obtained Cas9 mutant "enhanced" SpCas9 or eSpCas9. This enzyme could significantly reduce off-target effects without compromising on-target cleavage efficiency (Verma et al. 2022).

Application of CRISPR/Cas Technology in Medicine, Biology, and Industry

With the discovery of the CRISPR/Cas system as a gene-editing tool, biology and biotechnology research have been revolutionized. This is mainly due to its cost, effectiveness, specificity, user-friendliness, and ease of use in any laboratory, regardless of molecular biology skills. In contrast to ZFNs and TALENs, there is no requirement for protein engineering for each targeted gene in the CRISPR/Cas system. The CRISPR/Cas system merely depends upon a DNA construct that codes the target-specific gRNA and Cas9, and in the case of knock-in experiments, a donor template for HDR. Furthermore, using this system, multiple genes would be manipulated simultaneously and precisely, increasing the efficiency of the experiment (Xu et al. 2021).

The CRISPR/Cas9 system was first applied to cause single and multiple gene targeting, which would reveal the gene function or its regulation (Hsu et al. 2014). However, single and multiple gene targeting is only one of the CRISPR/Cas applications. This system has also been adopted for other purposes, one of which is genome-wide screening. Genome-wide screen assays are performed to discover genes whose inhibition or aberrant activation would drive phenotypes implicated in disease, development, or other biological processes. This application has been used for humans and mice to knock out their libraries (Xu et al. 2021). Other central CRISPR/Cas system applications include chromatin immunoprecipitation, transcriptional activation and repression, epigenetic editing, live imaging of DNA or mRNA, and therapeutic applications (Chira et al. 2017, Zhang et al. 2017, Adeli 2018).

Chromatin Immunoprecipitation

To characterize proteins and RNAs associated with chromatin, one should purify specific genomic loci. Flexible targeting and isolation of these loci can be efficiently conducted using the CRISPR/Cas9 system. To achieve this aim, we should introduce a nuclear localization signal and epitope tag into dCas9 (catalytically inactive Cas9) to create a DNA-binding protein, which gRNAs can target. The complex would be later purified with traditional chromatin immunoprecipitation (CHiP) techniques to obtain mass spectrometry for further investigation (Zhang et al. 2017).

Transcriptional Activation and Repression

In the last decade, several studies focused on the specificity and re-programming of the CRISPR/Cas9 system to develop targetable CRISPR/Cas9 ribonucleoprotein complexes that would activate (CRISPRa) or interfere (CRISPRi) with transcription of any type of coding region within a genome. By fusing dCas9 to a known-featured transcription-regulatory domain, this system utilizes gRNAs to direct the complex upstream of the transcription site. With the dCas9 being deactivated, this system can target the specific loci without causing cuts or changing the genomic DNA. Binding dCas9 to the targeted DNA sequence would recruit repressive or activating effectors to modify gene expression (Kampmann 2018).

Epigenetic Editing

Recent studies have shown that epigenetic modification to the genomic DNA and histone proteins play a crucial role in the biological process. For instance, methylation or acetylation are epigenetic marks at specific loci; these marks, together with histone residues, would either be inherited or acquired, influencing gene expression. Using the CRISPR/Cas9 genome editing system, it would be feasible to study the roles and targets of these epigenetic marks. This strategy would introduce epigenetic modifications at the desired genomic loci to model disease and examine the potential therapeutic approaches, including finding a method to control cancerous cells (Khan et al. 2016, Sun et al. 2016).

Live Imaging of DNA/mRNA

One should conduct DNA visualization to study cellular processes, including replication, transcription, recombination, and the interactions between DNA and associated proteins and RNA. Two main approaches have been proposed to achieve this goal: fluorescence *in situ* hybridization (FISH) and fluorescent tagging of DNA-binding proteins. The first approach utilizes fluorescently tagged nucleic acid probes to bind and visualize DNA. However, it would be impossible to carry out live imaging using this method due to the sample fixation. On the other hand, while the other method, i.e., fluorescent tagging of DNA-binding proteins, could be used for live imaging, it is only possible to use this approach for repetitive DNA sequences, such as telomeres (Verma et al. 2022).

Recent studies concentrated on using the CRISPR/Cas9 system to obtain live imaging and target any specific sequences with high efficiency and flexibility. To achieve this goal, inactivated dCas9 was tagged with fluorophores for imaging repetitive DNA elements and the genes encoding proteins which would result in the observation of chromatin organization throughout the cell cycle (Verma et al. 2022). Moreover, this system could also be used for live RNA imaging (Gaj et al. 2016, Verma et al. 2022).

Therapeutic Applications

CRISPR/Cas is one of the most valuable tools for gene therapy. This is mainly due to its high speed, being less expensive than other methods, and being safer because of its significant specificity. Since this method, in most cases, shows no off-target effects during genome modification, it can be used for transplant therapies that require donor matching, cancer immunotherapy, tissue regeneration, gene therapy, insect-borne diseases, viral diseases, as well as obesity and metabolism regulation (Khan et al. 2016).

Application of CRISPR/Cas System in the Industry

The CRISPR/Cas system application has expanded from medicine and biotechnology to industry. Several investigations focus on using CRIPSR activator (CRISPRa) and CRISPR interference (CRISPRi) systems on bacteria, yeasts, and fungi (Vyas et al. 2015, Min et al. 2016). For instance, several bacterial genes were modified using this system to produce mutant strains to obtain biofuels and biochemicals from *Clostridium beijerinckii* (Wang et al. 2015). Also, due to the importance of *Escherichia coli* in biotechnology, CRISPR/Cas9 was used to produce a common strain of this bacterium (Pyne et al. 2015). Furthermore, CRISPR/Cas9 technology was applied to make mutations in *Streptomyces coelicolor* to use the mutants as a source of pharmacologically active and industrially relevant secondary metabolites (Cobb et al. 2015, Tong et al. 2015). Other types of this system were mostly used to obtain strains capable of producing ethanol from synthesis gas in *Clostridium ljungahii,* capable of converting waste glycerol to butanol in *Clostridium pasteurianum,* and modifying *Bacillus subtilis* to get mutants as producers of industrial enzymes and valuable low molecular weight substrates (Liao et al. 2022).

In the case of yeasts, various CRISPR/Cas systems have also been utilized to create mutants based on the research interest. For example, CRIPSRa, CRISPRi, and CRISPR/Cas9 were used for common production strains of *Candida albicans, Kluyveromyces lactis, Pichia pastoris,* and *Saccharomyces cerevisiae* (Zhang et al. 2014, Vyas et al. 2015, Min et al. 2016, Liao et al. 2022). Also, mutant strains capable of phenol and formaldehyde catabolism were obtained from *C. albicans* using this approach (Liao et al. 2022). Most filamentous fungi, especially *Aspergillus* spp., were transformed to create mutants as sources and producers of enzymes, such as *A. nidulans, A. aculeatus, A. carbonarius, A. brasiliensis, N. crassa,* and many more species, using this approach (Vyas et al. 2015). Using CRISPR/Cas9, industrial researchers could also create ß-lactam antibiotics from *Penicillium chrysogenum* mutants (Pohl et al. 2016).

CRISPR/Cas vs. Other Genetic Modification Methods (Advantages and Disadvantages)

As for the RNAi, it is mostly impossible to knock out a gene of interest completely. In addition, simultaneously multiple targeting of the gene of interest is also not feasible using this system. Also, the knock-in of a specific region to the location of interest is a great challenge using this method. One of the major challenges of using RNAi to manipulate the gene of interest is the production of unstable mutants, which lose their genetic modifications over time (Barrangou et al. 2015, Baulcombe 2015).

As mentioned before, apart from RNAi, there are four main approaches for genome manipulation using nucleases, namely meganucleases, ZFNs, TALENs, and CRISPR/Cas9. Meganucleases cannot be routinely used for genome editing due to the lack of a clear correlation between meganuclease protein residues and target DNA sequence specificity (Verma et al. 2022). Given these challenges, new genome manipulation technologies focused on specificity, reducing off-target effects, and feasibility, which can all be found in the CRISPR/Cas technology (Hsu et al. 2014).

Zinc-finger nucleases also have some limitations, such as exhibiting context-dependent binding preferences. This occurs due to the cross-talk between adjacent modules assembling into a larger array. While some attempts have been conducted to decrease this limitation, the assembly of functional ZFNs with high specificity to bind the desired DNA is still a major challenge demanding a considerable screening process (Gaj et al. 2016).

Similar to ZFNs, TALENs would also encounter the same problem from context-dependent specificity, while the TALE DNA-binding monomers are, for the most part, modular. Furthermore, this method is costly and requires experts to construct TALENs repetitive sequences of new TALE arrays (Gaj et al. 2016).

Compared to the complex design of TALENs and ZFNs, which requires a customized protein for each gene sequence and has low delivery efficiency, CRISPR/Cas9, a genome engineering method, uses only a single nuclease and gRNA (Hsu et al. 2014). The simplicity of the designed gRNAs only based on Watson-Crick's rule of hybridization is a sufficient reason to choose this method from the others. Unlike the specific polypeptide-DNA binding of ZFNs and TALENs, designing a 20 nt sequence of gRNA is not time-consuming, and it is not tedious to optimize this system in contrast to ZFNs and TALENs (Zhang et al. 2014, Verma et al. 2022). One of the most significant characteristics of CRISPR/Cas9 genome manipulation is its stability compared to the RNAi approach (Zhang et al. 2014, Verma et al. 2022). Regardless of the type of organism to be manipulated. CRISPR/Cas9 can be applied efficiently to humans, animals, insects, plants, fungi, bacteria, viruses, and oomycetes with just a simple bacterial transformation or lentiviral approach. Consequently, genome editing using site-specific nucleases would be a valuable tool to understand the transgene experiments efficiently and accurately. Since CRISPR/Cas9 is an RNA-based guiding technique, it is much cheaper and easier to design to obtain the manipulation of a wide range of target sequences precisely (Liao et al. 2022).

While CRISPR/Cas9 is a valuable tool for genome engineering, it still faces challenges. Off-target effects are one of the most important challenges for CRISPR/Cas9 experts. This challenge could be resolved with a suitable selection of gRNAs, using variants of Cas9, such as dCas9, and using smaller-sized gRNAs (Zhang et al. 2014, Verma et al. 2022).

Limitation in choosing a PAM sequence is another major obstacle regarding using CRISPR/Cas9. Several Cas9 mutants have been engineered and applied to deal with this problem. For example, Cpf1 (known as Cas12) is another variant for Cas9 which recognizes T-rich PAM sequences with higher frequency than G-rich ones. Cpf1 is also smaller, making it a fabulous choice than Cas9 in genetic modification. Also, using other variants, including dCas9 and deactivated dCas9, would decrease the problem of PAM limitation (Zhang et al. 2014, Gaj et al. 2016).

Designing gRNA is another challenge for applying CRISPR/Cas9 to an individual organism. Even though many online databases have tried to solve this problem, this is still a challenge since one should also manually double-check the specificity and secondary structures only after confirming that the designed gRNA would be activated during the process (Verma et al. 2022).

One of the major concerns of the biotechnologist and molecular biologist trying to generate CRISPR/Cas9 mutants is the transformation approach and selection of a suitable vector for carrying the CRISPR/Cas complex (Verma et al. 2022). The plasmid vectors, lentiviral vectors, adeno-associated virus (AAV) vectors, Cas9 mRNA and gRNA, and crRNA/Cas9 RNP complex have been tested successfully according to the target organism (Gaj et al. 2016).

Plasmid vectors are the most commonly used for many organisms, including fungi and oomycetes. These vectors carry compatible or inducible Cas9 and gRNA, along with a reporter or selection marker, and are used to express Cas9 and gRNA (Fang et al. 2017). Lentiviral vectors' components are like plasmid vectors and are also used to express Cas9 and gRNA. Furthermore, using lentiviral vectors for the infection of those cell types, which are difficult to transfect, has also been reported several times (Verma et al. 2022). Having the same elements, AAV vectors are used for stable or transient expression of a Cas9 derived from *Staphylococcus aureus* (known as SaCas9) and gRNA and for non-toxic infection of dividing and non-dividing cells (Boddu et al. 2021). Cas9 mRNA and gRNA expression systems have transcription reactions *in vitro* to generate Cas9 mRNA and gRNA, delivered with microinjection or electroporation. The Cas9 mRNA and gRNA systems are used for only the transient expression of CRISPR gene editing components (Verma et al. 2022). As for the last-mentioned expression system, i.e., crRNA/Cas9 RNP complexes, purified Cas9 protein, and *in vitro* transcribed gRNA should be delivered with microinjection or electroporation to serve the target of transient expression of the CRISPR gene editing components (Verma et al. 2022).

Different delivery approaches have been proposed for the CRISPR/Cas9 system depending on the host organism. For plant pathologists, the most widely used delivery systems are the transformation of plasmids into competent cells, electroporation of plasmids and galactose induction of Cas9, and Agrobacterium-mediated transformation of gRNA/Cas9 vector (Verma et al. 2022). Transformation of plasmids into competent cells has been successful for bacteria, fungi, and oomycetes (Fang et al. 2017, Schuster and Kahmann 2019, Liao et al. 2022). Electroporation of plasmids and galactose induction of Cas9 protein has been conducted extensively in yeasts (Verma et al. 2022). Agrobacterium-mediated transformation of the gRNA/Cas9 vector has been directed to plants, fungi, and oomycetes (Wang et al. 2018, Liao et al. 2022).

CRISPR/Cas Application in Fungal Plant Pathogens

In recent decades, the utilization of CRISPR/Cas systems has developed significantly. Production of resistant plant mutants to a variety of diseases caused by devastating pathogens, such as *Xanthomonas* spp., *Pseudomonas syringae,* and *P. capsici,* as well as non-pathogenic stresses such as drought and salinity resistant plants, are some examples of the application of CRISPR/Cas9 systems in plant pathology (Verma et al. 2022). These studies tried to find the resistant genes in plants themselves by knocking them out using CRISPR/Cas9 or enhancing the transcription of resistant genes using the CRIPSRa method to protect the plants (Kampmann 2018). The application of the CRISPR/Cas systems has been conducted on plants and pathogens to find their genes responsible for pathogenicity. Filamentous fungi play significant roles in biotechnology as producers of enzymes and compounds used in medicine, biofuel, and the food industry. Due to the production of these valuable chemicals, these organisms have always been the main target of genetic modification, whether to increase or decrease the amount of chemicals they produce (Shi et al. 2017). Apart from being important in industry and medicine, filamentous fungi are also infamous as important plant, animal, and human pathogens. In plant pathology, numerous genera cause devastating diseases on various plants, some of which are *Fusarium, Botrytis, Alternaria, Colletotrichum, Ustilago, Puccinia,* and several other fungi. These fungal pathogens cause a substantial economic loss in agriculture yearly. Although several attempts were conducted to establish a CRISPR/Cas9 system in filamentous fungi, only the most important ones are discussed in this chapter. Information about the genetically well-studied

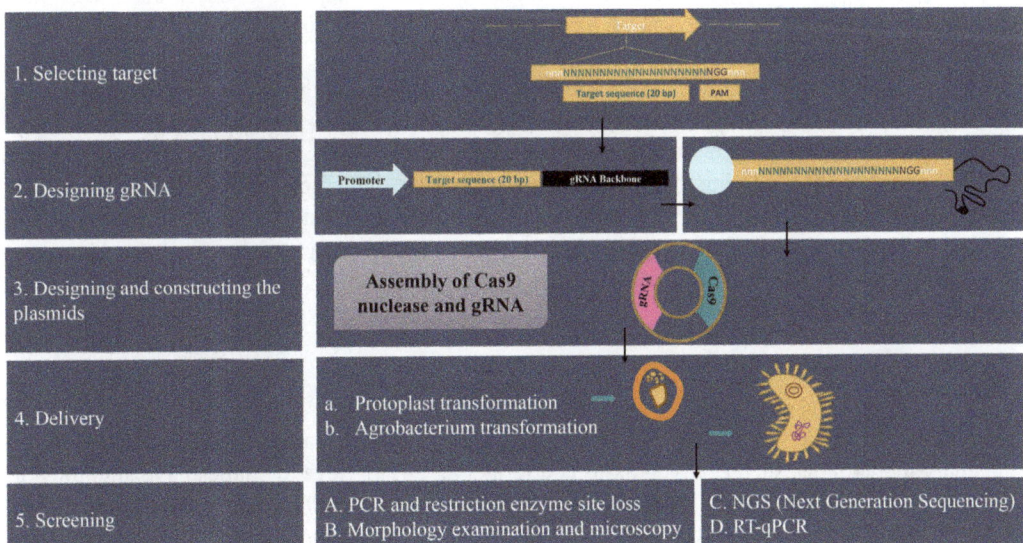

Figure 1. CRISPR/Cas9 system experiment in summary.

fungal plants' pathosystems would be a launchpad for other similar systems, such as diseases caused by oomycetes. Figure 1 illustrates a typical workflow of the CRISPR/Cas9 system. Depending on the microorganism type, different approaches would be applied for transformation.

Ustilago maydis

The basidiomycete fungus, *Ustilago maydis* (DC.) Corda, is the causal agent of corn smut disease. In the last two decades, researchers have been interested in this pathogen as a model genetic organism to investigate the plant-pathogen interactions with this biotrophic agent due to its small genome size compared to other eukaryotic pathogens (Schuster et al. 2018). There have been several successful attempts to modify the *U. maydis* genome for various purposes. These include the identification of a particular gene's function or its regulation. The first establishment of the CRISPR/Cas9 system to disrupt target genes in *U. maydis* was in 2016 (Schuster and Kahmann 2016). The protocol gave detailed information on conducting efficient genome editing using the CRISPR/Cas9 approach in fungi. Using this protocol, production of the mutants was mediated via a single-step transformation of a self-replicating plasmid that could express the *U. maydis* codon-optimized Cas9 gene with a U6 controlling promoter that would control gRNA. It was shown that approximately 70% of the progeny of an individual transformant could be successfully disrupted within the respective *b* gene (Schuster et al. 2018). This gene encodes a divergently transcribed pair of homeodomain proteins involved in the mating-type process. The self-replicating plasmid was lost rapidly without selection to avoid the adverse effects of Cas9. It was also shown that through the simultaneous disruption of functionally redundant genes and gene families, it is possible to study their contribution to the virulence of *U. maydis*. Using an all-in-one plasmid approach, researchers successfully established CRISPR/Cas9 into the smut fungus, so both the Cas9 cassette and gRNA cassette are located on a single plasmid (Schuster et al. 2018). Some other studies focused on the plant-side of the *U. maydis* pathosystems. For example, in a *Zea mays* L. study, the *ZmACD6* genes were successfully knocked out using CRISPR/Cas9 to determine their role in *U. maydis* (Zhang et al. 2019). This gene involves the salicylic acid pathway triggered by infection with the smut fungus.

Furthermore, in a successful study focusing on the multiplexing version of CRISPR/Cas9 (Zuo et al. 2020), the system was built using tRNA promoter-based expression cassettes, which facilitated the expression of several gRNAs from the same plasmid. This multiplexed version of CRISPR/Cas9 was successfully used to describe the contribution of oligopeptide transporters to the pathogen's virulence. The CRISPR/Cas9 technique was also applied to obtain indel mutations in the nitrate transporter gene, *um03864*, to discover its actual functions (Schroeder et al. 2020). It was believed that the gene would be involved in growth ability, mating efficiency, and pathogenesis. However, the mutants showed that while this gene has no effects on the pathogen mating behavior, it significantly reduces pathogen growth and pathogenicity.

Several strategies for genetic manipulation of *U. maydis* using CRISPR/Cas9 system were tested (Wege et al. 2021). Using this system, one could delete the targeted gene via HDR, introduce the point mutations via NHEJ, perform the heterologous complementation at the genomic locus, and conduct endogenous N-terminal tagging with the fluorescent protein mCherry on the smut causal agent. The precise protocol of the transformation of *U. maydis* using CRISPR/Cas9 for each genetic modification was also described.

Fusarium spp.

Fusarium species are cosmopolitan plant pathogens mostly known to cause wilt diseases and produce secondary metabolites, such as mycotoxins (Ferrara et al. 2019). Several studies have attempted to establish CRISPR/Cas systems in this pathogen. For example, *F. proliferatum* is a well-known plant pathogen worldwide and produces numerous mycotoxins. The fumonisins are the most important due to their fetal toxicity (Ferrara et al. 2019). It was possible to knock out the polyketide synthase gene (*FUM1*) from *F. proliferatum* (Matsush.) Nirenberg using the CRISPR/Cas9 system. This gene is required for fumonisin biosynthesis (Ferrara et al. 2019).

The CRISPR/Cas system application was also conducted for the *F. oxysporum* species complex. The fungus is known as one of the most important plant pathogens worldwide (Wang et al. 2018). However, due to the low efficiency and high off-target effects of other genetic modification approaches, it was difficult to obtain mutants with desired features. By developing the Cas9 ribonucleoprotein complex, it was possible to obtain knock-out transformants of *F. oxysporum* that were uracil auxotroph mutants. Furthermore, by analyzing the *bik1* mutants, it was revealed that this polyketide synthase is involved in synthesizing the reg pigment called bikaverin (Wang et al. 2018).

Alternaria alternata

Alternaria alternata (Fr.) Keissl. is one of the most well-known filamentous fungi for producing secondary metabolites and one of the most devastating post-harvest plant pathogens (Wenderoth et al. 2017). Despite its importance as a plant pathogen, genetic modification of this fungus is quite risky. This is mainly due to gene deletions which can lead to heterokaryotic strains. Consequently, the gene-function analysis would be rather tedious using such approaches. Moreover, the sexual cycle of this fungus is unknown, which makes it difficult to combine classical genetic methods with molecular biological ones (Wenderoth et al. 2017). However, it was recently shown that the CRISPR/Cas9 system could effectively be used for gene knock-out in *A. alternata*. Several white transformants were obtained by targeting two main genes associated with melanin biosynthesis, i.e., *pksA* and *brm2*.

Furthermore, the successful application of the CRISPR/Cas9 approach in generating uracil auxotrophic strains was reported (Wenderoth et al. 2017). Using this system, it was revealed that one of the secondary metabolites produced by *A. alternata*, polyketide-derived alternariol, is also responsible for the virulence and colonization of the fungus (Wenderoth et al. 2019). It was also shown that germination, sporulation, and secondary metabolism are regulated by light in *A. alternata* by targeting phosphoreceptors using CRISPR/Cas9 (Igbalajobi et al. 2019).

Pyricularia oryzae

Reports on the rice blast causal agent, *Pyricularia oryzae* Cavara (syn. *Magnaporthe oryzae* B. C. Coach), and genetic modification using CRISPR/Cas9 have arisen in the last decade. The rice blast agent is one of the most significant pathogens in rice paddy fields, causing a severe threat to global food security (Arazoe et al. 2015). Even though most studies focused on the rice plants themselves for genetic manipulations using CRISPR/Cas9 to get the desired mutants resistant to this disease, some studies also focused on the pathogen to characterize its pathogenicity-related genes and find a way to control the pathogen. Attempts to develop a rapid plasmid-free CRISPR/Cas9 approach successfully accelerated targeted transient mutations and specific gene editing in *P. oryzae* (Foster et al. 2018). A single crossover-mediated homology-directed repair approach of genome editing using CRISPR/Cas9 was also carried out for *M. oryzae* (Yamato et al. 2019).

Botrytis cinerea

The gray mold fungus, *Botrytis cinerea* Pers., is an ascomycete plant pathogen infecting more than 1000 species and causing a considerable loss in the fruits, vegetables, and flowers. The importance of the pathogen would lead the researchers to find a proper way to manage its negative effects on the plants (Hahn and Scalliet 2021). Recently, some studies reported the successful application of CRISPR/Cas9 in this plant pathogen. The introduction of optimized Cas9-gRNA RNPs into the protoplasts of this pathogen and utilization of a novel approach to combine RNP delivery with the co-transformation of transiently stable telomer-containing vectors allowed researchers to temporarily choose and screen for marker-free editing events (Leisen et al. 2020). This method excelled over other CRISPR/Cas9 methods because it provided no off-target effects. Besides, plant cell death-inducing proteins and the related metabolites for necrotrophic infection by *B. cinerea* were evaluated using the CRISPR/Cas9 system (Leisen et al. 2022). The results revealed that this fungus releases a highly redundant cocktail of proteins to acquire necrotrophic infection of a wide variety of hosts.

CRISPR/Cas and Oomycetes

Oomycetes are considered among the most important plant pathogens in agriculture, horticulture, aquaculture, and natural ecosystems. Due to the annual economic loss caused by these organisms, oomycetes have always been a subject of genetic manipulation (Schuster and Kahmann 2019). Even though there are not so many reports of the successful transformation of oomycetes using CRISPR/Cas systems, recent studies have focused on generating stable transformants using this system. Some of the most successful examples are discussed below.

Application of CRISPR/Cas9 in Gene Function Studies

Phytophthora sojae

Phytophthora is a fungus-like microorganism belonging to the kingdom *Stramenopila* and the most well-known devastating plant pathogen worldwide. The species assigned to this genus cause substantial economic losses to agriculture, horticulture and natural ecosystems each year. Due to the exceptionally low rate of homologous recombination, genome editing was not successful in the oomycetes (Cao et al. 2022).

The pathogen *P. sojae* is responsible for damping-off and root rot in soybean seedlings and established plants (Erwin and Ribeiro 1996). While DNA transformation of *P. infestans* and *P. sojae* was conducted a long time ago, targeted gene mutations and gene replacement were not possible since the insertion of transgenes happens solely by the NHEJ repairing system (Fang and Tyler 2016). Despite all these facts, the first application of the CRISPR/Cas9 system in the oomycetes was conducted for *P. sojae*. The system was designed to be compatible with the pathogen with an oomycete-specific nuclear localization signal and gRNA-specific transcription cassette (Fang and Tyler 2016). CRISPR/Cas9 targeting an endogenous gene, i.e., *Avr4/6,* involved as the RXLR effector in *P. sojae* infection, revealed that this system could be utilized by NHEJ and HDR pathways. Small indels occur in the homozygous mutants without the homologous template, which results from gene conversion triggered by cleavage of non-mutant alleles. Using the homologous template could successfully replace the *Avr4/6* gene with the *nptII* gene, an antibiotic resistance gene. It was also confirmed that *Avr4/6* is recognized by the *R* gene of soybean (Fang and Tyler 2016, Cao et al. 2022).

One of the largest conserved families of lipid transfer proteins in eukaryotes is oxysterol-binding proteins. These proteins could be targeted by the oxathiapiprolin oomycide (Miao et al. 2018). Two of these proteins, PsORP1 and PsORP2, were selected to study biological functions. It was previously proposed that *PsORP2* has a higher expression level in germinated cysts and late infection. Attempts to knock out this gene using CRISPR/Cas9 showed that this gene has no effects on sporangial or zoospore production, cyst germination, oospore production, virulence, or oxathiapiprolin sensitivity. Hence, it was concluded that PsORP2 would not be an essential protein for the development of virulence of *P. sojae* (Miao et al. 2018). As for PsORP1, using the CRISPR/Cas9 gene-editing tool, 16 mutations have been generated from HDR mutants. All these transformants showed high oxathiapiprolin resistance with enhanced or similar virulence, oospore production, and cyst germination, whereas their sporangial and zoospore production was significantly decreased (Miao et al. 2020).

Phytophthora infestans

Phytophthora infestans causes potato and tomato late blight and is infamous for causing the Irish famine. This pathogen is difficult to control and causes substantial economic loss annually (Erwin and Ribeiro 1996). The first attempts to establish the CRISPR/Cas9 RNP system in this pathogen were unsuccessful (van der Hoogen and Govers 2018).

Nevertheless, further studies successfully generated CRISPR/Cas9 mutants of *P. infestans.* This system was established in *P. infestans* using CRISPR/Cas12a (Ah-Fong et al. 2021). The vectors containing the nuclease and gRNA were expressed from a single transcript to achieve this goal. Using the variant of Cas9, i.e., Cas12a (formerly known as Cpf1), enabled the researchers to express the CRISPR/Cas system in *P. infestans.* Former systems of CRISPR/Cas9 were difficult to express, and the active nuclease was toxic to the pathogen. The CRISPR/Cas12a system has both DNase and RNase activity and fewer off-targets. With the RNAse activity of Cas12a, editing an elicitin in *P. infestans* was possible, which was the first report of the utilization of Cas12a in oomycetes (Ah-Fong et al. 2021).

Also, recent studies of the CRISPR/Cas9 system in *P. infestans* revealed that a colorimetric approach could be used to detect specific nucleotide sequences in plant pathogens. This approach was based on CRISPR/Cas9-triggered isothermal amplification and gold nanopores as optical probes (Chang et al. 2019). After the cleavage of the target DNA by the Cas9-gRNA complex, isothermal amplification would occur. Consequently, the product would be hybridized with the oligonucleotide-functionalized gold nanoparticles, resulting in the aggregation of gold nanoparticles and color change to purple. This method has a significantly higher specificity than the convenient CRISPR/Cas9 system and was applied to identify *P. infestans* genomic DNA (Chang et al. 2019). However, due to its costly nature, it would not be affordable for large-scale identification of plant pathogens in field studies.

Aphanomyces invadans

Species of *Aphanomyces, Achlya* and *Saprolegnia* are reported to be the most destructive pathogens of aquatic organisms. Even though less than 50 species are assigned to *Aphanomyces,* they are mostly specialized plant and animal pathogens, whereas others are considered saprotrophic or opportunists observed on animal and plant debris (Iberahim et al. 2018). The application of CRISPR/Cas9 has not yet been carried out on *Aphanomyces* species that infect plants; however, in a recent study using this system, the genome of *A. invadans* has been manipulated. The targeted gene was the serine protease to study the effects of this gene on the pathogenicity and virulence of the pathogen. *Aphanomyces invadans* are a primary cause of the epizootic ulcerative syndrome, a lethal disease, in cultured and wild fish worldwide (Iberahim et al. 2018). It was previously stated that extracellular proteins, mainly proteases, are produced by *A. invadans,* which might have a key role in destroying fish immunoglobulin. Serine proteases have been reported as a significant virulence factor in many bacteria and fungi as well (Majeed et al. 2018). Hence, targeting the encoding genes would reveal their roles in the pathogenicity of *A. invadans.* Using the CRISPR/Cas RNP complexes results in the effective mutation in the targeted gene. It confirms the hypothesis that serine proteases have a major role in the pathogenicity of this oomycete (Majeed et al. 2018).

Phytophthora litchii

The downy blight caused by *Phytophthora litchii* (C.C. Chen ex W.H. Ko, H.S. Chang, H.J. Su, C.C. Chen and L.S. Leu) Voglmayr, Göker, Riethm. and Oberw. (formerly *Peronophythora litchii* C.C. Chen ex W.H. Ko, H.S. Chang, H.J. Su, C.C. Chen and L.S. Leu) is one of the most important diseases of lychee (*Litchi chinensis* Sonn.). The pathogen can attack the host at almost all growth stages, i.e., early leaf emergence, flowering, fruit production, and postharvest (Kraturisha et al.

1995). Almost all plant pathogenic oomycetes have pectin acetylesterases involved in the enzymatic deacetylation of pectin, a major component of the higher plants' primary cell wall (Kong et al. 2019). Pectin acetylesterases show sequence and transcriptional polymorphism among the plant pathogenic oomycetes (Kong et al. 2019). In a recent study, to investigate the role of these components in oomycete pathogenicity, two genes, i.e., *plpae4* and *plpae5,* were knocked out using the CRISPR/Cas9 system. It was shown that *plpae5* mutants have less virulence and could not invade the hosts than the wild-type *P. litchii.* Hence, it was concluded that only *plpae5* plays a vital role in *P. litchii* pathogenicity (Kong et al. 2019).

Pythium spp.

The oomycete *Pythium ultimum* is considered one of the most dangerous plant pathogens with a broad host range. Like other pathogenic *Pythium* species, it causes damping-off and root rot, resulting in severe yield loss. This pathogen has two varieties, i.e., *Py. ultimum* var. *ultimum* and *Py. ultimum* var. *sporangiferum,* which are considered the most common and rare pathogen, respectively (Mostowfizadeh-Ghalamfarsa 2016). There are only a few reports of the genes involved in the sexual reproduction of oomycetes so far. For instance, PoStr proteins in *Py. oligandrum* plays an important role in oospore formation. The role of PoStr proteins was discovered by silencing the encoding gene via RNAi (Yang et al. 2013).

Another example is the knock-out *P. sojae*'s *PsYPK1* gene via CRISPR/Cas9. This gene codes a serine/threonine protein kinase and is involved in oospore formation in *P. sojae* (Feng et al. 2021). The last example is the utilization of RNAi to silence a gene that encodes a loricrin-like protein. This gene was also believed to confer with oospore formation in *P. infestans* (Feng et al. 2021). The mechanism of oospore formation regulation in *Py. ultimum* was described recently using CRISPR/Cas9 (Feng et al. 2021). Researchers have developed an adapted CRISPR/Cas9 transformation system for the genus *Pythium.* Oospore formation occurs under the regulation of *PuM90*, a stage-specific Puf family RNA-binding protein. Knocking out the responsible gene, i.e., PuM90, results in the formation of the mutants unable to produce oospores or in retarded oospores with empty oogonia (Feng et al. 2021).

The most recent successful attempt to establish the CRISPR/Cas9 system was reported for *Py. oligandrum.* This oomycete is a strong antagonist of fungal and other oomycete plant pathogens while promoting plant growth. Its antagonistic ability is hypothesized to be due to the existence of an elicitin-like protein called oligandrin. Knock-out mutants of oligandrin have been generated using the CRISPR/Cas9 technology to test this ability (Salmaninezhad et al. 2022). However, further investigations are in progress to enhance oligandrin's transcription using a donor template containing a strong promoter to induce the HDR pathway utilizing the CRISPR/Cas9 system.

Application of CRISPR/Cas9 in Gene Regulation Studies

Phytophthora sojae

The importance of transcriptional regulation for plant pathogen development and virulence is known. It has been shown that MADS-box transcription factors are members of a highly conserved transcriptional regulator family in eukaryotic organisms. Hence, these factors are involved in different biological processes (Leesutthiphonchai and Judelson 2018). There has been only one MADS-box gene, i.e., *PsMAD1,* identified in *P. sojae.* This gene is highly expressed during sporangia production and during infection (Lin et al. 2018). Therefore, its function remained unknown until recently. Using the CRISPR/Cas9 system, knock-out mutants of this gene were generated. It was revealed that the *PsMAD1* gene is a major regulator of *P. sojae* and contributes to zoosporogenesis and pathogenesis (Lin et al. 2018). It was also shown that these mutants have similar vegetative growth, oospore production, and sensitivity to various abiotic stresses to the wild-type strains. However, no zoospores formed in the mutant strains (Lin et al. 2018).

The avirulence effectors of *P. sojae* interact with the corresponding soybean resistance (R) proteins, resulting in robust plant immune responses (Ochola et al. 2020). To clarify how *Avr* gene expression patterns affect plant-microbe interactions, CRISPR/Cas9 system was used to knock in the promoters for early and late expression of *P. sojae*. With the assistance of RNA sequencing (RNA-Seq) results, it was revealed that the precise setup of *Avr* gene expression impacts the fitness of plant disease and could be utilized to improve crop resistance in disease control management (Ochola et al. 2020).

One of the most widely used approaches to manage diseases caused by *P. sojae,* is to use the resistance genes in soybean cultivars. This method is based on effector-triggered immunity in which resistance to *P. sojae* protein from soybean recognizes a specific effector from this pathogen that has been coded by the *Avr* gene (Białas et al. 2018). Many *Avr* genes have been recognized using the CRISPR/Cas9 system to knock out a specific gene to reveal its function. Using the CRISPR/Cas9 system showed that the RXLR effector gene, *Avr3a*, from *P. sojae* can be recognized by the *Rps8* gene in soybean (Arsenault-Labrecque et al. 2022). The *Avr3a* gene is responsible for the triggering immunity derived from the host plant resistance gene *Rps8*. Oomycetes can rapidly adapt to environmental changes, such as sudden host resistance, with their transcriptional plasticity (Gu et al. 2021). *Phytophthora sojae* can survive and overcome the host resistance gene *Rps1b* via silencing its corresponding effector gene, *Avr1b-1*. Generating knock-out and knock-in mutants of *P. sojae* using the CRISPR/Cas9 construct resulted in the deletion of the *Avr1b-1* gene and replacing this gene with the *mCherry* fluorescent coding gene. The results also showed that knock-out mutants obtained virulence on soybean, and by replacing the *Avr1b-1* gene with *mCherry,* the transformants were unable to infect the host plant. Consequently, the *Avr1b-1* gene may be remarkably susceptible to transcriptional variation (Gu et al. 2021).

Phytophthora capsici

Phytophthora capsici is an important and cosmopolitan plant pathogen causing enormous economic loss on various vegetable crops. This pathogen has been a model organism to study gene functions, plant-pathogen interactions, and fungicide resistance mechanisms. This is mainly due to its wide range of host plants and remarkable genetic diversity (Lamour et al. 2012). The first CRISPR/Cas9 gene-editing protocol for *P. capsici* was released using two plasmid and all-in-one plasmid systems. This protocol was similar to that of the *P. sojae* transformation protocol previously presented (Fang et al. 2017); however, it had slight modifications, such as the duration and temperature of centrifugation at protoplast isolation steps were mentioned (Wang et al. 2018).

Point mutations were conducted using CRISPR/Cas9 in *P. capsici* and *P. sojae* on the oxysterol binding protein-related protein 1 (ORP1), and the detection of ORP1 in oxathiapiprolin-resistant *P. capsici* isolates was conducted. The effects of the point mutations were also investigated on *P. capsici* phenotypes. It was confirmed that transformants containing heterozygous point mutations in PcORP1 have high levels of oxathiapiprolin resistance. While some transformants showed similar morphological behavior and characteristics as the wild-type strain, some could not produce sporangia (Wang et al. 2018).

Another study using the same transformation and gene editing method showed that an oxathiapiprolin resistance gene, called *PcMuORP1,* functions as a new selection marker for any species associated with *Phytophthora* spp. transformation and the CRISPR/Cas9 mediated genome manipulation (Wang et al. 2019). Researchers demonstrated that the gene *PcMuORP1* can be introduced into the CRISPR/Cas9 system. By inducing both NHEJ and HDR pathways, both knock-out and knock-in transformants could be obtained. The complementation of a deleted gene in *P. capsici* was gained using it as a selection marker. It was also shown that the use of *PcMuORP1* as a selection marker resulted in significantly higher efficiency of screening than the previously used neomycin phosphotransferase II marker. Furthermore, subsequent culturing and asexual reproduction without the selection marker resulted in losing the selection marker from the transformants. The loss of selection marker during subculturing indicates that this gene, i.e., *PcMuORP1,* has a minor long-term effect on the transformants' fitness (Wang et al. 2019).

Also, the sensitiveness of *P. capsici* isolates to the fluoxapiprolin fungicide was determined utilizing CRISPR/Cas9. Nine stable transformants were achieved by fungicide adaption that were all fluoxapiprolin resistant. It was also concluded that *P. capsici* resistance risk to this fungicide is moderate, and applying several point mutations in *PcORP1* results in various resistance levels to fluoxapiprolin (Miao et al. 2021).

Reports of the characterization of RXLR effecter *PcAvh1* from *P. capsici* are available (Chen et al. 2019). This effector is promptly induced at the early stages of infection and only during the infection (Chen et al. 2019). Knocking out of the gene responsible for coding using CRISPR/Cas9 resulted in the demolition of the pathogen virulence, whereas its overexpression led to the enhancement of disease development in the tested plants, i.e., *Nicotiana benthamiana* Domin and bell pepper (*Capsicum annuum* L.). These results confirmed that *PcAvh1* is a critical virulence factor (Chen et al. 2019).

The introduction of a new member of the piperidinyl thiazole isoxazoline class of fungicides, namely Ro34-1, was an important attempt to prohibit several developmental stages of most oomycete plant pathogens, especially *P. capsici* (Miao et al. 2020). The potential resistance risk of this fungicide was evaluated in *P. capsici,* and 12 resistant transformants were generated using CRISPR/Cas9. These mutants performed with lower fitness than the wild-type strain suggesting that the resistance risk in Ro34-1 is low in *P. capsici.* This study also validated site-directed mutagenesis of the targeted gene in *P. capsici* utilizing CRISPR/Cas9 (Lin et al. 2020).

Even though small cysteine-rich proteins contain fungal avirulence proteins and play key roles in plant-pathogen interactions (Zhang et al. 2021), their role and function in oomycetes were not clarified. These proteins have previously been identified in *Phytophthora* species (Templeton et al. 1994); however, the first official report of their characterization in *P. capsici* was revealed using CRISPR/Cas9. This protein is called SCR82 and is similar to the *P. cactorum* phytotoxic protein PcF. The homologs of this protein are only found in the species assigned to *Phytophthora*. Knocking out of the encoding *scr82* gene in *P. capsici* via CRISPR/Cas9 severely destroyed its virulence on the host plant and substantially reduced its resistance to oxidative stress. In contrast, the overexpression of *scr82* led to the enhancement of virulence and tolerance to oxidative stress. Hence, it was concluded that *scr82* is a conserved region across all *Phytophthora* species and plays a significant role both as a virulence factor and plant defense elicitor (Zhang et al. 2021).

Phytophthora palmivora

Phytophthora palmivora is a destructive plant pathogen infecting several plant species. This pathogen infects many economically important hosts, such as cacao and papaya (Miao et al. 2018). It is tetraploid and heterothallic. Its oospore germination rate is relatively low. Due to all these facts, functional genomic studies of this pathogen have been challenging (Miao et al. 2018). However, with the advent of CRISPR/Cas9 technology, simultaneous mutations of multiple alleles in the first generation of the transformants would be possible. The first attempt to create CRISPR/Cas9 mutants in *P. palmivora,* was through Agrobacterium-mediated transformation. This method was more convenient to perform in *P. palmivora,* providing single-copy integration in the pathogen that would facilitate the gene function analyses (Tian et al. 2020).

Since *P. palmivora* can infect all parts of papaya plants, it was hypothesized that it had evolved cysteine protease inhibitors produced by papaya. By targeting a cysteine protease inhibitor, PpalEPIC8, homozygous transformants were generated using CRISPR/Cas9 gene editing via Agrobacterium-mediated transformation (Gumtow et al. 2018). Since the mutants showed reduced pathogenicity on papaya, it was revealed that PpalEPIC8 plays a significant role in the pathogen virulence by inhibiting papain produced by papaya (Gumtow et al. 2018).

A secreted 15 kDa glycoprotein was previously isolated from *P. palmivora* strains obtained from papaya (Pettongkhao et al. 2020). However, its function was unknown until the study in which the CRISPR/Cas9 system was utilized. Generating knock-out mutants of the gene responsible for coding this 15 kDa glycoprotein resulted in the complete loss of *P. palmivora* ability to infect the host plant. Furthermore, it was revealed that the knock-out mutants of this glycoprotein had smaller sporangia, shorter germ tubes, and fewer appressoria. Hence, this glycoprotein plays an important role in the pathogenicity and development of *P. palmivora* (Pettongkhao et al. 2020).

Prospects and Challenges of using CRISPR/Cas in Oomycetes

CRISPR/Cas9 has been the most robust gene-editing tool in the last two decades compared to other genome engineering tools. This technique's simplicity, ease of use, accuracy, and time-saving nature highlight its value compared to other methods. Despite all these advantages, utilization of CRISPR/Cas9 in oomycetes is quite challenging due to some reasons.

Transformation Approach

The generation of CRISPR/Cas9 mutants has been carried out mainly via the polyethylene glycol (PEG-) transformation approach. This method works fine for several oomycetes, including *P. capsici, P. infestans, P. megakarya, P. parasitica, P. sojae, Py. oligandrum, Py. ultimum,* and several other oomycetes (Fang et al. 2017, Wang et al. 2018, Tian et al. 2020, Feng et al. 2021, Salmaninezhad et al. 2022). However, protoplast isolation and regeneration of protoplasts can be quite tricky. While some oomycetes require only a limited time for protoplast isolation, others need more time. For instance, it has been reported that *Py. oligandrum* strains require 2 to 3 hours of enzymatic digestion to release the protoplasts (Horner et al. 2012), while the time for *P. capsici* and *P. sojae* was reported to be only 40 minutes (Fang et al. 2017, Wang et al. 2018). Although the longer the hyphae stay in the enzyme solution, the more protoplasts would be obtained; nevertheless, long time periods has a negative effect on the regeneration of the protoplasts afterward. Besides, the protocol needs to be optimized for each species and each strain. For example, in a recent protocol of *P. sojae* transformation, glucose was added to the washing buffer of the protoplasts (Fang et al. 2017, Wang et al. 2018). However, adding glucose to this buffer resulted in the isolation of practically no protoplasts for *P. oligandrum* (Salmaninezhad et al. 2022).

On the other hand, the PEG-mediated transformation of oomycetes results in multi-copy integration, consequently disrupting multiple loci of the genome and complicating the gene function analyses. In contrast, Agrobacterium-mediated transformation produces single-copy integration that would conquer this obstacle (Tian et al. 2020). The CRISPR/Cas system was successfully established in *P. palmivora* using the Agrobacterium-mediated transformation (Tian et al. 2020).

Selection Markers

Although screening selection markers seems to cause no problems in CRISPR/Cas9 experiments, it has been proven to be one of the most challenging obstacles in oomycetes. The availability of selection markers for oomycetes, especially for *Phytophthora,* is limited. This limitation results in restraining transgenic manipulation in some cases (Wang et al. 2019). The most commonly used selection markers for oomycetes are reported to be fluorescent proteins, such as enhanced green fluorescent protein (eGFP), mCherry, as well as antibiotic resistance markers, such as neomycin phosphotransferase II (resistance to Geneticin), and hygromycin (Fang et al. 2017). However, using these markers would be somewhat difficult. For example, using fluorescent microscopy would be challenging because it requires special microscopy facilities, and is time-consuming. Moreover, removing these selection markers would be difficult to regenerate the strains that express the markers (Fang and Tyler 2016). According to almost all oomycetes CRISPR/Cas9-based transformation protocols, the mutants lose their desired features after three to four generations without using the selection markers (Fang and Tyler 2016).

Guide RNA Design

Even though several useful online tools are now available for gRNA designing, some bioinformatics software is available to help users identify the best possible gRNAs among several. The best proposed online tool to design gRNAs for oomycetes is EupaGDT (http://grna.ctegd.uga.edu/). This website includes many oomycete genomes imported from FungiDB (http://fungidb/org/) and many detailed instructions for designing and choosing candidate gRNAs (Fang et al. 2017). When designing a gRNA, three main factors should be considered, i.e., gRNA activity, gRNA specificity, and gRNA secondary structures (Fang et al. 2017, Wang et al. 2018, Tian et al. 2020). The activity can be checked with the EupaGDT website, in which any gRNA with a total score of ≥ 0.5 would have good activity. Based on the type of Cas9, one can choose different search options on this website; as for the oomycetes, the most commonly used Cas9 is SpCas9 which is derived from *S. pyogenes*. By selecting the PAM sequence in "additional option settings", the server searches for PAM off-targets. Moreover, since the oomycetes do not use RNA polymerase III for gRNA transcription, the 5' end of the gRNA does not need to start with G (Fang et al. 2017). Recently, a plug-in tool in Geneious (www.geneious.com) software has also been developed to design the gRNAs for CRISPR/Cas9 based on the gene of interest. One of the problems with the EupaGDT gRNA designing tool is that it does not have the option for an HDR transformation approach for the oomycetes. Another important note is that this website classifies non-target genomic DNA sequences with less than 3 nt mismatches. If a gRNA contains more than 3 nt mismatches, there would be a considerable problem because large variations have been reported across target sites, cell types, and species regarding the significance of base pairing at each position. Therefore, it is difficult to establish a reliable rule for gRNA design. As a result, it is necessary to utilize different strategies, such as deep sequencing or whole-genome sequencing, to confirm the absence of off-target mutations (Fang et al. 2017).

It is important to double-check the gRNA sequence with FungiDB and NCBI to check gRNA specificity. With the NCBI, a BLASTn searching machine would be useful. The BLAST search is the most critical step since it directly points out the off-target effects of gRNAs (Fang et al. 2017).

The last important characteristic of the gRNAs is their secondary structures which can be checked via the RNA structure (http://rna.urmc.rochester.edu/RNAstrucureWeb/Servers/Predict1/Predict1.html) website. In practice, it would be impossible to find a gRNA without any secondary structures; hence, it would be logical only to select the best ones with no more than three consecutive paired bases. Furthermore, the secondary structures should be checked at different temperatures to see whether the final gRNA would work at room temperature or not; this option is also included in the search engine of the *RNA structure* website. It is wise to design more than one gRNA to manipulate the gene of interest to get the best possible results (Fang et al. 2017, Wang et al. 2018).

In addition, it is important to confirm whether the mutation generated by CRISPR/Cas9 is genuinely responsible for the observed phenotype or not. Hence, it is also necessary to reintroduce the targeted gene back into the mutant to conduct a complementation experiment (Fang et al. 2017).

Plasmid Construction

Fortunately, with several new plasmid designing tools, it is possible to design any plasmid within only a few days. The plasmids should have multiple cloning sites containing blunt cutter enzymes at the location of each selection marker.

There are two main plasmids required for the expression of Cas9 and gRNA. Some studies use an all-in-one plasmid that expresses both Cas9 and gRNA simultaneously. Each 20 nt gRNA sequence should be flanked by two ribozymes, HDV and Hammerhead (HH) ribozymes. The first six nucleotides of HH ribozyme must be complementary to the first nucleotides of the target sequence. A conserved 80 nt gRNA scaffold is also required to form the complex with the Cas9 protein (Fang et al. 2017).

The construction of the plasmid is considered a laborious process, and in the case of ordering them from a company, it would be costly. However, one should double-check the plasmids ordered from another company with the specific primers to evaluate the existence of the desired fragments. If the plasmids are constructed manually, it is important to add *NheI-* and *BsaI* compatible overhangs to the end of the two oligonucleotides (gRNAs) correctly since oligo annealing-based cloning is very sensitive. Even a minor error may result in ligation failure (Fang et al. 2017).

Additionally, in the case of performing a knock-in experiment, it is very complicated to assemble a homologous repair template. This is mainly because four fragments, i.e., linearized plasmid carrier, left homology arm, right homology arm, and the donor DNA should be combined (Fang et al. 2017, Wang et al. 2018). Although the use of the In-Fusion cloning kit (Clontech) has been proposed for this purpose, it was proved to be problematic, and the sequential ligation of each fragment was proven to be more effective (Fang et al. 2017).

Homology-directed Repair Assays

To conduct accurate point mutations using the HDR pathway, providing an HDR template with the modified gRNA recognition site is necessary. It has been proposed that replacing degenerate codons would be the most straightforward approach to generating the required mutations to avoid cutting the HDR donor sequence (Fang et al. 2017).

Transformants' Screening

A fast way to detect and screen the efficiency of designed gRNAs is through the transient expression assays using the pooled transformants. For HDR assays, it would be wise to use a high-fidelity DNA polymerase to avoid mutations during PCR amplification. If the genomic DNA background is high and masks the mutations, it is recommended to design a nested PCR strategy to reduce the background (Fang et al. 2017). On the other hand, screening the mutants with the selection markers, such as antibiotic resistance, should be conducted carefully. For example, after protoplasts isolation, it is recommended to wait for the first regeneration of the protoplasts for about 18 h and then add the antibiotics. This technique applies to almost all of the oomycetes. Growing oomycetes on the regeneration medium before adding antibiotics allows the protoplasts to regenerate properly and adding the antibiotic markers after 18 h would significantly increase the number of positive transformants.

Toxic Nature of Cas9 for the Oomycetes

Several reports state that the Cas9 protein negatively affects oomycetes' growth rate. Hence, it can be concluded that it is important to optimize the concentration of the Cas9 plasmid when trying to generate CRISPR/Cas9 mutants. Previous studies suggest that the amount of pCas9 used for the oomycetes must not exceed 50 μg/mL (Wang et al. 2018). For *Phytophthora* and *Pythium* species, 30 μg/mL of the pCas9 was recommended (Fang et al. 2017, Salmaninezhad et al. 2022).

Single Zoospore Isolation

Single zoospore isolation is the most critical step in purification and obtaining the transformants with the desired mutations in oomycetes. This is mainly due to the coenocytic nature of the oomycete mycelium. In other words, each protoplast and mycelium are multinucleate, harboring nuclei with a diversity of genotypes (Lamour and Kamoun 2009, Fang et al. 2017). Hence, isolation of the single zoospores, which are more than 95% haploid and mononucleate, would assist the researchers in obtaining the homokaryotic mutants (Fang et al. 2017, Wang et al. 2018). However, knowing that a large group of oomycetes, such as half of the *Globisporangium* species or some of the downy-mildew causing *Peronosporales*, do not produce zoospores at all makes them difficult to manipulate using CRISPR/Cas9 system.

Conclusion

This chapter highlights the importance of the CRISPR/Cas system as the most robust gene-editing tool developed in the last two decades and discusses its application in the plant pathogenic oomycetes. It was previously suggested that further investigations of the oomycete transformation should consider two critical factors: consistency and easiness protoplast production in large numbers that are able to regenerate readily, and the development of more efficient vectors with more effective promoters for driving selectable marker gene expression when selecting stable transformants (Fang et al. 2017). Promoters derived from *Bremia lactuca* Regel and another complex promoter for *P. sojae* have been used in several studies (Fang et al. 2017, Wang et al. 2018, Salmaninezhad et al. 2022).

So far, the only applications of the CRISPR/Cas9 technique in oomycetes are limited to gene regulation and gene function. As discussed before, CRISPR/Cas systems are utilized for various purposes. Regarding the application of CRISPR/Cas9 in oomycetes, it would be worth optimizing other variants of Cas9, such as Cas9n (nickase Cas9) and dCas9 (deactivated Cas9) for other applications of this system, such as interference or activation of transcription of the targeted genes.

Furthermore, designing other transformation methods, such as microinjection of the RNP complex directly into the mycelium, would also facilitate the transformation process, especially in oomycetes such as *Saprolegnia* and *Achlya* species, where the size of mycelium would allow the researcher to perform the direct injection of CRISPR/Cas components to the hyphae. Using alternative transformation approaches provides advantages for the researchers since the protoplast isolation and transformation would be time-consuming and could not be applied to all oomycetes. Nevertheless, the increasing number of studies that, step by step, contribute to improving the prospects of the application of the CRISPR/Cas system in oomycete genetic studies is promising.

References

Adli, M. 2018. The CRISPR tool kit for genome editing and beyond. Nat. Commun. 9(1): 1–13.

Ah-Fong, A.M., A.M. Boyd, M.E. Matson and H.S. Judelson. 2021. A Cas12a-based gene editing system for *Phytophthora infestans* reveals monoallelic expression of an elicitor. Mol. Plant Pathol. 22(6): 737–752.

Alfonzo, J. D. 2015. RNAi, the guiding principle and keeping family happy. RNA 21: 555–556.

Aramayo, R. and E.U. Selker. 2013. *Neurospora crassa* a model system for epigenetics research. Cold Spring Harb. Perspect. Biol. 5: a017921.

Arazoe, T., K. Miyoshi, T. Yamato, T. Ogawa, S. Ohsato, T. Arie et al. 2015. Tailor-made CRISPR/Cas system for highly efficient targeted gene replacement in the rice blast fungus. Biotechnol. Bioeng. 112: 2543–2549.

Armas-Tizapantzi, A. and A.M. Montiel-González. 2016. RNAi silencing: a tool for functional genomics research on fungi. Fungal Biol. Rev. 30: 91–100.

Arsenault-Labrecque, G., P. Santhanam, Y. Asselin, B. Cinget, A. Lebreton, C. Labbé et al. 2022. RXLR effector gene Avr3a from *Phytophthora sojae* is recognized by Rps8 in soybean. Mol. Plant Pathol. 23: 693–706.

Asman, A. K., J. Fogelqvist, R. R. Vetukuri and C. Dixelius. 2016. *Phytophthora infestans* Argonaute 1 binds microRNA and small RNAs from effector genes and transposable elements. New Phytol. 211: 993–1007.

Białas, A., E. K. Zess, J. C. De la Concepcion, M. Franceschetti, H. G. Pennington, K. Yoshida et al. 2018. Lessons in effector and NLR biology of plant-microbe systems. Mol. Plant Microbe Interact. 31(1): 34–45.

Barrangou, R., A. Birmingham, S. Wiemann, R. L. Beijersbergen, V. Hornung and A.V.B. Smith. 2015. Advances in CRISPR-Cas9 genome engineering: lessons learned from RNA interference. Nucleic Acids Res. 43(7): 3407–3419.

Baulcombe, D. 2015. VIGS, HIGS and FIGS: small RNA silencing in the interactions of viruses or filamentous organisms with their plant hosts. Curr. Opin. Plant Biol. 26: 141–146.

Boddu, P.C., A.K. Gupta, J.S. Kim, K.M. Neugebauer, T. Waldman and M.M. Pillai. 2021. Generation of scalable cancer models by combining AAV-intron-trap, CRISPR/Cas9, and inducible Cre-recombinase. Commun. Biol. 4(1): 1–9.

Cao, J., M. Qiu, W. Ye, and Y. Wang. 2022. *Phytophthora sojae* Transformation Based on the CRISPR/Cas9 System. Bio-protocol. 12(6): e4352–e4352.

Capriotti, L., E. Baraldi, B. Mezzetti, C. Limera and S. Sabbadini . 2020. Biotechnological approaches: Gene overexpression, gene silencing, and genome editing to control fungal and oomycete diseases in grapevine. Int. J. Mol. Sci. 21(16): 5701–5730.

Chang, W., W. Liu, Y. Liu, F. Zhan, H. Chen, H. Lei et al. 2019. Colorimetric detection of nucleic acid sequences in plant pathogens based on CRISPR/Cas9 triggered signal amplification. Microchim. Acta. 186(4): 1–8.

Chen, X.R., Y. Zhang, H.Y. Li, Z.H. Zhang, G.L. Sheng, Y.P. Li et al. 2019. The RXLR effector PcAvh1 is required for full virulence of *Phytophthora capsici*. Mol. Plant Microbe Interact. 32(8): 986–1000.

Chira, S., D. Gulei, A. Hajitou, A.A. Zimta, P. Cordelier and I. Berindan-Neagoe. 2017. CRISPR/Cas9: transcending the reality of genome editing. Mol. Ther. Nucleic Acids. 7: 211–222.

Christian, M., T. Cermak, E.L. Doyle, C. Schmidt, F. Zhang, A. Hummel et al. 2010. Targeting DNA double-strand breaks with TAL effector nucleases. Genetics. 186(2): 757–761.

Cobb, R.E., Y. Wang and H. Zhao. 2015. High-efficiency multiplex genome editing of *Streptomyces* species using an engineered CRISPR/Cas system. ACS Synth. Biol. 4(6): 723–728.

Erwin, D.C. and O. K. Ribeiro. 1996. *Phytophthora: diseases worldwide* (No. 632.4 E73p). Minnesota, US: APS Press.

Fang, Y. and B.M. Tyler. 2016. Efficient disruption and replacement of an effector gene in the oomycete *Phytophthora sojae* using CRISPR/C as9. Mol. Plant Pathol. 17(1): 127–139.

Fang, Y., L. Cui, B. Gu, F. Arredondo and B.M. Tyler. 2017. Efficient genome editing in the oomycete *Phytophthora sojae* using CRISPR/Cas9. Curr. Protoc. Microbiol. 44(1): 21A–1.

Fang, L., W. Wang, G. Li, L. Zhang, J. Li, D. Gan et al. 2020. CIGAR-seq, a CRISPR/Cas-based method for unbiased screening of novel mRNA modification regulators. Mol. Syst. Biol. 16(11): e10025.

Feng, H., C. Wan, Z. Zhang, H. Chen, Z. Li, H. Jiang et al. 2021. Specific interaction of an RNA-binding protein with the 3′-UTR of its target mRNA is critical to oomycete sexual reproduction. PLoS Pathog. 17(10): e1010001.

Ferrara, M., M. Haidukowski, A.F. Logrieco, J.F. Leslie and G. Mulè. 2019. A CRISPR-Cas9 system for genome editing of *Fusarium proliferatum*. Sci. Rep. 9(1): 1–9.

Fire, A., S. Xu, M.K. Montgomery, S.A. Kostas, S.E. Driver and C.C. Mello. 1998. Potent and specific genetic interference by double-stranded RNA in *Caenorhabditis elegans*. Nature. 391: 806–811.

Foster, A.J., M. Martin-Urdiroz, X.Yan, H.S. Wright, D.M. Soanes and N.J. Talbot. 2018. CRISPR-Cas9 ribonucleoprotein-mediated co-editing and counterselection in the rice blast fungus. Sci. Rep. 8(1): 1–12.

Fulci, V. and G. Macino. 2007. Quelling post-transcriptional gene silencing guided by small RNAs in *Neurospora crassa*. Curr. Opin. Microbiol. 10: 199–203.

Gaj, T., S.J. Sirk, S.L. Shui and J. Liu. 2016. Genome-editing technologies: principles and applications. Cold Spring Harb. Perspect. Biol. 8(12): a023754.

Grenville-Briggs, L.J., V.L. Anderson, J. Fugelstad, A.O. Avrova, J. Bouzenzana, A. Williams et al. 2008. Cellulose synthesis in *Phytophthora infestans* is required for normal appressorium formation and successful infection of potato. Plant Cell. 20(3): 720–738.

Gu, B., G. Shao, W. Gao, J. Miao, Q. Wang, X. Liu et al. 2021. Transcriptional variability associated with CRISPR-Mediated gene replacements at the *Phytophthora sojae Avr1b-1*Locus. Front. Microbiol. 12: 338.

Gumtow, R., D.Wu, J. Uchida and M. Tian. 2018. A *Phytophthora palmivora* extracellular cystatin-like protease inhibitor targets papain to contribute to virulence on papaya. Mol. Plant Microbe Interact. 31(3): 363–373.

Gupta, R.M. and K. Musunuru. 2014. Expanding the genetic editing tool kit: ZFNs, TALENs, and CRISPR-Cas9. J. Clin. Invest. 124(10): 4154–4161.

Hahn, M. and G. Scalliet. 2021. One cut to change them all: CRISPR/Cas, a groundbreaking tool for genome editing in *Botrytis cinerea* and other fungal plant pathogens. Phytopathology. 111(3): 474–477.

Horner, N.R., L.J. Grenville-Briggs and P. Van West. 2012. The oomycete Pythium oligandrum expresses putative effectors during mycoparasitism of *Phytophthora infestans* and is amenable to transformation. Fungal Biol. 116(1): 24–41.

Hsu, P.D., E.S. Lander and F. Zhang. 2014. Development and applications of CRISPR-Cas9 for genome engineering. Cell. 157: 1262–1278.

Iberahim, N.A., F. Trusch and P. Van West. 2018. *Aphanomyces invadans*, the causal agent of epizootic ulcerative syndrome, is a global threat to wild and farmed fish. Fungal Biol. Rev. 32(3): 118–130.

Iberahim, N.A., N. Sood, P.K. Pradhan, J. van den Boom, P. van West and F. Trusch. 2020. The chaperone Lhs1 contributes to the virulence of the fish-pathogenic oomycete *Aphanomyces invadans*. Fungal Biol. 124(12): 1024–1031.

Jiang, F. and J. Doudna. 2017. CRISPR-Cas9 structure and mechanisms. Annu. Rev. Biophys. 46: 505–529.

Kampmann, M. 2018. CRISPRi and CRISPRa screens in mammalian cells for precision biology and medicine. ACS Chem. Biol. 13(2): 406–416.

Khan, F.A., N.S. Pandupuspitasari, H. Chun-Jie, Z. Ao, M. Jamal, A. Zohaib et al. 2016. CRISPR/Cas9 therapeutics: a cure for cancer and other genetic diseases. Oncotarget. 7(32): 52541.

Khurshid, H., S.A. Jan, Z.K. Shinwari, M. Jamal and S.H. Shah. 2017. The era of CRISPR/Cas9-mediated plant genome editing. Curr. Issues Mol. Biol. 26: 47–54.

Kim, S.C., P.M. Skowron and W. Szybalski. 1996. Structural Requirements for FokI-DNA interaction and oligodeoxyribonucleotide-instructed cleavage. J. Mol. Biol. 258(4): 638–649.

Kong, G., L.Wan, Y.Z. Deng, W. Yang, W. Li, L. Jiang et al. 2019. Pectin acetylesterase PAE5 is associated with the virulence of plant pathogenic oomycete *Peronophythora litchii*. Physiol. Mol. Plant Pathol. 106: 16–22.

Krappmann, S. 2016. CRISPR/Cas9 the new kid on the block of fungal molecular biology. Med. Mycol. 55: 16–23.

Kraturisha, C., M. Tospol, V. Rukvidhyasatra and K. Bhavakul. 1995. *Peronophythora litchii* associated with root rot of litchi. In *2. National Plant Protection Conference, Chiang Mai (Thailand), 9–11 Oct 1995.*

Lamour, K. and S. Kamoun. 2009. *Oomycete genetics and genomics: diversity, interactions and research tools.* New Jersey, USA: John Wiley and Sons.

Lamour, K.H., R. Stam, J. Jupe and E. Huitema. 2012. The oomycete broad-host-range pathogen *Phytophthora capsici*. Mol. Plant Pathol. 13(4): 329–337.

Latijnhouwers, M., P.J. de Wit and F. Govers. 2003. Oomycetes and fungi: similar weaponry to attack plants. Trends Microbiol. 11(10): 462–469.

Leesutthiphonchai, W. and H.S. Judelson. 2018. A MADS-box transcription factor regulates a central step in sporulation of the oomycete *Phytophthora infestans*. Mol. Microbiol. 110(4): 562–575.

Leisen, T., F. Bietz, J. Werner, A. Wegner, U. Schaffrath, D. Scheuring et al. 2020. CRISPR/Cas with ribonucleoprotein complexes and transiently selected telomere vectors allows highly efficient marker-free and multiple genome editing in *Botrytis cinerea*. PLoS Pathog. 16(8): e1008326.

Leisen, T., J. Werner, P. Pattar, N. Safari, E. Ymeri, F. Sommer et al. 2022. Multiple knockout mutants reveal a high redundancy of phytotoxic compounds contributing to necrotrophic pathogenesis of *Botrytis cinerea*. PLoS Pathog. 18(3): e1010367.

Li, L., S.S. Chang and Y. Liu. 2010. RNA interference pathways in filamentous fungi. Cell. Mol. Life Sci. 67: 3849–3863.

Liao, B., X. Chen, X. Zhou, Y. Zhou, Y. Shi, X. Ye et al. 2022. Applications of CRISPR/Cas gene-editing technology in yeast and fungi. Arch. Microbiol. 204(1): 1–14.

Lin, D., Z. Xue, J. Miao, Z. Huang and X. Liu. 2020. Activity and resistance assessment of a new OSBP inhibitor, R034-1, in *Phytophthora capsici* and the detection of point mutations in PcORP1 that confer resistance. J. Agric. Food Chem. 68(47): 13651–13660.

Lin, L., W. Ye, J. Wu, M. Xuan, Y. Li, J. Gao et al. 2018. The MADS-box transcription factor PsMAD1 is involved in zoosporogenesis and pathogenesis of *Phytophthora sojae*. Front. Microbiol. 2259.

Mahfouz, M.M., A. Piatek and C.N. Stewart Jr. 2014. Genome engineering via TALENs and CRISPR/Cas9 systems: challenges and perspectives. Plant Biotechnol. J. 12(8): 1006–1014.

Majeed, M., H. Soliman, G. Kumar, M. El-Matbouli and M. Saleh. 2018. Editing the genome of *Aphanomyces invadans* using CRISPR/Cas9. Parasites Vectors. 11(1): 1–10.

Malzahn, A., L. Lowder and Y. Qi. 2017. Plant genome editing with TALEN and CRISPR. Cell Biosci. 7(1): 1–18.

Matsu-ura, T., M. Beak, J. Kwon and C. Hong. 2015. Efficient gene editing in *Neurospora crassa* with CRISPR technology. Fungal Biol. Biotechnol. 2: 4.

Miao, J., Y. Chi, D. Lin, B.M. Tyler and X. Liu. 2018. Mutations in ORP1 conferring oxathiapiprolin resistance confirmed by genome editing using CRISPR/Cas9 in *Phytophthora capsici* and *P. sojae*. Phytopathol. 108(12): 1412–1419.

Miao, J., X. Liu, G. Li, X. Du and X. Liu. 2020. Multiple point mutations in PsORP1 gene conferring different resistance levels to oxathiapiprolin confirmed using CRISPR–Cas9 in *Phytophthora sojae*. Pest Manag. Sci. 76(7): 2434–2440.

Miao, J., C. Li, X. Liu, X. Zhang, G. Li, W. Xu et al. 2021. Activity and resistance-related point mutations in target protein PcORP1 of fluoxapiprolin in *Phytophthora capsici*. J. Agric. Food Chem. 69(13): 3827–3835.

Michielse, C.B., P.J. Hooykaas, C.A. van den Hondel and A.F. Ram. 2005. Agrobacterium-mediated transformation as a tool for functional genomics in fungi. Curr. Genet. 48(1): 1–17.

Miller, J.C., S. Tan, G. Qiao, K.A. Barlow, J.Wang, D. F. Xia et al. 2011. A TALE nuclease architecture for efficient genome editing. Nat. Biotechnol. 29(2): 143–148.

Min, K., Y. Ichikawa, C.A. Woolford and A.P. Mitchell. 2016. *Candida albicans* gene deletion with a transient CRISPR-Cas9 system. mSphere. 1: e00130–16.

Morange, M. 2015. What history tells us XXXVII. CRISPR-Cas: The discovery of an immune system in prokaryotes. J. Biosci. 40: 221–223.

Moon, S.B., D.Y. Kim, J.H. Ko and Y.S. Kim. 2019. Recent advances in the CRISPR genome editing tool set. Exp. Mol. Med. 51(11): 1–11.

Mostowfizadeh-Ghalamfarsa, R. 2016. Pythium *species in Iran*. Shiraz, Iran: Shiraz University Press.

Mostowfizadeh-Ghalamfarsa, R. and F. Salmaninezhad. 2020. Taxonomic challenges in the genus *Pythium*. In: M. Rai, K.A. Abd-Elsalam, A.P. Ingle. (Eds.) *Pythium: diagnosis, diseases and management* (pp. 179–199). New York, USA: CRC Press.

Mostowfizadeh-Ghalamfarsa, R., F. Salmaninezhad and G.I. Hardy. 2020. Species complexes within the genus *Phytophthora, the plant destroyer*. In: Agricultural Research Update. Nova Publications. USA.

Ochola, S., J. Huang, H. Ali, H. Shu, D. Shen, M. Qiu et al. 2020. Editing of an effector gene promoter sequence impacts plant-*Phytophthora* interaction. J. Integr. Plant Biol. 62(3): 378–392.

Paul, S.K., T. Akter and T. Islam. 2021. Gene editing in filamentous fungi and oomycetes using CRISPR-Cas technology. In Abd-Elsalam, K. A., and Lim (Eds.) *CRISPR and RNAi Systems: Nanobiotechnology approaches to plant breeding and protection* (pp. 723–753). The Netherlands: Elsevier.

Pettongkhao, S., N. Navet, S. Schornack, M. Tian and N. Churngchow. 2020. A secreted protein of 15 kDa plays an important role in *Phytophthora palmivora* development and pathogenicity. Sci. Rep. 10(1): 1–15.

Pirrello, C., G. Magon, F. Palumbo, S. Farinati, M. Lucchin, G. Barcaccia et al. 2023. Past, present, and future of genetic strategies to control tolerance to the main fungal and oomycete pathogens of grapevine. J. X. B. 74(5): 1309–1330.

Pohl, C., J.A. Kiel, A.J. Driessen, R.A. Bovenberg, and Y. Nygard. 2016. CRISPR/Cas9 based genome editing of *Penicillium chrysogynum*. ACS Synth. Biol. 5: 754–764.

Pyne, M.E., M. Moo-Young, D.A. Chung, and C.P. Chou. 2015. Coupling the CRISPR/Cas9 system with lambda red recombineering enables simplified chromosomal gene replacement in *Escherichia coli*. Appl. Environ. Microbiol. 81(15): 5103–5114.

Salmaninezhad, F., R. Mostowfizadeh-Ghalamfarsa and M. Thines. 2022. Transformation of *Pythium oligandrum* using CRISPR/Cas9. Online conference of the DPG working groups „Mycology" and „Host-Parasite Relationships". March 2022. Frankfurt, Germany: 13.

Saraiva, M., I. De Bruijn, L. Grenville-Briggs, D. McLaggan, A. Willems, V. Bulone et al. 2014. Functional characterization of a tyrosinase gene from the oomycete *Saprolegnia parasitica* by RNAi silencing. Fungal Biol. 118(7): 621–629.

Sattar, M.N., S. Iftikhar, I. El-Masri, N.J. Aldine, Z. El-Sebaaly and Y. N. Sassine. 2021. Breeding of Agaricus bisporus: strains, spawns, and impact on yield. In *Mushrooms: Agaricus bisporus* (pp. 190–239). Wallingford UK: CABI.

Schmidt, F.R. 2005. About the nature of RNA interference. Appl. Microbiol. Biotechnol. 67: 429–435.

Schroeder, L.A., S. Khanal and M.H. Perlin. 2020. CRISPR-Cas9 editing of nitrate transporter gene, um03849, in *Ustilago maydis*. Undergraduate Arts and Research Showcase. 6.

Schuster, M. and R. Kahmann. 2019. CRISPR-Cas9 genome editing approaches in filamentous fungi and oomycetes. Fungal Genet. Biol. 130: 43–53.

Schuster, M., C. Trippel, P. Happel, D. Lanver, S. Reißmann and R. Kahmann. 2018. Single and multiplexed gene editing in *Ustilago maydis* using CRISPR/Cas9. Bio-protocol. 8(14).

Shi, T.Q., G.N. Liu, R.Y. Ji, K. Shi, P. Song, L. J. Ren et al. 2017. CRISPR/Cas9-based genome editing for the filamentous fungi: the state of art. Appl. Microbiol. Biochemist. 101: 7435–7443.

Sun, L., B.M. Lutz and Y.X. Tao. 2016. The CRISPR/Cas9 system for gene editing and its potential application in pain research. Trans. Perioper. Pain Med. 1: 22–33.

Templeton, M.D., E.H. Rikkerink and R.E. Beever. 1994. Small, cysteine-rich proteins and recognition in fungal-plant interactions. MPMI-Mol. Plant Microbe Interact. 7(3): 320–325.

Tian, M., N. Navet and D. Wu. 2020. CRISPR-Cas9-mediated gene editing of the plant pathogenic oomycete *Phytophthora palmivora*. In Islam, M.T., Bhowmik, P.K., and Molla, A.K. (Eds.) *CRISPR-Cas Methods* (pp. 87–98). New York, NY: Humana.

Tong, Y., P. Charusanti, L. Zhang, T. Weber and S.Y. Lee. 2015. CRISPR-Cas9 based engineering of actinomycetal genomes. ACS Synth. Biol. 4(9): 1020–1029.

van den Hoogen, J. and F. Govers. 2018. Attempts to implement CRISPR/Cas9 for genome editing in the oomycete *Phytophthora infestans*. bioRxiv. 274829.

Verma, A.K., D. Chettri and A.K. Verma. 2022. Potential of CRISPR/Cas9-based genome editing in the fields of industrial biotechnology: Strategies, challenges, and applications. In Verma, P. (Ed.) *Industrial Microbiology and Biotechnology* (pp. 667–690). Singapore, Singapore: Springer.

Vijn, I. and F. Govers. 2003. *Agrobacterium tumefaciens* mediated transformation of the oomycete plant pathogen *Phytophthora infestans*. Mol. Plant Pathol. 4(6): 459–467.

Vyas, V.K., M.I. Barrasa and G.R. Fink. 2015. A *Candida albicans* CRISPR systerm permits genetic engineering of essential genes and gene families. Sci. Adv. 1: e1500248.

Wang, Y., Z.T. Zhang, S.O. Seo, K. Choi, T. Lu, Y. S. Jin et al. 2015 Marker less chromosomal gene deletion in *Clostridium beijerinckii* using CRISPR/Cas9 system. J. Biotechnol. 200: 1–5 85.

Wang, S., H. Chen, X. Tang, H. Zhang, W. Chen and Y.Q. Chen. 2017. Molecular tools for gene manipulation in filamentous fungi. Appl. Microbiol. Biotechnol. 101(22): 8063–8075.

Wang, W., Z. Xue, J. Miao, M. Cai, C. Zhang, T. Li et al. 2019. PcMuORP1, an oxathiapiprolin-resistance gene, functions as a novel selection marker for *Phytophthora* transformation and CRISPR/Cas9 mediated genome editing. Front. Microbiol. 2402.

Wang, Z., B.M. Tyler and X. Liu. 2018. Protocol of *Phytophthora capsici* transformation using the CRISPR-Cas9 system. In Ma W., and Wolpert, T. (Eds.) *Plant pathogenic fungi and oomycetes: Methods and protocols* (pp. 265–274). New York, USA: Humana Press.

Wege, S.M., K. Gejer, F. Becker, M. Bölker, J. Freitag and B. Sandrock. 2021. Versatile crispr/cas9 systems for genome editing in *Ustilago maydis*. J. Fungi. 7(2): 149.

Wenderoth, M., C. Pinecker, B. Vob and R. Fischer. 2017. Establishment of CRISPR/Cas9 in *Alternaria alternata*. Fungal Genet. Biol., 101: 55–60.

Weiland, J. J. 2003. Transformation of *Pythium aphanidermatum* to geneticin resistance. Curr. Genet. 42(6): 344–352.

Whisson, S.C., A.O. Avrova, P. Van West and J.T. Jones. 2005. A method for double-stranded RNA-mediated transient gene silencing in *Phytophthora infestans*. Mol. Plant Pathol. 6(2): 153–163.

Xu, T., M.L. Kempher, X. Tao, A. Zhou and J. Zhou. 2021. CRISPR. In Green, L.H., and Goldman, E. (Eds.) *Practical handbook of microbiology* (pp. 147–158). New York, USA: CRC Press.

Yamato, T., A. Handa, T. Arazoe, M. Kuroki, A. Nozaka, T. Kamakura et al. 2019. Single crossover-mediated targeted nucleotide substitution and knock-in strategies with CRISPR/Cas9 system in the rice blast fungus. Sci. Rep. 9(1): 1–8.

Yang X, Zhao W, Hua C, Zheng X, and Wang Y. 2013. Chemotaxis and oospore formation in *Phytophthora sojae* are controlled by G-protein-coupled receptors with a phosphatidylinositol phosphate kinase domain. Mol. Microbiol. 88(2):382–94.

Zhang, F., Y. Wen and X. Guo. 2014. CRISPR/Cas9 for genome editing: progress, implications and challenges. Hum. Mol. Genet. 23: 40–46.

Zhang, Y., J.F. Hu, H. Wang, J. Cui, S. Gao, A.R. Hoffman et al. 2017. CRISPR Cas9-guided chromatin immunoprecipitation identifies miR483 as an epigenetic modulator of IGF2 imprinting in tumors. Oncotarget. 8(21): 34177.

Zhang, Z., J. Guo, Y. Zhao and J. Chen. 2019. Identification and characterization of maize ACD6-like gene reveal ZmACD6 as the maize orthologue conferring resistance to *Ustilago maydis*. Plant Signal. Behav. 14(10): e1651604.

Zhang, Z.H., J.H. Jin, G.L. Sheng, Y.P. Xing, W. Liu, X. Zhou et al. 2021. A small cysteine-rich phytotoxic protein of *Phytophthora capsici* functions as both plant defense elicitor and virulence factor. Mol. Plant Microbe Interact. 34(8): 891–903.

Zhao, W., X. Yang, S. Dong, Y. Sheng, Y. Wang and X. Zheng. 2011. Transient silencing mediated by in vitro synthesized double-stranded RNA indicates that PsCdc14 is required for sporangial development in a soybean root rot pathogen. *Sci. China Life Sci.* 54(12): 1143–1150.

Ziv, C. and O. Yarden. 2010. Gene silencing for functional analysis: assessing RNAi as a tool for manipulation of gene expression. In Sharon, A. (Ed.), Molecular and cell biology methods for fungi, 1st ed. (pp. 77–100). New Jersey, USA: Humana Press.

Zuo, W., J.R. Depotter and G. Doehlemann. 2020. Cas9HF1 enhanced specificity in *Ustilago maydis*. Fungal Biol. 124(3–4): 228–234.

CHAPTER 19

Genetics of Oomycetes
A Gate for Systematic and Ecological Studies

Hossein Masigol,[1] *Reza Mostowfizadeh-Ghalamfarsa*[2,*]
and *Giles E. St. J. Hardy*[3]

Introduction

Oomycetes, Jack of all Trades

Oomycetes constitute one of the most ecologically diverse phylogenetic lineages within the group of osmotrophic fungus-like stramenopiles (Adl et al. 2012). Supported by their filamentous structure and some eco-physiological similarities, oomycetes were considered fungi. However, oomycetes' phylogenetic detachment from animals and their distinctive characteristics (e.g., swimming stages, diploid/polyploid vegetative stages, and cell walls made of cellulose) separated them from the kingdom Fungi (Beakes et al. 2012). As the heading of this section has tried to articulate idiomatically, oomycetes are indeed versatile microorganisms with the ability to survive in a broad spectrum of habitats and niches, adopt dissimilar lifestyles, and communicate ecologically with other living organisms.

Starting with habitats and niches, we see that oomycetes populate aquatic and terrestrial environments no matter how extreme they might get. For example, in addition to more common habitats, the presence of oomycete communities has been confirmed in extreme environments such as sediment at a deep-sea methane cold seep (Takishita et al. 2007), Arctic and Antarctic (Tojo et al. 2012), and even aerosol particles in the atmosphere (Lang-Yona et al. 2018).

The high ecological prevalence of oomycetes is mainly due to their ability to adopt a diverse range of lifestyles. The genus *Phytophthora* is probably one of the most well-known and diverse plant pathogenic taxa that have had devastative impacts on a wide range of crops and natural ecosystems (Yoshida et al. 2013, Scanu et al. 2021). The genera *Plasmopara* and *Albugo* have also been the causal agents of downy grapevine mildew and white rust in the *Brassicaceae* family, respectively. Oomycetes are notorious aquatic animal pathogens too. For instance, the genus *Saprolegnia* is responsible for infections in fish in the aquaculture and fish industry. As an endemic microorganism to all freshwater habitats, *Saprolegnia parasitica* Coker is responsible for declining natural populations of salmonids and other freshwater fish, causing ecological and economic losses (van West 2006). Another notorious oomycete animal pathogen is *Aphanomyces* spp. The genus includes two wide-spread pathogenic species called *A. astaci* Schikora and *A. invadans* Willoughby,

[1] Plankton and Microbial Ecology, Leibniz Institute for Freshwater Ecology and Inland Fisheries (IGB), 16775 Neuglobsow, Germany.
[2] Department of Plant Protection, School of Agriculture, Shiraz University, Shiraz, Iran 7144167186.
[3] Phytophthora Science and Management, Harry Butler Institute, Murdoch University, Murdoch, WA 6150, Australia.
* Corresponding author: rmostofi@shirazu.ac.ir

R.J. Roberts and Chinabut causing crayfish plague and epizootic granulomatous aphanomycosis (EGA), respectively. *Aphanomyces astaci*, listed among the 100 of the World's Worst Invasive Alien Species, has been transferred from the US to European countries and put endemic populations of native European crayfish species on the verge of extinction (Svoboda et al. 2017). In some extreme cases, some oomycete taxa like *Pythium insidiosum* De Cock, L. Mend., A.A. Padhye, Ajello and Kaufman have caused infection in mammals such as horses and dogs, and even humans through small wounds via contact with water containing the propagules (Gaastra et al. 2010).

Oomycetes' Genetics, Bridging the Gap between Systematic and Ecological Studies

Ecologists often challenge taxonomists' work as if trying to characterize and then categorize taxa as redundant, and it does not contribute so much to revealing their functions in different terrestrial and aquatic ecosystems. However, such a notion has shown to be wrong as, in many cases, the incorrect understanding of one group of organisms' taxonomy and diversity has the potential to impact all upstream ecological studies negatively. The importance of well-supported systematics is intensified for oomycetes, as many are destructive pathogens. The weaker their taxonomic and phylogenetic status is, the more confusing other aspects of their threats become. Indeed, appreciating the innate relationships between oomycetes has its own merits, and understanding the great diversification of taxa in many different habitats requires such an appreciation. Therefore, it is safe to say that genetic data in oomycetes is an adequate tool for extracting systematics information based upon a natural system derived from relationships of taxa.

In addition to application in the systematics of oomycetes, genetic data have clarified knowledge about the origin of the taxa, their distribution patterns, and reasons for adopting a specific lifestyle. For example, genetics has helped elevate knowledge about freshwater pathogens by tracking them down at the beginning of their emergence. It also can detect changes the pathogenic taxa have been through, thanks to genetic markers and microsatellite genotyping. The missing pieces of ecological studies also depend greatly on genetic markers for metabarcoding approaches. The necessity of metabarcoding lies in the fact that culture-dependent methods very negatively disturb the actual diversity of oomycetes in a given environment which could significantly impact our assumed ecological implications of oomycetes' diversity. For example, the order *Saprolegniales* is considered the most dominant oomycetes' lineage in freshwater ecosystems. However, the same is said about *Peronosporales* and *Pythiales* and the terrestrial environment. However, all these findings have been heavily culture-dependent, causing a significant fraction of oomycetes' real diversity to be missed, since many are obligate biotrophs. This is where metabarcoding approaches can help us study not just single isolates but oomycetes' communities.

Systematics

Classic Identification of Oomycetes

Oomycetes or phylum *Oomycota* have been assigned to the lineage of biflagellate heterokont eukaryotic organisms (Stramenopiles) within the *Chromalveolata* superkingdom (Tsui et al. 2009). Similar to fungi, morphological, morphometric and physiological-based identification of oomycetes have lasted since the last decade of the 19th century. The observation of too many similarities in sexual and/or asexual organs was probably the main reason for considering defining oomycetes as one of the fungal groups. The late 19th and early 20th century publications reveal a high level of sophistication in depicting and recording the finest morphological details of oomycetes. The genesis and structure of sporangia (asexual organs), and oogonia and antheridia (female and male sexual organs) were essential features to separate different taxa. Other distinctive features were: the way zoospores develop and release from sporangia, the radial growth rate at different temperatures, and possible pathogenicity toward some general and specific plant and animal hosts.

Reviewing the potential of classical identification in separating different orders of oomycetes reveals that it has been relatively practical up to the genera level. It is still possible to isolate and place oomycete stains in a specific order and sometimes genus according to their purely morpho-physiological features. For example, the orders *Saprolegniales* and *Peronosporales*, two of the most well-studied lineages of oomycetes, can be conclusively recognized by observing two or more features such as the shape of sporangia, zoospore formation, the size of mycelia, the number of oospores in oospheres, the origin of oogonia and antheridia, the possible substrates, and radial growth. However, the practicality and reliability of morpho-physiological features start falling apart from the genus level downward. Although the latest available dichotomous keys separate genera and species of oomycetes solely based on morpho-physiological features, scholars have had good reasons to doubt such framing and called for substantial changes, as has been the case for many fungal taxa as well.

There are three main challenges as we try to assign oomycete isolates to a genus or species. Firstly, there is too much variability among isolates. For instance, the length of a sporangium or gemma in one isolate could be at least 10 times shorter or longer than in another. The same is true with the number of oospores inside an oosphere. So, they cannot be solely investigated as reliable characteristics. As an illustration, in *Achlya androgyna* (W. Archer) T.W. Johnson and R.L. Seym., sporangial and oogonial length variations are 77–988 and 105–200 μm, respectively. Secondly, sexual organs, which happen to be very important in delineating genera and species, are often absent and cannot be observed under lab conditions. Therefore, many isolates must be defined as sterile even though they might produce sexual organs in nature. Thirdly, morphologic descriptions provided by species identification keys are sometimes either confusing or uncertain. In fact, they often do not exclude anything. This is an example from the *Saprolegnia* key (Johnson et al. 2002) to species where five different structures are assigned at the same time to describe sporangia of *Saprolegnia subeccentrica* (M.W. Dick) Milko and its renewal methods as follows: "...sporangia abundant in young and old cultures; cylindrical, clavate, or long-fusiform, often curved, or slightly irregular; proliferating internally; or renewed sympodially or in a basipetalous fashion...". Or, here, the key is describing morphology of gemmae in *Achlya apiculata* de Bary as follows: "...Gemmae sparse or abundant; fusiform, cylindrical, clavate, globose, infrequently irregular or branched; terminal or intercalary, single or catenulate...". In addition, superfluous use of relative adverbs such as "predominantly", "rarely", "extremely", "generally", "exclusively", as examples makes distinction more skeptical as it is not clear what proportion they are referring to.

Eventually, with the invention of PCR in the 1980s, genetics opened its way to oomycetes' taxonomy and caused many improvements in many taxa. Classical taxonomy of oomycetes was accompanied and supported by DNA-based approaches resulting in a better resolution in all taxonomic ranks.

Barcoding of Oomycetes

The application of molecular tools for identifying oomycetes dates back to the second half of the 20th century. Earlier, with all the biochemical and morphological similarities between fungi and oomycetes, some scholars speculated that, evolutionarily speaking, oomycetes should be separated from fungi. They cited amoebas, heterotrophic flagellates, algal groups, and even chytrid fungi. Baldauf et al. (2000) were some of the first researchers who used DNA sequences (the small subunit ribosomal RNA =SSU rRNA) to separate oomycetes from fungal lineages and establish the evolutionary relationship between oomycetes and other close eukaryotic organisms. They considered oomycetes as heterokonts within the chromalveolate "superkingdom," which was then confirmed by Cavalier-Smith and Chao (2006) and Tsui et al. (2009) using the same DNA regions. Therefore, the small subunit ribosomal RNA gene (known as 18S rDNA) became one of the most practical barcode markers to provide evidence for separating oomycetes from fungi. They were placed in the chromalveolate lineage developed from a common biflagellate (mastigonate ancestor) (Beakes et al. 2012). Additionally, regardless of its limitations in the lower taxonomic levels, the 18S rRNA gene assisted scholars in having a more organized picture of the diversity of oomycetes.

According to the 18S rDNA-based phylogeny, oomycetes were separated into two mega-groups, namely "Basal oomycetes" and "Crown oomycetes". The Basal oomycetes, mainly marine holocarpic species, form the deepest early-diverging clades like *Haptoglossales/ Eurychasmales, Haliphthorales/ Olpidiopsidales*, and *Atkinsiellales*. Species of these clades are predominantly marine obligate parasites of seaweeds, crustaceans, and nematodes. On the other hand, Crown oomycetes constituted more recent evolutionary clades. In contrast to Basal oomycetes, Crown oomycetes were more diverse in lifestyle, preferred habitats, and speciation. There were at least six different orders, namely *Albuginales, Leptomitales, Rhipidiales, Saprolegniales, Peronosporales*, and *Pythiales*. The observed phylogenetic diversification of these orders was more or less matched with their ecological roles confirming the practicality of the 18S rRNA gene in separating high taxonomic ranks of oomycetes. Some clades got separated mainly based on their preferred habitats. A good example is the order *Saprolegniales*, known as water molds, which live primarily in freshwater bodies and infect aquatic animals. Another ecological lineage is *Peronosporales*, known as downy mildews, which are considered mainly obligate biotrophic pathogens of plants. In contrast to fungi, in which the SSU rRNA gene was not that much help due to its very low variability among fungal lineages, it led to the discovery of evolutionary patterns in oomycetes (Lara and Belbahri 2011). The study illustrated that oomycetes, by their nature, are more animal pathogens as plant pathogenesis occurred only rarely in oomycete evolution. Also, they revealed that oomycetes distinguish themselves from other eukaryotic groups as they have experienced several transitions from soil/freshwater to the marine environment (and vice versa). With a better resolution regarding the evolutionary relationships between different lineages of oomycetes and revealing their ecological diversification, other genetic barcodes started to be used for lower taxonomic levels, too, as we will explore in the following sections.

Phylogenetic Studies and Their Taxonomic Applications

Following the success of the 18S rRNA gene in delineating oomycete-related taxa, other gene markers were gradually implemented to delineate genera and species of oomycetes. Logically, associated internal spacer region (ITS) sequences were one of the best options as they help resolve lower taxonomic levels, as confirmed in many fungal species. With more than 1200 citations until 2022, the study by Cook et al. (2000) is considered one of the most influential turning points in the study of oomycetes. For the first time, this study resolved the genus *Phytophthora* de Bary and related oomycetes based on the ITS sequences of genomic rDNA. Additionally, the study was a breakthrough because it tried to fit key morphological and physiological properties with the novel categorization achieved from the analyses of ITS sequences (see the next section for more information). Following the application of ITS sequences for inferring phylogenetic relationships between different genera and species of oomycetes, the boom phase of introducing new taxa began. Consequently, the number of new species multiplied during a short period (see the next section for more information).

Despite considering ITS sequences as de facto DNA barcodes for identification and taxonomic positioning of *Phytophthora* and other genera such as *Pythium* and *Plasmopara,* there were limitations. These included large amounts of insertion and deletions, the presence of indels due to differences in alleles or among the multiple copies of the ITS (Kageyama et al. 2007), and highly similar sequences for different species, which made scholars search for novel barcodes. Such concerns were reflected in another groundbreaking study by Robideau et al. (2011) that attempted to improve DNA barcoding of oomycetes by comparing ITS with other available markers such as cytochrome c oxidase subunit I (*cox*1, COI) or the D1–D3 region of the nuclear large subunit (LSU) rDNA. This study happened after a series of publications starting in 2000 that attempted to explore the diversity of some oomycete genera using both mitochondrial and nuclear loci, as we will explore in the next section. Supported by such publications, Robideau et al. (2011) concluded that *cox*1 and ITS sequencing could be used as complementary mitochondrial and nuclear DNA barcodes for the systematics of oomycetes. The study also discouraged using LSU sequences as they are highly conserved and did not vary between some closely related species that were distinctly separated with *cox*1 and ITS.

The next milestone in searching for suitable DNA barcodes was established by Choi et al. (2017), who compared *cox*1 and *cox*2 loci efficiency as a universal barcode of oomycetes. They highlighted some disadvantages of ITS rDNA and *cox*1 and proposed *cox*2 as an alternative for some taxa. Regarding ITS rDNA, they argued that ITS-based studies are categorically lacking in many oomycete taxa, especially those with economic importance such as *Bremia* and *Plasmopara*, due to repetitive insertions in their sequences (Thines 2007). They also pointed out the insufficiency of the ITS region for phylogenetic distinction in closely related species in some genera such as *Phytophthora* and *Peronospora* (Cook et al. 2000, Volgmayer et al. 2014). The study revised the conclusion obtained from Robideau et al. (2011) by introducing *cox*2, instead of *cox*1, together with ITS rDNA as universal oomycete DNA barcodes. The superiority of *cox*2 over *cox*1 has confirmed its ability to amplify all oomycete genera tested in the study. Higher PCR efficiency for historical herbarium specimens, availability of sequence data for several historical type specimens, more successful species identification, and higher interspecific and lower intraspecific divergences were some of the advantages of *cox*2.

The abovementioned investigations tried to create a framework in which all oomycete taxa can be studied using similar molecular toolboxes. Nevertheless, reviewing the current status of systematics shows that oomycete orders have stepped into somewhat different directions as scholars have faced specific obstacles. For instance, supported by the availability of various gene markers, multigene phylogenetic investigations of some taxa have reached their mature status. In contrast, the opposite has happened in other taxa as delineating genera and species is not satisfactory yet (Masigol et al. 2018, 2020). In the following section, we will discuss which gene markers have been employed in different orders of oomycetes and point to shortcomings associated with each of them. In particular, the orders *Peronosporales*, *Pythiales*, and *Saprolegniales* containing well-studied genera will be central as there are lessons to be learned for other oomycete orders regardless of all deficiencies.

Terrestrial Orders

Peronosporales

The order *Peronosporales* is one of the good examples of showing gradual improvements in oomycetes' systematics. Before the application of gene markers, evolutionary relationships between *Peronosporales* and closely related orders were unclear. However, Riethmüller et al. (2002) used nuclear large subunit rDNA sequences containing the D1 and D2 regions. They separated the order from the *Pythiales*, *Saprolegniales*, *Leptomitales*, and *Rhipidiales*, with the genus *Albugo* as the most basal lineage. The study also confirmed the divergence of the genus *Phytophthora* from the order *Pythiales* previously shown by Riethmüller et al. (2000), Cooke et al. (2000) and Petersen and Rosendahl (2000). A review of the rapid growth of *Phytophthora* species will reveal the success of applying various gene markers in *Peronosporales*' systematics.

As we said earlier, Cooke et al. (2000) used ITS sequences for inferring the molecular phylogeny of oomycetes, *Phytophthora* in particular. The study introduced the genus *Phytophthora* as a paraphyletic clade with at least 47 species (at the time) that are phylogenetically grouped into eight clades. These eight clades were more or less verified by key morphological features of *Phytophthora*, including sporangium type, reproductive behavior, the temperature in which the oomycete grows in its optimized status, and pathogenicity, even though some discrepancies were observed. Despite observed inconsistencies, trying to infer the taxonomy of *Phytophthora* with the help of both morphological features and gene markers was a breakthrough. Nevertheless, the more isolates associated with *Phytophthora* were isolated, the more contradictions were formed between morphological characterization and constructed phylogenetic relationships. Accordingly, it implied that more gene markers must be used to fit new isolates into one of those ITS-based clades. Following Cooke et al. (2000)'s study, Martin and Tooley (2003) inferred the phylogenetic relationships between *Phytophthora* species using sequence information of mitochondrially encoded cytochrome c oxidase subunit I and II genes. It was concluded that the groupings of *Phytophthora* species by *cox*1 and *cox*2 generally agreed with what was observed in the ITS tree, with several notable exceptions. This study was the first in which two gene markers were simultaneously used to understand the phylogenetic relationships of this genus. In Kroon et al. (2004), *Phytophthora* species were phylogenetically studied using a set of both mitochondrial (*cox*1 and NADH dehydrogenase subunit I) and nuclear (translation elongation factor 1α and β–tubulin) gene markers. The study was considered another breakthrough at the time for several reasons. Firstly, it conducted a phylogenetic analysis of *Phytophthora* without the ITS regions. Secondly, supported by the obtained multigene phylogeny, the mating system in *Phytophthora* species (as one of the important features in their systematics) was inferred to be a homoplasious trait, with at least eight independent transitions from homothallism to heterothallism. Thirdly, two new clades were proposed, and the number of *Phytophthora* clades reached 10, which were then supported by subsequent publications. Kroon et al. (2004)'s study was then followed by Blair et al. (2008), in which 234 isolates from 82 species of *Phytophthora* were phylogenetically analyzed using a set of seven nuclear gene markers, including 28S rDNA, 60S ribosomal protein L10, β-tubulin, elongation factor 1α, enolase, heat shock protein 90, and TigA gene fusion protein. The results confirmed the 10-clade phylogeny of *Phytophthora* species obtained from ITS and *cox*1 and *cox*2 regions. Yang et al. (2017) used a similar set of gene markers in another comprehensive multigene marker-based study on the phylogeny of 142 described and 43 provisionally named *Phytophthora* species, leading to new hypotheses regarding the evolutionary history of sporangial papillation of the species.

Another success story in the application of several gene markers is the development of two web-based, searchable, and user-friendly databases for rapid identification of *Phytophthora* species known as Phytophthora-ID.org (Grünwald et al. 2011), which rely on ITS or *cox*1 and *cox*2 regions and contain the relevant information from 700 *Phytophthora* isolates. The databases currently distinguish themselves from other sources such as GenBank as only valid ITS and *cox* (1 and 2) sequences associated with published *Phytophthora* species are included. These are fed by regular updates as soon as new combinations are published in the literature. Such efforts paved the way for subsequent publications, mainly starting in 2010. Although the introduction of novel *Phytophthora* species was still on the agenda, other movements became dominant. For example, thanks to the expansion of multigene phylogeny and the availability of sequences for many taxa, the phylogeny of many cryptic or complex species were re-evaluated without introducing new combinations. Taxonomic re-evaluation of *P. cinnamomi* Rands and *P. cinnamomi* var. *parvispora* Kröber and Marwitz (Scanu et al. 2014) (analysing ITS, β-tubulin, *cox*1 and *cox*2), and *P. cryptogea* Pethybr. and Laff. species complex (Safaiefarahani et al. 2015) (analysing ITS, β-tubulin, heat shock protein, *cox*1, and NADH dehydrogenase subunit I) are some of the examples of improving delineation of *Phytophthora* species using several gene markers. Therefore, applying several gene markers in improving the systematics of *Phytophthora* species has successfully led to a mature and progressive picture of their taxonomy.

Pythiales

The systematics of *Peronosporales* is deeply connected to another major order in oomycetes, namely *Pythiales*. Implementation of various gene markers, including SSU and LSU rDNA

(Dick et al. 1999, Leclerc et al. 2000) and *cox*2 (Hudspeth 2000), strongly supported the phylogenetic position of both orders in the class *Peronosporomycetes,* subclass *Peronosporomycetidae*. Hudspeth et al. (2003) concluded that phylogenetic analyses of *cox*2 cannot infer separate monophyletic orders for *Pythiales* and *Peronosporales*, even though the lifestyle associated with their members is different. *Peronosporales* are mainly obligate biotrophs responsible for downy mildew, while *Pythiales* possess facultative saprotrophy and hemibiotrophy. Like *Phytophthora*, the systematics of the genus *Pythium,* one of the most well-known genera in *Pythiales,* has embraced multigene phylogenetic-based taxonomy, leading to a rapid expansion of known species as follows in the next paragraph.

The credit for applying gene markers in the systematics of *Pythium* goes to Levesque and Cock (2004). They re-introduced 116 species and varieties of *Pythium* using phenetic analysis of ITS rDNA marker. *Pythium* species were divided into two major clades with 11 smaller ones, often correlated with host-type and morphology of sporangia (the clades were named A-K). After the establishment of ITS marker-based taxonomy, as new combinations were introduced, other regions such as β-tubulin and *cox*2 were used to improve the discriminative capacity of phylogenetic investigations (Belbahri et al. 2005, 2006). The application of several gene markers not only resolved the boundaries within *Pythium* species but clarified their phylogenetic relationships with *Phytophthora* species as well. This is where Villa et al. (2006)'s study postulated the intriguing possibility of considering several *Pythium* species (*Pythium helicoides* Drechsler, *P. ostracodes* Drechsler, *P. oedochilum* Drechsler, and *P. vexans* de Bary) as elusive intermediate species in the *Pythium*-to-*Phytophthora* evolutionary line. This notion can be reflected in the introduction of the novel genus named *Phytopythium* as an intermediate between the genera *Phytophthora* and *Pythium* (de Cock et al. 2015).

Two more recent good examples of giving resolution to the *Pythium* species concept are Bala et al. (2010) and Bahramisharif et al. (2013)'s studies which reported several new species of *Pythium* using three (ITS, LSU, *cox*1) and four (ITS, β-tubulin, *cox*1 and *cox*2) gene markers. Hyde et al. (2014) followed these studies and confirmed the 11-clade phylogeny of *Pythium* species. Also, they recommended 18S and 28S nuclear rRNA genes for gene-generic level phylogenies within *Pythium sensu lato*, ITS, and *cox*2 for sub-generic, inter- and intra-specific level phylogenies, and finally, ITS and *cox*1 for non-phylogenetic species identification. In addition to new species, *Pythium* is believed to be an aggregation of several lineages and needs to be phylogenetically revisited in the species and at the genus level. In one major study, Uzuhashi et al. (2010) proposed that the genus *Pythium sensu lato* must be segregated into five genera, including *Pythium sensu stricto, Ovatisporangium, Globisporangium, Elongbisporangium,* and *Pilasporangium* based on both phylogeny (LSU and *cox*2) and morphology. De Cock et al. (2015) re-evaluated the genus *Ovatisporangium* (using SSU, LSU rDNA and *cox*1 markers) and confirmed its position as a separate phylogenetic entity they named *Phytopythium*. Considering ever-increasing multigene phylogenetic efforts to unravel cryptic species and delimit the boundaries of each taxon, we are likely to witness more new combinations in the study of *Pythiales* in the future.

The Freshwater Order

Saprolegniales

The order *Saprolegniales* is mainly associated with freshwater ecosystems (Masigol et al. 2019, 2021a, b); a few terrestrial taxa such as plant pathogenic *Aphanomyces* species have also been recorded. *Saprolegniales'* systematics has been influenced by the implementation of gene markers but not as practical as *Peronosporales* and *Pythiales*, as stated above. Studying systematics of *Saprolegniales* has been solely dependent on morpho-physiological features for nearly 150 years. Although introducing the ITS and LSU gene markers was pivotal in delineating the families and genera, it failed to separate lower ranks. According to the LSU rDNA gene marker and morphology of oospores, the genera which were morphologically associated with the family *Saprolegniaceae sensu lato* were moved into the order *Saprolegniales* including *Saprolegniaceae sensu stricto*, *Verrucalvaceae*, and *Achlyaceae* (Beakes et al. 2014). The separation of these three families was also confirmed when Rocha et al. (2018) used both LSU and ITS rDNA to determine phylogenetic relationships between them. Nevertheless, the lack of non-ITS rDNA sequences (like LSU rDNA) from globally distributed isolates hinders such studies. Similar to other orders in oomycetes, efforts have also been made to fit the new arrangement of recently introduced families in *Saprolegniales* (obtained from the analysis of ITS and LSU markers) with their key morphological characteristics. Therefore, considering both gene markers and morphology for studying the diversity of *Saprolegniales* resulted in the emergence of three families, as stated above.

Along with applying mainly rDNA markers and introducing new families, the taxonomy of different genera (belonging to *Saprolegniaceae sensu lato*) was revisited, leading to several novel combinations and phylogenetic repositioning of taxa. For example, the genus *Achlya sensu lato* was shown to be polyphyletic according to the analysis of SSU rDNA and a part of isolates was transferred to two new combinations in *Newbya* gen. nov. (Spencer et al. 2002). The genus *Newbya* is defined morphologically by having centric and sub-centric oospores and laterally attached antheridia. Interestingly, the separation between *Newbya* and *Achlya sensu stricto* was confirmed by ITS and LSU rDNA markers (Beakes et al. 2014, Rocha et al. 2018), where *Newbya* isolates were separated from *Achlyaceae* (the genera: *Achlya sensu stricto*, *Thraustotheca*, and *Dictyuchus*) and closely related to *Saprolegniaceae sensu stricto*. Motivated by the relative success of rDNA markers in delineating families and genera within the order *Saprolegniales*, Steciow et al. (2014) achieved the most compelling multiple barcode assessment of *Saprolegniales* to date. This is currently the only study which has discussed phylogenetic relationships of major genera using a consensus tree of all three rDNA markers (SSU, ITS, and LSU). According to the multigene phylogenetic analysis, the genera of *Saprolegniales* are categorized into nine clades encompassing eleven documented genera and a new clade that has yet to be morphemically characterized. Also, zoospores' discharge mode, sporangia, and flagellated zoospores' type were considered key discriminant morphological features.

Marine Orders

Most marine lineages are associated with "Basal oomycetes" as predominant marine obligate parasites of marine microorganisms. Recent molecular investigations using various genomic regions have confirmed that these orders are among early-diverging branches of oomycetes with the ability to infect many organisms such as seaweed, crustacean, nematodes, and algae; it will be discussed later in the current chapter. In other words, studies on marine oomycetes have suggested that oomycete evolution started from marine environments. Generally speaking, the phylogeny of early-diverging oomycetes has been based on 18S/SSU (Sekimoto et al. 2007) and *cox*1 (Gachon et al. 2009), and *cox*2 (Hudspeth et al. 2000) and constantly updated ever since. In one of the latest updates, marine oomycetes were divided into eight monophyletic orders. The study has shown that the phylogenetic grouping of these eight orders is largely associated with their lifestyle. For example, orders such as *Eurychasmatales*, *Haptoglossales*, *Miraculales*, *Olpidiopsidales*, and *Pontismatales* are parasites of brown algae, invertebrate animals, diatoms, aquatic oomycetes, and red algae, respectively (Buaya et al. 2020). In the following paragraph, a review of important milestones in understanding the phylogeny of marine oomycetes will be presented.

The notion of marine-related taxa as the deepest lineages of oomycetes was examined by Kupper et al. (2006), who studied an oomycete pathogen of brown algae, namely *Eurychasma dicksonii* (E.P. Wright) Magnus, and showed it is separated from terrestrial and freshwater oomycetes using 18S rRNA. In 2007, the SSU rRNA, LSU rRNA, and *cox*2 genes were used to identify the *Haliphthoros*-like marine oomycete and *Haliphthoros milfordensis* Vishniac (as parasites of a wide range of marine crustaceans and some other marine animals) (Sekimoto et al. 2007). This study was important because it confirmed the phylogenetic separation of marine oomycetes from saprolegnian and peronosporalean clades. Such a separation was also recognized by Sekimoto et al. (2008), who used both SSU rRNA and *cox2* genes and introduced *Olpidiopsis porphyrae* Sekimoto, Yokoo, Y. Kawam. and D. Honda as a red algae parasite and a phylogenetically separated lineage that had branched out before the emergence of the abovementioned clades. Strittmatter et al. (2013) also showed the basal position of *Eurychasma dicksonii* (a pathogen of filamentous brown algae) within the oomycete lineage using LSU rRNA gene sequences. Later, Buaya et al. (2017) used partial nrSSU sequences to reconstruct the phylogeny of *Miracula helgolandica* Buaya, Hanic and Thines and *Olpidiopsis drebesii* Buaya and Thines, two basal oomycete parasitoids of marine diatoms and showed they are placed, as expected, among "Basal oomycetes". *Lagenisma coscinodisci* Drebes, *Miracula moenusica* A. Buaya and Thines, and *Olpidiopsis gillii* (de Wildeman) Friedmann are other examples of diatom pathogens whose phylogeny has been revealed by SSU and *cox2* sequences (Thines et al. 2015, Buaya and Thines 2019, Buaya et al. 2019b).

Interestingly, some investigations have shown that the presence of oomycetes in non-freshwater ecosystems is not limited to only basal lineages. *Halophythophthora* spp. is a perfect example of how marine oomycetes could arise from more recently evolved orders such as *Peronosporales* (Maia et al. 2021). In fact, in this extensive study, eight well-established *Halophytophthora* species from marine and brackish-water ecosystems in Portugal were characterized based on multigene phylogeny and detailed morphological and physiological descriptions. Such findings might suggest intriguing recent evolutionary adaptations of freshwater taxa to non-freshwater ecosystems.

Challenges

As shown above, the application of different gene markers is practical in improving the systematics of oomycetes. However, the current picture is far from complete as there are still inconsistencies among various lineages. Several arguments can be made to explain such shortcomings: I) one important obstacle is the unavailability of DNA sequences of gene markers for a large number of taxa. For instance, we are already aware of the high efficiency of some nuclear and mitochondrial gene markers in delineating *Peronosporales* and *Pythiales*. However, they have been rarely tested in *Saprolegniales* taxa. Therefore, gene markers that are proved to be operative in one or several groups of oomycetes must be eventually examined in others. That is how a more unified and reliable systematic containing all, or at least, major clades will be reachable; II) coalescing multigene markers-based phylogeny with the classical taxonomy of oomycetes is another old-age challenge oomycetologists face. If successful, this will integrate the polyphasic approach, which combines morphological, physiological, and ecological species concepts and hopefully excites the Consolidated Species Concept (CSC) in oomycete systematics; III) A momentum needs to be created for all scientists who study oomycetes to reclaim their contribution in the field of fungal systematics more independently (Masigol et al. 2021a, b, 2023).

Ecological Studies

Genetic Markers for Ecological Studies of Oomycetes

Genetic markers have been used for various purposes in oomycetes. Genetic markers have been a discriminative tool for some freshwater genera such as *Saprolegnia* to separate pathogenic isolates from non-pathogenic ones. The same is true with terrestrial pathogens, e.g., *Phytophthora*. Researchers have also applied genetic markers to trace the distribution patterns of some other genera, such as *Aphanomyces,* which can be found in both ecosystems. Therefore, in the following sections, we will discuss the application of genetic markers in both freshwater and terrestrial-associated taxa.

The Application of Genetic Markers in Terrestrial Ecosystem Studies of Oomycetes

Like their freshwater counterparts, studying oomycetes associated with terrestrial ecosystems (*Aphanomyces*, *Phytophthora*, and *Pythium* species, in particular) also benefits from applying genetic marker-based techniques such as RAPD, AFLP, and restriction fragment length polymorphism (RFLP). For instance, although researchers have tried to explain phenotypic characters in *Phytophthora* species populations in the absence of sexual reproduction, it has not been feasible for most species due to the limited number of genetic markers available for population genetic studies. In one study, Whisson et al. (1995) used both RAPD and RFLP markers to study the inheritance of virulence/avirulence in two populations of *Phytophthora sojae* Kaufm. and Gerd. They illustrated an initial genetic linkage map comprising 10 and 12 major and minor linkage groups, respectively. RAPD markers have also been used to study the distribution of *Phytophthora megakarya* Brasier and M.J. Griffin in Central and West Africa (Nyassé et al. 1999). The analysis of RAPD markers highly differentiated isolates inside Central Africa, reflecting an ancient evolution of *P. megakarya* in this part of Africa. Supported by studying 726 isolates of *P. infestans* in Canada during 1994-96, eight distinct genotypes of *P. infestans* were described using metalaxyl sensitivity and allozyme markers which implied rapid changes in the regional populations of the species (Peters et al. 1999). In another study, no close correlation among RAPD, AFLP and virulence groups could be established among 32 single-zoospore isolates of *Phytophthora infestans* (Mont.) de Bary (Abu-El Samen et al. 2003). Following the investigation of *P. infestans* lineages, Mahuku et al. (2003) used RAPD analysis to detect genetic variation in such a period. They concluded that migration and sexual recombination were important factors in the rapid changes of *P. infestans*' population in Canada, resulting from a low level of genetic diversity within populations.

Additionally, genetic markers such as RAPD and RFLP have had more diverse applications in *Pythium*, such as the differentiation of some *Pythium* species and varieties from others. For example, supported by RAPD data, it was revealed that three varieties of *P. ultimum*, including var. *ultimum*, var. *sporangiferum*, and group HS were not genetically distinct, even though they have morphological differences (Francis et al. 1994). In another study, Herrero et al. (1998) could separate *Pythium aphanidermatum* (Edson) Fitzp. isolates from *P. delicense* Meurs, *P. ultimum* Trow, *P. irregular* Buisman, and *P. paroecandrum* Drechsler isolates. Also, Matsumoto et al. (2000) divided geographically and morphologically diverse *Pythium irregulare* Buisman isolates into four groups based on RFLP and RAPD analyses. There were some correlations in their study: for example, seven isolates in groups III and IV were isolated only from sugar beet and sugar beet fields. Finally, RAPD methods have been used for separating pathogenic *P. insidiosum* De Cock, L. Mend., A.A. Padhye, Ajello and Kaufman from naturally occurring isolates (Pannanusorn et al. 2007).

Genetic markers such as RAPD and AFLP have become obsolete due to low reliability and being cost-ineffective. Therefore, one better, more practical, and cheaper genetic marker-based tool is inter-simple sequence repeat (ISSR)-PCR which studies fungal intraspecific genetic variation (Lindblom and Ekman 2005). Similar to fungi, ISSR has been used to identify repeated motifs in the genome of some oomycete taxa, such as *Pythium* and *Phytophthora* (Vasseur et al. 2005, Li and Liu 2021) (Figure 1). For example, Haghi et al. (2020) explored the genetic diversity of *Pythium oligandrum* Drechsler in Iran and observed no significant diversity among isolates of *P. oligandrum* isolated from nine provinces. This study has both practical and theoretical applications in studying the genetic diversity of the *Pythium* species. Firstly, its practicality lies in the fact that it develops a tool to screen *P. oligandrum* isolates for their biological control potential. Secondly, it investigates whether evolutionary mechanisms such as gene flow and genetic drift have influenced populations of *P. oligandrum* separated by kilometers. In another study, the applicability of ISSR markers has been extended to *Phytophthora capsici* Leonian to determine whether the genetic diversity of isolates obtained from sexual and asexual populations differs significantly. These findings are important as the epidemic regularity of the diseases is influenced by the genetic diversity of isolates (Li and Liu 2021).

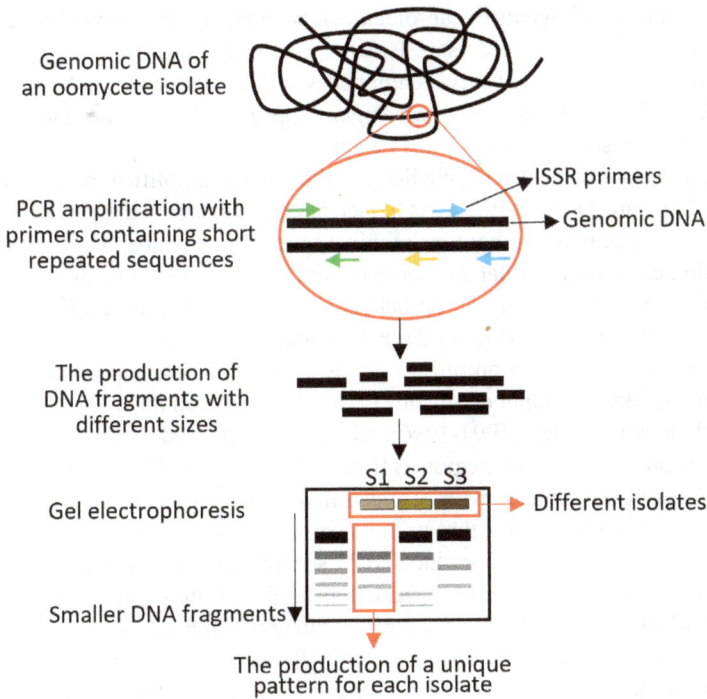

Figure 1. The application of inter-simple sequence repeat (ISSR)-PCR in studying the genetic diversity of the oomycetes' isolates.

Additionally, unprecedented growth in our understanding of genome content and organization has been practical for developing advanced management strategies such as gene silencing against mainly terrestrial oomycete pathogens. For instance, the applicability of transient gene silencing in determining gene function in *Phytophthora infestans* de Bary by introducing *in vitro* synthesized dsRNA into protoplasts derived from mycelium was examined using pairs of primers for the selected genes (Whisson et al. 2005). The study showed transient gene silencing as a practical tool for studying *P. infestans* functional genomics. In another study, van West et al. (2008) used genetic markers to detect the potential DNA methylation in genomic DNA of *P. infestans* and amplify the endogenous *infl* gene sequence. They concluded that the transcriptional gene silencing (TGS) phenomenon in *P. infestans* is mediated by changes in histone deacetylation.

Moreover, in a comprehensive review, Vetukuri et al. (2013) highlighted the importance of gene silencing in controlling gene expression at transcriptional and post-transcriptional levels, chromatin organization, adaptability, and diversification of *P. infestans*. Also, genetic primers were effectively used for northern blot hybridization, western blotting, and quantitative real-time PCR methods (Jahan et al. 2015). They evaluated an RNA silencing strategy and confirmed the functionality of a host-induced gene-silencing approach against *P. infestans* (Jahan et al. 2015).

The Application of Genetic Markers in Freshwater Ecosystem Studies of Oomycetes

Genetic markers have been practical to bridge our knowledge gap about the pathogenicity of *Saprolegnia* spp. isolates as well as the deep distributional patterns of *Aphanomyces* spp. isolates. *Saprolegnia parasitica* and *S. diclina* Humphrey (to a lesser extent) have always been the most common cause of saprolegniosis in many fish and amphibian species. However, we argue that this might not be the case everywhere. The notion of these species as the main pathogens in almost every corner of the world is problematic and not convincing due to the lack of effort to characterize them correctly. Reviewing the literature, we quickly understand that most studies have claimed the two species to be the main causal agents of saprolegniosis without any morphological and phylogenetic evidence. In other words, *Saprolegnia* spp. Isolates, considered a fish/amphibian pathogen, could have belonged to other *Saprolegnia* species, not necessarily *S. parasitica* and *S. diclina*. Therefore, since taxonomic studies on pathogenic *Saprolegnia* spp. have been either lacking or unreliable, other genetic markers-based techniques must be applied.

In response to the abovementioned challenges, the random amplification of polymorphic DNA polymerase chain reaction (RAPD-PCR) has helped researchers assess the genetic distance between different isolates and then reveal subgroups within a species. In essence, the RAPD-PCR technique divides morphologically very similar *Saprolegnia* spp. isolates into several subgroups. Then, we can assess whether there is any correlation between subgroups with isolates' geographical origin, pathogenicity, and other features (Figure 2). For example, three genetically distinct groups were inferred from the analysis of 686 amplified products in 67 different positions from 19 isolates isolated from catfish (Bangyeekhun et al. 2001). The study showed that all group 1 is composed of isolates obtained many years ago (1991-1996). Also, it was revealed that isolates in groups 2 and 3 were not limited to any geographical region as they had been isolated from various ponds. Although the study tried to associate isolates with *S. parasitica* or *S. diclina*, it suggested that they might belong to other *Saprolegnia* species that are not considered pathogens. In another study, Ghiasi et al. (2013) analyzed 385 amplified products in 21 separated positions from 23 isolates recovered from hatcheries in Mazandaran Province, Iran; and divided them into four genetically distant groups where only group 1 may represent a highly virulent clone of *Saprolegnia*. Similar to Bangyeekhun et al. (2001), although isolates were associated with *S. parasitica* and *S. diclina*, it does not rule out the possibility of a close relationship with other *Saprolegnia* species. As a result, studying the genetic diversity of *Saprolegnia* isolates using the RAPD-PCR technique became obsolete for three reasons. Firstly, the results of RAPD-PCR are often criticized due to their low reproducibility (i.e., the production of different patterns on agarose gel electrophoresis by a unique isolate under the same condition). Secondly, the method is highly time-consuming as several primers must be used to compensate low reproducibility of the method. Thirdly, such studies demand RAPD banding patterns of all possible pathogenic species of *Saprolegnia*, enabling effective comparison among isolated isolates. This information, however, is not available.

The same interest as explained above exists for *Aphanomyces* spp. isolates that cause crayfish plague and Epizootic ulcerative syndrome (EUS) in crayfish and fish species, respectively. In contrast to *Saprolegnia* spp. isolates, *Aphanomyces* spp. isolates are much harder to isolate, due to their slower growth rate than other fungal and oomycete isolates. For instance, even if *Aphanomyces* spp. isolates do exist on the internal parts and surfaces of their hosts (mainly fish and crayfish species), it will be challenging to isolate and keep them in the lab. Using culture-dependent methods, cultures were quickly dominated by fast-growing oomycetes such as *Pythium* and *Achlya* species. Also, most *Aphanomyces* spp. isolates are saprophytes (with a faster growth rate than the pathogenic

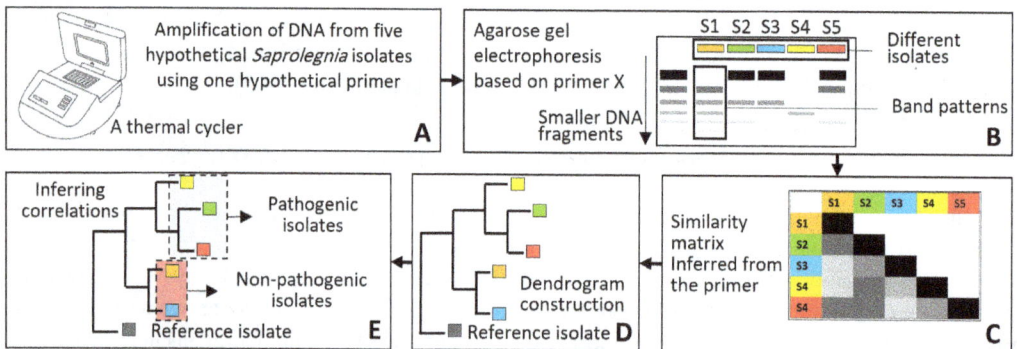

Figure 2. The application of random amplification of polymorphic DNA polymerase chain reaction (RAPD-PCR) in studying the genetic diversity of the *Saprolegnia* species isolates.

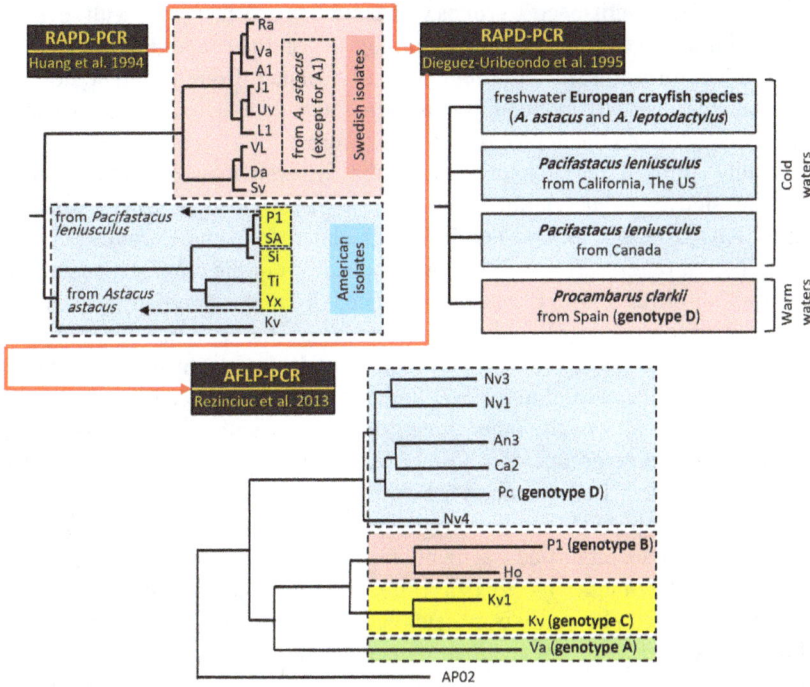

Figure 3. A review of the application of RAPD- and AFLP-PCR techniques for understanding the genetic diversity of *Aphanomyces astaci* worldwide.

ones) which frequently end up in cultures, stopping pathogenic *Aphanomyces* species from growing. However, there has been a great tendency to understand the genetic diversity of *Aphanomyces* spp. (*A. astaci* in particular) because isolates have been robustly disseminated from North America to the continents of Europe, Asia, and Australia over the last few decades. The question is how to justify such an enormous global invasion with respect to the genetic diversity of *Aphanomyces* spp. isolates. In contrast to *Saprolegnia* spp., more sophisticated methods (than RAPD-PCR techniques) have been used to understand the genetic diversity of *Aphanomyces* spp. isolates as discussed in the following paragraphs (Figure 3).

Anecdotally, we know that *A. astaci* was originally associated with crayfish species in North America (Oidtmann et al. 2006). Therefore, due to a long evolutionary arms race between *A. astaci* and endemic populations, some American crayfish species were more frequently farmed because they were less susceptible to crayfish plague. However, it became evident that these species are vectors of *A. astaci*, causing a significant threat to native European crayfish species. As soon as American crayfish species were transferred to European territory for commercial purposes, the crayfish plague became very serious as native European crayfish species were susceptible to *A. astaci*. The realization of the global distribution of crayfish plague causing a life-threatening condition in different crayfish species suggested high inter-species diversity of *A. astaci*. At that time, the RAPD-PCR had been used in fungi to assess the degree of genetic distance between different isolates. Therefore, Huang et al. (1994) used this method to determine how diverse *A. staci* isolates are (Figure 3). Gradually, determining the presence/absence of *A. astaci* in different crayfish species filled the gap in our knowledge regarding its origin and global dissemination pattern. For example, Huang et al. (1944) realized that Swedish isolates of *A. astaci* were divided into two sub-specific groups. One group constituted isolates isolated from two crayfish species (*Astacus astacus* and *A. leptodactylus*). The other group included isolates from *A. astacus* and North American crayfish (*Pacifastacus leniusculus*). The importance of this work lies in the fact that it presents the first contribution to understanding the genetic diversity of *A. astaci*. Using RAPD-PCR to expand *A. astaci* isolates' sub-specific groups was critical since each group with several isolates is often associated with different ecological, epidemiological, and physiological features. Previously, it was impossible to discriminate *A. astaci* isolates isolated from different hosts. However, RAPD-PCR became a sensitive method for assessing the degree of genetic distance between different *A. astaci* isolates and the degree of their relatedness.

This eventually allowed researchers to track the dissemination of isolates more robustly. In another similar study, Dieguez-Uribeondo et al. (1995) divided *A. astaci* isolates into four groups based on RAPD-PCR: (1) *A. astaci* isolates isolated from the European crayfish species (*Astacus astacus* and *A. leptodactylus*), (2) isolates isolated from the crayfish species *Pacifastacus leniusculus* from California, USA, (3) isolates isolated from the crayfish species *Pacifastacus leniusculus* from Canada, and (4) isolates isolated from the crayfish *Precambarus clarkii*. Interestingly, these genetic groups were confirmative of isolates' physiological properties too. For example, Dieguez-Uribeondo et al. (1995) found that isolates placed in the fourth group are subtropical taxa, while isolates of the other three groups are usually found in cold environments. Finally, five distinct genotype groups of *A. astaci* were identified as follows: group A (known as "As-genotype") isolated from European crayfish species, the groups B ("PI") and C ("PsII") from *P. leniuculus* of Californian and Canadian origin, respectively, and the groups D ("Pc") and E ("Or") from *O. limosus*. Also, amplified fragment length polymorphism (AFLP) analysis confirmed RAPD-based grouping (Rezinciuc et al. 2014). Although these genotypes were originally identified in the crayfish species listed above, their mosaic picture (Ungureanu et al. 2020) results from the multitude of invasive crayfish species and their dynamic spread.

With the announcement of *A. astaci* as one of the 100 worst invasive species in the world by the Global Invasive Species Specialist Group of the IUCN (International Union for Conservation of Nature; Lowe et al. 2004), researchers became even more committed to identifying *A. astaci* genotypes in mass mortalities of crayfish species and finally trace the sources of infection by improving knowledge on the diversity of this parasite. Although RAPD and AFLP approaches were practical to assign isolates to genetic groups, they suffered from two main challenges. Firstly, both RAPD and AFLP approaches require axenic cultures of the pathogen, whose isolation and maintenance in the lab are challenging. Secondly, they are not appropriate for large-scale and retrospective investigations. That is why other tools such as microsatellite markers emerged to study the diversity and epidemiology of *A. astaci* genotypes. Similar to RAPD and AFLP-based groupings, two lineages and four haplogroups (A, B, D, and E) were classified based on 14 developed microsatellite loci (Aast1- Aast14) (Grandjean et al. 2014). However, lack of amplification specificity, possible mixed infections, and difficulties in isolating DNA from pathogenic isolates of *A. astaci* from crayfish samples were some of the challenges faced by microsatellite markers (Maguire et al. 2016). Therefore, amplifying the mtDNA of ribosomal subunits was used to overcome the abovementioned difficulties and detect the pathogens in acute disease outbreaks in the wild, directly from the infected crayfish tissue samples (Makkonen et al. 2018). To summarize, *A. astaci* is a successful recipient of different molecular toolboxes. Currently, attempts to map the dissemination of *A. astaci* isolates and resolve their genetic variations worldwide is an ongoing quest.

The Application of Genetic Markers in Marine and Saltwater Ecosystem Studies of Oomycetes

Traditionally, oomycete orders that populate marine and saltwater ecosystems are less studied than their freshwater counterparts. However, topics such as their phylogenetic relationships, parasitism toward non-freshwater-associated organisms (diatoms, algae, and others), and other ecological implications in non-freshwater environments have gained attention over the last decades. Concerning pathogenic marine genera, *Lagenisma* spp., as important parasitoids of centric marine diatoms, are responsible for regulating diatom blooms which happen to be crucial in up to 40% of the global photosynthesis (Buaya et al. 2019a). Other pathogenic genera are *Eurychasma* and *Haptoglossa*, which can infect marine brown algae and nematodes, respectively (Grover and Barkoulas 2021).

The importance of such findings lies in the fact that these host organisms contribute to food webs and other geochemical processes. Therefore, any disturbance in the population structure by an oomycete pathogen might lead to great consequences for the entire habitat they are living in. However, marine oomycetes' depth of impact on natural communities and their potential host range is still unknown.

In addition to their pathogenic interactions, marine oomycetes are thought to harbor more complicated relationships. For instance, oomycete-virus interactions are a newly discovered phenomenon, with ecological implications for entire marine ecosystems, as viruses are the most abundant pathogens in the oceans and are involved in infecting bacteria, controlling microbial abundance, releasing dissolved organic matter, and manipulating global biogeochemical cycles (Breitbart et al. 2012). Supported by the fact that the mycoviruses had previously been reported from both fungi and terrestrial oomycetes, Botella et al. (2020) examined their presence in the genus *Halophytophthora* which is considered an exceptional group of marine microorganisms with an unknown role as a potential phytopathogen on coastal and estuarine grasses. For the first time, they confirmed the presence of harmful ssRNA viruses, including eight lineages of Bunya-like mycoviruses in one single *Halophytophthora* species. Although the virulence and the life cycles of these viruses are yet to be studied in detail, their close relationship with the genus *Halophytophthora* points to the crucial role of oomycetes in balancing other organisms' populations in marine habitats.

High-throughput Sequencing (HTS) and Metabarcoding

Studying Oomycete Communities

High-throughput sequencing (HTS-(and metabarcoding-based investigations have been a poorly studied topic in oomycetes. This has resulted in a distorted picture about the abundance and distribution in oomycetes' taxa. Results from most oomycete-related investigations are mainly based on culture-dependent methods, which leave most of the slow-growing and low-abundant taxa unexplored. Cultures are quickly colonized by common oomycetes such as *Pythium* spp. in terrestrial and *Saprolegnia* and *Achlya* species in aquatic landscapes. Consequently, oomycete communities have been poorly studied. Only in the last decade researchers have applied HTS and metabarcoding techniques to shed light on the hidden diversity of oomycetes, especially in aquatic ecosystems.

Before reviewing studies conducting large-scale surveys on oomycete communities using culture-independent methods, we need to articulate the importance of such findings: a general notion states that oomycete communities, unlike fungi, vary significantly in terrestrial and aquatic ecosystems (Põlme et al. 2020, Grossart et al. 2021). However, this notion quickly loses its validity as oomycete communities are constantly circulating from their releasing sources among natural and human-made habitats (terrestrial and aquatic). These sources could range from plant debris on the surface of the water and aquatic animals such as fish, crayfish, and amphibian species to agricultural fields and sediments. Therefore, the diversity of such turbulent environments for oomycete communities can only be understood by HTS and metabarcoding studies. These methods enable scientists to examine whether/how/to what extent oomycete communities are responsive toward biotic and abiotic changes. We can also test whether oomycetes' HTS-derived diversity matches traditional culture-dependent methods.

Understanding the real diversity of oomycete communities explained above has important ecological implications. These can be all addressed within the following lines of inquiry: I) the difference between oomycetes populating the shoreline and deeper parts of freshwater ecosystems, II) shifting oomycete communities as we move from terrestrial to the aquatic environment and vice versa, III) the possibility of aquatic environments to act as a vector for dissemination of both plant and animal pathogens, IV) the impact of water drainage from agricultural fields to natural water supplies, V) parasitic/saprophytic nature of oomycetes, and VI) their relationship with fungal taxa. High-throughput sequencing and metabarcoding methods have a great capacity to make scientists address and examine the ecological implications of oomycetes.

High-throughput sequencing and metabarcoding techniques have been effectively used to uncover the diversity of the fungal taxa for at least a decade resulting in the discovery of a tremendous number of uncultured and unknown taxa. Such a discovery was so striking that the term "Dark matter fungi" was coined by Grossar et al. (2016) to highlight our ignorance of the actual diversity of fungi. Although the abovementioned HTS and metabarcoding-related studies were mainly focused on fungi, taxa assigned to Stramenopiles occasionally appeared in the final community compositions as predators and ecosystem stabilizers. However, it was unclear what fraction of Stramenopiles belong to oomycetes (Debroas et al. 2017, Rojas-Jimenez et al. 2017). In more oomycete communities-orientated studies, advanced HTS techniques such as Illumina and Nanopore Sequencing have revealed taxa composition in oomycete communities. Reviewing the literature, we can observe two significant trends regarding applying advanced sequencing techniques. Firstly, all studies have been conducted in terrestrial environments and left the diversity of aquatic oomycete communities utterly unexplored. Secondly, these metagenomic analyses have been practical in deriving the ecological implications of oomycete communities in various habitats.

In one study, the type and abundance of oomycete communities with pea root rot in the Canadian prairies were determined by the Illumina MiSeq HTS technique. The genus *Pythium* and *Pythium heterothallicum* W.A. Campb. and F.F. Hendrix were the most prevalent taxa in three provinces during 2013-2014. Interestingly, *Aphanomyces euteiches* Drechsler, usually considered the most important and common oomycete pathogen of pea plants, had the lowest abundance. Also, multivariate analysis showed a difference in the relative abundance of species in oomycete communities regarding asymptomatic and diseases sites among years and provinces (Esmaeili Taheri et al. 2017). In another study, Sapp et al. (2019) used cytochrome c oxidase subunit II metabarcoding to examine whether there was a site-specific distribution of oak rhizosphere-associated oomycetes. They observed that oomycete communities were significantly correlated with ecological features of the sampling sites, including altitude, crown foliation, slope, soil skeleton, and soil nitrogen. These findings are crucial for predicting the spread of oomycete communities in forests and facilitating forest ecosystem management accordingly. Therefore, habitat-specificity (i.e., fluctuations in communities of microorganisms concerning environmental parameters of any given ecosystem) might not be limited to fungal taxa, as oomycetes have shown a similar trend, even though the number of studies is deficient. The practicality of using HTS tools was also demonstrated by Rossmann et al. (2021), where they introduced environmental DNA and metabarcoding techniques as a phytosanitary assessment tool for the detection of the pathogenic oomycete communities such as *Pythium* and *Phytophthora* species. Finally, Fiore-Donno and Bonkowski (2021) conducted an extensive landscape-scale metabarcoding survey and effectively examined any positive/negative influences of the most influential ecological and edaphic parameters in forests. They concluded that hemibiotrophs such as *Pythium* and *Phytophthora* species are mainly correlated with spruce beech forests, whose soil is sandier, their pH and C/N ratio are higher, and their forest intensity management is lower, compared to biotrophs such as *Lagena* and *Peronospora* species that prefer contrasting situations. The authors pointed out that understanding the community composition of oomycetes and the effects of abiotic and biotic parameters on them has been hampered due to a lack of systematic basic information. Therefore, HTS and metabarcoding tools make linking ecology, distribution, and diversity of oomycete communities possible.

Conclusion

The application of genetics to study the taxonomy and ecology of oomycetes has been reviewed in this chapter. DNA sequences successfully separate oomycetes from fungi and clarify their distant evolutionary relationship regarding the taxonomy. Additionally, oomycetes are phylogenetically categorized into two mega-groups, namely "Basal" and "Crown" oomycetes. In the current chapter, orders such as *Peronosporales*, *Pythiales*, and *Saprolegniales* received more attention due to their commercial and ecological relevance. We discussed that phylogenetic studies inferred from numerous genome regions lead to a significant revision and expansion in well-described genera such as *Phytophthora, Pythium*, and *Saprolegnia* (a destructive plant pathogen, cosmopolitan saprophyte, and notorious animal pathogen, respectively). Concerning ecology, genetics have been assisting in revealing distributional patterns of both terrestrial and freshwater taxa. Some DNA marker analyses such as RAPD and AFLP are among traditional markers whose use has been discouraged owing to their low reliability. In contrast, scientists have used ISSR markers as more reliable and cost-effective tools to study the genetic diversity of oomycete isolates more recently. Aside from culture-dependent studies, high-throughput sequencing and metabarcoding techniques are shedding light on unexplored ecological aspects of oomycetes. Although we have far from complete knowledge regarding the distribution of oomycete communities in various habitats, current and previously published investigations point to a habitat-specificity of oomycete communities in various habitats with different biological and limnological features (i.e., the presence/absence of some taxa in a given habitat). At the same time, the interconnectivity of terrestrial and aquatic landscapes might suggest that oomycete communities are constantly being circulated among these landscapes resulting in a more complicated picture of their diversity.

References

Abu-El Samen, F.M., G.A. Secor and N.C. Gudmestad. 2003. Genetic variation among asexual progeny of *Phytophthora infestans* detected with RAPD and AFLP markers. Plant Pathol. 52(3): 314–325.

Adl, S.M., A.G. Simpson, C.E. Lane, J. Lukeš, D. Bass, S.S. Bowser et al. 2012. The revised classification of eukaryotes. J. Eukaryoti. Microbiol. 59(5): 429–514.

Bahramisharif, A., S.C. Lamprecht, C.F. Spies, W.J. Botha and A. McLeod. 2013. *Pythium cederbergense* sp. nov. and related taxa from *Pythium* clade G associated with the South African indigenous plant *Aspalathus linearis* (rooibos). Mycologia 105(5): 1174–1189.

Bala, K., G.P. Robideau, N. Désaulniers, A.W.A.M. De Cock and C.A. Lévesque. 2010. Taxonomy, DNA barcoding and phylogeny of three new species of *Pythium* from Canada. Persoonia 25: 22–31.

Beakes, G.W., D. Honda and M. Thines. 2014. 3 Systematics of the Straminipila: Labyrinthulomycota, Hyphochytriomycota, and Oomycota. In D.J. McLaughlin, and J.W. Spatafora (Eds.), *Systematics and evolution* (pp. 39–97)., Berlin, Heidelberg: Springer.

Breitbart, M. 2012. Marine viruses: truth or dare. Annu. Rev. Mar. Sci. 4: 425–448.

Belbahri, L., G. Calmin, E. Sanchez-Hernandez, T. Oszako and F. Lefort. 2006. *Pythium sterilum* sp. nov. isolated from Poland, Spain and France: its morphology and molecular phylogenetic position. FEMS Microbiol. Lett. 255(2): 209–214.

Belbahri, L., G. Calmin, J. Pawlowski and F. Lefort. 2005. Phylogenetic analysis and real time PCR detection of a presumably undescribed *Peronospora* species on sweet basil and sage. Mycol. Res. 109(11): 1276–1287.

Beakes, G.W., S.L. Glockling and S. Sekimoto. 2012. The evolutionary phylogeny of the oomycete "fungi". Protoplasma 249(1): 3–19.

Blair, J.E., M.D. Coffey, S.Y. Park, D.M. Geiser S. and Kang. 2008. A multi-locus phylogeny for *Phytophthora* utilizing markers derived from complete genome sequences. Fungal Genet. Biol. 45(3): 266–277.

Botella, L., J. Janoušek, C. Maia, M.H. Jung, M. Raco and T. Jung. 2020. Marine oomycetes of the genus *Halophytophthora* harbor viruses related to Bunyaviruses. Front. Microbiol. 11: 1467.

Breitbart, M. 2012. Marine viruses: truth or dare. Annual Review of Marine Science 4: 425–448.

Buaya, A.T., S. Ploch, A. Kraberg, and M. Thines. 2020. Phylogeny and cultivation of the holocarpic oomycete *Diatomophthora perforans* comb. nov., an endoparasitoid of marine diatoms. Mycol. Prog. 19(5): 441–454.

Buaya, A., A. Kraberg and M. Thines. 2019a. Dual culture of the oomycete Lagenisma coscinodisci Drebes and Coscinodiscus diatoms as a model for plankton/parasite interactions. Helgol. Mar. Res. 73(1): 1–6.

Buaya, A.T., S. Ploch and M. Thines. 2019b. Rediscovery and phylogenetic placement of *Olpidiopsis gillii* (de Wildeman) Friedmann, a holocarpic oomycete parasitoid of freshwater diatoms. Mycoscience 60(3): 141–146.

Buaya, A.T. and M. Thines. 2019. *Miracula moenusica*, a new member of the holocarpic parasitoid genus from the invasive freshwater diatom *Pleurosira laevis*. Fun. Syst. Evo. 3(1): 35–40.

Buaya, A.T., S. Ploch, L. Hanic, B. Nam, L. Nigrelli, A Kraberg et al. 2017. Phylogeny of *Miracula helgolandica* gen. et sp. nov. and *Olpidiopsis drebesii* sp. nov., two basal oomycete parasitoids of marine diatoms, with notes on the taxonomy of Ectrogella-like species. Mycol. Prog. 16(11): 1041–1050.

Cooke, D.E.L., A. Drenth, J.M. Duncan, G. Wagels, and C.M. Brasier. 2000. A molecular phylogeny of *Phytophthora* and related oomycetes. Fungal Genet. Biol. 30(1): 17–32.

Dick, M.W., M.C. Vick, J.G. Gibbings, T.A. Hedderson and C.C.L. Lastra. 1999. 18S rDNA for species of *Leptolegnia* and other Peronosporomycetes: justification for the subclass taxa Saprolegniomycetidae and Peronosporomycetidae and division of the Saprolegniaceae *sensu lato* into the Leptolegniaceae and Saprolegniaceae. Mycol. Res. 103(9): 1119–1125.

Debroas, D., I. Domaizon, J.F. Humbert, L. Jardillier, C. Lepère, A. Oudart et al. 2017. Overview of freshwater microbial eukaryotes diversity: a first analysis of publicly available metabarcoding data. FEMS Microbiol. Ecol. 93(4): p.fix023.

De Cock, A.W.A.M., A.M. Lodhi, T.L .Rintoul, K. Bala, G.P. Robideau, Z.G. Abad et al. 2015. *Phytopythium*: molecular phylogeny and systematics. Pers.: Mol. 34: 25–39.

Esmaeili Taheri, A., S. Chatterton, B.D Gossen and D.L. McLaren. 2017. Metagenomic analysis of oomycete communities from the rhizosphere of field pea on the Canadian prairies. Can. J. Microbiol. 63(9): 758–768.

Fiore-Donno, A.M. and M. Bonkowski. 2021. Different community compositions between obligate and facultative oomycete plant parasites in a landscape-scale metabarcoding survey. Biol. Fertil. Soils 57(2): 245–256.

Francis, D.M., M.F. Gehlen, and D.A. St Clair. 1994. Genetic variation in homothallic and hyphal swelling isolates of *Pythium ultimum* var. *ultimum* and *P. utlimum* var. *sporangiferum*. MPMI 7(6): 766–775.

Gaastra, W., L.J. Lipman, A.W. De Cock, T.K. Exel, R.B. Pegge, J. Scheurwater et al. 2010. *Pythium insidiosum*: an overview. Vet. Microbiol. 146(1–2): 1–16.

Gachon, C.M., M. Strittmatter, D.G. Müller, J. Kleinteich and F.C. Küpper. 2009. Detection of differential host susceptibility to the marine oomycete pathogen *Eurychasma dicksonii* by real-time PCR: not all algae are equal. Appl. Environ. Microbiol. 75(2): 322–328.

Grandjean, F., T.Vrålstad, J. Diéguez-Uribeondo, M. Jelić, J. Mangombi, C. Delaunay et al. 2014. Microsatellite markers for direct genotyping of the crayfish plague pathogen *Aphanomyces astaci* (Oomycetes) from infected host tissues. Vet. Microbiol. 170(3-4): 317–324.

Grossart, H.P., E.A. Hassan, H. Masigol, M. Arias-Andres and K. Rojas-Jimenez. 2021. Inland Water Fungi in the Anthropocene: Current and Future Perspectives. The Encyclopedia of Inland Waters, Second Edition, Ed. Kendra Cheruvelil.

Grossart, H.P., C. Wurzbacher, T.Y. James and M. Kagami. 2016. Discovery of dark matter fungi in aquatic ecosystems demands a reappraisal of the phylogeny and ecology of zoosporic fungi. Fungal Ecol. 19: 28–38.

Grover, M. and M. Barkoulas. 2021. *C. elegans* as a new tractable host to study infections by animal pathogenic oomycetes. PLoS Pathog. 17(3): p.e1009316.

Grünwald, N.J., F.N. Martin, M.M. Larsen, C.M. Sullivan, C.M. Press, M.D. Coffey et al. 2011. Phytophthora-ID. org: a sequence-based *Phytophthora* identification tool. Plant Dis. 95(3): 337–342.

Haghi, Z., R. Mostowfizadeh-Ghalamfarsa and V. Edel-Hermann. 2020. Genetic diversity of Pythium oligandrum in Iran. J. Plant Pathol. 102(4): 1197–1204.

Herrero, M.L. and S.S. Klemsdal. 1998. Identification of *Pythium aphanidermatum* using the RAPD technique. Mycol. Res. 102(2): 136–40.

Hudspeth, D.S., D. Stenger and M.E. Hudspeth. 2003. A *cox*2 phylogenetic hypothesis for the downy mildews and white rusts. Fungal Divers. 13(4): 47–57.

Hudspeth, D.S., S.A. Nadler and M.E. Hudspeth. 2000. A COX2 molecular phylogeny of the Peronosporomycetes. Mycologia 92(4): 674–684.

Hudspeth, D.S.S. 2000. A *cox*2 phylogeny of the Peronosporomycetes (Oomycetes). Mycologia 92: 674–684.

Hyde, K.D., R.H. Nilsson, S.A. Alias, H.A. Ariyawansa, J.E. Blair, L. Cai et al. 2014. One stop shop: backbones trees for important phytopathogenic genera: I. Fungal Divers. 67(1): 21–125.

Jahan, S.N., A.K. Åsman, P. Corcoran, J. Fogelqvist, R.R. Vetukuri and C. Dixelius. 2015. Plant-mediated gene silencing restricts growth of the potato late blight pathogen *Phytophthora infestans*. J. Exp. Bot. 66(9): 2785–2794.

Johnson Jr, T.W., R.L. Seymour and D.E. Padgett. 2005. Systematics of the Saprolegniaceae: New combinations. Mycotaxon 92: 11–32.

Johnson, T.W., R.L. Seymour and D.E. Padgett. 2002. Biology and systematics of the Saprolegniaceae. The US: University of North Carolina Wilmington.

Kageyama, K., M. Senda, T. Asano, H. Suga and K. Ishiguro. 2007. Intra-isolate heterogeneity of the ITS region of rDNA in *Pythium helicoides*. Mycol. Res. 111(4): 416–423.

Kroon, L.P.N.M., F.T. Bakker, G.B.M. Van Den Bosch, P.J.M. Bonants and W.G. Flier 2004. Phylogenetic analysis of *Phytophthora* species based on mitochondrial and nuclear DNA sequences. Fungal Genet. Biol. 41(8): 766–782.

Kupper, F.C., I. Maier, D.G. Muller, S.L.D. Goer and L. Guillou. 2006. Phylogenetic affinities of two eukaryotic pathogens of marine macroalgae, *Eurychasma dicksonii* (Wright) *Magnus* and *Chytridium polysiphoniae* Cohn. Cryptogam., Algol. 27(2): 165–184.

Leclerc, M.C., J. Guillot and M. Deville. 2000. Taxonomic and phylogenetic analysis of Saprolegniaceae (Oomycetes) inferred from LSU rDNA and ITS sequence comparisons. Antonie van Leeuwenhoek 77(4): 369–377.

Levesque, C.A. and A.W. De Cock. 2004. Molecular phylogeny and taxonomy of the genus *Pythium*. Mycol. Res. 108(12): 1363–1383.

Lang-Yona, N., D.A. Pickersgill, I. Maurus, D. Teschner, J. Wehking, E. Thines et al. 2018. Species richness, rRNA gene abundance, and seasonal dynamics of airborne plant-pathogenic oomycetes. Front. Microbiol. 9: 2673.

Li, P. and D. Liu. 2021. Genetic diversity among asexual and sexual progenies of *Phytophthora capsici* detected with ISSR markers. Plant Prot. Sci. 57(4): 271–278.

Lindblom, L. and S. Ekman. 2005. Molecular evidence supports the distinction between *Xanthoria parietina* and *X. aureola* (Teloschistaceae, lichenized Ascomycota). Mycol. Res. 109(2): 187–199.

Mahuku, G., R.D. Peters, H.W. Platt and F. Daayf. 2000. Random amplified polymorphic DNA (RAPD) analysis of *Phytophthora infestans* isolates collected in Canada during 1994 to 1996. Plant Pathol. 49(2): 252–260.

Maguire, I., M. el Jelić, G. Klobučar, M. Delpy, C. Delaunay and F. Grandjean. 2016. Prevalence of the pathogen *Aphanomyces astaci* in freshwater crayfish populations in Croatia. Dis. Aquat. Org. 118(1): 45–53.

Maia, C., M. Horta Jung, G. Carella, I. Milenković, J. Janoušek, M. Tomšovský et al. 2021. Eight new *Halophytophthora* species from marine and brackish-water ecosystems in Portugal and an updated phylogeny for the genus. Pers.: Mol. 48: 54–90.

Makkonen, J., J. Jussila, J. Panteleit, N.S. Keller, A. Schrimpf, K. Theissinger et al. 2018. MtDNA allows the sensitive detection and haplotyping of the crayfish plague disease agent *Aphanomyces astaci* showing clues about its origin and migration. Parasitology 145(9): 1210–1218.

Masigol, H., P.van West, S.R. Taheri, J.M. Fregeneda-Grandes, L. Parvulescu, D. McLaggen et al. 2023. Advancements, deficiencies, and future necessities of studying Saprolegniales: a semi-quantitative review of 1073 published papers. Fungal Biol. Rev. (In Press).

Masigol, H., R. Mostowfizadeh-Ghalamfarsa and H.P. Grossart. 2021a. The current status of *Saprolegniales* in Iran: Calling mycologists for better taxonomic and ecological resolutions. Mycol. Iran. 8(2): 1–13.

Masigol, H., J.N. Woodhouse, P. van West, R. Mostowfizadeh-Ghalamfarsa, K. Rojas-Jimenez, T. Goldhammer et al. 2021b. Phylogenetic and Functional Diversity of Saprolegniales and Fungi Isolated from Temperate Lakes in Northeast Germany. J. Fungi. 7(11): 968.

Masigol, H., S.A. Khodaparast, R. Mostowfizadeh-Ghalamfarsa, K. Rojas-Jimenez, J.N. Woodhouse, D. Neubauer et al. 2020. Taxonomical and functional diversity of Saprolegniales in Anzali lagoon, Iran. Aquat. Ecol. 54(1): 323–336.

Masigol, H., S.A. Khodaparast, J.N. Woodhouse, K. Rojas-Jimenez, J. Fonvielle, F. Rezakhani et al. 2019. The contrasting roles of aquatic fungi and oomycetes in the degradation and transformation of polymeric organic matter. Limnol. Oceanogr. 64(6): 2662–2678.

Masigol, H., S.A. Khodaparast, R. Mostowfizadeh-Ghalamfarsa, S. Mousanejad, K. Rojas-Jimenez and H.P. Grossart. 2018. Notes on *Dictyuchus* species (Stramenopila, Oomycetes) from Anzali lagoon, Iran. Mycol. Iran. 5(2): 79–89.

Matsumoto, C., K. Kageyama, S.U. Haruhisa and M. Hyakumachi. 2000. Intraspecific DNA polymorphisms of *Pythium irregulare*. Mycol. Res. 104(11): 1333–1341.

Nyassé, S., L. Grivet, A.M. Risterucci, G. Blaha, D. Berry, C. Lanaud et al. 1999. Diversity of *Phytophthora megakarya* in Central and West Africa revealed by isozyme and RAPD markers. Mycol. Res. 103(10): 1225–1234.

Oidtmann, B., S. Geiger, P. Steinbauer, A. Culas and R.W. Hoffmann. 2006. Detection of Aphanomyces astaci in North American crayfish by polymerase chain reaction. Dis. Aquat. Org. 72(1): 53–64.

Pannanusorn, S., A. Chaiprasert, C. Prariyachatigul, T. Krajaejun, N.Vanittanakom, A. Chindamporn et al. 2007. Random amplified polymorphic DNA typing and phylogeny of *Pythium insidiosum* clinical isolates in Thailand. Southeast Asian J. Trop. Med. Public Health 38(2): 383.

Peters, R.D., H.W. Platt and R. Hall. 1999. Use of allozyme markers to determine genotypes of Phytophthora infestans in Canada. Can. J. Plant Pathol. 21(2): 144–153.

Petersen, A.B. and S. Rosendahl. 2000. Phylogeny of the Peronosporomycetes (Oomycota) based on partial sequences of the large ribosomal subunit (LSU rDNA). Mycol. Res.104(11): 1295–1303.

Põlme, S., K. Abarenkov, R. Henrik Nilsson, B.D. Lindahl, K.E. Clemmensen, H. Kauserud et al. 2020. FungalTraits: a user-friendly traits database of fungi and fungus-like stramenopiles. Fungal Divers. 105(1): 1–16.

Rezinciuc, S., J. Galindo, J. Montserrat and J. Diéguez-Uribeondo. 2014. AFLP-PCR and RAPD-PCR evidences of the transmission of the pathogen *Aphanomyces astaci* (Oomycetes) to wild populations of European crayfish from the invasive crayfish species, *Procambarus clarkii*. Fungal Biol. 118(7): 612–620.

Riethmüller, A., H. Voglmayr, M. Goker, M. Weiß and F. Oberwinkler. 2002. Phylogenetic relationships of the downy mildews (*Peronosporales*) and related groups based on nuclear large subunit ribosomal DNA sequences. Mycologia 94(5): 834–849.

Riethmüller, A., M. Weiß and F. Oberwinkler. 2000. Phylogenetic studies of Saprolegniomycetidae and related groups based on nuclear large subunit ribosomal DNA sequences. Can. J. Bot. 77(12): 1790–1800.

Robideau, G.P., A.W. De Cock, M.D. Coffey, H. Voglmayr, H. Brouwer, K. Bala et al. 2011. DNA barcoding of oomycetes with cytochrome c oxidase subunit I and internal transcribed spacer. Mol Ecol Resour. 11(6): 1002–1011.

Rocha, S.C., C.C. Lopez-Lastra, A.V. Marano, J.I. de Souza, M.E. Rueda-Páramo and C.L. Pires-Zottarelli. 2018. New phylogenetic insights into Saprolegniales (Oomycota, Straminipila) based upon studies of specimens isolated from Brazil and Argentina. Mycol. Prog. 17(6): 691–700.

Rojas-Jimenez, K., C. Wurzbacher, E.C. Bourne, A. Chiuchiolo, J.C. Priscu and H.P. Grossart. 2017. Early diverging lineages within *Cryptomycota* and *Chytridiomycota* dominate the fungal communities in ice-covered lakes of the McMurdo Dry Valleys, Antarctica. Sci. Rep. 7(1): 1–11.

Rossmann, S., E. Lysøe, M. Skogen, V. Talgø and M.B. Brurberg. 2021. DNA metabarcoding reveals broad presence of plant pathogenic oomycetes in soil from internationally traded plants. Front. Microbiol. 12: 645.

Safaiefarahani, B., R. Mostowfizadeh-Ghalamfarsa, G.S.J. Hardy and T.I. Burgess. 2015. Re-evaluation of the *Phytophthora cryptogea* species complex and the description of a new species, *Phytophthora pseudocryptogea* sp. nov. Mycol. Prog. 14(11): 1–12.

Sapp, M., N. Tyborski, A. Linstädter, A. Lopez Sanchez, T. Mansfeldt, G. Waldhoff et al. 2019. Site-specific distribution of oak rhizosphere-associated oomycetes revealed by cytochrome c oxidase subunit II metabarcoding. Ecol. Evol. 9(18): 10567–10581.

Scanu, B., T. Jung, H. Masigol, B.T. Linaldeddu, M.H. Jung, A. Brandano et al. 2021. *Phytophthora heterospora* sp. nov., a New Pseudoconidia-Producing Sister Species of *P. palmivora*. J. Fungi. 7(10): 870.

Scanu, B., G.C. Hunter, B.T. Linaldeddu, A. Franceschini, L. Maddau, T. Jung et al. 2014. A taxonomic re-evaluation reveals that *Phytophthora cinnamomi* and *P. cinnamomi* var. *parvispora* are separate species. Pathol. 44(1): 1–20.

Sekimoto, S., K. Yokoo, Y. Kawamura and D. Honda. 2008. Taxonomy, molecular phylogeny, and ultrastructural morphology of *Olpidiopsis porphyrae* sp. nov. (Oomycetes, straminipiles), a unicellular obligate endoparasite of *Bangia* and *Porphyra* spp. (Bangiales, Rhodophyta). Mycol. Res. 112(3): 361–374.

Sekimoto, S., K. Hatai and D. Honda. 2007. Molecular phylogeny of an unidentified Haliphthoros-like marine oomycete and *Haliphthoros milfordensis* inferred from nuclear-encoded small-and large-subunit rRNA genes and mitochondrial-encoded cox2 gene. Mycoscience 48(4): 212–221.

Spencer, M.A. 2002. Revision of *Aplanopsis*, *Pythiopsis*, and 'subcentric' *Achlya* species (Saprolegniaceae) using 18S rDNA and morphological data. Mycol. Res. 106(5): 549–560.

Steciow, M.M., E. Lara, C. Paul, A. Pillonel and L. Belbahri. 2014. Multiple barcode assessment within the *Saprolegnia-Achlya* clade (*Saprolegniales*, Oomycota, Straminipila) brings order in a neglected group of pathogens. IMA Fungus 5(2): 439–448.

Strittmatter, M., C.M. Gachon, D.G. Müller, J. Kleinteich, S. Heesch, A. Tsirigoti et al. 2013. Intracellular eukaryotic pathogens in brown macroalgae in the Eastern Mediterranean, including LSU rRNA data for the oomycete *Eurychasma dicksonii*. Dis. Aquat. Org. 104(1): 1–11.

Svoboda, J., A. Mrugała, E. Kozubíková-Balcarová and A. Petrusek. 2017. Hosts and transmission of the crayfish plague pathogen *Aphanomyces astaci*: a review. J. Fish Dis. 40(1): 127–140.

Takishita, K., N.Yubuki, N. Kakizoe, Y. Inagaki and T. Maruyama. 2007. Diversity of microbial eukaryotes in sediment at a deep-sea methane cold seep: surveys of ribosomal DNA libraries from raw sediment samples and two enrichment cultures. Extremophiles 11(4): 563–576.

Thines, M. 2007. Characterisation and phylogeny of repeated elements giving rise to exceptional length of ITS2 in several downy mildew genera (Peronosporaceae). Fungal Genetics and Biology 44(3): 199–207.

Thines, M., B. Nam, L. Nigrelli, G. Beakes and A. Kraberg. 2015. The diatom parasite *Lagenisma coscinodisci* (Lagenismatales, Oomycota) is an early diverging lineage of the Saprolegniomycetes. Mycol. Prog. 14(9): 1–7.

Tojo, M., P. Van West, T. Hoshino, K. Kida, H. Fujii, A. Hakoda et al. 2012. *Pythium polare*, a new heterothallic oomycete causing brown discolouration of *Sanionia uncinata* in the Arctic and Antarctic. Fungal Biol. 116(7): 756–768.

Tsui, C.K., W. Marshall, R. Yokoyama, D. Honda, J.C. Lippmeier, K.D. Craven et al. 2009. Labyrinthulomycetes phylogeny and its implications for the evolutionary loss of chloroplasts and gain of ectoplasmic gliding. Mol. Phylogenet. Evol. 50(1): 129–140.

Uzuhashi, S., M. Kakishima and M. Tojo. 2010. Phylogeny of the genus *Pythium* and description of new genera. Mycoscience 51(5): 337–365.

Van West, P., S.J. Shepherd, C.A. Walker, S. Li, A.A. Appiah, L.J. Grenville-Briggs et al. 2008. Internuclear gene silencing in *Phytophthora infestans* is established through chromatin remodelling. Microbiology 154(5): 1482–1490.

Van West, P. 2006. *Saprolegnia parasitica*, an oomycete pathogen with a fishy appetite: new challenges for an old problem. Mycologist 20(3): 99–104.

Vasseur, V., P. Rey, E. Bellanger, Y. Brygoo and Y. Tirilly. 2005. Molecular characterization of Pythium group F isolates by ribosomal-and intermicrosatellite-DNA regions analysis. Eur. J. Plant Pathol. 112(4): 301–310.

Vetukuri, R.R., A.K. Åsman, S.N. Jahan, A.O. Avrova, S.C. Whisson, and C. Dixelius. 2013. Phenotypic diversification by gene silencing in *Phytophthora* plant pathogens. Commun. Integr. Biol. 6(6): p.e25890.

Villa, N.O., K. Kageyama, T. Asano and H. Suga. 2006. Phylogenetic relationships of *Pythium* and *Phytophthora* species based on ITS rDNA, cytochrome oxidase II and β-tubulin gene sequences. Mycologia, 98(3): 410–422.

Whisson, S.C., A.O. Avrova, P. Van West and J.T. Jones. 2005. A method for double-stranded RNA-mediated transient gene silencing in *Phytophthora infestans*. Mol. Plant Pathol. 6(2): 153–163.

Whisson, S.C., A. Drenth, D.J. Maclean and J.A. Irwin. 1995. *Phytophthora sojae* avirulence genes, RAPD, and RFLP markers used to construct a detailed genetic linkage map. MPMI 8(6): 988–995.

Yang, X., B.M. Tyler and C. Hong, 2017. An expanded phylogeny for the genus *Phytophthora*. IMA Fungus 8(2): 355–384.

Yoshida, K., V.J. Schuenemann, L.M. Cano, M. Pais, B. Mishra, R. Sharma et al. 2013. The rise and fall of the *Phytophthora infestans* lineage that triggered the Irish potato famine. eLife 2: e00731.

CHAPTER 20

Molecular-Genetic Approaches to Protozoa and Their Pathogenicity

Gleice Ribeiro Orasmo[1],* and *Mariluce Gonçalves Fonseca*[2]

Introduction

Protozoa are unicellular eukaryotic organisms, in which all vital activities take place within the confines of a single plasma membrane; therefore, a single cell is a complete organism. The Protozoa are considered to be a subkingdom of the kingdom Protista, most of which are free-living organisms, although many organisms are parasites (Yaeger 1996). In 2003, Cavalier-Smith confirmed that Protozoa are accepted in this classification: Sarcomastigophora, Labyrinthomorpha, Apicomplexa, Microspora, Ascetospora, Myxospora and Ciliophora; however, the high genetic variation presented by these organisms makes taxonomic approaches very difficult. Single-celled eukaryotes are found wherever life exists and because they are highly adaptable, they spread easily from one place to another (Yaeger 1996).

Protist includes of approximately 65,000 known species, of which 10,000 are parasites of different animals (Levine et al. 1980, Vavra and Wallace 1980), with only a few dozen species infecting humans and can cause various diseases that affect the health of the population. Species pathogenic to humans mainly include the genera Acanthamoeba, which causes amoebiasis and keratitis (Khan 2009), Leishmania, which causes Leishmaniasis (Cuppolilo et al. 2003, Abeijon et al. 2020, Kumar et al. 2020, Azevedo et al. 2021), Plasmodium, agent's malaria (Lacorte et al. 2013, Gilabert et al. 2018, Chookajorn and Billker 2023), Trypanosoma, which causes Chagas' Disease (Cestari and Ramirez 2010, Kessler et al. 2013) and Toxoplasma, which causes toxoplasmosis (Vilares et al. 2018).

Protozoan diseases range from very mild to life-threatening. Individuals whose defenses are able to control but not eliminate a parasitic infection become carriers and constitute a source of infection for others (Yaeger 1996). Understanding the global distribution and genetic diversity of a given species of protozoan parasite is of fundamental importance to better understand the diseases these organisms cause and to find control methods that are more efficient, robust and long-lasting (Clark et al. 2016).

Despite immense genetic variation, molecular-level characterization of intestinal protists necessitates designing primers and probes for improved diagnostics and more accurate PCRs to detect all genetic variants or specifically differentiate between such variants (Stensvold et al. 2011). The occurrence of cases of human leishmaniasis in Brazil reaches very expressive numbers, with cases having been described in the five regions of the country. On average, with about 3500 cases registered and the incidence coefficient being 2.0 cases/100,000 inhabitants (Brasil 2023) makes

---

[1] Department of Biology, Natural Sciences Center, Federal University of Piauí, Teresina, Piauí, Brazil.
[2] Department of Biology, Federal University of Piauí, Picos, Piauí, Brazil.
* Corresponding author: gleice@ufpi.edu.br

Leishmania visceral fatality index in Brazil (2016 – 2020)

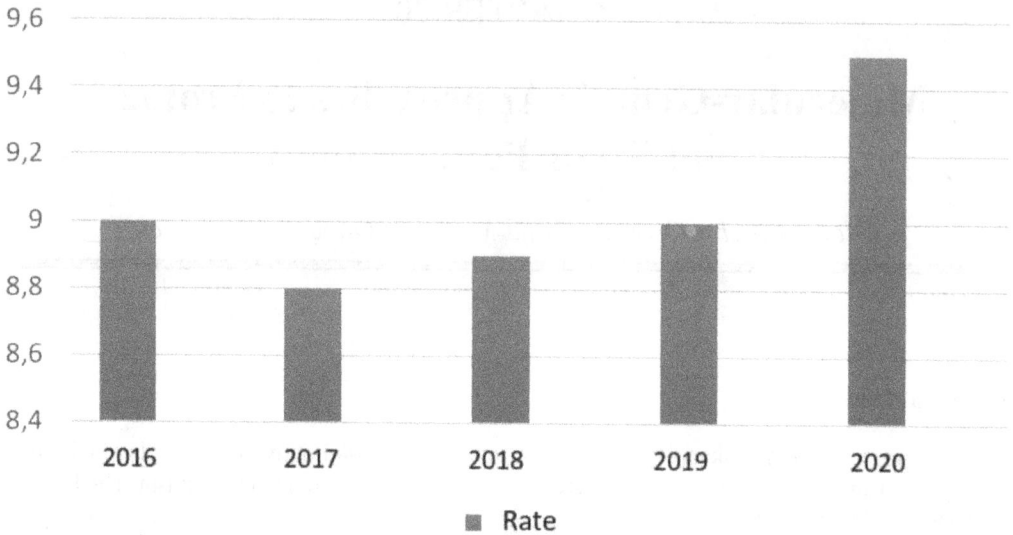

Figure 1. Leishmania visceral fatality index in Brazil (2016–2020). Source: the authors.

Brazil the country with the highest incidence of visceral leishmaniasis in the world (Azevedo et al. 2021). Despite the range of modern molecular tools for the diagnosis of visceral Leishmaniasis being available today, in Brazil, the official recommendation for the tests is to be determined by means of direct parasitological examination or culture, or in immunochromatography or immunofluorescence assays, which make limited deeper knowledge about the parasite and weakens the effectiveness of the treatment. Thus, the genetic diversity of *Leishmania infantum* in Brazil is not yet known (Azevedo et al. 2021). The visceral leishmaniasis lethality rate in Brazil between 2016 and 2020 is shown in Figure 1 - a significant increase in lethality is noted.

Thus, in this review, we intend to present some studies on molecular genetics in Protozoa, and verify how the use of molecular biology tools have contributed to understanding the diseases caused by these parasites. It is expected that these more in-depth analyses will bring more accuracy in diagnoses and can also provide greater efficiency in the treatment of these diseases.

Pathogenic Protozoa, Their Vectors and the Diseases they Affect in Humans

Among the important pathogenic microorganisms in human health that occur mainly in countries with precarious socioeconomic conditions are the representatives of the Protozoan group. This group comprises the Amoebae, the Sarcomastigophora, the Cilliophora and the Apicomplexa.

Human infection with gastrointestinal parasites is common worldwide, but higher morbidity and higher mortality is observed in developing countries, particularly among children, many of whom live in communities without access to sanitary facilities (Harhay et al. 2010). Infection with gastrointestinal parasites can cause acute or chronic gastroenteritis, but a considerable number of infected individuals do not have symptoms. According to David et al. (2015) among the most common species of protozoa that cause intestinal infections are *Cryptosporidium, Blastocystis* and *Entamoeba, Dientamoeba fragilis, Giardia duodenalis*, in addition to *G. lamblia* (Figure 2).

These amoebae were later identified as belonging to the genders *Acanthamoeba,* described so far as non-pathogenic (Culberton et al. 1959, Marciano-Cabral and Cabral 2003). In 1977, Pussard and Pons proposed the division of the genders *Acanthamoeba* into three morphological groups based

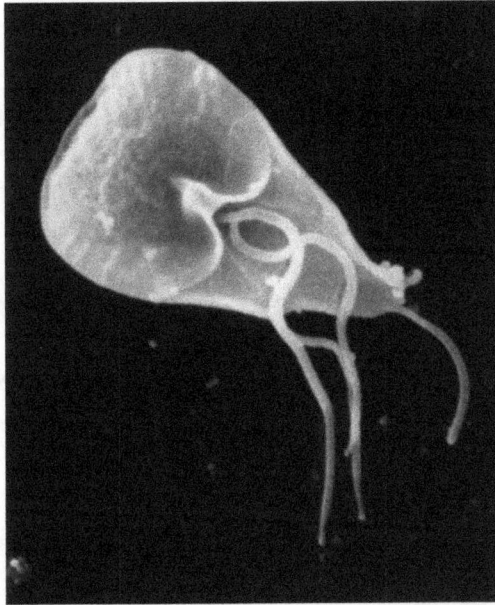

Figure 2. *Giárdia lamblia* cell. Source: https://pt.wikipedia.org/wiki/Gi%C3%A1rdia#/media/Ficheiro:Giardia_lamblia_SEM_8698_lores.jpg.

on the size and shape of the cysts. *Acanthamoeba* is a free-living amoeba of the gender amoeba and consists of many species that are naturally pathogenic (Khan 2009).

Naegleria fowleri, a type of free-living amoeba responsible for causing primary amoebic meningoencephalitis, is characterized by a rare and fatal infection of the central nervous system (brain and spinal cord). These amoebas can enter the brain through the nose when people swim in contaminated warm fresh water (Page 1988). The course of primary amoebic meningoencephalitis can be rapid, ranging from changes in smell or taste, headaches, neck stiffness, nausea and vomiting to confusion and death. To detect amoebas, doctors perform a lumbar puncture to obtain a sample of cerebrospinal fluid and sometimes take a small piece of tissue from the brain (biopsy) and then examine and analyze the sample obtained. Determining the best treatment is difficult, but doctors usually use a combination of drugs, including miltefosine. Free-living amoebas are protozoa that live in soil or water and do not need to live in people or animals. Although they rarely cause infections in humans, certain types of these amoebas can cause serious, life-threatening illness (Laoprasert et al. 2009).

N. fowleri is also known as the "brain-eating amoeba", is present all over the world. In the United States, most infections have occurred in southern states after exposure to warm fresh water from lakes and rivers in the summer. These amoebas can also live in hot springs or hot water discharged from industrial factories, poorly maintained swimming pools with little or no chlorination, and in soil (Laoprasert et al. 2009). It can even grow in water heaters at temperatures up to 46°C and survives for short periods at higher temperatures, however, *N. fowleri* does not live in salt water. When people, usually children or young people, swim in contaminated water, amoebas can enter the central nervous system through the nose. When they reach the brain, they cause inflammation and tissue destruction, which usually quickly leads to death (Culberton et al. 1959).

Different species of free-living amoebas can also cause granulomatous amoebic encephalitis, which is another very rare and usually fatal infection of the central nervous system by species of *Acanthamoeba* sp. or *Balamuthia mandrillaris*. Granulomatous amoebic encephalitis usually occurs in people with weakened immune systems or in poor general health, and usually progresses more slowly than primary amoebic meningoencephalitis (Tootla et al. 2022). Diagnosis is mainly performed by lumbar puncture and cerebrospinal fluid analysis, brain tissue biopsy. Imaging tests,

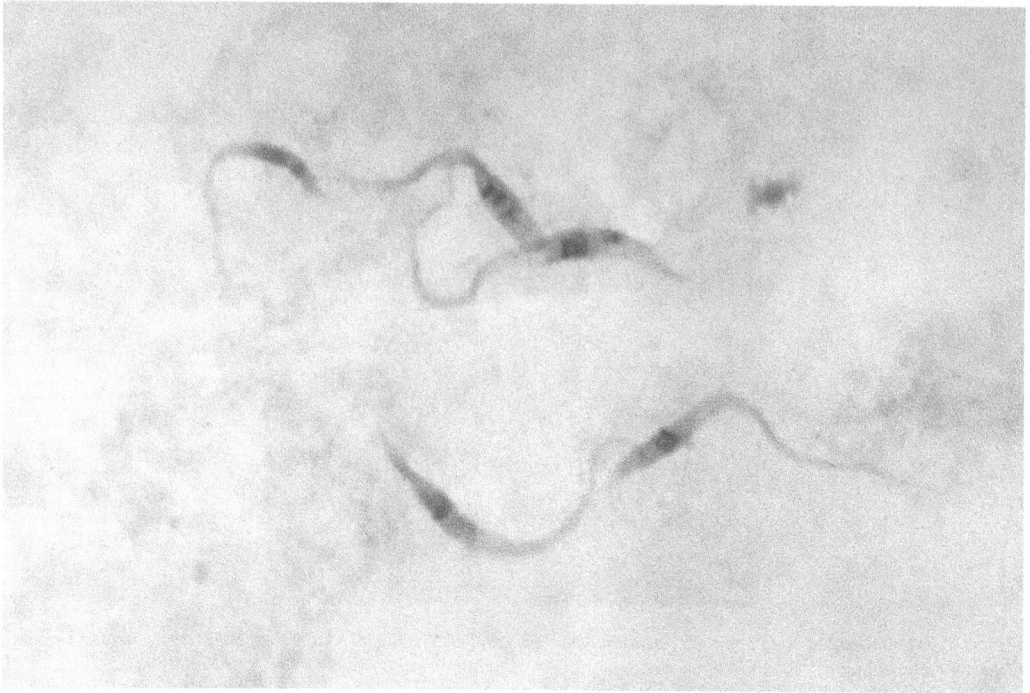

Figure 3. *Trypanosoma cruzi.* Source: https://pt.wikipedia.org/wiki/Tripanossoma#/media/Ficheiro:Trypanosoma_cruzi_crithidia.jpeg.

such as computed tomography (CT) and magnetic resonance imaging (MRI), are done to exclude other possible causes of brain infection, but these tests cannot confirm that amoebas are the cause. Other techniques are available from specialized laboratories and are more likely to detect amoebas. They include the following: Culture (growing microorganisms in the laboratory until they are in sufficient numbers to be identified); Polymerase chain reaction (PCR) to analyze amoeba genetic material or a biopsy of brain tissue that is stained and examined under a microscope or analyzed using PCR.

Amoebic keratitis is a rare infection of the cornea (the clear layer in front of the eye's iris and pupil) also caused by *Acanthamoeba* species. It usually occurs in people who wear contact lenses and causes painful ulcerations on the cornea and often vision is affected. Free-living amoebas are protozoa that live in soil or water and do not need to live in people or animals. Although they rarely cause infections in humans, certain types of these amoebae can cause serious, life-threatening infections in addition to amoebic keratitis. Infection is more likely if the lenses are worn while swimming or if the lens cleaning solution is not sterile. Some infections develop after an accidental scratch of the cornea. Symptoms of amoebic keratitis include redness of the eyes, excess tear production, foreign body sensation, and pain when the eyes are exposed to bright light. Thus, the use of molecular biology techniques for characterization and identification in environmental and clinical samples has been an important alternative in the verification of free-living amoeba isolates in severe cases (Lorenzo-Morales et al. 2015).

Steverding (2014) reported that Chagas disease or American trypanosomiasis affects six to seven million people worldwide, mainly in Latin America. The total estimated number of Chagas patients outside Latin America is more than 400,000 with the USA being the most affected country accounting for three-fourths of all cases. *Trypanosoma cruzi* (Figure 3) is the etiological agent of Chagas disease, for which there is no effective treatment or vaccine (Kessler et al. 2013).

The etiologic agent of Chagas disease, *Trypanosoma cruzi*, is a flagellate protozoan. Its evolutionary cycle includes the obligatory passage through hosts of several classes of mammals, including man, and hemipterous, hematophagous insects, commonly called kissing bugs, belonging to the Triatomidae family (Cestari and Ramirez 2010). In vertebrates, *T. cruzi* circulates in the blood and multiplies in tissues. In kissing bugs, it multiplies in the digestive tract; the infective forms are eliminated with their feces and urine. The transmission of the infection occurs mainly through the deposition of the vector's feces on human skin and mucous membranes.

The process of adaptation of triatomines to the human home depended on two factors that complemented each other: the kissing bug's need for food and its genetic mutations over time. With deforestation and the scarcity of wild animals (their natural sources of food), triatomines began to feed on domestic animals and humans, adapting to the peridomicile and home. Everything indicates that *T. cruzi* and its vectors have been present on this continent for a long time. However, human disease, at least in its endemic form, appears relatively recent (Steverding 2014).

The geographic distribution of endemic Chagas disease occurs in all areas where there are anthropophilic triatomines adapted to the human home, from Mexico to southern Argentina. For chagasic infection to occur under natural conditions, it is first necessary for susceptible people to meet triatomines infected with *T. cruzi*. The infection is transmitted through the feces and urine of triatomines, which defecate immediately after or during the bite (Hecht et al. 2010). Chagas infection has two very distinct phases: the acute or initial phase, asymptomatic or with few symptoms (the majority), or symptomatic, with fever, adenomegaly, hepaesplenomegaly, unilateral conjunctivitis (Romaña's sign), myocarditis and meningoencephalitis. It can be fatal in up to 10% of severe cases, the vast majority with meningoencephalitis, almost always fatal in children under two years of age, according to observations by Carlos Chagas himself and his contemporaries.

This phase is characterized by the presence of *T. cruzi* in the direct blood test. Approximately two months after the start of the acute phase, *T. cruzi* disappears from the bloodstream and can only be detected by special tests (xenodiagnosis, blood culture or PCR) (Li et al. 2005). After a latency period of 10 to 15 years, called the indeterminate form, patients can progress to 3 main types of the

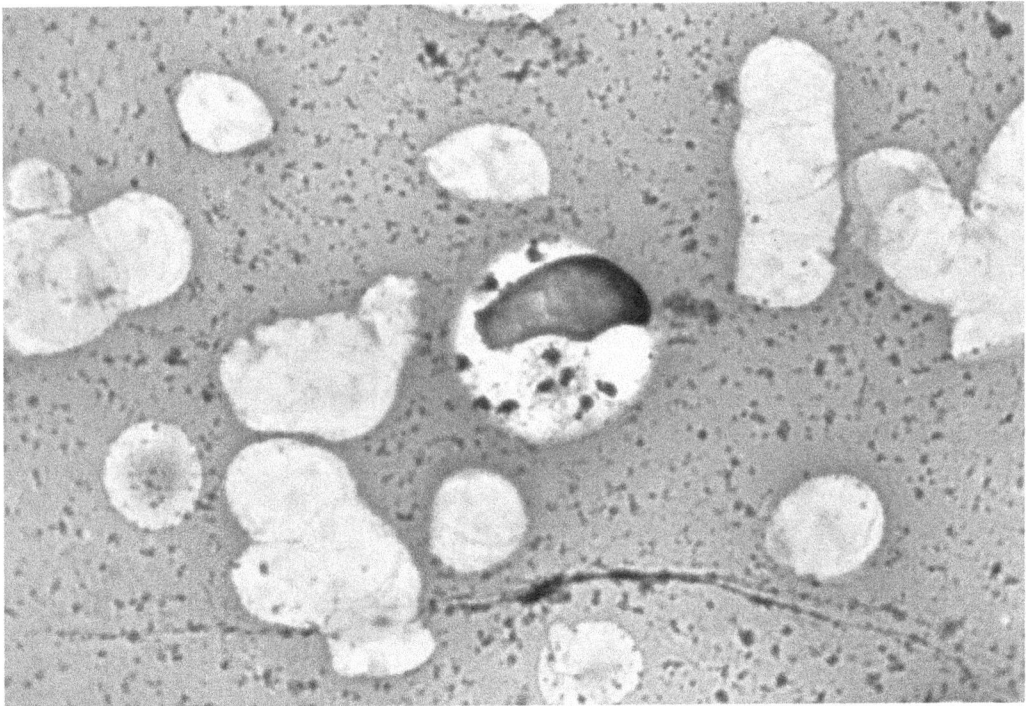

Figure 4. Amastigote form of *Leishmania donovani* in a bone marrow cell. Source: https://pt.wikipedia.org/wiki/Leishmania#/media/Ficheiro:Leishmania_donovani_01.png.

disease: the cardiac form, with chronic myocarditis, heart failure and eventually sudden death due to cardiac arrhythmia; digestive form, with megaesophagus and megacolon (exaggerated enlargement of the esophagus or colon by contraction of the corresponding sphincters); mixed form, with heart disease and "megas" simultaneously. About 50% of cases, depending on the endemic area, remain in the indeterminate form, without cardiac or digestive manifestations (Li et al. 2005).

According to Lainson and Shaw (1987), the genus *Leishmania* is divided into two subgenera: *Viannia* and *Leishmania*. *Leishmania (Viannia)* is the agent of the cutaneous-mucosal form, which is the most severe form of the disease, which occurs mainly in the countries of South and Central America represented by eight species, among them *Leishmania* (*Viannia*) *braziliensis* has emerged in Northeast of Brazil (Cuppolilo et al. 2003). *Leishmania* has three species distributed mainly in Central America, where it is predominant, and in South America responsible for the benign form of the disease (Nweze et al. 2021). In the visceral form of Leishmaniasis, the causative species is *Leishmania donovani* (Family Trypanosomatidae) which identifies three subspecies responsible for the disease in humans: *L. chagasi*; *L. infantum* and *L. donovani* (Figure 4). In the taxonomic position, it belongs to the phylum Sarcomastigophora and order Cinetoplastidae. They are mainly reported for Latin America, Asia, Africa, the Middle East, and some European countries (Nweze et al. 2021).

Leishmaniasis is caused by twenty or more species of *Leishmania* including several disorders that affect the skin, the mucous membranes of the nose, mouth or throat, or internal organs including the liver, spleen, and bone marrow. In cutaneous leishmaniasis, the first symptom is usually a well-defined lump at the site of a sand fly bite. It usually appears after several weeks or months and contains parasites inside white blood cells known as macrophages. As the infection spreads, more lumps may appear near the initial lump (Cuppolilo et al. 2003). The initial lump slowly enlarges and often becomes an open ulceration that may ooze or form a crust. Sores are usually painless and do not cause other symptoms unless a secondary bacterial infection develops at the site, characterized by redness in adjacent areas of skin, pain and sometimes fever. The sores eventually heal on their own after several months but may persist for more than a year. They leave permanent scars like those caused by burns. In rare cases, skin sores spread over the entire body. When this happens, the person is evaluated for HIV infection and other causes for the weakened immune system (Nweze et al. 2021).

In mucosal leishmaniasis, symptoms begin with a skin ulcer that heals on its own. Ulcers and tissue destruction may develop on mucous membranes inside the nose, mouth, or throat while the skin ulcer is present for months to years after it heals. The first sign may be nasal congestion, nasal discharge, or nosebleeds. Over time, people can become severely deformed (Nweze et al. 2021).

Visceral leishmaniasis can start suddenly but usually develops gradually over weeks to months after being bitten by an infected sand fly. People may have irregular episodes of fever (Azevedo et al. 2021). They may lose weight, have diarrhea, and experience general tiredness. The liver, spleen and sometimes lymph nodes increase in size. The number of white blood cells decreases, causing anemia, making people more susceptible to other infections. Left untreated, visceral leishmaniasis can result in death (Nweze et al. 2021). People who respond to treatment and those who are infected, but do not have symptoms are unlikely to have symptoms later unless their immune systems are weakened (for example, by AIDS or by drugs used to suppress the immune system, such as those used for preventing rejection of a transplanted organ). In people with AIDS, visceral leishmaniasis often recurs, and cutaneous leishmaniasis can cause ulcers all over the body (Azevedo et al. 2021).

After treatment of visceral leishmaniasis, plaques or lumps (nodules) may appear on the skin as other symptoms of visceral leishmaniasis go away. When sand flies bite people who have these areas of abnormal skin, the mosquitoes become infected and thus can spread the infection. Whether plaques and bumps appear and how long they remain on the skin depends on where people were infected like Sudan (located south of the Sahara) in Africa: plaques and lumps typically remain for a few months to a year; India and nearby countries: plaques and lumps can persist for years; Southern

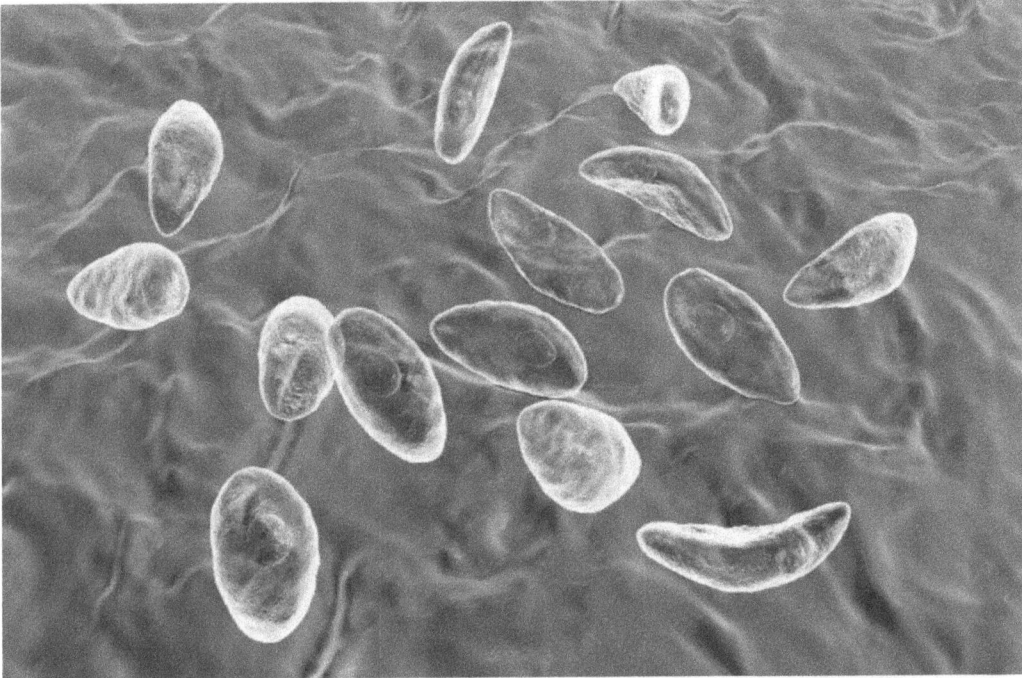

Figure 5. *Toxoplasma gondii.* Source: https://static.mundoeducacao.uol.com.br/mundoeducacao/2019/07/toxoplasma.jpg.

Europe, North Africa, Middle East and Latin America: plaques and lumps do not appear on the skin after treatment of visceral leishmaniasis (Engels and Zhou 2020).

Toxoplasma gondii is a sporozoan that infects many warm-blooded vertebrates, including humans (Dubey 2010, Nazir et al. 2018). The unicellular parasites of *T. gondii* (Figure 5) invade the host's cells, where they multiply (Alves et al. 2022, Silva et al. 2018). In adults, it causes asymptomatic chronic infection, which can reach 15% to 60% or more of the population, as well as generate an acute febrile condition with lymphadenopathy. In children, it produces a subacute infection with encephalopathy and chorioretinitis which, in congenital cases, is particularly severe. Immunosuppressed patients with positive serology develop encephalitis. The obligate endocellular parasite, *T. gondii,* preferentially invades cells of the mononuclear phagocytic system, leukocytes, and parenchymal cells. Once endocytosed, it remains in the parasitophorous vacuole without being digested and there it multiplies through a process of internal budding or endogeny. The parasite cycle has a sexual phase in the intestinal mucosa of definitive hosts and an asexual one in intermediate hosts. The transmission cycle includes cats where the sexual phase of toxoplasmas occurs; oocysts eliminated by them in the feces can also contaminate other animals, such as rodents and cattle, which become infected.

Malaria has always been, since ancient times, one of the main scourges of humanity. Currently, at least 300 million people contract malaria each year worldwide. Of these, about 1.5 million to 2 million die (Nweze et al. 2021). In Africa, nearly 3,000 children die of malaria each day. It is present in more than 90 countries, although with different prevalences. The most affected are India, Brazil (about 300,000 cases/year), Afghanistan and other Asian countries, including China. Malaria is typically a disease of the underdeveloped world (Engels and Zhou 2020). Also known as tertian or fourth fever, malaria has typical, almost unmistakable symptoms. It is manifested by episodes of chills followed by high fever, which last 3 to 4 hours. These episodes are usually accompanied by profound malaise, nausea, headaches, and joint pain. After the crisis, the patient can resume his usual life. But, after a day or two, the chills/fever condition returns and is repeated for weeks until the patient, untreated, heals spontaneously or dies amidst renal, pulmonary complications and cerebral coma. Treated in time, only exceptionally do people die of malaria. The interval between

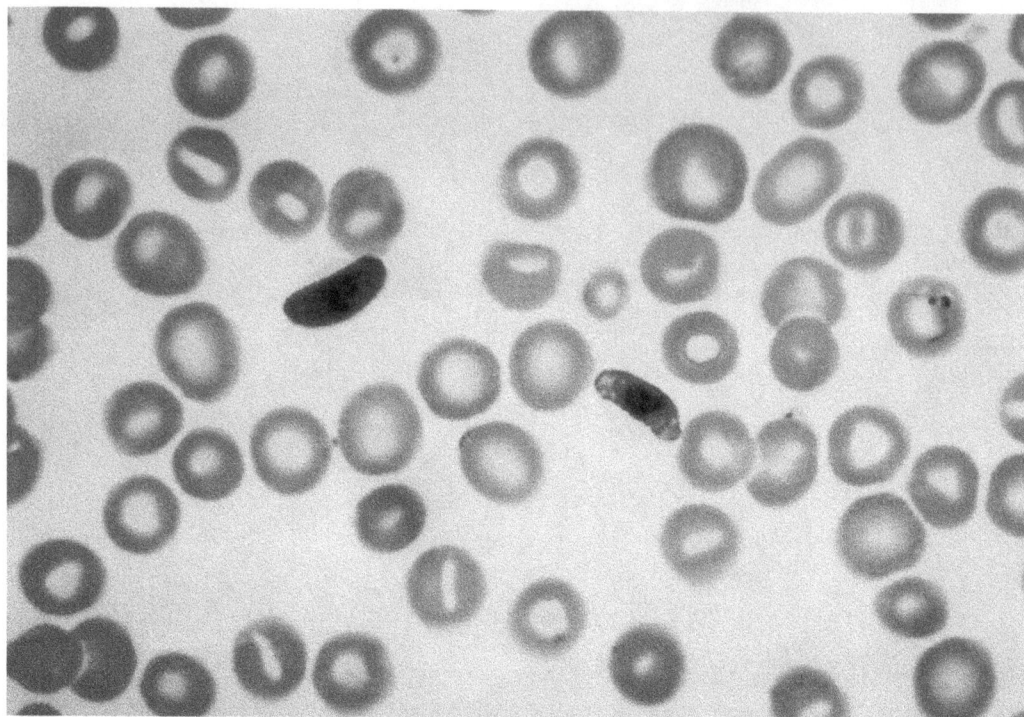

Figure 6. *Plasmodium falciparum* in human blood. Source: https://pt.wikipedia.org/wiki/Plasmodium_falciparum#/media/Ficheiro:Plasmodium_falciparum_01.png.

Table 1. Main pathogenics Protozoas, their vectors and the diseases that affect humans. Source: the authors.

Species*	Vector	Disease
Naegleria fowleri RF Carter, 1970 *Acanthamoeba* Volkonsky, 1931	Direct cycle of free-living species found in freshwater	- Primary and Granulomatous Amoebic Meningocephalitis-Keratitis
Trypanosoama cruzi Chagas, 1909	Hemiptera Triatominae	Chagas disease
Leishmania danovani Laveran and Mesnil, 1902 *Leishmania chagasi* Cunha and Chagas, 1937 *Leishmania infantum* Nicolle, 1908	Phlebotomine, Diptera Genders *Lutzomya* and *Phlebotomus*	Visceral leishmaniasis Cutaneous leishmaniasis
Toxoplasma gondii Nicolle and Manceaaux,1909	Cats (definitive hosts) Humans (intermediate hosts)	Toxoplasmosis
Plasmodium malariae Marchiafava and Celli, 1885 *Plasmodium vivax* Grassi e Feletti, 1890 *Plasmodium falciparum* Welch, 1897	Diptera, Culicidae Gender *Anopheles*	Malaria

*Source: Global Biodiversity Information Facility (GBIF). https://www.gbif.org/species/search acesso em: 01/06/2023.

episodes, the severity of the disease and its degree of mortality depend on many factors, but mainly on the species of parasite that causes malaria.

There are four species that cause human malaria: *Plasmodium vivax, P. falciparum, P. malariae* and *P. ovale. Falciparum* is responsible for a very serious form of malaria, formerly called malignant tertian (Lacorte et al. 2013). Of the annual deaths due to malaria, more than 95% are caused by *P. falciparum* (Figure 6).

Vivax causes a milder disease, benign tertian, which has the inconvenience of returning after apparently being cured, and this is because in the liver cells of the infected man, some hibernating forms may remain (Gilabert et al. 2018). The human malaria cycle is human-anopheline-human.

Generally, it is the female that attacks because she needs blood to ensure the maturation and laying of eggs. After biting an infected individual, the parasite develops part of its cycle in the mosquito and when it reaches the insect's salivary glands, it is ready to be transmitted to another person (Gilabert et al. 2018).

Table 1 shows the main species of pathogenic protozoa, as well as the diseases caused by these species and their vectors.

Molecular Tools Applied to Genetic Studies in Protozoan Parasites

Genetic-molecular analyses are valuable tools that provide advances in the understanding of Protozoa parasites, helping in the detection of parasites (Alves et al. 2022), in taxonomic studies, in the understanding of evolutionary processes (Gilabert et al. 2018, Van den Broeck et al. 2020), parasite-host co-evolution, genetic (Vilares et al. 2018) and genomic characterization (Chookajorn and Billker 2023, Kimmel et al. 2023), species identification (Clark et al. 2016, Srichaipon et al. 2019, Oliveira et al. 2021, Wang et al. 2021, Perrucci et al. 2021, Lynn et al. 2023, Maani et al. 2023, Xu et al. 2023), in more accurate detection of the diagnosis (Silva et al. 2018, Abeijon et al. 2020), understanding the mechanisms of infection (Cestari and Ramirez 2010) and the disease caused (Kumar et al. 2020), as well as helping with treatments.

Understanding the global distribution and genetic diversity of a given species of protozoan parasite is of fundamental importance to finding efficient, robust and long-lasting control methods (Clark et al. 2016). Genome-wide screening of an important species of Apicomplexa, the human malaria parasite, has transformed our understanding of protozoan parasites. Kimmel et al. (2023) reported a system that provided a powerful tool to enable gene-by-gene screening in *Plasmodium falciparum*. The authors systematically studied 33 proteins encoded on chromosome 3, one of the smallest chromosomes of *P. falciparum*. These new molecular toolkits are capable of transforming parasitology studies; however, new screening techniques, with different experimental nuances, are still needed to understand the biological uniqueness especially of malaria parasites (Chookajorn and Billker 2023). To unravel the evolutionary history and adaptation of *Plasmodium vivax* in different hosts, we used long and short reading frame technologies, in two *P. vivax*-like genomes and nine additional *P. vivax*-like genotypes, its closest genetic relative. The analyses show that the genomes of *P. vivax* and *P. vivax*-like are highly similar and collinear in the central regions. In addition, the data reveal that the evolution of *P. vivax* in humans did not occur at the same time as the other agents of human malaria, thus suggesting that transfer of parasites *Plasmodium* for humans happened several times independently throughout the history of the Homo genus. Two of these genes were also identified under positive selection in the main agent of human malaria, *P. falciparum*, thus suggesting their key role in the evolution of the ability of these parasites to infect humans or their anthropophilic vectors (Gilabert et al. 2018).

Using molecular tools, Lacorte et al. (2013) characterized the diversity of *Plasmodium* and *Haemoproteus* lineages in bird communities from three different habitats in southeastern Brazil based on prevalence, richness, and composition of lineages. The overall prevalence was 35.3%, ranging from 17.2% to 54.8% for each location evaluated. No significant association was observed between prevalence and habitat type. Eighty-nine *Plasmodium* and 22 *Haemoproteus* strains were identified, 86% of which were described for the first time, including an unusual infection of a non-columbiform host by *Haemoproteus* parasite. The composition of the parasitic communities of the Brazilian Cerrado and the dry tropical forest was similar, but different from the Atlantic Forest. No significant effects of habitat type on lineage richness were observed. The results also revealed that the samples with greater diversity of bird species presented a greater diversity of parasite lineages. Thus, these findings point to the importance of the Neotropical region (southeastern Brazil) as an important reservoir of new hemosporid lineages.

Different techniques have been used in molecular studies of the parasite *Toxoplasma gondii* and protozoa of the phylum Apicomplexa, which involve amplification via PCR, such as RFLP markers (Silva et al. 2018) and multiplex microsatellites, as well as techniques based on sequencing, Sanger Sequencing associated with Next Generation Sequencing (Sanger/NGS) (Vilares et al. 2018). Such techniques have been used for various purposes, such as parasite detection, in the characterization and analysis of genetic diversity, among others. To genetically characterize the parasite *Toxoplasma gondii*, 71 birds and 34 wild and captive mammals were evaluated in the State of Pernambuco, Brazil, using twelve RFLP PCR markers. Of the total number of animals evaluated, 32 samples were submitted to bioassays in mice, of which one isolate was positive (TgButstBrPE1), being detected in a socozinho (*Butorides striata*), a free-living bird; and 73 primary samples were submitted to biomolecular diagnosis using the PCR technique, targeting the repetitive fragment of 529 bp of *T. gondii* DNA, of which seven animals were positive. The seven primary samples and the positive isolate were subjected to RFLP PCR using the markers SAG1, 5'3'SAG2, alt. SAG2, SAG3, BTUB, GRA6, c22–8, c29–2, L358, PK1, Apico and CS3. The ToxoDB-RFLP #13 genotype of the socozinho isolate and the Type BrIII genotype of a captive otter (*Lontra longicaudis*) (PS-TgLonloBrPE1) were obtained. Therefore, the PCR-RFLP technique was effective in characterizing the parasite in Brazilian birds and mammals (Silva et al. 2018). Microsatellite analysis by multiplex PCR increased the discriminatory power by only one more identified type. However, when classical sequencing (Sanger sequencing) and Next Generation Sequencing (NGS) methodologies were added, the discriminatory power increased to 36 different types. With the aim of validating new methodologies for the genetic characterization of strains of *Toxoplasma gondii* of human origin isolated in Portugal. Vilares et al. (2018) estimated the genetic diversity of 68 strains of *T. gondii*, 51 strains being isolated through classical genotyping (Sag2). 5-microsatellite multiplex PCR, and a combination of Sanger and NGS sequencing of eight loci responsible for virulence were performed. The combination of these methodologies (Sanger/NGS) allowed the identification of mosaicism in *T. gondii* isolates in Portugal, previously not achieved by any of the other technologies used (Vilares et al. 2018).

Alves et al. (2022) aimed to detect at the molecular level the parasites *Toxoplasma gondii*, *Neospora caninum* and *Sarcocystis* spp., which are widespread protozoa capable of infecting a variety of domestic and wild animals (Dubey 2010, Darwich et al. 2012, Nazir et al. 2018). A total of 65 naturally infected birds were classified into 33 species. Tissue samples from the brain and heart of birds were subjected to DNA extraction and amplification, via PCR, of the 18S rRNA gene for *Sarcocystis* spp., of the NC5 gene for *N. caninum* and the 529 bp repetitive gene for *T. gondii*. The DNA of *N. caninum* was detected in two birds, corresponding to 3.07% and the DNA of *Sarcocystis* spp. was detected in three birds (4.62%). *T. gondii* DNA was not detected in any analyzed tissue. According to the authors, this is the first report of detection of DNA from *N. caninum*, *Sarcocystis* spp. in tissue samples for the analyzed bird species, thus extending the list of intermediate hosts.

The occurrence of DNA from Apicomplexan protozoa (*Theileria equi, Neospora caninum* and *Toxoplasma gondii*) was evaluated by Perrucci et al. (2021) in blood and milk samples collected from 33 healthy dairy females. A total of 73 blood and 73 milk samples were used for DNA extraction and analysis. In the blood samples, 11 females (33%) were positive for *T. equi*, while the milk samples were negative. *T. gondii* DNA was detected in the blood and milk of 3 dairy cows, whereas *N. caninum* DNA was found in 4 milk samples and 5 blood samples. According to the authors, this study is the first report on the presence of *N. caninum* DNA in the milk of naturally infected donkeys. The presence of *N. caninum* DNA in some of these females may suggest the possible occurrence of an endogenous cycle, while that of *T. gondii* in the collected milk may indicate a discontinuous excretion. Birds are important intermediate hosts of the protozoan parasites *Toxoplasma gondii* and *Neospora caninum*; however, little is known about the contamination of bird eggs by these parasites. Thus, Maani et al. (2023) aimed to investigate, by molecular detection, contamination by *T. gondii* and *N. caninum* in chicken (*Gallus domesticus*), duck (*Anas platyrhynchos*) and quail (*Coturnix coturnix*) eggs in three different geographic regions of Iran. The *T. gondii* and *N. caninum* parasites were detected by PCR by the RE and Nc5 genes, respectively. For the three species, the contamination rates with *T. gondii* and *N. caninum* were 10.7 and 5.9%, respectively. *T. gondii* contamination among chickens, ducks and quails were 12.2, 15.5 and 4.4%, respectively, while *N. caninum* was detected in 11.1, 3.3 and 1.1% of the same samples, respectively. Contamination rates increased with increasing humidity in three different regions. The authors concluded that the possibility of transmission of *T. gondii* eggs should not be overlooked by consumption of cooked or raw eggs.

Cryptosporidium is a protozoan that belongs to the phylum Apicomplexan; it is a parasite that causes diarrhea in humans and animals, with at least 46 species having been described, with more than 120 genotypes (Ryan et al. 2021a) and 23 species/genotypes of *Cryptosporidium* have been identified in humans, two of which, *C. hominis* and *C. parvum*, account for more than 95% of cases of cryptosporidiosis (Ryan et al. 2021b). A total of 621 fresh fecal samples were used to investigate the presence of *Cryptosporidium* spp. in the feces of racehorses in China and carry out the molecular characterization of the parasite. All DNA was analyzed for the presence of species/ genotypes and subtypes of *Cryptosporidium* through amplification of the ribosomal RNA small subunit and 60 kDa glycoprotein genes. PCR analysis revealed that 11 samples (1.8%) were positive for *Cryptosporidium* spp.; of these, seven were identified as *C. parvum* and four as *C. hominis*. Although the infection rate was relatively low, the isolates were identified as subtypes IIdA14G1 and IIdA15G1 (from *C. parvum*) and subtype IkA18G1 (from *C. hominis*), suggesting that these animals are a potential source of *Cryptosporidium* in humans (Xu et al. 2023). With the aim of investigating the presence of *Cryptosporidium*, as well as identifying the species, in cattle on farms in eight Brazilian states, a total of 408 fecal samples from calves were collected and screened using the Ziehl-Neelsen technique, positive samples being submitted to nested PCR. *Cryptosporidium* species were identified by PCR-RFLP, using SSPI, ASEI and MBOII enzymes. The Ziehl-Neelsen technique showed that 89.7% (35/39) of the evaluated farms and 52.9% (216/408) of the samples were positive. Protozoa were detected by nested PCR in 54.6% of the samples and the 56 samples submitted to PCR-RFLP showed *C. parvum*. These findings demonstrate the high level of *Cryptosporidium* spp. in circulation in cattle herds and the predominance of the species *C. parvum* (Oliveira et al. 2021).

To avoid the subjectivity associated with microscopic methods and with the evaluation of macroscopic pathological lesions, Clark et al. (2016) adopted non-quantitative PCR-based screening for *Eimeria* spp. and the occurrence of the OTU genotype. *Eimeria* spp. causes intestinal coccidiosis in all major livestock animals and is the most important parasite of domestic chickens considering economic impact and animal welfare. The study provided a comprehensive map of *Eimeria* occurrence for domestic chickens, confirming that all known species (*Eimeria acervulina, E. brunetti, E. maxima, E. mitis, E. necatrix, E. praecox, E. tenella*) are present on all six continents where chickens are found (including 21 countries). Analysis of 248 transcribed internal spacers, from sequences derived from 17 countries, provided evidence of possible allopatric diversity for species such as *E. tenella*, but not for *E. acervulina* and *E. mitis*, and highlighted a tendency for spread according to variance genetics. Three genetic variants previously described only in Australia and southern Africa have a wide distribution in the south, but not the northern hemisphere. The authors concluded that while the genotypes of the taxonomic unit are still unclear, the occurrence of the genetic variant of *Eimeria* could pose a risk to food security and animal welfare in Europe and North America, should these parasites spread to the northern hemisphere (Clark et al. 2016).

The DNA microarray approach has been widely used in parasitology research to facilitate understanding of disease mechanisms and identification of drug targets and biomarkers for diagnostic and therapeutic development (Kumar et al. 2020). Several studies have demonstrated the wide genetic diversity of the genus *Leishmania*, mainly considering *L. braziliensis* (Cuppolilo et al. 2003). Kumar et al. (2020) reviewed the DNA microarray approach, a technique that may be a new hope in the diagnosis, treatment and control of visceral leishmaniasis, one of the main infectious diseases that affect the poorest regions of the world. According to WHO 2016 weekly epidemiological record (World Health Organization 2016), 90% of global visceral leishmaniasis cases occurred in 6 countries (Bangladesh, Brazil, Ethiopia, India, South Sudan and

Sudan), and every year many deaths are reported due to visceral leishmaniasis (Torres-Guerreiro et al. 2017). According to Azevedo et al. (2021), Brazil is the country with the highest incidence of visceral leishmaniasis in the world. Abeijon et al. (2020) developed a conventional ELISA test associated with specific monoclonal antibodies (mAbs) for the diagnosis of visceral leishmaniasis, simultaneously detecting six biomarkers (Li-isd1, Li-txn1, Li-ntf2, Ld-mao1, Ld-ppi1 and Ld-mad1). Clinical validation of this new mAb-based multiplexed capture ELISA showed a sensitivity of 93%; thus, these results strongly support the hypothesis of its possible usefulness.

Aiming to evaluate a qPCR detection system for the diagnosis of visceral leishmaniasis (VL) in dogs, Cavalcanti et al. (2009) developed sets of specific primers for the *Leishmania donovani* complex, amplifying a 132 bp fragment of kDNA from *L. infantum*. The reaction was performed using the ABI PRISM 7000 system with the ABI PRISM software used to perform the analysis. When canine blood samples were evaluated using this system, the limit of detection of the method was 0.07 parasites per reaction, the efficiency was 94.17% (R2 = 0.93, slope = 3.47) and the sensitivity and specificity were 100% and 83.33% respectively. The use of such sensitive, reproducible and fast qPCR will be useful in the diagnosis and control of *L. infantum* infection in endemic areas, where serological surveys often underestimate the true prevalence of the disease.

Phylogenetic analyses based on single nucleotide polymorphisms (SNPs) using 637,821 SNP markers revealed the evolutionary history of parasites of the *Leishmania braziliensis* species complex, based on whole-genome sequencing of 67 isolates from 47 localities in Peru. Van den Broeck et al. (2020) first reported the origin of *Leishmania* Andina as a clade of quasi-clonal lineages that diverged from mixed Amazonian ancestors, with significant reduction in genome diversity and large structural variations resulting from parasite-host interactions; in addition, the patterns of population structure of these species were strongly associated with biogeographical origin. Changes in afforestation over the past 150,000 years have influenced the speciation and diversity of these Neotropical parasites. Secondly, the authors found that genomic-scale analyses provided evidence of meiotic-like recombination between Andean and Amazonian *Leishmania* species, resulting in a fully hybrid genome. By comparing complete ancestral nuclear and mitochondrial genomes, our data expand our assessment of the genetic consequences of diversification and hybridization in parasitic protozoa.

Blastocystis spp. are the most common protozoa in humans and animals worldwide. The genetic diversity of *Blastocystis* spp. can be associated with a wide range of symptoms. However, the prevalence of each subtype is different in each country. Until now, there is no standard method for subtyping *Blastocystis* spp. A sequential analysis using restriction fragment length polymorphism (RFLP) was developed for the rapid differentiation of subtypes of *Blastocystis* and the polymerase chain reaction (PCR) of ribosomal DNA small subunit (SSU rDNA). Of the 1,025 students participating in the survey, 416 (40.6%) were positive for *Blastocystis* spp., with subtype 3 being the most common (58.72%) among Thai students (Srichaipon et al. 2019).

Ciliates (Phylum Ciliophora), as a diverse group of protozoa, have a wide geographic distribution and play a vital ecological role as trophic links in microbial food webs (Lynn 2008). Sequence comparisons of nuclear small subunit ribosomal RNA (SSU rDNA) and mitochondrial genes have been used to identify known species and discriminate closely related species. Thus, genetic diversity within the ciliated order Euplotida was revealed by mitochondrial CO1 and nuclear ribosomal genes. According to the authors, CO1 is a suitable marker for the study of genetic diversity within Euplotes (Wang et al. 2021). Species identification in microbial eukaryotes generally requires morphological analysis combined with a molecular approach and, not infrequently, mating tests in distant populations to confirm whether they constitute a species. These three approaches were used to test hypotheses of the biogeography of the protozoan *Paramecium biaurelia* (one of the 15 cryptic species of the *P. aurelia* complex) collected worldwide from 92 sampling points over 62 years. The results indicated that despite the great distance between them, most of the *P. biaurelia* populations studied do not differ from each other (rDNA fragment) or differ only slightly (mtDNA COI fragment). The results may suggest that the current distribution is the result of recent dispersal by natural or anthropogenic factors, or that the low level of genetic diversity may be the result of a slow mutation rate of the studied DNA fragments (Tarcz et al. 2018).

The ciliates belonging to the *Paramecium aurelia* complex are formed by 15 species, and it is a good model for systematic and evolutionary studies. One member of the complex is *P. sonneborni*, of which only two lineages are known and studied (from the US and Cyprus) which show low viability in the F1 and F2 generations of cross-lineage hybrids and may be an example of ongoing allopatric speciation. Despite its molecular distinctiveness, the authors decipher that *P. sonneborni* must remain in the *P. aurelia* complex, making it a paraphyletic taxon. Morphological studies revealed that some characteristics of the nuclear apparatus of *P. sonneborni* correspond to *P. aurelia* spp. complex, while others are similar to *P. jenningsi* and *P. schewiakoffi*. According to the authors, the observed discordance indicates rapid division of *P. aurelia–P. jenningsi–P. schewiakoffi*, in which the genetic, morphological and molecular boundaries between species are not congruent (Przyboś et al. 2007).

For the protozoan *Trypanosoma cruzi*, the causative agent of Chagas disease, subverting the complement system and invading host cells is crucial for successful infection. C3b and C4b deposition assays revealed that *T. cruzi* primarily activates lectin and alternative complement pathways in non-immune human serum. These results establish that the complement system recognizes metacyclic *T. cruzi* trypomastigotes, resulting in the death of susceptible strains. The complement system, therefore, acts as a physiological barrier that resistant strains need to bypass for successful host infection (Cestari and Ramirez 2010). Chagas disease is transmitted by hematophagous insects known as "kissing bugs" (Hemiptera, Triatominae), with *Triatoma infestans* and *Rhodnius prolixus* being the two most important vector species. BLAST searches of repetitive DNA sequences, comparing the genomes of the two species, revealed that only 4 of the 42 satellite DNA families are shared between the two species, suggesting a large differentiation between the genomes of *Triatoma* and *Rhodnius*. Regarding the Y chromosome, fluorescence *in situ* hybridization (FISH) showed that the common satellite DNAs are absent in *T. infestans*, but are present in the Y chromosome of *R. prolixus*. These results support a different origin and/or evolution on the Y chromosome of both species (Pita et al. 2018). DNA isolation is the first stage of any work whose analysis is at the molecular-genetic level. Thus, it is crucial in molecular biology techniques that DNA extraction be successful, since the subsequent steps depend on the quality of the DNA, aiming at a satisfactory quantity and DNA free of proteins, RNAs, polysaccharides, among other molecules found in the worked fabric. Rotureal et al. (2005) developed a non-toxic and versatile method for extracting nuclear DNA from trypanosomatids by protein salting-out, resulting in satisfactory and rapid quantity and quality. Compared to commercial kits and phenol-chloroform procedures, the DNA yields obtained were similar. Furthermore, the samples were suitable for PCR and subsequent analyses. The reduction of manual work was noticed by this method in the routine use of medical diagnoses, in large-scale taxonomic and ecoepidemiological studies of trypanosomatids.

Congenital Chagas disease is a growing concern for the World Health Organization (WHO) and although El Salvador has the highest rates of Chagas disease in the Americas, pregnancy screening remains neglected. Lynn et al. (2023) carried out a maternal surveillance study on infection by *Trypanosoma cruzi* in western El Salvador. Of the 198 participating pregnant women, 6% were positive for *T. cruzi* by serology or molecular diagnosis. For molecular detection of the *T. cruzi* parasite, DNA was extracted from whole blood samples, and the presence of parasite DNA and quantification of parasite load was performed using a next-generation digital polymerase chain reaction (dPCR) assay. Half of the children born to *T. cruzi* positive women were admitted to the NICU due to neonatal complications. Older women and those who knew an infected relative or close friend were significantly more likely to test positive for *T. cruzi* infection at the time of delivery. The authors report that there is an urgent need to add maternal *T. cruzi* infections to mandatory pregnancy screening programs. Molecular methods were used to investigate a focus of *Trypanosoma cruzi* infection involving triatomines and Neotropical primates in a zoo located in the Brazilian Cerrado. Quantitative PCR (qPCR) was used to examine blood samples from 26

Table 2. Molecular techniques applied to the study of parasitic Protozoa.

Tools	Goals	Species	Authors
PCR[1]			
	Species identification	*Toxoplasma gondii*	Maani et al. 2023
		Neospora caninum	
	Species identification	*Toxoplasma gondii*	Perrucci et al. 2021
		Neospora caninum	
		Theileria equi	
	Species identification /	*Eimeria* spp.	Clark et al. 2016
	Infection mechanisms	*Trypanosoma cruzi*	Cestari and Ramirez 2010
	Species identification	*Cryptosporidium* spp.	Xu et al. 2023
dPCR[2]			
	Species identification	*Trypanosoma cruzi*	Lynn et al. 2023
PCR-RFLP[3]			
	Genotyping	*Toxoplasma gondii*	Silva et al. 2018
Nested PCR[4]			
	Investigate infection	*Trypanosoma cruzi*	Minuzzi-Souza et al. 2016
Nested PCR + RFLP			
	Species identification	*Cryptosporidium* spp.	Oliveira et al. 2021
RFLP + SSU rDNA[5]	Species identification	*Blastocystis* spp.	Srichaipon et al. 2019
SSU rDNA			
	Species identification	Ciliophora	Wang et al. 2021
SNPs[6]			
	Phylogenetics	*Leishmania* spp.	Van den Broeck et al. 2020
DNA Microarrays			
	Biomarkers identification	*Leishmania infantum*	Kumar et al. 2020
Sanger[7] **+ NGS**[8]			
	Genetical diversity	*Toxoplasma gondii*	Vilares et al. 2018
ELISA + mAbs[9]			
	Species identification	*Leishmania* spp.	Abeijon et al. 2020
Molecular Tool Kit			
	Genomic	*Plasmodium falciparum*	Kimmel et al. 2023 Chookajorn and Billker 2023
Genome sequencing	Adaptation	*Plasmodium vivax*	Gilabert et al. 2018
FISH[10]			
	Genomic	*Trypanosoma cruzi*	Pita et al. 2018
rDNA + mtDNA COI[11]			
	Species identification	*Paramecium biaurelia*	Tarcz et al. 2018

[1] PCR - Polymerase Chain Reaction
[2] dPCR - PCR digital
[3] RFLP - Restriction Fragment Length Polymorphism
[4] Nested PCR - Nested-Polymerase Chain Reaction
[5] SSU rDNA - Nuclear ribosomal gene sequences
[6] SNPs - Single-nucleotide Polymorphisms
[7] Sanger - Sanger sequencing
[8] NGS - Next Generation Sequencing
[9] mAbs - Specific monoclonal antibodies
[10] FISH - Fluorescence *in situ* hybridization
[11] mtDNA COI - fragment of mitochondrial DNA gene

primates and necropsy samples from two primates that died during the study. Infection in vectors was evaluated using nested PCR. Parasite lineages were determined in five vectors and two primates by comparing glucose-6-phosphoisomerase (G6pi) gene sequences. *Trypanosoma cruzi* was found in 44 vectors and 17 primates (six genera and eight species), being detected in three primates born to qPCR-negative mothers at the zoo and in two dead specimens. All G6pi sequences corresponded to the TcI lineage of *T. cruzi*. According to the authors, these findings strongly suggest vectorial transmission of *T. cruzi* within a small primate unit in the zoo and parasites, presumably coming from the nearby riparian forest (Minuzzi-Souza et al. 2016).

Kessler et al. (2013) constructed a fluorescent clone of *Trypanosoma cruzi*, in which the GFP gene is integrated into the chromosome that carries the Dm28c ribosomal cistron in *T. cruzi*. This fluorescent *T. cruzi* produces detectable amounts of GFP only in the replicative stages (epimastigote and amastigote). The fluorescence signal was also strongly correlated with the total number of parasites in *T. cruzi* cultures, providing a simple and rapid means of determining the growth-inhibiting dose of anti-*T. cruzi* in epimastigotes. This *T. cruzi* fluorescent clone is, therefore, an interesting tool for impartial detection of the parasite's proliferative stages, with multiple applications in the genetic analysis of *T. cruzi*.

In order to identify lateral DNA transfer events from the parasite to the host, Hecht et al. (2010) modified a PCR using additional targeted primers along with Southern blots, fluorescence and bioinformatics techniques. Natural human infections by *Trypanosoma cruzi* are documented, where mitochondrial minicircles integrated mainly in the retrotransposable LINE-1 of several chromosomes. The founders of five families show integrations in minicircles that were transferred vertically to their progeny. The final microhomology joining of 6 to 22 AC-rich nucleotide repeats in the minicircles and the host DNA mediates the integration of the foreign DNA. Mosaic recombination and hitchhiking retrotransposition events for different loci were more prevalent in germline compared to somatic cells. A pathway for the integration and maintenance of minicircles in the host genome is suggested. Thus, *T. cruzi* infection has the unexpected consequence of increasing human genetic diversity, and Chagas disease may be a fortuitous part of negative selection.

The use of molecular techniques and their applications, as well as the studies presented in this review are described in Table 2.

Molecular-genetic Contributions in Protozoa Parasites

The demarcation of boundaries between protist species is often problematic due to the absence of a uniform species definition, added to the abundance of diversity and the occurrence of convergent morphology (Przyboś et al. 2007). Thus, molecular-genetic studies are essential for a more refined evaluation.

Numerous studies have suggested a greater diversity of protozoa than originally proposed (Hausmann et al. 2003), so that protozoan taxonomy relies heavily on high-specialization techniques (Wang et al. 2021). In order to be able to detect all genetic variants or specifically differentiate between such variants, the molecular characterization of intestinal protists increasingly required increasingly effective tools, since it is necessary to design primers and probes for improved diagnoses, current and future. Thus, Stensvold et al. (2011) detected interspecific and intraspecific genetic diversity in single-celled intestinal parasites and its implications for nucleic acid-based diagnostics.

Silva et al. (2018) described the first isolation and genotypic characterization of *Toxoplasma gondii* in the wild bird (*Butorides striata*) and the first genotypic characterization of *T. gondii* in a captive otter using PCR-RFLP. In recent years, different researchers have adopted Next Generation Sequencing (NGS) to the detriment of classic sequencing (Sanger sequencing). The combination of these methodologies (Sanger/NGS) allowed obtaining a high discriminatory power in *T. gondii* isolates in Portugal (Vilares et al. 2018). Similar to advances in molecular epidemiology of other infectious diseases, the NGS methodology allows for a better understanding of the genetic profile of circulating strains, with gains in the prevention of congenital infection and the implementation of control measures. Next Generation Sequencing (NGS) is present and indispensable in genomic research (Vilares et al. 2018). Current molecular genetic studies in populations of *T. gondii* in Brazil have shown great genetic variability; in contrast, samples from various parts of the world in populations of the freshwater ciliate, *Paramecium biaurelia* (*P. aurelia* species complex, Ciliophora, Protozoa) revealed low genetic variability (Tarcz et al. 2018).

Visceral leishmaniasis is a serious, fatal disease caused by the parasites *Leishmania infantum* and *L. donovani* (Abeijon et al. 2020). The disease is endemic in 76 countries and, in the American continent, it is described in at least 12 countries. Of the cases registered in Latin America, 90% occur in Brazil (Brasil 2023). The disease has been described in several Brazilian municipalities, presenting important changes in the pattern of transmission, initially predominating in wild and rural environments and more recently in urban centers. However, the genetic diversity of *Leishmania infantum* in Brazil is not yet known, and the moleculars tools are not used in official recommended diagnostic tests (Azevedo et al. 2021), which limits disease control and treatment actions. Studies comparing the complete ancestral nuclear and mitochondrial genomes of Andean and Amazonian ancestors of *Leishmania* expand knowledge about the genetic consequences of diversification and hybridization in parasitic protozoa (Van den Broeck et al. 2020). Several studies have demonstrated the wide genetic diversity of the Protozoa, genus *Leishmania*, mainly considering *L. braziliensis* (Cavalcanti et al. 2009). Recent research has demonstrated the high sensitivity of real-time PCR (qPCR) in the diagnosis of *Leishmania infantum* infection (Cavalcanti et al. 2009).

In order to investigate the presence of *Cryptosporidium* spp. and perform the molecular characterization of the parasite in racehorses in China, Xu et al. (2023) analyzed 621 fresh fecal samples, of which 11 were positive (1.8%). *Cryptosporidium* infection was relatively minor in racehorses from parts of China. The identification of the human pathogen *C. hominis* and the zoonotic *C. parvum* in horses supports the zoonotic nature of the two species. The zoonotic subtypes IIdA14G, 1IIdA15G1 and IkA18G1 identified in racehorses and previously in humans suggest that racehorses may be a potential source of *C. parvum* in humans (Xu et al. 2023). These studies indicate the need for periodic verification of protozoal parasite infections.

Minuzzi-Souza et al. (2016) periodically checked for vectors and parasites to help eliminate pockets of *Trypanosoma cruzi* transmission in captive animal facilities. This should be of particular importance for captive breeding programs involving endangered mammals and would reduce the risk of accidental transmission of *T. cruzi* to zookeepers and veterinarians.

Transgenic parasites that express reporter genes are interesting tools to investigate parasite biology and parasite-host guests, with a view to developing new strategies for disease prevention and treatment. Kessler et al. (2013) developed a fluorescent *T. cruzi* clone, an interesting tool for impartial detection of parasite proliferative gains, with multiple applications in *T. cruzi* genetic analysis, including parasite-host parasite analyses, regulation of gene expression and development drugs. Understanding the global distribution and genetic diversity of a given species of protozoan parasite is of fundamental importance to finding efficient, robust and long-lasting control methods (Clark et al. 2016).

DNA transfer between species is an important biological process that leads to the accumulation of mutations inherited by sexual reproduction among eukaryotes. This demonstration of the contemporary transfer of eukaryotic DNA into the human genome and its subsequent inheritance by descendants introduces a significant change in the scientific concept of evolutionary biology and medicine (Hecht et al. 2010).

In this way, we can inquire that the studies that make use of molecular tools allow greater speed and precision of the genetics of the protozoa, not limited only in studies that aim at the genotyping of organisms, the elucidation of taxonomic uncertainties, or of the evolutionary processes and parasite-host relationship. In addition, these molecular tools have helped in understanding infections, and disease cycles, thus collaborating with more effective treatments.

Conclusions

The great genetic variability and diversity presented by the Protozoa parasites requires more refined techniques so that analysis and diagnosis can be more accurate. The breadth of genetic variation found in these organisms is almost limitless and requires a great deal of effort on the part of researchers in order to achieve success. Although this is something seemingly unattainable, the molecular approach has leveraged in recent years, and has shown great advances in studies, especially in accurate diagnoses. The use of molecular biology techniques has also helped in understanding the mechanisms of action of parasites and, in this way, contributed to the control of infections and guidelines for the treatment of diseases caused by these parasites. Thus, understanding the genetic diversity of a particular species of protozoan parasite is of fundamental importance to find efficient, consistent and definitive control methods.

References

Abeijon C., F. Alves, S. Monnerat, J. Mbui, A.G. Viana, R.M. Almeida et al. 2020. Urine-based antigen detection assay for diagnosis of visceral leishmaniasis using monoclonal antibodies specific for six protein biomarkers of *Leishmania infantum/Leishmania donovani*. PLoS Negl Trop Dis 14(4): e0008246.

Alves, M.E.M., F.D. Fernandes, P. Bräunig, L. Murer, C.E. Minuzzi, H.F. dos Santos et al. 2022. *Toxoplasma gondii*, *Neospora caninum* and *Sarcocystis* spp. in species of naturally infected birds. Pesq. Vet. Bras. 42: e07026.

Azevedo, R.C.F., R.E.da Silva, J.de O.J. Costa, I.P. Pesenato, S.P.J Souza, G.S.N. Castelli et al. 2021. Leishmaniose Visceral no Brasil: o que é preciso saber. BJGH. 01: 03.

Brasil. Ministério da Saúde. Situação Epidemiológica da Leishmaniose visceral.[internet]. Brasília, Brasil. [Acesso em maio 2023]. Disponível em: https://www.gov.br/saude/pt-br/assuntos/saude-de-a-a-z/l/leishmaniose-visceral/situacao-epidemiologica-da-leishmaniose-visceral.

Cavalcanti, M.de P., M.E.F.de Brito, W.V.de Souza, Y.de M. Gomes and F.G.C. Abath. 2009. The development of a real-time PCR assay for the quantification of *Leishmania infantum* DNA in canine blood. The Veterinary Journal 182: 356–358.

Cavalier-Smith, T. 2003. Protist phylogeny and the high-level classification of Protozoa. Eur. J. Protistol. 39(4): 338–348.

Cestari, I. and M.I. Ramirez. 2010. Inefficient complement system clearance of *Trypanosoma cruzi* Metacyclic Trypomastigotes Enables Resistant Strains to Invade Eukaryotic Cells. PLoS ONE 5(3): e9721.

Chookajorn, T. and O. Billker. 2023. Sideways: road to gene-by-gene functional screening in malaria parasites. Trends in Parasitol., 39(5): 317–318.

Clark, E.L., S.E. Macdonald, V. Thenmozhi, K. Kundu, R. Garg, S. Kumar, et al. 2016. Cryptic *Eimeria* genotypes are common across the southern but not northern hemisphere. Int. J. Parasitol. 46: 537–544.

Culbertson, C.G., J.W. Smith, H.K. Cohen and J.R. Minner. 1959. Experimental infection of mice and monkeys by *Acanthamoeba*. Am. J. Path. 35: 185–197.

Cupolillo, E., L.R. Brahim, C.B. Toaldo, M.P.de Oliveira-Neto, M.E.F.de Brito, A. Falqueto et al. 2003. Genetic Polymorphism and molecular epidemiology of *Leishmania (Viannia) braziliensis* from different hosts and geographic areas in Brazil. Jour. Clin. Microbiol. 41(7): 3126–3132.

Darwich L., O. Cabezón, I. Echeverria, M. Pabón, I. Marco, R. Molina-López et al. 2012. Presence of *Toxoplasma gondii* and *Neospora caninum* DNA in the brain of wild birds. Vet. Parasitol. 183(3/4): 377–381.

David, E.B., S. Guimarães, A.P.d. Oliveira, T.C.G.d. Oliveira-Sequeira, G.N. Bittencourt, A.R.M. Nardi et al. 2015. Molecular characterization of intestinal protozoa in two poor communities in the State of São Paulo, Brazil. Parasites and Vectors, 8: 103.

Dubey J.P. 2010. *Toxoplasma gondii* infections in chickens (*Gallus domesticus*): Prevalence, clinical disease, diagnosis and public health significance. Zoonoses Publ. Health 57(1): 60–73.

Engels, D. and X.N. Zhou. 2020. Neglected tropical diseases: an effective global response to local poverty-related disease priorities. Infect Dis Poverty. 9:10.

Gilabert A., T.D. Otto, G.G. Rutledge, B. Franzon, B. Ollomo, C. Arnathau et al. 2018. *Plasmodium vivax*-like genome sequences shed new insights into *Plasmodium vivax* biology and evolution. PLoS Biol 16(8): e2006035.

Harhay, M.O., J. Horton and P.L. Olliaro. 2010. Epidemiology and control of human gastrointestinal parasites in children. Expert Rev Anti Infect Ther. 8: 219–34.

Hausmann, K., N. Hülsmann, R. Radek. Protistology, 3rd ed.; E. Schweizerbart'sche Verlagsbuchhandlung: Berlin, Germany 2003.

Hecht, M.M., N. Nitz, P.F. Araujo, A.O. Sousa, A.dC. Rosa et al. 2010. Inheritance of DNA Transferred from American Trypanosomes to Human Hosts. PLoS ONE 5(2): e9181.

Kessler, R.L., D.F. Gradia, R.dC. Pontello-Rampazzo, É.E. Lourenço, N.J. Fidêncio, L. Manhaes et al. 2013. Stage-Regulated GFP Expression in *Trypanosoma cruzi*: Applications from Host-Parasite Interactions to Drug Screening. PLoS ONE 8(6): e67441.

Khan, N.A. 2009. Acanthamoeba - Biology and Pathogenesis. Caister Academic Press, Norfolk, 290p.

Kimmel, J., M. Schmitt, A. Sinner, P.W.T.C. Jansen, S. Mainye, G. Ramón-Zamorano et al. 2023. Gene-by-gene screen of the unknown proteins encoded on *Plasmodium falciparum* chromosome 3. Cell Syst. 14(9): 23–27.

Kumar A, S.C. Pandey and M. Samant. 2020. DNA-based microarray studies in visceral leishmaniasis: identification of biomarkers for diagnostic, prognostic and drug target for treatment. Acta Trop [Internet]. 208: 105512.

Lacorte, G.A., G.M.F. Félix, R.R.B. Pinheiro, A.V. Chaves, G. Almeida-Neto et al. 2013 exploring the diversity and distribution of neotropical avian malaria parasites – a molecular survey from Southeast Brazil. PLoS ONE 8(3): e57770.

Laoprasert, T., T. Nualchan, S. Chinabut and K. Hatai. 2009. Amoebae Isolated from Fish and Rearing Water at Oscar *Astronotus ocellatus* Farm. Aquaculture Sci. 57(2): 265–270.

Lainson, R. and J.J. Shaw. 1987. Evolution, classification and geographical distribution. pp. 12–120. *In*: Peters, W. and R Killick-Kendrick (eds.). The Leishmaniases in Biology and Medicine. London: Academic Press.

Levine, N.D., J.O. Corliss, F.E.g. Cox, G. Deroux, J. Grain, B.M. Honingberg et al. 1980. A newly revised classification of the Protozoa. J. Protozoo. 27(1): 37–58.

Li, F.-J., R.B. Gasserb, J.-Y. Zhenga, F. Claesc, X.-Q. Zhud and Z.-R. Luna. 2005. Application of multiple DNA fingerprinting techniques to study the genetic relationships among three members of the subgenus Trypanozoon (Protozoa: Trypanosomatidae). Mol. Cell. Probes 19: 400–407.

Lorenzo-Morales, J., N.A. Khan and J. Walochnik. 2015. An update on *Acanthamoeba keratitis*: diagnosis, pathogenesis and treatment. Parasite. 22: 10.

Lynn, D.H. 2008. The Ciliated Protozoa: Characterization, Classification and Guide to the Literature, 3rd ed.; Springer: Dordrecht, Germany.

Lynn, M.K., M.S. Rodriguez Aquino, P.M. Cornejo Rivas, M. Kanyangarara, S.C.W. Self, B.A. Campbell et al. 2023. Chagas disease maternal seroprevalence and maternal-fetal health outcomes in a Parturition cohort in Western El Salvador. Trop. Med. Infect. Dis. 8: 233.

Maani, S., K. Solhjoo, M.A. Bagherzadeh, A. Bazmjoo, H.G. Barnaaji, H. Hamed Mir et al. 2023. Prevalence of *Toxoplasma gondii* and *Neospora caninum* contaminations in poultry eggs: molecular surveillance in three different geographical regions of Iran. Food Safety and Risk 10: 3.

Marciano-Cabral, F. and G. Cabral. 2003. *Acanthamoeba* spp., as Agents of Disease in Humans. Clinical Microbiology Rev. 16: 273–307.

Minuzzi-Souza, T.T.C., N. Nitz, M.B. Knox, F. Reis, L. Hagström, C.A. Cuba Cuba et al. 2016. Vector-borne transmission of *Trypanosoma cruzi* among captive Neotropical primates in a Brazilian zoo. Parasites and Vectors, 9: 39.

Nazir M.M., M.M. Ayaz, A.N. Ahmed, A. Maqbool, K. Ashraf, M. Oneeb et al. 2018. Prevalence of *Toxoplasma gondii*, *Neospora caninum*, and *Sarcocystis* species DNA in the heart and breast muscles of Rock pigeons (*Columbia livia*). J. Parasitol. Res. 2018:1–4.

Nweze, J.A., F.N. Mbaoji1, Y-M. Li, L.-Y. Yang, S.-S. Huang, V.N. Chigor,et al. 2021. Potentials of marine natural products against malaria, leishmaniasis, and trypanosomiasis parasites: a review of recent articles. Infect Dis Poverty, 10:9.

Oliveira, J.S., F.D.C. Martins, W.A. Ladeia, I.B. Cortela, M.F. Valadares, A.M.R.N. Matos. et al. 2021. Identification, molecular characterization and factors associated with occurrences of *Cryptosporidium* spp. in calves on dairy farms in Brazil. Braz J. Vet. Parasitol., 30(4): e009621.

Page, F.C. 1988. A New Key to Freshwater and Soil Gymnamoebae. Freshwater Biological Association, Ambleside, Cumbria, pp.122.

Perrucci, S., L. Guardone, I. Altomonte, F. Salari, S. Nardoni, M. Martini et al. 2021. Apicomplexan protozoa responsible for reproductive disorders: Occurrence of DNA in Blood and Milk of Donkeys (*Equus asinus*) and Minireview of the Related Literature. Pathogens 10(111): 1–9.

Pita, S., P. Mora, J. Vela, T. Palomeque, A. Sánchez, F. Panzera et al. 2018. Comparative analysis of repetitive DNA between the main vectors of chagas disease: Triatoma infestans and Rhodnius prolixus. Int. J. Mol. Sci. 19: 1277–1288.

Przyboś, E., M. Prajer, M. Greczek-Stachura, B. Skotarczak, A. Maciejewska and S. Tarcz. 2007. Genetic analysis of the *Paramecium aurelia* species complex (Protozoa: Ciliophora) by classical and molecular methods. Syst. Biodivers., 5(4): 417–434.

Pussard, M. and Pons, R. 1977. Morphologies de la paroi kystique et taxonomie du genre *Acanthamoeba* (Protozoa, Amoebida). Protistol. 13: 557–610.

Rotureau, B., A. Gego and B. Carme. 2005. Trypanosomatid protozoa: A simplied DNA isolation procedure. Experimental Parasitology 111: 207–209.

Ryan, U.M., Y. Feng, R. Fayer and L. Xiao. 2021a. Taxonomy and molecular epidemiology of *Cryptosporidium* and *Giardia* - a 50 year perspective (1971–2021). Int. J. Parasitol. 51(13–14): 1099–119.

Ryan, U., A. Zahedi, Y. Feng and L. Xiao. 2021b. An update on zoonotic *Cryptosporidium* species and genotypes in humans. Animals (Basel). 11: 3307.

Silva, M.A., H.F.J. Pena, H.S. Soares, J. Aizawa, S. Oliveira, B.F. Alves et al. 2018. Isolation and genetic characterization of *Toxoplasma gondii* from free-ranging and captive birds and mammals in Pernambuco state, Brazil. Braz. J. Vet. Parasitol. 27(4): 481–487.

Srichaipon, N., S. Nuchprayoon, S. Charuchaibovorn, P. Sukkapan and V. Sanprasert. 2019. A simple genotyping method for rapid differentiation of blastocystis subtypes and subtype distribution of *Blastocystis* spp. in Thailand. Pathogens, 8(38): 1–15.

Stensvold, C.R., M. Lebbad and J.J. Verweij. 2011. The impact of genetic diversity in protozoa on molecular diagnostics. Tren in Parasit. 27(2): 53–58.

Steverding, D. 2014. The history of Chagas disease. Parasites & Vectors. 7: 317.

Tarcz, S., N. Sawka-Gądek and E. Przyboś. 2018. Worldwide sampling reveals low genetic variability in populations of the freshwater ciliate *Paramecium biaurelia* (*P. aurelia* species complex, Ciliophora, Protozoa). Org. Divers. Evol. 18: 39–50.

Tootla, H.D., B.S. Eley, J.M.N. Enslin, J.A. Frean, C. Hlela, T.N. Kilborn et al. 2022. Balamuthia mandrillaris Granulomatous Amoebic Encephalitis: The First African Experience. J Pediatric Infect Dis Soc. 28; 11(12): 578–581.

Torres-Guerrero, E., M. R. Quintanilla-Cedillo, J. Ruiz-Esmenjaud and R. Arenas. 2017. Leishmaniasis: A review. 6, 750.

Van den Broeck, F., N.J. Savill, H. Imamura, M. Sanders, I. Maes, S. Cooper et al. 2020. Ecological divergence and hybridization of Neotropical *Leishmania* parasites. PNAS 117(40): 25159–25168.

Vavra, J. and F.G. Wallace. 1980. A newly revised classification of the protozoa. J. Protozoo. 27(1): 37–58.

Vilares, A., V. Borges, D. Sampaio, L.Vieira, I. Ferreira, S. Martins et al. 2018. Sequenciação de nova geração: o paradigma dos parasitas (New generation sequencing: the parasite paradigm). Instituto Nacional de Saúde Doutor Ricardo Jorge (INSA), Boletim Epidemiológico, 10. Artigos Breves n.7.

Wang, C., Y. Hu, A. Warren and H.X. Xiaozhong. 2021. Genetic diversity and phylogeny of the genus euplotes (Protozoa, Ciliophora) revealed by the Mitochondrial CO1 and Nuclear Ribosomal Genes. Microorganisms 9: 2204.

Xu, C., Z. Wei, F. Tan, A. Liu, F. Yu, A. Zhao et al. 2023. Molecular detection and genetic characteristics of *Cryptosporidium* spp. in Chinese racehorses. Equine Vet J. 55: 474–480.

Yaeger, R.G. 1996. Protozoa: Structure, Classification, Growth, and Development. In: Medical Microbiology. 4th edition. Galveston (TX): University of Texas Medical Branch at Galveston; Chapter 77.

Index

About the Editors

Dr. hab. Sylwia Okoń is an Professor at the Institute of Genetics, Breeding and Plant Biotechnology, Faculty of Agrobioengineering at the University of Life Sciences in Lublin. She has twelve years of teaching and fifteen years of research experience. Her main research interests focus on the interactions between the host plant and the pathogen and increasing plant resistance to fungal pathogens. She is also involved in research in the field of genetic differentiation of fungal pathogens causing plant diseases, and their identification and classification. She conducts research related to phylogenetic and taxonomic analyses. She cooperates with University of Sydney (Australia), U.S. Department of Agriculture (USA), Leibniz Institute of Plant Genetics and Crop Plant Research (Germany), Julius Kühn Institute (JKI) (Germany). Dr. Okoń is the co-author of more than 60 scientific papers, she participated in numerous scientific conferences.

Dr. Beata Zimowska is Associate Professor at Department of Plant Pathology, Subdepartment Phytopathology and Mycology, Faculty of Horticulture and Land Scape, University of Life Sciences in Lublin, Poland. She has published more than 70 research publications, 30 popular articles in Polish and foreign. She is a member of several scientific societies. She has been awarded several awards: grade I team from HM Rector the Academy of Agriculture in Lublin – 2000, grade II team from HM Rector the University of Life Sciences in Lublin – 2009, individual grade I from HM Rector the University of Life Sciences for scientific activity in the years 2010–2012–2014, individual grade II from HM Rector the University of Life Sciences for scientific activity in the years 2014–2016–2017. She was co-organizer and member of Scientific Board 1st Eurasian Mycological Congress 3–5.07.2016, Manisa, Turkey and XIII Congress of Ecology and Environment with International Participation 12–15.09.2017, Edirne, Turkey. She serves as a referee for 15 international journals and is a member of the editorial board of 2 national and international journals. Since 2014 she has been cooperating with Bulgarian Institute of Genetics in Sofia, since 2015 with Department of Agriculture, University of Naples Federico II, Portici and Institute for Sustainable Plant Protection (IPSP)/National Research Council (CNR) Portici, Italy, since 2015 with Department of Biology Trakya University, Edirne, Turkey. Her research activities are aimed at: Phoma sensu lato group of fungi, endophytic fungi, fungal symbiont and their relationships with gall midges, mutualistic interactions between the gall causing species of dipterans from Asphondylia genus (Diptera: Cecidomyiidae) and on fungi developing in the inner wall of galls, etiology and epidemiology of infectious diseases of MAPs (medicinal and aromatic plants), with special regards of new and emerging infectious diseases (EIDs).

Dr. Mahendra Rai is presently a visiting Professor at the Department of Microbiology, Nicolaus Copernicus University, Torun, Poland. Formerly he was Professor and Head of the Department of Biotechnology, SGB Amravati University, Maharashtra State, India. He was a visiting scientist at the University of Geneva, Debrecen University, Hungary; University of Campinas, Brazil; VSB Technical University of Ostrava, Czech Republic, National University of Rosario, Argentina, and the University of Sao Paulo. He has published more than 425 research papers in national and international journals, with 81 h-index. In addition, he has edited/authored more than 70 books and 6 patents. Recently, he has been featured in Stanford's list of the top 2% of scientists in Nanoscience and Nanotechnology.

For Product Safety Concerns and Information please contact our EU
representative GPSR@taylorandfrancis.com
Taylor & Francis Verlag GmbH, Kaufingerstraße 24, 80331 München, Germany

www.ingramcontent.com/pod-product-compliance
Lightning Source LLC
Chambersburg PA
CBHW080713220326
41598CB00033B/5407

* 9 781032 358420 *